BIOGEOCHEMICAL CYCLING OF MINERAL-FORMING ELEMENTS

Studies in Environmental Science

Volume 1 **Atmospheric Pollution 1978**
Proceedings of the 13th International Colloquium, held in Paris, April 25–28, 1978
edited by M.M. Benarie

Volume 2 **Air Pollution Reference Measurement Methods and Systems**
Proceedings of the International Workshop, Bilthoven, December 12–16, 1977
edited by T. Schneider, H.W. de Koning and L.J. Brasser

Volume 3 **Biogeochemical Cycling of Mineral-Forming Elements**
edited by P.A. Trudinger and D.J. Swaine

Volume 4 **Potential Industrial Carcinogens and Mutagens**
by L. Fishbein

Studies in Environmental Science 3

BIOGEOCHEMICAL CYCLING OF MINERAL-FORMING ELEMENTS

Edited by

P.A. Trudinger

Baas-Becking Geobiological Laboratory, P.O. Box 378, Canberra City, A.C.T. 2601, Australia

D.J. Swaine

C.S.I.R.O., Fuel Geoscience Unit, P.O. Box 136, North Ryde, N.S.W. 2113, Australia

ELSEVIER SCIENTIFIC PUBLISHING COMPANY

Amsterdam — Oxford — New York — 1979

ELSEVIER SCIENTIFIC PUBLISHING COMPANY
335 Jan van Galenstraat
P.O. Box 211, 1000 AE Amsterdam, The Netherlands

Distributors for the United States and Canada:

ELSEVIER/NORTH-HOLLAND INC.
52, Vanderbilt Avenue
New York, N.Y. 10017

Library of Congress Cataloging in Publication Data
Main entry under title:

Biogeochemical cycling of mineral-forming elements.

(Studies in environmental science ; v. 3)
Includes bibliographical references and index.
1. Mineral cycle (Biogeochemistry)
I. Trudinger, P. A. II. Swaine, D. J.
III. Series.
QH344.B56 574.5'2 78-21297
ISBN 0-444-41745-1

ISBN 0-444-41745-1 (Vol. 3)
ISBN 0-444-41696-X (Series)

Printed in The Netherlands

CONTENTS

PREFACE

The term *mineral* in the title of this book is defined as "a homogeneous, naturally-occurring phase, ... restricted to inorganic crystalline phases" (Glossary of Geology and Related Sciences, American Geological Institute, 2nd edn, 1960, p. 186).

Minerals, so defined, dominate the world around us. They make up the bulk of the earth's crust and the skeletal structures of organisms, and they are used extensively by Man in his industrial, agricultural, artistic and cultural activities.

The elements from which minerals are formed undergo continual cycling within the environment. The cycles are influenced by a variety of factors not the least of which are, in many instances, biological in character.

It is the purpose of this book to review current knowledge of the major biological processes which are involved in these geochemical cycles and which influence, directly or indirectly, the formation, dissolution and transformation of minerals.

Chapter 1 outlines some general aspects of the biogeochemical cycling of elements. The chapters in sections 2—8 relate to specific classes of minerals selected on the basis of their quantitative or economic significance and the extent to which biogeochemical data are available. The last two chapters make recognition of Man as an organism which is making a profound impact on the mineral status of the earth. Chapter 9 deals with the use of minerals in agriculture and chapter 10 provides an insight into the future consequences of mineral utilization.

The book is not intended to provide a complete coverage of the biogeochemistry of minerals and the choice of topics largely reflects the editors' interests. We do regret, however, being unable to find authors to discuss carbonate deposition by protozoa and biological silicification to complete the sections carbon and silicon, respectively.

Since, as discussed in Chapter 1, biogeochemical cycles are interlinked, there is inevitably a degree of overlap between the subjects discussed in this book. Straight duplication has been avoided as much as possible but differing viewpoints on particular topics have been included to provide the non-specialist reader with an appreciation of the complexities surrounding hypotheses which are often not amenable to rigorous scientific proof.

Many colleagues, too numerous to mention specifically, have assisted in various ways in the planning and preparation of this book. Our particular thanks, however, must go to:

Mrs. Shirley Driessen, Miss Winnie Wong and Mrs. Robyn Raison who bore the brunt of secretarial and typing work associated with the editing,

the publishers, Sigma Xi, and Professor B.J. Skinner for permission to reproduce the article in Chapter 10,

all the authors for their time, thought and efforts, and

our publishers, Elsevier, for their patience during the book's lengthy gestation period.

P.A.T.
D.J.S.

Chapter 1

BIOGEOCHEMICAL CYCLING OF ELEMENTS — GENERAL CONSIDERATIONS

P.A. TRUDINGER [1], D.J. SWAINE [2] and G.W. SKYRING [1]

[1] *Baas Becking Geobiological Laboratory, P.O. Box 378, Canberra City, A.C.T. 2601 (Australia)*
[2] *C.S.I.R.O., Fuel Geoscience Unit, P.O. Box 136, North Ryde, N.S.W. (Australia)*

CONTENTS

INTRODUCTION

The concept of geochemical cycles is fundamental to a proper understanding of the status of an element whether it be solid, liquid or gas (Garrels et al., 1975). Changes in the state of an element depend on chemical and biological factors, and living matter is an important stage in the cycle of most elements (Ehrlich et al., 1977). A realistic appraisal of the role of an element and of the relevance of its place in a particular part of the geochemical cycle depends on the fact that the system is *dynamic*, not static. Hence, the simple statement of the total content of an element in a soil or water is but the starting point, and must be seen in the context of the cycle and the factors that may modify the value and change the form of the element. Not only is it necessary to ascertain changes in the form and amount of an element at the various stages of the cycle, but it is also necessary to find out how changes occur and the relevant reaction rates. Geochemical cycles are natural phenomena, but agricultural and industrial activities may modify and influence some stages of the cycles of certain elements. This may mean increases or decreases in the amount of the element at some stages of its cycle. Pollution should be seen as something imposed on the natural background. These consequences of human activity can be viewed as particular examples of the wide-ranging influences of the biosphere on the geochemical transformations of elements which are covered by the term "biogeochemical cycling".

"The chemical elements, including all the essential elements of protoplasm, tend to circulate in the biosphere in characteristic paths from environment to organisms and back to the environment. These more or less circular paths are known as *biogeochemical cycles*" (Odum, 1971, p. 86).

Odum also distinguished two basic groups of biogeochemical cycle:

(1) gaseous types in which the main element reservoir is the atmosphere or hydrosphere and

(2) sedimentary types in which the main reservoir is the earth's crust.

The *reservoir* is here defined as the "large, slow, moving, generally non-biological component" of the earth as distinct from the *cycling pool* which exchanges "rapidly between organisms and their immediate environments". There is, of course, not necessarily a clear-cut distinction between the two groups and many biogeochemical cycles involve all three reservoirs.

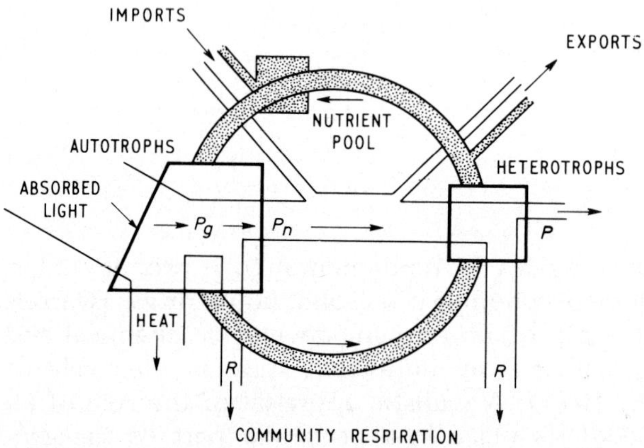

Fig. 1.1. The integration of a biogeochemical cycle (stippled) with an energy-yielding circuit shown in a simplified diagrammatic form. Note the contrast between the cycling of material and the one-way flow of energy. Pg = gross production, Pn = net primary production, which may be consumed within the system by heterotrophs or exported from the system, P = secondary production, R = respiration. (Reproduced from Odum, 1971, with permission of W.B. Saunders Co., Philadelphia).

A biogeochemical cycle is, overall, an endergonic process which relies ultimately on solar energy. This is illustrated in Fig. 1.1 where a generalized biogeochemical cycle (shaded area) is superimposed on a simplified one-way, energy-flow diagram. The nutrient pool is the reservoir(s) from which the cycling elements are derived. The reservoir also provides a sink for the products of biogeochemical reactions which become, in the short-term, unidirectional.

All organisms are constructed from elements and it follows, therefore, that all organisms are involved in element cycling. As will be obvious from the discussions in this book, however, many of the biogeochemical processes of significance in mineral turnover are the preserve of microorganisms. There are a number of reasons why this should be so:

(1) microorganisms make up the bulk of the mass of the biosphere and their rates of growth are generally several orders of magnitude greater than those of higher organisms,

(2) the microbial world embraces a wider range of environments than the plant and animal spheres,

(3) microorganisms carry out many unique reactions of geochemical significance, and

(4) the period over which microorganisms have colonized earthly environments is 4-5 times that occupied by higher organisms.

CONTENTS OF TRACE ELEMENTS IN SOME EARTH MATERIALS

The contents of some trace elements in the continental crust, shales, soils, bituminous coals and plankton are given in Table 1.1 to provide some perspective when considering other aspects of these elements. In each of these situations, organic matter is associated with the elements to a greater or a lesser degree. This is not usually very marked with crustal rocks except shales, but may be a major factor for some elements in surface soils and coals. The data in Table 1.1 show that, for some elements, e.g. beryllium, cadmium, cobalt and molybdenum, the contents of the various reservoirs are similar, while for others, there may be enrichments relative to the crust, e.g. boron and sulfur in many shales, soils and coals, mercury, nickel and selenium in many shales, and germanium in some coals.

There is a good deal of information on the inorganic forms of several elements in many rocks, soils and coals, but much remains to be done on the organic associations of trace elements. For example copper, lead and zinc are associated with humic acids, probably through carboxyl or phenolic groups (Saxby, 1969; Nissenbaum and Swaine, 1976). Vanadium porphyrins occur in petroleum (Davis, 1967), but the form of vanadium in coal has not been established. In most shales and coals, trace elements probably occur partly inorganically and partly organically bound.

TABLE 1.1.

Contents of trace elements (values as μg element g^{-1} of dry material)

	Crust [a]	Shale [b]	Soil [c]	Coal [d]	Plankton [e]
As	(1.8)	13	1—50	<1—13	—
Ba	700	580	100—3000	<40—500	17—323
Be	(2.8)	3	up t0 10	<0.4—5	—
B	(10)	100	2—100	4—200	—
Cd	(0.2)	0.3	up to 0.5	up to 0.2	0.4—6.4
Cl	(130)	180	mean 100	<100—400	—
Cr	35	90	5—1000	<1.5—20	(3.5)
Co	10	19	1—40	<0.6—2.0	(5)
Cu	25	45	2—100	6—30	6.5—57.5
F	625	740	mean 200	50—500	—
Ga	15	19	up to 60	1.5—10	(1.5)
Ge	1.5	1.6	up to 5	1—30	—
Pb	15	20	2—200	2—40	2.1—31
Mn	(950)	850	200—3000	4—600	3.6—15.0
Hg	(0.08)	0.4	up to 1	<0.03—0.2	0.08—0.71
Mo	(1.5)	2.6	0.2—5	0.3—4	(1)
Ni	13	68	5—500	3—50	2.0—11.6
P	(1050)	700	mean 650	40—1000	(4250)
Sc	10	13	up to 20	1—10	—
Se	(0.05)	0.6	0.1—2	0.3—1.6	—
Ag	(0.07)	0.07	up to 5	<0.2—0.6	<0.1—0.26
Sr	350	300	50—1000	<20—400	125—8900
S	(260)	2400	mean 700	Mostly up to 13000	<u>6000</u>
Th	10.5	12	up to 10	<0.2—6	—
Sn	(2)	6	up to 10	<0.9—7	(35)
Ti	3600	4600	1000—10000	500—3000	(80)
U	2.5	3.7	1—6	<0.4—3.8	—
V	60	130	20—500	10—60	(5)
Zn	52	95	10—300	<15—200	69—970
Zr	240	160	60—2000	15—300	(20)

[a] Mean values for Upper Post-Archean Continental Crust (Taylor, 1979); values in parentheses are from Taylor (1964).
[b] Mean values (Turekian and Wedepohl, 1961).
[c] Ranges for normal soils (Swaine, 1955); mean values are from Bowen (1966).
[d] Ranges for 90% of values for bituminous coals from New South Wales and Queensland, Australia (Swaine, 1977).
[e] Median values for 7 groups of plankton from the coast of Oregon, USA (Martin and Knauer, 1973); values in parentheses (Vinogrodova and Kovalsky, 1962) and underlined (Vinogradov, 1953) are mainly for diatoms.

BIOGEOCHEMICAL PROCESSES

The chemical environment may be influenced by organisms in several ways which may be arbitrarily classified under two general headings:

Primary processes — those involving specific metabolic reactions which are generally, but not always, essential for the organism's growth and activity, and

Secondary processes — those which are brought about as the result of biological activities but which are not an integral part of the organism's physiology.

Primary processes

Two of the most important primary biogeochemical processes are the accumulation of elements into the cellular structure of organisms and metabolic changes in the oxidation state of elements leading to the accumulation of oxidized or reduced forms of elements in the external milieu.

A third process, biomethylation, has been the subject of interest in recent years in view of its implications with respect to the entry of toxic metals into the food chain. It is also a potentially important reaction for mobilizing and transporting certain elements through the environment.

Accumulation of elements. The following chemical elements are known, or suspected, to be physiologically essential in the biosphere *: Al, B, Ba(?), Br(?), C, Ca, Cl, Co, Cr, Cu, F(?), Fe, H, I, K, Mg, Mn, Mo, N, Na, O, P, Rb(?), S, Se, Si, Sr(?), V, Zn. Many of these elements exist in trace amounts in the lithosphere and hydrosphere and are concentrated by organisms as illustrated in Table 1.2 which lists some enrichment factors exhibited by marine organisms: the data are from Bowen (1966) who also gives a detailed account of the element requirements for different classes of organisms and of the physiological functions of elements. In general, there are examples of specific biological concentration effects in all organisms from bacteria to higher plants and animals. Sometimes the products of element accumulation are minerals of which some examples are given in Table 1.3 and many more described in other chapters.

It is evident from Tables 1.1 and 1.2 that the biosphere is selective with respect to element accumulation and it may thus play a significant role in determining the chemical characteristics of the environment. This selectivity is particularly pronounced in so-called "accumulator organisms" which are characterized by an unusual ability to absorb and accumulate large amounts of specific elements (Table 1.4). In some instances the abnormal accumulation is associated with a specific physiological function — e.g., Ca and Si with calcareous and siliceous organisms and Cu with organisms containing copper-protein respiratory molecules — but in others the reasons for the elevated element levels are obscure.

* There are some reports that Sn, Ni and As are essential in human nutrition (see Wolf, 1978).

TABLE 1.2

Typical enrichments of elements in marine plants (Calculated from data of Bowen, 1966)

A Essential elements		B Non-essential elements	
Element	Enrichment factor (relative to sea water)	Element	Enrichment factor (relative to sea water)
Al	6,000	Ag	830
B	24	As	10,000
C	12,300	Au	1,200
Ca	25—750	Ba	1,000
Co	2,300	Be	1,700
Cu	3,700	Bi	3,530
Fe	70,000	Cd	4,000
I	500—25,000	Cr	20,000
K	140	Cs	1,400
Mg	4	F	3,5
Mn	26,500	Ga	17,000
Mo	45	Hg	1,000
Na	3	Li	170
P	50,000	Ni	600
Rb	70	Pb	267,000
S	14	Sn	330
Se	8,900	Ti	12,000—80,000
Si	500—6,700	W	35
Sr	33—175		
V	1,000		
Zn	15,000		

TABLE 1.3

Inorganic constituents reported in hard parts of marine organisms (from Lowenstam, 1974)

Carbonates	Calcite Aragonite Vaterite Monohydrocalcite Amorphous	Iron oxides	Magnetite Goethite Lepidocrocite Amorphous hydrates
Phosphates	Dahllite Francolite Amorphous calcium phosphate hydrogel Amorphous ferric phosphate hydrogel	Sulfates	Celestite Barite Gypsum
		Halides	Fluorite
		Oxalates	Weddellite Whewellite
Silica	Opal		

The apparent accumulation of yttrium by ferns is also unusual in that no physiological function is known for this element. Nevertheless, it is now clear that element concentration by organisms is not restricted to essential elements (Table 1.2B). Despite the many large enrichment factors the actual concentrations of the listed non-essential elements are generally in the ng g^{-1} to $\mu g\ g^{-1}$ range. Their accumulation may be due to non-specific combination with complexing and other sites in the organism. In this respect, it is of interest that the cellular concentrations of many elements reflect the relative amounts of these elements in the environment, a fact which provides the basis for biogeochemical prospecting where geochemical anomalies are sought by analysis of the elemental composition of plants (Brooks, 1972). This subject is complex, however, since the concentrations of elements in plants, or parts of plants depend on several factors including species, pH and drainage of soils.

Oxidations and reductions. The synthesis of the overall biosphere is largely a reduction of CO_2, water, sulfate and nitrate (or N_2) to provide the major element components (C, H, O, N, S) of living organic matter. Oxidized forms of these elements are regenerated by subsequent respiratory, fermentative and other degradative processes.

Aside from this general organic oxidation-reduction cycle, a number of electron-transfer reactions, catalysed mainly by microorganisms, lead to

TABLE 1.4

Examples of accumulator organisms (from Bowen, 1966)

Element	Accumulator organisms
Al	Club-mosses or *Lycophyta*
As	Brown algae, coelenterates
B	Brown algae, sponges
Br	Brown algae, sponges, coelenterates, molluscs
Ca	Protozoa, sponges, coelenterates, echinoderms, molluscs, vertebrates
Cl	Soft coelenterates
Cu	Annelids, arthropods, most molluscs
F	Vertebrates (in skeletons)
Fe	Some bacteria and plankton, horsetails
I	Diatoms, brown algae, sponges, coelenterates, marine annelids
Mn	Ferns (?), marine crustacea
Na	Soft coelenterates
Si	Horsetails, diatoms, some protozoa and sponges
Sr	Brown algae (in preference to Ca)
V	Some ascidians
Y	Ferns
Zn	Coelenterates

Note: All organisms are accumulators of C, H, N, P, S.

TABLE 1.5

Some biological oxidations and reductions of inorganic compounds (from Silverman and Ehrlich, 1964, unless otherwise stated)

Element	Known or presumed reaction	Relevant chapter in this book
	Oxidations	
As	$AsO_3^{3-} + 0.5\ O_2 \rightarrow AsO_4^{3-}$	—
Fe	$Fe^{2+} - e^- \rightarrow Fe^{3+}$	4
I [a]	$2\ I^- - 2\ e^- \rightarrow I_2$	—
Mn	$Mn^{2+} + 0.5\ O_2 + 2\ OH^- \rightarrow MnO_2 + H_2O$	5
N	see p. 12	—
P	$HPO_3^{2-} + H_2O - 2\ e^- \rightarrow HPO_4^{2-} + 2\ H^+$	3.1
S	$H_2S + 0.5\ O_2 \rightarrow S^0 + H_2O$	6.1, 6.2, 6.4
	$H_2S + 2\ O_2 \rightarrow SO_4^{2-} + 2\ H^+$	
	$S^0 + H_2O + 1.5\ O_2 \rightarrow SO_4^{2-} + 2\ H^+$	
Se	see Table 1.8	—
	Reductions	
As	$AsO_4^{3-} + 2\ H^+ + 2\ e^- \rightarrow AsO_2^- + 2\ OH^-$	—
Cu	$Cu(OH)_2 + H^+ + e^- \rightarrow Cu(OH) + H_2O$	—
Fe	$Fe(OH)_3 + H^+ + e^- \rightarrow Fe(OH)_2 + H_2O$	4
Hg [b]	$Hg^0 + 2\ d^- \rightarrow Hg^{2-}$	—
I [c]	$IO_3^- + 6\ H^+ + 6\ e^- \rightarrow I^- + 3\ H_2O$	—
Mo	$Mo_6O_{21}^{6-} + 6\ H^+ + 3\ e^- \rightarrow 6\ MoO_{2.5} + 6\ OH^-$	—
Mn	$MnO_2 + 2\ H^+ + 2\ e^- \rightarrow Mn^{2+} + 2\ OH^-$	5
N	see p. 12	
P	$HPO_4^{2-} + 2\ H^+ + 2\ e^- \rightarrow HPO_3^{2-} + H_2O$	3.1
S	$SO_4^{2-} + 10\ H^+ + 8\ e^- \rightarrow H_2S + 4\ H_2O$	6.1, 6.3
Se	see Table 1.8	
U	$UO_2(OH)_2 + 2\ H^+ + 2\ e^- \rightarrow U(OH)_4$	8
V	$VO_4^{3-} + 3\ H^+ + e^- \rightarrow VO(OH)_2 + OH^-$	—

[a] Gozlan and Margalith, 1973, 1974.
[b] Tsunogai and Sase, 1969.
[c] Summers and Sugarman, 1974.

changes in the oxidation state of elements in the extracellular environment (Table 1.5). Some of the reactions are physiologically significant — oxidations of Fe^{2+} and reduced sulfur compounds provide energy for growth, while sulfate and nitrate are electron acceptors for anaerobic oxidative processes. Other reactions may be fortuitous side reactions of physiological processes, e.g., the metabolism of selenium appears to parallel that of sulfur and both may involve the same enzymes (Shrift, 1973). It is also not yet proven that reductions of compounds of As, Cu, Fe, Mo, Mn, U and V are other than non-specific consequences of the reducing activities of organisms. Several of the reactions listed in Table 1.5 provide the basis of discussions in other chapters throughout the book.

Biomethylation. Methylation and methyl transfer are important reactions in the organic metabolism of organisms (Mudd, 1973) and it has been known for some time that dimethyl selenide, dimethyl telluride and di- and trimethyl arsine are biosynthetic products of the metabolism of inorganic compounds selenium, tellurium and arsenic, respectively, by microorganisms (Challenger, 1951).

Biomethylation of metal ions is a relatively new discovery. It appears to be a means by which some organisms cope with toxic elements although the products may be inhibitory to other organisms. Methylation of mercuric ions in sediments was demonstrated by Jensen and Jernelöv (1969). Hg^{2+} is converted to methyl mercury (CH_3Hg^+) by *Clostridium cochlearium* (Yamanada and Tonomura, 1972a, b), and to both methyl mercury and dimethyl mercury (CH_3-Hg-CH_3) by extracts of a methanogenic bacterium (Wood et al., 1968). The biologically-mediated methylation of mercury has now been confirmed in several laboratories (see reviews by Alexander, 1974, Jernelöv and Martin, 1975) and appears to involve both enzymic processes and chemical reactions between Hg^{2+} and methyl corrinoids. The latter are biological methylating agents in organisms (Wood, 1974). Decomposition of organic mercury compounds has been demonstrated in sediments (Spangler et al., 1973; Jernelöv and Martin, 1975) and by bacteria (Furakawa and Tonomura, 1972a, b). The product is metallic mercury (Hg^0) and the reaction appears to supply a mechanism for mercury resistance in bacteria (Summers and Sugarman, 1974).

Although most research has been carried out with mercury, the possibility of a wider role for biomethylation of metals is suggested by reports of microbial formation of tetramethyl lead [$(CH_3)_4Pb$] in a number of lake sediments (Wong et al., 1975) and of methyl tin (CH_3Sn) by *Pseudomonas* sp. (Jernelöv and Martin, 1975).

Secondary processes

Secondary biogeochemical processes arise as the result of non-specific modifications of the physical and chemical characteristics of the environment due to metabolic and other activities of organisms.

Some of the more important chemical modifications, many examples of which are noted in other chapters, are:

(1) Production and consumption of acids and bases (e.g. organic acids, CO_2, OH^-, NH_4^+) which may have profound effects on environmental pH and alkalinity.

(2) Production and consumption of O_2 which control the oxygen fugacity and Eh of the milieu.

Since the stability of minerals is markedly influenced by Eh and pH (Garrels and Christ, 1965), the potential importance of these two secondary processes in the formation and dissolution of minerals is clear.

A third type of secondary process results from the liberation, during life and after death, of organic compounds which play important roles in metal transport, mobilization and fixation (e.g., Manskaya and Drosdova, 1968; Saxby, 1969; 1976), in the formation of petroleum (Gallegos, 1976) and coal (Given et al., 1973), and in providing the energy-yielding substrates for other primary biogeochemical processes.

Physical effects of organisms may be due to secondary chemical processes (e.g. weathering) or may result from their direct involvement with the environment. The upper regions of sediments (the zones of bioturbation) are frequently disturbed by the burrowing of benthic organisms and fossil tracks have been recognized in ancient rocks (Crimes and Harper, 1970). The boring and abrasive activities of algae are discussed in Chapter 2.4.

Plants are instrumental in consolidating terrestrial surfaces and, by the same token, the removal of plant cover leads to an acceleration of weathering and erosion. Motile organisms transport elements from one place to another. One of the most potent biogeochemical agents affecting the physical structure of the environment is Man himself who, in the space of a few centuries, has produced more change than other agents have accomplished over millions of years.

INTEGRATION OF BIOGEOCHEMICAL PROCESSES

A biogeochemical cycle results from the integration of several biological, chemical and, in many cases, physical processes. In this section, we discuss the *nitrogen cycle* to illustrate the ways in which organisms cooperate to effect the cyclic turnover of an element, and the *selenium cycle* to demonstrate the interaction between the biosphere and geosphere.

Nitrogen cycle

Nitrogen is a major component of living cells and the biosphere contains about 14 Pg N (Table 1.6). Although this amount is small compared with

TABLE 1.6

Global inventories of nitrogen in Pg (from Söderlund and Svensson, 1976)

Biospheric	
Terrestrial plants	11—14
Terrestrial animals	0.2
Terrestrial microorganisms	0.5
Oceanic plants	0.3
Oceanic animals	0.17
Total	12.17—15.17
Non-biospheric	
Litter	1.9—3.3
Soil (non-living organic-N + inorganic-N)	316
Rocks	1.9×10^8
Sediments	4×10^5
Coal	120
Oceans (non-living organic-N)	557
(inorganic: N_2, N_2O, NO_3^-, NO_2^-, NH_4^+)	2260
Atmosphere (N_2, N_2O, NH_3, NH_4^+, NO_x, NO_3^-, organic-N)	$>3.9 \times 10^6$
Total	$>1.9 \times 10^8$

the total crustal N, biospheric nitrogen is rapidly recycled, mainly through the agencies of microorganisms. A simplified version of the biological nitrogen cycle, which omits the contribution of nitrogenous gases from industry, is shown in Fig. 1.2.

Ammonium ions are central to the biological nitrogen cycle being the precursors of the organic-N constituents (amino acids, nucleic acids, etc.) of all living organisms. The main processes leading to the formation of ammonium ions are (1) the reduction of atmospheric N_2 (dinitrogen fixation)

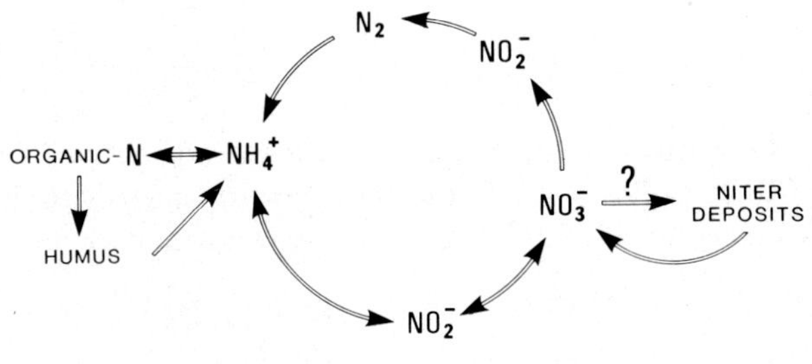

Fig. 1.2. A simplified version of the biological nitrogen cycle.

catalysed by cyanophytes * and many bacteria, and by the symbiotic consortia of rhizobia with legumes, and (2) the reduction of nitrate, with the intermediate formation of nitrite, catalysed by essentially all microorganisms and plants. Organic-N provides a secondary source of NH_4^+ which is released as the result of post mortem degradation of organic matter by autolytic enzymes and bacteria. Recycling of organic-N occurs more rapidly than that of organic-C since the C/N ratio of freshly formed organic matter is less than that of humus. Humus is also slowly degraded and the C/N ratio of the remaining organic material continues to increase until refractory kerogens are formed.

Atmospheric N_2 is regenerated from NH_4^+ by a series of biological oxidative and reductive reactions. *Nitrosomonas* spp. and *Nitrobacter* spp. oxidize NH_4^+ to NO_2^- and NO_2^- to NO_3^-, respectively, with molecular O_2 and the overall process is sometimes referred to as nitrification. Both *Nitrosomonas* and *Nitrobacter* are autotrophs which derive energy from their respective oxidations. The reduction of NO_3^- to N_2 (denitrification) is carried out by a range of anaerobic and facultative bacteria. These organisms utilize NO_3^- as an oxidant for the anaerobic oxidation of organic compounds. Many organisms carry out only the first step of denitrification, the reduction of NO_3^- to NO_2^-, a fact which might account for the significant amounts of nitrite (500 Tg N) present in oceanic waters (Söderlund and Svensson, 1976).

Niter deposits [including KNO_3, Chilean saltpeter, $NaNO_3$, and wall saltpeter, $Ca(NO_3)_2$] represent localized reservoirs of terrestrial nitrogen. Their entry into the biogeochemical cycle is catalysed by Man, but a role for organisms in their formation, while possible (Breger, 1911), has yet to be proven.

Many of the reactions given in Fig. 1.2 are difficult to quantify but recent estimates suggest that the present-day global interchange of biogenic N_2 between the atmosphere and lithosphere/hydrosphere is in the order of 170-270 Tg y^{-1} (Söderlund and Svensson, 1976). It should be emphasized that Fig. 1.2 shows only the biological components of the geochemical nitrogen cycle and gives no description of the movement of nitrogen between the various terrestrial reservoirs. A recent account of the overall biogeochemical nitrogen cycle is given by Söderlund and Svensson (1976).

Selenium cycle

The selenium contents of the important natural reservoirs are shown in Table 1.7. Although about 50 selenium minerals are known, selenium is usually found associated with heavy metal sulfides either as selenides or, to a

* Cyanophytes are commonly called blue-green algae but some consider that they should be classified as bacteria — the cyanobacteria (e.g. Stanier and Cohen-Bazire, 1977).

TABLE 1.7

Selenium contents (as μg Se g^{-1})

Igneous rocks (basalt, granite)	0.05	Taylor (1964)
Sandstones	0.05	Turekian and Wedepohl (1961)
Carbonates	0.08	Turekian and Wedepohl (1961)
Shales	0.6	Turekian and Wedepohl (1961)
Volcanic ash	0.1—2.6	Wells (1967)
Fumarole deposit, Iceland	265	Mroz and Zoller (1975)
Sediments — NW part of Pacific Ocean	0.1—1.7	Sokolova and Pilipchuk (1973)
Iron-manganese concretions	0.4—0.8	Sokolova and Pilipchuk (1973)
Soils	0.1—2; some up to 1200	Swaine (1955)
Plants	mostly 0.1—15; some up to 5000	Anderson et al. (1961)
Freshwaters	0.0001—0.4, mean 0.0002	Louderback (1975)
Sea water	0.00009	Bowen (1966)
Coal: U.S.A. (mostly Illinois Basin)	0.4—8.1, mean 2.3	Gluskoter et al. (1977)
Australia	0.21—2.5, mean 0.79	Porritt and Swaine (1976)
Crude oil various	0.01—1.1	Filby and Shah (1975)
Phosphate rocks	up to 95, mostly <20	Swaine (1962)
Superphosphate	up to 25, mostly <4	Swaine (1962)

limited extent, as a substituent for S in crystal lattice S of sulfides. Basaltic and granitic rocks, the major constituents of the continental crust, are low in selenium with a mean content of 0.05 μg Se g^{-1} compared with 0.6 μg Se g^{-1} for shales. The higher contents in shales may be partly the result of biological concentration. Biological factors may also contribute to the high Se contents of deep-sea sediments relative to sandstones and carbonates. Deep-sea sediments are relatively rich in iron-manganese concretions and there is some evidence that the formation of these concretions is related to microbial activity (see Chapters 4 and 5). Unlike sulfur, selenium is retained in marine sediments, mainly by adsorption. Selenium is emitted from volcanoes in gaseous form and volcanic ash and tuffaceous volcanic material are often high in selenium.

In sediments and soils, the chemistry of selenium differs from that of sulfur in that the stability of selenite is similar to that of sulfate. The reduction of selenite to elemental selenium, which tends to immobilise selenium in soils and water, is an important process in the natural environment. Selenates are only stable under alkaline oxidising conditions and have been found, for example, in the Chilean nitrate deposits.

The cycling of selenium between major reservoirs is shown in Fig. 1.3. It is assumed that volcanic emanations and igneous and metamorphic rocks are

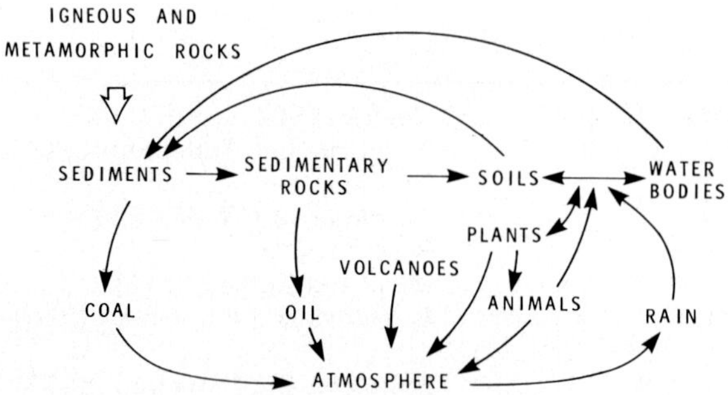

Fig. 1.3. The geochemical cycle of selenium.

the ultimate sources of selenium which then circulates through the more mobile hydrospheric, lithospheric and atmospheric reservoirs. Biochemical reactions which operate within this cycle are listed in Table 1.8. Most of our information on these reactions relate to terrestrial biota but it is probable that similar reactions take place in oceanic waters and sediments.

Selenium is essential in animal, including human nutrition (Allaway, 1968; Gunn et al., 1976) although its precise physiological function is not known

TABLE 1.8

Biological reactions involved in the selenium cycle (Silverman and Ehrlich, 1964; Shrift, 1964, 1973; Martin, 1973)

Reaction	Organisms
$SeO_4^{2-} \rightarrow SeO_3^{2-}$	?
$SeO_3^{2-} \rightarrow Se^0$	Bacteria
$Se^0 \rightarrow H_2Se$	Bacteria
$SeO_4^{2-} \rightarrow$ Organic-Se	Plants, bacteria, fungi
$SeO_3^{2-} \rightarrow$ Organic-Se	Animals, plants, fungi, bacteria
SeO_4^{2-}, SeO_3^{2-}, Se^0, Organic-Se $\rightarrow Se(CH_3)_2$	Bacteria, fungi, plants, animals
$Se^0 \rightarrow SeO_4^{2-}$	Bacteria
$H_2Se \rightarrow SeO_4^{2-}$	Bacteria
Organic-Se $\rightarrow SeO_4^{2-}$	?

(Scott, 1973). However, since the dietary requirement is in the order of 1 μg g^{-1} or less (Scott, 1973), utilization of Se by animals is probably of minimal geochemical significance. The major known biological reservoir for selenium is plant material and it has been estimated that the annual uptake of this element by grasses and crop plants is in the same order as the annual production of selenium by mining. Certain plants, such as species of *Astragalus*, *Xylorhiza*, *Stanleya* and *Oonopsis*, are selenium accumulators and their consumption can lead to chronic selenosis (e.g., blind staggers) in animals (Martin, 1973; see also Chapter 9). Accumulator plants differ from non-accumulator plants in their organic Se metabolism (Shrift, 1973); in particular much of the total selenium in accumulator plants is non-proteinaceous in contrast to that in non-accumulators (Peterson and Butler, 1967).

Despite the clear-cut demonstrations of selenium accumulation by plants and metabolism by both plants, fungi and bacteria (Table 1.8), evidence for a nutritional selenium requirement by these groups of organisms is equivocal (Shrift, 1973). It is possible, however, that formation of organic-selenium compounds may be a mechanism for counteracting selenium toxicity (Doran and Alexander, 1977).

A biogeochemical link between the hydrosphere/lithosphere and the atmosphere is provide by the formation of volatile, methylated derivatives of selenium. The production of dimethyl selenide, $(CH_3)_2Se$, has been reported for fungi (Challenger, 1935, 1951), a soil bacterium (Doran and Alexander, 1977) and animals (Shrift, 1973), while volatile selenium compounds, including dimethyl diselenide, $(CH_3)_2Se_2$, can be liberated by plants (Lewis, 1976). In lake sediments, microorganisms produce volatile dimethyl selenide, dimethyl diselenide and an unknown selenium compound (Chau et al., 1976) and the evolution of $(CH_3)_2Se$ from soils was reported by Francis et al. (1974). A major unresolved question is the means by which reduced organic selenium compounds are reoxidized. Bacteria capable of utilizing $(CH_3)_2Se$ and $(CH_3)_2Se_2$ as sole carbon sources have been isolated from soil (Doran and Alexander, 1977) but the fate of Se was not determined.

Secondary biogeochemical processes play an important role in selenium cycling. During weathering and the formation of soils, there are often increases in selenium, and in some cases, the decay of accumulator plants builds up high selenium concentrations in surface soils. Much selenium is retained in soils where it occurs as selenite, elemental selenium and organically combined selenium, and is adsorbed on clays and iron and manganese oxides. There is, however, some removal by leaching, which is a source of selenium for rivers and lakes and ultimately for the sea. Selenium may be present in water in true solution, adsorbed on particulate matter or complexed with organic matter. Humic material is the main form of dissolved organic matter in water, probably because of preferential utilization of polysaccharides and other biopolymers by microorganisms (Reuter and Perdue, 1977). Chemical changes in the movement of selenium in the soil-plant-animal cycle have been reviewed by Allaway, (1973).

Man contributes to the Se cycle through his industrial activity and his use of fertilisers (see Chapter 9). During the combustion of coal and oil (Fig. 1.3) some selenium is released to the atmosphere, dispersed and returned to the earth mainly in rain. The full extent of Man's influence has, however, yet to be evaluated.

INTERDEPENDENCE OF BIOGEOCHEMICAL CYCLES

Since a large number of elements are essential for all, or at least some, of the components of the biosphere (see p. 6), it is obvious that the biogeochemical cycles of these elements will be interdependent. This is illustrated at a biochemical level by Fig. 1.4. Photosynthesis results in the evolution of O_2 and the incorporation of CO_2 into the organic matter of living cells which, at the same time, incorporate nitrogen, sulfur and phosphorus. Organic matter and O_2 are used to drive independent cycles of sulfur, nitrogen and carbon each of which requires the participation of phosphorus. The latter three cycles also regenerate carbon, sulfur and nitrogen in the form required for the initial photosynthetic cell production. The five element cycles are thus clearly interdependent and any change in one cycle will in the long term have a profound influence on the operation of the other four.

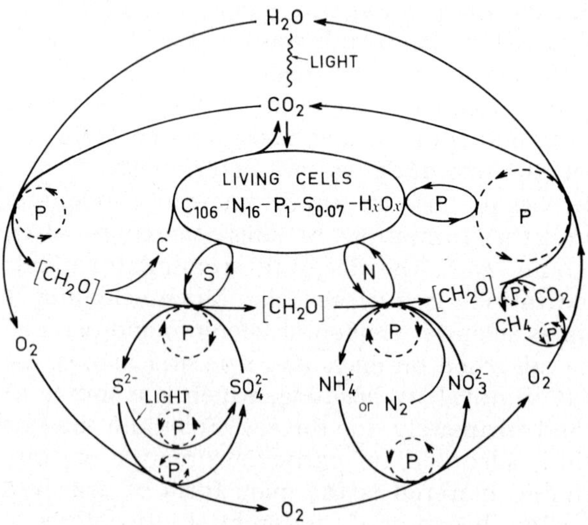

Fig. 1.4. Interrelations between the cycles of carbon, sulfur, phosphorus, nitrogen and oxygen.

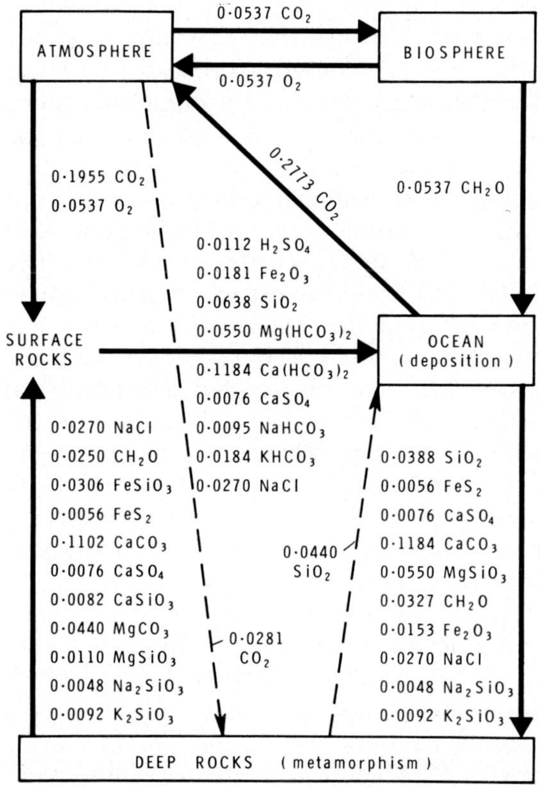

Fig. 1.5. A steady-state cycle involving organic carbon and minerals of calcium, magnesium, iron, sodium and potassium. All material fluxes are in units at 0.1 P mol y^{-1}. Note the large number of processes involved that have the same order of magnitude of importance. (Reproduced from Garrels and Perry, 1974, with permission of John Wiley and Sons, Inc., New York).

Figure 1.5 extends this notion to the geochemical level which shows an estimate of the influence of the global biosphere carbon and oxygen cycles on the fluxes of major elements through the terrestrial reservoirs, and includes the effects of both primary and secondary biogeochemical processes.

Both Figs. 1.4 and 1.5 also serve to highlight, as foreshadowed in Fig. 1.1, the dominant role of photosynthesis in controlling biogeochemical cycles.

Biogeochemical successions

The foregoing discussions have dealt with general aspects of the long-term, global cycling of elements. The various stages in a biogeochemical cycle are, however, separated in space and/or time and, at any particular time, one process may dominate in any particular environment and may, later, be succeeded by another. This situation arises because the environment is in a continual state of change caused not only by geological processes (climatic changes, crustal movements, weathering, etc.) but also by organisms themselves which, through consumption of nutrients, excretion of products and other activities, create the conditions necessary for the development of other groups of organisms with different biogeochemical characteristics.

The history of biochemical evolution provides a notable example of biogeochemical succession. It is now widely believed that primitive organisms were anaerobes which fermented prebiotic organic matter formed in a primordial oxygen-free environment (Oparin, 1961). The geochemical effects of such organisms would have been limited in kind — to production of organic acids and CO_2 — and extent, due to the limited availability of organic matter. Further development of living organisms depended on the emergence of photosynthesis, the means of trapping unlimited amounts of radiant energy to support synthetic reactions. Early photosynthesis may have been linked to anaerobic oxidation of sulfides (Peck, 1966) or ferrous iron (Cloud, 1974), but the major event was undoubtedly the development of the oxygen-evolving photosynthetic procaryotes.

The formation of atmospheric oxygen has been the subject of numerous discussions (e.g. Berkner and Marshall, 1964; Brinkman, 1969; Van Valen, 1971; Walker, 1975; Schidlowski, 1975; Dimroth and Kimberly, 1976; Cloud, 1976). Abiological mechanisms have been proposed, but it is now widely held that photosynthetic organisms were responsible for the initial development of atmospheric oxygen. This set the stage for subsequent evolution of oxygen-linked respiration which not only completed the biological oxygen cycle but also added oxidative stages to the environmental transformations of carbon, sulfur, nitrogen, iron and many other elements. Indeed it is arguable that, aside from the origin of life, the development of photosynthesis is, geochemically, the most profound biological event which has yet taken place on earth. The geological period at which atmospheric oxygen reached significant levels, however, is still a matter of controversy.

In the modern environment, examples of regional biogeochemical successions are numerous. As an illustration, we may consider the consequences of a transition from oxygenic to anaerobic conditions. The capacity of a biotic system in a particular environment to oxidize photosynthetically-derived organic matter to CO_2 and water is a function of the supply of oxygen. In most terrestrial and marine environments, oxygen is plentiful and most organic matter is rapidly recycled. Oxygen consumption, however, may exceed that rate of its diffusion in, for example, water-logged soils and enclosed water bodies, particularly when an abundance of nutrients is available to support the vigorous growth of photosynthetic organisms. In these situations anaerobic reactions such as fermentation, and the reductions of nitrate, sulfate and CO_2 become predominant (Claypool and Kaplan, 1974; see also Fig. 2.2.1, p. 50).

Within the anaerobic environment itself, biogeochemical successions may develop. Several authors (Nissenbaum et al., 1972; Martens and Berner, 1974, 1977; Claypool and Kaplan, 1974; Whelan et al., 1978) have noted that, in anaerobic sediments, methane accumulates to the greatest extent below the zone of sulfate reduction. That this phenomenon represents a biogeochemical succession was shown by Cappenberg (1974a) who reported stratification of sulfate-reducing and methanogenic bacteria in the bottom deposits of the freshwater Lake Vechten in The Netherlands. In an investigation employing ^{14}C-labelled substrates and specific inhibitors, Cappenberg et al. (Cappenberg, 1974b; 1975; Cappenberg and Prins, 1974) demonstrated the biochemical basis for the succession of these organisms whereby the products of organic degradation by the sulfate reducers become the substrates for methanogenesis (Fig. 1.6). Although not shown directly, it was implicit that fermentation processes supplied the simple organic compounds and hydrogen required for sulfate reduction and methane formation.

The precise reasons for the spatial separation of sulfate reduction and methanogenesis are nevertheless not known with certainty. Nissenbaum et al. (1972) and Claypool and Kaplan (1974) suggested that methane generation and sulfate reduction are necessarily mutually exclusive metabolic processes and it has been shown that small amounts of sulfate (down to 0.2 mM) may inhibit methanogenesis in sediments (MacGregor and Keeney, 1973; Winfrey and Zeikus, 1977). On the other hand, Cappenberg (1975) demonstrated commensal growth of a sulfate-reducing bacterium and a

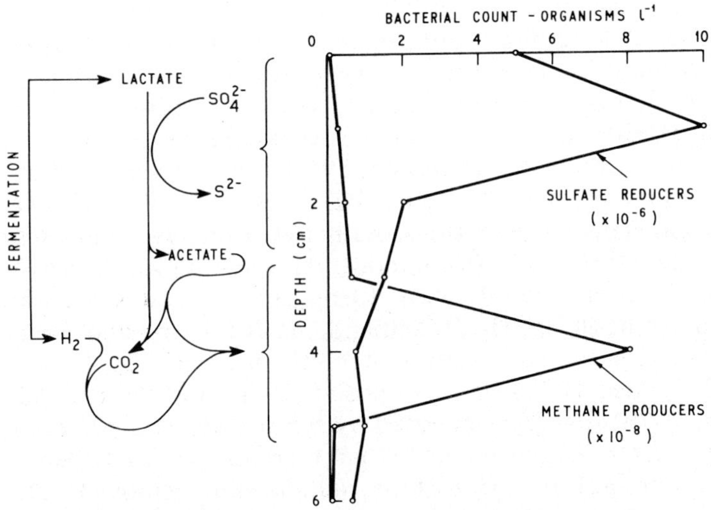

Fig. 1.6. Bacterial and biochemical stratification in sediments of Lake Vechten, The Netherlands. Data on bacterial populations from Cappenberg (1974a).

methanogenic bacterium in continuous cultures, and the simultaneous reduction of sulfate and methanogenesis in sediments incubated under H_2 was reported by Oremland and Taylor (1978). The co-existence of sulfate reducers and methane producers in sewage sludge digesters had been earlier recorded by Lawrence et al. (1964) and Lawrence and McCarty (1964).

In the Lake Vechten sediments referred to above, stratification of the two groups of organisms appeared to be due largely to the sensitivity of methane producers to H_2S which became inhibitory above about 0.1 mM (Cappenberg, 1975). In many sediments, however, H_2S does not accumulate to a significant extent due to its fixation as iron sulfides and its diffusion from the sediment. Winfrey and Zeikus (1977), for example, reported that, in Lake Mendota sediments, the amount of sulfide which could be added before free H_2S appeared in the pore waters was about 25 times the concentration of sulfate required to inhibit methanogenesis.

Several investigators (Reeburgh, 1976: Barnes and Goldberg, 1976; Martens and Berner, 1977; Reeburgh and Heggie, 1977; Whelan et al., 1978) have suggested that, in some environments, methane is produced throughout the biologically-active, anaerobic zones of sediments but is consumed in the sulfate-reducing zones. The reaction between methane and sulfate (eqn 1) is energetically favourable;

$$CH_4 + SO_4^{2-} + 2\ H^+ \rightarrow H_2S + CO_2 + 2\ H_2O\text{: } \Delta G^\circ = -95\ \text{kJ} \quad (1)$$

and there is a brief, but unconfirmed, report of the isolation of a methane-oxidizing, sulfate-reducing bacterium (Panganiban and Hanson, 1976). Nevertheless, most attempts to demonstrate methane utilization by sulfate-reducing organisms have been unsuccessful (Sorokin, 1957; Oremland and Taylor, 1978; see also Chapter 6.1) and further verification is required before the methane consumption hypothesis is acceptable.

The notion of competition between organisms for substrates was suggested by Stumm (1966) who argued that successions of biochemical energy-yielding processes would be governed by thermodynamics, those yielding large amounts of energy, e.g. $2\ CH_2O + SO_4^{2-} \rightarrow S^{2-} + 2\ CO_2 + 2\ H_2O$: $\Delta G^\circ = 920$ kJ, taking precedence over those with lower energy yields, e.g. $2\ CH_2O \rightarrow CH_4 + CO_2$: $\Delta G^\circ = -414$ kJ (Claypool and Kaplan, 1974). Observations which support the concept of competition for substrates are the stimulation of methanogenesis by H_2 (Oremland, 1975; Ferry and Peck, 1977) and the reversal of sulfate inhibition of methanogenesis by H_2 and acetate (Winfrey and Zeikus, 1977). It is proposed (Ferry and Peck, 1977, Oremland and Taylor, 1978) that H_2, generated by fermentation, is preferentially utilized during sulfate reduction and that methanogenesis becomes prominent when the competition of sulfate reduction is removed. The possibility that sulfate reducers themselves may supply H_2 for methanogenesis is suggested by the experiments of Bryant et al. (1977) who demonstrated commensal growth of *Desulfovibrio* sp. and an H_2-utilizing methanogenic bacterium at low concentration, or in the absence, of sulfate.

There are then several possible explanations for the stratification of sulfate reduction and methane accumulation in nature and indeed many factors may contribute to this phenomenon and different mechanisms may apply in different environments. One complicating factor is that methane is produced not only by CO_2 reduction but also by fermentations (Mechalas, 1974; Mah et al., 1977) and the conditions required for these reactions are not necessarily the same. Nevertheless, the stratification of methane and sulfide serves to further illustrate the close interdependence of organisms with different metabolic specialities.

CONCLUDING REMARKS

A biogeochemical cycle is a complex system of interacting biochemical, chemical and physical processes. The various cycles are interlinked so that one cycle may be influenced by processes which are mechanistically unconnected with that cycle. It also follows that all organisms have a direct or indirect influence on most geochemical element cycles while the biosphere itself is dependent on the chemical and physical properties of the environment (Trudinger and Bubela, 1967; Brock; 1966). It is not surprising, therefore, that for many element cycles we have qualitative information on the roles that organisms *may* play but little or no data on their quantitative significance. Recent attempts to partly redress this situation are the project on Biogeochemical Cycles initiated by the Scientific Committee on Protection of the Environment (SCOPE: Svensson and Söderlund, 1976; Stumm, 1977) and the International Symposia on Environmental Biogeochemistry (Nriagu, 1976; Krumbein, 1978).

Projection of biogeochemical cycling through geological time relies on interpretation of geological and palaeontological evidence which becomes increasingly limited beyond the Cambrian. Stable isotopes and the fossil evidence of microstructures and stromatolites provide some insight into the early biogeochemistry of, for example, carbon and sulfur, but for the majority of elements, there appears to be no geological evidence which can be readily equated with biological activity. In these instances, our notions of past biogeochemical events rely entirely on extrapolation from the observed processes occurring in the modern environment. These extrapolations, in turn, require assumptions to be made on the chemical and physical conditions which prevailed on earth in past geological eras.

Nevertheless, despite these limitations, it will be evident from the succeeding chapters that great advances have already been made even though most of the research spans only a few decades. Modern techniques and instrumentation such as satellite imagery, scanning electron microscopy, X-ray analysis, electron and ion probes and computer assisted analysis are providing powerful tools for geoscientists. It is to expected, therefore, that the next few decades will see major advances in our understanding of the geochemical role of the biosphere.

ACKNOWLEDGEMENTS

The Baas Becking Laboratory is supported by the Commonwealth Scientific and Industrial Research Organization, the Bureau of Mineral Resources and the Australian Mineral Industries Research Association, Ltd. We extend our appreciation to Mrs. V.M. Barhatov for the preparation of illustrations.

REFERENCES

Alexander, M., 1974. Microbial formation of environmental pollutants. Adv. Appl. Microbiol., 18: 1—73.

Allaway, W.H., 1968. Agronomic controls over the environmental cycling of trace elements. Adv. Agron., 20: 235—274.

Allaway, W.H., 1973. Selenium in the food chain. Cornell Vet., 63: 151—170.

Anderson, M.S., Lakin, H.W., Beeson, K.C., Smith, F.F. and Thacker, E., 1961. Selenium in Agriculture. Agriculture Handbook No. 200, Agricultural Research Service, United States Department of Agriculture, 65 pp.

Barnes, R.O. and Goldberg, E.D., 1976. Methane production and consumption in anoxic marine sediments. Geology, 4: 297—300.

Berkner, L.V. and Marshall, L.C., 1964. The history of growth of oxygen in the earth's atmosphere. In: P.J. Brancazio and A.G.W. Cameron (Editors), The Origin and Evolution of Atmospheres and Oceans. John Wiley, New York, NY, pp. 102—126.

Bowen, H.J.M., 1966. Trace Elements in Biochemistry. Academic, New York, NY, 241 pp.

Breger, C.L., 1911. Origin of some mineral deposits by bacteria. Min. Eng. World, 35: 289—291.

Brinkman, R.T., 1969. Dissociation of water vapor and evolution of oxygen in the terrestrial atmosphere. J. Geophys. Res., 74: 5355—5368.

Brock, T.D., 1966. Principles of Microbial Ecology. Prentice-Hall, New Jersey, 306 pp.

Brooks, R.R., 1972. Geobotany and Biogeochemistry in Mineral Exploration. Harper and Row, New York, NY, 290 pp.

Bryant, M.P., Campbell, L.L., Reddy, C.A. and Gabill, M.R., 1977. Growth of *Desulfovibrio* in lactate or ethanol media low in sulfate in association with H_2-utilizing methanogenic bacteria. Appl. Environ. Microbiol., 33: 1162—1169.

Cappenberg, T.E., 1974a. Interrelations between sulfate-reducing and methane-producing bacteria in bottom deposits of a fresh water lake. I. Field observations. Antonie van Leeuwenhoek; J. Microbiol. Serol., 40: 285—295.

Cappenberg, T.E., 1974b. The interrelations between sulfate-reducing and methane-producing bacteria in bottom sediments of a fresh water lake. II. Inhibition experiments. Antonie van Leeuwenhoek; J. Microbiol. Serol., 40: 297—306.

Cappenberg, T.E., 1975. A study of mixed continuous cultures of sulfate-reducing and methane-producing bacteria. Microb. Ecol., 2: 60—72.

Cappenberg, T.E. and Prins, R.A., 1974. Interrelations between sulfate-reducing bacteria and methane-producing bacteria in bottom deposits of a fresh water lake. III. Experiments with ^{14}C-labelled substrates. Antonie van Leeuwenhoek; J. Microbiol. Serol., 40: 457—469.

Challenger, F., 1935. The biological methylation of compounds of arsenic and selenium. Chem. Ind., 54: 657—662.

Challenger, F., 1951. Biological methylation. Adv. Enzymol., 12: 429—491.

Chau, Y.K., Wong, P.T.S., Silverberg, G.A., Luxon, P.L. and Bengert, G.A., 1976. Methylation of selenium in the aquatic environment. Science, 192: 1130—1131.

Claypool, G.E. and Kaplan, I.R., 1974. The origin and distribution of methane in marine sediments. In : I.R. Kaplan (Editor), Natural Gases in Marine Sediments. Plenum, New York, NY, pp. 99—139.

Cloud, P., 1974. Evolution of ecosystems. Am. Sci., 62: 54—66.

Cloud, P.E., 1976. Beginnings of biospheric evolution and their biogeochemical consequences. Paleobiology, 2: 351—387.

Crimes, T.P. and Harper, J.C. (Editors), 1970. Trace Fossils, Geological Journal Special Issue, No. 3, 547 pp.

Davis, J.B., 1967. Petroleum Microbiology. Elsevier, Amsterdam, 604 pp.

Dimroth, E. and Kimberley, M.M., 1976. Precambrian atmospheric oxygen: evidence in the sedimentary distributions of carbon, sulfur, uranium and iron. Can. J. Earth Sci., 13: 1161—1185.

Doran, J.W. and Alexander, M., 1977. Microbial transformation of selenium. Appl. Environ. Microbiol., 33: 31—37.

Ehrlich, P.R., Ehrlich, A.H. and Holdren, J.P., 1977. Ecoscience: Population, Resources, Environment. Freeman, San Francisco, CA, 1053 pp.

Ferry, J.G. and Peck, H.D., 1977. Relationship between sulfate reduction and methane production in salt marsh sediments. Abstracts of the Annual Meeting of the American Society for Microbiology, p. 236; N49.

Filby, R.H. and Shah, K.R., 1975. Neutron activation methods for trace elements in crude oils. In: T.F. Yen (Editor), The Role of Trace Metals in Petroleum. Ann Arbor Science, Ann Arbor, pp. 89—110.

Francis, A.J., Duxbury, J.M. and Alexander, M., 1974. Evolution of dimethylselenide from soils. Appl. Microbiol., 28: 248—250.

Furakawa, K. and Tonomura, K., 1972a. Metallic mercury releasing enzyme in mercury-resistant *Pseudomonas.* Agric. Biol. Chem., 36: 217—226.

Furakawa, K. and Tonomura, K., 1972b. Induction of a metallic mercury releasing enzyme in mercury-resistant *Pseudomonas.* Agric. Biol. Chem., 36: 2441—2448.

Gallegos, E.J., 1976. Biological fossil hydrocarbons in shales. Dev. Petroleum Sci., 5: 149—180.

Garrels, R.M. and Christ, C.L., 1965. Solution, Minerals, and Equilibria. Harper and Row, New York, NY, 450 pp.

Garrels, R.M. and Perry, E.A., 1974. Cycling of carbon, sulfur and oxygen through geologic time. In : E.D. Goldberg (Editor), The Sea, Vol. 5. Wiley-Interscience, New York, NY, pp. 303—336.

Garrels, R.M., Mackenzie, F.T. and Hunt, C., 1975. Chemical Cycles and the Global Environment. Kaufmann, Los Altos, CA, 206 pp.

Given, P.H., Casagrande, D.J., Imbalzano, J.F. and Lucas, A.J., 1973. Biochemical aspects of early stages of coal formation. Proceedings of Symposium on Hydrogeochemistry and Biogeochemistry (E. Ingerson, Editor), 2: 240—263.

Gluskoter, H.J., Ruch, R.R., Miller, W.G., Cahill, R.A., Dreher, G.B. and Kuhn, J.K., 1977. Trace Elements in Coal: Occurrence and Distribution. Circular No. 499, Illinois State Geological Survey, 154 pp.

Gozlan, R.S. and Margalith, P., 1973. Iodide oxidation by a marine bacterium. J. Appl. Bacteriol., 36: 407—417.
Gozlan, R.S. and Margalith, P., 1974. Iodide oxidation by *Pseudomonas iodooxidans*. J. Appl. Bacteriol., 37: 493—499.
Gunn, S.A., Harr, J.R., Levander, O.A., Olson, O.E., Schroeder, H.J., Allaway, W.H., Lakin, H.W. and Boaz, T.D., 1976. Selenium. National Academy of Sciences, Washington, DC, 203 pp.
Jensen, S. and Jernelöv, A., 1969. Biological methylation of mercury in aquatic organisms. Nature, 223: 753—754.
Jernelöv, A. and Martin, A.L., 1975. Ecological implications of metal metabolism by microorganisms. Annu. Rev. Microbiol., 29: 61—77.
Krumbein, W.E. (Editor), 1978. Environmental Biogeochemistry and Geomicrobiology, 3 Vols. Ann Arbor Science, Ann Arbor, 1052 pp.
Lawrence, A.W. and McCarty, P.L., 1964. The role of sulfide in preventing heavy metal toxicity in anaerobic treatment. Water Pollut. Control Fed. J., 37: 393—409.
Lawrence, A.W., McCarty, P.L. and Guerin, F.J.A., 1964. The effects of sulfides on anaerobic treatment. 19th Industrial Waste Conference, Purdue University. Proceedings, pp. 343—347.
Lewis, B.G., 1976. Selenium in biological systems, and pathways for its volatilization in higher plants. In: J.O. Nriagu (Editor), Environmental Biogeochemistry, Vol. 1. Ann Arbor Science, Ann Arbor, pp. 389—409.
Louderback, T., 1975. Selenium and the environment. Miner. Ind. Bull., Colo. Sch. Mines, 18: 1—14.
Lowenstam, H.A., 1974. Impact of life on chemical and physical processes. In: E.D. Goldberg (Editor), The Sea, Vol. 5, Wiley-Interscience, New York, NY, pp. 715—796.
Macgregor, A.N. and Keeney, D.R., 1973. Methane formation by lake sediments during in vitro incubation. Water Res. Bull., 9: 1153—1158.
Mah, R.A., Ward, D.M., Baresi, L. and Glas, T.L., 1977. Biogenesis of methane. Annu. Rev. Microbiol., 31: 309—341.
Manskaya, S.M. and Drosdova, T.V., 1968. Geochemistry of Organic Substances. Pergamon, Oxford, 345 pp.
Martens, C.S. and Berner, R.A., 1974. Methane production in the interstitial waters of sulfate-depleted marine sediments. Science, 185: 1167—1169.
Martens, C.S. and Berner, R.A., 1977. Interstitial water chemistry of anoxic Long Island Sound sediments. I. Dissolved gases. Limnol. Oceanogr., 22: 10—25.
Martin, J.H. and Knauer, G.A., 1973. The elemental composition of plankton. Geochim. Cosmochim. Acta, 37: 1639—1653.
Martin, J.L., 1973. Selenium assimilation in animals. In: D.L. Klayman and W.H.H. Günther (Editors), Organic Selenium Compounds: Their Chemistry and Biology. Wiley-Interscience, New York, NY, pp. 633—691.
Mechalas, B.J., 1974. Pathways and environmental requirements for biogenic gas production in the ocean. In: I.R. Kaplan (Editor), Natural Gases in Marine Sediments. Plenum, New York, NY, pp. 11—25.
Mroz, E.J. and Zoller, W.H., 1975. Composition of atmospheric particulate matter from the eruption of Heimaey, Iceland. Science, 190: 461—464.
Mudd, S.H., 1973. Biochemical mechanisms of methyl transfer. Metabol. Conjugation Metabol. Hydrol., 3: 297—350.
Nissenbaum, A. and Swaine, D.J., 1976. Organic matter-metal interactions in Recent sediments: the role of humic substances. Geochim. Cosmochim. Acta, 40: 809—816.
Nissenbaum, A., Presley, B.J. and Kaplan, I.R., 1972. Early diagenesis in a reducing fjord, Saanich Inlet, British Columbia — 1. Chemical and isotopic changes in major components of interstitial water. Geochim. Cosmochim. Acta, 36: 1007—1027.

Nriagu, J.O. (Editor), 1976. Environmental Biogeochemistry, 2 Vols. Ann Arbor Science Publ., Ann Arbor, 773 pp.

Odum, E.P., 1971. Fundamentals of Ecology, 3rd edn. W.B. Saunders Co., Philadelphia, 574 pp.

Oparin, A.I., 1961. Life, Its Nature, Origin and Development. Academic, New York, NY, 182 pp.

Oremland, R.S., 1975. Methane production on shallow water, tropical marine sediments. Appl. Microbiol., 30: 602—608.

Oremland, R.S. and Taylor, B.E., 1978. Sulfate reduction and methanogenesis in marine sediments. Geochim. Cosmochim. Acta, 42: 209—214.

Panganiban, A. and Hanson, R.S., 1976. Isolation of a bacterium that oxidizes methane in the absence of oxygen. Annual Meeting of the American Society for Microbiology, Abstract I 59, p. 121.

Peck, H.D., 1966. Some Evolutionary Aspects of Inorganic Sulfur Metabolism. Lectures on theoretical and applied aspects of modern microbiology. University of Maryland, 22 pp.

Peterson, P.J. and Butler, G.W., 1967. Significance of selenocystathionine in an Australian selenium-accumulating plant, *Neptunia amplexicaulis*. Nature, 213: 599—600.

Porritt, R.E. and Swaine, D.J., 1976. Mercury and selenium in some Australian coals and fly-ash. Preprint, Institute of Fuel Conference Sydney, Paper 18, 9 pp.

Reeburgh, W.S., 1976. Methane consumption in Cariaco Trench waters and sediments. Earth Planet. Sci. Lett., 28: 337—344.

Reeburgh, W.S. and Heggie, D.T., 1977. Microbial methane consumption reactions and their effect on methane distributions in fresh water and marine environments. Limnol. Oceanogr., 22: 1—9.

Reuter, J.H. and Perdue, E.M., 1977. Importance of heavy metal-organic matter interactions in natural waters. Geochim. Cosmochim. Acta, 41: 325—334.

Saxby, J.D., 1969. Metal-organic chemistry of the geochemical cycle. Rev. Pure Appl. Chem., 19: 131—150.

Saxby, J.D., 1976. The significance of organic matter in ore genesis. In: K.H. Wolf (Editor), Handbook of Strata-bound and Stratiform Ore Deposits, Vol. 1. Elsevier, Amsterdam, pp. 111—133.

Schidlowski, M., 1975. Archean atmosphere and evolution of the terrestrial oxygen budget. In: B.F. Windley (Editor), The Early History of the Earth. Wiley-Interscience, New York, NY, pp. 525—535.

Scott, M.L., 1973. Nutritional importance of selenium. In: D.L. Klayman and W.H.H. Günther (Editors), Organic Selenium Compounds: Their Chemistry and Biology. Wiley-Interscience, New York, NY, pp. 629—661.

Shrift, A., 1964. A selenium cycle in nature? Nature, 201: 1034—1035.

Shrift, A., 1973. Metabolism of selenium by plants and microorganisms. In: D.L. Klayman and W.H.H. Günther (Editors), Organic Selenium Compounds: Their Chemistry and Biology. Wiley-Interscience, New York, NY, pp. 763—814.

Silverman, M.P. and Ehrlich, H.L., 1964. Microbial formation and degradation of minerals. Adv. Appl. Microbiol., 6: 153—206.

Söderlund, R. and Svensson, B.H., 1976. The global nitrogen cycle. In: B.H. Svensson and R. Söderlund (Editors), Nitrogen, Phosphorus and Sulphur — Global Cycles. SCOPE Report 7. Ecol. Bull. (Stockholm), 22: 23—73.

Sokolova, E.G. and Pilipchuk, M.F., 1973. Geochemistry of selenium in sediments in the NW part of the Pacific Ocean. Geochem. Int., 10: 1152—1160.

Sorokin, Yu.I., 1975. On the ability of sulfate-reducing bacteria to utilize methane for the reduction of sulfate to hydrogen sulfide. Dokl. Akad. Nauk SSSR, 115: 816—818.

Spangler, W.J., Spigarelli, J.L., Rose, J.M. and Miller, H.M., 1973. Methylmercury: bacterial degradation in lake sediments. Science, 180: 192—193.

Stanier, R.Y. and Cohen-Bazire, G., 1977. Phototrophic prokaryotes: the cyanobacteria. Annu. Rev. Microbiol., 31: 225—274.
Stumm, W., 1966. Redox potential as an environmental parameter; conceptual significance and operational limitation. Third International Conference on Water Pollution Research, Munich, Germany, pp. 1—16.
Stumm, W. (Editor), 1977. Global Cycles and Their Alterations by Man. Dahlem Konferenzen, Abakon, Berlin, 347 pp.
Summers, A.O. and Sugarman, L.I., 1974. Cell-free Mercury (II) — reducing activity in a plasmid-bearing strain of *Escherichia coli*. J. Bacteriol., 119: 242—249.
Svensson, B.H. and Söderlund, R., (Editors), 1976, Nitrogen, Phosphorus and Sulphur — Global Cycles, SCOPE Report 7. Ecol. Bull. (Stockholm), 22, 192 pp.
Swaine, D.J., 1955. The Trace-Element Content of Soils. Technical Communication No. 48, Commonwealth Bureau of Soil Science. Commonwealth Agricultural Bureaux, England, 157 pp.
Swaine, D.J., 1962. The Trace-Element Content of Fertilizers. Technical Communication No. 52, Commonwealth Bureau of Soils. Commonwealth Agricultural Bureaux, England, 306 pp.
Swaine, D.J., 1977. Trace elements in coal. In: D.D. Hemphill (Editor), Trace substances in Environmental Health-XI, University of Missouri, pp. 107—116.
Taylor, S.R., 1964. Abundance of chemical elements in the continental crust: a new table. Geochim. Cosmochim. Acta, 28: 1273—1285.
Taylor, S.R., 1979. Chemical composition and evolution of the continental crust: the rare earth element evidence. In : M.W. McElhinny (Editor), The Earth Its Origin, Structure and Evolution, Academic, New York, NY, in press.
Trudinger, P.A. and Bubela, B., 1967. Microorganisms and the natural environment. Miner. Deposita, 2: 147—157.
Tsunogai, S. and Sase, T., 1969. Formation of iodide-iodine in seawater. Deep Sea Res., 16: 489—496.
Turekian, K.K. and Wedepohl, K.H., 1961. Distribution of the elements in some major units of the earth's crust. Bull. Geol. Soc. Am., 72: 175—191.
Van Valen, L., 1971. The history and stability of atmospheric oxygen. Science, 171: 439—443.
Vinogradov, A.P., 1953. The Elementary Chemical Composition of Marine Organisms. Memoir No. 2, Sears Foundation. Sears Foundation for Marine Research, Yale University, New Haven, 647 pp.
Vinogradova, Z.A. and Kovalsky, V.V., 1962. Elemental composition of the Black Sea plankton. Dokl. Akad. Nauk, 147: 1458—1460.
Walker, J.G.C., 1975. Implication for atmospheric evolution of the inhomogeneous accretion model of the origin of the earth. In: B.F. Windley (Editor), The Early History of the Earth. Wiley-Interscience, New York, NY, pp. 537—546.
Wells, N., 1967. Selenium content of soil-forming rocks. N.Z. J. Geol. Geophys., 10: 198—208.
Whelan, T., Bernard, B.B. and Brooks, J.M., 1978. Carbon isotope variations in total carbon dioxide and methane from interstitial waters of nearshore sediments. In: B.W. Robinson (Editor), Stable Isotopes in Earth Sciences. New Zealand Department of Scientific and Industrial Research, Bulletin 220, pp. 39—47.
Winfrey, M.R. and Zeikus, J.G., 1977. Effect of sulfate on carbon and electron flow during microbial methogenesis in fresh water sediments. Appl. Environ. Microbiol., 33: 275—281.
Wolf, W.R., 1978. Nutrient trace element composition of foods: analytical needs and problems. Anal. Chem., 50: 190A—194A.
Wong, P.T.S., Chau, Y.K. and Luxon, P.L., 1975. Methylation of lead in the environment. Nature, 253: 263—264.

Wood, J.M., 1974. Biological cycles for toxic elements in the environment. Science, 183: 1049—1052.
Wood, J.M., Kennedy, F.S. and Rosen, C.G., 1968. Synthesis of methylmercury compounds by extracts of a methanogenic bacterium. Nature, 220: 173—174.
Yamanada, M. and Tonomura, K., 1972a. Formation of methylmercury compounds from inorganic mercury by *Clostridium cochlearium*. J. Ferment. Technol., 50: 159—166.
Yamanada, M. and Tonomura, K., 1972b. Further studies of formation of methylmercury from inorganic mercury by *Clostridium cochlearium* T-2. J. Ferment. Technol., 50: 893—900.

Chapter 2.1

THE CARBON CYCLE

S. GOLUBIĆ

Department of Biology, Boston University, 2 Cummington Street, Boston, MA 02215 (U.S.A.)

W. KRUMBEIN

Environmental Laboratory, AVZ-3-145 Universität Oldenburg, P.O. Box 2503, D-2900 Oldenburg (F.R.G.)

J. SCHNEIDER

Geology-Palaeontology, University of Göttingen, Göttingen (F.R.G.)

CONTENTS

INTRODUCTION — INORGANIC AND ORGANIC CARBON CYCLES

Within its global biogeochemical cycle, the element carbon occurs in the form of various chemical compounds that are continuously being transformed and moved within two major, mutually interconnected cycles: (1) the inorganic carbon, or carbonate cycle in which carbon passes through a series of chemical equilibria; and (2) the organic carbon cycle in which carbon passes through the processes of biosynthesis and mineralization of organic matter (Fig. 2.1.1).

Carbon occurs in the atmosphere predominantly as gaseous CO_2, and in the lithosphere as rocks, mostly $CaCO_3$ and fossil organic carbon. The atmospheric and lithospheric carbon pools are linked by the hydrospheric carbonate buffer system:

$$H_2CO_3 \rightleftharpoons HCO_3^- \rightleftharpoons CO_3^{2-}$$

The relative amounts of atmospheric carbon, hydrospheric inorganic carbon and lithospheric carbonate carbon are about 1 : 60 : 10^5. The amount of

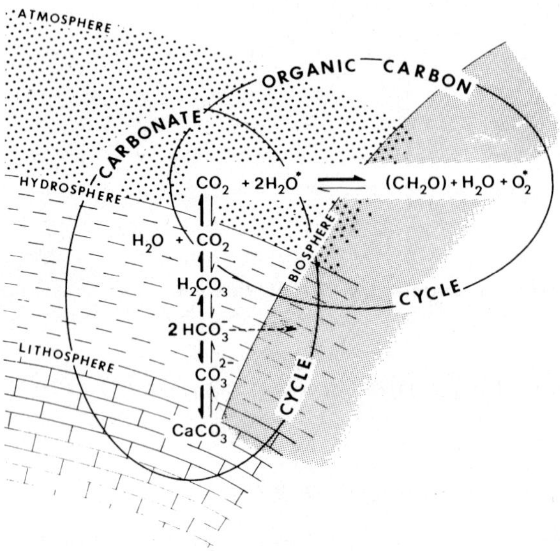

Fig. 2.1.1. Interaction between the organic and inorganic (carbonate) carbon cycles. Processes promoting precipitation, bold arrows; processes promoting dissolution, fine arrows; incorporation of bicarbonate by plants, dashed arrow.

organic carbon in the biosphere is nearly equal to that of the atmospheric carbon. The ratios of marine to terrestrial organic carbon are 1 : 66 for living biomass and 1 : 1 for dissolved and particulate organic carbon. Most of the reduced carbon on earth is in modern and ancient sediments (mostly shales and limestones) as coal, petroleum and dispersed organic particles in the ratio of 1 : 32 : 16,000. Total sedimentary reduced carbon constitutes about 18,000 times the amount of atmospheric carbon in CO_2 (see Garrels et al., 1976; Holland, 1978).

Photosynthesis, as the main driving force of the organic carbon cycle, strongly affects the carbonate cycle. Incorporation of CO_2 by green plants (including procaryotic blue-green algae *) may locally exhaust its supply, so that CO_2 becomes (at least temporarily) a limiting factor for photosynthetic activity. This has been shown in salt marshes and sugar cane fields, in plants which have the relatively newly discovered C-4 photosynthetic pathway. This pathway, in which CO_2 is fixed via phosphoenolpyruvate, is found in certain green land plants and, unlike the universal Calvin C-3 pathway, where CO_2 is fixed via ribulose diphosphate, permits the utilization of carbon dioxide at very low partial pressures. Hence, the presence of this pathway, which is prevalent in plants of arid, sunny, and hot environments, permits the depletion of local atmospheric carbon dioxide down to parts per million from its usual 0.03% (lately approaching 0.04%). Both free CO_2 in the atmosphere and hydrosphere, and the bicarbonate ion from the carbonate buffer

* See footnote on p. 12.

system of the hydrosphere are incorporated during photosynthesis. Bicarbonate assimilation has been known since Ruttner's works (1921, 1947). For a detailed review see Raven (1970). Depletion of the free CO_2 causes a shift in the equilibria between the participating carbonate species (Fig. 2.1.1 bold arrows) resulting in precipitation of $CaCO_3$. However, the rates of exchange between species may be too slow for an effective CO_2 replenishment. Plants that can use bicarbonate ions in photosynthesis, such as aquatic macrophytes and many algae, are able to overcome such limitations (Fig. 2.1.1, dashed arrow). The environmental consequence of bicarbonate incorporation is a rise in pH, which further reduces the amount of dissolved CO_2 and initiates more carbonate precipitation.

Aerobic respiration, on the other hand, releases CO_2 owing to the oxidation of organic matter, and thus, increases the concentration of CO_2 in the system. The equilibria of the carbonate buffer system are shifted in the opposite direction and a dissolution of $CaCO_3$ results (Fig. 2.1.1, fine arrows). In addition to oxidative respiration, partial degradation (fermentaton) of organic matter may result in formation of organic acids which also contribute to the dissolution of carbonates. Furthermore, carbonate may be dissolved by the activity of chelating organic compounds, or by the enzyme carbonic anhydrase.

The interaction between the carbonate cycle and the organic carbon cycle takes place under a variety of circumstances. At one end of the spectrum, carbonate chemistry may be under direct enzymic control (see Chapter 2.2). It may take place within cells, within organisms, or within micro-environments in close contact with living tissues (e.g., molluscan mantle). At the other extreme, where products of metabolic activities modify the overall chemistry of the environment, carbonate dissolution or precipitation may be influenced indirectly. The closer the contact between the organism and the substrate, the more specific are the biogenic dissolution and crystallization patterns that remain as traces of biological activity in sediments.

At present, the carbonate cycle is largely maintained and regulated by life processes (see Garrels et al., 1976). This raises the question whether any carbonate dissolution or precipitation in natural waters can be regarded as strictly inorganic. Biological control over carbonate dissolution and precipitation covers a wide range of micro- and macro-environments (see Chapters 2.2 and 2.4). A wealth of information concerning calcification and carbonate dissolution in biological systems has been recently published in form of symposium volumes edited by Sognaes (1963), Carriker et al. (1969), Crimes and Harper (1970), and Frey (1975).

CHEMICAL BASIS OF CARBONATE EQUILIBRIA

Dissolution or precipitation of carbonates depends on the availability of CO_2 which is dissolved in the aqueous environment. The thermodynamic equilibria in the simple inorganic system H_2O-CO_2-$CaCO_3$ are the following:

$$CO_2 \text{ gas} \underset{}{\overset{\text{water}}{\rightleftharpoons}} CO_2 \text{ dissolved} \tag{1}$$

$$CO_2 + H_2O \rightleftharpoons H_2CO_3 \tag{2}$$

$$H_2CO_3 \rightleftharpoons H^+ + HCO_3^- \qquad K_1 = \frac{[H^+]\cdot[HCO_3^-]}{[H_2CO_3]} \tag{3}$$

$$HCO_3^- \rightleftharpoons H^+ + CO_3^{2-} \qquad K_2 = \frac{[H^+]\cdot[CO_3^{2-}]}{[HCO_3^-]} \tag{4}$$

Reaction (2) is slow ($T^{1/2}$ = 15 s at 15°C), reactions (3) and (4) are very fast. Above a pH of 8, the following reactions are dominant:

$$CO_2 + OH^- \rightleftharpoons HCO_3^- \tag{5}$$

$$HCO_3^- + OH^- \rightleftharpoons CO_3^{2-} + H_2O \tag{6}$$

These reactions are pH- and temperature-dependent. The pH-dependence is illustrated in a diagram (Fig. 2.1.2) from Wattenberg (1936, p. 174) which describes the relative abundance of H_2CO_3, HCO_3^-, and CO_3^{2-} in pure water and in sea water (35‰) as a function of pH. The inorganic carbonate system has been known since the works of Buch et al. (1932) and Wattenberg (1930, 1936), and was discussed more recently by Buch (1960a, b), Harvey (1960), Berner (1971) and Bathurst (1975).

The solubility of $CaCO_3$ in the oceanic reservoirs is influenced by the formation of inorganic complexes of CO_3^{2-} with Mg^{2+} (Pytkowicz, 1965, 1969, 1973) and with Na^+. Garrels et al. (1961) estimated that 75% of the total CO_3^{2-} in sea water is complexed as dissolved $MgCO_3$ and 15% as

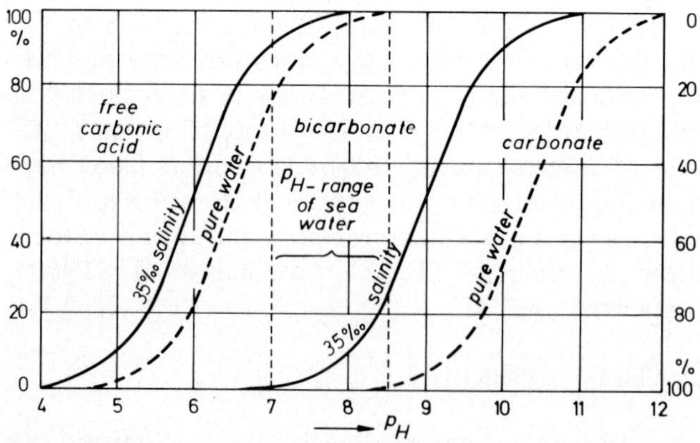

Fig. 2.1.2. The relative abundance of H_2CO_3, HCO_3^-, and CO_3^{2-} in pure water and in sea water depending on pH (after Wattenberg, 1936).

$NaCO_3^-$. This means that in sea water only 10% of the CO_3^{2-} is available for the formation of $CaCO_3$. Further factors inhibiting the precipitation of $CaCO_3$ are dissolved or particulate organic substances which build metal organic complexes with Ca^{2+} or are adsorbents for Ca^{2+} (e.g. Suess, 1970, 1973; Chave and Suess, 1970, Pytkowicz, 1973). Carbonate particles in nature are frequently coated with thin films of organic matter, thus hindering the exchange of $CaCO_3$ with the undersaturated water around the particle. Therefore, both dissolution and precipitation of $CaCO_3$ are strongly influenced by a variety of inorganic and organic substances in the surrounding environment.

Generally, the solubility of $CaCO_3$ is positively correlated with (i.e., increases with an increase of) the solubility and partial pressure of CO_2, hydrostatic pressure, Mg^{2+} incorporated in crystal lattices, and organic complexing; it is negatively correlated (i.e., decreases with an increase of) temperature, pH, concentration of CO_3^{2-} and grain size (see also Bathurst, 1975, p. 242; Schneider, 1976, pp. 21—26).

Addition or removal of CO_2 in an aqueous environment is regulated by life processes, primarily by respiration and photosynthesis. Respiration increases the amount of CO_2, causing a drop in pH and CO_3^{2-} concentration and therefore a higher solubility of $CaCO_3$, thus promoting carbonate dissolution. Photosynthesis decreases the amount of CO_2 in waters causing an increase in pH and CO_3^{2-} concentration (and therefore a lower solubility of $CaCO_3$) and thus promotes carbonate precipitation.

GLOBAL CYCLE OF CARBON

The cycling of the element carbon depends primarily on its chemical properties. Like nitrogen and sulfur, but unlike phosphorus, silica, and iron, carbon forms volatile, soluble, and insoluble compounds and can thus circulate through the atmosphere, hydrosphere and lithosphere. Since carbon is the basic element of life its circulation through the biosphere is of particular importance.

From the global viewpoint, two interlocking carbon cycles have been recognized: the *exogenic* cycle (including the carbonate and the organic carbon cycles) with carbon cycled by the atmosphere, hydrosphere and biosphere, and the *endogenic* cycle, with carbon cycled by the crust of the earth, the lithosphere and, in part, by the external mantle (Fig. 2.1.3). The exogenic cycle includes the processes of weathering and erosion, transport, sedimentation and early diagenesis (see Krumbein, 1972). It is powered by solar energy as an external energy source. The endogenic cycle, on the other hand, includes the processes of diagenesis, metamorphism, orogenesis and epeirogenesis. It is powered by the earth's internal heat flow, maintained by radioactive decay in the mantle and crust, and by the residual energy from

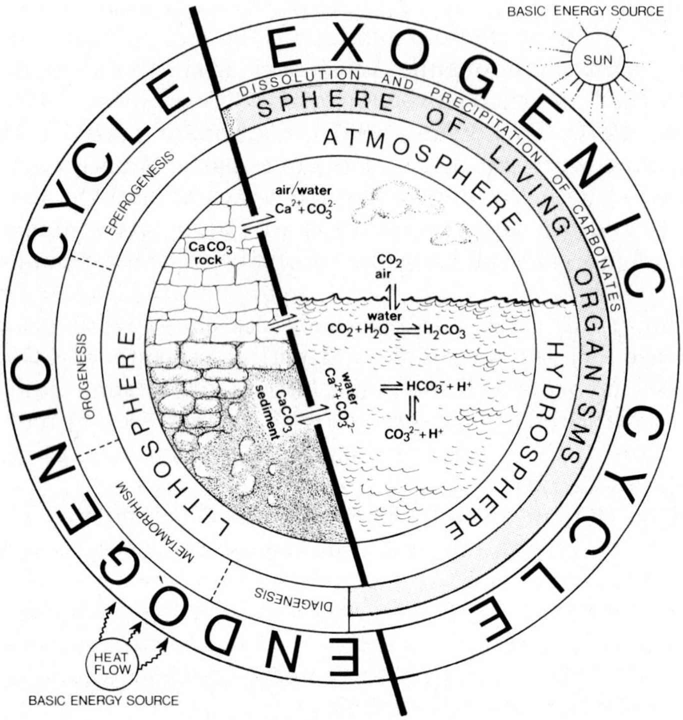

Fig. 2.1.3. Schematic presentation of the exogenic and endogenic cycling of inorganic carbon.

the accretion phase of the planet that is presumably derived from the conversion of gravitational to thermal energy. Carbon enters the exogenic cycle as CO_2 through volcanic emanations, as fossil organic carbon in sediments and fossil fuels and as $CaCO_3$ through uplift of carbonate rocks. Its exogenic circulation starts with the weathering process.

As a part of the exogenic *carbonate* cycle, atmospheric CO_2 enters the waters, becoming dissolved and partially hydrated. The dissolving power of CO_2-enriched water contributes to weathering: thus carbonate ions from calcareous rocks go into solution in a 1 : 1 ratio with atmospheric CO_2. In addition, Ca^{2+} and Mg^{2+} ions from silicate rocks combine with CO_2, forming bicarbonates. Although more carbonate than silicate is weathered out of the rocks, silicate weathering requires two mol of CO_2 for each mol of Ca^{2+} or Mg^{2+} and thus takes more CO_2 from the atmosphere. An equilibrium is then established between gaseous CO_2, dissolved CO_2, carbonic acid, bicarbonate and carbonate ions, as well as between bicarbonate and silicate in solution. This carbon is transported by rivers in the oceanic pool and is ultimately buried in sediments. Significantly more of the atmospheric CO_2, about 10%

of its total content, is drawn yearly into the organic carbon cycle (Holland, 1978).

As a part of the exogenic *organic carbon* cycle, CO_2 from the atmosphere and hydrosphere (including that in bicarbonate ions) enters the biosphere via enzymic reactions in organisms, mainly through the processes of photosynthesis and heterotrophic CO_2-assimilation (see Chapter 2.2). The carbonate cycle harbors the inorganic carbon needed for biosynthesis, and therefore serves as a nutrient pool for the organic cycle. This nutrient pool is tapped and depleted by photosynthesis and becomes recharged by respiration and other, mostly microbial, degradational processes. Carbon becomes biochemically reduced in the process of biosynthesis; it becomes oxidized in the process of biological mineralization (i.e., complete degradation of organic matter). Rates of incorporation and exchange within the organic carbon cycle are roughly 200 times higher than those in the carbonate cycle. Thus, exogenic cycling is dominated and regulated by life processes and can be interpreted as biogeochemical cycling (Fig. 2.1.4).

The combined active biomass of all primary producers acts as a gigantic pump that concentrates inorganic carbon dispersed in both the atmosphere and hydrosphere and deposits it in the reduced organic form, $(CH_2O)_n$. A comparison between the average contents (%w/w) of carbon in the atmosphere (0.03), hydrosphere (0.0014) and lithosphere (0.16), as opposed to that of the biosphere (24.9), impressively illustrates the efficiency of the biological carbon pump (see Deevey, 1970).

The major part of the organic carbon continues to cycle through food chains and webs, to become ultimately mineralized through microbial degradation. A minor part remains in the environment in the form of dissolved and particulate organic matter and becomes transported to the oceans and captured in sediments. On land the oxidation-reduction cycle of carbon is nearly closed: the fixed CO_2 is almost totally recycled. Vigorous biological release of CO_2 during microbial degradation of organic matter in soils significantly intensifies the weathering of carbonate (and other) rocks by lowering of pH. In the oceans, most of the photosynthetically-fixed CO_2 is recycled in the surface layers. A much smaller part reaches the ocean floors in form of dead particulate or dissolved organic matter, which is then oxidized in sediments with high oxidation potential (e.g., those of the deep sea) or incorporated in sediments with low oxidation potential (e.g., shelf areas).

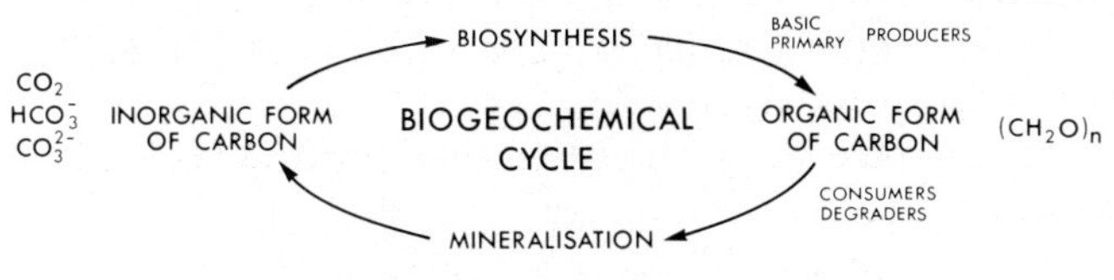

Fig. 2.1.4. The biogeochemical cycling of carbon.

Biological degradation and mineralization continues in newly deposited sediments. In general, about 30 to 50% of the organic carbon trapped in sediments is released as CO_2. This release contributes to dissolution of the deposited carbonate, and a substantial portion of both organic and inorganic carbon returns to the water column at the sediment-water interface (Bordovski, 1965). Another part may be fixed within the sediment as carbonate via sulfate reduction, or released by methane formation (Curtis, 1977; Krumbein et al., 1977). The remaining carbon contributes to the formation of sedimentary rocks.

As a part of the endogenic cycle, sedimentary rocks are modified by the processes of diagenesis and metamorphism and become uplifted millions of years later by the forces of diastrophism that are expressed through orogenic and epeirogenic movements. Orogenic movements result in folding, thrusting and metamorphism of rocks, while epeirogenic movements are predominantly vertical and may occur independently of orogenesis. From our understanding of plate tectonics, i.e., large scale recycling of the earth's crust through plate consumption and sea floor spreading, we may conclude that part of the carbon in sediments will leave the lithosphere and enter the mantle, returning to the atmosphere via volcanic degassing. This carbon then joins the lithospheric carbon in the process of weathering and thus another turn of the exogenic cycle begins. Active plate tectonics in conjuction with an extensive hydrosphere and biosphere is responsible for carbon cycling on the planet Earth in contrast with other interior planets of our solar system (Degens, 1977).

The amount of carbon in the atmosphere is approximately 0.055 E mol. About 3.23 E mol of carbon (primarily carbonate and bicarbonate in the oceans) are present in the hydrosphere, and the biosphere contains an estimated amount of 0.04 E mol of carbon. The bulk of the earth's carbon, however, is stored in the lithosphere: 4,000 E mol as $CaCO_3$, and 1,000 E mol as organic carbon incorporated in sediments (Garrels et al., 1975, 1976). Newer results on the subject of the biogeochemical cycling of carbon were discussed at the SCOPE workshop in Hamburg/Ratzeburg (Stumm, 1977).

If we assume that the entire carbon cycle is maintained in a steady state condition (Siever, 1968; Garrels and Mackenzie, 1972; Garrels et al., 1975, 1976), it follows that the amounts of carbon present at any time in the biosphere, atmosphere and hydrosphere remain relatively small. Thus this carbon circulates at relatively high rates, and has a short residence time within each of these three pools. Most of the carbon is stored in sediments and is thus removed from the exogenic circulation for millions of years. This carbon returns slowly via the endogenic cycle (Fig. 2.1.5).

The following residence times have been calculated for carbon (Garrels et al., 1975): 11 y in the biosphere, 4 y in the atmosphere, 385 y in the hydrosphere, and 342 M y in the lithosphere. These calculations represent, in a sense, the statistical probability of positioning an "orbiting average

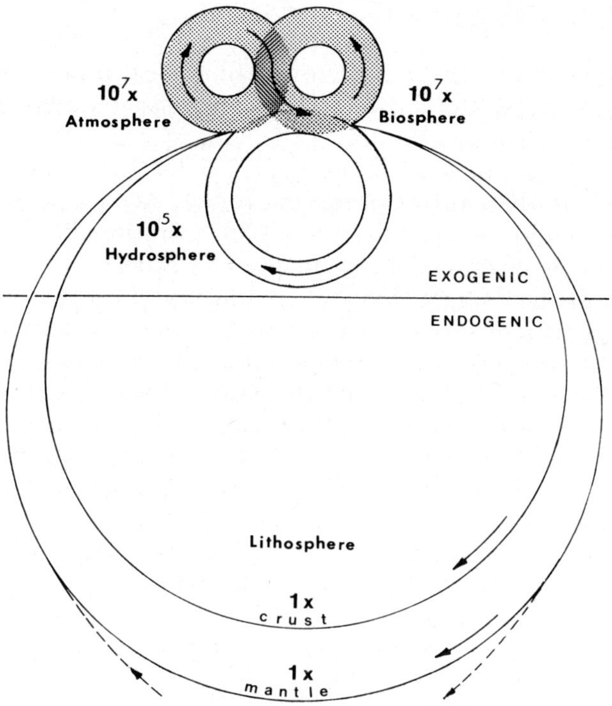

Fig. 2.1.5. The relative cycling frequency of an average carbon atom within the atmosphere, biosphere and hydrosphere (exogenic cycles) relative to the cycling in the lithospere (endogenic cycles).

C-atom'' on a global scale. However, the actual cycling and turn-over rates of carbon in biological systems within various micro-environments have to be considered ''short-circuited'' and much more rapid, as numerous radiocarbon labeling experiments show. Life processes also stimulate and regulate multiple subcycles of carbonate dissolution and reprecipitation (see Chapters 2.2 and 2.4). On the other hand, some carbon deposited in stable continental interiors (cratons) remains protected from dissolution (or oxidation) and recycling from the time of deposition billions (10^9) of years ago up to the present.

We may conclude that the overall carbon cycling depends on exchange rates between endogenic and exogenic cycles which, in turn, are determined by geothermal powers and crustal activity (Holland, 1978). The rapid and complex interplay of various exogenic carbon cycles, however, is homoeostatically regulated by life processes (Lovelock and Margulis, 1974).

Maintenance and regulation of the carbon cycle by biological processes implies that, through Earth's history, the entire carbon cycle has undergone evolutionary changes that paralleled the evolution of living systems. There is a general agreement that life originated earlier than 3.5 Gy B.P., under much more reducing conditions than those prevailing today. According to this view, free oxygen was present in extremely low concentrations in the earth's atmosphere prior to the evolution of oxygenic photosynthesis.

We can only speculate about the distribution and circulation of carbon on the early earth. Taking into consideration the absence of a powerful draining of the atmospheric CO_2 by photosynthesis, in conjunction with the same or higher volcanic CO_2 output, we may conclude that the amount of CO_2 in the early atmosphere was probably several times higher than today (Holland, 1968). The observation that most anaerobic bacteria, thought to be primitive organisms, grow optimally at concentrations of CO_2 several times higher than those of the present-day atmosphere is consistent with Holland's concept. The amount of dissolved carbonate in the hydrosphere was then adjusted to an elevated CO_2-pressure and thus was also higher. Therefore, in comparison with present conditions, more carbon was relocated from the lithosphere into the atmosphere and hydrosphere although this did not necessarily affect the overall rates of carbon cycling.

The overall cycling rates of carbon have always depended upon geothermal forces that cause volcanic CO_2 exhalation and carbonate uplift, independently of life processes. The hydrosphere's carbonate buffer system kept the oceans at near saturation, with respect to $CaCO_3$, at all times (Holland, 1972). Rates of deposition and accumulation of carbonate rocks per unit of geological time have also been within the range of fluctuation observed for the Phanerozoic (Garrels et al., 1976). The form of carbonate deposits, however, was certainly different in the Precambrian from those of today, i.e. it was a predominantly chemical precipitate, rather than biogenic skeletal carbonate (Monty, 1973).

Complex organic compounds are not stable under oxidizing conditions and can be generated abiotically only under relatively reducing conditions. Sources of energy, thought to have been available on the early earth (e.g. electrical discharge, UV), can drive abiotic organic synthesis under laboratory conditions (Miller and Orgel, 1974; Ponnamperuma, 1972). The stability of such organic compounds is considered essential for the origin of life on earth. However, nothing is known about the cycling of the pre-biotic organic carbon or of its possible accumulation rates. Early "oil slicks", possibly several meters in thickness, or organic rich sediments may have accumulated in the primordial ocean (e.g. Lasaga et al., 1971). We may wonder whether the spheroidal organic microfossils found in the oldest Precambrian sediments resulted from chemical or biological evolution (Schopf, 1972). Upon the origin of life, such organic matter was possibly the initial food source that allowed the evolution of a variety of heterotrophic metabolic systems, starting with simple anaerobic fermentation.

Replication, and the consequent exponential population growth which is intrinsic to all living systems, must have exhausted any accumulated resource within a relatively short geological time span and have caused a concurrent accumulation of metabolic products. It is, therefore, reasonable to expect that the selective pressures of early organic evolution favored successions of metabolic types where each step used the previous metabolic product as a resource. Exhaustion of all resources as the result of such unidirectional metabolic sequences must have induced an evolution of cyclic metabolic schemes. Since heterotrophs use the organic matter both as a source of carbon and as an energy source, it is probably the energy crisis that prompted the next selective pressure, because energy cannot be recycled. The evolution of photosynthetic energy-generating systems (and of usable organic compounds) probably solved this early energy crisis.

The evolutionary stages of photosynthesis presumably progressed through the following steps: photo-organotrophy, autotrophic reduction of CO_2 using a series of hydrogen donors (e.g., organic compounds, H_2S), and finally the autotrophic reduction of CO_2 using H_2O as the hydrogen donor. Oxygenic photosynthesis represented a significant qualitative jump in the evolution of life. Vast, previously untapped resource reservoirs, i.e. the carbonate cycle as the source of carbon, and the hydrosphere as the source of hydrogen thus became available to organisms. The consequence was an accumulation of metabolic products, specifically photosynthesized organic matter, and of molecular oxygen.

Considering again the nature of exponential population growth we can visualize the effects of early perpetual "algal blooms". The accumulation of free oxygen in the atmosphere up to levels equal to or higher than the present one, and of corresponding amounts of organic matter in sediments, probably occurred in a relatively short geological time period. This conclusion is at variance with that of Berkner and Marshall (1965) who proposed a rise of free O_2 up to a 10% of the present level at the end of the Precambrian (600 My ago). Our conclusion also differs from the models of Fischer (1972) which propose a slow and gradual rise of oxygen levels throughout most of the Precambrian. An early and sudden rise of oxygen levels is, however, more consistent with the sedimentary evidence in the oxidation state of both Precambrian and Phanerozoic marine sediments (Dimroth and Kimberley, 1976).

Stromatolites have been found in Precambrian strata dating from about 3.1 Gy B.P. onwards (Gebelein, 1976, p. 500), and became widespread about 2.3 Gy ago (Awramik et al., 1976, p. 166). Most recent stromatolites are built by the oxygen-releasing photosynthetic organisms, cyanophytes. However, non-biogenic and bacterial stromatolites have been discussed by Walter

(1976a, b), Monty (1971) and Doemel and Brock (1974) which implies that the presence of stromatolitic structures does not necessarily indicate oxygen release. Thus the calendar date for the origin of photosynthesis remains obscure. This date may be difficult to determine if the transition from the reducing to the oxidizing atmosphere indeed took place early and within a geologically short time. Schidlowski et al. (1974) and Junge et al. (1975) have calculated carbon and oxygen isotope values for sedimentary rocks from the early Precambrian to recent times. They concluded that the ratio of 1 : 4 between organic carbon and carbonate carbon, which is typical for the present-day sedimentary rocks, was already established 3.3 Gy ago, and that the build-up of the organic carbon reservoir must have been established earlier. Their calculations indicate that the build-up of this organic carbon, and thus a significant oxygenic photosynthesis, must have started 3.7 Gy ago. Taking into account the calculations of Garrels et al. (1975) and Li (1972) for the global mass balance of rocks, it can be derived that the oxygen equivalent of the organic carbon reservoir has remained stable since that time.

Oxygenic photosynthesis first took place in the Precambrian hydrosphere (Cloud and Giber, 1970). The oxygen produced was initially bound by hydrospheric ferrous iron derived from a previous anoxic atmospheric weathering. In an oxygenated hydrosphere, ferric iron precipitated as banded iron ore formations (see Chapter 4). In this way, the oxygen content of the hydrosphere was held low (reviewed by Gebelein, 1976, pp. 501—502). The first terrigenous red bed formation, which indicates an oxygenated atmosphere, has been dated at 1.8 Gy B.P. (Schidlowski et al., 1974).

Free O_2 acted as a poison for most anaerobic organisms and they needed evolutionary adjustments in order to survive (Margulis, 1972). It is interesting to note that certain cyanophytes isolated from recent stromatolites and thermal springs are able to switch from oxygenic photosynthesis to anoxygenic photosynthesis depending on the levels of ambient O_2 and H_2S (Cohen et al., 1975, 1977; Castenholz, 1977; Krumbein and Cohen, 1976). Apart from its toxic effect, the presence of free oxygen in the atmosphere offered new metabolic possibilities. Oxygen is the ideal hydrogen acceptor for energy-yielding biological redox systems. The evolution of cytochrome-operated oxidative respiration combined with enzymic CO_2 release (e.g. Krebs cycle) provided the mechanism for an effective decomposition and mineralization of the photosynthesized organic matter, and thus for a completion of the modern organic carbon cycle.

The build-up of the original, photosynthetically-generated oxygenated atmosphere resulted from an evolutionary lag between the event of photosynthesis and that of the oxidative respiration that followed. Its maintenance, however, throughout the latter eons was essential to keep the organic cycle operative. Because the residence time of O_2 in the present atmosphere is relatively short in the geological sense (6000 y; Holland,

1978), a slight malfunction of its homeostatic regulation would have dramatic repercussions on the environment, including a complete anoxia. The continuity of aerobic life and the fossil record assure us that global anoxic conditions have not occurred. Indeed, the homeostasis of the carbon cycle with the active participation of life is one of the many mutually-adjusted regulatory mechanisms that maintain optimal conditions for life on Earth (Lovelock and Margulis, 1974; Margulis and Lovelock, 1974).

Speculations on the future of the carbon cycle are even more vague. The burning of fossil fuels by civilized man takes advantage of the cumulative effect of past photosynthesis and works clearly in the direction of imbalance. The increase in atmospheric CO_2 is the most immediate, and proportionally greatest consequence of this activity. It is not yet clear whether its accumulation will produce a greenhouse effect that will modify the future climate by temperature increase, or whether an increase in the earth's albedo, by the accompanying production of smoke and dust, will lead to temperature decrease (Bach, 1976).

The annual increase of CO_2 in the atmosphere is at present about 388 T mol C y^{-1}. In the past 100 y the atmospheric concentration of CO_2 has been increased by 15% and there is no reason to believe that fossil fuel burning will stop in the years to come. Only a part (40—50% according to Garrels et al., 1976) of man's net CO_2-output remains airborne: a part of it goes to the ocean where it may cause increased dissolution of carbonate (Broecker, 1973). This change may be sufficient to destroy the coral reef belt of the tropical seas (Alexandersson, 1976). On the other hand, from thermodynamic models it seems improbable that all CO_2, which has been produced by fuel burning but did not accumulate in the atmosphere, has gone to the sea (Broecker, 1973). According to Garrels et al. (1975, 1976) about 280 T mol of CO_2 does not accumulate in the atmosphere or in the oceans. The sink for this enormous yearly amount is still unknown (Baes et al., 1977; Kerr, 1977). Latest calculations by Woodwell and Houghton (1977), Bolin (1977) and Reiners and Wright (1977) show that the problem might be even more serious. Burning of wood and deforestation for agricultural purposes add about the same amount of CO_2 per annum to the atmosphere as fossil fuel burning.

It has been suggested (Garrels et al., 1975; Junge, 1977; Stumm, 1977) that a net increase in photosynthesis may account for the "missing" CO_2. Unfortunately, a possible worldwide productivity increase is very difficult to determine. The accuracy of CO_2 measurement in the atmosphere is about 1‰; the accuracy in determining biological productivity (even after successful IBP-programs) is considerably lower. Global extrapolations are, therefore, much too uncertain. In addition, the forest areas of the world, where such a potential production increase might be expected, are presently decreasing.

We have to look for other potential reservoirs which are large enough to accommodate the annually "missing" amount of some 4 T mol of carbon. Major reservoirs of organic carbon, where a net increase in organic production could be stored, and which seem to be presently increasing, are recent sediments of lakes, rivers, estuaries and oceans. Man's use of phosphorus and nitrogen as fertilizers and detergents leads to eutrophication of aquatic systems with consequent increases in productivity (see Hutchinson, 1973). The photosynthetically fixed carbon is rapidly turned over through aquatic biota and transferred to sediments. Phosphorus and nitrogen, however, do not accumulate in the sediment in the same proportion as carbon. This indicates that these nutrients are recycled, returning to the water column where they stimulate another carbon-fixing cycle. An observed consequence is a net increase in anoxic sediments and water bodies throughout the world, which in turn provides an increase in traps for incompletely oxidized organic carbon. Photosynthetically precipitated carbonate, and sulphate reduction in sediments may, contribute to an increase of carbonate deposit and thus work toward a restoration of the long term ratio (1 : 4) of organic carbon to carbonate.

At this point, we may find comfort in the thought that our environment is homeostatically well buffered: for example, the oceanic carbonate-silicate buffer system and anoxic sediments may absorb substantial CO_2-addition. On the other hand, we really do not know the capacity of nature's homeostasis, particularly when it is exposed to multiple stresses of the present environmental deterioration (e.g. Holland, 1972).

ACKNOWLEDGMENTS

We thank Barry Cameron, H.D. Holland and Lynn Margulis for their valuable suggestions and unpublished data. Malcolm Walter and Susan Campbell critically read the manuscript. The work was supported by the NSF Grants Ga-43391 and GA-31168 to S. Golubić.

REFERENCES

Alexandersson, E.T., 1976. Actual and anticipated petrographic effects of carbonate undersaturation in shallow sea water. Nature, 262: 653—657.

Awramik, S.M., Margulis, L. and Barghoorn, E.S., 1976. Evolutionary processes in the formation of stromatolites. In: M.R. Walter (Editor), Stromatolites. Developments in Sedimentology, No. 20, Elsevier, Amsterdam, pp. 149—162.

Bach, W., 1976. Global air pollution and climatic change. Rev. Geophys. Space Phys., 14: 429—474.

Baes, C.F., Goeller, H.E., Olson, J.S. and Rotty, R.M. 1977. Carbon dioxide and climate: the uncontrolled experiment, Am. Sci., 65: 310—320.

Bathurst, R.G.C., 1975. Carbonate Sediments and Their Diagenesis. 2nd edn, Developments in Sedimentology, No 12, Elsevier, Amsterdam, 658 pp.

Berkner, L.V. and Marshall, L., 1965. On the origin and rise of oxygen concentrations in the earth's atmosphere. J. Atmos. Sci., 22: 225—226.
Berner, R.A., 1971. Principles of Chemical Sedimentology. McGraw Hill, New York, NY, 240 pp.
Bolin, B., 1977. Changes of land biota and its importance for the carbon cycle. Science, 196: 613—615.
Bordovski, O.K., 1965. Sources of organic matter in marine basins. Mar. Geol., 3: 5—31.
Broecker, W.S., 1973. Factors controlling CO_2 content in the oceans and atmosphere. In: C.M. Woodwell and E.V. Pecan (Editors), Carbon and the Biosphere, U.S. At. Energy Comm., Washington, DC, pp. 32—50.
Buch, K., Harvey, H.W., Wattenberg, H. and Gripenberg, S., 1932. Über das Kohlensäuresystem im Meerwasser. Rapp. Proc. Verb. Cons. Int., 79: 1—70.
Buch, K., 1960a. Dissoziation der Kohlensäure. Gleichgewicht und Puffersystems. Ruhlands Handbuch der Pflanzenphysiologie, Vol. 5. Springer, Berlin, pp. 1—11.
Buch, K., 1960b. Kohlendioxid im Meerwasser. Ruhlands Handbuch der Pflanzenphysiologie. Vol. 5, Springer, Berlin, pp. 47—61.
Carriker, M.R., Smith, E.H. and Wilce, R.T. (Editors), 1969. Penetration of calcium carbonate substrates by lower plants and invertebrates. Am. Zool., 9: 629—1020.
Castenholz, R.W., 1977. The effect of sulphide on the blue-green algae of hot springs. II. Yellowstone National Park. Microb. Ecol., 3: 79—105.
Chave, K.E. and Suess, E., 1970. Calcium carbonate saturation in seawater: Effects of dissolved organic matter. Limnol. Oceanogr., 15: 633—637.
Cloud, P. and Gibor, A., 1970. The Oxygen Cycle. Sci. Am., 223: 110—123.
Cohen, Y., Padan, E. and Shilo, M., 1975. Facultative anoxygenic photosynthesis in the Cyanobacterium *Oscillatoria limnetica*. J. Bacteriol., 123: 855—861.
Cohen, Y., Krumbein, W.E. and Shilo, M., 1977. Solar Lake (Sinai). 2. Distribution of photosynthetic microorganisms and primary production. Limnol. Oceanogr., 22: 609—620.
Crimes, T.P. and Harper, J.C. (Editors), 1970. Trace Fossils. Geol. J., Spec. Issue, 3: 1—547.
Curtis, C.D., 1977. Sedimentary geochemistry: environments and processes dominated by involvement of an aqueous phase. Phil. Trans. R. Soc. London, Ser. A, 286: 235—237.
Deevey, E.S., 1970. Mineral cycles. Sci. Am., 223: 148—158.
Degens, E.T., 1978. The Protobiosphere, Chem. Geol., 22: 177—187.
Dimroth, E. and Kimberley, M.M., 1976. Precambrian atmospheric oxygen: evidence in the sedimentary distribution of carbon, sulfur, uranium, and iron. Can. J. Earth Sci., 13: 1161—1185.
Doemel, W.N. and Brock, T.D., 1974. Bacterial stromatolites: Origin of laminations. Science, 184: 1003—1085.
Fischer, Alfred G., 1972. Atmosphere and the evolution of life. Main Currents in Modern Thought, 285: 1—6.
Frey, R.W. (Editor), 1975. The Study of Trace Fossils. Springer, Berlin, 562 pp.
Garrels, R.M. and Mackenzie, F.T., 1972. A quantitative model for the sedimentary rock cycle. Mar. Chem., 1: 27—41.
Garrels, R.M., Lerman, A. and Mackenzie, F.T., 1976. Controls of atmospheric O_2 and CO_2: Past, present, and future. Am. Sci., 64: 306—315.
Garrels, R.M., Mackenzie, F.T. and Hunt, C., 1975. Chemical Cycles and the Global Environment. Kaufmann, Los Altos, CA, 206 pp.
Garrels, R.M., Thompson, M. and Siever, R., 1961. Control of carbonate solubility by carbonate complexes. Am. J. Sci., 259: 24—45.
Gebelein, C.D., 1976. The effects of the physical, chemical and biological evolution of the earth. In: M.R. Walter (Editor), Stromatolites. Developments in Sedimentology, No. 20. Elsevier, Amsterdam, pp. 499—515.

Harvey, H.W., 1960. The Chemistry and Fertility of Seawaters. Cambridge University Press, 2nd edn, 240 pp.
Holland, H.D., 1968. The abundance of CO_2 in the Earth's atmosphere through geologic time. In: L.H. Ahrens (Editor), Origin and Distribution of the Elements. Pergamon, Oxford, pp. 949—954.
Holland, H.D., 1972. The geologic history of sea water — an attempt to solve the problem. Geochim. Cosmochim. Acta 36: 637—651.
Holland, H.D., 1978. Chemistry of the Atmosphere and the Oceans. Wiley, New York, NY, 351 pp.
Hutchinson, G.E., 1973. Eutrophication. Am. Sci., 61: 269—279.
Junge, C.E., Schidlowski, M., Eichmann, R. and Pietrek, H., 1975. Model calcualtions for the terrestrial carbon cycle — Carbon isotope geochemistry and evolution of photosynthetic oxygen. J. Geophys. Res., 80: 4542—4582.
Junge, C.E., 1977. Stable isotope fractionation in geochemical and environmental cycles. In: W. Stumm (Editor), Global Chemical Cycles and Their Alteration By Man. Dahlem Konferenzen, Berlin, pp. 33—44.
Kerr, R.A., 1977. Carbon dioxide and climate: carbon budget still unbalanced. Science, 197: 1352—1353.
Krumbein, W.E., 1972. Role des microorganismes dans la génèse, la diagénèse et la dégradation des roches en place. Rev. Ecol. Biol. Sol., 9: 283—319.
Krumbein, W.E. and Cohen, Y. 1976. Primary production, mat formation and lithification: Contribution of oxygenic and facultative anoxygenic cyanobacteria. In: E. Flügel (Editor) Fossil Algae, Springer, Berlin, Heidelberg—New York, pp. 37—56.
Krumbein, W.E., Cohen, Y. and Shilo, M., 1977. Solar Lake, (Sinai) 4. Stromatolitic cyanobacterial mats. Limnol. Oceanogr., 22: 635—656.
Lasaga, A.C., Holland, H.D. and Dwyer, M.J., 1971. Primordial oil slick. Science, 174: 53—55.
Li, Y.H., 1972. Geochemical mass balance among lithosphere, hydrosphere, and atmosphere. Am. J. Sci., 272: 119—137.
Lovelock, J.E. and Margulis, L., 1974. Atmospheric homeostasis by and for the biosphere: The gaia hypothesis. Tellus, 26: 1—9.
Margulis, L., 1972. Early cellular evolution. In: C. Ponnamperuma (Editor), Exobiology. North-Holland, Amsterdam, pp. 342—368.
Margulis, L. and Lovelock, J.E., 1974. Biological modulation of the earth's atmosphere. Icarus, 21: 471—489.
Miller, S.L. and Orgel, L.E., 1974. The Origins of Life on the Earth. Prentice-Hall, Englewood Cliffs, NJ, 229 pp.
Monty, C.L.V., 1971. An autoecological approach of intertidal and deep water stromatolites. Ann. Soc. Geol. Belg., 94: 265—276.
Monty, C., 1973. Precambrian background and phanerozoic history of stromatolitic communities, an overlook. Ann. Soc. Geol. Belg., 96: 585—624.
Ponnamperuma, C., 1972. The Origins of Life. Dutton, New York, NY, 215 pp.
Pytkowicz, R.M., 1965. Rates of inorganic calcium carbonate precipitation. J. Geol., 73: 196—199.
Pytkowicz, R.M., 1969. Chemical solution of calcium carbonate in seawater. Am. Zool., 9: 673—679.
Pytkowicz, R.M., 1973. Calcium carbonate retention in supersaturated seawater. Am. J. Sci., 273: 515—522.
Raven, J.A., 1970. Exogenous inorganic carbon sources in plant photosynthesis. Biol. Rev., 45: 167—221.
Reiners, W.A. and Wright, H.E., 1977. Impact of prehistoric and present fire patterns on the carbon dioxide content of the atmosphere. In: W. Stumm (Editor), Global Chemical Cycles and their Alterations by Man. Dahlem Konferenzen, Akabon, Berlin, pp. 121—135.

Ruttner, F., 1921. Das elektrolytische Leitvermögen verdünnter Lösungen unter dem Einflusse submerser Gewächse. I. Sitz. Ber. Akad. Wiss. Wien. Math., Naturw. Kl. Abt. I., 130: 71—108.

Ruttner, F., 1947. Zur Frage der Karbonatassimilation der Wasserpflanzen, I. Teil: Die beiden Haupttypen der Kohlenstoffaufnahme. Oesterr. Bot. Z., 94: 265—294.

Schidlowski, M., Eichmann, R. and Junge, C.E., 1974. Evolution des irdischen Sauerstoff — Budgets und Entwicklung der Erdatmosphäre. Umschau, 74: 703—707.

Schneider, J., 1976. Biological and Inorganic Factors in the Destruction of Limestone Coasts. Contributions to Sedimentology, No. 6, E. Schweizebartsche Verlagsbuchhandlung, Stuttgart, 112 pp.

Schopf, J.W., 1972. Precambrian paleobiology. In: C. Ponnamperuma (Editor), Exobiology. North-Holland, Amsterdam, pp. 6—61.

Siever, R., 1968. Sedimentary consequences of a steady state ocean-atmosphere. Sedimentology, 11: 5—29.

Sognnaes, R.F. (Editor), 1963. Mechanisms of Hard Tissue Destruction. Amer. Assoc. Advancement Sci. Publ., 75, 764 pp.

Stumm, W., (Editor), 1977. Global Chemical Cycles and Their Altertion By Man. Dahlem Konferenzen, Akabon, Berlin, 347 pp.

Suess, E., 1970. Interaction of organic compounds with calcium carbonate. I. Association phenomena and geochemical implications. Geochim. Cosmochim. Acta, 34: 157—168.

Suess, E., 1973. Interaction of organic compounds with calcium carbonate. II. Organocarbonate association in recent sediments. Geochim. Cosmochim. Acta, 37: 2435—2447.

Walter, M.R., 1976a. Geyserites of Yellowstone National Park: An example of abiogenic "stromatolites" In: M.R. Walter, (Editor), Stromatolites. Developments in Sedimentology, No. 20 Elsevier, Amsterdam, pp. 87—112.

Walter, M.R. (Editor), 1976b. Stromatolites. Developments in Sedimentology, No. 20. Elsevier, Amsterdam, 790 pp.

Wattenberg, H., 1930. Kalziumkarbonat und Kohlensäuregehalt des Meerwassers. Wiss. Erg. Dtsch. Atlant. Expedit. FVS Meteor, 1925—1927 VIII, 33.

Wattenberg, H., 1936. Kohlensäure und Kalziumkarbonat im Meere. Fortschr. Miner. Kristallogr., 20: 168—195.

Woodwell, G.M. and Houghton, R.A., 1977. Biotic influences on the world carbon budget. In: W. Stumm (Editor), Global Chemical Cycles and Their Alterations by Man. Dahlem Konferenzen, Akabon, Berlin, pp. 61—72.

Chapter 2.2

CALCIFICATION BY BACTERIA AND ALGAE

W.E. KRUMBEIN

Environmental Laboratory, AVZ-3-145 Universität Oldenburg, P.O. Box 2503, D-2900 Oldenburg (F.R.G.)

CONTENTS

INTRODUCTION

The role of calcification in the cycling of carbon has been discussed in Chapter 2.1, and its importance can be gauged from calculations by Li (1972) that the ratios, by weight, of sedimentary rocks are 71 : 12 : 15 : 2 for shales, sandstones, carbonates and evaporites, respectively. This implies that the carbonate cycle accounts for 15% of the total sediment cycle which has been estimated to have a turnover time of 300 My.

Although purely inorganic precipitation of carbonates may take place in the natural environment, biologically controlled precipitation appears to predominate as evidenced by the fact that the ratio of reduced carbon to carbonate rocks has remained almost constant at 1 : 4 since the Precambrian (Eichmann and Schidlowski, 1975; Garrels et al., 1975, 1976; Junge, 1977; Junge et al., 1975; Schidlowski et al., 1976).

This chapter deals with those biological factors which promote calcification and gives a short account of environments favouring biogenic or inorganic carbonate deposition.

BIOLOGICAL FACTORS IN CALCIFICATION

By far the most important biological controls on calcification are those which influence the equilibrium between the inorganic carboxy species (Chapter 2.1, p. 32). This may result from the consumption of CO_2 in photosynthesis, chemosynthesis and heterotrophic CO_2 fixation; the release of CO_2 during respiration and fermentation; and from the production of bases and acids due to varied metabolic activities.

Photosynthesis

Photosynthesis is the fundamental biochemical process which distinguishes plants, algae and certain bacteria from all other organisms. The process involves a complex series of reactions whereby radiant energy is converted to chemical energy and cell material is synthesized, generally from CO_2.

The underlying unity of oxygenic (algal and plant) and non-oxygenic (bacterial) photosynthesis was demonstrated by van Niel (1930) and is expressed by eqn (1):

$$CO_2 + 2\ H_2A \xrightarrow{\text{Light}} (CH_2O) + H_2O + 2\ A \qquad (1)$$

In oxygenic photosynthesis H_2A is water and A, oxygen (eqn (2)):

$$CO_2 + 2\ H_2O \xrightarrow{\text{Light}} (CH_2O) + H_2O + O_2 \qquad (2)$$

Tracer studies with ^{18}O demonstrated that all oxygen is derived from water.

In non-oxygenic photosynthesis H_2A is an inorganic sulfur compound (Eqns (3-5)) or an organic compound (e.g. eqn (6)) and A an oxidized product:

$$CO_2 + 2\ H_2S \xrightarrow{\text{Light}} (CH_2O) + H_2O + 2\ S \qquad (3)$$

$$3\ CO_2 + 2\ S + 5\ H_2O \xrightarrow{\text{Light}} 3(CH_2O) + 2\ H_2SO_4 \qquad (4)$$

$$\text{Sum: } 4\ CO_2 + 2\ H_2S + 4\ H_2O \xrightarrow{\text{Light}} 4(CH_2O) + 2\ H_2SO_4 \qquad (5)$$

$$CO_2 + 2\ \text{isopropanol} \xrightarrow{\text{Light}} (CH_2O) + H_2O + 2\ \text{acetone} \qquad (6)$$

It is interesting to note that bacterial photosynthesis may not necessarily result in a net CO_2 uptake; a net CO_2 evolution may occur in some circumstances. Species of purple photosynthetic bacteria, for example, can use organic compounds instead of CO_2, as the major carbon source for cell synthesis. When the exogenic carbon source is less oxidized than cell material oxidation of the exogenic carbon to the level of cell material is coupled to reduction and assimilation of CO_2. However if the carbon source is more oxidized than cell material, part of the substrate is necessarily oxidized anaerobically to CO_2.

According to Olson (1970) bacterial reactions, exemplified by eqns (3)-(6), may represent the most primitive forms of photosynthesis. They may, therefore, have had a major influence on calcification in the early Precambrian, prior to the development of the oxygen-evolving algae (see Chapter 6.1).

Carbonate precipitation controlled by photosynthetic CO_2 assimilation is a major process in lakes and in the formation of travertines. However, in the marine environment carbonate precipitation is influenced by many processes and in only a few instances has been shown to be the direct result of photosynthetic CO_2 assimilation. On the other hand, photosynthesis is a major factor in the natural carbonate cycle and controls the total carbonate equilibrium to the extent that a constant ratio between organic-C and carbonate-C is maintained in sediments even though little direct precipitational relation can be established between the two.

Dark CO_2 fixation

Fixation of CO_2 is not exclusively restricted to photosynthetic organisms. Chemolithotrophic microorganisms such as thiobacilli, nitrifying bacteria and iron-oxidizing bacteria incorporate CO_2 as their sole, or at least major carbon source. Moreover, heterotrophic microorganisms, which require organic carbon sources, can incorporate large amounts of CO_2. Under certain conditions, heterotrophic CO_2 fixation may even exceed photosynthetic assimilation in lakes and the marine environment (Cohen et al., 1977; Geyh et al., 1974; Romanenko, 1964); Thus carbonate precipitation may be caused by several types of organisms during certain periods in the productivity cycle.

Carbon dioxide is also fixed in the dark by photosynthetic organisms by the so-called Wood-Werkman reaction (Wood and Stjernholm, 1962). The CO_2 assimilated, however, rarely exceeds that formed by dark respiration; i.e. there is no net CO_2 uptake. On the other hand, the amount of organic carbon derived from CO_2 may be as high as 30% in heterotrophic bacteria and 90% in mixotrophic organisms. In the natural environment, non-photosynthetic CO_2 fixation by these organisms, together with the above-mentioned dark fixation by photosynthetic organisms, may under some conditions account for 90 to 100% of organic C-production.

The most important mechanism of heterotrophic CO_2 fixation is β-carboxylation of phosphoenol pyruvate (PEP) by PEP carboxylase. The products of β-carboxylation are regulators of intracellular pH and influence the fate of OH^- in the cell. β-carboxylation yields fairly strong acids from neutral molecules (e.g. carbohydrates) and the weak acid, H_2CO_3. This reaction is promoted in many plants in response to cytoplasmic alkalinity, when, for example, OH^- generated by nitrate reduction is not excreted. In algae, however, the OH^- resulting from nitrate reduction or photo-incorporation of HCO_3^- is excreted (Raven, 1974). The enzyme, ribulose diphosphate carboxylase, an important component of the CO_2-fixing mechanism in photo- and chemolithotrophs, can also function as an oxidase and be involved in the excretion of glycolate and glycolic acid. Thus the mechanisms of CO_2 assimilation in heterotrophs and lithotrophs are linked to those controlling cellular pH and ion exchange which, in turn, are associated with bicarbonate assimilation and Ca^{2+}-uptake.

Anaerobic respiration and fermentation

Diverse physiological activities of heterotrophic bacteria may create conditions favouring carbonate deposition. In anaerobic environments, fermenta-

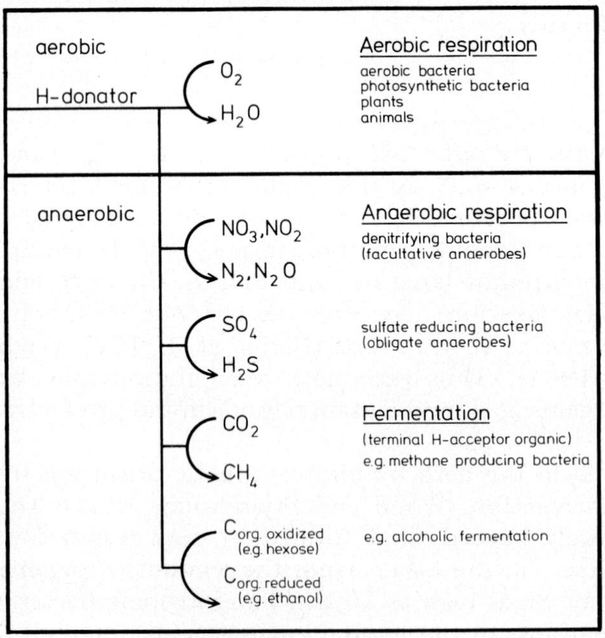

Fig. 2.2.1. Respirative pathways which are related to CO_2 exchange and carbonate precipitation by bacteria and algae.

tion and anaerobic respiration are of major importance (See Fig. 2.2.1).

In fermentation, relatively reduced organic compounds are oxidized at the expense of reduction of more oxidized organic matter. Simple organic acids and alcohols, CO_2, and sometimes H_2, are the usual products.

The term "anaerobic respiration" describes oxidation of organic matter by inorganic electron acceptors other than O_2. Two of the most important inorganic electron acceptors are nitrate and sulfate and their reduction may be accompanied by calcium carbonate formation, In nitrate reduction (denitrification) by heterotrophs, either N_2 or N_2O are formed and the overall reduction-calcification processes are expressed by eqns (7) and (8) where 'C' represents organic carbon:

$$Ca(NO_3)_2 + 3H_2 + 'C' \rightarrow CaCO_3 + 2\,H_2O + N_2 \qquad (7)$$

$$Ca(NO_3)_2 + 2\,H_2 + 'C' \rightarrow CaCO_3 + 2\,H_2O + N_2O \qquad (8)$$

Sulfate reduction is carried out by *Desulfovibrio* spp. and certain other bacteria (see Chapter 6.1). The reaction may yield $CaCO_3$ according to eqn (9):

$$CaSO_4 + 2(CH_2O) \rightarrow CaCO_3 + H_2S + CO_2 + H_2O \qquad (9)$$

Ammonia production

Precipitation of $CaCO_3$ is also promoted by ammonia which may be produced by organisms under both aerobic and anaerobic conditions. The most common reaction is oxidative deamination of amino acids which liberates ammonia according to eqn (10):

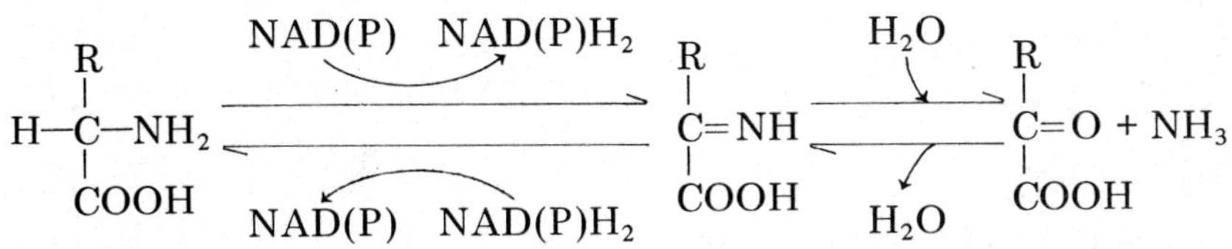

(10)

The general reactions by which $CaCO_3$ is formed, are shown in eqns (11) and (12):

$$2\,NH_4OH + Ca(HCO_3)_2 \rightarrow CaCO_3 + (NH_4)_2CO_3 + 2\,H_2O \qquad (11)$$

$$(NH_4)_2CO_3 + CaSO_4 \rightarrow CaCO_3 + (NH_4)_2SO_4 \qquad (12)$$

Uptake and transfer of CO_2

In most organisms, CO_2 diffuses passively into the cell. At high pH values, however, active transport of HCO_3^- has been demonstrated or assumed for many organisms (Raven, 1970, 1972, 1974; Kessler, 1974). Determination of whether bicarbonate assimilation takes place or not in a certain situation is complicated by the fact that the assumed external pH changes during bicarbonate assimilation may be influenced by pH changes caused by other processes occurring at the same time. Ammonia uptake, for example, lowers the external pH. Proton fluxes as a consequence of different net inflows of cations and anions can also modify external pH values. In the case of non-axenic cultures or measurements in nature, the pH of the milieu will be influenced by the total microbiological flora.

Once inside the cell, HCO_3^- is converted to CO_2 by the enzyme, carbonic anhydrase. CO_2 is then fixed by carboxydismutase and OH^- is excreted to maintain ionic balance. Carbonic anhydrase is also associated with the extracellular carbonate dissolution by boring organisms (Schneider, 1976) and with the CO_2-transfer system for intracellular calcification. It represents a key enzyme in the biological cycling of carbonate (Degens, 1976; Raven, 1974).

Alkalinity, pH and carbonate deposition and solution

Carbonate precipitation and solution is a function of pH and alkalinity, which is the capacity of water to neutralize hydrogen ions. The neutralizing capacity of carboxy ions is termed "carbonate alkalinity" (CA) while "total alkalinity" (TA) includes CA and that due to other anions (mainly boron species and OH^-) *minus* the offsetting effect of hydrogen ions present in solution.

Photosynthesis decreases CA but has no effect on TA. This is illustrated by eqns (13)-(16), for the formation of organic matter from the various carboxy species.

$$CO_3^{2-} + 2\,H^+ \rightarrow CH_2O + O_2 \qquad (13)$$

$$HCO_3^- + H^+ \rightarrow CH_2O + O_2 \qquad (14)$$

$$H_2CO_3 \rightarrow CH_2O + O_2 \qquad (15)$$

$$CO_2 + H_2O \rightarrow CH_2O + O_2 \qquad (16)$$

(*Editorial comment:* It should be noted that eqns (13)-(16) are summations and do not reflect the fact that O_2 is derived from water; see p. 48).

The influence of organisms on carbonate precipitation and dissolution cannot, however, be simply explained in terms of these photosynthetic reactions (Davies and Kinsey, 1973; Smith and Key, 1975; Brewer and Goldman, 1976). Brewer and Goldman (1976) also found that the system, $CO_2 \rightleftharpoons HCO_3^- \rightleftharpoons CO_3{}^{2-}$ cannot be adequately described in terms of pH and alkalinity. Thus the equilibria between carbonate formation and solution in aqueous solutions associated with organisms cannot be simply measured from TA, CA and pH.

Excretion of OH^- during bicarbonate assimilation is variable and depends largely on the differences between intracellular and external pH. These differences in turn are influenced by whether or not calcification takes place. *Sargassum* excretes OH^- in the absence of bicarbonate assimilation showing that the two processes may be uncoupled. Schneider (1976) described environments in which bicarbonate uptake was not coupled to OH^- excretion at high pH values. The dark precipitation of carbonate by Characeae and Zooxanthellae (Goreau, 1963) is an example of HCO_3^- uptake — OH^- excretion uncoupling.

Raven (1974) has given an account of pH changes, and, therefore, possible changes in the calcifying environment, of cells during H^+ uptake or release (see also Schneider 1976; Chapter 6.3). According to Raven and Smith (1974, 1976) an increase in pH of the solutions surrounding the cells can be observed during CO_2 fixation, HCO_3^- fixation, organic acid assimilation and NO_3^- assimilation and when the anion influx exceeds that of cations. These processes may then produce precipitation of carbonates. The pH of the environment decreases during NH_4^+ assimilation and the excretion of organic acids like lactate and glycolate, and when there is an excess of cation influx over anion influx. These reactions all modify the environment in favour of dissolution of carbonates. Since many of the pH-increasing reactions are directly or indirectly coupled with the pH-decreasing reactions, it is sometimes difficult to decide whether, for example, HCO_3^- uptake associated with NH_4^+ assimilation or glycolate excretion will, in fact, cause calcification in the environment.

Calcium uptake and deposition

The uptake of Ca^{2+} by organisms has been summarized by Wasserman and Kallfelz (1970) who concluded that Ca^{2+} may be either actively or passively transported through the cell membrane depending on the relative pH differences, calcium concentrations and binding sites. In most cells, calcium is actively transported outwards, resulting in a relatively low intracellular concentration of calcium. The endoplasmic reticulum (or in animal cells, the sarcoplasmic reticulum) acts as the calcium "pump" to maintain the calcium level in nerve cells or for intracellular calcification. On a microscopic scale, different types of calcification and Ca^{2+}-uptake or absorption have been demonstrated in Characeae and *Nitella* species depending on pH and the stage of development of the cells, or parts or a single cell (Arnott and Pautard, 1970; Pautard, 1970).

The Golgi apparatus and calcification

Calcification in eucaryotic organisms is frequently related to the Golgi apparatus (see review by Darley, 1974 and Degens, 1976). The Golgi and/or dictyosomes are defined areas of the cytoplasm and endoplasmic reticulum occurring exclusively in eucaryotes. The endoplasmic reticulum forms a continuum through the cytoplasmic region of eucaryotic cells and is continuous with the cell membrane. Many ribosomes — less abundant than in procaryotic cells — are aligned on the surfaces of the endoplasmic reticulum. The Golgi apparatus is much larger than the ribosomes and often extends from the nucleus to the anterior part of the cells. The Golgi apparatus, though different in shape and structure from the endoplasmic reticulum, seems to have similar structural units. The numerous membranes of the Golgi lie closer to one another than those of the endoplasmic reticulum. On one surface, the Golgi forms vacuoles. It is possible that cisternal membranes arise from the endoplasmic reticulum, become smoother, and change to smooth endoplasmic reticulum and later to Golgi bodies. Since the Golgi undergoes a continuous change in form, and since it connects the membrane system with the cell interior, it has been suggested, and later proved, that the Golgi serves as a packaging system in which several forms of cellular products are enclosed in a particular chemical environment within a particular morphological unit and then transferred to the cell surface. This mechanism is responsible for such diverse functions as the transfer of stinging organelles or nematocysts of coelenterates to the surface and the transfer of either carbonate plates, or organic matrices for carbonate plates, in many algae. It also provides the transfer system for removing unwanted or detrimental cellular products to the sites of excretion.

The Golgi may also be regarded as a membrane-interface between endoplasmic reticulum and plasmalemma.

It consists of three typical components:

(1) Membranes, lamellate and closely packed, which are connected at their ends and thus form a cavity that surrounds the vacuoles.

(2) Small vesicles which group together and, in some way, bud from the lamellate cisternae, but not the flat cisternae. As a result both rough and smooth surfaces are found in the Golgi apparatus.

(3) Large vacuoles which are grouped and usually surrounded by the membrane packages.

The vesicles derived from the Golgi cisternae have polysaccharide coatings which, after transfer, become located on the outside of the plasmalemma. The surfaces of the Golgi apparatus (forming and maturing faces) are involved in an intricate transfer system. The forming face is responsible for the replication of the Golgi by budding from the endoplasmic reticulum. At a later stage, the maturing face becomes responsible for the formation of vacuoles and secretory granules or plates. The other end of the Golgi apparatus is involved in the formation of plasmalemma.

According to Arnott and Pautard (1970) and Degens (1976) the organization of calcification, either intracellularly or on excreted organic matrices, may be specified by the genetic code. Degens (1976) assumed that a "carrier protein" serves as a template for an acidic polypeptide (the "mineralization matrix") which has a strong affinity for calcium ions (see Schiffmann et al., 1970 for a discussion of "calcifying matrices"). Combination of the carrier protein and mineralization matrix initiates crystallization of the mineral in question.

The carrier protein is formed in the Golgi apparatus on the inside of the cisternae and may either be transported outside the cell to combine with the mineralization matrix or begin calcifying within the cisternae and be subsequently extruded.

Other biological controls on calcification

Cellular constituents, including organic matter excreted by living organisms or released after death, may have profound effects on calcification. For example, Suess and Fütterer (1972) have shown that humic acid accelerates ooid growth. Chave and Suess (1970) and Suess (1970, 1973) have shown and discussed the fact, that organic substances form thin, sometimes monomolecular coatings on carbonate particles, thus influencing the solubilization and precipitation of carbonates in the marine environment. Similar results have recently been reported by Ferguson et al. (1978) and Davies et al. (1978). Phosphates have also often been mentioned as inhibitors of carbonate crystallization (e.g. Simkiss, 1964). Mitterer (1968, 1972) demonstrated that organic compounds such as citric acid or polypeptides in marine sediments may in some instances favour carbonate precipitation while in other circumstances they inhibit the reaction. Mitterer (1972) described calcified protein matrices in ooids and carbonate mud, which presumably resulted from extracellular activity of proteinaceous material.

CARBONATE PRECIPITATION BY BACTERIA AND ALGAE

In general, carbonate precipitation by bacteria can be regarded as environmentally-influenced precipitation not involving calcifying matrix systems or cellular organization. Calcification in and by algae is on the borderline between both environmental calcification and calcification controlled by the cellular apparatus, respectively.

Precipitation by bacteria

Pytkowicz (1965) concluded that the complete precipitation of calcium from seawater by inorganic mechanisms would require about 100 ky. The precipitation may be either hindered (Suess, 1970, 1973) or enhanced (Mitterer, 1968, 1972) by biological activities in seawater.

Bavendamm (1932) and, later, Krumbein (1971, 1972) summarized the evidence for $CaCO_3$ precipitation due to bacterially-induced environmental changes, and reviewed attempts to differentiate between biogenic and inorganic processes. Calcium carbonate precipitation in organic-rich anaerobic milieux has been demonstrated in the laboratory by Berner (1969, 1971), Krumbein (1969, 1973, 1974) and Morita et al. (1975).

In many of these experiments, the pH rose in the early stages of the experiments. In others, the pH was stable or at least rose only slowly. Once (Krumbein, 1974), the overall pH fell from the initial value. These contradictory pH effects, and several other field observations, indicate that $CaCO_3$ precipitation in nature under anaerobic conditions is rarely a single process connected to bacterial sulfate reduction. Acid formation (the end or intermediate products of organic degradation), ammonia production (see Eqn (10)) or saponification (Berner 1968, 1971) may have an influence on the pH, Eh, and partial pressure of CO_2 and other gases in the environment. Krumbein and Cohen (1977), Krumbein et al. (1977) and Jørgensen and Cohen (1977) have shown that, in some recent stromatolitic microbial mats, sulfate reduction and $CaCO_3$ formation are stoichiometrically related, organic carbon compounds being quantitatively transformed into carbonates throughout a depth profile of 90 cm. The minerals found during the different stages of lithification included monohydrocalcite, aragonite, high magnesium calcite, and dolomite. It is not yet clear why bacterial carbonate precipitation yields such a large variety of minerals, it appears not to depend on the relative abundances of Mg and Ca. Krumbein (1973, 1974, and unpublished data) found that, in laboratory experiments only aragonite, high-Mg calcite and monohydro-calcite were formed. Dolomite was never observed, even though in some experiments Mg was enriched about 5 times in comparison to the Mg : Ca ratios in seawater. However, it has been observed, both in laboratory and field experiments (Greenfield, 1963; Krumbein, 1973, 1974, unpublished data; Owen, 1973; Krumbein and Cohen, 1977; Krumbein et al., 1977) that Mg becomes bound to the organic matter. The subsequent decay and transformation of Ca- and Mg-organic complexes, such as fatty acid complexes or simple Mg salts of acetic acid, may, perhaps, result in the formation of dolomite.

Further examples of precipitation of $CaCO_3$ as the result of bacterial activities have been noted by Krumbein (1972, 1974), McCallum and Guhathakurta (1970), Pautard (1970), Morita et al. (1975) and Oppenheimer (1961). The reaction was observed in decaying algae by Adolphe and Billy (1974) and Billy (1967, 1975), and in peptone media with seawater by Shinano and Sakai (1969). $CaCO_3$ precipitation by heterotrophic bacteria was reported by Lalou (1957), Shinano (1972) and Shinano and Sakai (1969). Heterotrophic bacteria were also believed to be responsible for $CaCO_3$ formation in some fresh water environments (Billy, 1975; Krumbein, 1968). Mineral aggregates formed as the results of these processes were described by Krumbein (1973, 1974), McCallum and Guhathakurta (1970), Pochon et al. (1964), Puri and Collier (1967). Monty (1967), Trichet (1972) and Purdy (1963) assumed that pellets and pellet-like structures which are found in algal mats and algal decay environments may be of bacterial origin. Many of the oncoidal structures found in carbonate sediments and carbonate rocks have been shown to be of bacterial origin (Krumbein and Cohen, 1974; 1977). The chemolithotrophs include acid-forming bacteria which are generally involved in $CaCO_3$ solution rather than deposition. In some instances, however, *Thiobacillus denitrificans* may contribute to carbonate precipitation due to its denitrification activity. In summary then, there is a great deal of evidence that in many environments bacterial processes described by eqns (7) to (12) may be responsible for extracellular carbonate precipitation.

Cyanobacteria (see footnote on p. 12)

Carbonate precipitation by cyanobacteria has been extensively reviewed by Golubić (1973) and the role of these organisms in the formation of stromatolites is discussed in Walter (1976).

Precipitation may be due to an increase in environmental pH and/or a shift in the carbonate bicarbonate equilibrium system, as the result of photosynthetic CO_2 incorporation. However, deposition of the carbonates in the immediate vicinity of the cyanobacteria has also been demonstrated by Friedman (1955) for *Geitleria* and by Jaag (1945) for *Rivularia*. Golubić (1973) concluded that incrustation of cyanobacteria with carbonates is not species-specific and not exclusively dependent on photosynthetic activity. Potts and Krumbein (1978) demonstrated that the cyanobacteria *Pleurocapsa minor* and *Plectonema gloeophilum*, caused the precipitation of different kinds of carbonate particles at the same site in desert stromatolites.

Cyanobacterial lacustrine crusts in central Europe were first described by Kann (1941) and Pia (1934) and later re-evaluated by Schneider (1977). The ecology of the encrusting cyanobacteria has been described by Golubić and Fischer (1975) who reported that CO_2 may be repeatedly cycled over in the algal balls before it is finally fixed as $CaCO_3$. However, the dynamics of carbonate precipitation and dissolution in the "Furchensteine" or "furrowed stones" of freshwater environments is still controversial and different opinions on this subject were summarized by Golubić (1973) and Schneider (1977). In particular, it may be noted that deposition of irregular $CaCO_3$ deposits on cyanobacterial filaments of a recent stromatolitic environment (Krumbein and Cohen, 1977) appears to involve also the associated bacterial flora.

Precipitation by algae

Calcification in algae is a very common phenomenon and has been extensively reviewed by Arnott and Pautard (1970) and Pautard (1970). All algae are capable of precipitating $CaCO_3$ in the environment as the result of photosynthetic CO_2 incorporation but certain organisms play a more specific role in calcification. Two models of calcification have recently been proposed, which relate physiological control on interspace chemistry to assimilatory and respiratory mechanisms. Borowitzka and Larkum (1976) suggested that, in green algae, carbonate precipitation is due to the removal of both CO_2 and HCO_3^- from intracellular spaces. They calculated that the rate of calcification was 10% of the CO_2 assimilation rate. On the other hand Digby (1977) advanced a model for coralline algae in which excretion of acid occurs at the *outer* surface of a certain compartment while carbonate deposition takes place in intracellular spaces due to a spatial separation of the acidic and basic surfaces of the cell walls. This hypothesis relates carbonate deposition more to the splitting of water than the uptake of CO_2 from the external medium.

Chrysophyta

Detailed morphological and chemical descriptions of the process of calcification in the coccolithophorid members of the Chrysophyta have been given by Manton and Leedale (1969) and Paasche (1966, 1968). The best studied organisms are the marine *Cricosphaera carterae* (Manton and Leedale, 1969) and the freshwater *Hymenomonas roseola* (Manton and Peterfi, 1969). Milliman (1974) gives an account of the distribution of coccolithophorid muds and Loreau and Purser (1973) describe members from the Cretaceous. The chemical composition of coccolithophorids and coccolithophorid oozes has been described by Thompson (1972) who found that most of the recent marine coccolithophorid oozes are mainly made up from calcite. The calcite is very low in Mg and Sr but high in Mn, a rather surprising result for carbonate of marine origin.

Calcification involving the Golgi system in eucaryotic cells has been studied best in the *Haptophyceae*. Some of the *Haptophyceae* form organic surface scales of a very distinct structural pattern. In addition several of these scale-forming algae, the coccolithophorid members, produce calcified scales. The calcified organic matrix or "coccolith" and its formation from the Golgi have been described for *Coccolithus huxleyi*, *C. pelagicus*, and *Hymenomonas* sp. (Paasche, 1965, 1968; Paasche and Klaveness, 1970), for *Prymnesium* (Manton, 1966) and for *Chrysochromulina* (Manton, 1967).

Calcified scales formed inside the Golgi and then extruded have so far only been found in coccolithophorids. In other forms, the matrix is formed in the Golgi but calcification takes place in the cytoplasm. In some cases, the plates of coccolithophorids calcify inside the T-shaped cisternae where the stem is associated with the Golgi stack while the scale is already forming with its future distal surface facing the stack. The scales of coccolithophorids always consist of two layers of a cellulose network of microfibrils radially and concentrically arranged. The outer layer of the scale is made up of acidic polysaccharides and small amounts of protein, probably containing hydroxyproline. Whether this matrix calcifies or not is under genetic control.

Calcification in coccolithophorids is stimulated by light even when CO_2 fixation is inhibited. On the other hand, under normal conditions, there is a direct relation between scale formation and photosynthesis. It has been suggested therefore (Paasche, 1965) that light provides an additional energy source through photophosphorylation. It is interesting in this context that calcification in corals by symbiotic zooxanthellae is also stimulated by light. Coccolithophorids sink faster in the presence of calcified scales than in the naked state and it has been suggested that the light-scattering effect of the scales may enhance light trapping in deeper water (Blankley, 1971).

Rhodophyta (red algae)

The red algae are the most prominent and widely distributed carbonate-depositing algae. They occur in the northern seas as well as in the tropical waters and become encrusted in the intertidal zone as well as in water depths exceeding 50 m.

The coralline algae consist of vegetative and reproductive parts called thalli and conceptacles. They have been variously classified in two major groups, the Melobesoideae and Corallineae, or three groups, the Lithophylloideae, Mastophorideae, and Melobesoideae.

In general, encrusting species need hard ground for growth and many form algal balls or oncoids in both agitated and non-agitated environments. Several of the encrusting coralline algae are particularly rich and diversified in northern seas (e.g. *Lithothamnium* or *Lithophyllum*). Encrustation and calcification is generally an extracellular and intra-utricular process (see p. 58) and occurs in the dark as well as in the light. Nevertheless inhibitors of photosynthesis and carbonic anhydrase depress calcification in coralline algae. Moreover, the particular types of incrustation (high slender branches, tightly packed branches in round sometimes cauliflower-like heads or massive crustose growth) are now known to be species-dependent and are the result of photoactivated reactions rather than environmental energy as was once thought.

The importance of the encrusting and branching red algae in certain environments lies not so much in their contribution to calcification but rather in the cementation and structural development of reefs. Outstanding examples of the contribution of the coralline algae to carbonate deposition in recent environments are the algal cup reefs of the Bermudas (Schroeder, 1972), the terraces of encrusting algae of the intertidal zone in the Mediterranean, the massive incrustations of the rocky littoral of northern seas (e.g. Heligoland, North Sea), and the incrustation and oncoidal development in deep waters of tropical seas where there is little or no coral reef growth under heavy clastic sedimentation.

The main minerals produced by coralline algae are aragonite and some high-Mg calcite. There may be enrichment of Sr (Lowenstam, 1964). Calcification is mainly linked to metabolic CO_2 since most of the red algae are enriched in ^{13}C and ^{16}O. Baily and Bisalputra (1970) suggested that vesicles are involved in intra-utricular crystallization in coralline algae which implies that the activity of the Golgi apparatus is also involved (see p. 54).

Chlorophyta (green algae)

Among the chlorophyta, the most prominent calcifying groups are the Codiaceae and the Dasycladaceae. The Dasycladaceae were more important in past geological eras than at present but some of its members have been subjected to intensive study since they are relatively easy to cultivate and sufficiently large for single-cell experiments to be performed.

Among the Codiaceae are some of the best studied examples in terms of growth pattern, calcification speed, ecology, and distribution pattern in certain environments. *Penicillus*, *Udotea*, *Halimeda*, and several other species belong to this group. Growth and calcification of *Halimeda* have been studied extensively by Goreau (1963) and Wilbur et al. (1969): a plant can mature within 24 h, thus facilitating experimental studies. Calcification in these algae occurs first on fibril structures which are attached to the wall. There is no well-defined organic matrix recognizable although the intra-utricular liquid is separated completely from the outer environment.

Penicillus precipitates aragonite extracellularly (Perkins et al., 1972) and intracellularly, as well as in the intra-utricular spaces (Wilbur et al., 1969). The crystals vary considerably in size and form and two different mechanisms must be involved in their formation. The biochemistry of associated polysaccharides has been studied extensively by Böhm (1969, 1972, 1973). He found that 66-83% of excess calcium can be accounted for by the polysaccharide fraction. This contrasts with the situation in coralline algae where there are indications (Böhm, 1972) that brucite and silicate may be involved in cation excess. The calcareous green algae usually deposit aragonite although some calcite has been detected by scanning electron microscopy. The presence of rhombs is not conclusive evidence for calcite. The green algae tend to be enriched in Sr, and ^{13}C is enriched in the heads in relation to the stems. Many of the green calcareous algae produce carbonate in the form of small aragonite needles on and in organic material such as the fibrous material on the cell wall or the intra-utricular slime. They may contribute to carbonate sediments large amounts of carbonate which is difficult to identify and relate to the original plant.

Dinoflagellata

Several calcifying dinoflagellates have been described. In many cases, the cysts or resting cells are calcified while the vegetative cell is not (Fütterer

1976; Wall and Dale 1968). Sarjeant (1974) has summarised information on living and fossil dinoflagellates, and calcareous cyst formation. Calcification seems to follow a matrix controlled pattern. The contribution of dinoflagellates to the formation of carbonate sediments and rocks seems to be of minor importance.

CALCIFICATION ENVIRONMENTS

Mountains, deserts, rocks and soils are carbonate depositional environments of different magnitude and importance. In general, calcification in these environments is at a minimum at an average temperature of around 15°C and increases as the average temperature increases or decreases from this minimum. Thus, mountains, arctic and antarctic cold and dry areas, as well as subtropical and tropical areas are depositional environments, while humid areas have less calcification. Decreasing rainfall in continental environments also favours calcification. Desert areas (cold and warm), desert soils, subtropical soils, and dry mountain ridges are carbonate depositional environments. Many of these environments show special examples of biogenic carbonates (Potts and Krumbein, 1978). Evaporites in extremely hot environments are examples of completely abiogenic carbonate deposition.

Caves are important environments of carbonate deposition but, although there is some evidence for biogenic calcification within them (Friedman, 1955; Pochon et al., 1964; Schneider, 1977), it is probable that, here, inorganic carbonate deposition is the major factor.

Calcification and deposition of carbonates in rivers are uncommon but may be found in some stagnant parts, and in the upper reaches, when the deposits may occur as calcareous tufas, travertines, and sinters (Golubić, 1967, 1973). In some rivers, oncoids and algal balls are formed.

Many lakes have calcifying phytoplankton and contain varved carbonate sediments. The organisms are restricted in species but abundant in numbers. Precipitation of carbonates may be governed mainly by the cycle of productivity in the water body of the lake (Müller, 1966). Salt lakes, like the Great Salt Lake, and salt pans are calcifying environments in which evaporitic carbonates of abiogenic origin may form (Füchtbauer and Müller, 1970). Many lacustrine crusts occur in lakes of temperate regions (Kann, 1941; Schneider, 1977).

Shallow marine environments include coral and algal reefs as well as other bioherms and many favour calcification by benthic fauna. Stromatolites and stromatolitic environments are also typical shallow marine formations. The shallow marine carbonate environment may be subdivided into more or less agitated waters with dominantly benthic fauna, calm shallow areas with carbonate muds (e.g. Bahama Banks) with ooids as typical forms of deposits and reef areas with their complicated patterns of calcification and deposition (Bathurst, 1975; Kinsey and Davies, Chapter 2.5).

In supratidal and intertidal zones, many calcifying environments are encountered. Bacteria, algae, grasses and calcifying metazoans contribute to the calcification. According to some authors, only inorganic calcification occurs in beachrock environments, although others have presented evidence for biologically controlled calcification (Bricker, 1971; Krumbein, 1969, 1979).

Carbonates are found in almost all deep-sea sediments, the majority containing more than 30%. Nevertheless the deep sea is not generally regarded as a calcifying environment. In general, the "carbonate snow" originating from the productive areas redissolves below the lysocline. However, since the settling velocity exceeds the rate of dissolution, the surface sediments of many deep-sea areas still contain high percentages of biogenic carbonates. Almost no carbonate is found below a critical water depth of 5 km. Pteropods are found down to 3.5 km and globigerines at 4 km depth. Coccoliths may dominate the deep-sea carbonate fraction in specific areas such as the Black Sea (Degens and Ross, 1974) and the Mediterranean (Milliman, 1974). The last author, in an account of recent carbonate sediment distribution, concludes that deep-sea sediments contain 89% of the carbonates of the surface sediments of the world. Shallow water carbonates and slopes of the continental shelves account for only 4.8% and 5.8%, respectively. Only a negligible amount of terrestrial carbonates is of recent origin.

CONCLUDING REMARKS

Biological calcification and the biological cycle of carbonates control the availability of CO_2 and its storage in sediments and rocks. These in turn have a strong influence on climatic and other properties of the natural environment.

An example of the importance of this relationship is the burning of fossil fuel. This may increase CO_2 levels to such an extent as to endanger life on earth unless balanced by increased calcification and carbonate deposition.

In general, calcification is governed by photosynthetic and non-photosynthetic CO_2 fixation and, one may recognize an evolutionary sequence from simple chemical precipitation caused by the environmental effects of procaryotes to the highly organized membranous calcification systems found in eucaryotic algae and animals.

Since calcification depends strongly on environmental factors which in turn are subject to biological influences, it is probable that few instances of carbonate deposition in nature can be regarded as completely abiogenic.

ACKNOWLEDGEMENTS

I wish to acknowledge valuable discussions with J.A. Raven, critical typing by G. Koch and financial support of research work by Deutsche Forschungsgemeinschaft grants Kr 333/11 and 12.

REFERENCES

Adolphe, J.-P. and Billy, C., 1974. Biosynthèse de calcite par une association bactérienne aérobie. C.R. Acad. Sci. Paris, 278, Ser. D: 2873—2875.

Arnott, H.J. and Pautard, F.G.E., 1970. Calcification in plants. In: H. Schraer (Editor), Biological Calcification. Appleton-Century-Crofts, New York, NY, pp. 375—446.

Bailey, A. and Bisalputra, T., 1970. A preliminary account of the application of thinsectioning, freeze-etching, and scanning electron microscopy to the study of coralline algae. Phycologia, 9: 83—101.

Bathurst, R.G.C., 1975. Carbonate Sediments and Their Diagenesis. Developments in Sedimentology, 12, Elsevier, Amsterdam, 658 pp.

Bavendamm, W., 1932. Die mikroskopische Kalkfällung in der tropischen See. Arch. Mikrobiol., 3: 205—276.

Berner, R.A., 1968. Calcium carbonate concretions formed by decomposition of organic matter. Science, 159: 195—197.

Berner, R.A., 1969. Chemical changes affecting dissolved calcium during the bacterial decomposition of fish and clams in sea water. Mar. Geol., 7: 253—274.

Berner, R.A., 1971. Bacterial processes effecting the precipitation of calcium carbonate in sediments, In: O.P. Bricker (Editor), Carbonate Cements. Johns Hopkins University Studies in Geology, No. 19, Johns Hopkins, Baltimore, pp. 247—251.

Billy, C., 1967. Alginolyse et association bactérienne en milieu marin. Vie et Milieu, Ser. A, 18: 1—6.

Billy, C., 1975. Isolement des constituants d'une bactérienne productrice de calcite. C.R. Acad. Sci., Paris, 281: 621—623.

Blankley, W.F., 1971. Autotrophic and Heterotrophic Growth and Calcification in Coccolithophorids. Doctoral thesis, University of California, San Diego, CA, pp.

Böhm, E.L., 1969. Calcification in the Calcareous Alga *Halimeda opuntia* (L) *(Chlorophyta, Udoteaceae)*. Ph.D. thesis, University of the West Indies.

Böhm, E.L., 1972. Concentration and distribution of Al, Fe and Si in the calcareous alga *Halimeda opuntia*. Int. Rev. Ges. Hydrobiol., 57: 631—636.

Böhm, E.L., 1973. Composition and Ca-binding properties of the water soluble polysaccharides in the calcareous alga *Halimeda opuntia* (L) *(Chlorophyta, Udoteaceae)*. Int. Rev. Ges. Hydrobiol., 58: 117—126.

Borowitzka, M.A. and Larkum, A.W.D., 1976. Calcification in the green alga Halimeda III. The sources of inorganic carbon for photosynthesis and calcification and a model of the mechanism of calcification. J. Exp. Bot., 27 : 879—893.

Brewer, P.G. and Goldman, J.C., 1976. Alkalinity changes generated by phytoplankton growth. Limnol. Oceanogr., 21: 109—117.

Bricker, O.P. (Editor), 1971. Carbonate Cements. Johns Hopkins University Studies in Geology, No. 19, Johns Hopkins, Baltimore, 376 pp.

Chave, K.E. and Suess, E., 1970. Calcium carbonate saturation in sea water, effects of dissolved organic matter. Limnol. Oceanogr., 15: 633—638.

Cohen, Y., Krumbein, W.E. and Shilo, M., 1977. Solar Lake (Sinai). 2. Distribution of photosynthetic microorganisms and primary production. Limnol. Oceanogr., 22: 609—620.

Darley, W.M., 1974. Silifications and calcification. In: W.D.P. Stewart (Editor), Algal Physiology and Biochemistry. Blackwell Scientific Publications, Oxford, pp. 655—675.

Davies, P.J. and Kinsey, D.W., 1973. Organic and inorganic factors in beach rock formation, Heron Island, Great Barrier Reef. J. Sediment. Petrol., 43: 59—81.

Davies, P.J., Bubela, B., and Ferguson, J., 1978. The formation of ooids. Sedimentology, 25: 703—730.

Degens, E.T., 1976. Molecular mechanisms on carbonate, phosphate, and silica deposition in the living cell. Top. Curr. Chem., 64: 1—112.

Degens, E.T. and Ross, D.A. (Editors), 1974. The Black Sea. Geology, Chemistry, and Biology. Amer. Ass. Petrol. Geol., Tulsa, Oklahoma, 633 pp.

Digby, P.S.B., 1977. Photosynthesis and respiration in the coralline algae. *Clathromorphum circumscriptum* and *Corallina officinalis*, and the metabolic basis of calcification. J. Mar. Biol. Ass., 57: 1111—1124.

Eichmann, R. and Schidlowski, M., 1975. Isotopic fractionation between coexisting organic carbon-carbonate pairs in Precambrian sediments. Geochim. Cosmochim. Acta, 39: 535—595.

Ferguson, J., Bubela, B., and Davies, P.J., 1978. Synthesis and possible mechanism of formation of radial carbonate oids. Chem. Geol., 22: 285—308.

Friedmann, I., 1955, *Geitleria calcarea*, n. gen. n. sp., a new athmophytic lime-encrusting blue-green alga. Bot. Not., 108: 439—445.

Füchtbauer, H. and Müller, G., 1970. Sedimente und Sedimentgesteine. Schweizerbarth, Stuttgart, 726 pp.

Fütterer, D., 1976. Kalkige Dinoflagellaten ("*Calciodinelloideae*") und die systematische Stellung der *Thoracosphaeroideae*. Neues Jahrb. Geol. Palaeontol. Abh., 151: 119—141.

Garrels, R.M., Mackenzie, F.T. and Hunt, C., 1975. Chemical Cycles and the Global Environment. Kaufmann, Los Altos, CA, 206 pp.

Garrels, R.M., Lerman, A. and Mackenzie, F.T., 1976. Controls of atmospheric O_2 and CO_2: Past, present, and future. Am. Sci., 64: 306—315.

Geyh, M.A., Krumbein, W.E. and Kudrass, H.-R., 1974. Unreliable ^{14}C dating of long-stored deep-sea sediments due to bacterial activity. Mar. Geol., 17: M45—M50.

Golubić, S., 1967. Algenvegetation der Felsen. Eine ökologische Algenstudie im dinarischen Karstgebiet. Binnengewaesser, 1: 211—214.

Golubić, S. 1973. The relationship between blue-green algae and carbonate deposits. In: N.G. Carr and B.A. Whitton, The Biology of Blue-green Algae. Bot. Monogr. 9, Blackwell Scientific Publications, Oxford, pp. 434—472.

Golubić, S. and Fischer, A.G., 1975. Ecology of calcareous nodules forming in Little Conestoga Creek near Lancaster, Pennsylvania. Verh. Int. Verein. Limnol., 19: 2315—2323.

Goreau, T.F., 1963. Calcium carbonate deposition by coralline algae and corals in relation to their roles as reed-builders. Ann. N.Y. Acad, Sci., 109: 127—167.

Greenfield, L.H., 1963. Metabolism and concentration of Ca and Mg and precipitation of $CaCO_3$ by a marine bacterium. Ann. N.Y. Acad. Sci., 109: 23—45.

Jaag, O., 1945. Untersuchungen über die Vegetation und Biologie der Algen des nackten Gesteins in den Alpen, im Jura und im schweizerischen Mittelland. Beitr. Kryptogamenflora Schweiz, 9: 560 pp.

Jørgensen, B.B. and Cohen, Y., 1977. Solar Lake (Sinai), 5. The sulfur cycle of the benthic cyanobacterial mats. Limnol. Oceanogr., 22: 657—666.

Junge, C.E., 1977. Stable isotope fractionation in geochemical and environmental cycles. In: W. Stumm (Editor), Dahlem Conference on Global Chemical Cycles and their Alterations by Man. Abakon. Berlin, pp. 33—44.

Junge, C.E., Schidlowski, M., Eichmann, R. and Pietrek, 1975. Model calculations for the terrestrial carbon cycle: Carbon isotope geochemistry and evolution of photosynthetic oxygen. J. Geophys. Res., 80: 4542—4582.

Kann, E., 1941. Krustensteine in Seen. Arch. Hydrobiol., 37: 504—532.

Kessler, E., 1974. Hydrogenase, photoreduction and anaerobic growth. In: W.D.P. Stewart (Editor), Algal Physiology and Biochemistry. Blackwell Scientific Publications, Oxford, pp. 456—473.

Krumbein, W.E., 1968. Geomicrobiology and geochemistry of Nasi-Lime—Crust in Israel. In: G. Müller and G.M. Friedman (Editors), Recent Developments in Carbonate Sedimentology in Central Europe. Springer, Heidelberg, pp. 138—147.

Krumbein, W.E., 1969. Über den Einfluss der Mikroflora auf die exogene Dynamik (Verwitterung und Krustenbildung). Geol. Rundsch., 58: 333—363.

Krumbein, W.E., 1971. Sedimentmikrobiologie und ihre geologischen Aspekte. Geol. Rundsch., 60: 438—471.

Krumbein, W.E., 1972. Rôle des microorganismes dans la génèse, la dégradation et la diagénèse des roches en place. Rev. Ecol. Biol. Sol., 9: 283—319.

Krumbein, W.E., 1973. Mikrobiologische Untersuchungen zur Fällung von Kalziumkarbonat aus Meerwasser. In: Biologische Anstalt Helgoland, Jahresberichte, Hamburg, pp. 50—54.

Krumbein, W.E., 1974. On the precipitation of aragonite on the surface of marine bacteria. Naturwissenschaften, 61: 167.

Krumbein, W.E., 1979. Photolithotrophic and chemoorganotrophic activity of bacteria and algae as related to beachrock formation and degradation.(Gulf of Aqaba, Sinai). Geomicrobiol. J., in press.

Krumbein, W.E. and Cohen, Y., 1974. Biogene, klastische und evaporitische Sedimentation in einem mesothermen monomiktischen ufernahen See (Golf von Aqaba). Geol. Rundsch. 63: 1 035—1965.

Krumbein, W.E. and Cohen, Y., 1977. Primary production, mat formation and lithification: contribution of oxygenic and facultative anoxygenic cyanobacteria. In: E. Flügel (Editor), Fossil Algae. Springer, Berlin, pp. 38—56.

Krumbein, W.E., Cohen, Y., and Shilo, M., 1977. Solar Lake (Sinai). 4. Stromatolitic cyanobacterial mats. Limnol Oceanogr., 22: 635—656.

Lalou, C., 1957. Studies on the bacterial precipitation of carbonate in sea water. J. Sediment. Petrol., 27: 190—197.

Li, Y.-H., 1972. Geochemical mass balance among lithosphere, hydrosphere, and atmosphere. Am. J. Sci., 272: 119—137.

Loreau, J.P. and Purser, B.H., 1973. Distribution and ultrastructure of holocene ooids in the Persian Gulf. In: B.H. Purser (Editor), The Persian Gulf, Springer, Berlin, pp. 297—327.

Lowenstam, H.A., 1964. Coexisting calcites and aragonites from skeletal carbonates of marine organisms and their strontium and magnesium contents. In: Recent Researches in the Fields of Hydrosphere, Atmosphere and Nuclear Geochemistry. Maruzen and Co., Tokyo, pp. 373—404.

Manton, I., 1966. Observations on scale production in *Prymnesium parvum.* J. Cell Sci., 1: 375—380.

Manton, I., 1967. Further observations on scale formation in *Chrysochromulina* chiton. J. Cell Sci., 2: 411—418.

Manton, I. and Leedale, G.F., 1969. Observations on the microanatomy of *Coccolithus pelagicus* and *Cricosphaera carterae,* with special reference to the origin and nature of coccoliths and scales. J. Mar. Biol. Ass. U.K., 49: 1—16.

Manton, I. and Peterfi, L.S., 1969. Observations on the fine structure of coccoliths, scales, and the protoplast of a freshwater coccolithophorid, *Hymenomonas roseola*

Stein, with supplementary observations on the protoplast of *Cricosphaera carterae*. Proc. R. Soc. B., 172: 1—15.

McCallum, M.F. and Guhathakurta, K., 1970. The precipitation of calcium carbonate from seawater by bacteria isolated from Bahama Bank sediments. *J. Appl. Bacteriol.*, 33: 649—655.

Milliman, J.D., 1974. Marine Carbonates. Springer, Berlin, 375 pp.

Mitterer, R.M., 1968. Amino acid composition of organic matrix in calcareous oolites. Science, 162: 1498—1499.

Mitterer, R.M., 1972. Calcified proteins in the sedimentary environment. In: H.R. von Gaertner and H. Wehner (Editors), Advances in Organic Geochemistry, Pergamon, Oxford, pp. 441—452.

Monty, C.L.V., 1967. Distribution and structure of recent stromatolitic algal mats, Eastern Andros Island, Bahamas. Ann. Soc. Geol. Belg., 90: 55—99.

Morita, R.Y., Cuff, C., and Patrick, J., 1975. Anaerobic microbial precipitation of calcite: A preliminary study. Abstract of Ann. Meeting, 1975. Am. Soc. Microbiol., p. 190.

Müller, G., 1966. Die Sedimentbildung im Bodensee. Naturwissenschaften, 53: 237—247.

Olson, J.M., 1970. The evolution of photosynthesis. Science, 168: 438—446.

Oppenheimer, C.H., 1961. Note on the formation of spherical aragonitic bodies in the presence of bacteria. Geochim. Cosmochim. Acta, 23: 295—296.

Owen, R.W., 1973. Red Sea Algal Sediments and the Hoyt Limestone of New York: A Comparison of Recent and Cambrial Algal Deposition. Thesis, Rensselaer Polytechnic Institute, Troy, New York, 117 pp.

Paasche, E., 1965. The effect of 3-(p-chorophenyl)-1, 1-dimethylurea (CMU) on photosynthesis and light-dependent coccolith formation in *Coccolithus huxleyi.* Physiol. Plant., 18: 138—145.

Paasche, E., 1966. Adjustment to light and dark rates of coccolith formation. Physiol. Plant., 19: 271—278.

Paasche, E., 1968. Biology and physiology of coccolithophorids. Ann. Rev. Microbiol., 22: 71—86.

Paasche, E. and Klaveness, D., 1970. A physiological comparison of coccolith-forming and naked cells of *Coccolithus huxleyi.* Arch, Mikrobiol., 73: 143—152.

Pautard, F.G.E., 1970. Calcification in unicellular organisms. In: H. Schraer (Editor), Biological Calcification. North Holland, Amsterdam, pp. 105—201.

Perkins, R.D., McKenzie, M.D. and Blackwelder, P.L., 1972. Aragonite crystals with Codiacean algae: distinctive morphology and sedimentary implications. Science, 175: 624—626.

Pia, J., 1934. Die Kalkbildung durch Pflanzen. Beih. Bot. Zentralbl., 52, Abt. A: 1—72.

Pochon, J., Chalvignac, M. and Krumbein, W.E., 1964. Recherches biologiques sur le Mondmilch, C.R. Acad. Sci. Paris, 258: 5113—5115.

Potts, M. and Krumbein, W.E., 1978. Desert stromatolites; genetic control of calcification in calcrete by blue-green algae. Abstract X. International Congress on Sedimentology, Jerusalem: 521—522.

Purdy, E.G., 1963. Recent carbonate facies of the Great Bahama Bank. J. Geol., 71: 334—355 and 472—497.

Puri, H.S. and Collier, A., 1967. Role of microorganisms in formation of limestones. Trans. Gulf. Coast. Assoc. Geol. Soc., 17: 355—367.

Pytkowicz, R.M., 1965. Rates of inorganic calcium carbonate nucleation. J. Geol., 73: 196—199.

Raven, J.A., 1970. Exogenous inorganic carbon sources in plant photosynthesis. Biol. Rev., 45: 167—221.

Raven, J.A., 1972. Endogenous inorganic carbon sources in plant photosynthesis. II. Comparison of total CO_2 production with measured CO_2 evolution in the light. New Phytol., 71: 995—1014.

Raven, J.A., 1974. Carbon dioxide fixation. In: W.D.P. Stewart (Editor), Algal Physiology and Biochemistry. Blackwell Scientific Publications, Oxford, pp. 434—455.

Raven, J.A. and Smith, F.A., 1974. Significance of hydrogen ion transport in plant cells. Can. J. Bot., 52: 1035—1048.

Raven, J.A. and Smith, F.A., 1976. Cytoplasmic pH regulation and electrogenic H^+ extrusion. Curr. Adv. Pl. Sci., 24: 649—660.

Romanenko, W.I., 1964. Heterotrophic assimilation of CO_2 by the aquatic microflora. Mikrobiologiya, 33: 679—683.

Sarjeant, W.A.S., 1974. Fossil and Living Dinoflagellates. Academic, London, 182 pp.

Schidlowski, M., Eichmann, R. and Junge, C.E. 1976. Carbon isotope geochemistry of the Precambrian Lomagundi carbonate province, Rhodesia. Geochim. Cosmochim. Acta, 40: 449—455.

Schiffmann, E., Martin, G.R. and Miller, E.J., 1970. Matrices that calcify. In: H. Schraer (Editor), Biological Calcification, Cellular and Molecular Aspects. North Holland, Amsterdam, pp. 27—67.

Schneider, J., 1976. Biological and Inorganic Factors in the Destruction of Limestone Coasts. Contributions to Sedimentology, No. 6, Springer, Berlin, 112 pp.

Schneider, J., 1977. Carbonate construction and decomposition by epilithic and endolithic micro-organisms in salt- and freshwater. In: E. Flügel (Editor), Fossil Algae, Springer, Berlin, pp. 248—260.

Schroeder, J.H., 1972. Fabrics and sequences of submarine carbonate cements in Holocene Bermuda cup reefs. Geol. Rundsch., 61: 708—730.

Shinano, H. and Sakai, M., 1969. Studies of marine microorganisms taking part in the precipitation of calcium carbonate. I. Calcium carbonate precipitation in peptone medium prepared with seawater and artificial seawater. Bull. Jpn. Soc. Sci. Fisheries, 35: 1001—1969.

Shinano, H., 1972. Studies of marine microorganisms taking part in the precipitation of calcium carbonate. V. A taxonomic study of marine bacteria taking part in the precipitation of calcium carbonate. Bull. Jpn. Soc. Sci. Fisheries, 38: 833—838.

Simkiss, K., 1964. Phosphates as crystal poisons in calcification. Biol. Rev., 39: 487—505.

Smith, S.V. and Key, G.S., 1975. Carbon dioxide and metabolism in marine environments. Limnol. Oceanogr., 20: 493—495.

Suess, E., 1970. Interaction of organic compounds with calcium carbonate. I. Association phenomena and geochemical implications. Geochim. Cosmochim. Acta, 34: 157—168.

Suess, E., 1973. Interaction of organic compounds with calcium carbonate. II. Organo-carbonate association in recent sediments. Geochim. Cosmochim. Acta, 37: 2435—2447.

Suess, E. and Fütterer, D., 1972. Aragonitic ooids: experimental precipitation from seawater in the presence of humic acid. Sedimentology, 19: 129—139.

Thompson, G., 1972. A geochemical study of some lithified carbonate sediments from the deep sea. Geochim. Cosmochim. Acta, 36: 1237—1253.

Trichet, J., 1972. Etude du mecanisme de la nucléation du carbonate de calcium dans les dépôts dus à des Cyanophycées (Polynésie Française). C.R. Congr. Géologique Internationale, Montréal, Sect. 7, pp. 631—638.

Van Niel, C.B., 1930. Photosynthesis of bacteria. In: Contributions to Marine Biology. Stanford University Press, Palo Alto, CA, p. 158.

Wall, D. and Dale, B., 1968. Modern dinoflagellate cysts and evolution of the Peridiniales. Micropaleontology, 14: 265—304.

Walter, M.R. (Editor), 1976. Stromatolites. Developments in Sedimentology, 20, Elsevier, Amsterdam, 790 pp.

Wassermann, R.H. and Kallfelz, F.A., 1970. Transport of calcium across biological membranes, In: H. Schraer (Editor), Biological Calcification, Cellular and Molecular Aspects, North Holland, Amsterdam, pp. 313—345.

Wilbur, K.M., Colinvaux, L.H. and Watabe, N., 1969. Electron microscope study of calcification in the alga *Halimeda* (order *Siphonales*). Phycologia, 8: 27—35.

Wood, H.G. and Stjernholm, R.L., 1962. Assimilation of carbon dioxide by heterotrophic microorganisms. In: I.C. Gunsalus and R.Y. Stanier (Editors), The Bacteria, Vol. 3, Academic, New York, NY, pp. 41—117.

Chapter 2.3

CARBONATE TURNOVER AND DEPOSITION BY METAZOA

K.M. WILBUR

Department of Zoology, Duke University, Durham, NC 27706 (U.S.A.)

K. SIMKISS

Department of Zoology, The University of Reading, Whiteknights, Reading RG6 2AJ (England)

CONTENTS

INTRODUCTION

Although there are potentially a large number of carbonate minerals that might be formed by metazoa, the only important ones are the polymorphic forms of calcium carbonate: high and low magnesian calcites, aragonite, and much less commonly, vaterite. The reasons for this have been discussed elsewhere (Wilbur and Simkiss, 1968). The occurrence of calcium carbonate as the major mineral of invertebrate skeletons has two important consequences for the relationship of biogenic mineralization to inorganic processes of the environment.

The first concerns natural cycles. The natural cycles distributing calcium and carbonate ions on the earth's surface are well known and have been identified and summarized by Skirrow (1975). Estimates of the sizes of the calcium and carbon dioxide compartments in these schemes and of the fluxes between them are given by Pytokowicz (1973). It is therefore possible to identify the importance of the metazoa in these schemes. Terrestrial metazoa such as annelids and molluscs clearly play an important role in the redistribution of calcareous minerals over the surface of the land. In aquatic systems, the function of metazoa is even more important for, in their absence, many of the natural cycles would be disrupted. The reason for this is the capacity of invertebrates to form relatively insoluble calcium carbonate from solutions from which they would not precipitate spontaneously. It is now generally agreed that the oceans are supersaturated with respect to Ca^{2+} and CO_3^{2-} to depths of 0.1—1 km. Below these levels, the effects of increasing pressure and increasing CO_2 concentrations result in undersaturation even though the sea water in those regions may be in contact with calcareous sediments (Skirrow, 1975). Minerals from the sea bed may be carried to the surface by upwelling but, because they remain there as a supersaturated solution, they are dependent upon some form of biomineralization to return them as sediments. A somewhat similar phenomenon occurs in freshwaters (Otsuki and Wetzel, 1974), so it is possible in a very simplistic way to envisage organisms as catalyzing the crystallization of minerals from their surroundings.

A second effect of metazoa on inorganic processes of the environment relates to their capacity for *rapid* deposition of calcium carbonate as compared to inorganic precipitation in natural waters. Many body fluids of metazoa, like sea water, are supersaturated with calcite (Potts, 1954) but, whereas Pytkowicz (1965) estimated that it would take 0.1 My to nucleate crystals of calcium carbonate from sea water, most metazoa are able to do this very rapidly and at precise sites. In addition, the type of calcium carbonate that is normally precipitated from sea water is aragonite, apparently because of the influence of magnesium ions (Simkiss, 1964), whereas many marine metazoa form skeletons of calcite, suggesting that they might control the ionic composition of the fluids at the sites of mineralization. Some, such as the arthropods, may deposit skeletons which are very poorly crystalline.

This comparison between inorganic processes and the more complex systems found in organisms will serve as a framework in our discussion of biological mineralization, and we shall make frequent reference to these processes in speaking of carbonate deposition by metazoa. The bases for differences between living and inanimate systems to which we have referred will be approached in three ways: first, as descriptions of the methods by which skeletons are formed in several invertebrate phyla; second, as a summary of the major kinds of cellular systems that are involved; and third, as an analysis of the chemical and physiological parameters of these systems.

SKELETAL FORMATION OF INVERTEBRATE PHYLA

Porifera

The mineral of modern calcareous sponges occurs as magnesian calcite spicules which take a great variety of forms with lengths from 40 to 1,200 μm (Jones, 1970). The spicules may be free in the sponge tissue or organized into a loose or articulated supporting framework, and the orientation of the spicule appears to be determined by the cells which form it (Jones, 1967). Each spicule is a single birefringent crystal. Spicule analysis has indicated a content of 5.2—12.9 mol% $MgCO_3$ and the presence of SO_4, Fe_2O_3, Sr, Li, Na, Ba, and Mn (Jones and James, 1969; Jones, 1970; Jones and Jenkins, 1970).

The developing spicules are enclosed in a thin retractile organic membrane which, in larger spicules, occurs as a thicker sheath. Formation of the sheath and the deposition of carbonate are accomplished extracellularly by single cells called sclerocytes. The spicules have one to four rays and each ray is

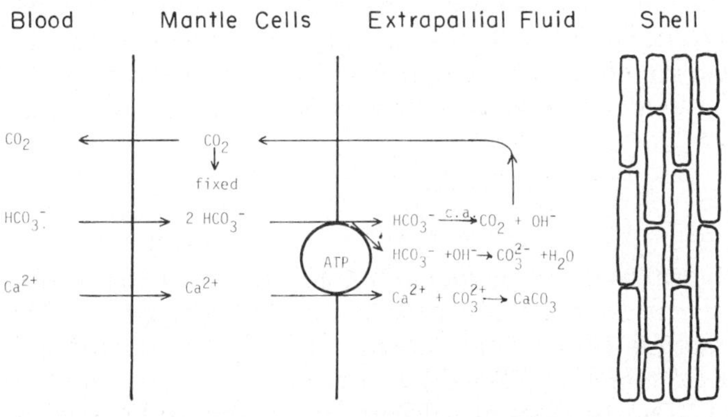

Fig. 2.3.1. Proposed pathways and reactions of ions in the formation of $CaCO_3$ of shell. The scheme includes active transport of Ca^{2+} (Wood, 1973) and HCO_3^- (Wheeler, 1975), the action of carbonic anhydrase (c.a.) (Wheeler, 1975), and CO_2 fixation (Hammen and Wilbur, 1959).

formed by sclerocytes moving along it and adding $CaCO_3$ to the crystal lattice. The number of sclerocytes which produce a spicule will depend upon the type of spicule. For example, each needle-shaped monaxon is formed by two sclerocytes whereas three-rayed spicules are formed by six sclerocytes. Details of the association of sclerocytes and spicules have been described by Jones (1970) and Ledger and Jones (1977). It appears that in depositing $CaCO_3$, the sclerocytes take in Ca^{2+} and HCO_3^- and remove H^+ or produce OH^- to the degree that crystalline $CaCO_3$ is formed (see Fig. 2.3.1). The deposition rate will be a function of the rate of transport of Ca^{2+} and HCO_3^- and has been measured by microscopic observation of the increase in length of individual spicules (Jones, 1964). In sponge juveniles, it has been found that the amount of spicule formation is related to the calcium concentration of the medium (Ledger and Jones, 1977). Information on the initiation of crystal nucleation in calcareous sponge spicules is completely lacking. In fact, it is uncertain whether organic material which might function in nucleation is present within calcareous spicules (Ledger and Jones, 1977), although there is evidence that siliceous spicules do contain organic compounds (see Wilbur, 1976). Interestingly, siliceous sponge spicules are formed within cells rather than extracellularly as in calcareous spicule deposition (Simpson, 1968; Simpson and Vaccaro, 1974).

The skeleton of the coralline sponges (Sclerospongia) in contrast to the calcareous sponges (Calcispongia) is compact rather than spicular and may be several centimetres in size (Hartman and Goreau, 1970). The coralline sponges also differ in that they may play an important role in reef formation, especially below depths of about 80 m (see Milliman, 1974). Their unit of structure is the sclerodermite which consists of a small organic centre from which aragonitic crystals radiate toward other centres which together make up the crystalline mass. Long siliceous spicules are trapped between the aragonitic crystals. Neither the details of carbonate deposition nor the cells responsible are known, although it appears that the crystals of the sclerodermites develop in association with organic material around the heads of the siliceous spicules.

Coelenterata

The coelenterates which play a major role in marine carbonate deposition include the Hydrozoa, specifically the Stylasterina and Milleporina, the Octocorallia and the Scleractinia (Zooantharia). The mineral composition and crystal type of the skeletal material formed by these groups have been tabulated by Milliman (1974). Our discussion will largely concern skeletogenesis in the Scleractinia since these corals have received primary attention in experimental studies.

Calcification in Scleractinia

Anatomical aspects. The skeletons of corals are formed of crystals of aragonite in an organic matrix. Calcification begins when the free-swimming larva attaches to the substratum. The larva then metamorphoses into a polyp which continues to deposit extracellularly $CaCO_3$ and organic matrix through the secretory activity of the calicoblastic epithelium. The epithelium is a layer of cells at the base of the polyp adjoining the skeletal area at which deposition takes place.

In the scleractinian corals, the calicoblastic epithelium forms a basal plate and, through tissue growth, the edges of the plate are extended as a crystalline sheet of $CaCO_3$ called the epitheca (Barnes, 1972). The form of the epitheca varies with species. It may serve as the substratum over which the calicoblastic epithelium deposits $CaCO_3$, thus thickening the skeleton. In order for the epitheca to form extracellularly, the site of calcification facing the calicoblastic epithelium must be isolated from the sea-water medium, and this is accomplished by a lappet of tissue adjoining the calicoblastic body wall and enclosing the site of secretion (Barnes, 1972). Within this lappet cavity, the primary epithecal layer is deposited. Once deposited, the primary layer itself provides a closed environment in which further deposition can take place.

Calcification mechanisms. The mechanisms of $CaCO_3$ deposition have not been clearly defined. Two major questions are involved: first, the relation of the organic portion of the skeleton to the initiation and control of crystal growth and, second, the role of the symbiotic photosynthetic zooxanthellae, the algae which grow within the tissue of hermatypic corals (Chapman, 1974). Another persistent question has been the site of crystal initiation, whether it is entirely extracellular or whether it is partly intracellular as well (Muscatine, 1971).

Site of crystal initiation. Recent evidence for intracellular calcification comes from observations of crystals forming in association with protein within Golgi vesicles (Hayes and Goreau, 1976, 1977). The crystals are thought to be released from the cells and to initiate extracellular crystal formation (Hayes and Goreau, 1977). An examination of the problem in recently settled *Pocillopora* (Vandermeulen, 1975) has indicated that intracellular storage is insufficient to account for the observed amount of skeletogenesis. The other view, that of extracellular crystallization, considers that crystals form external to the calicoblastic epithelium in an organic material containing mucopolysaccharide secreted by the epithelium (Goreau, 1959; Vandermeulen, 1975). It has been suggested that the organic material may play a role in crystal nucleation and in the control of crystal size, shape, and orientation (Goreau, 1959, 1963; Wainwright, 1963). The presence in coral skeletons and alcyonarian spicules of a high content of aspartic acid residues, in addition to those which may be in the amide form, has led to the suggestion that carboxyl groups on protein may possibly be involved in calcium binding and calcification (Mitterer, 1978). Extracellular calcification could occur by either of two mechanisms: (1) If sea water supersaturated with calcium and carbonate ions had access to the secreted organic matrix *temporarily*, the calicoblast cells could create an environment in which crystal nucleation and growth could occur by removing H^+ and any inhibitors of crystal nucleation from the extracellular space; (2) Crystals could also be formed by the transport of Ca^{2+} and HCO_3^- from the calicoblast cells to the site of crystal formation isolated from sea water (see also Barnes, 1970) with a subsequent removal of H^+.

Sources of skeletal carbonate. Metabolic CO_2 contributes to the formation of skeletal carbonate. Crossland and Barnes (1974) have proposed that the hydrolysis of urea by urease may be one of the sources of CO_2 which is converted to carbonate. The NH_3 produced from the urea would neutralize H^+ resulting from $CaCO_3$ formation and so favor $CaCO_3$ deposition (Campbell and Speeg, 1969; Crossland and Barnes, 1974).

The $CaCO_3$ of the coral skeleton comes from Ca^{2+} and HCO_3^- of sea water, as shown by the uptake of radioisotopes (Goreau, 1963; Crossland and Barnes, 1974), in addition to carbonate derived from metabolic CO_2 (Pearse, 1970; Crossland and Barnes, 1974). The origin of skeletal carbonate can accordingly be considered as being derived from two compartments, the external medium and the "milieu interieur" of the coral (Weber and Woodhead, 1970). A part of the metabolic CO_2 is utilized by the zooxanthellae and may be incorporated subsequently into $CaCO_3$ and the organic matrix. This pathway will not occur in ahermatypic corals which lack zooxanthellae. The relative contributions to skeletal carbonate of carbon originating within the organism and from sea water can be determined by measuring the $^{13}C/^{12}C$ ratio of the carbonate. For example, the ratio of $^{13}C/^{12}C$ in photosynthate produced by zooxanthellae is lower than that in HCO_3^-, CO_3^{2-}, and H_2CO_3 in sea water (Land et al., 1975). If the coral utilizes both photosynthetically-fixed carbon and that from sea water in depositing skeletal material, the $^{13}C/^{12}C$ ratio will be intermediate, and its value will depend upon the proportions of carbon from the two sources. In addition to photosynthesis, the lighter, smaller isotope (^{12}C) will become enriched relative to its concentration in media in processes involving diffusion because it diffuses faster and has a lower bond energy that its heavier counterpart.

The $^{13}C/^{12}C$ ratios are commonly expressed as $\delta^{13}C$ values in which

$$\delta^{13}C = \left[\frac{(^{13}C/^{12}C)\ \text{sample} - (^{13}C/^{12}C)\ \text{standard}}{(^{13}C/^{12}C)\ \text{standard}}\right] \times 1000.$$

The $\delta^{13}C$ values of hermatypic corals from widely separated localities were found to be in the range of approximately +1 to −2 when expressed as the mean ±1 standard deviation, whereas ahermatypic corals fell in the range of approximately −5 to −9 (Weber, 1974, Figs. 2.3.2 and 2.3.3). However, there are exceptions to the narrow range for hermatypic species (Land ct al., 1975).

The concept of two compartments, the external medium and the milieu interieur, which undergo mixing (Weber and Woodhead, 1970) is almost certainly an oversimplification and has been brought into question by Land et al. (1977) and by Goreau (1977). Many factors, both environmental and internal, affect the relative amounts of ^{13}C and ^{12}C fixed in the coral skeleton with the result that precise interpretations of the $^{13}C/^{12}C$ ratios are not possible. The complexity of interpreting the ratios is indicated by the following list of factors which have been put forward as possible influences on the ratios: (1) physiological activities and membrane permeabilities which may influence movement of isotopes within the organism and across its outer covering; (2) growth rate (Land et al., 1975); (3) seasonal fluctuations in $\delta^{13}C$ which may be related to differences in growth rate (Goreau, 1977); (4) linked carbon pools within the coral relating to biochemical pathways and the translocation of intermediates and synthesized compounds (Goreau, 1977); (5) isotopic composition of the zooxanthellae which may contribute photosynthetic products to the skeleton (Land et al., 1975); (6) diurnal variation in the synthesis and utilization of compounds; (7) isotopic concentration of food taken in by the coral (Smith, 1972; Land et al., 1977); and (8) rate of diffusion of metabolic CO_2 from coral tissue which in turn may be influenced by ambient conditions in the microenvironment of the coral

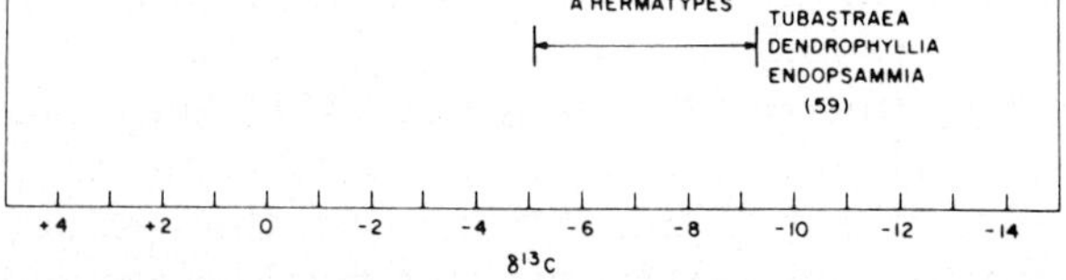

Fig. 2.3.2. Comparison of the range in carbon isotopic composition for hermatypic corals, ahermatypic corals, and marine carbonate precipitated in isotopic equilibrium with the $CO_3^{2-} - HCO_3^- - CO_2$ system of ambient seawater. All of the corals (111 hermatypes and 59 ahermatypes) were collected from shallow-water, reef surface environments at Heron Island, Australia. The length of the horizontal bars is equivalent to the mean ± one standard deviation (Weber, 1974).

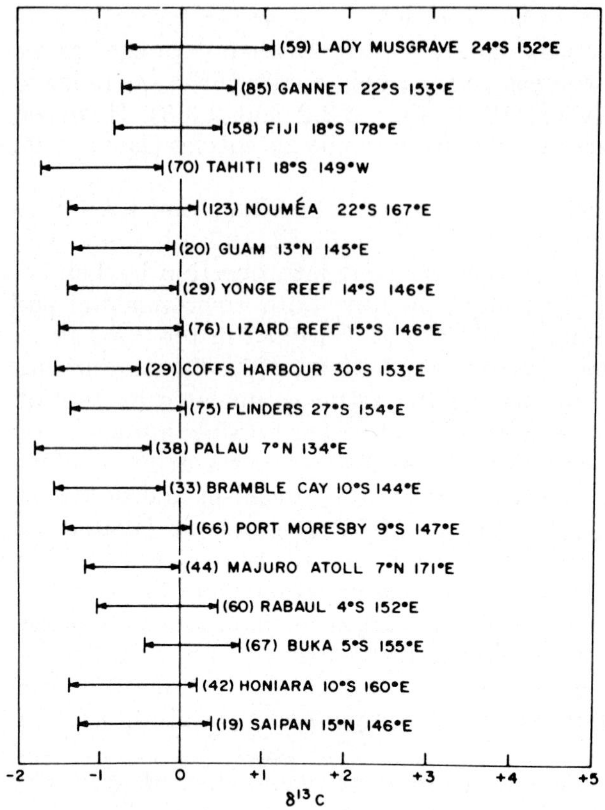

Fig. 2.3.3. Dispersion of the carbon isotopic composition of the hermatypic scleractinian genera *Acropora*, *Porites*, and *Montipora*, all living in shallow-water, surface-reef environments at 18 different, widely-scattered localities. The length of the horizontal bar corresponds to the mean ± one standard deviation. Figures in parentheses indicate the number of individual coral specimens of the three genera analyzed (Weber, 1974).

polyps (Weber, 1974). A further complication in the analysis of $^{13}C/^{12}C$ ratios arises from the fact that the ratio may be different within different parts of a single coral colony, the more rapidly growing regions commonly, but not invariably, showing a depletion of ^{13}C relative to ^{12}C (Land et al., 1975). This finding indicates: (1) an unequal distribution of compounds resulting from metabolic processes which distinguish between the two isotopes and; (2) a change in the ratio as the rapidly growing portions age and other processes with a lesser discrimination between the isotopes assume relatively more importance.

Goreau (1977b) has included several of the factors mentioned above in a model which shows the interrelationships of 9 carbon pools. The major factors controlling isotope composition of skeletal carbon and the tissues are considered to be: (1) the degrees of openness and exchange between pools; (2) the existence of large fixed external pools; (3) isotope fractionation dependent on the growth rate; and (4) the extent of consumption of tissue pools.

Crystals. The coral skeleton is a complex structure of crystals, many of which develop from spherulitic clusters (Bryan and Hill, 1940; Jell, 1974). Details of crystal organization within various portions of the skeleton as seen in the scanning electron microscope are given by Sorauf (1970, 1972) and Jell (1974). The crystals are grouped in various ways and commonly, but not always, have their major axis perpendicular to the calicoblastic epithelium (Sorauf, 1970; Vandermeulen and Watabe, 1973; Jell, 1974). The growing crystals, termed needles or crystallites, may form fibres, which in turn may be joined in lamellae or sclerodermites radiating from so-called calcification centres. The lamellae may form still larger aggregates called trabeculae (Barnes, 1970; Jell, 1974). Crystals may also be fused into lath-like units (Sorauf, 1970; Vandermeulen and Watabe, 1973). The complexity of crystalline structures of considerable size present in coral skeletons indicates strongly that the calicoblast cells do not exert control of the pattern of crystal growth. Rather, the pattern is governed by conditions within the extracellular space which are influenced by the cells. This does not preclude the participation of intracellular crystal formation in $CaCO_3$ deposition, however, and both intracellular and extracellular crystal growth may play a role in skeletogenesis.

Calcification rates *. The rate of deposition of $CaCO_3$ in corals can be measured in various ways: (1) by linear measurements of growth in colonies marked by staining (Barnes, 1970; Macintyre and Smith, 1974; Land et al., 1975; see Stoddart, 1969 for literature on linear growth); (2) by linear measurements with a laser during periods of 12 h (Strömgren, 1976); (3) by the incorporation of ^{45}Ca (Goreau, 1959); (4) by incorporation of alizarin red (which is proportional to calcium deposition) (Lamberts, 1974); (5) by increase in mass in situ (Bak, 1973); (6) by x-radiography (Buddemeier et al., 1974); and (7) by combining x-radiography of growth bands and radioisotope decay to determine age of regions within coral branches (Moore and Krishnaswami, 1974).

Examples of $CaCO_3$ deposition rates for several species of corals as determined by ^{45}Ca are given in Table 2.3.1. Two estimates for coral reefs are included for comparison. The considerable variation between species and within a single species is clearly evident. The figures represent mean values for the branch tips or outer edge where deposition rates are highest rather

* Calcification rates in corals are discussed also in Chapter 2.5.

than for the colony as a whole. In fact, the rates between different regions of a colony may differ by 5- and 6-fold (Goreau and Goreau, 1959). Species of massive corals were found to have lower calcium deposition rates than smaller species. Most rates for individual species in Table 2.3.1 are based on measurements of deposition occurring during periods of hours in light. This circumstance, together with differences in rates between regions of a colony, means that the rates in Table 2.3.1 cannot be extrapolated to monthly or annual rates. However, monthly deposition rates of colonies have been followed by underwater measurements of changes in mass (Bak, 1974) and annual rates have been estimated from band widths due to cyclic skeletal

TABLE 2.3.1

Calcium deposition rates in corals

The calcium deposition rates for individual species represent the highest mean in situ deposition and standard deviation. These rates are found in the branch tips or the outer edge, depending upon the species. Superscripts: 1, Jamaica; 2, Hawaii; 3, Australia. In some cases, the deposition rates have been recalculated to conform to the units at the heads of the columns. In doing this, it was assumed that the sea water contained 10 mM Ca, that protein is 16% N, and (in one case) that counting efficiency was 60%.

Species	μg Ca cm^{-2} h^{-1}	μg Ca mg N^{-1} h^{-1}	Reference
Acropora palmata [1]	69 ± 9.5	29.5 ± 0.78	Goreau and Goreau, 1959
Acropora cervicornis [1]	20 ± 1.7	41.7 ± 11.7	Goreau and Goreau, 1959
Acropora cervicornis [1] Incubated tips	—	24.2 ± 7.56	Pearse and Muscatine, 1971
Millepora complanata [1]	38 ± 5.7	35.7 ± 3.89	Goreau and Goreau, 1959
Porites furcata [1]	36 ± 6.7	15.4 ± 3.70	Goreau and Goreau, 1959
Colpophyllia natans [1]	80 ± 1.2	10.4 ± 1.85	Goreau and Goreau, 1959
Montastrea annularis [1]	54 ± 11.9	7.32 ± 1.71	Goreau and Goreau, 1959
Montastrea annularis [1] During 24 h, calculated	3.89	—	Barnes and Taylor, 1973
Pocillopora damicornis [2] Calculated	—	37.8 ± 13.1	Vandermeulen and Muscatine, 1974
Acropora acuminata [3] Incubated tips, calculated	—	234 ± 36.5	Crossland and Barnes, 1977
Coral reefs (corals and algae) Pacific, during 24 h, calculated	18 ± 1.3	—	Smith and Kinsey, 1976
Coral reef Enewatok, during 24 h, calculated	6.3	—	Smith and Harrison, 1977

density variations made evident by x-ray transmission of coral sections (Buddemeier et al., 1974). Band dimensions have indicated that during any given year linear growth rates of a specimen at Eniwetok Atoll may vary by a factor of nearly two. Also, rates between colonies showed considerable variability. Because of this variability, measurements of linear growth appear to be of limited value in correlating growth with environmental factors.

From comparison of laboratory (Goreau, 1959; Clausen and Roth, 1975) and field (Goreau and Goreau, 1959) measurements, it would appear that laboratory measurements of ^{45}Ca deposition under appropriate conditions can provide a reasonably good index of deposition in the field during short periods.

Effects of light. Coral calcification is accelerated by light as demonstrated by the rate of deposition of ^{45}Ca, the highest rates occurring in sunlight with marked reduction on cloudy days and very low rates in darkness (Goreau, 1959; Pearse and Muscatine, 1971; Vandermeulen et al., 1972). Goreau and Goreau (1959) found an average light/dark ratio of 9.02 in 11 species of hermatypic corals. In *Acropora cervicornis*, the calcification rate was found to be directly proportional to light intensity over the range 0 to 10,500 lux (Chalker and Taylor, 1975). The effect of light is related to the requirement of light for photosynthesis by the zooxanthellae within the coral tissues, and the rate of photosynthesis will be proportional to light intensity up to a maximum rate. Colonies from which zooxanthellae were removed by maintenance in darkness (Goreau and Goreau, 1960), or exposed to 3-(3,4-dichlorophenyl)-1,1-dimethylurea (DCMU) which inhibits photosynthesis, had markedly reduced calcification rates (Vandermeulen et al., 1972; Chalker and Taylor, 1975). Although light greatly enhances calcification rate, calcification *per se* is not dependent on light since deposition occurs in deep-sea ahermatypic corals which lack zooxanthellae. Also, hermatypic corals deposit calcium slowly in total darkness (Goreau, 1963).

The mechanisms of linkage between photosynthesis and calcification remain to be clarified. However, several explanations have been advanced (Simkiss, 1964; Chapman, 1974; Chalker and Taylor, 1975) including the following:

(1) The zooxanthellae increase the calcification rate by removing H_2CO_3 and CO_2 resulting from the formation of $CaCO_3$ (Goreau, 1959).

(2) The zooxanthellae may provide photosynthate used in the formation of the organic matrix, and the rate of matrix synthesis may determine the rate of $CaCO_3$ deposition (Wainwright, 1963; Muscatine and Cernichiari, 1969).

(3) Photosynthesis is related to oxidative phosphorylation which may provide AT as an energy source for the active transport of calcium and bicarbonate and for the synthesis of organic matrix (Chalker and Taylor, 1975).

(4) Photosynthesis by zooxanthellae may facilitate the growth of $CaCO_3$

crystals by removing organic and inorganic phosphates which are crystal poisons (Simkiss, 1964).

It will be apparent that these suggested explanations are not mutually exclusive and all could apply.

Mollusca

The shell-forming system. The molluscs deposit oriented crystals of $CaCO_3$ surrounded and penetrated by a protein matrix and arranged in a variety of patterns and layers to form the shell. Commonly, two or more crystal patterns are present within a single shell. Detailed accounts of the mineral architecture of shell will be found in MacClintock (1967), Taylor et al. (1969, 1973), Kennedy et al. (1970), Grégoire (1972) and in earlier works cited by these authors. The crystals are of 3 types: calcite, aragonite, and much less frequently, vaterite. Two of these, calcite and aragonite, may occur in separate layers of the shell. The well-ordered, and often complex, crystalline structure of shell is assembled extracellularly by the mantle, a thin organ which lines the inner shell surface and is bounded by a single layer of epithelial cells. The cells secrete the organic components of the shell and transport Ca^{2+} and HCO_3^- into the extrapallial fluid between the mantle and shell. Here, $CaCO_3$ is deposited in association with the organic material on the inner shell surface.

The deposition of $CaCO_3$ crystals can be viewed as occurring progressively within 4 compartments in linear arrangement (Wilbur, 1972): (1) the outer medium; (2) the mantle containing blood with its outer layer of epithelial cells; (3) the extrapallial fluid compartment between mantle and shell; and (4) the shell (Fig. 2.3.1). Although a net movement of ions occurs inward from the outer medium to the extrapallial fluid in $CaCO_3$ deposition, an interchange in both directions can take place. Thus, when a freshwater gastropod is placed in Ca-free medium, calcium may be lost from the animal without a decrease in blood and tissue calcium, indicating replacement of calcium of the blood and tissue compartment by shell calcium (Greenaway, 1971).

Mechanisms of calcium carbonate deposition. The carbonate of shell has its origin in the bicarbonate of the medium (Hammen and Wilbur, 1959), metabolic CO_2 (Campbell and Speeg, 1969), or in some cases from calcium carbonate spherules present in the tissues (Watabe et al., 1976). In marine bivalves, bicarbonate probably comes chiefly from the medium (Wheeler et al., 1975). The principal source of shell calcium is the medium with tissue spherules providing a secondary supply. The $CaCO_3$ of molluscs is thought to be deposited in isotopic equilibrium with the medium (Milliman, 1974). However, discrimination in favour of ^{13}C and ^{18}O has been observed (Keith et al., 1964; Tourtelot and Rye, 1969), and this discrimination may be different in aragonitic and calcitic layers.

In order for $CaCO_3$ to be deposited in shell formation, the concentrations of Ca^{2+} and CO_3^{2-} must exceed the solubility product. The pH of the extrapallial fluid is crucial, for it must be maintained at a level which will promote the dissociation of HCO_3^- and the precipitation of $CaCO_3$. The protons generated by the precipitation of carbonate must be removed from the extrapallial fluid for $CaCO_3$ deposition to continue. The first requirement, that of ionic concentrations in the extrapallial fluid exceeding the solubility product, appears to be facilitated by concentrations of blood Ca^{2+} and CO_3^{2-} that reach saturation or super-saturation levels (Potts, 1954; Burton, 1970, 1976). A favorable diffusion gradient of these ions occurs toward the extrapallial fluid as $CaCO_3$ is crystallized at that site. The extrapallial fluid itself is also probably saturated in marine bivalves (Crenshaw, 1972a). The pH range of the extrapallial fluid of most marine molluscs is 7.1—7.8 and 8.0—8.3 for freshwater molluscs (Wilbur, 1964; Crenshaw, 1972a; Wada and Fujinuki, 1976). Freshwater molluscs maintain a degree of saturation in the extrapallial fluid roughly equivalent to that in marine molluscs by actively concentrating calcium from the medium (Potts, 1954) and by maintaining a higher carbonate concentration (Wada and Fujinuki, 1976). The direct application of analyses of ion concentrations of the extrapallial fluid to $CaCO_3$ deposition must be made with circumspection (Wilbur, 1972). The double compartmentalization of the extrapallial space and can lead to complications, since shell deposition takes place principally in the compartment at the periphery of the shell, whereas analyses usually reflect concentrations in the central compartment. Also, shell deposition may not be proceeding at the time of sampling so that the pH and ion concentrations recorded may be different from those during active deposition. The pH and Ca^{2+} concentrations in *Mercenaria* undergo cyclic changes which are not necessarily associated with shell deposition (Crenshaw and Neff, 1969; Crenshaw, 1972b).

The mechanisms by which the animal regulates pH to produce a sufficiently alkaline environment for calcification are not known. The following are three suggested mechanisms. The first proposes that carbonate forms from CO_2 by the secretion of hydroxyl ions (Simkiss, 1976). The second advocates the removal of protons by ammonia to form ammonium ions (Campbell and Boyan, 1976). In the third proposed mechanism, HCO_3^- is actively transported by the mantle cells into the extrapallial fluid where conversion to carbonate is favored by the action of carbonic anhydrase which catalyzes the formation of OH^- and CO_2 from HCO_3^- (Wheeler, 1975) (Fig. 2.3.1). Loss of resulting CO_2 from the extrapallial fluid is by diffusion along the concentration gradient into the tissues and to the outer medium. This concentration gradient is maintained in part by CO_2 fixation by the tissues (Hammen and Wilbur, 1959). Calcium could move into the extrapallial fluid as a counter-ion of HCO_3^- or be transported (Wood, 1973).

Organic components of shell. The shell of molluscs contains organic material called conchiolin which comprises from as little as 0.01% in certain neogastropod species to more than 5% in cephalopods (Wilbur and Simkiss, 1968). This organic matrix is largely protein with soluble and insoluble fractions, the proportions of the fractions depending upon the species. The insoluble fraction surrounds and separates crystals and crystalline layers. This fraction contains crosslinked proteins which may be quinone-tanned as indicated by a high tyrosine content. The soluble protein fraction, on the other hand, appears to be present within the $CaCO_3$ crystals and is demonstrable only after pulverized shell is decalcified (Meenakshi, personal communication; Crenshaw, 1972).

The deposition of molluscan shell is an orderly sequence in which crystal nucleation is followed by crystal growth within an organic matrix to form multiple layers which are both separated and bound together by the matrix. Five functions have been proposed for the organic matrix: (1) initiation of crystal nucleation; (2) control of crystal orientation; (3) limitation of crystal growth; (4) bonding between crystals and crystal layers; and (5) inhibition of shell fracture by virtue of separation of crystals and crystal layers (Currey and Taylor, 1974). Another organic component of shell is a sclerotized protein layer, the periostracum, which covers the outer surface. It is the substratum on which the outermost layer of crystals is deposited and has two additional functions which we shall mention below. Further information on composition and properties of the organic components of shell will be found in Wilbur and Simkiss (1968), Grégoire (1972), Degens (1976), and Wilbur (1976).

Crystal nucleation. A glycoprotein which specifically binds calcium in the presence of excess sodium ions has been isolated from the water-soluble fraction of the matrix of the clam *Mercenaria* (Crenshaw, 1972b) and another Ca-binding protein has also been isolated from a similar fraction of the shell of the snail *Nassa* (Krampitz et al., 1976). Also, Ca-binding sulfated polysaccharide sites have been identified in the matrix covering the crystals in *Nautilus* shell (Crenshaw and Ristedt, 1976). Crenshaw and Ristedt have suggested that Ca-binding sites may bring about a local increase in Ca^{2+} and CO_3^{2-} by ionotropy resulting in the formation of crystal nuclei. Degens (1976) has proposed that a polypeptide fraction of shell matrix is activated on attachment to another protein fraction, thus initiating calcium fixation and the formation of CaO_6 and CaO_9 polyhedra.

Crystal orientation. The orientation of crystals in any given region of shell is commonly, but not invariably, uniform. These crystals form on or within the organic matrix which may therefore be a factor influencing their orientation. In the bivalves *Pinctada* and *Pteria*, the b-axis of the crystals does in fact correspond to the long axis of matrix fibrils (Wada, 1970) but it is not clear which component is orienting the other. Crystals growing within the matrix could conceivably orient the matrix covering them. If the matrix in turn were to influence the early growth of crystals of the layer next deposited, the orientation of successive crystal layers could be maintained. At the present time, the evidence does not clearly indicate the role of the matrix in crystal orientation.

Layer formation. Nacreous shell is a layered structure in which single layers of crystals of approximately uniform thickness are separated by sheets of organic matrix. Crystal layering could occur as a sequence of crystal growth on a sheet of matrix followed by secretion of a layer of matrix which limits crystal thickness (Wilbur, 1972; Crenshaw and Ristedt, 1976; Degens, 1976). While this may be true, it is clear that the mechanisms in the structuring of shell are complex, involving factors in the control of matrix secretion, its polymerization, and crystal deposition (Wilbur, 1972, 1976). Layering in cross lamellar shell involves an additional factor in that crystal orientation is changed with the deposition of each major layer but is the same in alternate layers. This very interesting feature of crystal orientation remains unexplained.

Isolation of shells from the medium. The organic components of shell function in enclosing the site of shell formation and in retarding crystal solubilization and ion exchange with the environment. The periostracum forms a curtain between the mantle edge and the shell edge and so acts as a barrier between the outer medium and the extrapallial fluid. This makes possible the creation of a microenvironment in which shell deposition can take place. The periostracum also provides an outer sclerotized hydrophobic layer on the shell surface and this protects against solubilization of the $CaCO_3$ in acidic freshwater environments. The insoluble fraction of the organic matrix which surrounds the crystals may serve to retard solubilization by acids produced during anaerobic periods in living animals, and exchange between shell crystals and the environment after death. However, this protection is limited as evidenced by erosion and loss of shell weight in living animals during periods not favorable to growth (e.g., Wada and Fujinuki, 1976) and by exchange when isolated shells are placed in a medium containing ^{45}Ca (Wheeler et al., 1975).

Annelida

Both the polychaetes and the oligochaetes produce mineralized secretions. Many polychaetes live within organic tubes which they secrete and which may contain varying amounts of calcareous debris. It is, however, only among a few members of the Serpulidae that a true calcareous tube is secreted, and worms of this group may be major contributors to marine coral ecosystems (Hubbard, 1974). The tubes are formed by a variety of intracellular and extracellular processes involving special calcium glands and general epithelial mucus cells. The minerals of the tube are high magnesium calcite or aragonite: the composition of the tubes has been reviewed by Defretin (1971). In contrast to the serpulids, oligochaetes deposit minerals mainly in terrestrial habitats where they may contribute significantly to redistributing calcareous minerals. The mineral deposits of the oligochaetes are mainly produced by the calciferous glands of the oesophagus which are particularly well developed in the Lumbricidae (Stephenson, 1930).

The serpulid tube is initiated at the time when the larva settles. In *Spirorbis*, there is a single medio-dorsal gland in the wall of the posterior intestine which discharges during a period of 5—8 s and produces the initial site of attachment (Quiénvreux, 1963). The calcareous tube grows from this site by the intermittent activity of a pair of subepithelial glands which open beneath a fold of epidermis which forms the collar at the anterior end of the worm. In *Pomatoceros*, two calcium glands secrete a viscous white fluid consisting of calcium carbonate in mucoprotein. According to Hedley (1956b), there may also be some mineral secreted by the mucus cells of the so-called ventral shield area into which the calcium glands open. These secretions are moulded onto the anterior end of the tube by the collar and soon harden to form the tube (Hedley, 1956a,b). Undoubtedly, the two ventro-lateral glands are the main source of mineral but additional calcium-secreting glands are present in *Serpula* (Hedley, 1956b).

Much of the interest in the secretion of the serpulid tube relates to the activity of the columnar cells which line the calcium glands. The relationship of the ultrastructure of these cells to carbonate deposition was investigated by Neff (1971a) who showed that, in *Pomatoceros caeruleus*, calcite was precipitated intracellularly in multivesicular bodies prior to its secretion into the lumen of the gland. These glands were capable of producing calcite equal to their volumes within 3 h and it might be thought, therefore, that the production of intracellular mineral is the basis of production of tube secretion in all polychaetes. In the closely related species of *Eupomatus*, however, an entirely different mechanism appears to exist and aragonite is deposited extracellularly in the lumen of the duct (Neff, 1967). In *Spirorbis spirorbis*, Nott and Parkes (1975) proposed that there is a mechanism for concentrating calcium in the intercellular spaces of the collar. Thus, mucus with a high calcium content is secreted into the lumen of the gland and crystals of calcite only form when this comes into contact with the carbonate in sea water.

The main source of ions for tube formation appears to be sea water and the rate of secretion is markedly influenced by the ambient calcium concentration (Neff, 1969). Swan (1950), in some rather crude pulse labelling experiments with Sr, showed that it took many hours for the ion to pass through the worm and become incorporated in the secretions of the calcium glands. This is in keeping with the fact that ions may be taken through the general body surface (Fretter, 1953).

The epithelial cells of the collar of *Spirorbis* have numerous microvilli which project beyond the cuticle. Electron probe studies have led Nott and Parkes (1975) to propose that they are involved in actively taking up calcium from the surrounding sea water for use in tube formation. In addition, calcium ions are stored in a variety of tissues. Thus, numerous calcium-rich cells occur in the stomach of *Spirorbis* (Nott and Parkes, 1975) and in the stomach, intestine, chloragocytes (containing chloragosomes) and nephridea of *Pomatoceros triqueter* (Vovelle, 1956). Also, Neff (1971b) observed hydroxyapatite and calcium magnesium phosphate deposits in vacuoles of the epithelial cells on the anterior surface of the collar of *Pomatoceros caeruleus*. The rate of turnover of these deposits is not known.

The region of tube secretion in serpulids is normally restricted to its anterior edge. If the posterior end of the tube is damaged, it is repaired by a different mechanism. Under these circumstances, special mucus cells in the ventral epithelium of the epidermis of the abdomen produce a calcareous secretion which is applied to the damaged region of the tube and builds up a substantial mineral deposit.

The calciferous glands of oligochaetes occur as pouches of the oesophagus. Detailed studies have been made of their histology (van Gansen-Semal, 1959) and ultrastructure (Nakahara and Bevelander, 1969). ^{45}Ca injected into the coelomic fluid of *Lumbricus terrestris* is rapidly taken up by the basal parts of the epithelial cells of the calciferous glands. During the next few hours, it can be traced passing to the distal regions of the cell and is then found in the calcite spherules of the glands (Nakahara and Bevelander, 1969). According to Crang et al. (1968), the deposition of mineral occurs cyclically and involves the intracellular formation of calcite crystals within mitochondria which then discharge into the lumen of the glands. The functions of the calciferous glands are not well understood but they are thought to be involved in acid-base balance or in the excretion of calcium. They discharge into the alimentary tract so that, in certain areas, worm casts are rich in these mineral deposits.

Arthropoda

The arthropods represent over 75% of all known animals and thus cover a wide diversity of forms and habitats. Most of the work on calcification in arthropods has concentrated on the exoskeleton, which is particularly well mineralized in the Crustacea, but there are also some limited studies on intracellular calcification (Waku and Sumimoto, 1974) and the formation of calcareous tubes of insect larvae (Marshall and Chueng, 1973). Among the Crustacea, there has been much interest in the cyclical changes in mineralization associated with the moulting of the exoskeleton. Information is available on crabs (*Uca, Carcinus, Searima*), shrimps (*Crangon, Paleonectes*), lobsters (*Panulirus, Homarus*), crayfish (*Orconectes, Cambarus*) and isopods

(*Ligia*) (see McWhinnie et al., 1969). Barnacles also moult but retain their calcified skeleton and increase in size throughout life (Costlow, 1959; Bubel, 1975).

The main histological changes in the decapod cuticle during moult were described by Travis (1963, 1965, 1970) and are summarized in the beautiful ultrastructural study of the integument of the fiddler crab *Uca* by Green and Neff (1972).

The crustacean cuticle is made of chitin and protein and has an outer tanned epicuticle within which lies a tanned and calcified exocuticle. The thickest layer is the inner endocuticle which is heavily calcified but contains no tanned problems. The inside of the exoskeleton is formed by a membranous layer which is neither tanned nor mineralized but which covers the cellular epidermis. This layer contains a variety of cell types including melanophores and epidermal cells but the secretion of the exoskeleton is mainly performed by the cuticle-secreting cells and the associated intra-epidermal or reserve cells. The exoskeleton is penetrated by tegumental ducts and by pore canals. The first of these carry secretions from glands through the thickness of the cuticle while the latter may consist of 50—90 fine protoplasmic extensions of each cuticle-secreting cell penetrating the exoskeleton at a density of about 4×10^6 pore canals mm^{-2}.

During premoult, the exoskeleton is resorbed into the animal (Travis, 1965) and there is an increase in the protein-bound calcium of the blood although the ionic calcium remains constant (Greenaway, 1974a,b). At this time, an increase in the rate of unidirectional calcium flux across the isolated epidermis toward the haemolymph is seen (R. Roer, personal communication). Most of the resorbed calcium leaves the animal in the urine (Travis, 1951) or across the gills (Greenaway, 1974b), although some of it is stored as granules in the hepatopancreas (Becker et al., 1974) or in the gastroliths which develop at this time (Travis and Friberg, 1963a). The crayfish loses 83% of its total body calcium during these few days and about 60% of that not excreted is retained in the gastroliths (Greenaway, 1974b).

The synthesis of the new epicuticle and exocuticle commences during premoult, requiring the resorption of the old exoskeleton across its surface. In fact, high rates of calcium flux occur across the pre-exuvial cuticle-epidermis complex in vitro (R. Roer, personal communication). The exocuticle is not mineralized until after moult and then it is calcified simultaneously with the formation and mineralization of the endocuticle. Thus, the exocuticle exists before it is mineralized but the endocuticle is secreted and calcified simultaneously (Travis, 1965).

The epidermal cells appear to be involved in calcifying both the exocuticle and the endocuticle. Early in postmoult, "strings of granules" containing inorganic calcium were observed to be extruded from the epidermis into the endocuticle. At the same time, the calcification of the exocuticle occurred as a wave of hardening moved from its outer edge inwards (Travis, 1960). The calcium in this case appears to have been transported by way of the pore canals. The calcium carbonate which is deposited in the exocuticle develops without much orientation around the lamellar fibres of the chitin-protein cuticle. It is poorly crystalline throughout most of the exoskeleton but, in the pore canals and the inner part of the endocuticle, well-developed calcite crystals have been detected (Travis and Friberg, 1963b). In addition to carbonate, 4.8 to 9.4% of the calcium in the cuticle of *Astacus fluviatilis* may be calcium phosphate (Welinder, 1975).

During moulting, crustaceans rarely feed. Therefore, the calcium used for forming the exoskeleton can only come from the body stores and by the uptake of environmental calcium. According to Porcella et al. (1969), the calcium uptake in *Daphnia* occurs shortly before moult and can be described as a first-order chemical reaction as is also the case in crayfish and lobster (Schurr and Stamper, 1962). The cuticle is normally impermeable to calcium over most of its surface (Chaisemartin, 1965), and it is thought that calcium is therefore absorbed across the specialized gill epithelium. In the crayfish, calcium uptake starts 15 to 30 min after moult and reaches a rate of 2 μmol g^{-1} body wt h^{-1}. Uptake appears to involve a calcium pump since it occurs against an electrochemical gradient and is a saturable process (Greenaway, 1974c). Bicarbonate ions appear to be involved in the process, calcium uptake falling by 60% in their absence. This has led Greenaway (1974c) to suggest that the carbonate component of the exoskeleton is derived from both environmental bicarbonate and from metabolic CO_2 as in the corals.

The uptake of calcium falls off rapidly as moulting is completed. McWhinnie (1962) found that during intermoult the crayfish *Orconectes virilis* still removed about 2.5 mg Ca in 20 days. When animals were injected with 1 μCi of ^{45}Ca, left for 12 h and then stimulated with hormone injections to resorb the skeleton, McWhinnie et al. (1969) found that calcium was removed from the cuticle but that the specific activity of the exoskeleton rose. They interpreted this as indicating that sites of calcium deposition and resorption are anatomically separated and that, based on the distribution of ^{45}Ca, deposition occurs in the outer layers while resorption occurs from the inner part of the endocuticle. These results are similar to those of Hayes and Armstrong (1961) who kept lobsters in ^{85}Sr for 24 h and then showed that the exocuticle was the most heavily mineralized part of the skeleton and also contained a thick band of radioisotope. If it is assumed that the isotope did not enter through the cuticle, then this experiment would support the concept that there is continual secretion of mineral from the pore canals of the cuticle and resorption of mineral from the innermost part of the endocuticle. The observation of Drach (1939), that calcification was most intense around the pore canals and proceeded from the outer layers inward, would also fit this scheme. These suggestions differ somewhat from the method of mineralization envisaged by Travis and Friberg (1963b). They noted that the exocuticle was radio-opaque before it was calcified and interpreted this as showing that ions existed in a metastable state when the exocuticle was secreted. They suggested, therefore, that mineralization was induced by such simple changes in the physicochemical environment as a modification of the pH or enzyme content of the fluids. This scheme is difficult to assess since it appears likely that the old exoskeleton is resorbed across the surface of the newly secreted exocuticle, so that some retention of mineral might be expected in such studies. All of these hypotheses agree, however, as to the importance of the pore canals in inducing mineralization.

DECALCIFICATION AND EROSION

We now turn from the deposition of skeletal carbonate by invertebrates to their activities in decalcifying and eroding skeletal material and rock substrata.

The solubilizing of $CaCO_3$ of shell occurs as a normal process in living molluscs and may well take place in other invertebrate skeletons as well. In bivalves, decalcification on the inner shell surface may result from an increase in organic acids and metabolic CO_2 during short periods of anaerobiosis or long-term adverse environmental conditions (Wilbur, 1972 for references). During winter months, the shell weight of marine bivalves may show a perceptible weight loss (Wada and Fujinuki, 1976). Within a single day, calcium deposition and shell decalcification may alternate several times (Crenshaw, 1972a). When the valves are closed, anaerobiosis probably results in acid accumulation and shell etching. Then on opening, the acids disappear by aerobic metabolism, and $CaCO_3$ can be deposited.

$CaCO_3$ is also solubilized in all types of plant and animal skeletons and in calcareous rock through the chemical action of burrowing and boring organisms (Smith and Carriker, 1969; Carriker and Smith, 1969). Results on substrate penetration by microorganisms and invertebrates have been discussed in a symposium by Carriker et al. (1969) and, more recently, in a review by Milliman (1974). Taxa containing organisms known to dissolve carbonate materials * are bacteria, algae (Edwards and Perkins, 1974; Perkins and Tsentas, 1976), lichens, fungi (Edwards and Perkins, 1974), Porifera, Annelida, and Mollusca. Burrows are also formed by Bryozoa, Platyhelminthes, Phoronida, Brachiopoda, Sipunculoidea, Crustacea, including barnacles, and Echinodermata. In these groups, there is uncertainty as to whether the burrowing is carried out by chemical or mechanical action or both. Even in those cases in which dissolution is clearly evident, the mechanisms are not well known. The suggested agents which may be involved include acids. chelating compounds, carbonic anhydrase and other enzymes (Carriker and Smith, 1969).

---

* See also Chapter 2.4.

Following the dissolution of skeletons and substrata, Ca^{2+} and CO_3^{2-} are of course returned to the medium, but more important ecologically is the fact that some organisms weaken reef structures which favors their further breakdown through wave action (Milliman, 1974). Activity by borers, grazers, browsers and predators also directly increases the rate of sediment formation from corals, other skeletal material, and rock substrata through dissolution and fragmentation. The amount of sediment formed by organisms such as in the siliceous boring marine sponges may be considerable. These animals produce fine grained sediments from carbonate rocks and skeletons of corals, red algae, and molluscs as a result of the dissolving action of certain amoebocytes which remove small chips from the substratum. Fütterer (1974) estimated that the chips may constitute 2—3% of the total sediment in a region of the Persian Gulf and as much as 30% of the sediment in the lagoon at Fanning Island in the Pacific. Rützler (1975) found that the sponge *Cliona* through chemical etching produced 16 mg $CaCO_3$ as sediment per mg dry wt of sponge per year. The rate of excavation of Bermuda rock by this organism may amount to 256—3,000 g m^{-2} y^{-1}, depending upon the density of the sponges. The clam *Tridacna crocea* in Australian waters penetrates unfixed coral heads by mechanical and chemical burrowing. The resulting sediment production has been estimated at 140 g m^{-2} y^{-1}. This estimate has not been corrected for chemical action, a factor which is difficult to quantify. The erosion rate may be increased 20-fold in areas of high density of clams. Two estimates of erosion from grazing by littorinid molluscs indicate that 0.1 mm (Emery, 1946) or 0.6 mm (North, 1954) of intertidal sandstone may be removed per year. The amount at the higher rate would be 3300 kg removed from an area of 3000 m^2.

CELLULAR SYSTEMS OF CALCIUM CARBONATE DEPOSITION

It appears that, in most cases, the deposition of $CaCO_3$ by invertebrates is accomplished by one of three morphologically distinct systems which have common properties. The systems are: (1) calcification within vesicles or vacuoles; (2) extracellular calcification by single cells; and (3) extracellular calcification by epithelia. We now briefly consider each of these systems in summarizing the detailed information on various invertebrate groups given on pp. 71—87.

Intracellular calcification within vesicles and vacuoles

Intracellular calcification within vesicles, or in vacuoles within syncytia, is present in a wide variety of tissues in every phylum examined thus far. Commonly, the calcium deposited takes the form of amorphous calcium spherules formed in vesicles associated with the endoplasmic reticulum or Golgi systems (Pautard, 1976). The spherules are of variable composition and seem to function as mineral stores, as excretory products, and for buffering against metabolic acids (Simkiss, 1976; Watabe et al., 1976; Sminia et al., 1977). In a number of organisms, the intracellular or vacuolar system of calcification is also capable of producing skeletal structures. Thus the spicules of the alcyonarian *Renilla* are formed in this way (Dunkelberger and Watabe, 1974). The formation of tooth plates (Kniprath, 1974) and larval spicules of echinoderms (Okazaki, 1965; Tilney and Porter, 1967; Millonig, 1970) also fall in this category.

The formation of the calcified structures involves movement of organic material from its presumed site of synthesis in the endoplasmic reticulum and Golgi regions into the vesicle or vacuole where it is joined by inorganic ions which may be pumped into this site. Some of these ions are presumably bound to the organic material. The organic material normally becomes incorporated into the calcified structures, although this is uncertain in skeletal structures of echinoderms (Wilbur, 1976).

Extracellular calcification by single cells

Single cells may deposit $CaCO_3$ extracellularly. This occurs in the formation of the test in protozoa (Pautard, 1970) and the spicules of calcareous sponges (Jones, 1970). Single cells also have the capacity to form portions or entire skeletons of echinoderm larvae in vitro (Okazaki, 1975); and cells which do not form distinct epithelia regenerate spines of echinoderms extracellularly (Heatfield and Travis, 1972). These results suggest the possibility that cells which are not in an epithelial layer may also deposit the meshwork of the echinoderm test and spines extracellularly.

A mechanism which would accomplish extracellular mineralization by single cells may involve the entrance of Ca^{2+} and HCO_3^- from the medium and their movement through the cell and onto the site of deposition of $CaCO_3$ on the crystal surface. The details and metabolic costs of these movements remain obscure.

Extracellular calcification by epithelia

The extracellular formation of crystals of $CaCO_3$ by epithelia is the most common type of skeleton-forming system and occurs in corals (Vandermeulen, 1975), molluscs (Wilbur, 1964), some annelids (see p. 84) brachiopods (Williams, 1971), and arthropods (Travis, 1970). In the arthropods, elaborate cellular extensions penetrate the mineralized carapace and are important in the calcification process. Formation of a mineralized skeleton in these taxa involves the movement of Ca^{2+} and HCO_3^- across a layer of cells from the body fluid (absent in corals). Nucleation and crystal growth take place in an organic milieu secreted by the epithelium.

This system of classifying the cellular basis of calcification in three systems is clearly not absolute. There are situations in which the arrangement of cells is such that it is not clear whether the mineral is strictly intra- or extracellular. There are also situations in which the single cell-syncytial or the single cell-epithelial distinction is not absolute. However, the system of classification is both convenient and important for two reasons. First, it reduces an extensive amount of information on many taxonomic groups to reasonable proportions; and, second, by simplifying the morphological aspects of calcification, it emphasizes certain concepts. Thus, in this simplified form, it is apparent that the basis of invertebrate calcification is an oriented membrane which deposits mineral on or near the surface facing away from the cytoplasmic contents. If we now start with this simplified concept, we can rebuild — and reduce — the complexities of calcification in various invertebrate groups described earlier with some new insights on the sites and energetics of the mechanisms of mineralization.

Chemical and physiological parameters of invertebrate calcification

Intracellular and extracellular fluids

Mineralization may occur outside cells in extracellular fluids or inside cells within a vesicle or vacuole, as we have mentioned. If mineralization takes place extracellularly, the ions used in mineralization may pass from a general extracellular fluid compartment (Fig. 2.3.4) across a cellular compartment and into a mineralizing fluid compartment. The ions may be transported from the general extracellular fluid compartment through the cells (route a) or by way of an intercellular route (route b). The energy relationships between routes (a) and (b) may be quite different and are probably important in the mineralization process. Ions which form mineral in an intracellular vesicle or vacuole will also come from extracellular fluid, move into the cell and then into the vesicle (route c). If the ions of the extracellular fluid which form the extracellular mineral are derived by pumping from intracellular sources (route a), then there is no critical distinction between intracellular mineralization within a vesicle or vacuole (route c) and extracellular mineralization except the arrangement of the membranes. If, however, the general extracellular fluid and the mineralization fluid are in communication by intercellular channels (route b), then there is a major difference in energy requirements. The distinction may be quantified in relation to calcium. Route (a) involves transfer from an intracellular ionic Ca^{2+} concentration of 1 μmol l^{-1} up to the level of 1 mmol l^{-1} ionic Ca^{2+} which is probably necessary for biomineralization (Borle, 1973). Route (b) starts with a Ca^{2+} concentration of 1 mmol l^{-1}. Thus, a difference of three orders of magnitude between the two routes defines the essential difference between these two types of mineralization.

It is generally assumed that route (a) is the one involved in forming the

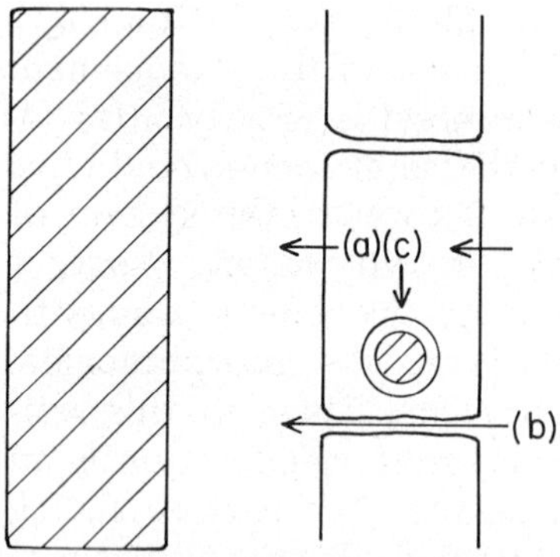

Fig. 2.3.4. Pathways of ions across cells in the formation of $CaCO_3$. (a) Ion movement through a cell to an extracellular space where mineral deposition occurs (large area marked diagonally). (b) Ion movement through intercellular space. (c) Ion movement into an intracellular vesicle where mineral deposition occurs (smaller circle with diagonal markings).

exoskeleton of crustaceans. In the molluscs, some authors favour the same view but, since the extrapallial fluid is so similar to blood (Crenshaw, 1972a), it has been suggested that the calcium diffusion potentials recorded by Istin and Maetz (1964) may, in fact, represent route (b) (Simkiss, 1976).

The CO_2/HCO_3^- system involved in calcification can be considered in a somewhat similar manner. The question as to whether the carbonate ion of the mineral is derived from extracellular fluids originating in the external medium or whether it arises from intracellular metabolic CO_2 has been examined in corals (Pearse, 1970), molluscs (Campbell and Speeg, 1969; Wheeler et al., 1975), and arthropods (Greenaway, 1974c) (see pp. 74, 80, 87). It may be that both sources contribute to the skeletal carbonate. The significance of determining the intracellular and extracellular sources of carbonate is not simply that it enables one to draw up a balance sheet of input or output but that it serves as an indication of the participation of cellular function in mineralization processes.

Membrane transport

The morphological classification adopted in an earlier section (see pp. 89—91) emphasizes the important role of membrane systems in calcification. Membranes, both in intracellular and extracellular calcification, are thought to be involved in an active transport of calcium to the site of calcification. They may also be involved in facilitating the availability of bicarbonate ions and in removing protons released during calcification. Thus, all the main ion species involved in biological calcification may be controlled by membrane processes. The ions are related according to the empirical equation

$$Ca^{2+} + HCO_3^- \rightarrow CaCO_3 + H^+$$

The evidence in support of this concept of active transport in invertebrate calcification comes mainly from comparative studies which show that Ca^{2+} and H^+ pumps, and in some cases HCO_3^- pumps, are involved in normal cellular homeostasis. Some of these pumps may be exchange systems, e.g., Ca^{2+}/H^+ exchange, adding calcium ions and removing protons from the site of calcification. Alternatively, there may be counterion flow, e.g., HCO_3^- following a Ca^{2+} pump as a counterion (see p. 81, Mollusca). These are two possibilities currently attracting much interest but, at the present time, the transporting functions of membranes at sites of calcification have not been satisfactorily characterized.

Effects of other ions

The deposition of mineral on one side of a cell membrane is affected by both the availability of calcium and carbonate ions and also by their interaction with other ions. Interfering ions not only impede the effective collisions of calcium and carbonate but they may also form ion pairs with the calcium and carbonate ions (Skirrow, 1975). Mg^{2+} and PO_4^{3-} may also interfere with the growth of the crystal lattice so that, in their presence, the rate of calcification may be reduced and a particular crystal type may be favoured (Kitano et al., 1976). One of the functions of cellular membranes is probably to control the ionic composition of invertebrate skeletons and that of some intracellular mineral deposits.

Energy of activation

Although many biological fluids appear to be supersaturated with calcite, spontaneous mineralization does not occur. A number of explanations have been proposed, including the effects of crystal poisons and the high solubility of small crystal nuclei. An alternative approach to the problem is to suggest that there is an "energy barrier" which has to be overcome by the

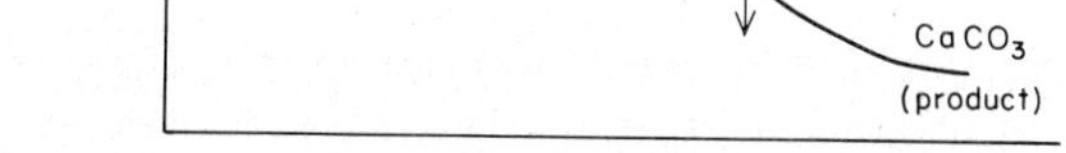

Fig. 2.3.5. Calcium and carbonate ions may not react spontaneously but require an "energy of activation" before they form the calcium carbonate product and release their "energy of reaction". The broken line indicates that certain materials (e.g. organic matter) may reduce the energy of activation and thus facilitate calcification.

energy of activation before calcification proceeds (Fig. 2.3.5) (see Neuman and Neuman, 1958).

In enzyme kinetics the rate of reaction (k) is related to the energy of activation (E) by the Arrhenius equation:

$$k = C\,e^{-E/RT}$$

where C is a constant for the particular system, R is the gas constant and T is the absolute temperature.

The energy of activation of a system can be easily derived by plotting ln k against $1/T$ giving a slope of $-E/R$. This has been done for many enzyme systems and the approach has also been applied to a number of physiological processes in what has been referred to as the "Crozier concept". The evidence relating to this and the whole question as to its possible validity in a heterogeneous system have been discussed at length by Dawes (1964). The concept has, however, been applied to a number of calcifying systems and for echinoderms Heatfield (1970) obtained a value of 64.9 kJ mol^{-1}: for crustaceans, Porcella et al. (1969) calculated a value of 62.8 kJ mol^{-1} while Weber and White (1974) found 86.5 kJ mol^{-1} for corals. These values are of the same magnitude as other biological phenomena such as nerve discharge and the flashing of fireflies (Stearn, 1949).

It should be realized that, in a complex physiological process such as calcification, an estimate of the energy of activation will not identify any particular reaction. According to the Crozier concept the rate of the overall process will be controlled by the slowest step in the series so that if this changes, a different energy of activation will be found. This is obviously an oversimplified approach but it is interesting to note that the calcification of the cuticle of *Daphnia* normally has an energy of activation of 62.8 kJ mol^{-1} but that this rises to 71.8 kJ mol^{-1} during moult (Porcella et al., 1969).

Rate of calcification

The rate of calcification is strongly influenced by the availability of calcium ions in the environment (for annelids, Neff, 1969; crustacea, Greenaway, 1974a, b, c; echinoderms, Okazaki, 1965). It is suggested by Porcella et al. (1969) that, in many crustacea, calcium deposition follows first-order reaction kinetics. Thus, the rate of calcification of the exoskeleton of *Daphnia magna* after a moult can be described by the equation

$$dC/dt = k(C_{max} - C)$$

where dC/dt = rate of mineral deposition, k = deposition coefficient, C_{max} = maximum quantity of calcium accumulated, C = quantity of calcium accumulated at time t.

An alternative explanation of these data is the suggestion originally proposed by Schurr and Stampier (1962) for the crayfish (*Cambarus longulus longerostris*) that the exponential relationship exists because the rate of mineralization is dependent upon the progressive filling of a fixed number of free sites in the exoskeleton.

Factors influencing mineral deposition

The organic matrix. In corals, annelids, arthropods, and molluscs, the matrix permeates the calcified deposits. However, there is uncertainty as to whether matrix material exists within the mineral of calcareous sponges and echinoderms (Wilbur, 1976). In all probability, the matrix has a significant role both in initiating and terminating calcium carbonate deposition during skeletal formation (see pp. 81, 82). Crenshaw and Ristedt (1976) have suggested that the nucleation of $CaCO_3$ crystals in molluscs may be mediated by negatively charged groups of sulfated polysaccharide which may be a part of a sulfated calcium-binding glycoprotein molecule (Crenshaw, 1972b) associated with the crystals of molluscan shell. Adsorption of protein on crystals may interrupt crystal growth. The adsorption may involve negatively charged carboxyl groups fixed to carbonate surfaces (Mitterer, 1972). Substances including phosphate (Simkiss, 1964) and lipids (Suess, 1970) may also inhibit growth of carbonate crystals by covering their surfaces.

Enzymes. An obvious additional factor in biomineralization is enzyme activity. Enzymes are involved in the movements of ions by membrane pumps, the synthesis of organic matrix and sclerotized skeletal material, photosynthesis of coral zooxanthellae, and reactions involving CO_2. The best studied of these is carbonic anhydrase which catalyses the reactions

$$CO_2 + H_2O \rightleftharpoons H_2CO_3$$
$$CO_2 + OH^- \rightleftharpoons HCO_3^-$$

The enzyme has been found in corals (Goreau, 1959), annelids (Clark, 1975), crustaceans (Costlow, 1959), molluscs (Wilbur, 1972, for references), and echinoderms (Heatfield, 1970). In all cases, sulfanilamide inhibitors of the enzyme added to the medium reduced the rate of calcification at least 50%, indicating that catalysis by carbonic anhydrase is required for the normal rate of mineralization. In theory, the reactions should occur without enzyme catalysis and this appears to be true. Various mechanisms of action have been suggested (e.g., Goreau, 1959; Istin and Girard, 1970b) but the exact role of carbonic anhydrase remains uncertain.

The microenvironment of calcification. In extracellular calcification, crystallization occurs within a volume of fluid which is isolated from the external medium. The fluid in most cases is a thin layer between the cells responsible for calcification and the skeletal surface on which $CaCO_3$ is deposited. Cells can alter the composition and concentration of the ions in the fluid by creating a barrier between the external medium and the site of mineralization. In this way, any inhibitory substances in the medium may be excluded from the calcifying surface while organic material associated with mineralization may be retained. Exclusion of the medium is accomplished in various ways: (1) by envelopment of the calcification site by a layer of cells which enclose a large calcifying area, as in the corals and molluscs; (2) by individual cells attached to crystal surfaces, as in the calcareous sponges; or (3) cellular secretion of material through which the medium cannot penetrate, as in the periostracum of molluscs.

REMODELING

Skeletal remodeling in which the form of the skeleton is altered during its development has been observed to occur in some invertebrates and may be a general feature of invertebrate mineralization. It appears likely that the remodeling is brought about by the cells and the cellular membranes which form the skeleton and which retain an association with the mineral after it has been deposited. This is clearly the case in the arthropods during the moult cycle when the entire exoskeleton is resorbed (Travis, 1965) and the mineral components excreted (Greenaway, 1974a, b, c) or stored (McWhinnie et al., 1969) prior to mineralizing the new skeleton. In this case, the same epidermis appears to be capable of depositing and resorbing the exoskeleton, and according to McWhinnie et al. (1969), a similar process may continue throughout intermoult. In the molluscs, the shell may be resorbed during periods of anaerobic metabolism (Dugal, 1939; Crenshaw and Neff, 1969) and during egglaying of snails with calcified eggshells (Tompa, 1976). Shell remodeling differs from these in that the shell is reshaped as its grows. The following are examples of remodelling: (1) removal of the shell apex in gastropods when a certain size has been reached (Hyman, 1967); (2) resorption of spines and apertural teeth; and (3) thinning and resorption of the columella of gastropods (Meglitsch, 1964).

Remodeling without macroscopic changes may occur in echinoderm skeletal material. Towe (1967) has suggested that in skeletal formation oriented polycrystalline growth first occurs and is followed by recrystallization by continual fusion and coalescence. In regenerating spines, Davies et al. (1972) showed that the first material to be deposited was calcite but that this became modified to a high magnesium calcite as it aged. Clearly, some process of solution and redeposition or exchange normally occurs in the echinoid skeleton and it would be of considerable interest to know whether part of this process was also cell-mediated. The concept of cell-mediated resorption has been invoked by Nichols and Currey (1968) as a way of removing imperfections from the surface of echinoid skeletons and thereby increasing their mechanical strength.

The fracture or removal of a piece of skeleton initiates modeling of a different type from those discussed. Cell processes are activated resulting in secretion of $CaCO_3$ and repair of the injured area. The processes of repair have been investigated in annelids (see p. 84), molluscs, and echinoderms (see Wilbur, 1972).

RHYTHMIC AND INCREMENTAL GROWTH

Invertebrate growth is characteristically incremental rather than continuous. The increments are indicated by markings or changes in conformation, commonly microscopic, in the skeleton. Growth increments are present in corals, serpulids, molluscs, barnacles, insects, echinoderms and probably other groups as well (Neville, 1967; Losada, 1974; Pearse and Pearse, 1975). The markings result from an interruption of deposition or a change in ratio of mineral to organic material deposited by the cells which form the skeleton. The conformational changes are due to the form assumed by the secreting epithelium and the amount of material deposited. In bivalve shells the markings are lines of organic material deposited. In bivalve shells, the markings are lines of organic materials on the internal surface parallel to the growing edge (Pannella, 1975) or ridges on the outer surface (Clark, 1974; Wheeler et al., 1975). The markings and ridges recording the rate of growth of individual increments within periods which are subdaily, daily or longer (Pannella, 1976). Annual banding is present in corals (Knutson et al., 1972; Buddemeier et al., 1974).

Growth by increments appear to be the result of two types of effects: (1) environmental fluctuations, including tides, temperature, light, and food supply; and (2) endogenous rhythms (Palmer, 1973; Pannella, 1975). Of these, Pannella (1976) considers tides to be of major importance in the formation of increments in bivalves. Exposure at low tide induces shell closure and appears to increase the organic/inorganic ratio of shell deposition by limiting $CaCO_3$ deposition, producing an organic marking or layer. A more pronounced deposition of $CaCO_3$ occurs at high tide, extending the area of the shell edge (MacClintock and Pannella, 1969). A new increment is formed on the next tidal cycle. The pattern of markings resulting from tidal influences will follow a lunar periodicity. Also, photoperiodicity has been shown to be important in triggering the formation of a growth increment at a particular time of day in the bivalve *Argopecten* (Wrenn, 1972). Such growth increments regulated by light will conform to solar periodicity which will be out of phase with lunar periodic effects by 50 min d^{-1}. A solar periodicity in the formation of ridges in the epitheca of corals was formerly thought to occur, but this is now uncertain (Pannella, 1975). The growth pattern of an organism may reflect more than one kind of cycle and, in the bivalve *Mercenaria*, lunar, solar, and endogenous cycles are considered to have an influence (Pannella, 1975).

Growth increments which extend the area of the skeleton are formed by epithelia which may contract and extend, as in corals (Barnes, 1972), molluscs (Clark, 1974), and serpulids. An increment will be formed only if the epithelium covers the site of deposition, secretes, and provides a microenvironment favoring crystal formation. Withdrawal of the epithelium or cessation of secretion will terminate the increment. Increments in thickness in areas permanently covered by epithelia will result from intermittent secretion and control of the microenvironment.

From these considerations, incremental growth of the skeleton will clearly be governed by the form and metabolism of the contiguous epithelium. The mechanisms involved have been but little explored. However, it will be evident that environmental and endogenous stimuli affecting nerves and muscles controlling epithelial movement will have an influence. Thus, mechanical stimuli from tidal movements or responses of photoreceptors (Jacklett, 1973a,b; Strumwasser, 1973), which result in muscle contraction, can terminate and prevent increment formation. Similarly, environmental factors such as temperature and light can be expected to influence the size and the initiation of growth increments by acting upon ion pumps, synthesis of organic matrix and its secretion, and photosynthesis.

In summary, it appears that the pattern of frequency of formation and accordingly the size of growth increments may reflect one or more of three rhythms: lunar, solar, and endogenous (Pannella, 1975). Their expression will result from environmental and internal influences affecting the movement and metabolism of tissues which produce the skeleton.

ACKNOWLEDGEMENTS

We thank Dr. M.A. Crenshaw and Dr. C.S. Sikes for helpful suggestions in the preparation of the manuscript.

REFERENCES

Bak, R.P.W., 1973. Coral weight increment in situ. A new method to determine coral growth. Mar. Biol., 20: 45—49.

Bak, R.P.W., 1974. Available light and other factors influencing growth of stony corals through the year in Curacao. Proc. Second Internat. Coral Reef Symp., Brisbane. Great Barrier Reef Committee, pp. 229—233.

Barnes, D.J., 1970. Coral skeletons: An explanation of their growth and structure. Science, 170: 1305—1308.

Barnes, D.J., 1972. The structure and formation of growth-ridges in scleractinian coral skeletons. Proc. R. Soc. London, Ser. B., 182: 331—350.

Barnes, D.J. and Taylor, D.L., 1973. In situ studies of calcification and photosynthetic carbon fixation in the coral *Montastrea annularis*. Helgol. Wiss. Meeresunters., 24: 284—291.

Becker, G.L., Chen, C.H., Greenawalt, J.W. and Lehninger, A.L., 1974. Calcium phosphate granules in the hepatopancreas of the blue crab *Callinectes sapidus*. J. Cell. Biol., 61: 316—326.

Borle, A.B., 1973. Calcium metabolism at the cellular level. Fed. Proc., 32: 1944—1950.

Bryan, W.H. and Hill, D., 1940. Spherulitic crystallization as a mechanism of skeletal growth in the hexacorals. Proc. R. Soc. Queensl., 52: 78—91.

Bubel, A., 1975. An ultrastructural study of the mantle of the barnacle *Elminus modestus* in relation to shell formation. J. Exp. Mar. Biol. Ecol., 20: 287—324.

Buddemeier, R.W., Maragos, J.E. and Knutson, D.W., 1974. Radiographic studies of coral reef exoskeletons: Rates and patterns of coral growth. J. Exp. Mar. Biol. Ecol., 14: 179—200.

Burton, R.F., 1970. Tissue buffering in the snail, *Helix aspersa*. Comp. Biochem. Physiol., 37: 193—203.

Burton, R.F., 1976. Calcium metabolism and acid-base balance in *Helix pomatia*. In: S.P. Davies (Editor), Perspectives in Experimental Biology. I. Zoology. Pergamon, Oxford, pp. 7—16.

Campbell, J.W. and Boyan, B.D., 1976. On the acid-base balance of gastropod molluscs. In: N. Watabe and K.M. Wilbur (Editors). The Mechanisms of Mineralization in the Invertebrates and Plants. University of South Carolina Press, Columbia, pp. 109—133.

Campbell, J.W. and Speeg, K.V., Jr., 1969. Arginine biosynthesis and metabolism in terrestrial snails. Comp. Biochem. Physiol., 25: 3—32.

Carriker, M.R. and Smith, E.H., 1969. Comparative calcibiocavitology: Summary and conclusions. Am. Zool., 9: 1011—1020.

Carriker, M.R., Smith, E.H. and Wilce, R.T. (Editors), 1969. Penetration of calcium carbonate substrates by lower plants and invertebrates. Am. Zool., 9: 629—1020.

Chaisemartin, C., 1965. Voies de passage du calcium entre milieu extérieur et intérieur chez l'écrevisse postexuviale. C.R. Soc. Biol., 159: 1214—1217.

Chalker, B.E. and Taylor, D.L., 1975. Light enhanced calcification, and the role of oxidative phosphorylation in calcification of the coral *Acropora cervicornis*. Proc. R. Soc. London, Ser. B., 190: 323—331.

Chapman, G., 1974. The skeletal system. In: L. Muscatine and H.M. Lenhoff (Editors), Coelenterate Biology. Reviews and New Perspectives. Academic, New York, NY, pp. 93—128.

Clark, A.M., 1957. The distribution of carbonic anhydrase in the earthworm and snail. Aust. J. Sci., 19: 205—207.

Clark, G.R., 1974. Growth lines in invertebrate skeletons. Ann. Rev. Earth Planet. Sci., 2: 77—99.

Clausen, C.D. and Roth, A.A., 1975. Estimation of coral growth rates from laboratory ^{45}Ca-incorporation rates. Mar. Biol., 33: 85—91.

Costlow, J.D., 1959. Effect of carbonic anhydrase inhibitors on shell development and growth in *Balanus improvisus*, Darwin. Physiol. Zool., 32: 173—184.

Crang, R.E., Holsen, R.C. and Hitt, J.B., 1968. Calcite production in mitochondria of earthworm calciferous glands. Biol. Sci., 18: 299—301.

Crenshaw, M.A., 1972a. The inorganic composition of molluscan extrapallial fluid. Biol. Bull., 143: 506—512.

Crenshaw, M.A., 1972b. The soluble matrix from *Mercenaria mercenaria* shell. Biomineralization, 6: 6—11.

Crenshaw, M.A. and Neff, J.M., 1969. Decalcification at the mantle-shell interface in molluscs. Am. Zool., 9: 881—885.

Crenshaw, M.A. and Ristedt, H., 1976. The histochemical localization of reactive groups in septal nacre from *Nautilus pompilius* L. In: N. Watabe and K.M. Wilbur (Editors), The Mechanisms of Mineralization in the Invertebrates and Plants. University of South Carolina Press, Columbia, pp. 355—367.

Crossland, C.J. and Barnes, D.J., 1974. The role of metabolic nitrogen in coral calcification. Mar. Biol., 28: 325—332.

Crossland, C.J. and Barnes, D.J., 1977. Calcification in the staghorn coral *Acropora acuminata:* variations in apparent skeletal incorporation of radioisotopes due to different methods of processing. Mar. Biol., 43: 57—62.

Currey, J.D. and Taylor, J.D., 1974. The mechanical behavior of some molluscan hard tissues. J. Zool. Lond., 173: 395—406.

Davies, T.T., Crenshaw, M.A. and Heatfield, B.M., 1972. The effect of temperature on the chemistry and structure of echinoid spine regeneration. J. Paleontol., 46: 874—883.

Dawes, E.A., 1964. Enzyme kinetics (optimum pH, temperature and activation energy). In: M. Florkin and E.H. Stotz (Editors), Comparative Biochemistry, Vol. 12, Elsevier, Amsterdam, pp. 89—125.

Defretin, R., 1971. The tubes of polychaete annelids. In: M. Florkin and E.H. Stotz (Editors), Comprehensive Biochemistry. Elsevier, Amsterdam, 26 C: 713—745.

Degens, E.T., 1976. Molecular mechanisms on carbonate, phosphate, and silica deposition in the living cell. Top. Curr. Chem., 64: 1—112.

Drach, P., 1939. Mue et cycle d'intermue chez les crustacés décapods. Ann. Inst. Oceanogr., 19: 103—391.

Dugal, L.P., 1939. The use of calcareous shell to buffer the product of anaerobic glycolysis in *Venus mercenaria.* J. Cell. Comp. Physiol., 13: 235—251.

Dunkelberger, D.G. and Watabe, N., 1974. An ultrastructural study on spicule formation in the penatulid colony *Renilla reniformis.* Tissue Cell, 6: 573—586.

Edwards, B.D. and Perkins, R.D., 1974. Distribution of microborings within continental margin sediments of the Southeastern United States. J. Sediment. Petrol., 44: 1122—1135.

Emery, K.O., 1946. Marine solution basins. J. Geol., 54: 209—228.

Fretter, V., 1953. Experiments with radioactive Sr (Sr^{90}) on certain molluscs and polychaetes. J. Mar. Biol. Assoc. U.K., 32: 367—386.

Fütterer, D.K., 1974. Significance of the boring sponge *Cliona* for the origins of fine grained material of carbonate sediments. J. Sediment. Petrol., 44: 79—84.

Goreau, T.F., 1959. The physiology of skeleton formation in corals. I. A method for measuring the rate of calcium deposition by corals under different conditions. Biol. Bull., 116: 59—75.

Goreau, T.F., 1963. Calcium carbonate deposition by coralline algae and corals in relation to their roles as reef-builders. Ann. N.Y. Acad. Sci., 109: 127—167.

Goreau, T.F. and Goreau, N.I., 1959a. The physiology of skeleton formation in corals. I. A method for measuring the rate of calcium deposition by corals under different conditions. Biol. Bull., 116: 59—75.

Goreau, T.F. and Goreau, N.I., 1959b. The physiology of skeletal formation in corals. II. Calcium deposition by hermatypic corals under various conditions in the reef. Biol. Bull., 117: 239—250.

Goreau, T.F. and Goreau, N.I., 1960. The physiology of skeleton formation in corals. IV. On isotopic equilibrium exchanges of calcium between corallum and environment in living and dead reef-building corals. Biol. Bull., 119: 416—427.

Goreau, T.J., 1977a. Coral skeletal chemistry: physiological and environmental regulation of stable isotopes and trace metals in *Montastrea annularis.* Proc. R. Soc. London, Ser. B., 196: 291—315.

Goreau, T.J., 1977b. Carbon metabolism in calcifying and photosynthetic organisms: Theoretical models based on stable isotope data. Proc. Third Internat. Coral Reef Symp., II, 395—401.

Green, J.P. and Neff, M.R., 1972. A survey of the fine structure of the integument of the fiddler crab. Tissue Cell, 4: 137—171.

Greenaway, P., 1971. Calcium regulation in the freshwater mollusc *Limnaea stagnalis* (L.) (Gastropoda: Pulmonata). I. The effect of internal and external calcium concentrations. J. Exp. Biol., 53: 333—349.
Greenaway, P., 1974a. Total body calcium and haemolymph calcium concentration in the crayfish *Austropotamobius pallipes* (Lereboullet). J. Exp. Biol., 61: 19—26.
Greenaway, P., 1974b. Calcium balance at the premoult stage of the freshwater crayfish *Austropotamobius pallipes* (Lereboullet). J. Exp. Biol., 61: 27—34.
Greenaway, P., 1974c. Calcium balance at the premoult stage of the freshwater crayfish *Austropotamobius pallipes* (Lereboullet). J. Exp. Biol., 61: 35—45.
Grégoire, C., 1972. Structure of the molluscan shell. In: M. Florkin and B.T. Scheer (Editors), Chemical Zoology, Vol. 7. Mollusca. Academic, New York, NY, pp. 45—102.
Hammen, C.S. and Wilbur, K.M., 1959. Carbon dioxide fixation in marine invertebrates. I. The main pathway in the oyster. J. Biol. Chem., 234: 1268—1271.
Hartman, W.D. and Goreau, T.F., 1970. Jamaican coralline sponges: their morphology, ecology and fossil relatives. Symp. Zool. Soc. London, 25: 205—243.
Hayes, D.K. and Armstrong, W.D., 1961. The distribution of mineral in the calcified carapace and claw shell of the American lobster *Homarus americanus*, evaluated by means of microroentgenograms. Biol. Bull., 121: 307—315.
Hayes, R.L. and Goreau, N.I., 1976. Cellular mechanisms regulating coral skeletogenesis. J. Cell. Biol., 70: 408A.
Hayes, R.L. and Goreau, N.I., 1977. Intracellular crystal-bearing vesicles in the epidermis of scleractinian corals, *Astrangia danae* (Agassiz) and *Porites porites* (Pallas). Biol. Bull., 152: 26—40.
Heatfield, B.M., 1970. Calcification in echinoderms: effects of temperature and Diamox on incorporation of calcium-45 in vitro by regenerating spines of *Strongylocentrotus purpuratus*. Biol. Bull., 139: 151—163.
Heatfield, B.M. and Travis, D.F., 1975. Ultrastructural studies of regenerating spines of the sea urchin *Strongylocentrotus purpuratus*. Cell types without spherules. J. Morphol., 145: 13—50.
Hedley, R.H., 1956a. Studies on serpulid tube formation. I. The secretion of the calcareous and organic components of the tube by *Pomatoceros triqueter*. Q. J. Microsc. Sci., 97: 411—419.
Hedley, R.H., 1956b. Studies on serpulid tube formation. II. The calcium secreting glands in the peristomium of *Spirorbis, Hydroides* and *Serpula*. Q. J. Microsc. Sci., 97: 421—427.
Hubbard, J.A.E.B., 1974. Barnes technique amended for analysing fabric and cavity development in coral reef communities. J. Paleontol., 48: 769—777.
Hyman, L.H., 1967. The Invertebrates, Vol. 6, *Mollusca* I. McGraw-Hill, New York, NY, 720 pp.
Istin, M. and Girard, J.P., 1970a. Dynamic state of calcium reserves in freshwater clam mantle. Calcif. Tissue Res., 5: 196—205.
Istin, M. and Girard, J.P., 1970b. Carbonic anhydrase and mobilisation of calcium reserves in the mantle of lamellibranchs. Calcif. Tissue Res., 5: 247—260.
Istin, M. and Maetz, J., 1964. Permeabilité au calcium du manteau de lamellibranches d'eau douce étudiée à l'aide des isotopes ^{45}Ca et ^{47}Ca. Biochim. Biophys. Acta, 88: 225—227.
Jacklett, J.W., 1973a. Neuronal population interactions in a circadian rhythm in *Aplysia*. In: J. Salanki (Editor), Neurobiology of Invertebrates Mechanisms of Rhythm Regulation. Akadémiai Kiado, Budapest, pp. 363—380.
Jacklett, J.W., 1973b. The circadian rhythms of the eye of *Aplysia*. Effects of low calcium and high magnesium. J. Comp. Physiol., 87: 329—338.
Jell, J.S., 1974. The microstructure of some scleractinian corals. Proc. Second Internat. Coral Reef Symp., Great Barrier Reef Committee, Brisbane, 2: 301—320.

Jones, W.C., 1964. Photographic records of living oscular tubes of *Leucosolenia variabilis*. II. Spicule growth, form and displacement. J. Mar. Biol. Assoc., 44: 311—331.

Jones, W.C., 1967. Sheath and axial filament of calcareous sponge spicules. Nature, 214: 365.

Jones, W.C., 1970. The composition, development, form and orientation of calcareous sponge spicules. Symp. Zool. Soc. London, 25: 91—123.

Jones, W.C. and James, D.W.F., 1969. An investigation of some calcareous sponge spicules by means of electron probe micro-analysis. Micron, 1: 34—39.

Jones, W.C. and Jenkins, D.A., 1970. Calcareous sponge spicules: a study of magnesian calcites. Calcif. Tissue Res., 4: 314—329.

Keith, M.L., Anderson, G.M. and Eichler, R., 1964. Carbon and oxygen isotopic composition of mollusc shells from marine and freshwater environments. Geochim. Cosmochim. Acta, 28: 1757—1786.

Kennedy, W.J., Morris, N.J. and Taylor, J.D., 1970. The shell structure, mineralogy and relationships of the Chamacea (Bivalvia). Paleontol., 13: 379—413.

Kitano, Y., Kanamori, N. and Yoshioka, S., 1976. Influence of chemical species on the crystal type of calcium carbonate. In: N. Watabe and K.M. Wilbur (Editors), The Mechanisms of Mineralization in the Invertebrates and Plants. University of South Carolina Press, Columbia, pp. 191—202.

Kniprath, E., 1974. Ultrastructure and growth of the sea urchin tooth. Calcif. Tissue Res., 14: 211—228.

Knutson, D.W., Buddemeier, R.W. and Smith, S.V., 1972. Coral chronometers: seasonal growth bands in reef corals. Science, 177: 270—277.

Krampitz, G., Engels, J. and Cazaux, C., 1976. Biochemical studies on water-soluble proteins and related components of gastropod shells. In: N. Watabe and K.M. Wilbur (Editors), The Mechanisms of Mineralization in the Invertebrates and Plants. University of South Carolina Press, Columbia, pp. 155—173.

Lamberts, A.E., 1974. Measurement of alizarin deposited by coral. Proc. Second Internat. Symp. on Coral Reefs, Great Barrier Reef Committee, Brisbane, 2: 241—244.

Land, L.S., Lang, J.C. and Barnes, D.J., 1975. Extension rate: A primary control on the isotopic composition of West Indian (Jamaican) scleractinian reef coral skeletons. Mar. Biol., 33: 221—233.

Land, L.S., Lang, J.C. and Barnes, D.J., 1977. On the stable carbon and oxygen isotopic composition of some shallow water, ahermatypic, scleractinian coral skeletons. Geochim. Cosmochim. Acta, 41: 169—172.

Ledger, P.W. and Jones, W.C., 1977. Spicule formation in the calcareous sponge *Sycon ciliatum*. Cell Tissue Res., 8: 553—567.

Losada, F.J., 1974. Studies on growth and shell deposition in barnacles of the genus *Balunus*. Ph.D. Thesis, Duke University, Durham, NC, 120 pp.

Macintyre, I.G. and Smith, S.V., 1974. X-radiographic studies of skeletal development in coral colonies. Proc. Second Internat. Symp. on Coral Reefs, Great Barrier Reef Committee, Brisbane, 2: 277—287.

MacClintock, C., 1967. Shell structure of patelloid and bellerophontoid gastropods (Mollusca). Yale Univ. Peabody Mus. Nat. Hist. Bull., 22: 1—140.

MacClintock, C. and Pannella, G., 1969. Time of calcification in the bivalve mollusk *M. mercenaria* (L) during the 24 hour period. Abstract. Ann. Meeting Geol. Soc. Am., p. 140.

Marshall, A.T. and Chueng, W.W.K., 1973. Calcification in insects: the dwelling-tube and midgut of Machaerotid larvae (Homoptera). J. Insect Physiol., 19: 963—972.

McWhinnie, M.A., 1962. Gastrolith growth and calcium shifts in the freshwater crayfish *Oronectes virilis*. Comp. Biochem. Physiol., 7: 1—14.

McWhinnie, M.A., Cahoon, M.O. and Johanneck, R., 1969. Hormonal effects on calcium metabolism in crustacea. Am. Zool., 9: 841—855.

Meglitsch, P.A., 1967. Invertebrate Zoology. Oxford University Press, New York, 901 pp.

Milliman, J.D., 1974. Marine Carbonates. Springer, New York, pp. 253—261.
Millonig, G., 1970. A study of the formation and structure of the sea urchin spicule. J. Submicrosc. Cytol., 2: 157—165.
Mitterer, R.M., 1972. Biogeochemistry of aragonite mud and oolites. Geochim. Cosmochim. Acta, 36: 1407—1422.
Mitterer, R.M., 1978. Amino acid composition and metal binding capability of the skeletal protein of corals. Bull. Mar. Sci., 28: 173—180.
Moore, W.S. and Krishnaswami, S., 1974. Correlation of x-radiography revealed banding in corals with radiometric growth rates. Proc. Second Internat. Symp. on Coral Reefs, Great Barrier Reef Committee, Brisbane, 2: 269—276.
Muscatine, L., 1971. Calcification in corals. In: H.M. Lenhoff, L. Muscatine and L.V. Davis (Editors), Experimental Coelenterate Biology. University Hawaii Press, Honolulu, pp. 227—238.
Muscatine, L. and Cernichiari, E., 1969. Assimilation of photosynthetic products of zooxanthellae by a reef coral. Biol. Bull., 137: 506—523.
Nakahara, H. and Bevelander, G., 1969. An electron microscope and autoradiographic study of the calciferous glands of the earthworm *Lumbricus terrestris.* Calcif. Tissue Res., 4: 193—201.
Neff, J.M., 1967. Calcium carbonate tube formation by serpulid polychaete worms: Physiology and ultrastructure, Ph.D. Thesis, Duke University, 305 pp.
Neff, J.M., 1969. Mineral regeneration by serpulid polychaete worms. Biol. Bull., 136: 76—90.
Neff, J.M., 1971a. Ultrastructural studies of the secretion of calcium carbonate by the serpulid polychaete worm *Pomatoceros caeruleus.* Z. Zellforsch., 120: 160—186.
Neff, J.M., 1971b. Ultrastructure of calcium phosphate containing cells in the serpulid polychaete worm *Pomatoceros caeruleus.* Calcif. Tissue Res., 7: 191—200.
Neuman, W.F. and Neuman, M.W., 1958. The chemical Dynamics of Bone Mineral. University of Chicago Press, Chicago, Il, 200 pp.
Neville, A.C., 1967. Daily growth layers in animals and plants. Biol. Rev., 42: 421—441.
Nichols, D. and Currey, J.D., 1968. The secretion, structure and strength of echinoderm calcite. In: S.M. McGee-Russell and K.F.A. Ross (Editors), Cell Structure and Its Interpretation. E. Arnold, London, pp. 251—261.
North, W.J., 1954. Size distribution, erosive activities and gross metabolism efficiency of the marine intertidal snails *Littorina planaxis* and L. *scutulata.* Biol. Bull., 106: 185—197.
Nott, J.A. and Parkes, K.R., 1975. Calcium accumulation and secretion in the serpulid polychaete *Spirorbis spirorbis.* J. Mar. Biol. Assoc. U.K., 55: 911—923.
Okazaki, K., 1965. Skeleton formation of sea urchin larvae. Continuous observation of the process of matrix formation. Exp. Cell Res., 40: 585—596.
Okazaki, K., 1975. Spicule formation by isolated micromeres of the sea urchin embryo. Am. Zool., 15: 567—582.
Otsuki, A. and Wetzel, R.G., 1974. Calcium and total alkalinity budgets and calcium carbonate precipitation of a small hard-water lake. Arch. Hydrobiol., 73: 14—30.
Palmer, J.D., 1973. Tidal rhythms: The clock control of the rhythmic physiology of marine organisms. Biol. Rev., 48: 377—418.
Pannella, G., 1975. Palaeontological clocks and the history of the earth's rotation. In: G.D. Rosenberg and S.K. Runcorn (Editors), Growth Rhythms and the History of the Earth's Rotation. Wiley, London, pp. 253—284.
Pannella, G., 1976. Tidal growth patterns in recent and fossil mollusc bivalve shells: a tool for the reconstruction of paleotides. Naturwissenschaften, 63: 539—543.
Pautard, F.G.E., 1970. Calcification in unicellular organisms. In: H. Schraer (Editor), Biological Calcification. Cellular and Molecular Aspects. Appleton-Century-Crofts, New York, NY, pp. 105—201.
Pautard, F.G.E., 1976. Calcification in single cells: with an appraisal of the relationship between *Spirostomum ambiguum* and the osteocyte. In: N. Watabe and K.M. Wilbur

(Editors), The Mechanisms of Mineralization in the Invertebrates and Plants. University of South Carolina Press, Columbia, pp. 33—53.
Pearse, V.B., 1970. Incorporation of CO_2 into coral skeleton. Nature, 228: 383.
Pearse, V.B. and Muscatine, L., 1971. Role of symbiotic algae (zooxanthellae) in coral calcification. Biol. Bull., 141: 350—363.
Pearse, J.S. and Pearse, V.B., 1975. Growth zones in the echinoid skeleton. Am. Zool. 15: 567—582.
Perkins, R.D. and Tsentas, C.I., 1976. Microbial infestation of carbonate substrates planted on the St. Croix shelf, West Indies. Geol. Soc. Am. Bull., 87: 1615—1628.
Porcella, D.B., Rixford, C.E. and Slater, J.V., 1969. Molting and calcification in *Daphnia magna.* Physiol. Zool., 42: 148—159.
Potts, W.T.W., 1954. The inorganic composition of the blood of *Mytilus edulis* and *Anodonta cygnaea.* J. Exp. Biol., 31: 376—385.
Pytkowicz, R.M., 1965. Rates of inorganic calcium carbonate nucleation. J. Geol., 73: 196—199.
Pytkowicz, R.M., 1973. The carbon dioxide system in the oceans. Schweiz Z. Hydrol., 35: 8—28.
Quiénvreux, C., 1963. Secrétion tupibare des larves de Spirorbinae. (Annélides Polychètes). Cah. Biol. Mar., 4: 399—406.
Rützler, K., 1975. The role of burrowing sponges in bioerosion. Oecologia, 19: 203—216.
Schurr, J.M. and Stamper, M.N., 1962. Model for the accumulation of strontium and calcium by recently moulted crayfish (*Cambarus longulus longerostris*). Limnol. Oceanogr., 7: 474—477.
Simkiss, K., 1964. Phosphates as crystal poisons. Biol. Rev., 39: 487—505.
Simkiss, K., 1976. Cellular aspects of calcification. In: N. Watabe and K.M. Wilbur (Editors), The Mechanisms of Mineralization in the Invertebrates and Plants. University of South Carolina Press, Columbia, pp. 1—31.
Simpson, T.L., 1968. The structure and function of sponge cells: new criteria for the taxonomy of poecilosclerid sponges (Demospongiae). Yale Univ. Peabody Mus. Nat. Hist. Bull., 25: 1—141.
Simpson, T.L. and Vaccaro, C.A., 1974. An ultrastructural study of silica deposition in the freshwater sponge *Spongilla lacustris.* J. Ultrastruct. Res., 47: 296—309.
Skirrow, G., 1975. The dissolved gases-carbon dioxide. In: J.P. Riley and G. Skirrow (Editors), Chemical Oceanography, 2nd edn., Vol. II. Academic, New York, NY, pp. 1—143.
Sminia, T., De With, N.D., Bos, J.L., Van Nieuwmegen, M.E., Witter, M.P. and Wondergem, J., 1977. Structure and function of the calcium cells of the freshwater pulmonate snail *Lymnaea stagnalis.* Neth. J. Zool., 27: 195—208.
Smith, B.N., 1972. Natural abundance of the stable isotopes of carbon in biological systems. Bioscience, 22: 226—231.
Smith, E.H. and Carriker, M.R., 1969. Introduction to the symposium. Am. Zool., 9: 633—634.
Smith, S.V. and Harrison, J.T., 1977. Calcium carbonate production of the *Mare incognitum,* the upper windward reef slope, at Enewetak Atoll. Science, 197: 556—559.
Smith, S.V. and Kinsey, D.W., 1976. Calcium carbonate production, coral reef growth, and sea level change. Science, 194: 937—939.
Sorauf, J.E., 1970. Microstructure and formation of dissepiments in the skeleton of the recent Scleractinia (hexacorals). Biomineralization, 2: 1—22.
Sorauf, J.E., 1972. Skeletal Microstructure and microarchitecture in Scleractinia (Coelenterata). Palaeontol., 5: 88—107.
Stearn, A.E., 1949. Kinetics of biological reactions with special reference to enzymic processes. Adv. Enzymol., 9: 25—74.
Stephenson, J., 1930. The Oligochaeta, Oxford University Press, Oxford.
Stoddart, D.R., 1969. Ecology and morphology of recent coral reefs. Biol. Rev., 44: 433—498.

Strömgren, T., 1976. Skeletal growth of the hydrocoral *Millepora complanata* Lamarck in relation to light. Limnol. Oceanogr., 21: 156—160.

Strumwasser, F., 1973. Neural and humoral factors in the temporal organization of behavior. Physiologist, 16: 9—54.

Suess, E., 1970. Interaction of organic compounds with calcium carbonate. Geochim. Cosmochim. Acta, 34: 157—168.

Swan, J.E., 1950. The calcareous tube secreting glands of the serpulid polychaetes. J. Morphol., 86: 285—314.

Taylor, J.D., Kennedy, W.J. and Hall, A., 1969. The shell structure and mineralogy of the Bivalvia. Introduction, Nuculacea-Trigonacea. Bull. Br. Mus. Nat. Hist., Zool., Suppl., 3: 1—125.

Taylor, J.D., Kennedy, W.J. and Hall, A., 1973. The shell structure and mineralogy of the Bivalvia. II. Lucinacea-Clavagellacea. Conclusions. Bull. Br. Mus. Nat. Hist., Zool., 22: 255—294.

Tilney, L.G. and Porter, K.R., 1967. Studies on the microtubules in *Heliozoa*. II. The effect of low temperature on these structures in the formation and maintenance of the axopodia. J. Cell. Biol., 34: 327—343.

Tompa, A.S., 1976. Calcification of the egg of the land snail *Anguispira alternata* (Gastropoda: Pulmonata). In: N. Watabe and K.M. Wilbur (Editors), The Mechanisms of Mineralization in the Invertebrates and Plants. University of South Carolina Press, Columbia, pp. 427—444.

Tourtelot, H.A. and Rye, R.O., 1969. Distribution of oxygen and carbon isotopes in fossils of late Cretaceous age, western interior region of North America. Bull. Geol. Soc. Am., 80: 1904—1922.

Towe, K.M., 1967. Echinoderm calcite: Single crystal or polycrystalline aggregate. Science, 157: 1048—1050.

Travis, D.F., 1951. Physiological changes which occur in the blood and urine of *Panuliris argus* Latreille during the molting cycle. Anat. Rec., 111: 573.

Travis, D.F., 1960. Matrix and mineral deposition in skeletal structures of the decapod crustacea. In: R.F. Sognnaes (Editor), Calcification in Biological Systems, A.A.A.S., Pub. 64, pp. 57—116.

Travis, D.F., 1963. Structural features of mineralization from tissue to macromolecular levels of organization in the decapod crustacea. Ann. N.Y. Acad. Sci., 109: 177—245.

Travis, D.F., 1965. The deposition of skeletal structures in the crustacea. 5. The histomorphological and histochemical changes associated with the development and calcification of the branchial exoskeleton in the crayfish *Orconectes virilis* Hagen. Acta Histochem., 20: 193—222.

Travis, D.F., 1970. The comparative ultrastructure and organization of five calcified tissues. In: H. Schraer (Editor), Biological Calcification: Cellular and Molecular Aspects. Appleton-Century-Crofts, New York, NY, pp. 203—311.

Travis, D.F. and Friberg, U., 1963a. The deposition of skeletal structures in the Crustacea. IV. Microradiographic studies of the gastrolith of the crayfish *Orconectes virilis* Hagen. J. Ultrastruct. Res., 8: 48—65.

Travis, D.F. and Friberg, U., 1963b. The deposition of skeletal structures in the Crustacea. VI. Microradiographic studies of the exoskeleton of the crayfish *Orconectes virilis* Hagen. J. Ultrastruct. Res., 69: 285—301.

Trench, R.K., 1975. Deposition of calcium carbonate skeletons by corals: an appraisal of physiological and ecological evidence. In: E.A. Drew, J.N. Lythgoe and J.D. Woods (Editors), Underwater Research. Academic, London, pp. 381—394.

Vandermeulen, J.H., 1975. Studies on reef corals. III. Fine structural changes of calicoblast cells in *Pocillopora damicornis* during settling and calcification. Mar. Biol., 31: 69—77.

Vandermeulen, J.H. and Muscatine, L., 1974. Influence of symbiotic algae on calcification in reef corals: Critique and progress report. In: W.B. Vernberg (Editor), Symbiosis in the Sea. University of South Carolina Press, Columbia, pp. 1—19.

Vandermeulen, J.H., Davis, N.D. and Muscatine, L., 1972. The effect of inhibitors of photosynthesis on zooxanthellae in corals and other marine invertebrates. Mar. Biol., 16: 185—191.

Vandermeulen, J.H. and Watabe, N., 1973. Studies on reef corals. I. Skeleton formation by newly settled planula larva of *Pocillopora damicornis*. Mar. Biol., 23: 47—57.

Van Gansen-Semal, P., 1959. Structures des glandes calciques d'*Eisenia foetida Sav.* Bull. Biol. Fr. Belg., 93: 38—63.

Vovelle, J., 1956. Processus glandulaires impliques dans la reconstruction du tube chez *Pomatoceros triqueter* (L) Annelide Polychaete (Serpulidae). Bull. Lab. Marit. Dinard, 42: 10—32.

Wada, K., 1970. The structure and formation of pearl. In: H. Yukawa (Editor), Profiles of Japanese Science and Scientists. Kodansha, Tokyo, pp. 227—244.

Wada, K. and Fujinuki, T., 1976. Biomineralization in bivalve molluscs with emphasis on the chemical composition of the extrapallial fluid. In: N. Watabe and K.M. Wilbur (Editors), The Mechanisms of Mineralization in the Invertebrates and Plants. University of South Carolina Press, Columbia, pp. 175—190.

Wainwright, S.A., 1963. Skeletal organization in the coral, *Pocillopora damicornis*. Q. J. Microsc. Sci., 104, Pt. 2: 169—183.

Waku, Y. and Sumimoto, K., 1974. Metamorphosis of midgut epithelial cells in the silkworm (*Bombyx mori*) with special regard to the calcium salts deposits in the cytoplasm. II. Electron microscopy. Tissue Cell, 6: 127—136.

Watabe, N., Meenakshi, V.R., Blackwelder, P.L., Kurtz, E.M. and Dunkelberger, D.G., 1976. Calcareous spherules in the gastropod. *Pomacea paludosa*. In: N. Watabe and K.M. Wilbur (Editors), The Mechanisms of Mineralization in the Invertebrates and Plants. University of South Carolina Press, Columbia, pp. 283—308.

Weber, J.N., 1974. $^{13}C/^{12}C$ ratios as natural isotopic tracers elucidating calcification processes in reef-building and non-reef-building corals. Proc. 2nd Int. Symp. on Coral Reefs, Great Barrier Reef Committee, Brisbane, 2: 289—298.

Weber, J.N. and White, E.W., 1974. Activation energy for skeletal aragonite deposited by the hermatypic coral *Platygyra* spp. Mar. Biol., 26: 353—359.

Weber, J.N. and Woodhead, P.M.J., 1970. Carbon and oxygen isotope fractionation in the skeletal carbonate of reef-building corals. Chem. Geol., 6: 93—123.

Welinder, B.S., 1975. The crustacean cuticle II. Deposition of organic and inorganic material in the cuticle of *Astacus fluviatilis* in the period after moulting. Comp. Biochem. Physiol., 51B: 409—416.

Wheeler, A.P., 1975. Oyster mantle carbonic anhydrase: Evidence for plasma membrane-bound activity and for a role in bicarbonate transport. Ph.D. Thesis. Duke University, Durham, NC, 212 pp.

Wheeler, A.P., Blackwelder, P.L. and Wilbur, K.M., 1975. Shell growth in the scallop *Argopecten irradians*. I. Isotopic incorporation with reference to diurnal growth. Biol. Bull., 148: 472—482.

Wilbur, K.M., 1964. Shell formation and regeneration. In: K.M. Wilbur and C.M. Yonge (Editors), Physiology of Mollusca, Vol. I. Academic, New York, NY, pp. 243—282.

Wilbur, K.M., 1972. Shell formation in mollusks. In: M. Florkin and B.T. Scheer (Editors), Chemical Zoology, Vol. 7. Mollusca. Academic, New York, NY, pp. 103—145.

Wilbur, K.M., 1976. Recent studies of invertebrate mineralization. In: N. Watabe and K.M. Wilbur (Editors), The Mechanisms of Mineralization in the Invertebrates and Plants. University of South Carolina Press, Columbia, pp. 79—108.

Wilbur, K.M. and Simkiss, K., 1968. Calcified shells. In: M. Florkin and E.H. Stotz (Editors), Comprehensive Biochemistry, 26A. Elsevier, Amsterdam, pp. 221—295.

Williams, A., 1971. Comments on the growth of the shell of articulate brachiopods. Smithson. Contrib. Paleobiol., 3: 47—67.

Wood, D., 1973. Calcium movement and electrogenesis across the isolated clam mantle. Ph.D. Thesis, Washington State University, Pullman, 122 pp.

Wrenn, S., 1972. Daily increment formation and synchronization in the bay scallop. Am. Soc. Zool. Summer Meeting, Div. Comp. Physiol. Biochem. with Am. Physiol. Soc., Am. Zool., 12: xxxvii.

Chapter 2.4

CARBONATE DISSOLUTION

S. GOLUBIĆ

Department of Biology, Boston University, 2 Cummington Street, Boston, MA 02215 (U.S.A.)

J. SCHNEIDER

Geology-Palaeontology, University of Göttingen, Göttingen (F.R.G.)

CONTENTS

ABIOTIC AND BIOGENIC CARBONATE DISSOLUTION

Throughout Earth's history life has continuously modified the environment so that it is unusual today to encounter any chemical process that is not in some way influenced by it. Biogenically increased concentrations of CO_2, organic acids, and chelating agents all affect carbonate dissolution in natural waters.

Chemically pure water can solvate only about 13 mg l^{-1} $CaCO_3$. When rainwater at 15°C is in equilibrium with atmospheric CO_2 (0.03% of the total atmospheric volume), up to 0.6 mg l^{-1} enters into solution. If this hydrated CO_2 is in equilibrium with calcium carbonate, it can maintain up to 50 mg l^{-1} of $CaCO_3$ in solution (Cole, 1975, p. 187). It is important to note that half of the bicarbonate in solution derives from the atmospheric CO_2, and the other half from the available lithospheric carbonate. Since CO_2 solubility increases with decreasing temperature, so does the solubility of carbonate. Carbonate dissolution greater than 50 mg l^{-1} depends on an increased partial pressure of CO_2 which is usually biogenic. Oxidative respiration in soils, for example, is able to increase the ambient CO_2 content up to 500 times (Ruttner, 1960).

Carbonate dissolution in natural systems is often intimately tied with carbonate precipitation, and both processes need to be studied together. Extensive carbonate precipitation, such as travertine deposition in karstic waters, is usually preceded by intensive carbonate dissolution (Golubić, 1973). In terrestrial environments, ground water may serve as a transportation mechanism connecting areas of intensive carbonate dissolution with areas where the carbonate is redeposited. In aquatic environments, carbonate is precipitated in illuminated surface waters where there is an excess of photosynthesis (trophogenic layer); carbonate is dissolved in dark, deep waters where organic matter is decomposed and CO_2 released (tropholytic layer). However, conditions promoting carbonate dissolution are also maintained within the littoral zone micro-environments of well-aerated shallow waters. Such conditions are encountered just a few millimeters below the surface of algal mats, or beneath undisturbed fine grain sediments. Similar micro-environments exist in fissures and cavities of rocks and under carbonate crusts and travertine deposits. Extremely localized and directed carbonate dissolution is carried out by specialized boring organisms which leave specific traces of their activity in carbonate substrates.

In this chapter, we shall first discuss the prevalent mechanisms of carbonate removal, the relationship between carbonate-removing organisms and carbonate substrates, and then briefly review the principal taxonomic groups involved.

MECHANISMS OF CARBONATE REMOVAL

Various mechanisms by which organisms may affect carbonate substrates have been proposed (Pia, 1937), but few are proven or understood. These mechanisms may be categorized as follows:

(1) biological corrosion — chemical dissolution of carbonates by organisms, and

(2) biological abrasion — mechanical removal of carbonate by organisms.

Both mechanisms can be summarized as bioerosion. When biological corrosion and abrasion act together, a synergistic effect is often observed. The result of bioerosion is a highly profiled morphology of rock surfaces known as biokarst (Schneider, 1976). Two groups of organisms are mainly responsible for carbonate bioerosion (Milliman, 1974): borers — penetrating hard substrates, and grazers — scraping the substrate while feeding on epilithic and endolithic algae.

Biological corrosion

Biological corrosion may affect carbonate substrates directly, or indirectly by a modification of ambient water chemistry. Most commonly, biocorrosive activity involves an increase in partial pressure of CO_2 within the carbonate buffer system. The additional CO_2 usually stems from oxidative respiration of various organisms including bacteria, fungi, soil animals and plant roots. Hydration of CO_2 and the corrosive activity of carbonic acid in biological systems is often enzymically regulated by carbonic anhydrase, which enhances the following reaction (eqn. (1))

$$CO_2 + H_2O \xrightarrow{\text{carbonic anhydrase}} H^+ + HCO_3^- \tag{1}$$

Carbonic anhydrase is generally associated with photosynthesis and respiration. Its association with carbonate dissolution has been established in clionid sponges (Hatch, 1975) and predatory snails (Chetail and Binot, 1967; Chetail and Fournie, 1969; Smarsh et al., 1969), and is suspected in other invertebrates and carbonate-boring algae and fungi.

Any biological activity that results in lowering the pH of ambient waters improves conditions favorable for carbonate dissolution. For example, environments rich in decomposing organic matter are characterized by a high CO_2 output and thus by low pH. Furthermore, the production of organic acids, particularly of humic acids that complex calcium, represents a widespread, indirect carbonate dissolution mechanism in natural waters. An important mechanism of indirect carbonate dissolution originates from the interference between the carbonate cycle and the biogeochemical cycling of sulfur at the transition between anaerobic and aerobic environments in the free water or within the sediment. Several different metabolic types of bacteria may participate in the process. Organic sulfur is released as H_2S, mainly from S-containing amino acids, by various heterotrophic bacteria. By far the greater amount of H_2S in natural environments, however, is derived by anaerobic bacterial reduction of sulfate. Sulfate is abundant in the oceans although in fresh waters it may be a limiting factor in H_2S formation. Hydrogen sulfide is oxidized under anaerobic conditions by photosynthetic green and purple bacteria, and in transitional anaerobic-aerobic zones by non-photosynthetic sulfur bacteria (Beggiatoaceae, *Thiobacillus*; Pfennig, 1975). The most important sulfur-oxidizers are the thiobacilli, species of which oxidize sulfur compounds to sulfuric acid thus dissolving available carbonate. The oxidation and reduction of sulfur compounds is discussed in Chapters 6.1 to 6.4.

Mineralized tissues of skeletons and shells of molluscs, annelids, barnacles, and echinoderms may be affected by an enzymic dissolution (digestion) of the organic matrix which contains calcium carbonate crystallites (Carriker and Smith, 1969; Carriker et al., 1969). An ability to digest organic matter from within animal skeletons has been demonstrated for bacteria (DiSalvo, 1969) and fungi (Schneider, 1976). Loosened carbonate particles are then subject to either mechanical or chemical removal.

Carbonate particles embedded in the faecal pellets of planktonic crustaceans are protected from dissolution by an organic capsule *. This mechanism is largely responsible for the presence of remains of calcareous nannoplankton in deep ocean sediments below the compensation level for carbonate (Honjo, 1976). Similarly, carbonate particles suspended in sea water are apparently coated by a layer of organic matter. After this layer is experimentally removed by a strong oxidant (e.g. hypochlorite or H_2O_2), the rate of carbonate dissolution increases significantly (Chave and Suess, 1967, 1970).

Biological abrasion

Biological abrasion or mechanical removal of carbonate is common in animals. Carbonate-penetrating bivalves erode the substrate mainly by scraping with their shells (Ansel and Nair, 1969) or by a combination of mechanical and chemical activity (Kleemann, 1973). Herbivorous snails and chitons remove carbonate with their radulas while feeding on epilithic and endolithic algae (Craig et al., 1969; Schneider, 1976). Echinoderms and fishes remove carbonate with their teeth. Substantial amounts of carbonate are scraped away by parrot fish on coral reefs (Darwin, 1899). Mechanical rasping has been proposed as a means of carbonate penetration by endolithic barnacles (Tomlinson, 1969; Seilacher, 1969). Carnivorous gastropods that feed on other shell-bearing organisms are able to penetrate the shells of their prey by a combined chemical and mechanical action (Carriker, 1969). A complex chemical and mechanical removal of specifically sized carbonate chips is carried out by clinoid sponges (Cobb, 1969; Fütterer, 1974; Hatch, 1975; Rützler and Rieger, 1973; Rützler, 1975; see also Fig. 2.4.1).

Destruction of carbonates by organisms also involves translocation of material from the site of biological corrosion or abrasion, and its redistribution in the environment. Carbonate particles which are mechanically removed by gastropods and echinoderms are transported through their digestive tracts and distributed via faecal pellets. Sponges remove the chips of car-

---

* See also Chapter 2.3.

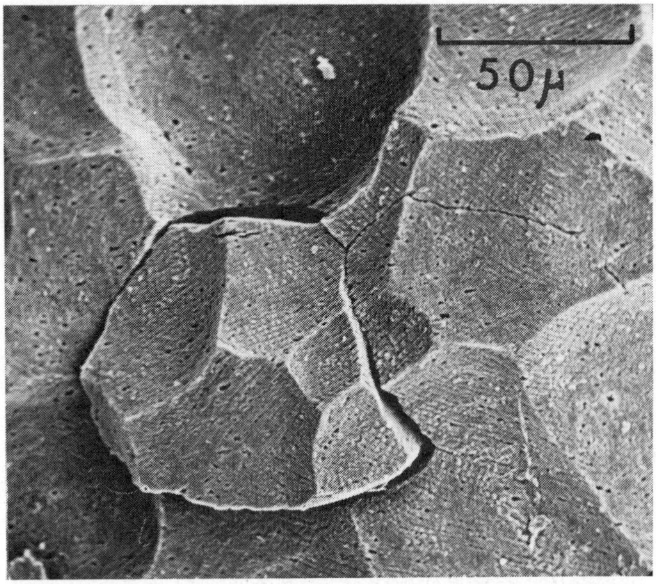

Fig. 2.4.1. Traces of clionid sponge boring and chip-removal in a shell of the bivalve *Venus* sp.; Mediterranean; SEM, courtesy of Dr. D. Fütterer, Kiel, W. Germany.

bonate rocks or shells and discard them through the oscula to the surface, thus contributing to the fine grain sediment fraction of the surrounding sea bottom (Fütterer, 1974; Rützler, 1975).

The highest rates of carbonate removal are encountered where biological corrosion and abrasion are acting synergistically. Biological abrasion may enhance biological corrosion by increasing the surface to volume ratio of carbonate particles and by removing loose carbonate particles together with a contact layer of newly carbonate-saturated ambient water, thus exposing fresh carbonate surfaces to dissolution. Conversely, biological corrosion of $CaCO_3$ and the enzymic removal of the associated organic matter may loosen a structure sufficiently to enhance biological abrasion.

Only a few determinations of destruction rates by bioerosional processes in the marine environment have been made, and the rates reported were determined in different ways and are not directly comparable. They range between 0.01 mm and 1.6 mm y^{-1}. The extremely high destruction rates reported by Jehu (1918) for pholads (12.7 mm y^{-1}) and Neumann (1966) for the boring sponge *Cliona lampa* (14 mm y^{-1}) represent maximum rates of the actual boring which does not necessarily continue after the organism has become established. Rützler (1975) determined the destruction rate of *Cliona lampa* as 0.1—1.0 mm y^{-1}. Lists of authors, localities, tidal zones, determination methods and destruction rates are tabulated by Milliman (1974, Table 66, p. 269) and Schneider (1976, Table 14, p. 76).

RELATIONSHIPS BETWEEN ORGANISMS AND CARBONATE SUBSTRATES

Rock surfaces exposed to weather in terrestrial environments, as well as the rocky shores and bottoms of lakes, rivers and oceans are colonized by a diversified flora and fauna. Specialized microorganisms, such as certain genera of cyanophytes, diatoms, fungi, lichens and associated bacteria, can survive extreme ecological changes such as sudden wetting and drying. Such adaptations enable these organisms to colonize bare, periodically wetted cliffs on mountain slopes (Jaag, 1945; Golubić, 1967). For these "plants of the rock", the term "lithophyta" was designated (Ercegović, 1925). Freshwater lithophytes colonize the spray zone of lakes. Characteristically colored zones of marine supratidal and intertidal lithophytes outline the coasts of the world's oceans (Ercegović, 1932, 1934; LeCampion-Alsumard, 1969, 1970; Schneider, 1976).

Lithophytes may extend into the interior of the rock, colonizing fissures and interstitial spaces (Friedmann et al., 1967; Friedmann, 1971), or they may penetrate actively into carbonate substrates (Golubić, 1969). In French literature, an algal division introduced by Chodat (1898) is often applied: "les algues cariantes" are specialized epiliths that affect the underlying rock surface by corrosive metabolic products; "les algues perforantes" penetrate the carbonate and extend into its interior. Prolonged activity of boring algae combined with secondary precipitation results in the formation of a highly-refractive micritic envelope around carbonate particles (Bathurst, 1966). In permanently submerged habitats, lithophytes are accompanied by animals and macroscopic algae. This diversified, yet specialized assemblage of organisms is generally termed *lithobionts*. The surface of the rock is inhabited by *epiliths*, and the interior of the rock by *endoliths*.

The different ways in which organisms colonize the interior of rock are summarized in Table 2.4.1.

Chasmoendoliths and cryptoendoliths can affect the substrate by water retention. Their metabolic products are released into the microenvironment and may contribute to carbonate dissolution (~algues cariantes). True endoliths, or euendoliths, are those that penetrate actively into the interior of the substrate forming tunnels which follow specific boring patterns (Golubić, Friedmann and Schneider, unpublished results).

Rock destruction by organisms is designated as microscopic or macro-

TABLE 2.4.1

Endolithic colonization of carbonate substrates

Chasmoendoliths	colonize fissures
Cryptoendoliths	colonize structural cavities
Euendoliths	penetrate the substrate

scopic depending on whether the organisms are smaller or larger than about 100 μm. Rock discoloration is mostly caused by dense populations of microscopic epilithic and endolithic algae rather than by "weathering". The abundance and ecological importance of microscopic boring organisms may exceed that of the macroscopic borers, although their presence may not be as obvious.

Chemical activity takes place in the microenvironment along the interface between the rock and the organism, and frequently leaves etching marks. The direction and morphology of biogenic etching is determined by the shape and movement of the organism, but the morphology of etching marks may also reflect the structure of the substrate. Arrangements of calcium carbonate molecules within the substrate, planes of crystal cleavage and twinning, size of carbonate grains in limestones and dolomites, and the arrangements of crystallites and organic lamellae in skeletal carbonates, can all be reflected in the morphology of the etching marks (Golubić, 1969; Golubić et al., 1975; Schneider, 1976).

Mechanical activity of boring organisms is also reflected in scratch marks (e.g. Warme, 1975). In general, boring patterns are determined both by the specificity of the boring organism as well as by the particular nature of the bored substrate, and both factors need to be considered in their interpretation (Golubić and Schneider, 1972).

Taking the influence of the substrate into account, the recognizable biological specificity of boring patterns is often sufficient to allow taxonomic determination. This is a significant property for the palaeontological interpretation of fossil borings of soft-bodied organisms which are not preserved after burial (Bromley, 1970; Golubić et al., 1975).

ORGANISMS THAT REMOVE CARBONATE

Mechanisms for chemical and mechanical mobilization of carbonate evolved independently in various taxonomic groups. Consequently, the endolithic ecological niche is occupied by a heterogenous community. Endolithic organisms range in size over several orders of magnitude, leaving a correspondingly wide spectrum of boring and scraping traces on carbonate substrates. Density of colonization and the capacity for carbonate removal are both size-dependent. Microscopically small procaryotes (cyanophytes) and eucaryotic algae comprise the principal primary producers among epiliths and endoliths alike. They range from 1 to 100 μm (mostly 5—15 μm) in diameter. Bacteria and fungi are the microbial heterotrophs of this community. They occupy the lower end of the same size range. Association between algae and fungi in carbonate rocks show all degrees of integration, from the occasional attack on algal cells by boring fungal hyphae to an establishment of several specialized carbonate-boring lichens (Schneider,

1976). Animals occupy various heterotrophic nutritional niches. They are generally larger, but overlap in the lower size range with microorganisms. Two different nutritional types affect carbonate substrates: grazing animals feed on primary producers that are on or within the substrate, while suspension-feeding animals bore into carbonate substrate for protection.

We shall discuss the importance and occurrence of the main representatives of the carbonate-removing organisms, their boring patterns, and means of carbonate removal. The contribution of various groups of organisms to carbonate bioerosion, with respect to destruction mechanisms, trace morphology, and habitat is summarized by Neumann (1968), Milliman (1974), Warme (1975), and Schneider (1976).

Bacteria

The presence of bacteria in every natural environment leads to the assumption that their effect on carbonate dissolution must be substantial. However, due to their small size, they hardly leave marks which can be specifically recognized as bacterial in origin. The potential of bacteria to cause carbonate dissolution can be deduced from published laboratory information on bacterial physiology, and from soil and ground-water chemistry. In nature, these effects are best referred to as indirect. The opposite effect, i.e. calcium carbonate precipitation, was demonstrated for bacteria that decompose proteins and release ammonia (Greenfield, 1963; see also Chapter 2.2). However, the major outcome of bacterial activity is decomposition of carbohydrates and other organic carbon compounds, resulting in their conversion into CO_2 and organic acids, both of which cause carbonate dissolution rather than precipitation. The ability of bacteria to weaken the skeletons of reef corals by decomposing structural organic matter was demonstrated by DiSalvo (1969).

We observed similar bacterial activity on gastropod shells in aquaria. Bacteria enter through damaged places under the periostracum and spread radially between periostracum and the palisade layer of $CaCO_3$ crystallites, forming circular colonies. As the intercrystallite organic matter (Kobayashi, 1969) is gradually degraded, a perfectly circular plug of loosened crystallites falls out, leaving a perforation similar to that produced by predatory snails. Bacterial perforations, however, are more varied in size and often fuse into larger holes.

Corrosion of concrete by bacteria that oxidize sulfur compounds into sulfuric acid was reported by Parker (1945, 1947) and has been examined more recently by Fjerdingstad (1969). The lowest pH is tolerated by *Thiobacillus concretivorus* Parker (pH 2.75 ± 0.42) and *T. thiooxidans* Waksman and Joffe (pH 2.35 ± 0.32). Within 100 days of incubation in a *T. thiooxidans* culture, 60% of a 3.6 g block of concrete were dissolved. Bacterial destructive activity has also been established on buildings and monuments (Krumbein and Pochon, 1964; Krumbein, 1968, 1973). *Thiobacillus* spp. and *Nitrosomonas-Nitrobacter* spp. oxidize sulfur and nitrogen compounds in air and water to produce H_2SO_4 and HNO_3 (Krumbein, 1968, 1972, 1973; Krumbein and Pochon, 1964). Bacteria are often associated with boring algae, but there is no evidence of their direct participation in carbonate penetration (Fig. 2.4.2-1d).

Cyanophytes (see footnote on p. 12)

Cyanophytes represent one of the most important groups of microbial endoliths. Only taxa with proven endolithic affinities will be mentioned here, reducing substantially the list of carbonate boring species published by Pia (1937). Coccoid cyanophytes (*Gloeocapsa* spp.) dominate the chasmoendolithic and cryptoendolithic niches in desert rocks from the Negev and Arizona to the Dry Valleys of Antarctica (Friedmann et al., 1967; Friedmann, 1971; Friedmann and Ocampo, 1976) and in the rocks along ocean coasts (LeCampion-Alsumard, 1969; Schneider, 1976). They do not actively penetrate the rock and no boring patterns originate from their activity. Some epilithic, chasmoendolithic and cryptoendolithic algae also associate with fungi, forming lichens that may actively penetrate carbonate substrates, both in freshwater and marine environments.

Freshwater mollusc shells and carbonate pebbles are frequently perforated by the cyanophytes *Hyella fontana* Huber et Jadin and *Schizothrix perforans* (Ercegović) Geitler. *Schizothrix lacustris* A. Braun is a facultative borer that participates both in lacustrine carbonate crust deposition and excavation of furrows in carbonate rocks in lakes (Schneider, 1977).

Marine wave and spray zones (supralittoral) and the intertidal zone (eulittoral) are exposed to drastic fluctuations in wetting and drying, salinity, temperature and insolation. These environments are dominated by cyanophytes, which are established in several characteristically coloured zones (Ercegović, 1932; LeCampion-Alsumard, 1969, 1970; Schneider, 1976). *Hyella balani* Lehmann and *Hormathonema paulocellulare* Ercegović dominate the intertidal zone of sun-exposed coasts. *Hyella* occupies well-drained surfaces, while *Hormathonema* prefers small depressions in the rock that retain moisture. *Hyella balani*, with yellow-brown gelatinous envelopes, bores shallow pits in the drying upper intertidal zones and deep conical pits with diverging tunnels in the lower intertidal zone and on barnacle shells. *Hormathonema paulocellulare*, with blue-grey gelatinous envelopes, forms parallel tunnels perpendicular to the surface with few lateral branchings.

Subtidal habitats are the domain of *Hyella caespitosa* Bornet et Flahault. *H. caespitosa* bores long individual tunnels (Fig. 2.4.2-1,2) rather than common conical pits and forms a dense network of intertwined borings several hundred microns deep. Both species of *Hyella* meet and overlap in the lower intertidal zone. Shaded areas of the intertidal zone are sometimes dominated

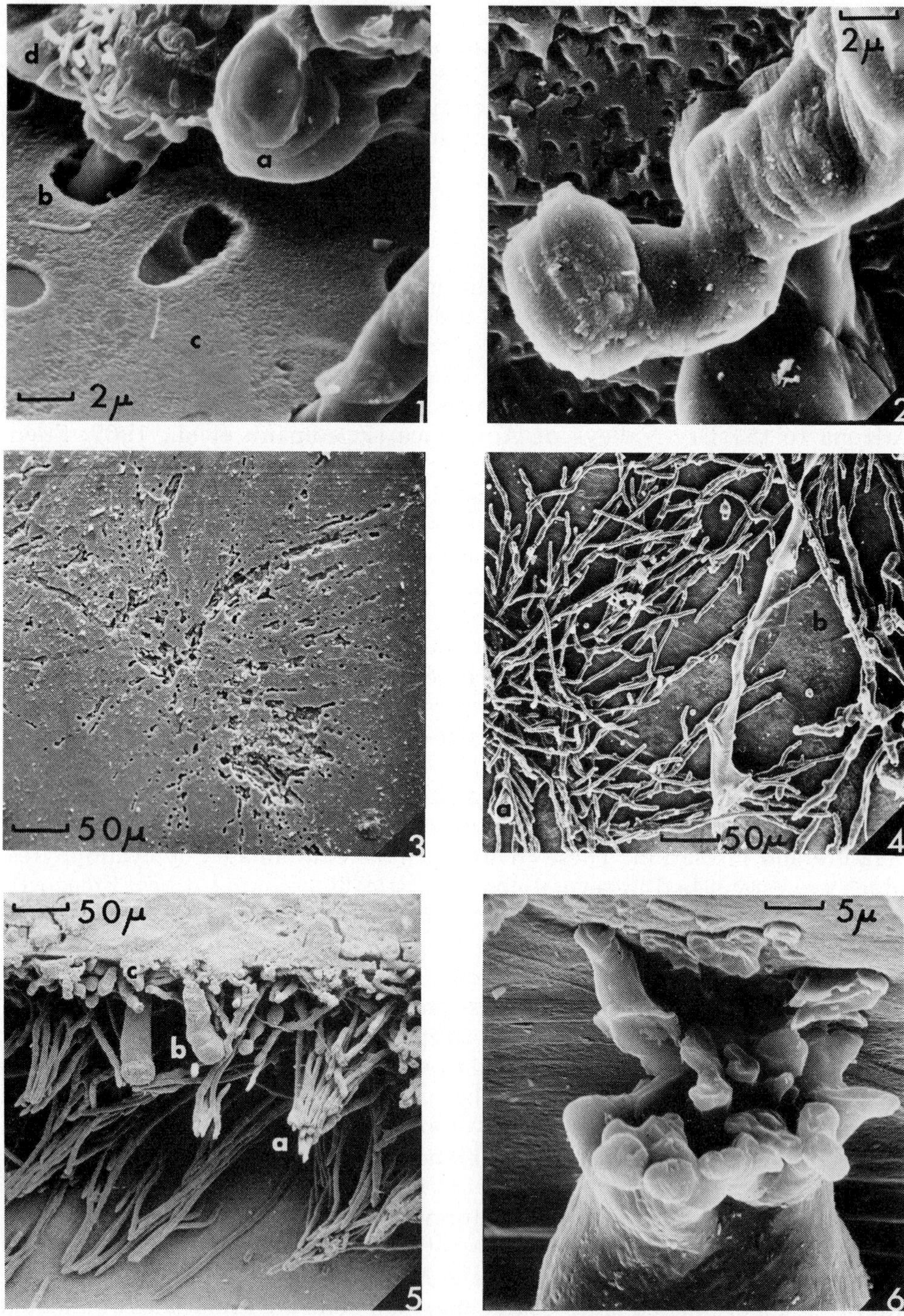
d
a
b
c
2μ
1
2μ
2
50μ
3
b
a
50μ
4
50μ
c
b
a
5
5μ
6

by *Kyrtuthrix dalmatica* Ercegović, a heterocyst-forming and thus nitrogen-fixing filamentous cyanophyte. The filaments of *Kyrtuthrix* penetrate with their central portions, forming loops that contain two trichomes. The tunnels are therefore elongated to oval in cross section. Due to its narrow, vertical range of distribution, *Kyrtuthrix* is an excellent ecological indicator.

Other filamentous cyanophytes, e.g. *Mastigocoleus testarum* Lagerheim and *Plectonema terebrans* Bornet et Flahault, have a wider range of distribution, including intertidal and subtidal zones. *Mastigocoleus* is a heterocyst-bearing cyanophyte with basket-shaped ramified filaments and borings. It has been found from the intertidal zone down to a depth of about 30 m. *Plectonema terebrans* has the widest range of distribution, from the intertidal zone to more than 250 m (strait of Florida, K. Lukas, unpublished). Together with the chlorophyte *Ostreobium*, *Plectonema* penetrates coral heads in tropical oceans. *P. terebrans* bores fine (1—2 μm), occasionally branched tunnels which can be easily confused with those formed by fungi. Unlike fungi, the ramifications of *Plectonema* are never dichotomous, and no formation of sporangia or fruiting bodies ever occurs.

Eucaryotic algae

Several taxa of eucaryotic algae have developed specialized adaptions for the endolithic mode of life. Among rhodophytes, only certain stages in the life cycles of the genera *Porphyra* and *Bangia* live endolithically. They were initially described as the separate genus and species *Conchocelis rosea* (Batters, 1892). The work of Drew (1949, 1958) established the identity of *Conchocelis* as growth stages in the life cycles of Bangiales.

Endolithic chlorophytes show a wide range of boring adaptations. Most belong to the order Chaetophorales, and a few to Siphonales. Recent work by Kornmann (1959, 1960) and Nielsen (1972) has clarified the systematics of several chaetophoracean borers. *Gomontia polyrhiza* (Lagerheim) Bornet et Flahault (sensu Nielsen) had been the subject of much taxonomic confusion. The boring form of *G. polyrhiza* is a sack attached to the surface by numerous rhizoidal appendages which sinks into the substrate forming a

Fig. 2.4.2. Microbial borings in calcite (Iceland spar, Plates 1 and 3), and their resin casts (araldite, Plates 2 and 4—6); 1. *Hyella caespitosa* Bornet et Flahault in culture. Surface colony embedded in gelatinous envelope (a) sends boring branches (b) into calcite substrate (c); d = associated bacteria. 2. Resin-cast of a boring of *Hyella caespitosa* in calcite, lower intertidal zone. 3. Surface view of shallow radiating borings of *Eugomontia sacculata* Kornmann in calcite spar. 4. Boring tunnels cast in araldite of *Eugomontia sacculata* (a) and *Entocladia testarum* Kylin (b). 5. Mixed community of algal borers. Tunnels cast in araldite: (a) = *Hyella balanii* Lehmann; (b) = *Gomontia* sp.; (c) = *Eugomontia sacculata* Kornmann. 6. *Gomontia polyrhiza* (Lagerheim) Bornet et Flahault detail with detached surface rhizoids (note change in magnification).

simple rounded pit (Fig. 2.4.2-6). *Eugomontia sacculata* Kornmann has been confused with *Gomontia* in the past; *Eugomontia* forms ramified filaments and boring tunnels that periodically widen in a series of sack-like swellings. Young filaments radiate from the point of initial penetration while adult thalli excavate an intertwined complex network of tunnels and turn the substrate into a light green, brittle, easily disintegrating mess (see Fig. 2.4.2-3/4). *Phaeophila dendroides* (Crouan) Batters and *Phaeophila tenuis* (Kylin) R. Nielsen produce shallow, radiating boring patterns. *P. tenuis* forms tunnels of about the same size as *Eugomontia* but differs from the latter by differentiating fine, undulating setae. This growth form is also reflected in boring casts which have a more spiny appearance than that of *Eugomontia. Phaeophila dendroides* has larger dimensions and characteristic cross-shaped ramifications. Bristles protrude out of the bored substrate so that they are not visible on the casts. *Entocladia testarum* Kylin is the least abundant and least known of the boring Chaetophorales. All these species are characterized by bag-like widenings of their tunnels and distinguishing them on the basis of boring patterns is not easy.

Ostreobium quekettii Bornet et Flahault is a highly specilized boring siphonal chlorophyte that universally bores coral skeletons. In colder seas, it is found in bivalve shells as well as in inorganic carbonate substrates. *Ostreobium* is well adapted to low light intensities, so that it extends into deeper waters and penetrates deeper into carbonate substrates than most endoliths. Together with *Plectonema terebrans* (cyanophyte), *Ostreobium* is the only boring organism capable of keeping pace with coral growth (Lukas, 1973). *Ostreobium* forms an array of different boring tunnels, most of which are very fine and intricately branched (1—2 μm in diameter). However, the main trunks can reach a diameter 20 times as wide. In addition to tubular borings, *Ostreobium* forms bags and widenings of irregular shapes. A slightly larger species *Ostreobium constrictum* Lukas has recently been discovered in Atlantic corals (Lukas, 1974). An example of a mixed community of algal borers is shown in Fig. 2.4.2-5.

Boring by brown algae has been described in the literature (Barnes and Topinka, 1969), but it refers mainly to the anchoring of macroscopic algal holdfasts.

Fungi and lichens

Since fungi are heterotrophic, their vertical distribution within the carbonate substrate does not directly depend on light. Boring fungi range in their distribution from freshwater environments, over supratidal and intertidal zones of the oceans, down to the abyssal depths. Yet various forms of endolithic fungi do have zonal distribution that is presumably a function of the available food source. Fungi in air-exposed habitats are specialized to withstand water shortages, and are most successful when associated with algae in the form of lichens. A large number of carbonate boring lichens are known from rock surfaces in freshwater realms where they represent the principal destructive agent; a few are found in the marine environment. A long list of endolithic lichens (more than 60 taxa) was published by Pia (1937); epi- and endolithic lichens from the marine inter- and supratidal zones were reported by Schneider (1976).

The fungus *Pharcidia balanii* (Winter) Bauch is common in the calcareous shells of barnacles and in intertidal carbonate rocks (Kohlmeyer, 1969; Schneider, 1976). It is frequently associated with microscopic algae, as in the lichen *Arthopyrenia sublittoralis* (Leight) Arnold, but this association is not obligatory. The endolithic fungi *Ostracoblabe implexa* Bornet et Flahault and *Lithopythium gangliiforme* Bornet et Flahault are common within the intertidal zone and in shallow marine waters. Both species have been found to parasitize endolithic algae. Their life cycles are not known. It is interesting to note that both fungi penetrate into carbonate substrata in the absence of algae, and explore the interior of the rock much deeper than the algae. Apparently, these fungi have been selected to intercept their hosts in an environment where algae are very frequent. *Ostracoblabe* bores thin tunnels (1—2 μm in diameter) and has irregular lateral branching. It can be easily confused with the cyanophyte *Plectonema terebrans.* The main distinguishing character of the fungus is spindle-shaped swellings along the tunnels. *Lithopythium* has irregular or dichotomous branchings and forms spherical fruiting bodies. Both fungi have been observed to penetrate carbonate in a straight line connecting algal filaments, indicating a chemotactic ability to detect the host.

Various species of fungi are encountered in mollusc or barnacle shells where they exploit the organic matter laid down by the animal between the crystallites of $CaCO_3$ (see e.g. Plate 8 of Schneider, 1976). Several of these fungi have been described by Zebrowski (1936) and Porter and Zebrowski (1937) only on the basis of excavations and thus must be considered trace fossil taxa, since the living organisms have not been described.

Animals

Bioerosion by animals is well documented in zoological, palaeontological and sedimentological literature. The overwhelming literature and annotated bibliographies prompted Warme (1975) to tabulate 9 principal reference sources with 25 to 4000 references each (see also Milliman, 1974). These works impressively document the significance of animal activity in carbonate destruction. Both biological corrosion and biological abrasion, or a combination of these processes, are carried out by animals. As a result, these activities may either increase the carbonate in solution, or contribute to production of more or less fine grain sediment. The following animal bioeroders can be distinguished (Milliman, 1974):

(i) Borers — penetrate hard substrates

(ii) Grazers — feed by scraping the substrate, usually in search of epi- and endolithic microorganisms

(iii) Browsers — deposit-feeders and sediment-ingestors that carry out a further destruction of substrate during digestion

(iv) Predators — actively hunt and consume their prey, cracking shells or carapaces.

Important taxonomic groups of animals that remove carbonate substrates are: Foraminiferida, Porifera, Coelenterata, Turbellaria, Nematoda, Phoronida, Bryozoa, Brachiopoda, Sipunculida, Polychaeta, Cirripedia, Decapoda, Isopoda, Pholadida, Mytilida, Gastrochaenidae, Gastropoda, Cephalopoda, Amphineura, Echinoidea, and Pisces.

Destruction mechanisms vary greatly among species and taxonomic groups, and in many cases are not yet understood. For description or discussion of the various mechanisms in different groups, see Milliman (1974, pp. 253—261), Warme (1975, pp. 194—219) and tabulated lists in Neumann (1968, pp. 78—79) and Schneider (1976, Tab. 11, p. 50).

CARBONATE REMOVAL MECHANISMS ALONG A HYPSOGRAPHIC PROFILE (see Fig. 2.4.3)

Terrestrial environments: Karst forms

The formation of karren (lapies) is initially an inorganic process whereby CO_2-saturated water films of rain, and snow-water runoff, dissolves carbonate rock superficially. The occasional wetting of rocks is sufficient to permit a settlement of endolithic lichens that continue to dissolve the rocks down to a depth of 1—3 mm. Lichen thalli spread horizontally below the rock surface and their fruiting bodies connect to the surface forming characteristic round holes. The fungal hyphae gradually disintegrate the rock surface into tiny particles which are then more susceptible to dissolution by rain- and snow-water runoff. Various endolithic lichen species that contribute to this process are described by Schneider (1976, pp. 33, 48, 56 ff, 61). Because of moisture retention, the karren floors are occupied by deeper boring lichens. As a consequence, their relief is enhanced biologically.

In addition to karstic surface forms, dissolution of carbonate continues beneath the soil cover, but results in different morphologies. Water percolating through a soil ecosystem becomes enriched with CO_2 produced by microorganisms, plant roots, animals and the decomposing organic matter. This aggressive water dissolves the underlying carbonate rock. Various organic substances such as humic acids which are effective chelating agents for Ca^{2+} ions aid the dissolving process, lowering the solubility product $[Ca^{2+}] \times$

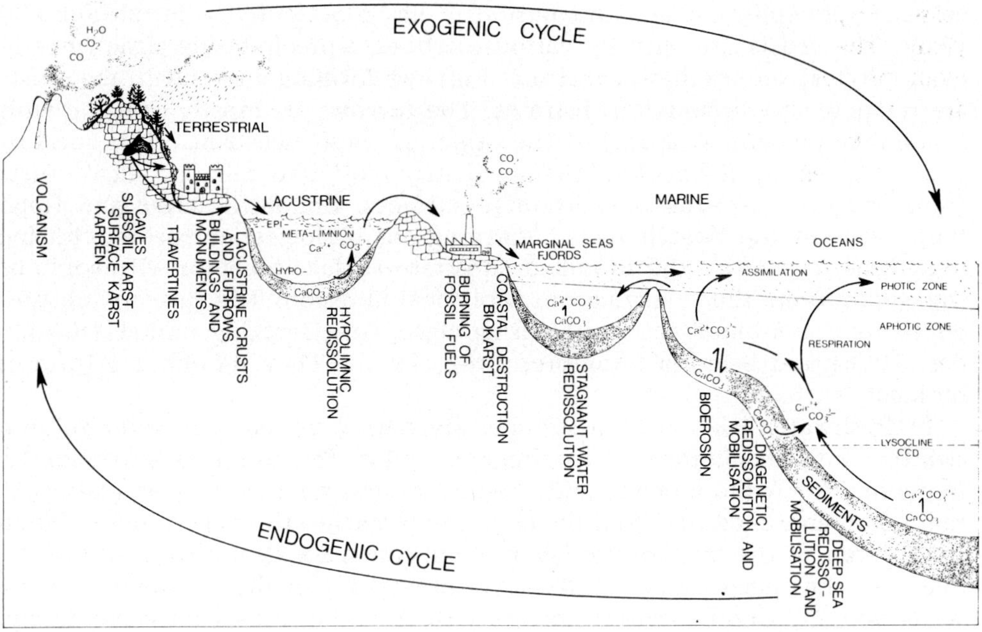

Fig. 2.4.3. Carbonate removal mechanisms along an idealized hypsographic profile from terrestrial, over lacustrine (freshwater) to marine environments.

$[CO_3^{2-}]$. Deep grooves, holes, pillars and caverns of the subsurface karst are characterized by rounded outlines. Sharp forms characteristic of surface karst are largely missing in the subsoil corrosion because capillary contact rather than runoff determines the corrosion site. The cave systems of karstic regions are witnesses of extensive past subsoil dissolution processes. Secondary precipitation of the dissolved carbonate is common in larger ventilated subterranen caves, where excessive CO_2 escapes from the seeping water into the cave atmosphere. The result is cavestone formation. Precipitation of $CaCO_3$ by fungal hyphae in the form of loose sinter crusts occurs in some caves (Schneider, 1977).

The entire cycle of corrosion and precipitation of carbonate is repeated on a smaller scale within travertine waterfalls. Here, the algae and mosses guide the primary precipitation of carbonate, while the underlying bacteria are responsible for its dissolution. Structural caverns below water parabolas are the site of secondary cavestone deposition (Golubić, 1969, 1973).

Lacustrine environments

Lacustrine crusts and furrows are common in the shallow sublittoral of carbonate-rich lakes. The co-occurrence of two chemically opposite processes — precipitation and dissolution — has represented a puzzle for 120 years. The crusts are built by various carbonate-precipitating algae (mostly cyanophytes) on any hard substrate. Furrows forming underneath the crusts are restricted to carbonate substrates. The furrows are mostly irregular with a brainlike pattern. A model of the origin of these lacustrine crusts and furrows is given by Schneider (1977). "Patch reef"-like calcified algal crusts form irregular patterns on carbonate surfaces. Endolithic algae and fungi bore between the "patch reefs", carrying out biological corrosion. Grazing organisms rasp away the loosened substrate surface together with some of the microborers thus carrying out biological abrasion. The furrows are produced by a combined action of boring algae, fungi, grazing animals (Schneider, 1976) and bacteria (Golubić, 1962, 1973). They represent a form of biokarst.

In freshwater lakes with a summer stagnation period, the water column stratifies into an illuminated trophogenic epilimnion and a dark tropholytic hypolimnion. A thermocline and chemocline characterize this separation. A rain of organic material from the epilimnion reaches the hypolimnion. When the oxygen of the tropholytic layer is consumed for degradation of organic material, the redox-discontinuity layer moves out of the sediment into the water column. CO_2 behaves reciprocally to O_2 and increases correspondingly. The resulting pH-drop causes a dissolution of carbonate material which has been biogenically precipitated in the epilimnion. This carbonate is recycled to the surface waters during the autumn and winter circulation and can be transported through the outflowing rivers to the oceans. A similar situation exists in temporary or permanently stratified waters of freshwater, brackish and marine environments. Examples of such environments are the monimolimnions of meromictic lakes, the Baltic Sea (see Wefer, 1976) and Norwegian fjords where a high CO_2 input is maintained by vigorous degradation of organic matter.

Some enclosed and stratified marginal seas do not exhibit a carbonate lysocline and can incorporate a substantial amount of carbonate into their sediments. In the Black Sea, for example (see Degens and Ross, 1974), layers of intact coccoliths are seasonally deposited, reflecting the production of coccolithophorids in the illuminated surface waters. The sediment contains 40% $CaCO_3$. The absence of a lysocline and the "survival" of carbonate particles while passing through the anoxic bottom waters can be explained by a discrepancy between the residence time of the deep versus surface water. The bottom waters of the permanently stratified Black Sea have a prolonged residence time (2000 y) during which microbial sulfate reduction maintains reducing conditions and high H_2S levels. Under reducing conditions the decomposition rate of organic carbon, and thus CO_2 release, is much lower than under oxidized conditions. In addition, some of the CO_2 is removed by an anaerobic reduction to methane. Higher turnover and primary production rates in the illuminated surface waters are accompanied by a correspondingly high production of coccoliths and other skeletal carbonates which sink into the anoxic waters and keep them permanently saturated with respect to $CaCO_3$. All excess carbonate then becomes incorporated into the sediment.

Marine environments

Various phenomena of coastal destruction have been discussed earlier in this paper (p. 108) to illustrate the principles involved and, therefore, will not be repeated here.

Only 20% of the mostly biologically precipitated $CaCO_3$ in the oceans is preserved in the form of carbonate sediments or as carbonates in clastic sediments; 80% of the marine $CaCO_3$ is redissolved. This dissolution takes place partly by bioerosion in shallow water and shelf areas, partly by diagenetic redissolution and mobilization within sediments at the redox-discontinuity layer, and is strongly influenced by microorganisms. At high latitudes, an early diagenetic dissolution takes place in oxygenated but $CaCO_3$-undersaturated bottom waters and pore waters (Alexandersson 1972, 1975, 1976). The most important part of carbonate dissolution takes place in the deep sea of all latitudes, below the carbonate compensation depth (or lysocline, e.g. Berger, 1970; Broecker, 1974, p. 36 ff; Peterson, 1966). These waters contain more CO_2 than cold surface waters. This surplus of total CO_2 results from the oxidation of organic matter. Because of the resulting pH-drop, the water is able to dissolve $CaCO_3$. There are different carbonate compensation depths in the Atlantic and in the Pacific and for calcite and aragonite, respectively. The deep Pacific shows a lower carbonate ion concentration than the deep Atlantic. Carbonate which is redissolved in the ocean depth is recycled to the surface waters within a mixing time of some 1000 y (see Garrels et al., 1975).

SYNERGISTIC EFFECTS IN BIOLOGICAL CARBONATE REMOVAL

It should be clear from the preceding discussions that no single biological factor affects the carbonate dissolution alone, and that the most significant rates of carbonate destruction result from a synergistic, mutual enhancement of various biological and abiotic factors. We shall now discuss two examples of such synergistic effects which may illustrate these points. One example deals with microbial endolithic colonization and reveals a selective advantage of the endolithic mode of life. The other concerns the synergistic effects in coastal destruction and their geological consequences.

Grazing pressure and endolith colonization

In an experiment performed in the intertidal zone of the Mediterranean Sea near Marseille, France, Iceland spar crystals were exposed to microbial colonization. Within a few weeks, the surface of the exposed crystals was completely covered by blue-green algae, and any further colonization ceased. *Dermocarpa* sp., an epilithic cyanophyte, comprised 95—98% of the colonizers (see Plate 1, Fig. 3 of LeCampion-Alsumard, 1975). The remaining 2—5% of the surface was colonized by spores that immediately penetrated the calcite after settling and later grew into thalli of the endolithic cyanophyte, *Hyella balani.* This contrasts with the high density of the established endoliths (over 90%) normally found in the intertidal zone. Soon after completion of the colonization, the exposed crystals were grazed by herbivorous snails, mostly by *Littorina* sp. Snails left clean tracks in removing the epilithic algae, but did not affect the endoliths. Subsequently, the grazed areas were recolonized by the same proportions of epiliths to endoliths.

Dermocarpa is characterized by a regular and high sporulation rate, and by a continuous and gradual cell growth. Consequently, exposed surfaces are shared by clonal populations of same age. By measuring the average cell sizes of *Dermocarpa,* it is possible to date the snail tracks relative to each other, and to determine the frequency of cell removal by snails. We observed that the areas frequently visited by snails correlate with a higher density of endoliths (T. LeCampion-Alsumard and Golubić, unpublished results). We concluded that epilithic algae with high reproduction rates are superior competitors in colonizing new rock surfaces, limiting the proportion of endolithic settlers. However, epiliths are vulnerable to snail grazing, while endoliths are able to escape by boring into substrate. By repeated cycles of grazing and recolonization, endoliths accumulate with time. The grazing pressure thus works in favor of the endolithic niche on two levels: it spares the endoliths, but removes their competitors.

The cumulative effect of biogenic carbonate removal on coastal destruction

The combined activity of coexisting destructive forces (micro- and macroorganisms and environmental energy factors, e.g. waves) exceeds significantly the destruction capacity of single species and has a cumulative effect (Schneider, 1977, p. 259).

The activity of endolithic boring algae as light-dependent photosynthetic organisms is limited to a relatively thin surface layer of carbonate rocks. As the algae penetrate into the dark interior of the rock, their boring activity slows down and finally ceases at a depth approaching their light compensation point (where photosynthesis equals respiration). Accordingly, algal borings represent a self-limiting surface phenomenon, that stabilizes with time. After they have reached their compensation point, left by themselves, boring algae would not cause any further destruction of the rock.

Both epilithic and endolithic algae are significant primary producers of organic matter on rocky substrates. They both serve as a food source for a number of grazers (gastropods, echinoderms, fishes, etc.) that are equipped with sharp, rasping radulas and teeth. While grazing on carbonate rock surfaces, the grazers remove a layer of algae together with the surface layer of carbonate rock which has been loosened by the endolithic microflora. Scratch marks left by grazers can often be observed on rocks and shells. Removal of a thin layer of the substrate results in a deeper light penetration, and the compensation depth for algae is thereby displaced towards the interior of the rock. Algal-boring activity resumes according to the changed light conditions. A continuation of algal penetration, therefore, depends on the continuous removal of the rock by grazers. Thus rates of algal borings and animal grazings are interdependent and mutually regulated.

When grazing is slower than algal boring, the latter becomes light limited and also slows down. If, on the other hand, the grazing rate exceeds the maximum boring rate, an area may become barren by overgrazing. Normally, there is a well-balanced ecological equilibrium between boring and grazing rates. In this way, an intrinsically static and self-stabilizing algal corrosion becomes a dynamic and cumulative destructive agent when combined with abrasion by grazers (Schneider, 1976). On a different scale, suspension-feeding endolithic animals and eroding environmental forces play a role similar to that of the microborers/grazers. Like algal endoliths, these endolithic animals bore the substrate for shelter and would by themselves remain a near-surface phenomenon, maintaining contact with their food source. However, their boring activity weakens the rocky substrate to a degree that it may collapse by its own weight, or become a victim of minor environmental energies such as waves and currents. The combined synergistic activity of biological and environmental factors has a cumulative effect on coastal carbonate destruction and therefore is of geological importance.

ACKNOWLEDGEMENTS

We thank Dr. T. Le Campion-Alsumard for the materials, SEM preparations and unpublished results made available for this publication. SEM-time and assistance, and photomicrographic services were made available by the University of Hamburg, West Germany. Drs. E.T. Degens, W. Krumbein and Lynn Margulis critically read the manuscript and gave valuable suggestions, and encouragement.

The work is supported by the Alexander von Humboltd foundation, West Germany and by the NSF Grants BO-2527-1 and GA-43391 to S. Golubić and by the Deutsche Forschungsgemeinschaft, West Germany (Az.:Schn 16/1-4) to J. Schneider.

REFERENCES

Alexandersson, E.T., 1972. Micritization of carbonate particles: Processes of precipitation and dissolution in modern shallow-marine sediments. Bull. Geol. Inst. Univ. Uppsala, N.S., 3: 201—236.

Alexandersson, E.T., 1975. Etch patterns on calcareous sediment grains: petrographic evidence of marine dissolution of carbonate minerals. Science, 189: 47—48.

Alexandersson, E.T., 1976. Actual and anticipated petrographic effects of carbonate undersaturation in shallow sea-water. Nature, 262: 653—657.

Ansell, A.D. and Nair, N.B., 1969. A comparative study of bivalves which bore mainly by mechanical means. Am. Zool., 9: 857—868.

Barnes, H. and Topinka, J.A., 1969. Effect of the nature of the substratum on the force required to detach a common littoral alga. Am. Zool., 9: 753—758.

Bathurst, R.G.C., 1966. Boring algae, micrite envelopes and lithification of molluscan biosparite. Geol. J., 5: 15—32.

Batters, E.A., 1892. On Conchocelis, a new genus of perforating algae. Phycol. Mem., 1: 25—29.

Berger, W.H., 1970. Planktonic foraminifera: selective solution and the lysocline. Marine Geol., 8: 111—138.

Broecker, W.S., 1974. Chemical Oceanography. Harcourt Brace Jovanovich, New York, N.Y., 214 pp.

Bromley, W.G., 1970. Borings as trace fossils and "Entobia cretacea" Portlock as an example. In: T.P. Crimes and J.C. Harper (Editors) Trace fossils. Geol. J., Spec. Issue, 3: 49—90.

Carriker, M.R., 1969. Excavation of boreholes by the gastropod, Urosalpinx: an analysis by light and scanning electron microscopy. Am. Zool., 9: 917—933.

Carriker, M.R. and Smith, E.H., 1969. Comparative calcibiocavitology: summary and conclusions. Am. Zool., 9: 1011—1020.

Carriker, M.R., Smith, E.H. and Wilce, R.T. (Editors), 1969. Penetration of calcium carbonate substrates by lower plants and invertebrates. Am. Zool., 9: 629—1020.

Chave, K.E. and Suess, E., 1967. Suspended minerals in seawater. Trans. N.Y. Acad. Sci., 2: 991—1000.

Chave, K.E. and Suess, E., 1970. Calcium carbonate saturation in seawater: effects of dissolved organic matter. Limnol. Oceanogr., 15: 633—637.

Chétail, M. and Binot, D., 1967. Mise en évidence et rôle de l'anhydrase carbonique dans l'organe accessoire de performation de *Purpura lappillus* L., Gastéropode Prosobranche C.R. Acad. Sci., Paris, 264: 946—948.

Chétail, M. and Fournié, J., 1969. Shell-boring mechanism of the gastropod, *Purpura* (Thaïs) *Lapillus*: a physiological demonstration of the role of carbonic anhydrase in the dissolution of $CaCO_3$. Am Zool., 9: 983—990.

Chodat, R., 1898. Sur les algues perforantes d'eau douce. Études de biologie lacustre. Bull. Herbier Boissier, 6: 431—476.

Cobb, W.R., 1969. Penetration of calcium carbonate substrates by the boring sponge, *Cliona*. Am. Zool., 9: 783—790.

Cole, G.A., 1975. Textbook of Limnology. C.V. Mosby Co., St. Louis, 283 pp.

Craig, A.K., Dobkin, S., Grimm, R.B. and Davidson, J.B., 1969. The gastropod, *Siphonaria pectinata*: a factor in destruction of beach rock. Am. Zool., 9: 895—901.

Darwin, Ch., 1899. Über den Bau die Verbreitung der Corallen-Riffe. 2nd Edn; J.V. Carus (Editor), E. Schweizerbart, Stuttgart, 231 pp.

Degens, E.T. and Ross, D.A., 1974. The Black Sea — Geology, Chemistry and Biology. Amer. Ass. Petr. Geol., Tulsa, OK, 633 pp.

DiSalvo, L.H., 1969. Isolation of bacteria from the corallum of *Porites lobata* (Vaughn) and its possible significance. Am. Zool., 9: 735—740.

Drew, K.M., 1949. Conchocelis-phase in the life-history of *Porphyra umbilicalis* (L) Kütz. Nature, 164: 748—749.

Drew, K.M., 1958. Studies in the Bangiophycidae. IV. The Conchocelis-phase of *Bangia fuscopurpurea* (Dillw) Lyngbye in culture. Publ. Stazione Zool. Napoli, 30: 358—372.

Ercegović, A., 1925. La végétation des lithophytes sur les calcaires et les dolomites en Croatie. Acta Bot. Croat., Zagreb, 1: 64—114.

Ercegović, A., 1932. Études écologiques et sociologiques des Cyanophycées lithophytes de la côte yougoslave de l'Andriatique. Bull. Int. Acad. Yougoslave Sci. A., Classe Sc. math. et nat., 26: 33—56.

Ercegović, A., 1934. Wellengang und Lithophytenzone an der ostadriatischen Küste. Acta Adriat., Split, 3: 1—20.

Fjerdingstad, E., 1969. Bacterial corrosion of concrete in water. Water Res., 3: 21—30.

Friedmann, I., 1971. Light and scanning electron microscopy of the endolithic desert algal habitat. Phycologia, 10: 411—428.

Friedmann, I., Lipkin, Y. and Ocampo-Paus, R., 1967. Desert algae of the Negev (Israel). Phycologia, 6: 185—200.

Friedmann, I. and Ocampo, R., 1976. Endolithic blue-green algae in the Dry Valleys: Primary producers in the Antarctic desert ecosystem. Science 193: 1247—1249.

Fütterer, D., 1974. Significance of the boring sponge "*Cliona*" for the origin of fine grained material of carbonate sediments. J. Sediment. Petrol., 44: 79—84.

Garrels, R.M., Mackenzie, F.T. and Hunt, C., 1975. Chemical Cycles and the Global Environment. W. Kaufmann Inc., Los Altos, CA, 206 pp.

Golubić, S., 1962. Zur Kenntnis der Kalkinkrustation und Kalkkorrosion im Seelitoral. Schweiz. Z. Hydrol., 24: 229—243.

Golubić, S., 1967. Algenvegetation der Felsen, eine ökologische Algenstudie im dinarischen Karstgebiet. Binnengewässer, 23, Schweizerbart, Stuttgart, 183 pp.

Golubić, S., 1969. Distribution, taxonomy and boring patterns of marine endolithic algae. Am. Zool., 9: 747—751.

Golubić, S., 1973. The relationship between blue-green algae and carbonate deposits. In: N. Carr and B.A. Whitton (Editors), The Biology of Blue-green Algae. Blackwell Scientific Publications, Oxford, pp. 434—472.

Golubić, S. and Schneider, J., 1972. Relationship between carbonate substrate and boring patterns of marine endolithic microorganisms. Geol. Soc. Am., Ann. Meet., Abstr. and Progr., 4: 518.

Golubić, S., Friedmann, I. and Schneider, J. (1979). The lithobiontic ecological niche with special reference to micro-organisms.

Golubić, S., Perkins, R.D. and Lukas, K.J., 1975. Boring microorganisms and microborings in carbonate substrates. In: R.W. Frey (Editor), The Study of Trace Fossils. Springer, New York, NY, pp. 229—259.

Greenfield, L.J., 1963. Metabolism and concentration of calcium and magnesium and precipitation of calcium carbonate by a marine bacterium. Ann. N.Y. Acad. Sci., 109: 23—45.

Hatch, W., 1975. The Implication of Carbonic Anhydrase in the Physiological Mechanism of Penetration of Carbonate Substrate by the Marine Burrowing Sponge *Cliona celata*. Unpubl. Ph.D. Diss., Boston University, 158 pp.

Honjo, S., 1976. Coccoliths: production, transportation and sedimentation. Mar. Micropaleontol., 1: 65—79.

Jaag, O., 1945. Untersuchungen über die Vegetation und Biologie der Algen des nackten Gesteins in den Alpen, im Jura und im Schwezerischen Mittelland. Beitr. Kryptogamenflora Schweiz, 9: 1—560.

Jehu, T.J., 1918. Rock boring organisms as agents in coast erosion. Scott. Geogr. Mag., 34: 1—10.

Kleemann, K., 1973. Der Gesteinsabbau durch Ätzmuscheln an Kalk-Küsten. Oecologia, 13: 377—395.

Kohlmeyer, J., 1969. The role of marine fungi in the penetration of calcareous substances. Am. Zool., 9: 741—746.

Kobayashi, I., 1969. Internal microstructure of the shell of bivalve molluscs. Am. Zool., 9: 663—672.
Kornmann, P., 1959. Die heterogene Gattung Gomontia. I. Der sporangiale Anteil. *Codiolum polyrhizum.* Helgol. Wiss. Meeresunters., 6: 229—238.
Kornmann, P., 1960. Die heterogene Gattung Gomontia. II. Der Fädige Anteil. *Eugomontia sacculata* nov. gen. nov. sp. Helgol. Wiss. Meeresunters., 7: 59—71.
Krumbein, W.E., 1968. Zur Frage der biologischen Verwitterung: Einfluss der Mikroflora auf die Bausteinverwitterung und ihre Abhängigkeit von edaphischen Faktoren. Z. Allg. Mikrobiol., 8: 107—117.
Krumbein, W.E., 1972. Rôle des microorganismes dans la genèse, la diagenèse et la dégradation des roches en place. Rev. Ecol. Biol. Sol, 9: 283—319.
Krumbein, W.E., 1973. Über den Einfluss von Mikroorganismen auf die Bausteinverwitterung — eine ökologische Studie. Deutsche Kunst- und Denkmalpflege, Jg. 1973, Deutscher Kunstverlag, München-Berlin, pp. 54—71.
Krumbein, W.E. and Pochon, J., 1964. Écologie Bactérienne des Pierres Altérées des Monuments. Ann. Inst. Pasteur, 107: 724—732.
Le Campion-Alsumard, T., 1969. Contribution à l'étude des cyanophycées lithophytes des étages supralittoral et médiolittoral (région de Marseille). Tethys, 1: 119—172.
Le Campion-Alsumard, T., 1970. Cyanophycées marines endolithes colonisant les surfaces rocheuses dénudées (Étages Supralittoral et Médiolittoral de la région de Marseille). Schweiz. Z. Hydrol., 32: 552—558.
Le Campion-Alsumard, T., 1975. Étude expérimentale de la colonisation d'éclats de calcite par les Cyanophycées endolithes marines. Cah. Biol. Mar., 16: 177—185.
Lukas, K.J., 1973. Taxonomy and Ecology of the Endolithic Microflora of Reef Corals, with a Review of the Literature on Endolithic Microphytes. Unpubl. Ph.D. Diss., Univ. Rhode Island, 159 pp.
Lukas, K.J., 1974. Two species of the chlorophyte genus *Ostreobium* from skeletons of Atlantic and Caribbean reef corals. J. Phycol., 10: 331—335.
Milliman, J.D., 1974. Recent Sedimentary Carbonates. Part 1: Marine Carbonates. Springer, Heidelberg, 375 pp.
Neumann, A.C., 1966. Observations on coastal erosion in Bermuda and measurements of the boring rate of the sponge *Cliona lampa.* Limnol. Oceanogr., 11: 92—108.
Neumann, A.C., 1968. Biological erosion of limestone coasts. In: R.W. Fairbridge (Editor), Encyclopaedia of Geomorphology. Reinhold Book Corp., New York, NY, pp. 75—81.
Nielsen, R., 1972. A study of the shell-boring marine algae around the Danish island Laeso. Bot. Tidsskr., 67: 245—269.
Parker, C.D., 1945. The corrosion of concrete I. The isolation of a species of bacterium associated with the corrosion of concrete exposed to atmospheres containing hydrogen sulphide. Aust. J. Exp. Biol. Med. Sci., 23: 81—90.
Parker, C.D., 1947. Species of sulphur bacteria associated with the corrosion of concrete. Nature, 159: 439—441.
Peterson, M.N.A., 1966. Calcite: rates of dissolution in a vertical profile in the central Pacific. Science, 154: 1542—1544.
Pfennig, N., 1975. The phototrophic bacteria and their role in the sulfur cycle. Plant Soil, 43: 1—16.
Pia, J., 1937. Die kalklösenden Thallophyten. Arch. Hydrobiol., 31: 264—328; 341—398.
Porter, C.L. and Zebrowski, G., 1937. Lime-loving molds from Australian sands. Mycologia, 29: 252—257.
Ruttner, F., 1960. Kohlendioxyd und Kohlensäure in Süsswasser. Ruhlands Handbuch der Pflanzenphysiologie, Vol. 5. Springer, Berlin, pp. 62—69.
Rützler, K., 1975. The role of burrowing sponges in bioerosion. Oecologia, 19: 203—216.

Rützler, K. and Rieger, G., 1973. Sponge burrowing: Fine structure of *Cliona lampa* penetrating calcareous substrata. Mar. Biol., 21: 144—162.

Schneider, J., 1976. Biological and Inorganic Factors in the Destruction of Limestone Coasts. Contributions to Sedimentology No 6. Eschweizerbart'sche Verlagsbuchhandlung, Stuttgart, 112 pp.

Schneider, J., 1977. Carbonate construction and decomposition by epilithic and endolithic microorganisms in salt-and-freshwater. In: E. Flügel, (Editor), Fossil Algae, Recent Results and Developments. Springer, Berlin, pp. 248—260.

Seilacher, A., 1969. Paleoecology of boring barnacles. Am. Zool., 9: 705—719.

Smarsh, A., Chauncey, H.H., Carriker, M.R. and Person, P., 1969. Carbonic anhydrase in the accessory boring organ of the gastropod, *Urosalpinx*. Am. Zool., 9: 967—982.

Tomlinson, J.T., 1969. Shell-burrowing barnacles. Am. Zool., 9: 837—840.

Warme, J.E., 1975. Borings as trace fossils and the processes of marine bioerosion. In: R.W. Frey (Editor), The Study of Trace Fossils. Springer, Berlin, pp. 181—227.

Wefer, G., 1976. Umwelt, Produktion und Sedimentation benthischer Foraminiferen in der westlichen Ostsee. Unpublished Doctoral thesis, Kiel, 103 pp.

Zebrowski, G., 1936. New genera of Cladochytriaceae. Ann. Missouri Bot. Garden, 23: 553—564.

Chapter 2.5

CARBON TURNOVER, CALCIFICATION AND GROWTH IN CORAL REEFS

D.W. KINSEY

University of Georgia Marine Institute, Sapelo Is., GA 31327 (U.S.A.)

P.J. DAVIES

Bureau of Mineral Resources, P.O. Box 378, Canberra, A.C.T. 2601 (Australia)

CONTENTS

INTRODUCTION

Coral reefs are unique among marine ecosystems in their ability to produce biogenic calcium carbonate and to retain it in the form of a wave-resistant structure. Such structures are best developed in the Pacific, Indian and western Atlantic Oceans, between latitudes 30° north and south. Until recent years, detailed coral-reef studies have been comparatively neglected, and many current ideas derive from scattered studies, largely on a few Pacific atolls, although headway has also been made in the western Atlantic and in the Indian Ocean. It is a paradox that the largest modern reef complex, the Great Barrier Reef off eastern Australia, has received less concerted study than most other reefal systems throughout the world. This chapter will therefore concentrate on work accomplished in the southern and northern Great Barrier Reef (Fig. 2.5.1A,B).

Trends in coral-reef research have developed models of reef growth at the

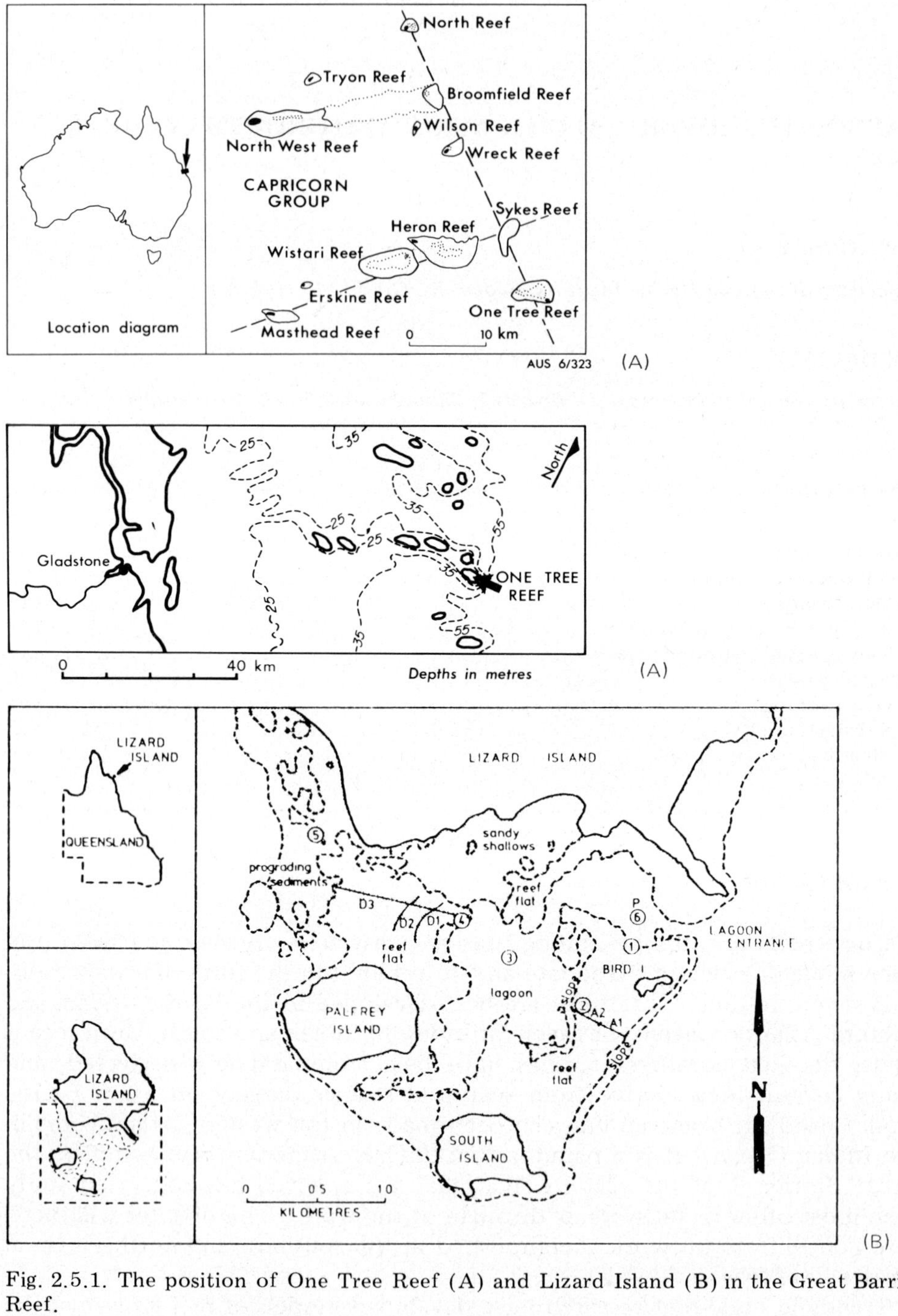

Fig. 2.5.1. The position of One Tree Reef (A) and Lizard Island (B) in the Great Barrier Reef.

zonal level, largely ignoring efforts devoted to models of the total system. While such major concepts as Darwin's subsidence theory or the role of Pleistocene sea-level fluctuations have been substantiated, it is probably too great a step to assert that the major problems in reef studies are at the individual assemblage or the zonal level. Our studies therefore emphasize the total system. In this paper, we summarize the present state of knowledge of overall processes and present analyses of the carbonate budget of a reef in terms of growth and destruction.

BACKGROUND CONSIDERATIONS

The main chemical component of a coral-reef system is calcium carbonate, which occurs either as high-Mg calcite, aragonite or low-Mg calcite. The mineralogy of the principal coral-reef inhabitants and the derived sediments is shown in Table 2.5.1A and B. In most reef systems, high-Mg calcite is most abundant close to the algal ridge, while aragonite is usually dominant seawards of the algal ridge, and in the lagoons. These various forms of calcium carbonate are created by, and for the direct benefit of the resident community. As the community develops, it modifies and then largely controls its own substratum, i.e., it can optimize its structure and orientation to changes in sea level provided that the rate of change does not exceed too much the metabolic growth capacity of the system.

Maximum potential for the translation of growth into "self-determined" morphology is achieved during a rising sea level, or during a stillstand following a rise. This is the present situation and differs substantially from past conditions of subaerial exposure of living reefs during the many Pleistocene fluctuations in sea level. Clearly, during exposure, the shape of reefs is controlled by weathering processes, but just as clearly, these weathered earlier reefs provided ideal sites for further reef growth as sea level rose above them. Thus, the shape of the new reef would initially be substratum controlled, but as growth continued, the original substratum control would become less influential as the reef communities developed their own substratum. Most if not all modern reefs therefore will consist of a series of superimposed "veneers" of new growth representing the periods of elevated sea level. Obviously in sites subject to maximum subaerial erosion, these veneers may erode away completely.

While general reef morphology is determined by a response to wind and waves and the need to optimize nutrient input, the upward growth of the reef to either reach or maintain itself at the sea surface is clearly a response to the need for solar energy. Not only is the input of carbon via the processes of photosynthesis totally dependent on light, but the processes of calcification and physical growth are linked to photosynthesis in all the organisms dominating the production of the carbonate mass of coral reefs.

TABLE 2.5.1

A. Mineralogy of principal reef constituents.
B. Percentage carbonate types in two reefs from the Great Barrier Reef.

(A)

Reef inhabitant	Mineralogy		
	Aragonite	High-Mg Calcite	Low-Mg Calcite
Scleractinian corals	X		
Alcyonarian corals		X	Few
Sponges		X	X
Echinoids		X	X
Asteroids		X	
Ophiuroids		X	
Molluscs	X	X	X
Foraminifera	X	X	Few
Green algae (*Halimeda*)	X		
Red coralline algae		X	

High-Mg calcite = 11—19 mol% $MgCO_3$ } After Chave (1954)
Low-Mg calcite = 0—5 mol% $MgCO_3$

(B)

Locality	Environment	Mineralogy		
		% Aragonite	% High-Mg Calcite	% Low-Mg Calcite
One Tree Reef [a]	Prograding sand sheet	50—60	40	10
	Windward lagoon	50	40	10
	Leeward lagoon	40—50	40	5
Heron Island [b]	Beach rock	35	65	Trace
	Beach sands	40	60	Trace

[a] Davies et al. (1976)
[b] Davies and Kinsey (1973).

(*Editorial note*: For a general discussion of the relationship between photosynthesis and calcification, see Chapter 2.2.)

Before exploring the intricacies of reef biogenic processes, we should consider some aspects of reef community structure and zonation. Reefs are comprised of large numbers of spectacular animals, particularly the corals themselves, but the far less conspicuous plants (mostly algae) have a standing crop which exceeds that of the animals. The name *coral reef* therefore is misleading and probably to some degree incorrect. Carbonates of coral origin, a major part of the reef mass, provide an unstable framework, which is stabilized by the calcifying algae (Yonge, 1930; Maxwell, 1968) aided by non-biological processes such as inorganic cementation (Maxwell, 1962; Ginsberg et al., 1971; Land and Goreau, 1971). Wave destruction, transportation and deposition are also important processes in developing a particular reef type (Maragos et al., 1973; Roberts, 1974; Davies et al., 1976; Baines and McLean, 1976; Roberts et al., 1977).

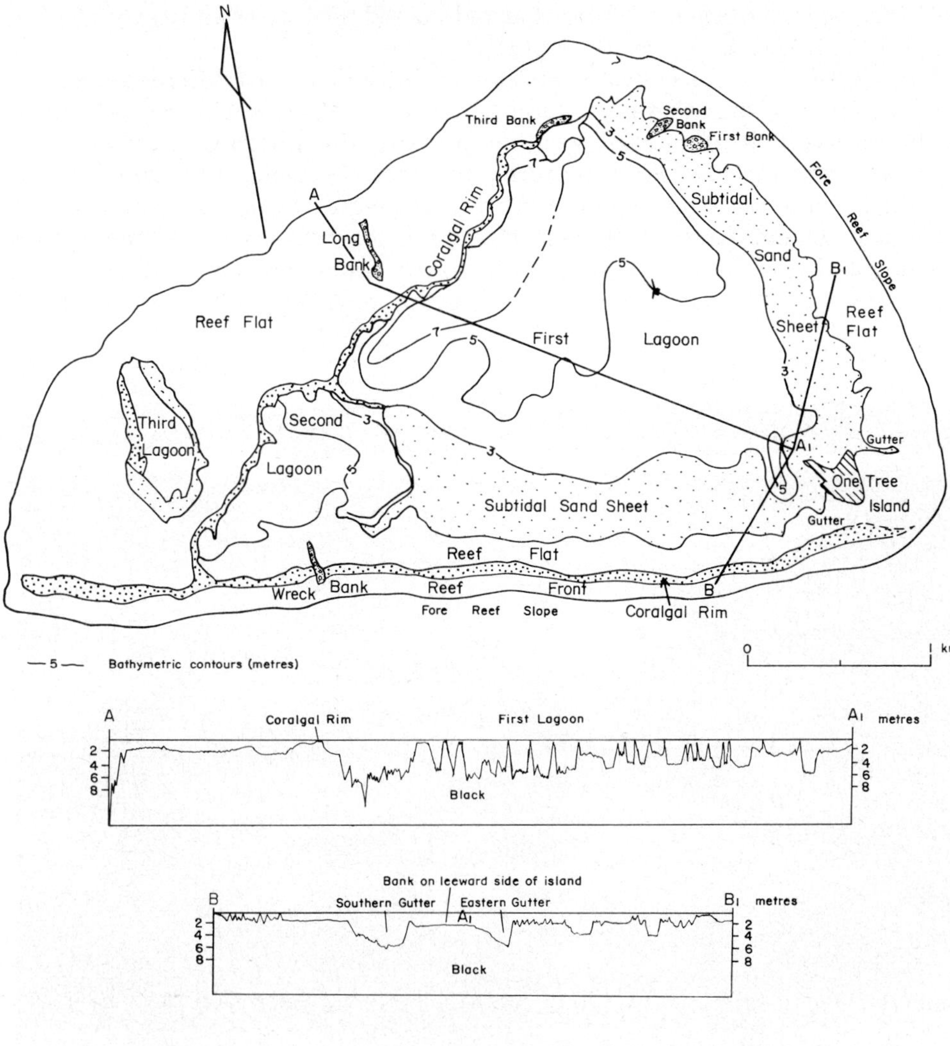

Fig. 2.5.2. The principal physical and biological zones of One Tree Reef, southern Great Barrier Reef.

Two groups of carbonate-secreting algae are of particular relevance: the foliose green algae (*Halimeda*) and red algae which, by their generation of fine carbonate sands or muds assume a passive infilling role in reef growth, and the encrusting coralline algae (*Porolithon*) which are the fundamental binding and cementing agents of the reef structure. Yonge (1930) describes the encrusting coralline algae as "the cement which binds the bricks into a homogeneous rampart" (1930, p. 67). It is regrettable that coralline algae, which are of enormous importance to the existence of coral reefs, have been the subject of less concerted research effort than any other major group of reef organisms. Notable exceptions are discussed by Littler and Doty (1975); Adey (1975), and Adey and Burke (1976).

The physical and biological zonation exhibited by most coral reefs is similar (Fig. 2.5.2) although their lateral and vertical extents will vary according to the variation in dominant weather pattern, the protection afforded by adjacent reefs, the local tidal patterns, the overall effects of the shape of the erosional platforms from which most modern reefs have grown, and possibly the impact of catastrophic storms. The upper seaward slope, and the distinctive serrated margin of the spur and groove zones, are areas of high coral cover (Fig. 2.5.3) and extensive encrusting coralline algal development.

Fig. 2.5.3. *Acropora* spp. on the upper part of the southern side of One Tree Reef.

(A)

(B)

Fig. 2.5.4. A. The stepped seaward margin of One Tree Reef. B. Consolidated pavement of the algal ridge, One Tree Reef.

(A)

(B)

Fig. 2.5.5. A. Extensive coral cover of the outer reef flat, together with encrusting calcifying algae; One Tree Reef. B. Coral/algal dominated reef flat, with sparse sand patches. One Tree Island in top left of photograph.

(A)

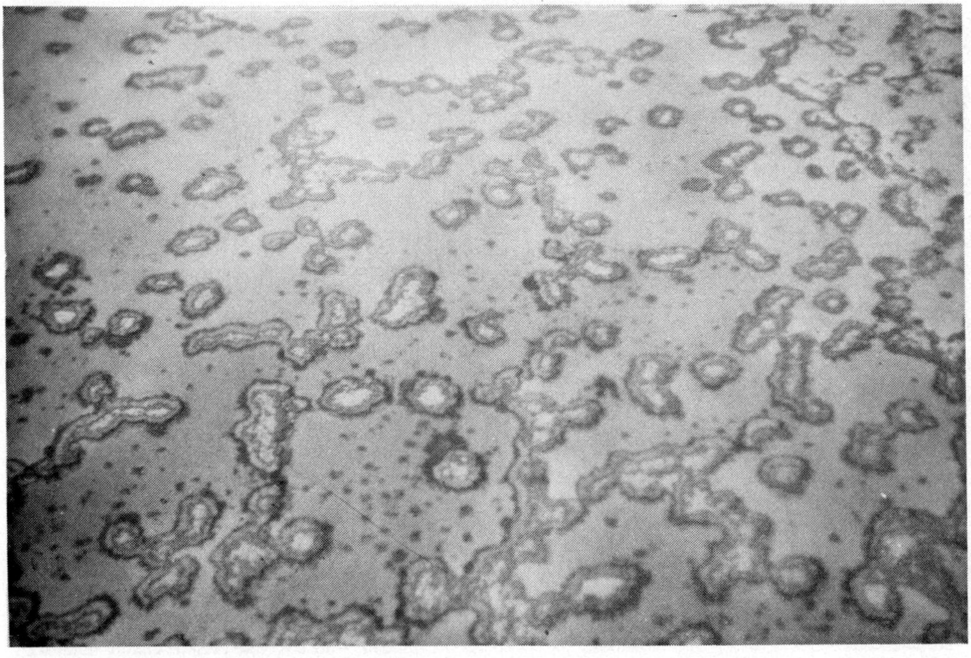

(B)

Fig. 2.5.6. A. Sand dominant zone of inner reef flat. Small (2—3 m diameter) *Pocillopora* colonies grow 1 m off the sandy bottom. One Tree Reef. B. Reticulate patch reef pattern in lagoon. One Tree Reef (aerial view).

(A)

(B)

Fig. 2.5.7. A. "Piecrust" (*Porolithon* covered) surface of the reticulate patch reefs. This is typical also of the leeward reef crest. One Tree Reef. B. Delicate and diverse coral development in 10—12 m of water on leeside of One Tree Reef.

Although calcification is thought to be rapid in this area there is little evidence of extensive carbonate build up, and Smith and Harrison (1977) suggest that the activity declines markedly with depth beyond the reef crest (at Enewetok). The origin of the spur and groove structure is still disputed by many and may prove to be complex (Shinn, 1963). Discussion has for many years revolved around a constructional origin (Munk and Sargent, 1954; Maxwell, 1968; Friedman, 1968), an erosional origin (Newell et al., 1951) and an origin resulting from inheritance of a pre-existing grooved surface (Emery et al., 1954; Kendall and Skipworth, 1969).

The seaward crest of the reef (Fig. 2.5.4A) is dominated by encrusting calcareous algae forming the algal ridge, a consolidated pavement usually devoid of corals and swept clean of fine sediments (Fig. 2.5.4B). The platform itself is bound together by encrusting algae and is surfaced by algal turf. Leeward of the pavement area is the outer reef flat composed of extensive coral communities interspersed with sand and rubble (Figs. 2.5.5A,B). This changes leeward into a zone where sand dominates over coral (Fig. 2.5.6A) before eventually giving way to a prograding sand sheet marking the edge of the lagoon (Fig. 2.5.2). The coral outcrops of the reef flat are extensively mixed with encrusting and foliose algae. Individual growth forms are compact. The prograding sand sheet between the reef flat and the lagoon is comprised of gravel and sand derived from the windward reef front and reef flats. Patch reef developments in lagoons show either linear, reticulate (Fig. 2.5.6B) or discrete patterns. The leeward reef flat is usually narrow (Fig. 2.5.7A) with shallow coral cover and little or no development of an algal pavement, though there is extensive development of coralline algae over all surfaces. The leeward slopes are areas of both fine and coarse sediments, commonly in large loosely bound structures which exhibit delicate and diverse coral development (Fig. 2.5.7B) because of the protection provided by the reef.

CARBON TURNOVER

Carbon turnover within a total reef community is a function of two distinct, biochemically interacting cycles. The first is the metabolic cycle consisting of the photosynthetic fixation of CO_2, and the release of CO_2 by respiration and decomposition processes. Superimposed on this are the direct incorporation of organic compounds (dissolved or particulate; living or non-living) which originate outside the reef systems (in the adjacent ocean waters), and the loss of organic compounds from the reef system into the out-flowing water. The second is the inorganic carbonate cycle involving the biological and non-biological precipitation and dissolution of carbonates. Superimposed on this is the loss of particulate carbonates in suspension in the out-flowing water (incoming suspended carbonates would normally be of very little consequence).

Most workers have accepted that exchange of dissolved organic matter with the ocean is insignificant relative to the rate at which CO_2 is fixed photosynthetically by reef communities. Marshall et al. (1975) established this to be valid for the Enewetok reef flat, although Kinsey (1972) presented some evidence from biological oxygen demand estimates on filtered samples that levels of dissolved organic matter were somewhat higher over the One Tree Island reef than in the surrounding ocean. By contrast, increases in particulate organic carbon may be very marked, and have been noted by a number of workers (Odum and Odum, 1955; Quasim and Sankaranarayanan, 1970; Tranter and George, 1972; Marshall and Talek, 1972; Johannes and Gerber, 1974; Gerber and Marshall, 1974; Marshall, 1972; Glynn, 1973). This material is predominantly algal detritus (Johannes and Gerber, 1974) from the reef front surf zone. It represents interzonal transport of amounts of material equivalent to up to 10% of the photosynthetic production of the seaward reef zones. Much of the material has been found to be deposited on the inner reef flats with some net transport into the lagoon (Odum and Odum, 1955; Marshall, 1965, 1968; Quasim and Sankaranarayanan, 1970; Johannes and Gerber, 1974; Marshall et al., 1975). However, complete reef systems have been found to exhibit a very small net loss to organics of the ocean at a rate equivalent to less than 0.5% of the total in situ photosynthetic turnover (Gordon, 1970; Smith and Pesret, 1974; Smith and Jokiel, 1975).

Thus the creation and reutilization of dissolved or suspended particulate organic matter is dominantly an internal function of reef systems with negligible net exchange with the ocean. Consequently, a considerable insight can be gained into the metabolic activity and carbon budget of a reef from a mass balance study of the CO_2 flux as this will indicate production and consumption within the system. Except in unusual and very enclosed reef environments (Smith and Pesret, 1974), it is unlikely that CO_2 is inherently a limiting nutrient. However, there is considerable evidence (Davies and Kinsey, 1973; Smith and Pesret, 1974; Smith and Jokiel, 1975; Smith and Kinsey, 1976) that inadequate water movement in lagoon environments may have the effect of limiting metabolic activity possibly by restricting access to the nutrients in the water. CO_2 is therefore, a very satisfactory monitoring parameter for determining the response of the system to all other variables such as more limiting nutrients or physical parameters. Fortunately, direct monitoring techniques are available which facilitate the determination of each of the components of the carbon flux in the presence of the others, without the need to modify or isolate the environment into any controlled experimental format.

The traditional parameter for monitoring CO_2 flux through the "organic" cycle in open community studies is oxygen (Sargent and Austin, 1954; Odum and Odum, 1955; Kohn and Helfrich, 1957; Gordon and Kelly, 1962; Kinsey, 1972).

In practice, the monitoring of rates of change of the parameters outlined above is carried out in the water overlying the reef community. This is done either by using upstream/downstream sampling where the current is predicted, or by marking particular water masses with soluble dyes so that a timed series of samples can be taken while the marked water remains over the community or zone being considered (Sargent and Austin, 1954; Smith and Marsh, 1973; Marsh and Smith, 1978; Kinsey, 1978). Together with data concerning volume transport rates and general bathymetry, it is possible to determine rates of net CO_2 change per unit area of the reef community. By determining the net CO_2 consumption through the organic cycle over the period of daylight, an estimate can be made of the net photosynthesis of the community. Similarly, by determining the net CO_2 production throughout the night, estimates can be made of community respiration and decomposition. On the assumption that the same respiration and decomposition rates apply during both day and night, estimates can then be made of gross photosynthetic production and gross respiration and decomposition. Because inorganic carbonate precipitation and dissolution both occur during daylight and at night, only net rates of calcification can be determined. The details of all the methods and field procedures outlined above are described in detail in the "Handbook of Coral Reef Research Methods" (Kinsey, 1978; Smith and Kinsey, 1978).

The organic carbon cycle

It is difficult to discuss the carbonate production of a coral reef system without first understanding the stoichiometric biological framework within which that activity is occurring (Kinsey and Domm, 1974; Marsh and Smith, 1978). The inverse relationship between O_2 and CO_2 flux is reasonably precise and predictable for most natural aerobic systems and oxygen can be measured with great accuracy. Furthermore, oxygen flux is totally unaffected by the CO_2 flux through the inorganic carbonate cycle. The main deficiency of the O_2-monitoring technique is that corrections must be applied to account for exchange between the water and the atmosphere via diffusion at the air/sea interface (Odum and Hoskin, 1958; Kinsey and Domm, 1974; Kinsey, 1978); however, O_2-monitoring is still a valuable tool for monitoring general community metabolism.

An alternative method is to determine changes in total CO_2 content of the sea water by monitoring pH, chlorinity, temperature and alkalinity (Park, 1969; Smith and Key, 1975; Smith and Kinsey, 1978) and to derive values for the flux of CO_2 through the "organic" cycle, by subtracting from the total flux that component attributable to the inorganic carbonate flux as determined by alkalinity measurements. These estimates are relatively free from errors caused by diffusion, as CO_2 exchange with the atmosphere is much slower than that of O_2. However, the absolute precision of the measurement is less than can be obtained with oxygen monitoring.

Broecker and Takahashi (1966) and Smith and Kinsey (1978) have developed a technique for monitoring the CO_2 flux through the inorganic carbonate cycle, based on the fact that removal of inorganic carbonates from sea water will decrease the total alkalinity of the water, and dissolution of inorganic carbonates will increase the total alkalinity. This relationship is stoichiometric within the ranges normally encountered in coral-reef systems, though it does have limitations in severely restricted environments (Davies and Kinsey, 1973; Smith, 1973; Brewer and Goldman, 1976).

Few studies to date have adequately estimated the overall metabolic activity of an entire reef system. Table 2.5.2 summarizes the available data, but even so, the estimates extend only from crest to crest and exclude the outer slopes. The estimates for One Tree Island Reef and Lizard Island Reef are the weighted means of a series of zonal studies. The Canton estimate is based on overall long-term changes. The principal conclusion to be gained from such data is that the reef systems reported clearly show little or no net

TABLE 2.5.2

Metabolic turnover of CO_2 by "total" reef systems (excluding calcification)

Note: All estimates exclude the activities of the outer slopes. However in the case of Canton and Fanning Islands, these do not contribute to the virtually closed systems. These two reefs also lack reef-flats subject to inflowing water because of the continuous islands along the reef crests. 1 = Mean water level; 2 = low water springs.

Reef	Description	Latitude	Metabolic rates (mmol CO_2 m^{-2} d^{-1})		
			Gross photo-synthesis	Gross respiration	Net gain
One Tree Reef [a]	5 × 7 km lagoonal reef with unbroken perimeter at MWL[1].	23°	190	195	−5
Lizard Island [b]	2.5 × 3 km lagoonal reef joining three small granitic islands. Broken perimeter at LWS[2].	13°S	270	270	0
Canton Island [c]	15 × 6 km atoll, landlocked except for single pass.	3°S	500	495	+5
Fanning Island [d]	Relatively enclosed atoll.	4°N	—	—	0

[a] Kinsey (1977).
[b] Kinsey, unpublished data.
[c] Smith and Jokiel (1975).
[d] Smith and Pesret (1974).

TABLE 2.5.3

Metabolic turnover of CO_2 by reef perimeter zones (excluding calcification)

	Dominant standing crop	Latitude	Metabolic rates (mmol CO_2 m^{-2} d^{-1})		
			Gross photo-synthesis	Gross respiration	Net gain
Reef-front areas					
Lizard Island pinnacle [a]	"Total" coral cover	14°S	800	615	185
One Tree Island surf zone pavement [b]	Algal pavement with some soft algal cover	23°S	170	45	125
Seaward reef-flats					
Lizard Island [a]	Mostly algal with some corals	14°S	560	580	−20
One Tree Reef [c]	35% coral cover	23°S	600	617	−17
Enewetak Tr. II [d]	Coral/algal	11°N	500	500	0
Enewetak Tr. III [d]	Algal turf and pavement	11°N	830	440	390
Enewetak Tr. II [e]	Coral/algal	11°N	830	830	0
Rongelap [f]	Algal turf and pavement (some poorly developed corals)	11°N	340	300	40

[a] Kinsey, unpublished data.
[b] Kinsey (1977).
[c] Kinsey and Domm (1974).
[d] Smith and Marsh (1973). (Tr. = transect.)
[e] Odum and Odum (1955). (Tr. = transect.)
[f] Sargent and Austin (1954).

gain of organic carbon. They are in virtually complete equilibrium consuming by respiration and decomposition all the organic material which they create by photosynthesis. The superimposed exchange of organic matter directly with the surrounding ocean is probably at a low level compared to this basic CO_2 flux (loc. cit.).

Table 2.5.3 shows the available data for the seaward perimeter zones of reefs other than fringing reefs. There are a number of differences between these results and those from the total reef systems (Table 2.5.2). With the exception of the bare algal pavement at One Tree Reef the results suggest a reasonably uniform gross activity for seaward shallow water zones with little or no latitudinal variation and, significantly, little correlation with variation in community structure. However, net gain is quite variable with

TABLE 2.5.4

One Tree Reef [a] — Zonation of metabolic CO_2 turnover (excluding calcification)

Zone	Metabolic rates (mmol CO_2 m^{-2} d^{-1})		
	Gross photosynthesis	Gross respiration	Net gain
Outer slopes — seaward	not known		
Surf-zone pavement	170 (25%)	45 (25%)	125
Reef-flat coral zone	600 (4%)	617 (7%)	−17
Sand flats (some filamentous algae and sparse corals)	75 (25%)	117 (25%)	−42
Lagoon with reticulated surface patch reefs	260 (6%)	260 (10%)	0
Lagoon with small submerged reefs	120 (10%)	145 (10%)	−25
Narrow leeward reef flat	not known		
Outer slopes — leeward	not known		
Overlying water (planktonic activity)	5	17	−12
Weighted mean	190	195	−5

[a] Median transect S-N. Dominant wind from SE. Data after Kinsey (1977) and Kinsey and Domm (1974). All figures are full seasonal means from data over several years. Figures in parenthesis are standard deviations as percentages, for data during any one month.

some sites requiring an input to achieve equilibrium, e.g., organic material from the outer seaward slopes and others showing substantial net gains which are probably lost as detrital and soluble organic matter towards the lagoon.

Considering now the detailed zonation of One Tree Reef (Table 2.5.4), it is apparent that, of those zones examined, the major site of metabolic activity is the reef flat. It is, however, a net consuming zone requiring an input of organic carbon. The algal pavement, which clearly exhibits net production is seaward of the reef flat and it is reasonable to propose that such a seaward zone is exporting its excess production of organic carbon, probably as algal detritus, to the inner zones and lagoons. Over the whole year, there appears to be a slight net loss from the total reef system. However, if the reef-front pinnacle at Lizard Island (Table 2.5.3) is generally representative of outer slope environments, i.e., they are significant net producers, then it is apparent that the One Tree Reef system may well be at "perfect" zero gain if the outer seaward slopes are taken into account.

One other facet of the organic activity of the reef which should be stressed is that activity measured at one time of the year may not be at all representative of the overall annual pattern of activity. Table 2.5.5 shows the seasonal variation in activity over the reef flat at One Tree Reef. The magnitude of the activity varies by a factor of 2.5 and the extent of gain or loss by an order of magnitude.

TABLE 2.5.5

Seasonal variation — One Tree Reef — coral zone of reef flat

Month	Extreme diurnal temp. range (°C)	Potential daily insolation ($kJ\ cm^{-2}\ d^{-1}$)	Metabolic rates ($mmol\ CO_2\ m^{-2}\ d^{-1}$)		
			Gross photosynthesis	Gross respiration	Net gain
June	18—22	1.05	300	440	−140
Sept.	20—25	2.1	600	615	−15
Dec.	25—35	2.7	750	700	50
April	23—27	1.7	750	875	−125

Note: Data after Kinsey (1977) and Kinsey and Domm (1974). All figures are based on data over several years.
The potential daily insolation is the calculated total insolation on a clear day assuming an atmospheric transmission coefficient of 0.7.

Thus the organic activity of a coral reef may be considered as a highly structured process with overall self-sufficiency, i.e., autotrophism, but with essentially no gain of fixed carbon. There is, however, considerable interzonal transfer of organic matter with the sediment zones acting as major zones of decomposition. There is little or no net import or export of organic carbon from the total reef system and therefore virtually all organic carbon is created and ultimately consumed within the system. These considerations imply a negligible rate of net biomass growth even though individual populations may increase and decrease substantially.

While these conclusions hold over a whole year, it is likely that there are periods well removed from steady-state during the cycle of the seasons. In many respects, it seems that zonation or the location of a community has a far greater bearing on its activity than the actual biological structure of the community. Perhaps the most important conclusions of all relate to the significance of the seaward perimeter zones as the sites of major activity, with the outermost of these (outer slopes and algal crest pavements) being probably net producers of organic matter which is then "fed" back to the remainder of the system (Smith and Marsh, 1973).

The inorganic carbon cycle

Even fewer investigations have been made on the quantitative fluxes of carbonates in coral reefs than have been made on the turnover of carbon through the "organic" cycles. Most of the data on carbonate fluxes are summarized in Smith and Kinsey (1976) and are repeated in modified form in Tables 2.5.6 and 2.5.7. These data are quoted only as net calcification, because independent estimates of production and dissolution are not possible.

TABLE 2.5.6

Net gains in carbonate CO_2 by "total" reef systems (Data sources as for Table 5.2.2)

	$mmol\ m^{-2}\ d^{-1}$	$kg\ CaCO_3\ m^{-2}\ y^{-1}$
Canton Atoll (Lat. 3°S)	14	0.5
Fanning Atoll (Lat. 4°N)	27	1.0
One Tree Reef (Lat. 23°S)	41	1.5
Lizard Island (Lat. 14°S)	50	1.8

All reefs examined exhibit a conspicuous daily gain of carbonate (expressed as $CaCO_3$, Table 2.5.6). The figures vary by a factor of 3—4 with little correlation with latitude. However, this gain implies a positive mass (inorganic) growth of the reefs in contrast to the virtually complete lack of organic growth reported for similar systems in the previous section (Table 2.5.2).

Calcification rates for reef flat and seaward zones are shown in Table 2.5.7 and may be compared with the results of metabolic CO_2 turnover for the same locations in Table 2.5.3. Whereas the "organic" activity varied to a moderate extent, the results for net carbonate deposition are remarkably similar for the different environments. The variations in community structure seem to have no effect on this activity and there is no correlation with latitude. While the number of reefs represented is small, it covers a rather wide range of reefal types and it is suggested (Smith and Kinsey, 1976) that there is some factor, as yet not understood, setting a "ceiling" on the calcification rates in these peripheral seaward zones regardless of the development of the community structure. The potential for positive physical

TABLE 2.5.7

Calcification on seaward reef flats

Locality	Seaward reef flats	$mmol\ CO_2\ m^{-2}\ d^{-1}$	$kg\ CaCO_3\ m^{-2}\ y^{-1}$
One Tree Reef [a]	Algal pavement (no coral)	110	4
One Tree Reef [a]	Coral reef flat zone	123	4.5
Lizard Island [b]	Seaward slope coral pinnacle	101	3.7
Lizard Island [b]	Reef flat (algal pavement)	104	3.8
Lizard Island [b]	Reef flat (coral/algal zones)	99	3.6
Enewatok Atoll [c]	Reef flat (coral/algal)	110	4
Enewatok Atoll [c]	Reef flat (algal)	110	4

[a] Kinsey (1977).
[b] LIMER (1976).
[c] Smith (1973).

TABLE 2.5.8

Calcification and potential vertical growth for various zones of the One Tree Reef system. Values are based on the means of data obtained during most months over the period 1969—1975. (Data after Kinsey, 1977.)

Zone	Net calcification		Vertical growth potential
	mmol CO_2 m^{-2} d^{-1}	kg $CaCO_3$ m^{-2} y^{-1}	($mm\ y^{-1}$)
Seaward slopes	not known	not known	not known
Algal pavement	110	4.0	2.8
Reef flat coral zone	123	4.5	3.1
Sand flats	8	0.3	0.2
Reticulated lagoon	41	1.5	1.0
Deep lagoon	14	0.5	0.3
Leeward flat (not contrib. to lagoon)	not known	not known	not known
Outer lee slopes	not known	not known	not known
Weighted means	41	1.5	1.0

growth of the reef in these seaward zones is considerable. Whether the precipitated carbonates accumulate in the zone of their formation or are redistributed, is the subject of a later section.

Considering now the zonation of the calcification activity across a complete reef system, and again using One Tree Reef (the only one for which such information is available) as an example, it is apparent that the general distribution pattern of net carbonate deposition (Table 2.5.8) follows the pattern of photosynthetic and respiratory activity (Table 2.5.4). A notable exception is the high level of calcification relative to the low level of "organic" activity in the algal pavement. Again there are no data for the outer slopes of this reef but, drawing once more from the seaward pinnacle data from Lizard Island (Table 2.5.7), we propose that the shallow parts of the outer slopes of a reef are likely to exhibit the same production as reef flat sites, i.e., approximately 4 kg $CaCO_3$ $m^{-2}y^{-1}$. It would be hard to imagine a reef environment supporting a higher standing crop of corals than the Lizard Island pinnacle or a lower standing crop than the One Tree Reef algal pavement. Yet both these zones have almost identical rates of carbonate deposition. The crux of the distinction between the respective roles played by these two zonal types, i.e., outer slope and algal pavement, relates to the nature of the sediments which each produces. Corals predominantly create large, movable building blocks whereas the algal pavement produces either total consolidation or fine sediments. The significance of these roles to the formation of a reef structure will be discussed later.

Numerical data for calcification at One Tree Reef are much more scant than those available for organic activity. However, the general seasonal trend indicated in Table 2.5.5 also applies to calcification activity. All zones exhibit significant to high net gain unlike the situation applying to the organic activity, and major inter-zonal transfer of accumulated materials is certain to occur. Based on the evidence so far available, positive inorganic growth is likely to be a feature of all modern coral reefs. As with the organic activity, the zonation of the reef has a much greater bearing on the calcification activity than do the obvious variations in biological structure of the community. The perimeter zones are clearly the sites of major activity the rates of which may be little affected by community type and latitude. The potential of these zones to accumulate carbonates is approximately 4 kg $CaCO_3$ $m^{-2}y^{-1}$.

PHYSICAL GROWTH

Most published works on reef growth stress or imply a two-dimensional outward expansion, with lateral growth of the living reef at sea level being the dominant process. Most attention has focussed on biological, chemical and physical forces which influence this expansion (Hubbard, 1974; Kinsey and Domm, 1974; Jaubert and Vasseur, 1974; Roberts, 1974; Scatterday, 1974; Shinn, 1963; Stoddart, 1969). Maximum expansion has long been considered to occur along windward margins, the earliest documentation of which was Darwin's hypothesis for the growth of barrier reefs from fringing reefs (Darwin, 1842). In more recent times, Maxwell (1968) invoked a similar mechanism for the growth of platform reefs in the Great Barrier Reef, and suggested as well that continued outward growth was accompanied by central degeneration and lagoon formation. We believe that inherent within such ideas are basic misinterpretations of the relations between rates of reef growth and rates of sea-level change, the possible effects of substratum on future colonization, and the acceptance without proof that the production of calcium carbonate exceeds that removed.

Ideas concerning mechanisms of reef growth in the Great Barrier Reef have been hampered by the paucity of data. However, it is known that reefs have existed since the middle Miocene (Maxwell, 1973), and that many of these reefs were exposed to subaerial erosion for long periods of time, especially during the last glacial period which lowered sea level to −100 m *. Purdy (1974) concluded that the shape of modern reefs in British Honduras reflects the shape of the surface from which they grow, which had been sculptured during the last sea level low. Adey et al. (1977), Halley et al. (1977) and Shinn et al. (1977) also accept substratum control for the growth of Caribbean reefs, but consider that subaerial erosion played very little part in determining substratum shape. Lewis (1968) and Rosen (1971) postulated some

* Sea levels are expressed relative to present-day levels.

substratum control for the growth of some Indian Ocean reefs. It is clear therefore that substratum is an important factor to be considered when analyzing controls on reef growth. Moreover, the depth of the growth substrate will establish a maximum age for the date of coral colonization. Initial colonization in the Great Barrier Reef started around 9 ky ago (Davies, 1975; Davies et al., 1976, 1977; Davies and Kinsey, 1977; Hopley, 1977; Thom et al., 1978) from a depth of between 20—25 m below present sea level (Davies, 1974), which accords with conclusions from most reef areas throughout the world (Lewis, 1968; Stoddart, 1971; Milliman, 1973; Broecker et al., 1968; Chappell and Polach, 1976; MacIntyre and Glynn, 1976). At about this time ($\simeq$9 ky B.P.), the rate of sea-level rise changed from about 10 mm y^{-1} to 6 mm y^{-1}; sea level in eastern Austalia stabilized around 6 ky B.P. (Thom and Chappell, 1975).

Growth rates

The mean calcification values in various environments at One Tree Reef are given in Table 2.5.8. These data may be converted to an implied vertical growth rate potential (Table 2.5.8) assuming that accrual is dominantly aragonite (density = 2.89 g cm^{-3}) and that there is 50% porosity after normal compaction. The latter assumption is supported by the porosity data for One Tree Reef shown in Table 2.5.9. Data in Table 2.5.8 show a maximum vertical growth rate of approximately 3 mm y^{-1} around the perimeter zones.

At the present-day sea level, it is obvious that little or none of the potential vertical growth is expressed and that accrued material must represent lateral growth. However, during the period of the Holocene transgression when the young reef was submerged by sea level rising at 6—10 mm y^{-1}, it seems reasonable to postulate that growth would be predominantly vertical at a rate similar to that currently exhibited by the active perimeter zones, i.e., 3 mm y^{-1}. Similar figures have been obtained by radiocarbon dating for the whole of Holocene reef growth in New Guinea (Chappell and Polach, 1976) and in the Caribbean (McIntyre and Glynn, 1976). Little of this

TABLE 2.5.9

Porosities in corals and sediments from One Tree Reef (Data from Davies and Kinsey, 1977.)

Lithology	Porosity %
Favites	55.5
Incipient cemented beach rock	45—49
Partially cemented beach rock	30.0
Porites	60.0
Medium sands	46—50
Fine to medium sands	33—41

growth material is likely to have been transferred to the lower-lying areas by physical destruction of the reef because of the protection offered by the increasing water cover. Similarly, we believe that most particulate material loosened by bio-erosion is likely to have remained virtually in situ, and would not invalidate the postulated vertical growth rate of 3 mm y^{-1}. However, it seems reasonable to suggest that the central areas of the reef would have been subjected to some of the obvious, but ill-defined factors which limit growth activity in such zones today (Smith and Pesret, 1974; Smith and Jokiel, 1975; Smith and Kinsey, 1976) and that growth of these areas, therefore, would have lagged behind the perimeter growth.

Substratum effects

The effects of substratum on reef growth in the Great Barrier Reef have been demonstrated by Harvey (1977), Davies et al. (1976, 1977), and Davies and Kinsey (1977). The reefs of the northern Great Barrier Reef have grown off substrates at a depth of 6—19 m (Harvey, 1977). Most data are however available from the southern Great Barrier Reef (Table 2.5.10). The most complete data are from One Tree Reef.

One Tree Reef is situated on the southeast extremity of a platform with a maximum depth of 20 to 25 m (Fig. 2.5.1). This is generally verified by echo profiles (Figs. 2.5.2, 2.5.8A) across the reef from north to south. Seismic refraction studies on the reef margins, in the lagoon, and on patch reefs, together with bathymetric and scuba studies have allowed the reconstruction shown in Fig. 2.5.8. On the south side of the reef, a cliff occurs at 10 to 20 m depth on which little coral grows. The top of the cliff at 10 m depth also corresponds with the maximum depth of the main lagoon (Fig. 2.5.8A) which is that part where sedimentation rates are lowest (Davies et al., 1976). Studies on the northern lee slope reveal it to be highly irregular with channels 7—15 m deep alternating with coral growth climbing to near present

TABLE 2.5.10

Depth of pre-Holocene growth surface beneath reefs of the southern Great Barrier Reef (Data are metres below reef flat level (from Davies et al., 1977).)

Reef	Depth of pre-Holocene growth surface			
	Windward side	Leeward side	Lagoon	Patch reefs
Wreck	7.9—12.8	7.3—12.0		
Sykes	9.5			
One Tree	10.6—19.1	7.3—18.0	16.1—16.6	11.9
Fitzroy	9.7—16.7	12.1—17.3		
Fairfax	8.2—12.5	13.2		

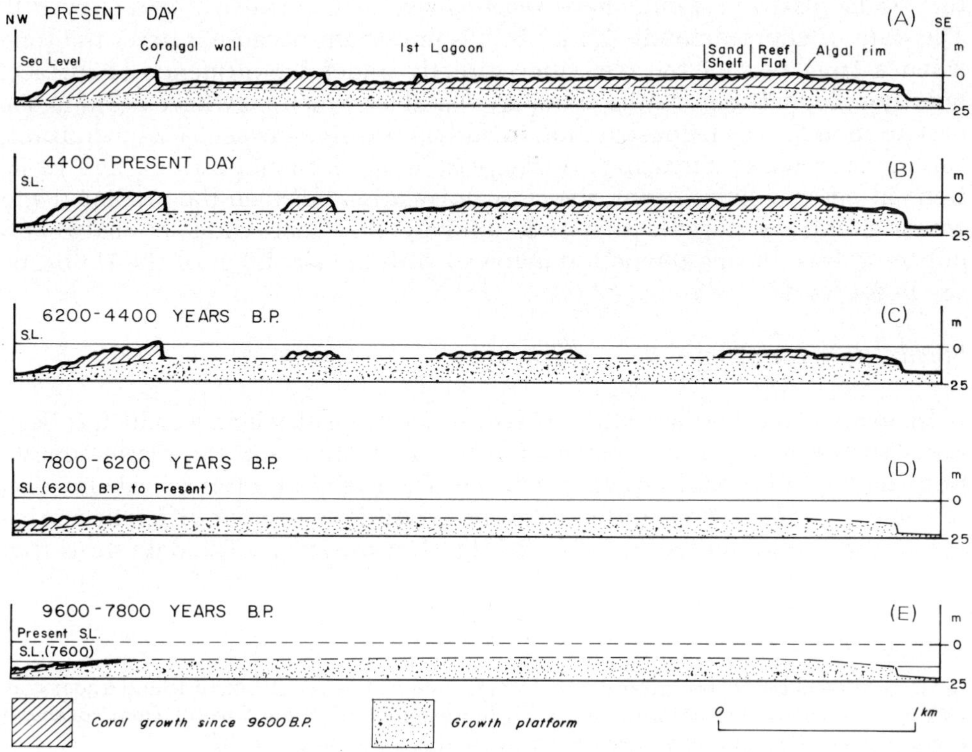

Fig. 2.5.8. Evolution of One Tree Reef since 9.6 ky B.P. The dark line is the top of the growing reef.

day low water and an extremely irregular debris-strewn bottom sloping down to depths of 20—25 m. We suggest therefore, that One Tree Reef has grown off a platform which varies in the manner shown in Fig. 2.5.8E. Its subsequent development accompanying the rise in sea level is shown in Figs. 2.5.8D,C,B. Sea level reached the 20 to 25 m depth platform about 9.0—9.6 ky B.P. (Fig. 2.5.8E), and reached the —10 m level by 7.8 ky B.P. Reef growth of about 3.5—5 m in that time would have been limited to the western lee side, the declivity of the southern cliff inhibiting growth on this side. The sea reached its present level by about 6.2 ky B.P. (Thom and Chappell, 1975). New growth between 7.8—6.2 ky B.P. would have occurred on the —10 m platform to a thickness of about 5 m with overlying water 5 m deep by 6.2 ky B.P. (Fig. 2.5.8D). The further growth between 7.8—6.2 ky B.P. on the western lee side would have given a total reef thickness of 8.5—10 m but here the overlying water would still be more than 10 m deep by 6.2 ky B.P. Coral growing from the —10 m platform would have reached sea level by 4.4 ky B.P. (Fig. 2.5.8C) while that which originally grew from the —20 to —25 m platform would have reached sea level between 2.0—1.2 ky B.P. The date of approximately 4.4 ky B.P. is important because it dates the time when a large part of the reef came directly under the influence of surface-modifying conditions. While there are variables which may have acted in the past to modify this reconstruction including possible lower sea temperatures, variations in water chemistry, differences in species composition, and variations in percent living cover, the close correlation between the predicted and actual situations suggests that it is unlikely that such variables had major impact at least during the period involved with the last 20 m of the Holocene rise in sea level.

Steady-state conditions

In many published accounts of reef growth, insufficient weight has been given to the relationships between reef calcification and the marine conditions under which that growth occurred. Data relating rates of calcification to removal on the windward reef flat at Lizard Island under different marine conditions are presented in Table 2.5.11 (Davies, 1977). The data show that

TABLE 2.5.11

Relationships between the production and removal of $CaCO_3$ at Lizard Island under conditions of slack water, no wind (Case I); flood tide and 10 knot winds from southeast (Case II), and ebb tide and 20 knot winds from the southeast (Case III)

$$\text{Theoretical time of removal of accrued carbonate (days)} = \frac{\text{Total potential annual accrual}}{\text{Suspension load} \times \text{flow-rate} \times 24\text{ h}}$$

In a 100 m transect, the total potential annual accrual = 400 kg of $CaCO_3$

Case I. Minimum load conditions
Suspension load = 0.5 g m^{-3}
Flow-rate = 2 m min^{-1}
Suspension load d^{-1} = 1.44 kg
Theoretical time of removal of accrued carbonate = 266 d

Case II. Intermediate load conditions
Suspension load = 0.3 g m^{-3}
Flow-rate = 12 m min^{-1}
Suspension load d^{-1} = 5.2 kg
Theoretical time of removal of accrued carbonate = 80 d

Case III. Maximum load conditions
Suspension load = 2.5 g m^{-3}
Flow-rate = 3.3 m min^{-1}
Suspension load d^{-1} = 11.4 kg
Theoretical time of removal of accrued carbonate = 35 d

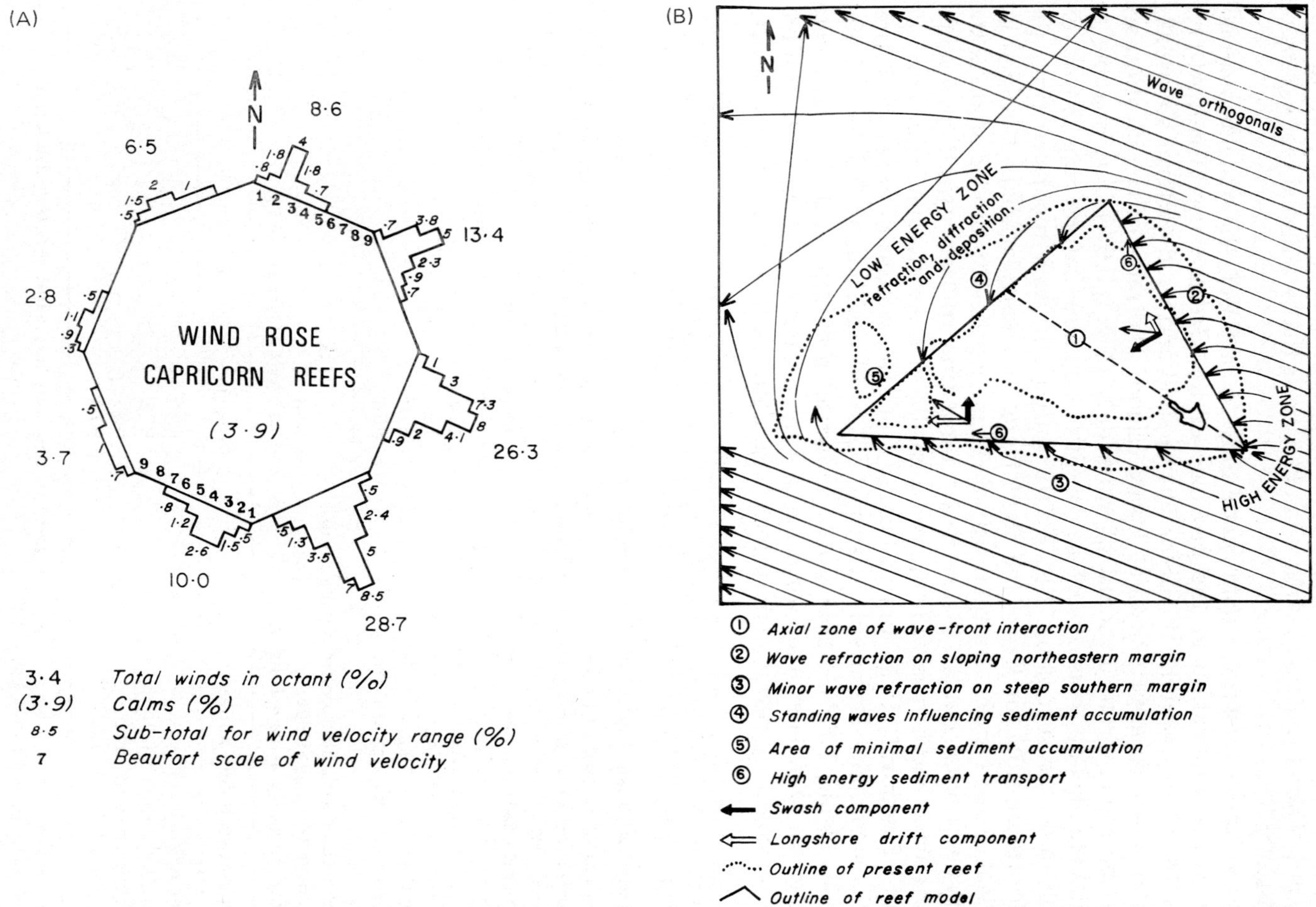

Fig. 2.5.9. A. Wind rose for area of Capricorn reefs. B. Model of wave patterns affecting One Tree Reef in east-southeasterly weather.

the theoretical maximum amount of calcium carbonate produced is easily removed within the time of its deposition, and the direction of removal is to leeward. The corollary to such a conclusion is that windward reef extension is unlikely.

While these numerical data relate only to one reef, the conclusions may have wider application in the province of the Great Barrier Reef. At One Tree Reef, for example, the dominant weather and wind patterns are usually similar to those experienced at Lizard Island. The rates of calcification are also the same. From a record of wind data (Fig. 2.5.9A) it is possible to model the likely effects of surface conditions on a reef after it has grown vertically into the surf zone. Fig. 2.5.9B shows one such reconstruction, the basic assumption for One Tree Reef being its triangular shape outlined by vertical growth from the platform at 10 m depth. The model predicts the northwest extension of the reef, the formation of lagoonward prograding sand sheets, and a massive leeward expansion of the reef, all of which are the result of removal of calcium carbonate from the windward edge. These predictions compare favourably with the observed situation at One Tree Island (Davies et al., 1976). Table 2.5.12 shows an attempted calcium carbonate budget for One Tree Reef for the past 4 ky, i.e. the time during which it has been affected by surface conditions. The data show that the overall growth

TABLE 2.5.12

Calcium carbonate budget of One Tree Reef.

The areas of contributing and receiving zones calculated by planimetric measurement. The contributing zones are assumed to calcify at a rate of 4 kg m^{-2} y^{-1}

Sediment contributing zones		Area ($m^2 \times 10^{-3}$)
1. Southern reef front		413
2. Southern algal ridge		190
3. Southern reef flat		670
4. Northern reef flat and front		850
Total area of sediment contributing zones excluding the algal ridge	=	1 933
All zones calcify at ca 4.0 kg m^{-2} y^{-1}	=	7 732 kg $CaCO_3$ y^{-1}
Sediment receiving zones		**Area ($m^2 \times 10^{-3}$)**
1. Lee edge		2 480
2. Southern sand sheet		822
3. Northern sand sheet		538
4. Lagoon (?)		2 670

Overall growth = 1.3 kg m^{-2} y^{-1}. Growth of lee edges and sand sheets = 2.0 kg m^{-2} y^{-1} = 6 m vertical growth in 4 ky.

of the sediment receiving areas in 1.3 kg m^{-2} y^{-1}. However, if the sand sheets and leeward edges are considered the major sites of deposition, then the rate of accumulation becomes 2 kg m^{-2} y^{-1}. This figure ignores any growth on the leeward side attributable to coral growing in that environment. However, Davies et al. (1976) noted a rich and branching coral assemblage, which because of its branching habit is readily susceptible to heavy weather from the northwest. If the rate of calcification of this leeward coral zone was only equivalent to that of the protected reticulate patch reefs of the lagoon (1.5 kg m^{-2} y^{-1}), then the total leeward growth would equal that produced and lost on the windward edge. Although we have not measured calcification rates on the leeward sides, the free water movement together with the abundant growth and wide variety of growth forms suggests a calcification rate which is likely to be higher than that exhibited by lagoonal situations. Such calcification would remain in the leeward environment so that the total calcification exhibited by the leeward edge would be greater than that measured at windward situations.

A major question arising from the conclusions at Lizard Island is why are coral reefs able to exist at all under such rigorous conditions? Possible answers are: (1) the data are totally unrepresentative; this seems unlikely; (2) the calcification figures of 4 kg m^{-2} y^{-1} for reef flat environments are much lower than rates on reef fronts which to date have rarely been measured. However, calcification data for a reef pinnacle at Lizard Island (Limer 1975 Expedition Team) closely resembling a reef front situation showed rates similar to reef flat environments, i.e. 4 kg m^{-2} y^{-1}. Further, Smith and Harrison (1977) indicate that the reef front at Enewetok has rates of calcification which fall rapidly with depth and never exceed those of the reef front; (3) the production rates are correct but the type of building block ultimately determines the potential leeward sediment, and the amount of visible erosion on windward sides. For example, corals are more easily broken than encrusting calcareous algae, serving to further emphasize the critical role of these algae in reef development. Coral reefs could not exist at sea level, under present-day equilibrium conditions, without them. It should be noted that in Table 2.5.12, the algal flat is omitted as a potential producer of calcium carbonate in estimating budgets because we consider that calcification in the algal zone is accrued in situ, which is why algal flats are topographically the highest parts of a reef system.

Two further points merit brief attention. If conditions in Table 2.5.12 approximate to the yearly norm, then a windward reef flat of about 400 m width would represent the equilibrium situation on which the production of calcium carbonate would exactly balance the amount lost. Most windward reef flats in fact vary in width between 100—400 m.

In view of the compelling evidence that windward reef environments have a capability of removing all that is produced, we believe that the origin of reef front morphological features should be re-examined. Although there has been much past discussion over the origin of the groove and buttress system, classical opinion generally invokes a constructional origin (Maxwell, 1968), the buttresses representing the tips of the windward growing reef front. Much more serious thought should now be given to the possibility that the buttress ends represent the original front of an eroding reef (Newell et al., 1951). The meagre 2 m of accretion in 10 ky shown by Land (1974) for Jamaican, and Buddemeier et al. (1975) for Enewetok buttress systems adds weight to this suggestion.

CONCLUSIONS

(1) In terms of organic carbon production, reef systems clearly exhibit little or no net gain, i.e. photosynthetic production equals oxidation and decomposition.

(2) The exchange of organic matter with the oceans is small compared with the basic carbon dioxide flux within the system.

(3) A coral reef may be considered as a self-supporting system, but with considerable interzonal transfer. The major site of metabolic activity is the reef flat, and probably the upper reaches of the seaward slope. The reef flat is a net consumer requiring an input of organic carbon. This gain is probably obtained from the algal pavement and the upper slope, which are seaward of the reef flat.

(4) In contrast to the carbon turnover, all reefs examined showed a net gain of calcium carbonate.

(5) Changes in carbonate flux show little conspicuous correlation with community structure, or with latitude.

(6) Net carbonate deposition over all the reefs examined shows an extraordinary degree of consistency.

(7) All zones exhibit significant net carbonate gains, with the perimeter zones clearly the sites of major activity. Maximum potential vertical growth rates of 3 mm y^{-1} occur at these sites.

(8) Such growth rates operating over the period of Holocene reef growth would give 25 m of vertical growth before surf action transformed the vertical potential into lateral accretion.

(9) Substrate has played a major role in defining the shapes of present-day reefs. At One Tree Reef, perimeter growth from a platform at 10 m depth defined both the shape of the reef and the central lagoon.

(10) Calcium carbonate deposited in windward perimeter zones at or near the surface is removed and transported in a leeward direction, where it forms prograding sand wedges infilling lagoons, and leeward extension of the reef. Windward reef zones are likely to be net destructional and not constructional sites.

(11) It is likely that the total accrual of carbonates on the leeward side,

i.e. coral growth plus derived detritus from windward sites, is greater than that deposited on windward perimeter zones.

(12) Reef growth during the Holocene on reefs studied by us has been in two directions, vertical, and laterally leeward.

ACKNOWLEDGEMENTS

Figures 2.5.1, 2.5.2, 2.5.8 and 2.5.9 are reproduced with permission of the Director, Bureau of Mineral Resources.

REFERENCES

Adey, W.H., 1975. The algal ridges and coral reefs of St. Croix: Their structure and Holocene development. Atoll Res. Bull., 187: 1—67.

Adey, W.H. and Burke, R., 1976. Holocene bioherms (algal ridges and bank-barrier reefs) of the Eastern Caribbean. Bull. Geol. Soc. Am. 87: 95—109.

Adey, W.H., Macintyre, I.G. and Stuckenrath, R., 1977. Relict barrier reef system at St. Croix. Proceedings of the Third International Symposium on Coral Reefs, Vol. 2, pp. 15—22.

Baines, G.B.K. and McLean, R.F., 1976. Resurveys of 1972 Hurricane Rampart of Funafuti Atoll, Ellice Islands. Search, 7: 36—37.

Brewer, P.G. and Goldman, J.C., 1976. Alkalinity changes generated by phytoplankton growth. Limnol. Oceanogr., 21: 108—117.

Broecker, w.S. and Takahashi, T., 1966. Calcium carbonate precipitation of the Bahama Banks. J. Geophys. Res., 71: 1575—1602.

Broecker, W.S., Thurber, L.D. and Goddard, J., 1968. Milankovitch hypothesis supported by precise dating of coral reefs and deep sea sediments. Science, 159: 297—300.

Buddemeier, R.W., Smith, S.V. and Kinzie, R.A., 1975. Holocene windward reef-flat history, Enewetok Atoll. Bull. Geol. Soc. Am., 86: 1581—1584.

Chappell, J. and Polach, H.A., 1976. Holocene sea level change and coral reef growth at Huon Peninsula, Papua New Guinea. Bull. Geol. Soc. Am., 87, 235—240.

Chave, K.E., 1954. Aspects of the biogeochemistry of magnesium. 1. Calcareous marine organisms. J. Geol., 62: 266—283.

Darwin, C., 1842. The Structure and Distribution of Coral Reefs. Smith, Elder and Co., London, 214 pp.

Davies, P.J., 1974. Sub surface solution unconformities at Heron Island, Great Barrier Reef. Proceedings of the Second International Symposium on Coral Reefs, Vol. 2, pp. 573—578.

Davies, P.J., 1975. Formation of the Great Barrier Reef. Habitat Aust., 3: 3—8.

Davies, P.J., 1977. Modern reef growth — Great Barrier Reef. Proceedings of the Third International Symposium on Coral Reefs, Vol. 2, pp. 325—330.

Davies, P.J. and Kinsey, D.W., 1973. Organic and inorganic factors in recent beach rock formation, Heron Island, Great Barrier Reef. J. Sediment. Petrol., 43: 59—81.

Davies, P.J. and Kinsey, D.W., 1977. Holocene reef growth — One Tree Island, Great Barrier Reef. Mar. Geol., 24: M1—M11.

Davies, P.J., Radke, B. and Robison, C., 1976. Geological and sedimentary development of One Tree Reef. BMR J. Aust. Geol. Geophys., 1: 231—240.

Davies, P.J., Marshall, J.F., Foulstone, D., Thom, B.G., Harvey, N., Short, A.D. and Martin, K., 1977. Reef growth, Southern Great Barrier Reef — Preliminary results. BMR J. Aust. Geol. Geophys., 2: 69—72.

Emery, K.O., Tracey, J.I. and Ladd, H.S., 1954. Geology of Bikini and Nearby Atolls. U.S. Geol. Surv. Prof. Pap., 260-A: 1—265.

Friedman, G.M., 1968. Geology and geochemistry of reefs, carbonate sediments, and waters, Gulf of Aqaba (Elat), Red Sea. J. Sediment. Petrol., 38: 895—919.

Gerber, R. and Marshall, N., 1974. Reef pseudoplankton in lagoon trophic systems. Proceedings of the Second International Symposium on Corals and Coral Reefs, Vol. 1, pp. 105—107.

Ginsberg, R.N., Marszalek, D.S. and Schneidermann, N., 1971. Ultrastructure of carbonate cements in a Holocene algal reef of Bermuda. J. Sediment. Petrol, 41: 472—482.

Glynn, P.W., 1973. Ecology of a Caribbean coral reef, the *Porites* reef flat biotype: Part II. Plankton community with evidence for depletion. Mar. Biol., 22: 1—21.

Gordon, D.C., 1970. Organic carbon budget of Fanning Island Expedition 1970. Hawaii Institute of Geophysics, Report HIG-70-23, pp. 23—29.

Gordon, M.C. and Kelly, H.M., 1962. Primary productivity of a Hawaiian coral reef: A critique of flow respimetry in turbulent waters. Ecology, 43: 473—480.

Halley, R.B., Shinn, E.A., Hudson, J.H. and Lidz, B., 1977. Recent and relict topography of Boo Bee Patch Reef, Bilize. Proceedings of the Third International Symposium on Coral Reefs, Vol. 2, pp. 29—35.

Harvey, N., 1977. The identification of subsurface solution disconformities on the Great Barrier Reef, Australia, between 14°S and 17°S, using shallow seismic refraction techniques. Proceedings of the Third International Symposium on Coral Reefs, Vol. 2, pp. 45—51.

Hopley, D., 1977. The age of the outer ribbon reef surface, Great Barrier Reef, Australia: Implications for hydro-isostatic models. Proceedings of the Third International Symposium on Coral Reefs, Vol. 2, pp. 23—28.

Hubbard, J.A.E.B., 1974. Scleractinian coral behaviour in calibrated current experiment; An index to their distribution patterns. Proceedings of the Second International Symposium on Coral Reefs, Vol. 2, pp. 107—126.

Jaubert, J.M. and Vasseur, P., 1974. Light Measurements: Duration aspect and the distribution of benthic organisms in an Indian Ocean coral reef (Tulear, Madagascar). Proceedings of the Second International Symposium on Coral Reefs, Vol. 2, pp. 127—142.

Johannes, R.E. and Gerber, R., 1974. Import and export of net plankton by an Eniwetok coral reef community. Proceedings of the Second International Symposium on Coral Reefs, Vol. 1, pp. 97—104.

Kendall, C.G.St.C. and Skipworth, P.A.D'E., 1969. Geomorphology of a recent shallow water carbonate province: Khor Al Bazam, Trucial Coast, South West Persian Gulf. Bull. Geol. Soc. Am., 80: 865—891.

Kinsey, D.W., 1972. Preliminary observations on community metabolism and primary productivity of the pseudo-atoll reef at One Tree Island, Great Barrier Beef. Proceedings of a Symposium on Corals and Coral Reefs (Mandapam Camp, India 1969). Marine Biological Association of India, pp. 13—32.

Kinsey, D.W., 1977. Seasonality and zonation in coral reef productivity and calcification. Proceedings of the Third International Symposium on Coral Reefs, Vol. 2, pp. 383—387.

Kinsey, D.W., 1978. Productivity and calcification estimates using slack-water periods and field enclosures. In: D.R. Stoddart and R.E. Johannes (Editors), Coral reefs: research methods. UNESCO pp. 439—468.

Kinsey, D.W. and Domm, A., 1974. Effects of fertilization on a coral reef environment —

Primary production studies. Proceedings of the Second International Symposium on Coral Reefs, Vol. 1, pp. 49—66.
Kohn, A.J. and Helfrich, P., 1957. Primary organic productivity of a Hawaiian coral reef. Limnol. Oceanogr., 2: 241—251.
Land, L.S., 1974. Growth rate of a West Indian (Jamaican) reef. Proceedings of the Second International Symposium on Coral Reefs, Vol. 2, pp. 409—412.
Land, L.S. and Goreau, T.F., 1971. Submarine lithification of Jamaican reefs. J. Sediment. Petrol., 40: 457—462.
Lewis, M.S., 1968. The morphology of the fringing coral reefs along the east coast of Mahe, Seychelles. J. Geol., 76: 140—153.
LIMER, 1976. Metabolic processes of coral reef communities at Lizard Island, Queensland. LIMER 1975 Expedition team. Search, 7: 463—468.
Littler, M.N. and Doty, M.S., 1975. Ecological components structuring the seaward edges of tropical Pacific reefs: The distribution, communities and productivity of *Porolithon*. J. Ecol., 63: 117—129.
Macintyre, I.G. and Glynn, P.W., 1976. Evolution of modern Caribbean fringing reef, Galeta Point, Panama. Bull. Am. Assoc. Petrol. Geol. 60: 1054—1072.
Maragos, J.E., Baines, G.B.K. and Beveridge, P.J., 1973. Tropical cyclone creates a new land formation on Funafuti Atoll. Science, 181: 1161—1164.
Marsh, J.A. and Smith, S.V., 1978. Productivity measurements of coral reefs in flowing water. In: D.R. Stoddard and R.E. Johannes (Editors), Coral reefs: research methods. UNESCO, pp. 361—378.
Marshall, N., 1965. Detritus over the reef and its potential contribution to adjacent waters of Eniwetok Atoll. Ecology, 46: 343—344.
Marshall, N., 1968. Observations on organic aggregates in the vicinity of coral reefs. Mar. Biol., 2: 50—53.
Marshall, N., 1972. Mucus and zooxanthellae from reef corals. Proceedings of the Symposium on Corals and Coral Reefs (Mandapam Camp, India 1969). Marine Biological Association of India, pp. 59—65.
Marshall, N. and Talek, G., 1972. Particulate and dissolved organic carbon in an atoll reef environment. Eniwetok Marine Biological Laboratory Annual Report, 1972, p. 16.
Marshall, N., Durbin, A.G., Gerber, R. and Talek, G., 1975. Observations on particulate and dissolved organic matter in coral reef areas. Int. Rev. Ges. Hydrobiol., 60: 335—345.
Maxwell, W.G.H., 1962. Lithification of carbonate sediments in the Heron Island Reef, Great Barrier Reef. J. Geol. Soc. Aust., 8: 217—238.
Maxwell, W.G.H., 1968. Atlas of the Great Barrier Reef. Elsevier, Amsterdam, 258 pp.
Maxwell, W.G.H., 1973. Geomorphology of eastern Queensland in relation to the Great Barrier Reef. In: O.A. Jones and R. Endean, (Editors), Biology and Geology of Coral Reefs. Academic, New York, NY, Vol. 1, pp. 233—272.
Milliman, J., 1973. Caribbean coral reefs. In O.A. Jones, and R. Endean (Editors), Biology and Geology of Coral Reefs. Academic, New York, NY, Vol. 1, pp. 1—50.
Munk, W.H. and Sargent, M.C., 1954. Adjustment of Bikini Atoll to ocean waves. U. S. Geol. Surv. Prof. Pap., 260-C: 275—280.
Newell, N.D., Rigby, J.K., Whiteman, A.J. and Bradley, J.S., 1951. Shoal water geology and environments, Eastern Andros Island, Bahamas. Bull. Am. Mus. Nat. Hist., 97: 1—29.
Odum, H.T. and Hoskin, C.M., 1958. Comparative Studies on the Metabolism of Marine Waters. Public Institute of Marine Science, Texas, Vol. 5, pp. 16—46.
Odum, H.T. and Odum, E.P., 1955. Trophic structure and productivity of a windward coral reef community on Enewetak Atoll. Ecol. Monogr., 25: 291—320.
Park, P.K., 1969. Oceanic CO_2 system: An evaluation of ten methods of investigation. Limnol. Oceanogr., 14: 179—186.
Purdy, E.G., 1974. Reef configurations: cause and effect. In: L.F. LaPorte (Editor), Reefs in Time and Space. Society of Economic Palaeontologists and Mineralogists, Special Publication, 18: pp. 9—76.

Quasim, S.Z. and Sankaranarayanan, V.N., 1970. Production of particulate organic matter by the reef on Kavaratti Atoll (Laccadives). Limnol. Oceanogr., 15: 574—578.

Roberts, H.H., 1974. Variability of reefs with regard to changes in wave power around an island. Proceedings of the Second International Symposium on Coral Reefs, Vol. 2, pp. 497—512.

Roberts, H.H., Murray, S.F. and Suhayda, J.N., 1977. Physical processes in a fore-reef shelf environment. Proceedings of the Third International Symposium on Coral Reefs, Vol. 2, pp. 507—515.

Rosen, B.R., 1971. Principal features of reef coral ecology in shallow water environments, Mahe, Seychelles. In: D.R. Stoddard and M. Yonge (Editors), Regional Variation in Indian Ocean Coral Reefs. Symposium of the Zoological Society of London, No. 28, pp. 163—183.

Sargent, M.C. and Austin, T.S., 1954. Biological economy of coral reefs. Bikini and nearby atolls, Marshall Islands. U.S. Geol. Surv. Prof. Pap., 260-E: 293—300.

Scatterday, J.W., 1974. Reefs and associated coral assemblages off Bonaire, Netherlands Antilles, and their bearing on Pleistocene and recent reef models. Proceedings of the Second International Symposium on Coral Reefs, Vol. 2, pp. 85—106.

Shinn, E., 1963. Spur and groove formation on the Florida Reef Tract. J. Sediment. Petrol., 33: 291—303.

Shinn, E.A., Hudson, J.H., Halley, R.B. and Lidz, B., 1977. Topographic control and accumulation rate of some Holocene Coral Reefs: South Florida and Dry Tortugas. Proceedings of the Third International Symposium on Coral Reefs, Vol. 2, pp. 1—7.

Smith, S.V., 1973. Carbon dioxide dynamics: A record of organic carbon production, respiration, and calcification in the Enewetak windward reef flat community. Limnol. Oceanogr., 18: 106—120.

Smith, S.V. and Harrison, J.T., 1977. Calcium carbonate production of the *Mare Incognitum,* the upper windward reef slope, at Enewetok Atoll. Science, 197: 556—559.

Smith, S.V. and Pesret, F., 1974. Processes of carbon dioxide flux in the Fanning Atoll lagoon. Pac. Sci., 28: 225—245.

Smith, S.V. and Key, G.S., 1975. Carbon dioxide and metabolism in marine environments. Limnol. Oceanogr., 20: 493—495.

Smith, S.V. and Jokiel, P.L., 1975. Water composition and biogeochemical gradients in the Canton Atoll Lagoon. 2. Budgets of phosphorus, nitrogen, carbon dioxide and particulate materials. Mar. Sci. Commun. 1: 165—207.

Smith, S.V. and Kinsey, D.W., 1976. Calcium carbonate production, coral reef growth, and sea level change. Science, 194: 937—939.

Smith, S.V. and Marsh, J.A., 1973. Organic carbon production and consumption on the Windward Reef Flat of Enewetok Atoll. Limnol. Oceanogr., 18: 953—961.

Smith, S.V. and Kinsey, D.W., 1978. Calcification and organic carbon metabolism as indicated by carbon dioxide. In: D.R. Stoddart and R.E. Johannes (Editors), Coral reefs: research methods. UNESCO, pp. 469—484.

Stoddard, D.R., 1969. Ecology and morphology of recent coral reefs. Biol. Rev., 44: 433—498.

Stoddard, D.R., 1971. Environment and history in Indian Ocean reef morphology. In: D.R. Stoddard and M. Yonge (Editors), Regional Variation in Indian Ocean Coral Reefs. Symposia of the Zoological Society of London, No. 28, pp. 3—38.

Thom, B.G. and Chappell, J., 1975. Holocene sea levels relative to Australia. Search, 3: 90—93.

Thom, B.G., Orme, G.R. and Polach, H., 1978. Drilling investigations of Bewick and Stapleton Islands. Phil. Trans. R. Soc. London, 291: 37—54.

Tranter, D.J. and George, J., 1972. Zooplankton abundance at Kavarati and Kalpeni atolls in the Laccadives. Proceedings of the Symposium on Corals and Coral Reefs (Mandapam Camp India, 1969), Marine Biological Association of India, pp. 239—256.

Yonge, M., 1930. A Year on the Great Barrier Reef. Putnam, London—New York, 246 pp.

Chapter 3.1

BIOGEOCHEMISTRY OF PHOSPHATE MINERALS

D. McCONNELL

Ohio State University, Columbus, OH 43210 (U.S.A.)

CONTENTS

THE PHOSPHORUS CYCLE

Before discussing the details of phosphate minerals, it seems essential to trace the movements of phosphates within the environment. Fig. 3.1.1 shows food in the uppermost position of the cycle, soil at the opposite pole of the diagram, and phosphorite in a more central position — because of its importance.

Phosphorite, for our purposes, is the term applied to calcium phosphate rocks, whether they are accumulations of bones, precipitates directly from sea water or replacements of calcareous rocks. They are very extensive on the earth, occurring on all continental land masses with the possible exception of Antarctica, where commercial deposits have not yet been found. Phosphorites comprise the principal geologic storage bin for inorganic phosphates.

The path of inorganic phosphates between phosphorite and human food comprises other intermediate stages: superphosphates (phosphatic fertilizers), soil minerals and plants. There may be — for non-vegetarians — also an animal intermediate between plants and human food.

The subcycle phosphorite to igneous rocks is accomplished through magmatic assimilation of sedimentary or metamorphic rocks that contain

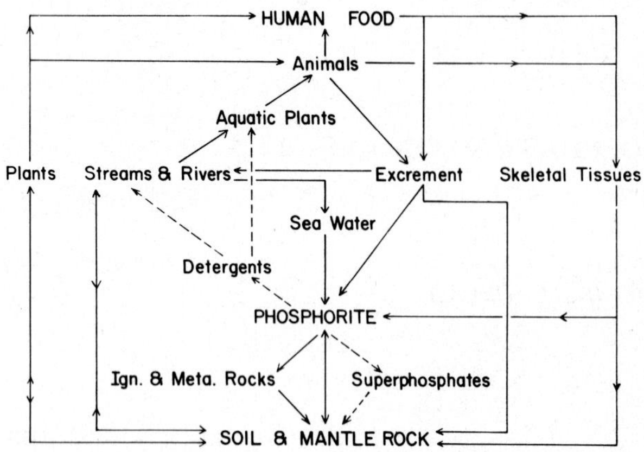

Fig. 3.1.1. Transitional paths of phosphates in nature. Dashed lines indicate paths instituted by man.

appreciable phosphorus. Weathering of such igneous rocks contributes to the unconsolidated mantle rocks, leaching of which ultimately produces dissolved phosphates in the runoff carried by streams and rivers to the ocean. In highly cultivated areas, there may be a significant contribution to surface waters from use of phosphatic fertilizers.

Considerable attention has been devoted in recent years to discharges of spent phosphatic solutions [detergents] with respect to their enhancement of growth of aquatic plants, particularly certain algae. This subcycle is shown by dashed lines in Fig. 3.1.1 inasmuch as it is not natural, and it is necessarily incomplete because it is not known to what extent these organic phosphates are ultimately disintegrated and to what extent they are incorporated in lacustrine sediments.

Organic phosphates which result from ingestion and metabolism of food may be passed on to soil or may be retained in the building of skeletal tissues of developing new animals. In mature animals, the weight of skeletal tissues is relatively stable, although metabolic processes provide for turnover of calcium and phosphorus. Under either circumstance, phosphates ultimately contribute to phosphorites as bone beds, or phosphatizing solutions which interact with limestones, or phosphatic cements of sandstones. Weathering accounts for the arrow connecting phosphorites with soil and mantle rock, but this direction is reversible to the extent that phosphatic solutions can enrich phosphorites through downward percolation. The path between streams and rivers and soil and mantle rock is also reversible for similar reasons.

The superphosphate path between phosphorites and soil and mantle rock

TABLE 3.1.1

Phosphatic minerals likely to form under the influence of organisms

Mineral	Chemical composition
Apatite	(see dahllite and francolite)
Brushite	$CaHPO_4 \cdot 2\ H_2O$ (monocl.)
Monetite	$CaHPO_4$ (tricl.)
Whitlockite	$Ca_9(Mg, Fe)H(PO_4)_7$
Dahllite	carbonate hydroxyapatite (pseudohex.)
Francolite	carbonate fluorapatite (pseudohex.)
Dehrnite	Na-containing member of apatite group
Lewistonite	K-containing member of apatite group
Bobierrite	$Mg_3(PO_4)_2 \cdot 8\ H_2O$ (monocl.)
Vivianite	$Fe_3(PO_4)_2 \cdot 8\ H_2O$ (monocl.)
Newberyite	$MgHPO_4 \cdot 3\ H_2O$ (orthorh.)
Biphosphammite	$NH_4H_2PO_4$ (tetr.)
Phosphammite	$(NH_4)_2HPO_4$ (?)
Archerite	$(K, NH_4)H_2PO_4$ (tetr.)
Struvite	$NH_4MgPO_4 \cdot 6\ H_2O$ (orthorh.)
Stercorite	$NH_4NaHPO_4 \cdot 4\ H_2O$ (tricl.)
Dittmarite	$NH_4MgPO_4 \cdot H_2O$ (orthorh.)
Phosphorrösslerite	$MgHPO_4 \cdot 7\ H_2O$ (monocl.)
Schertelite	$Mg(NH_4)_2H_2(PO_4)_2 \cdot 4\ H_2O$ (tricl.)
Hannayite	$Mg_3(NH_4)_2H_4(PO_4)_4 \cdot 8\ H_2O$ (tricl.)
Collinsite	$Ca_2(Mg, Fe)(PO_4)_2 \cdot 2\ H_2O$ (tricl.)
Messelite	$Ca_2Fe(PO_4)_2 \cdot 2\ H_2O$ (tricl.)
Crandallite	$CaAl_3(PO_4)_2(OH)_5 \cdot H_2O$ (hex.)
Millisite	$(Na, K)CaAl_6(PO_4)_4(OH)_9 \cdot 3\ H_2O$ (tetr.)
Wardite	similar to millisite (tetr.)
Taranakite	Highly hydrated phosphate of (Al, Fe) with (K, Na, NH_4, Ca) (hex.)
Francoanellite	similar to taranakite (hex.?)
Montgomeryite	$Ca_4MgAl_4(PO_4)_6(OH)_4 \cdot 12\ H_2O$ (monocl.)
Overite	$CaMgAl(PO_4)_2(OH) \cdot 4\ H_2O$ (orthor.)
Brazilianite	$NaAl_3(PO_4)_2(OH)_4$ (monocl.)
Minyulite	$KAl_2(PO_4)_2(OH, F) \cdot 4\ H_2O$ (orthorh.?)
Gordonite	$MgAl_2(PO_4)_2(OH)_2 \cdot 8\ H_2O$ (tricl.)
Davisonite	$Ca_3Al(PO_4)_2(OH)_3 \cdot H_2O$ (hex.)
Englishite	Ca, K, Al hydrated phosphate (orthorh.)
Lehiite	highly hydrated phosphate of Al, Ca and (Na, K) (?)
Kingite	highly hydrated phosphate of Al (tricl.)
Wavellite	$Al_3(PO_4)_2(OH)_3 \cdot 5\ H_2O$ (orthorh.)
Trolleite	$Al_4(PO_4)_3(OH)_3$ (?) (monocl.)
Variscite	$AlPO_4 \cdot 2\ H_2O$ (orthorh.)
Metavariscite	$AlPO_4 \cdot 2\ H_2O$ (monocl.)
Redondite	$(Al, Fe)PO_4 \cdot 2\ H_2O$ (orthorh.)
Barrandite	$(Al, Fe)PO_4 \cdot 2\ H_2O$ (orthorh.)
Clinobarrandite	$(Al, Fe)PO_4 \cdot 2\ H_2O$ (monocl.)
Senegalite	$Al_2PO_4(OH)_3 \cdot H_2O$ (orthorh.)
Strengite	$FePO_4 \cdot 2\ H_2O$ (orthorh.)
Phosphosiderite	$FePO_4 \cdot 2\ H_2O$ (monocl.)
Ludlamite	$Fe_3(PO_4)_2 \cdot 4\ H_2O$ (monocl.)

TABLE 3.1.1. (continued)

Mineral	Chemical composition
Koninckite	$FePO_4 \cdot 3\ H_2O$ (?) (tetr.)
Vauxite	hydrated ferrophosphate of Al (tricl.)
Metavauxite	hydrated ferrophosphate of Al (monocl.)
Paravauxite	hydrated ferrophosphate of Al (tricl.)
Beraunite	hydrated ferroferriphosphate (monocl.)
Cacoxenite	$Fe_9(PO_4)_4(OH)_{15} \cdot 18\ H_2O$ (hex.)
Tinticite	$Fe_6(PO_4)_4(OH)_6 \cdot 7\ H_2O$ (?)
Sigloite	hydrated ferroferriphosphate of Al (tricl.)
Barbosalite	$Fe^{2+}Fe_2^{3+}(PO_4)_2(OH)_2$ (monocl.)
Dufrenite	hydrated ferroferriphosphate (monocl.)
Rockbridgeite	hydrated ferroferriphosphate (orthorh.)
Cyrilovite	$NaFe_3(PO_4)_2(OH)_4 \cdot 2\ H_2O$ (tetr.)
Leucophosphite	hydrated ferriphosphate of K (monocl.)
Anapaite	$Ca_2Fe(PO_4)_2 \cdot 2\ H_2O$ (tricl.)
Mitridatite	hydrated ferriphosphate of Ca (monocl.)
Calcioferrite	$Ca_2Fe_2(PO_4)_3(OH) \cdot 7\ H_2O$ (monocl.)
Efremovite	hydrated ferriphosphate of Ca (?)
Ardealite	$CaHPO_4 \cdot CaSO_4 \cdot 4\ H_2O$ (monocl.)
Bradleyite	$Na_3Mg(PO_4)(CO_3)$ (?)
Viséite	hydrated Al phosphate silicate of Na, Ca (cubic)
Kribergite	hydrated phosphate sulfate of Al (?)
Azovskite	hydrated ferriphosphate (?)
Bolivarite	amorphous hydrated Al phosphate
Vashegyite	hydrated phosphate of Al (orthorh.)
Sasaite	$(Al, Fe)_{14}(PO_4)_{11}(OH)_7SO_4 \cdot 83\ H_2O$ (orthorh.?)
Evansite	amorphous hydrated phosphate of Al
Delvauxite	amorphous hydrated ferriphosphate
Richellite	hydrated ferriphosphate of Ca (?)

— although augmented by man, and thus dashed — is a very important one, and involves some complex minerals about which little was known until comparatively recently.

Among the subcycles shown in Fig. 3.1.1, all involve organisms other than man except the following: phosphorites → superphosphates, phosphorite → igneous and metamorphic rocks, and phosphorites → detergents. The path between igneous and metamorphic rocks → soil and mantle rock involves biochemical weathering as does phosphorites → soil and mantle rock. It has been known for many years that phosphorites are more soluble in humic acids, and phosphatic fertilizers are rated on their citric acid solubilities.

Other portions of the phosphorus cycle are self-explanatory; the only system of greater complexity would be one involving a gaseous phase (the atmosphere) and such processes as photosynthesis.

Poorly known, however, are the catalytic actions of organisms in the path from sea water → phosphorites, and some persons apparently still believe that precipitation of carbonate apatite takes place from sea water as a strictly inorganic reaction for which one can write a solubility product in the form of ionic activities. Such efforts are largely vitiated, nevertheless, by the failure of these same individuals to include the carbonate (bicarbonate) ion in their computations (McConnell, 1970a). The carbonate ion occurs in both the liquid and solid phases to an extent which cannot be ignored and, indeed, it may be the controlling factor which governs precipitation (McConnell et al., 1961).

The minerals presumed to form under biochemical or biogeochemical influences are listed in Table 3.1.1. Some of the omissions — turquoise, for example — will be explained below.

OXIDATION AND REDUCTION

Phosphate minerals, particularly those containing iron and manganese, are sensitive to oxidizing vs. reducing environments, although it has been suggested (Fisher, 1973) that only after all of the iron has been oxidized to the ferric condition does manganese take on a valence above two. Although minerals for which an essential cation is manganese are excluded from Table 3.1.1 — because manganese is not a prevalent component of the environmental paragenesis being considered — several of the minerals listed may have small amounts of divalent manganese substituting for iron. However, Mn(II) is not included in the formulas for such minerals as rockbridgeite and messelite, and such minerals as switzerite, faheyite, bermanite and frondelite are not included because they contain prominent amounts of manganese — in some cases Mn(III).

Van Wambeke (1971) has considered the oxidation and leaching among phosphate minerals with the type formula $A_xB_y(PO_4)_z(OH)_n$, where A is Li, Na, K, Ca, Ba, Sr, Pb, Fe^{2+}, Mn^{2+}, Cu, Zn, Bi, Re, Th or U; and B is Fe^{3+} or Al. He states the "deficiencies in A positions may reach 70% or more and 40 to 50% of PO_4 groups may be substituted by H_4O_4 with preservation of the original structure." Other changes observed were: (1) an abnormally high content of stable B cations, (2) only minor changes in the physical properties but an increase in unit-cell volumes, (3) a decrease in density, and (4) alteration of refractive indices, with a decrease when B is Al but an increase when B is Fe^{3+}.

Although this study (Van Wambeke, 1971) was concerned primarily with secondary processes more vigorous than weathering — starting with minerals which have compositions different from those likely to be formed by biogeochemical processes — nevertheless, it indicates the trend toward a greater degree of hydration and leaching [removal] of certain soluble cations (NH_4^+, Na^+, K^+ and Ca^{2+}), which will be discussed later, as a result of weathering processes.

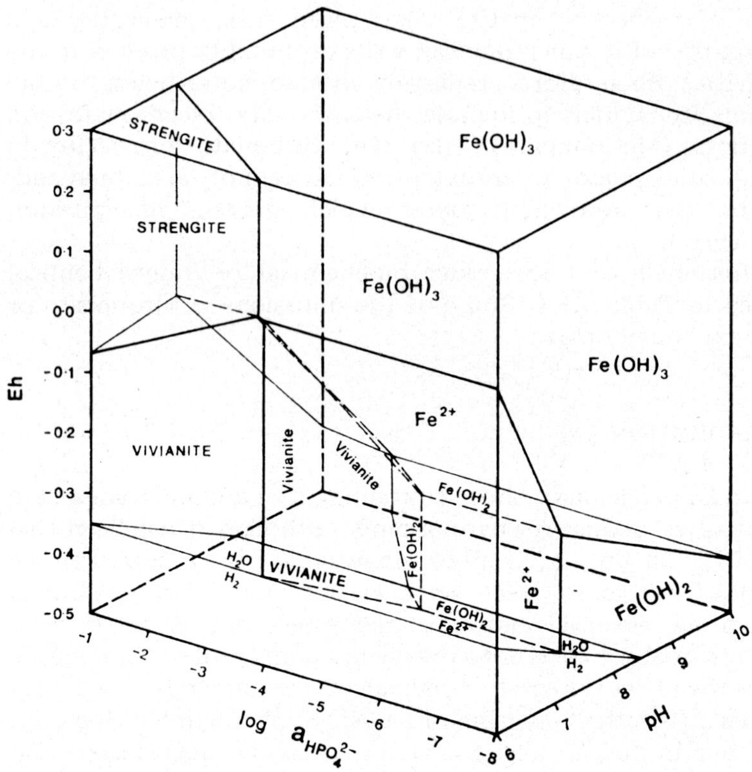

Fig. 3.1.2. Stability relations in the system $Fe(OH)_2$-H_3PO_4-H_2O. The chemical restraints are: $a_{HCO_3^-} \leqslant 10^{-3.5}$ and $a_{Fe^{2+}} = 10^{-4}$, $a_{HS^-} \sim 0$. Reproduced with permission from Nriagu (1972).

The relations involving pH, the oxidation potential [Eh] and the activity of HPO_4^{2-} have been indicated for $Fe(OH)_2$, $Fe(OH)_3$, vivianite and strengite by Nriagu (1972), who proposed the following transformations: vivianite → so-called kertschenite → amorphous ferric phosphate → strengite. His diagram is shown as Fig. 3.1.2, where phosphosiderite, of course, presumably occupies the same volume of the diagram as strengite, although the energy content must be slightly different in order for the dimorph to form. He states, interestingly, that although the diagram represents equilibrium conditions, "the reaction paths are not kinetically prohibited."

Kinetic considerations were suspected to be overriding factors in the case of crystallization of hydrous aluminum phosphates (McConnell, 1976b), and an attempt was made to relate the crystalline vs. the amorphous condition to the atomic products $(Al \cdot P^2)/H^2$ per 8 oxygens contained in the unit cell. The boundary between vashegyite, wavellite, kingite, etc. [crystalline] and bolivarite and evansite [amorphous] was approximately $0.01 = Al \cdot P^2/H^2$.

Stated in simple language: the tendency toward crystallinity is directly related to a product of functions of the aluminium and phosphorus atoms, but inversely related to a function of the amount of hydroxyls (water) associated with a structural unit consisting of 8 oxygens. The volumes of such units of 8 oxygens must be surmised for bolivarite and evansite, of course (McConnell, 1976b).

These studies (Van Wambeke, 1971; Nriagu, 1972; McConnell, 1976b) necessarily disregard the catalytic effects of organisms, which would be very difficult to evaluate in terms of existing knowledge concerning the conditions of formation of the minerals involved. Indeed, even in the case of precipitation of dahllite in vitro, the presence of a common metabolic catalyst [carbonic anhydrase] does not seem to be essential, although it greatly increases the rapidity with which this precipitation takes place, as well as reducing the required CO_3^{2-} activity (McConnell et al., 1961). However, the biochemical influence is indicated by Patrick et al. (1973), who conclude p. 565: "Microbial enzyme systems that function in a flooded [acid] soil are apparently effective in lowering the activation energy for strengite reduction with the result that there is a shift in the critical redox potential to a higher value." Again, the dimorphic form (phosphosiderite) is necessarily omitted from purely chemical considerations.

In Table 3.1.2 some of the relations are shown among iron-bearing

TABLE 3.1.2

State of oxidation of iron in several minerals

Fe(II)	Fe(II) and (III)	Fe(III)
Ludlamite	Barbosalite	Azovskite
Vivianite	Beraunite	Cacoxenite
	Dufrenite	Delvauxite
	Rockbridgeite	Koninckite
		Phosphosiderite
		Strengite
		Tinticite
	Including Al	
Metavauxite		Barrandite
Paravauxite	Sigloite	Clinobarrandite
Vauxite		Redondite
	Including Ca, Mg, K, Na	
Anapaite		Calcioferrite
Collinsite	Richellite	Cyrilovite
Messelite		Efremovite
		Leucophosphite
		Mitridatite

minerals, but this must not be taken to indicate any genetic relationship. The differences within any one group (contained within a column) are primarily in the degree of hydration — from the chemical viewpoint, at least. From the viewpoint of structural similarities, the transition vivianite → strengite, for example, seems no more probable than vivianite → phosphosiderite. One notices the sparsity of minerals containing both ferrous and ferric ions as well as other cations (below the dashed lines of Table 3.1.2), but this may depend on inadequate information.

With the exception of sigloite, which contains merely 2.76% of FeO, the other minerals containing Fe(II and III) are very dark in color, whereas those containing Fe(II) tend toward pale greenish; collinsite is light brown, however. Minerals containing Fe(III) tend to be yellow or reddish and light colored under the microscope. The strengite-phosphosiderite group may be essentially colorless. Mitridatite is either dark or light, probably depending upon the Fe(II) content as well as the Mn(II or III) content (Van Wambeke, 1971). Azovskite is dark brown probably for the same reasons.

Vivianite ($Fe_3(PO_4)_2 \cdot 8H_2O$) is very sensitive to oxidizing conditions; it is virtually colorless when all of the iron is in the reduced state, but becomes pale greenish-blue fairly quickly on exposure to air. Further exposure may produce dark blue or even bluish black. Rosenquist (1970) claims to have determined the solubility product of this mineral, but his analogy with "hydroxyl apatite" is inappropriate inasmuch as hydroxyapatite is not known to form in natural environments (McConnell, 1970a). The relations among this mineral, tinticite and strengite are discussed in some detail by Nriagu and Dell (1974).

Most weathering processes, to be sure, take place in an environment where aerobic bacteria are active, unless there are considerable quantities of nitrogenous organic matter present. Anaerobic conditions, which usually also imply pH values close to (or less than) 7, are encountered in certain marine muds, guano deposits, and elsewhere. The anaerobe, *Clostridium acidiurici* (Liebert), which is known to occur in soils, produces ammonia, carbon dioxide and acetic acid from uric acid, guanine and xanthine (Baker and Beck, 1942), while *Streptococcus allantoicus* forms — in addition to ammonia and carbon dioxide — urea, oxamic acid, etc. Oxalic acid is a common product in the decomposition of guano.

Although *Escherichia coli* and *Clostridium butyricum* are capable of reducing PO_4^{3-} to PO_3^{3-} and even PO_2^{3-} (Tsubota, 1959), the opposite reaction has been demonstrated for $PO_2^{3-} \rightarrow PO_4^{3-}$ in connection with *Bacillus caldolyticus* (Heinen and Lauwers, 1974). This suggests that P — to a limited extent, at least — is capable of cycling of the sort common to S and N.

Apatite apparently forms in vitro at pH values as low as 6.8 (possibly as low as 6.4) (Simpson, 1968) and seems to be free from dependence on the oxidation-reduction potential. At Christmas Island (Indian Ocean) a carbonate fluorhydroxyapatite is thought to form through aerobic decay of guano (Truman, 1965) whereas at et-Tabun cave (Israel) an apatitic substance may have formed under anaerobic conditions (Goldberg and Nathan, 1975).

The above-mentioned organic acids may be expected to produce conditions in which silica of silicate minerals is dissolved, and — assuming the phosphate concentration is adequate or sufficient time has elapsed at lower concentrations — the silicate minerals become replaced by phosphate minerals. Although little is known concerning the conditions of oxidation vs. reduction encountered in guano deposits, such a mode of replacement must be attributed to the conversion of feldspar laths to variscite-metavariscite at Malpelo Island (McConnell, 1943). Indeed, the conditions may be aerobic at the surface, whereas anaerobic conditions obtain at the lower portions in contact with the country rock.

However, the reverse situation — that is, replacement of apatitic substance by quartz — has been described (Horowitz, 1967) as a diagenetic process in sediments, so it becomes difficult to understand the conditions of Eh and pH which govern reactions of this sort without a knowledge of the entire history of events although Cook (1970, p. 2115) has concluded that "phosphatization and calcitization of silica, and the reverse reaction, may be explained by pH fluctuation, probably within the range of pH 7 to 10". Furthermore, laboratory study of the system CaO-P_2O_5-H_2O (Skinner, 1973) indicates that equilibrium obtains slowly even at temperatures about 300°C. While most minerals can be assumed to have formed under equilibrium conditions — in environments that existed at the time — it must be remembered that amorphous mineral substances [such as, bolivarite and evansite] have not attained a status of minimum energy during thousands to millions of years.

Thus we find the Eh, pH and organic catalysts are unknown variables in a system which may be well defined with respect to concentrations of components, temperature and pressure. The disentanglement of such knotty problems remains to challenge scientists of future generations. The whole story surely is not revealed by Krumbein and Garrels (1952).

MINERALS OF SOIL AND MANTLE ROCK

Although many soil scientists had considered the possible mechanisms which soils employ for the retention [fixation] of phosphorus, it remained for Haseman et al. (1950) to demonstrate that phosphorus could — and in experimental situations did — replace the silicon of micas and clay minerals in order to form crystalline hydrous aluminium phosphates of sodium, ammonium and potassium. Prior to experimentation by this group, associated with the laboratories of the Tennessee Valley Authority (TVA), most authors attributed the retention of phosphorus by soils to combination with calcium to produce fairly insoluble minerals; to adsorptive, exchangeable combination with silicate minerals; and to formation of phosphates of iron and/or aluminium with colloidal characteristics.

It was demonstrated in this work (Haseman et al., 1950) that a synthetic product essentially identical with taranakite (called palmerite by these authors) could be produced by treating illite or kaolinite with a solution of acidic (pH 3) potassium phosphate at 95°C during 100 h. Solutions of magnesium phosphates, on the other hand, gave products which yielded X-ray diffraction patterns "almost identical with that of barrandite," using illite, kaolinite and goethite as starting materials.

By 1967, investigators with the TVA * laboratories (Lehr et al., 1967) had produced "Crystallographic Properties of Fertilizer Compounds," which lists 17 of the 75 phosphate minerals shown in Table 3.1.1. To this list should be added octacalcium phosphate ($Ca_4H(PO_4)_3 \cdot 2.5H_2O$), which has not been discovered in a typical mineral occurrence — probably because of its tendency to hydrolyze to form apatite (Haseman et al., 1950). Their list included, in addition: hopeite, berlinite, chlorspodiosite and hilgenstockite. Hopeite, being a phosphate of zinc, was purposely omitted from his list of biominerals by McConnell (1973b) because of its composition. Berlinite, chlorspodiosite and hilgenstockite are not known to form at temperatures, pressures or concentrations of phosphatic solutions ordinarily encountered among biominerals. Although the compounds $NH_4H_2PO_4$ and $(NH_4)_2HPO_4$ are listed by Lehr et al. (1967), the mineral names (biphosphammite and phosphammite) were not assigned, as is true also for $MgHPO_4 \cdot 3H_2O$ (newberyite).

This compendium by the TVA group lists about 75 'orthophosphate' compounds which are not known as minerals — some of which contain barium, strontium and zinc. For each compound (or mineral), the following measurements are given: optical properties, density, X-ray powder diffraction pattern, and infrared absorption spectra. Probable space groups are given for more than 50 'orthophosphate' compounds and minerals.

'Pyrophosphates' are not considered in the present work because they do not seem to form naturally in soil; however, phyllophosphates (sheet structures) and tectophosphates (framework structures) have been described for viséite (McConnell, 1952) and kehoeite (McConnell and Foreman, 1974) of the latter group, and for kingite and vashegyite (McConnell, 1974) and taranakite (McConnell, 1976a) of the former group.

Omitted from both lists (Lehr et al., 1967; McConnell, 1973b) was the newly discovered hydrous aluminium phosphate, senegalite, as well as the recently recognized orthorhombic trimorph of $(Al,Fe)PO_4 \cdot 2H_2O$, redondite. Senegalite, according to Johan (1976), occurs in the oxidized zone of a magnetite deposit at Kouroudiako, Senegal; it is orthorhombic and has the composition $AlPO_4(OH)_3 \cdot H_2O$. Recently a new mineral, francoanellite, has

* TVA = Tennesee Valley Authority.

been described from the caves of Castellana, Puglia, Italy (Balenzano et al., 1976). In composition, this mineral closely resembles taranakite, with which it is associated, except that it contains less water. Deficiencies in Table 3.1.1 will undoubtedly arise with the discovery of new minerals and assignments of structures.

While the above six paragraphs deal primarily with inorganic interactions — that is, with the path phosphorite → superphosphates → soil and mantle rock — it must be remembered that the inorganic chemistry is much less complex than the organic chemistry, and only about half of the organic phosphate has been isolated and identified (Anderson, 1975). "Bacteria, molds and fungi, zooplankton, insects, higher plants, and animals continually excrete measurable amounts of organophosphate," according to Scharpf (1973, p. 395).

Salts of phytic acid (*myo*-inositol hexaphosphate) are common with respect to their occurrence in plant materials, but there is no evidence to show that iP_6 is either synthesized by microorganisms or reaches the soil through decomposition of plant material and animal excreta (Anderson, 1975). On the other hand, esters (containing glycerol, *myo*-inositol and *chiro*-inositol) of monophosphorylated carboxylic acids have been isolated (Anderson and Malcolm, 1974), in addition to the esters, iP_6 and iP_5, which frequently comprise more than half of the organic phosphate in soil. Compounds involving nucleic acids and phospholipids also have been identified in soils, and while these substances are undoubtedly of considerable importance to the propagation of plants, their importance within the reaction

$$\text{organic phosphates} \rightleftharpoons \text{phosphate ions} \rightleftharpoons \text{phosphatic minerals}$$

is yet to be evaluated except for very simple systems, particularly those which do not contain other mineral matter (silicates, carbonates, etc). It has been suggested that sugar phosphates and phosphoprotein may be present in small amounts (Anderson, 1975).

The production of a mineral substance as a direct consequence of the metabolic processes of a bacterium will be discussed later. However, the system involved here is a very simple one compared with one likely to be encountered in a soil, primarily in the absence of iron and aluminium.

MINERALS OF ROCK PHOSPHATES OF MAGNESIUM, ALUMINIUM AND IRON

Insular phosphates of two different types exist, depending on whether guano interacts with igneous rocks of intermediate or basic types, or whether such action is confined to calcareous accumulation — such as coral. The first type, which will now be discussed, has more diversified mineralogical compositions. Rock phosphates of magnesium, aluminium and iron comprise less extensive deposits than phosphorites and, consequently, are of less economic importance.

These deposits may be complex with respect to their mineralogy, as at Bomi Hill and Bambuta, Liberia, West Africa (Axelrod et al., 1952). On the other hand, their mineralogy may be comparatively simple, as at Malpelo Island, Colombia (McConnell, 1943). In Liberia, interaction with excreta of bats interacted with iron ore to produce strengite and phosphosiderite (both $FePO_4 \cdot 2H_2O$), leucophosphite and a mineral of the rockbridgeite group, while at Malpelo Island the interaction between the excreta of sea birds (chiefly the marked booby, *Sula dactylotra*) and an augite-andesite produced merely the dimorphic pair (strengite-phosphosiderite), and for reaction with an amygdaloidal scoria, the feldspar laths were either clear quartz or variscite with metavariscite (McConnell, 1943: see Figs. 3.1.3 and 3.1.4).

Redondite, according to Kato (1965), appears to be a trimorph of the variscite-metavariscite series to which the name "Messbach-type variscite" has been applied (Čech and Slánský, 1965). It was described in 1869 by Shepard (1869), and had not been re-examined in terms of topotype material until 1958 (McKie, 1958), at which time differences in the X-ray diffraction pattern, as well as differences in optical properties from those of

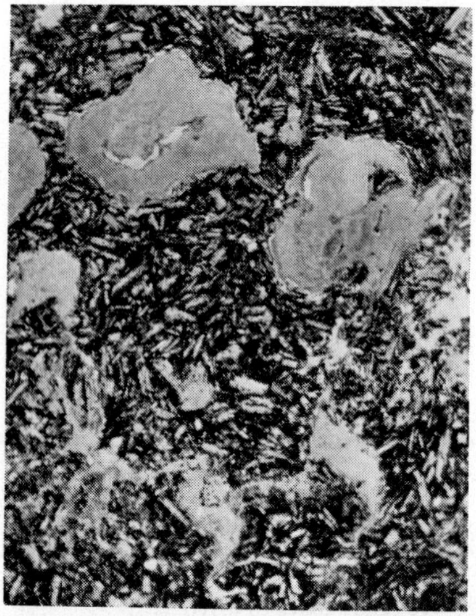

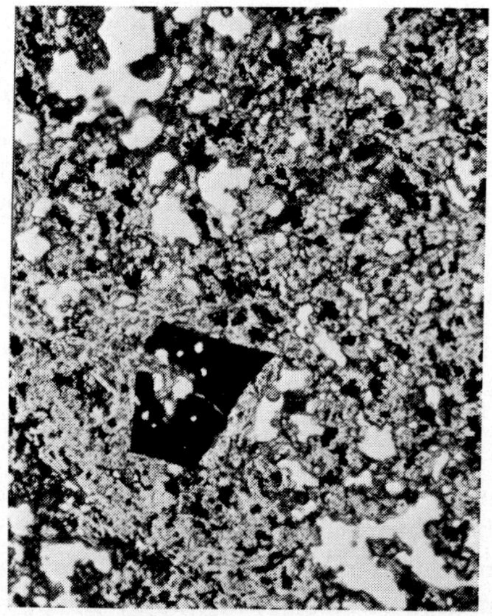

Fig. 3.1.3. Amygdaloid from Malpelo Island, D.F., Colombia. Some of the amygdules and some of the lathlike crystals of the mesostasis are phosphates of aluminium, whereas other amygdules are opal (or mixtures) and some of the lathlike crystals (originally feldspars) are now quartz. Magnification 33×. Reproduced with permission (McConnell, 1943).

Fig. 3.1.4. Rock phosphate from Malpelo Island, D.F., Colombia. The porous rock is essentially a mixture of strengite and phosphosiderite, but contains relict oxides (titaniferous magnetite). Magnification 33×. Reproduced with permission (McConnell, 1943).

variscite, caused the introduction of the name tangaite, which has not been accepted. Later work has indicated an orthorhombic structure and an approximate doubling of the *c* direction of variscite (Kato, 1965; Salvador and Fayos, 1972). Redondite occurs at Redonda Island, West Indies, in association with apatite.

A highly hydrated ferriphosphate of aluminium — recently described by Martini (1978) — believed to have formed through the action of guano on clay minerals in a dolomitic cave near Carlstonville, western Transvaal. Sasaite seems to be related structurally to vashegyite and kingite (McConnell, 1974) according to Martini (1978).

Other products of interaction of various rocks with bat excreta are taranakite (Sakae and Sudo, 1975), dittmarite (Mrose, 1971), mirabilite (Hutchinson, 1950), biphosphammite (Hutchinson, 1950; Pryce, 1972), phosphammite, struvite, newberyite, bobierrite, schertelite, hannayite, stercorite, monetite, whitlockite and brushite. It is noticeable (Table 3.1.1) that most of these mineral include the ammonium ion, which results from decomposition of urea and/or uric acid, and later the more stable minerals (Bobierrite, for example) persist after leaching of alkali ions (including NH_4^+) has been accomplished.

From a cave inhabited by humans at Mount Carmel, Israel, montgomeryite and crandallite have been reported (Goldberg and Nathan, 1975), in addition to dahllite and what is reported to be hydroxyapatite — said to be "almost devoid of CO_2" although no analysis is given. This represents a mixed type of deposit in which calcium is derived from limestone and aluminium from clay.

Although not products of interaction of guano, in this instance, millisite and crandallite are major constituents of the aluminium phosphate zone of the Bone Valley Formation of west-central Florida, where they are accompanied by wavellite (Owens et al., 1960). The phosphates comprise the cementing material of a silty sandstone containing kaolinite, in addition to quartz (silt and sand). According to Fisher (1966), other phosphate minerals likely to occur in sediments richer in iron are: beraunite, cacoxenite, laubmannite, phosphosiderite, rockbridgeite, strengite, turquoise, and possibly diadochite. (Turquoise is omitted from Table 3.1.1 because of its copper content, as are minerals which contain significant amounts of manganese.)

Similar aluminium-iron phosphates (barrandite [plus clinobarrandite?], crandallite, millisite, montgomeryite and wavellite) occur at Christmas Island (Indian Ocean), except that a carbonate fluorhydroxyapatite is dominant in deposits overlying carbonate rocks (Trueman, 1965). Barrandite masses preserve the texture of volcanic rocks (limbergites). Secondary veins of crandallite within the barrandite indicate the sequence of formation, and small vugs in the veins of crandallite contain apatite. Crandallite replaces carbonate apatite, however, in a sandstone of Florida (Blanchard, 1972). The

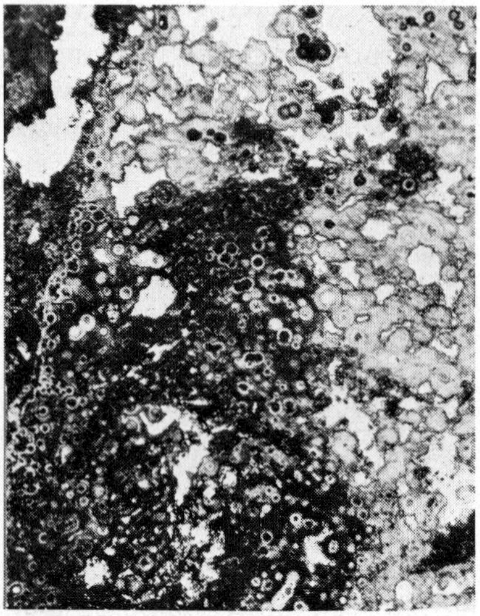

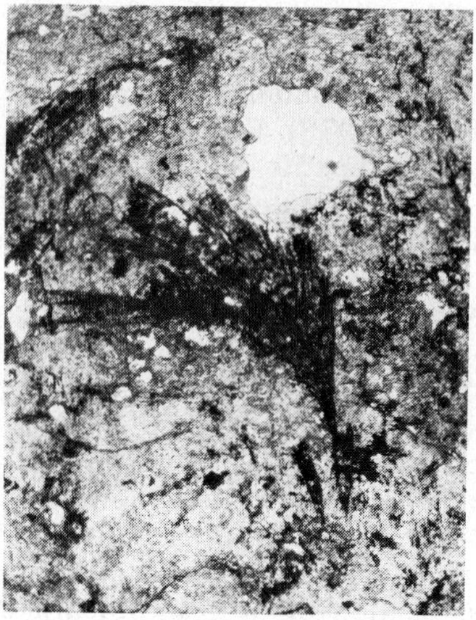

Fig. 3.1.5. Rock phosphate from Gran Roque, D.F., Venezuela. The clear spherulites are barrandite, whereas cloudy portions are microcrystalline mixtures of barrandite and a variety of apatite. Magnification 22×. Reproduced with permission (McConnell, 1950).

Fig. 3.1.6. Rock phosphate from Connétable Island, French Guiana. The aluminium-phosphate rock contains grains of residual quartz as well as relict organic structures. Magnification 22×. Reproduced with permission (McConnell, 1950).

association of a calcium-phosphate mineral with aluminium-iron phosphates is not unusual (McKie, 1958), although it is difficult to ascertain which of these minerals formed later through textural relationships at Gran Roque, D.F., Venezuela (Fig. 3.1.5). The dark centres of some of the otherwise clear spherulites of barrandite suggest that this mineral is replacing the apatitic mineral. The association of organic matter — presumably woody substance — is well illustrated in Fig. 3.1.6, where the essential mineral is barrandite (McConnell, 1941). The original rock type was not identifiable from specimens available, but was assumed to be an argillaceous sediment; it contained relict grains of quartz, as well as secondary dahllite in minor quantities.

Many of the minerals listed in Table 3.1.1, in addition to those mentioned above, are likely possibilities as a result of weathering or low-temperature metasomatism (McConnell, 1950). The principal difference between these secondary minerals and those produced from direct action of guano on various rocks is that the latter minerals tend to contain more alkali ions (including ammonium) and more water. Senegalite (Johan, 1976) has recently been described as occurring in the oxidation zone of a magnetite deposit — in association with turquoise, augelite, wavellite and crandallite — and this suggests that augelite ($Al_2PO_4(OH)_3$) might be added to Table 3.1.1, although other occurrences of this mineral supposedly indicate higher temperatures for its formation. Indeed, the phosphates which occur in pegmatites (Fisher, 1958) and those which occur in rock phosphates of magnesium, aluminium and iron (McConnell, 1950) are assumed to show some overlap because some of these minerals may form during a significant range of temperature. Vivianite, a particularly unstable mineral, occurs in clays where reducing conditions obtain, and has recently been reported from freshwater sediments of the Great Lakes (Superior, Erie and Ontario) by Nriagu and Dell (1974).

Hutchinson's monograph (1950) on "The Biogeochemistry of Vertebrate Excretion" is outstanding on all aspects (including production) of phosphatic deposits derived through interaction of various rocks and excreta of birds and bats. Unfortunately, with the exception of the cave deposits near Skipton, Victoria, Australia, Hutchinson's work is not very specific about the phosphate minerals containing magnesium, aluminium and iron. He gives some analyses (Table 3.1.3) of fresh excrement of birds, mammals and bats, where the greatest variation occurs for total nitrogen. The nitrogen, being contributed significantly by urea and/or uric acid, is low for the mammals because of loss at sea of the contribution by urine.

Although the exact biogeochemistry is not well understood, it is assumed that silica is preferentially leached from the feldspars, pyroxenes, amphiboles, etc. through partial (or complete) dissolution, and replacement of SiO_4 groups by PO_4 groups takes place. Some of the silica thus dissolved may be

TABLE 3.1.3

Analyses (%) of excrement of bats, birds and mammals (from Hutchinson, 1950)

	I	II	III	IV	V	VI
H_2O	9.40	43.96	29.40	22.28 }	83.65	82.63
Organic matter	81.75	18.94	17.74	56.03 }		
Total N	21.66	2.33	2.86	17.41	10.25	11.73
P_2O_5	4.30	16.34	16.80	7.14	6.95	7.42
Alkalies, etc.	3.70	20.36	27.84	—	—	—
Insol. (sand)	0.85	0.40	8.22	1.47	0.16	1.39
Na_2O				3.51	—	—
K_2O				2.51	3.85	1.57
CaO				3.67	2.36	4.56
MgO				0.50	1.40	1.03
Al_2O_3				—	0.00	0.49
Fe_2O_3				—	0.38	0.78
SO_3				0.30	3.00	3.80

I = Excrement of *Pelecanus occidentalis thagus*, dried.
II = Faeces of *Otaria byronica* (sea lion).
III = Guano from faeces of seal, Peru.
IV = Guano from birds, Peru.
V = Fresh guano from bats, Lares, Puerto Rico.
VI = Fresh guano from bats, San German, Puerto Rico.

reprecipitated as opal (forming amygdules) or quartz replacing the feldspar laths, as is the situation at Malpelo Island (Figs. 3.1.3 and 4) with respect to an amygdaloid, whereas other laths are converted to variscite and metavariscite (McConnell, 1943). A most interesting speculation arises in terms of taranakite, which can be readily synthesized from illite (Haseman et al., 1950). It has been assumed tentatively that the structural configuration of $n[X_2O_5]$ sheets has been preserved in taranakite in order to form a hydrous aluminium phyllophosphate (McConnell, 1976a). This hypothesis was predicated upon the supposed phyllophosphate structures of vashegyite and kingite (McConnell, 1974); possibly sasaite (Martini, 1978) is similar.

MINERALS OF PHOSPHORITES

Minerals of phosphorites are quite simple when compared with those of rock phosphates of magnesium, aluminium and iron. They consist essentially of apatites, whitlockite, monetite and brushite. However, francolite or dahllite (carbonate fluorapatite and carbonate hydroxyapatite) probably comprise more than 99% of the phosphatic constituent of the average phosphorite.

These two minerals (francolite and dahllite) are very deceptive in their crystallochemical properties, and this has led to the introduction of numerous synonyms, including: collophane, monite, nauruite, ornithite, and sombreite, with respect to insular phosphorites (Frondel, 1943). Formed at slightly higher temperatures were: grodnolite, kurskite, podolite, quercyite and staffelite (McConnell, 1938); all of these are synonyms for carbonate apatites. The names dehrnite and lewistonite are preserved for carbonate apatites containing sodium and potassium. The structure of apatite appears to have very little tolerance for magnesium; although other substitutions for calcium (Sr, Ba, Pb, etc.) are known to occur in natural apatites, they are of little practical importance with respect to phosphorites. Sr may be of geochemical interest with respect to the genesis of phosphorites; there is discrimination against Sr incorporation (compared with Ca) both for in vivo and in vitro systems (Nordin et al., 1962).

The low-temperature metasomatism of these minerals—probably with catalytic assistance of organisms or their metabolic products — is implied by the colloform textures displayed by some of them. The agate-like banding is well displayed for dahllite (Fig. 3.1.7), and the phosphatic cement of a glauconitic sandstone (Fig. 3.1.8) shows no evidence of elevated temperature. Fig. 3.1.9 shows hexagonal outlines of an apatitic mineral surrounded by later quartz from a spherulitic nodule from a Cenomanian phosphatic sandstone. Similar, although smaller, spherulites with a radial-plumose structure are found in the Thermopolis Formation (Upper Cretaceous) in the vicinity of Cody, Wyoming (McConnell, 1935; Mitchell and Sherwood,

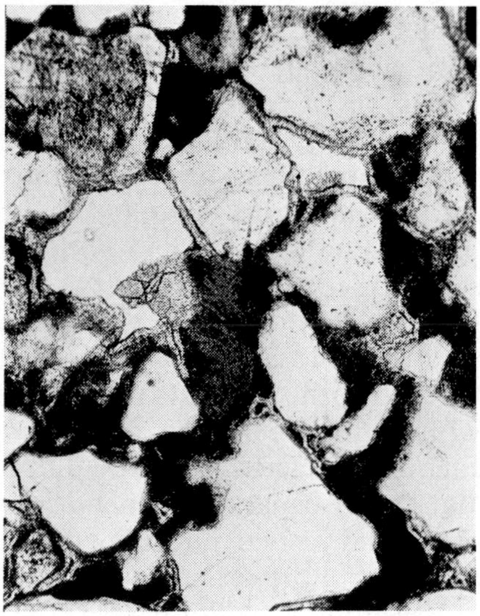

Fig. 3.1.7. Agate-like banding in dahllite (so-called quercyite) from Castillo de Belmez, Spain. Individual crystals have their *c* axes approximately normal to the bands, portions of which are cloudy because of inclusions (presumably clay). Magnification 37×. Reproduced with permission (McConnell, 1950).

Fig. 3.1.8. Sandy phosphorite from near Kursk, R.S.F.S.R. (so-called kurskite). The calcium-phosphate mineral forms both an anisotropic and apparently isotropic cement for the glauconitic sandstone. Magnification 67×. Reproduced with permission (McConnell, 1950).

Fig. 3.1.9. Phosphatic concretion from the valley of the Dniester River, Uk.R.S.S.R. showing euhedral crystals of a variety of apatite — presumably francolite — within a matrix of quartz. The concretions also contain glauconite, pyrite, etc. Magnification 67×. Reproduced with permission (McConnell, 1950).

1958). These examples are discussed here in order to indicate the lack of deep burial of sediments in which these carbonate apatites occur. Further discussion of concretions and nodules will appear later.

As already implied, phosphorites are of two general types: insular and continental. The first type, although locally significant, does not greatly add to the world's supply of phosphate rock. They are formed primarily by the interaction of guano with limestone (coral), and the stable mineral that results is dahllite-francolite. Kaneshima (1962) found for deposits at the Ryukyu Islands that there was progressive leaching of very small amounts of zinc by rain water and sea water; the principal mineral is dahllite which obtained both uranium and fluorine from interaction with sea water. Whitlockite seems to form in preference to apatite only when there is adequate magnesium to stabilize its structure, in excess of 2% by weight of MgO being usual (Jensen and Rowles, 1957). It occurs as a minor component of several insular phosphorites, and has the following synonyms: martinite, pyrophosphorite, and zeugite.

Brushite ($CaHPO_4 \cdot 2H_2O$) loses water readily at temperatures below 100°C, and thus is converted to monetite ($CaHPO_4$). Pseudomorphs of carbonate apatite after brushite have been reported. The name, metabrushite, has been applied to partially dehydrated material. While both minerals occur sparingly in insular phosphorites, with increasing proportions of calcium, apatite becomes the more stable (insoluble) mineral. Brushite is isostructural with gypsum ($CaSO_4 \cdot 2H_2O$), so it is not surprising that ardealite ($Ca_2HPO_4SO_4 \cdot 4H_2O$) should exist, particularly in view of their frequent association.

Cave deposits differ from insular deposits because organic matter accumulates and is subject to bacterial action — both aerobic and anaerobic. Comparatively little is known about the guano minerals that form through interaction with limestone or dolomite, but Bridge (1973) has examined such crustal deposits by microscopic and X-ray diffraction methods of material occurring in a bat cave in Western Australia. In addition to 7 phosphate minerals (Table 3.1.4), he found in the guano of Murra-el-elevyn cave four sulfates, an organic substance (guanine) and calcite. It is surprising that oxalic acid and/or oxalates were not found. Six chemical analyses are given that show total N ranging from 0.21 to 2.20% and P_2O_5 from 13.0 to 28.6%, and the minerals are indicated individually for the 6 samples.

Still another group of phosphatic deposits comprise nodular bodies which occur near (or on) the margins of the continental shelves. These are composed of francolite and numerous detrital minerals, including glauconite: many contain Globigerinae. Of even greater economic insignificance are lacustrine deposits, such as those of Eocene age in Wyoming (Love, 1964), and sodium phosphates found in saline lakes with high alkalinity (Fahey, 1962).

Vast continental phosphorites exist in parts of the Florida Peninsula,

TABLE 3.1.4

Minerals identified in guano in contact with limestone

Mineral	Composition	Occurrences (of 6)
Aphthitalite	$(K, Na)_3Na(SO_4)_2$	3
Biphosphammite [a]	$NH_4H_2PO_4$	2
"Biphosphammite (K)" [a]	$(K, NH_4)H_2PO_4$	3
Brushite [a]	$CaHPO_4 \cdot 2\ H_2O$	2(3?)
Calcite	$CaCO_3$	3
Dahllite [a]	carbonate hydroxyapatite	1
Guanine	$C_5H_5N_5O$	2
Gypsum	$CaSO_4 \cdot 2\ H_2O$	5
Hannayite [a]	hydrous NH_4, Mg phosphate	1
Mirabilite	$Na_2SO_4 \cdot 10\ H_2O$	1
Monetite [a]	$CaHPO_4$	5
Syngenite	$K_2Ca(SO_4)_2 \cdot H_2O$	6
Taylorite	$(K, NH_4)_2SO_4$	2
Whitlockite [a]	$(Ca, Mg)_3(PO_4)_2$	4
Unidentified	—	3

[a] Phosphate mineral. "Biphosphammite (K)" was designated as "ammonian KH_2PO_4" (Bridge, 1973), and was later described as archerite (Bridge, 1977).

Idaho (including parts of Montana, Wyoming and Utah), and Tennessee, with respect to the United States. Other comparable deposits are found in the U.S.S.R., Africa, Europe and Asia. South American deposits worthy of note occur in Brazil, Peru and Venezuela.

It has been supposed (Kazakov, 1937) that such deposits could precipitate directly from sea water as the result of inorganic processes, and many geologists, even today, fail to recognize the influence of organisms pointed out by McConnell (1965), Bushinskii (1967) and others. On the other hand, a profuse development of algal stromatolites (Banerjee, 1971) has been found in connection with Precambrian phosphorites on the Indian shield, and it has been concluded (Patwardhan and Ahluwalia, 1973) in conjunction with the Mussoorie phosphorites (Lower Himalaya, India) "that organisms could not be absent during deposition of the rocks of this region" (p. 385), although "the present phosphorite did not form entirely through the decay of accumulated hard parts of animal remains, containing calcium phosphate, in the basin of deposition" (p. 384). Indeed, the dark material (carbonaceous matter) of many phosphorites is believed to have its origin in organisms, and this suggests that many of these deposits formed under reducing conditions, such as those suggested for the shallow-water deposits (Miocene) of Beaufort County, North Carolina (Rooney and Kerr, 1967).

Further evidence that Kazakov's (1937) inorganic hypothesis is quite inadequate to account for the formation of modern phosphorite deposits off the coast of Peru is supplied by Manheim et al. (1975, p. 243), who state: "Four simultaneous requirements for formation of phosphorites are: (a) sediments rich in organic detritus blanketed by (b) water with low concentrations of dissolved oxygen, (c) low rates of inorganic (especially terrigenous) sedimentation and (d) low but not negligible concentrations of calcium carbonate in the sediment." Nevertheless, it should be pointed out that in experiments conducted by McConnell et al. (1961) a calcium carbonate precursor was not necessary. Precipitation of dahllite took place in air (at room temperature) provided an organic catalyst was present. The conditions for precipitation of francolite at oceanic temperatures, admittedly, might be quite different, however, and certain inhibitors must be kept in mind.

In vitro systems have been studied extensively and the common enzyme, carbonic anhydrase, has been found to function as a catalyst in the precipitation of dahllite (McConnell et al., 1961), whereas other substances act as inhibitors to this precipitation. Included in the latter category are pyrophosphate and polyphosphate ions (Fleisch and Neuman, 1961), sulfanilamide (McConnell et al., 1961) and the Mg^{2+} ion, which tends to favor formation of whitlockite (Trautz et al., 1964). One of the most interesting experiments is the formation of dahllite as an intracellular product of bacteria, which will be discussed later.

Between 1935 and 1950, Cayeux made important contributions emphasizing the role of organisms in the formation of "Phosphates de chaux" — particularly one notes his 1936 paper on bacteria. At the International Geologic Congress, Algiers, 1952, it is noteworthy that Charles (1953) and Willcox (1953) relied heavily upon Cayeux's data and hypotheses, whereas a presentation by McKelvey, Swanson and Sheldon (1953) placed emphasis on the hypothesis of Kazakov (1937), without any mention of Cayeux's work.

Parker (1975) proposes a somewhat more complex origin for the glauconitic, conglomeratic phosphorites from the continental margin of South Africa. However, there seems to be ample evidence that organisms are required for the precipitation of phosphorites. It is interesting to note that Whitton (1967) has found that *Nostoc verrucosum* is a phosphate accumulator, which suggests that related marine blue-green algae might bear further investigation.

Replacement deposits, which Braithwaite (1968) believes to be diagenetic and to indicate higher sea level, occur on Remire, Amirantes (island in the Indian Ocean). Four modes of emplacement are indicated: (1) as derived phosphate pebbles of carbonate origin; (2) as primary phosphate sediment, lacking evidence of a carbonate origin; (3) as a primary phosphate cement, lining voids within a calcarenite; and (4) as a result of in-place phosphatization of a calcarenite. Whitlockite was identified — in addition to a carbonate apatite — in at least one specimen by X-ray diffraction methods.

Reiterating, the phosphatic mineral of such phosphorites is essentially francolite, a carbonate fluorapatite of somewhat variable composition (McConnell, 1971; Rooney and Kerr, 1967). Although not proven to be contained within the apatitic phase through isomorphic substitution, some of the continental phosphorites are of considerable interest because of accumulations of uranium, thorium, yttrium, rare earths, scandium, and vanadium therein. These rarer components are thought to be related to diagenetic processes, in which case they were extracted from sea water during the early formative histories of the phosphorites.

Table 3.1.5 shows the comparative abundance of trace elements in crustal rocks, sea water and phosphorites, where it is noticeable that, with few exceptions, enrichment of elements in sea water shows comparable enrichment in phosphorites, whereas those elements depleted in sea water are similarly depleted in phosphorites. According to Tooms et al. (1969) this indicates that the depositional environment (sea water) must play a considerable part in contributing to the minor element contents of phosphorites. However, the composition of the phosphorite which forms during diagenetic processes in sea water must depend upon energy relations that are most complex; otherwise one would be at a loss to explain the very small amount of chlorine which enters into the francolite as a substitution for fluorine, because the concentration is greater than 10,000 to one in favor of chloride ions in sea water.It should be evident that during diagenesis the composition of the interstitial water will depart from that of sea water, and this fact may play an important role in the biogeochemical reactions which take place.

The question of where the uranium occurs in apatite — provided it does occur in the apatitic phase — can hardly be resolved in terms of the quantities present. In some circumstances, part of the uranium seems to be associated with organic substances. Furthermore, both tetravalent and hexavalent U occur (Altschuler et al., 1958), so that it could be replacing either calcium or phosphorus or both, according to the ionic radii of Shannon and Prewitt (1969) as shown in Table 3.1.6. One concludes, therefore, that although U^{6+} is somewhat larger than aluminium, it might substitute for P in apatite, and that U^{4+} can almost certainly substitute for Ca.

Some of the ions shown in Table 3.1.6 have been proved to enter the structure of fluorapatite, either as synthetic products or as naturally occurring minerals. For example, S and Si replace P in about equal amounts in ellestadite (McConnell, 1938), and Al replaces both Ca and P in heated morinite, when converted to apatite (Fisher and McConnell, 1969). Presumably, these same situations can obtain for biologic apatites also; that is, small amounts (traces or more) of the constituents shown in Table 3.1.6 are permissible theoretically. Whether they actually do substitute depends, among other factors, upon their relative availabilities within the biologic environment. Thus, the formation of an aluminium-rich apatite within the organic milieu seems highly improbable; one would expect instead

TABLE 3.1.5

Abundance of trace elements in crustal rocks, sea water and phosphorites (III and IV) *

	I ppm	II ppb	III ppm	IV ppm
Ag	0.07	0.28	1—50	3
As	1.8	2.6	0.4—188	40 [a]
B	10	4450	3—33	
Ba	425	21	1—1000	100
Be	2.8	0.0006	1—10	
Cd	0.2	0.11	1—10	
Ce	60	0.0013	9—85	
Co	25	0.39	0.6—11.8	
Cr	100	0.2	7—1600	1000
Cu	55	23.	0.6—394	100
Hg	0.08	0.15	10—1000	
I	0.5	64	0.15—280	
La	30	0.0029	7—130	300
Li	20	170	1—10	
Mn	950	1.19	0—10,000	30
Mo	1.5	10	1—138	30
Ni	75	6.6	1.9—30	100
Pb	13	0.03	0—100	
Rb	90	120	0—100	
Sb	0.2	0.33	1—10	7 [a]
Sc	22	0.004	10—50	10
Se	0.05	0.09	1—9.8	10 [a]
Sn	2	0.81	10—15	
Sr	375	8100	1800—2000	1000
Th	7.2	0.0015	5—100	
Ti	4400	1	100—3000	
U	1.8	3.3	8—1300	90 [a]
V	135	1.9	20—500	300
Y	33	0.003	0—50	300
Zn	70	11	4—345	300
Zr	165	0.026	10—500	30

I = Abundance in crustal rocks, according to Mason (1966).
II = Composition of sea water with 3.5% salinity, according to Turekian (1968).
III = Ranges of worldwide phosphorite analyses given in Swaine (1962) and Tooms et al. (1969).
IV = Modal values (except [a] means average) for the Phosphoria Formation, according to Gulbrandsen (1966).

crandallite if both Ca and Al were available.

Kaneshima (1962) has indicated that the deposits on Ryukyu Island contain very little uranium when compared with continental deposits, and this suggests that accumulation of uranium is a secondary or diagenetic process,

* Alternative values for crustal rocks are given on p. 4.

TABLE 3.1.6

Radii of ions capable of substituting in apatite in small amounts (Shannon and Prewitt, 1969)

Ion	CN [a]	Radius (Å)	Ion	CN [a]	Radius (Å)
For Ca^{2+}	VIII	1.12	*For* P^{5+}	IV	0.17
Mg^{2+}	VIII	0.89	Cr^{5+}	IV	0.35
Sr^{2+}	VIII	1.25	V^{5+}	IV	0.355
Ba^{2+}	VIII	1.42	Mo^{6+}	IV	0.42
Mn^{2+}	VIII	0.93	Se^{6+}	IV	0.29
Na^{+}	VIII	1.16	Si^{4+}	IV	0.26
K^{+}	VIII	1.51	S^{6+}	IV	0.12
Ce^{3+}	VIII	1.14	Sb^{5+}	VI [b]	0.61
Al^{3+}	VI [b]	0.53	Al^{3+}	IV	0.39
U^{3+}	VI [b]	1.06	U^{6+}	IV	0.48
U^{4+}	VIII	1.00	U^{5+}	VI [b]	0.76

[a] CN = coordination number.
[b] Al^{3+} (VI) and U^{3+} (VI) become larger when corrected to CN = VIII, but the values are uncertain, whereas Sb^{5+} (VI) and U^{5+} (VI) become smaller with CN = IV; probably U^{5+} (IV) is only slightly larger than U^{6+} (IV) — perhaps 0.49 Å.

requiring contact of the phosphorite with sea water. (These recent deposits were also quite low in fluorine because of lack of contact with sea water.)

Cook (1972) has recognized two different types of phosphorite in northwest Queensland: pelletal and non-pelletal. The lanthanide distribution in the former is normal for marine phosphorites, whereas it is depleted in the latter type with the exception of the heavy lanthanides which are relatively more abundant. Pelletal types contain greater concentrations of elements which are known to substitute in the structure of apatite, whereas the other types contain components which probably are of detrital origin or are derived from weathering.

Arsenic, in amounts between 3—15 $\mu g\ g^{-1}$, occurs in land-pebble phosphates of Florida, but does not seem to be associated with the apatitic phase (Stow, 1969; McConnell, 1970a). A direct correlation of As/Fe, however, suggests that part (or all) the As might be present as a substitute for P in an iron phosphate: as much as 1.7% of iron occurs in some of the samples, as well as 0.28% of organic carbon.

Typical contaminants of Florida phosphorites are the phosphates: wavellite, crandallite, barbosalite, rockbridgeite, dufrenite and vivianite (Swanson and Legal, 1967). Other than phosphates are: turgite (hydrated iron oxide), clayballs, chert and sandstone pebbles, and phosphatic limestone pebbles.

NODULES AND CONCRETIONS

Nodular bodies (including coprolites) and concretions are well known; they are typically composed of collophane (microcrystalline francolite) but frequently contain detrital minerals, as well as other secondary minerals. In some cases such concretions contain fossils (Fig. 3.1.10) which may have served as nucleation agents during diagenetic accumulation of the phosphatic material.

Such a concretion (shown in thin section in Fig. 3.1.10 and bisected as in Fig. 3.1.11) is believed to have existed initially as an organic gel in a quiet muddy marine sediment. The entrapment of a crab in such a slimy mess is believed to be coincidental, because a fossil crab is not always present (Stenzel, 1934). The irregular fractures are believed to represent subsequent shrinkage [desiccation (?)] cracks which are now partially filled with pyrite, suggesting initial anaerobic petrification. More or less contemporaneously with fracturing, the phosphatization supposedly took place.

Nodular phosphorites are well known from continental shelves, mostly at depths of from 100 to 500 m, which are believed to be replacements of fossiliferous limestone and other sediments by francolite, essentially in situ

Fig. 3.1.10. Phosphatic concretion from Brazos County, Texas. The large dark object is an appendage of a fossil crab, the tip of which is truncated by a calcite vein. The matrix material is essentially collophane (isotropic francolite), but contains glauconite, quartz, limonite, pyrite, gypsum, etc. in addition to fossil fragments. Magnification 27×. Reproduced with permission (McConnell, 1950).

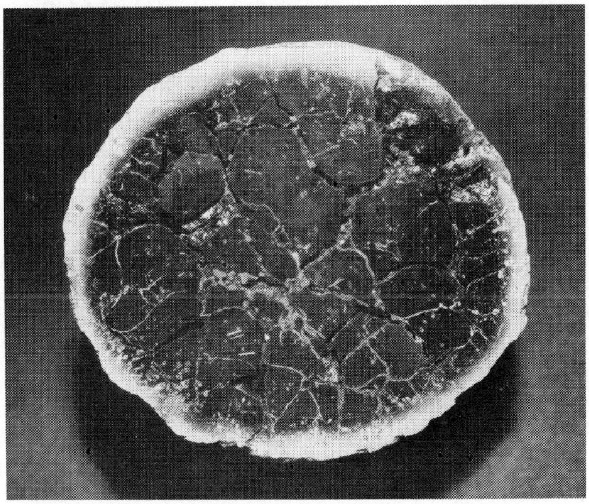

Fig. 3.1.11. Phosphatic concretion from the middle Eocene (Claiborne fauna), Brazos County, Texas (see also Fig. 3.1.10). Dense, almost opaque, collophane shows contraction fractures some of which now contain pyrite. The light-colored outer rim is the result of weathering. Reduced to about 1/2. Specimen courtesy of H.B. Stenzel.

(Parker, 1971, 1975). Similar replacement of biogenic carbonates is suggested by phosphatization of Recent foraminifera (D'Anglejan, 1967). According to Romankevich and Baturin (1972) "Lithification is accompanied by accumulation of phosphorus and loss of organic components, the retention of which increases in the sequence: carbohydrates, free lipids, nitrogen compounds, bound lipids. The CO_2 released by decomposition of the organic matter is in part absorbed by carbonate-fluorapatite". Nodular phosphorite off California and Mexico (Dietz and Emery, 1950) contains numerous foraminifera (Miocene to Recent) and is believed to represent accretion of additional phosphatic material on Miocene nodules, inasmuch as the present submarine nodules occur on an unconformity. The milieu in which these nodules are forming includes attached sponges, bryozoans and brachiopods, as is indicated by photographs taken at depths of 160—260 m. The spherulitic concretions that occur at the base of the Thermopolis Formation in Wyoming and Montana (McConnell, 1935) were mentioned previously. These spherulites, although smaller, are not dissimilar to worn concretions from the Kalyus sediments of Podoliya (Velikanov, 1975), except the Ukrainian ones (Fig. 3.1.9) have smoothed exterior surfaces which indicate mechanical weathering, whereas those from the Big Horn Basin have a rugose exterior. Those from the western U.S. average a 2.5 to 3.5 cm diam., whereas the others are 8 to 20 cm. Both are essentially carbonate apatites, but contain pyrite also, and it has been suggested that the dahllite of the Wyoming concretions is pseudomorphic after pyrite (Mitchell and Sherwood, 1958).

Concerning concretions found in the basal Colorado Shale of north-central Montana, Pecora et al. (1962, p. B33) made the following comments:

"The occurrence of spherulitic phosphate nodules poses a dual problem: (1) the origin of the original homogeneous discrete nodules, and (2) the development of the spherulitic structure. We believe that abundant information exists in the geologic literature to suggest that these nodules formed as concretions at numerous nucleation loci in unconsolidated mud on the sea floor by precipitation of microcrystalline carbonate-fluorapatite from sea water. We seek some hydrogeochemical process by which this phenomenon can form discrete concretions randomly distributed within 'blue mud' (= black shale) over hundreds of square miles, and conclude that geologic relations provide the best clues.

"Marine shale of Cretaceous age, stratigraphically higher than the spherulitic phosphate horizon in the Bearpaw Mountains, contains calcareous concretions with varying proportions of Ca, Mg, Fe, and Mn. Many of these concretions, septarian or homogeneous, contain unbroken delicate fossil shells, like *Baculites*, and were probably formed before compaction of the shale was completed (Clifton, 1957). An analogy in authigenic process for origin of the calcareous and phosphatic concretions at different horizons in the Colorado Shale is likely, although carbonate and phosphate concretions are not formed at the same horizon."

Remarkable preservation of coprolites in the Bridger Formation [Eocene] of southwestern Wyoming has been described (Bradley, 1946), including the contained microorganisms which consist of bacteria, desmids, freshwater flagellate algae (?), ostracods and possibly Radiolaria, although the last are not known as freshwater fossils. The analysis of a coprolite indicated francolite (87%), containing Ce_2O_3 0.28, La_2O_3 0.38, V_2O_5 0.01% and a trace of As_2O_3, plus magnesian calcite (4.7%), opaline silica (1.7%) and insoluble components consisting of barite, quartz, feldspar and clay.

Phosphatized wood, identified as a Mesozoic conifer *Cedroxylon*, is common in the deposits of the Dandaragan district of Western Australia (Simpson, 1912). Abundant wood (francolite) diam., and nodules up to 10 cm occur in a greensand considered to be Jurassic in age. The nodules consist of quartz (32.5%), "collophanite" [francolite] (46%), glauconite (12.5%), "iron ore" (5.5%) and feldspar (3.5%) (Matheson, 1948). Although attributed to inorganic processes (Matheson, 1948), the formation of francolite now composing the wood and nodules could have involved organic acids and catalysts. Phosphatized [apatized] wood has been reported from other localities, including the Pacific sea floor (400 m depth) where Goldberg and Parker (1960) resolve the matter of phosphatization essentially in terms of chemical concentrations — that is, without reference to organisms except for the contribution of the wood to "anaerobism."

Although there have been numerous attempts to show the presence of other minerals (brushite, whitlockite and 'octacalcium phosphate') as primary constituents of normal bones and teeth, there is no straightforward evidence that such tissues contain any mineral other than dahllite (McConnell, 1973a). This statement applies also to possible precursors within these tissues, and also to a so-called amorphous calcium phosphate, which has been assumed to be present on the basis of spurious, indirect evidence even in the case of nascent dental enamel. An electron diffraction pattern of non-deproteinized bone is shown as Fig. 3.1.12.

It is true, of course, that in vitro experiments will produce other mineral phases, but since these systems are not at equlibrium within a physiological system, there is no reason for accepting them as valid histochemical similitudes. Never has a diffraction pattern of normal bone (or dental enamel) been shown to contain interference maxima (lines) of any other

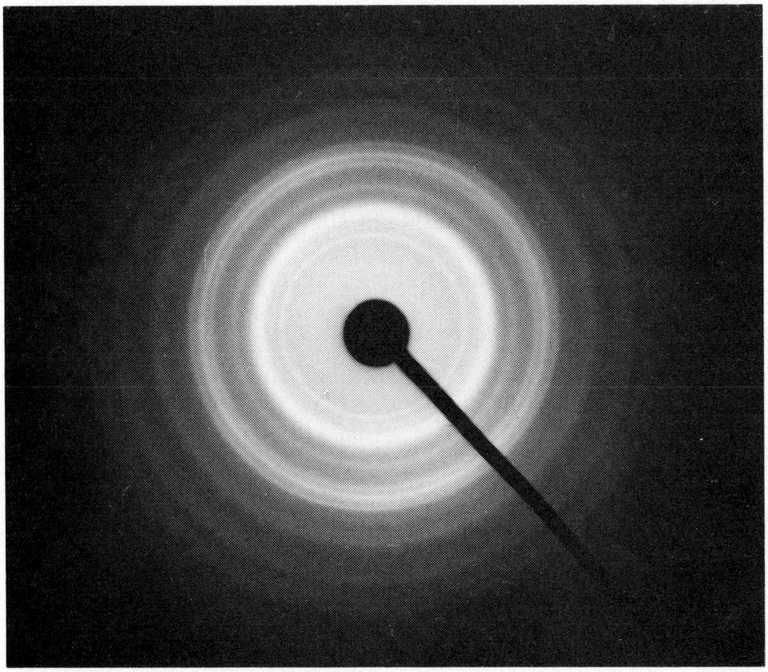

Fig. 3.1.12. Electron diffraction pattern of nondeproteinized bone, all interference maxima of which are attributable to dahllite. The relative intensities have been altered by photomanipulation in order to enhance the weaker maxima at larger angles (McConnell and Foreman, 1971). Copyright 1971 by the American Association for the Advancement of Science.

TABLE 3.1.7

Elemental analyses of bone and teeth

	Bovine bone	Dentin	Enamel	Enamel
Ca	26.70	26.2	37.0	36.41 [c]
P [a]	12.47	—	—	17.48 [c]
CO_2 [b]	3.48	—	—	2.24 [c]
Na	0.731	0.55	0.70	0.70 [c]
K	0.055	—	—	0.037 [c]
Mg	0.436	0.87	0.28	0.21 [c]
Sr	0.035	d	d	—
N	4.92	—	—	—
F	0.07	—	—	—
Cl	0.08	0.035	0.32	—

[a] Expressed as PO_4^{3-} by Armstrong and Singer (1965).
[b] Expressed as CO_3^{2-} by Armstrong and Singer (1965).
[c] Average of two samples (Little, 1961).
[d] See Table 3.1.8.

substance in addition to dahllite prior to heating, refluxing in ethylenediamine or similar drastic treatment. Two authors (Francis and Webb, 1971) became so entranced with a hypothesis relating brushite and 'hydroxyapatite' through epitaxy that they failed to note the *absence* of intense lines of brushite [at spacings 7.59 to 7.58 and 4.24 Å] from their diffraction pattern.

Indeed, in the light of current mineralogical theory (McConnell, 1973a) pertaining to carbonate apatites, there is no need to search for either a precursor or any other solid phase in bone — either crystalline or noncrystalline. The chemical composition is indicated for the principal constituents of bovine (dry, fat-free) bone, compared with dental enamel and dentin (Table 3.1.7). When converted to oxides, these values for bone become:

CaO	MgO	Na_2O	K_2O	SrO	P_2O_5	CO_2	F	Cl	Sum *
37.56	0.72	0.99	0.07	0.04	28.58	3.48	0.07	0.08	71.54

(* Corrected for F and Cl by −0.05)

Adding N and citric acid accounts for 77.32% of the substance; the remainder is undoubtedly organic carbon, chemically-combined water, and the moisture which would remain in "dry" bone. Herein lies the difficulty: no satisfactory method has been devised for separating these constituents. The water content cannot be assumed to be that of theoretical hydroxyapatite because carbonate apatites may have H_3O^+ substituting for Ca^{2+}, H_2O substituting for OH, and/or $H_4O_4^{4-}$ substituting for PO_4^{3-} (McConnell, 1960, 1970b).

While there has always been an implied relationship between metabolic processes far more complex than the activities of the inorganic ions, recent attempts have attained some success in identifying at least one agent which seems to act as an intermediate product or to exert a catalytic effect. For example, Ennever et al. (1974) have been able to show that fully decalcified and lipid-extracted bone matrix would not recalcify, whereas the phospholipid fraction of the extract would produce apatite [dahllite] after exposure to a "metastable calcium phosphate solution" for a week. This result is consistent with Irving's observations (1958a and b) that recalcifying sites of bone interact with lipid stains, suggesting that a lipid is one of the organic components capable of inducing mineralization. Nevertheless, as will be pointed out later, in vitro experiments with carbonic anhydrase indicate a similar ability on the part of this common enzyme to product microcrystalline dahllite.

Axiomatic is the fact that bone is produced in a very exactly controlled system with respect to temperature, pressure, pH, Eh and concentrations of various inorganic ions and organic complexes, as well as ionic strength. Attempts to relegate such an extremely complex system to simple calculations applicable to inorganic reactions will continue to be contraproductive, and many of these attempts are sterile inasmuch as they neglect the carbonate (or bicarbonate) ion which is present in both the liquid and the solid phase [dahllite]. Indeed, some of these calculations disregard the "genetic memory factor" to the extent that bone could be produced in any organ or tissue of the body — apparently being unmindful of the fact that dental enamel forms in different tissues from dentin and has a quite different microtexture.

Trace-element contents are shown for 'normal' human teeth in Table 3.1.8, according to Retief et al. (1971). It is unfortunate that Mo, V, Li and Se were not considered in this study because the first three are believed to have caries-inhibiting effects, whereas the last is supposed to increase incidence of caries (Pärkö, 1975). Small amounts of Li would be tolerated as a

TABLE 3.1.8

Some trace elements ($\mu g\ g^{-1}$) in human teeth (Retief et al., 1971) [a]

	Dentin	Enamel		Dentin	Enamel
Sr	94	111	Ag	2	0.6
Zn	174	263	Cr	2	1.0
Ba	129	125	Co	1	0.1
Al	69	86	Sb	0.7	1.0
Fe	93	118	Mn	0.6	0.6
Br	114	34	Au	0.07	0.1

[a] See Table 3.1.7 for major constituents found by Retief et al. (1971)

replacement for Ca in the same way that small amounts of Mo, V, and Se presumably could substitute for P (see Table 3.1.6). Uranium, which is commonly associated with fossil teeth and bones, may occur as a separate mineral phase, but may also be within the apatitic phase, substituting for Ca or even for P — depending upon its valence. In a series of fossil teeth, it was found (Seitz and Taylor, 1974) that although the U in dental enamel increased with age, there was a more rapid build up in dentin followed by a decrease in U content from a maximum of 750 $\mu g\ g^{-1}$ at 1.7 My to 350 $\mu g\ g^{-1}$ for older samples of dentin. This phenomenon is attributed to the greater organic content of the dentin, but may be related also to the greater porosity and permeability of dentin.

It has long been known that the inorganic component of the scales of bony fish is apatite. Carlström (1963) investigated the nature of "ganoin" of the scales of gar pike, for example, and found it to consist of a carbonate apatite [dahllite], as reported by Ørvig (1967). The scales of Permian fish, however, may be now essentially francolite (Konta, 1956). The fibrocartilage of the spinal column of the leopard shark contains numerous microcrystals of dahllite which are aligned with their *c* axes roughly parallel with the axis of the spinal column (McConnell et al., 1961).

The fossilization process of bones and teeth may involve organic reactions; it usually also involves introduction of fluorine. The fluorine content cannot be measured by methods involving X-ray diffraction, however, and has limited applicability for age determination (McConnell, 1962). Łoziński (1973, p. 433) found for some Polish specimens that "the increase of the rare earth content in fossil bones takes place simultaneously with the fluoridization, . . . (and) increases almost ten times from the Pleistocene to the Cretaceous." Uranium may be associated with fossil bones to the extent of 0.83% U, according to Altschuler et al. (1958), although it is usually much less abundant — perhaps a tenth as much. The organic acids found in fossil bones have been investigated by Wyckoff (1971), who found 19 amino acids to be present — chiefly aspartic, glutamic, glycine and alanine.

PATHOGENIC DEPOSITS IN VERTEBRATES

One of the most prevalent deposits is oral calculus, which occurs among humans, dogs, horses, cows, sheep and cats, as well as non-domesticated animals. In humans, such subgingival deposits originate as plaque, a milieu composed of organic debris that is not properly removed through oral hygiene. Entrapment of inorganic components, such as the scouring agent of a dentifrice, is also possible within plaque. Calculus per se is dahllite, but may include other mineral phases: whitlockite, brushite and possibly monetite (Westerden and Little, 1958). Schroeder and Baumbauer (1966) indicate that brushite and octacalcium phosphate occur most frequently in

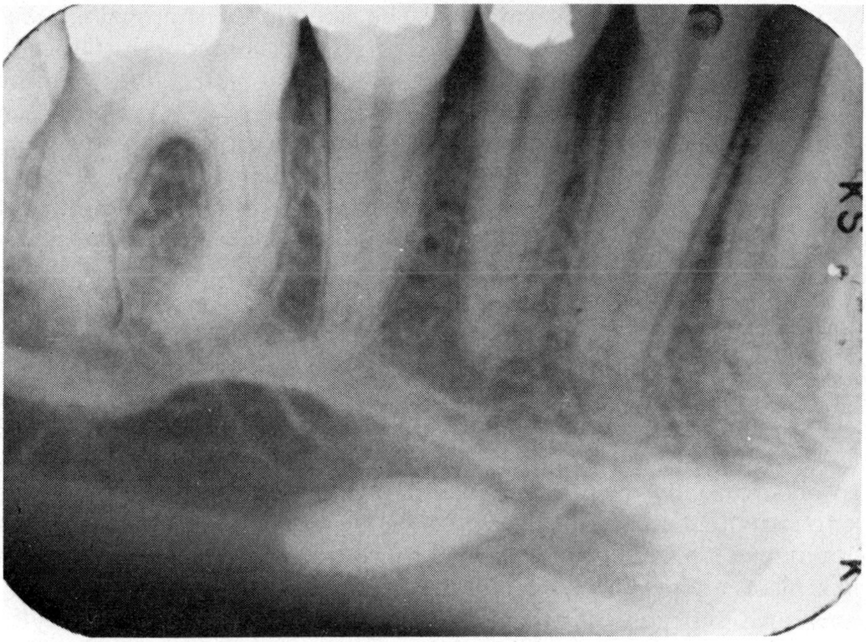

Fig. 3.1.13. Concretionary body (sialolith) within the submandibular gland. Photo by Gus C. Pappas (McConnell, 1973b). Copyright 1973 by John Wiley and Sons.

younger specimens, suggesting that dahllite and whitlockite are the more stable forms in normal physiological environments. The stabilization of the whitlockite structure by magnesium ions has already been mentioned.

Stones from salivary glands and ducts are less common and are believed to be related to the excretion of carbonic anhydrase by these glands (McConnell, 1973b). Based upon examination of a small number of such stones by X-ray diffraction, they are essentially dahllite (Fig. 3.1.13).

Although many studies have attempted to relate oral calculus to the microflora which exists in plaque, such efforts appear to be self-defeating when it is realized that similar deposits occur on teeth of rats raised under germ-free conditions (Fitzgerald and McDaniel, 1960). Thus it becomes apparent that, although such dahllite is surely related to organic processes, it is related to metabolic products of a vertebrate rather than microorganisms.

In vitro experiments (McConnell et al., 1961) indicated that a substance which was crystallochemically comparable with oral calculus could be produced both from pooled human saliva and from a calcifiable synthetic solution to which a few mg l^{-1} of carbonic anhydrase had been added. When a few mg l^{-1} of sulfanilamide was added also, such a precipitate did not form, clearly indicating an interrelation between the formation of dahllite and the presence of an organic catalyst. The activity of the enzyme was destroyed by the sulfanilamide, as well as by related compounds. Other in vitro experiments involving a bacterium (*Bacterionema matruchotii*) will be discussed later, and experiments by Ennever et al. (1974) have been described previously in connection with bone mineralization.

Uroliths of humans have been studied in some detail by Gibson (1974), Lonsdale et al. (1968) and Prien and Frondel (1947). While the predominant phosphate is dahllite, some uroliths are essentially oxalates; these are somewhat more frequent than apatite-oxalate mixed stones and are said to occur usually in acidic sterile urine (Prien and Frondel, 1947). Among 87 cases of urinary calculi (uroliths), Herman et al. (1958) found 11 to contain more than 0.1% of fluorine, presumably in combination with the dahllite.

Another frequent phosphatic stone — found for 90 of 600 cases — was a mixture of apatite and struvite, while those containing brushite comprised only about 1.6% (Prien and Frondel, 1947). These apatite-struvite mixtures are said to occur usually in infected alkaline urine, and occur with all ratios of apatite to struvite. Stones of this type have been equated with the activity of *E. coli*, which can produce ammonia — as also does *Proteus mirabilis*. Both of these organisms are pathogenic when present in the urinary tract.

Five cases have indicated "hannayite" (probably dittmarite or newberyite or a mixture of both) occurring in human uroliths; presumably this substance results from decomposition of struvite in a less alkaline environment. Monetite is rarely encountered in renal calculi where the pH is 5.1 or less (Gibson, 1974); it may represent the dehydration product of brushite under rather severe conditions of acidity. Brushite seems to occur as a transitional phase toward dahllite when the pH is 6.4 or higher. Although phosphorrösslerite has not been reported, it probably could form (like newberyte) as a decomposition product of struvite.

Bezoars (intestinal or stomach stones) are associated with lithofellic acid ($C_{20}H_{36}O_4$), which may be the principal component of some of them (Van Tassel, 1972). Such enteroliths are infrequent, but occur principally in ruminants and other herbivorous animals (deer, horses and other grazing mammals). Milton and Axelrod (1951) reported a stone found with the skeletal remains of a white-tailed deer which consisted essentially of brushite, whereas that from a horse was struvite with some newberyite. Another from the stomach of a deer was essentially newberyite. The tendency for struvite to alter to newberyite has been mentioned previously. An enterolith described by Hutton (1941) was composed of iron-containing bobierrite ($Mg_3(PO_4)_2 \cdot 8H_2O$, 57%; $Fe_3(PO_4)_2 \cdot 8H_2O$, 36% and $Mn_3(PO_4)_2 \cdot 8H_2O$, 7%). This stone had a diameter of about 20 mm and may represent fossilized material inasmuch as it was found on a raised beach. Another stone, secreted by an unknown animal, was essentially struvite (Hutton, 1945).

Ellis (1963) described phosphatic coatings — layers of brushite and whitlockite — which developed upon 'bullets' administered into the reticulorumen of Scottish lambs in order to relieve a nervous disorder known as phalaris staggers. The bullets were composed of cobalt oxide, bound together with bentonite, and baked at 1000°C. The formation of the coatings, which occurred in only a small percent of cases, seems to have been brought under control by adding citric acid to the bullets, although additions of calcium hydrogen phosphate or calcium carbonate to the diet of lambs did not induce such deposits on the bullets. Ellis (p. 606) concluded: "This suggests that saturation is not enough and that some idiosyncratic factor is involved."

Various other concretionary or nodular bodies occur in various tissues and organs of humans. Dahllite seems to be the principal mineral involved in calcification products occurring in the spleen, prostate gland, appendix, testes, the walls of the bronchi and in the lungs — the last in connection with histoplasmosis. However, chronic renal failure produces calcified visceral tissues which are either whitlockite or its immediate precursor (LeGeros et al., 1973). The "sand" of the pineal gland is dahllite, but may be normal rather than pathologic; it seems to increase in quantity with age, and its function is unknown.

Induced calcification of the cardiovascular system has been accomplished in cattle by administration of high levels of vitamin D; here the mineral substance found in the tissues was dahllite, as demonstrated by X-ray diffraction (Capen et al., 1966). Such disturbances of normal metabolic processes have been known for many years, but the mechanism of physiological response remains a mystery.

OTHER BIOLOGIC PRECIPITATES

Although the otoliths which occur in the labyrinths of the ears of most vertebrates are calcium carbonate [calcite, aragonite, vaterite and/or monohydrocalcite], there are two notable exceptions among the Cyclostomata, which are not true vertebrates: e.g. the lamprey and the hag-fish (Carlström, 1963). In the latter case, an apatitic diffraction pattern was obtained readily, whereas in the former a satisfactory pattern was obtained only after heating the specimen to 700°C. Both types are carbonate apatites inasmuch as effervescence was obtained with acid despite the absence of a carbonate phase. Carlström (1963) did not guess at the causes of such phylogenetic differences, although he suggested the use of otoliths (statoliths and statoconia) for taxonomic purposes. The axolotl, *Ambystoma mexicanum*, has a most peculiar type of otolith, consisting of densely packed statoconia within a thin shell. The shell is apatitic, whereas the statoconia are aragonite (Carlström, 1963; Hastings, 1935).

Fossils called conodonts (including neurodontiforms) occur from the lower Cambrian to the top of Triassic sediments. The earliest mineralogical designation of the substance composing these denticular plates was "collophane" (McConnell in Stauffer, 1938), which was then recognized as a carbonate apatite. Subsequently, analyses for fluorine have indicated that, in their present condition, such fossils are francolite (Hass and Lindberg, 1946), but it must be remembered that the fossilization process usually includes enrichment in fluorine. Thus these small fossils might have been dahllite (or possibly dehrnite) in their original condition. Surely they are *not* calcium metaphosphate ($Ca(PO_3)_2$), a composition erroneously assigned to *Archeognathus* by Rhodes and Wingard (1957). It is pertinent to note that calcium metaphosphate had never been previously reported as occurring in nature — nor has its occurrence been confirmed. The fluorine, strontium (0.4%) and traces of yttrium and rare earths are inhomogeneously distributed — decreasing inwardly — suggesting that these differences are postmortem changes (Pietzner et al., 1968). The organic matter associated with demineralized conodonts is probably amino acids (Pietzner et al., 1968). The zoological affinities of conodonts remain a mystery.

The inarticulate brachiopod, *Lingula*, has francolite as the inorganic component of its shell (McConnell, 1963b) that may be associated with oxidative metabolism (Hammen et al., 1962) inasmuch as this organism exceeded both a mussel (*Modiolus demissus*) and an oyster (*Crassostrea virginica*) in activities of several enzymes, including carbonic anhydrase. The activity of this enzyme was particularly high in the mantle (the organ associated with shell formation), and this fact may be related to the formation of oral calculus (Draus et al., 1962), although it is not known why *Lingula* is distinctive in formation of a phosphatic shell rather than a calcareous one (see Fig. 3.1.14.)

The gizzard plates of *Scaphander lignarius* contain fluorite, but also showed phosphorus as a major constituent by electron-probe analysis (Lowenstam and McConnell, 1968). The phosphorus-containing component appears to be amorphous with respect to X-ray diffraction prior to heating, after which a pattern of dahllite was obtained (in addition to fluorite) according to Lowenstam (1972). The renal concrements (uroliths) of *Nautilus pompilius* are similarly amorphous prior to heating, after which the X-ray diffraction pattern of whitlockite was obtained — presumably because of the high magnesium content (McConnell and Ward, 1978). Lowenstam (1972) reports a phosphatic component *after heating* the hard tissue for nine other marine invertebrates of five different phyla but only in the case of *Brachipoda* was a diffraction pattern of phosphate mineral obtained on untreated material. The brachiopod was *Pelagiodiscus atlanticus* and the mineral was determined to be francolite (see p. 158, Lowenstam, 1972) because of its fluorine content (ca. 3%), although "dahllite" is indicated in Lowenstam's Table 1. On heating, whitlockite was obtained for four specimens that were high in MgO and did not show calcite as a constituent after heating. Watabe (1956) found "dahllite" as a constituent of the first larval shell of the oyster *Pinctada martensii*. The denticles of two genera of chitons (*Chiton* and *Acanthopleura*) are composed of francolite (as well as

Fig. 3.1.14. X-ray powder diffraction patterns of (a) portion of the marginal shell of *Lingula* and (b) synthetic carbonate apatite prepared by Klement (1936).

being coated with lepidocrocite and magnetite) according to Lowenstam (1967).

Certain unicellular organisms — particularly *Bacterionema matruchotii* — are capable of inducing precipitates which yield X-ray diffraction patterns of dahllite or whitlockite on low-temperature (radio frequency) ashing (Ennever et al., 1971). Whether dahllite or whitlockite is obtained seems to depend upon the intensity of the current during ashing, the more intense treatment yielding whitlockite. The culture medium was a complex one, containing nine vitamins, pimelic and thioctic acids, casein hydrolysate, adenine, guanine, thymine, uracil, xanthine, inorganic components and a buffer. Isolation of the active component (Ennever et al., 1974) has led to the conclusion that a phospholipid is involved during in vitro calcification. Presumably the phospholipid acts as an intermediate product — rather than as a catalytic agent.

Although phosphorus is known to be present in many other living plants and animals (Clarke and Wheeler, 1922; Vinogradov, 1953; Rhodes and Bloxam, 1971), this phosphorus has not been identified with any particular mineral and probably is present as an organic phosphate. Apatite has been identified as the mineral component of Cambrian ostracods of Sweden, but it seems probable that the apatite is of postmortal origin, as is true of several other reported species and localities. Plants are not known to accumulate phosphatic hard parts, although phosphorus is an essential component of nucleic acids. P_2O_5 may reach as much as 7% of the ash for some marine algae (Vinogradov, 1953).

SUMMARY AND CONCLUSIONS

In this chapter, an attempt has been made to interrelate the organic and inorganic chemistries of phosphatic biominerals in view of the very meager information currently available. This has been done with whatever assistance could be obtained from geologists, mineralogists, chemists and crystallographers, on the one hand, and, on the other, biologists, biochemists, soil scientists and pathologists. Many of the data are more qualitative than quantitative, and thus many conjectures necessarily must be substituted for hypotheses.

One conclusion seems to emerge: organisms, or their metabolic products, influence the formation of biominerals to the extent that straightforward physical-chemical calculations are inadequate to account for what actually takes place in natural systems. This statement, while expressed in general terms, is particularly true for phosphatic biominerals. Equilibrium seems to obtain very slowly, and it is suspected that many of these systems do not attain equilibrium because of kinetic factors. That is, before equilibrium can be effected under a particular set of circumstances, there is a change in the circumstances (ionic activities, pH, Eh, temperature, catalytic agents, etc.), thus creating new phase boundaries if, indeed, not an entirely different system.

Some of the differences in energy content must be very subtle. For example, strengite has been mentioned frequently in systems involving FeO, P_2O_5 and H_2O, although the dimorphic form (phosphosiderite) occurs with strengite at some localities. During weathering — and/or 'biometasomatism', if one will tolerate the use of such a word — there may be some tendency toward preservation of structural configurations of the replaced mineral by its replacement. Such a hypothesis has been proposed for hydromica → taranakite, wherein the sheet structure composed of linked SiO_4 tetrahedra has been preserved by substitution of PO_4 tetrahedra, to some extent. This hypothesis is, in turn, predicated upon the supposition that vashegyite and kingite are phyllophosphates.

Then, there are the amorphous inorganic substances to contend with, such as those found by Lowenstam (1972) associated with fossil phyla. Richellite is another example of a substance which can be converted to a crystalline phase by heating (McConnell, 1963a). Perhaps the same situation exists for bolivarite, evansite and delvauxite.

Many questions have been raised, but few have been answered. The fascination of biomineralogy lies in the fact that for every answer sought, the partial attainment of an answer brings into focus several new questions which also require answers.

REFERENCES

Altschuler, Z.S., Clarke, Jr., R.S. and Young, E.J., 1958. Geochemistry of uranium in apatite and phosphorite. U.S. Geol. Surv. Prof. Pap., 314-D: 45—90.

Anderson, G., 1975. Other organic phosphorous compounds. In: J.E. Gieseking, Soil Components, vol. 1, Organic Components, Springer, New York, NY, pp. 305—331.

Anderson, G. and Malcolm, R.E., 1974. The nature of alkali-soluble soil organic phosphates. J. Soil Sci., 25: 282—297.

Armstrong, W.D. and Singer, L., 1965. Composition and constitution of the mineral phase of bone. Clin. Orthop., 38: 179—190.

Axelrod, J.M., Carron, M.K., Milton, C. and Thayer, T.P., 1952. Phosphate mineralization at Bomi Hill and Bambuta, Liberia, West Africa. Am. Mineral., 37: 883—909.

Baker, H.A. and Beck, J.V., 1942. *Clostridium acidiurici* and *Clostridium cylindrosporum*, organisms fermenting uric acid and some other purines. J. Bacteriol., 43: 291—304.

Balenzano, F., Dell'Anna, L. and DiPierro, M., 1976. Francoanellite, $H_6K_3Al_5(PO_4)_8 \cdot 13H_2O$, a new mineral from the caves of Castellana, Puglia, southern Italy. Neues Jahrb. Mineral., Monatsh., 1976, 49—57.

Banerjee, D.M., 1971. Precambrian phosphorite of Udaipur, Rajasthan, India. Bull. Geol. Soc. Am., 82: 2319—2330.

Blanchard, F.N., 1972. Physical and chemical data for crandallite from Alachua County, Florida. Am. Mineral., 57: 473—484.

Bradley, W.H., 1946. Coprolites from the Bridger Formation of Wyoming: Their composition and microörganisms. Am. J. Sci., 244: 215—239.

Braithwaite, C.J.R., 1968. Diagenesis of phosphatic carbonate rocks on Remire, Amirantes, Indian Ocean. J. Sediment. Petrol., 38: 1194—1212.

Bridge, P.J., 1973. Guano minerals from Murra-el-elevyn Cave, Western Australia. Mineral. Mag., 39: 467—469.

Bridge, P.J., 1977. Archerite, $(K, NH_4)H_2PO_4$, a new mineral from Madura, Western Australia. Mineral. Mag., 41: 33—35.

Bushinskii, G.I., 1967. Inhibitors and stimulators in lithogenesis. Lithology and Mineral Resources, July-Aug. 1967, No. 4: 495—499 (transl. of Litologia Poleznye i Iskopaemye).

Capen, C.C., Cole, C.R. and Hibbs, J.W., 1966. The pathology of hypervitaminosis D in cattle. Vet. Pathol., 3: 350—378.

Carlström, D., 1963. A crystallographic study of vertebrate otoliths. Biol. Bull., 125: 441—463.

Cayeux, L., 1936. Existence de nombreuses bactéries dans les phosphates sédimentaires de tout âge: consequences. C.R. Acad. Sci. (Paris), 203: 1198.

Čech, F. and Slánský, E., 1965. X-ray powder study and thermal investigation of $AlPO_4 \cdot 2H_2O$ minerals. Acta Univ. Carol. Geol., 1965, No. 1: 1—30.

Charles G., 1953. Sur l'origine des gisements de phosphates de chaux sédimentaires. Congres Geologique International, Alger, 1952. Comptes Rendus de la Dix-neuvième Session: Origine des Gisements de Phosphates de Chaux, 11: 163—184.

Clarke, F.W. and Wheeler, W.C., 1922. The inorganic constituents of marine invertebrates. U.S. Geol. Surv. Prof. Pap., 124, 62 pp.

Clifton, H.E., 1957. Carbonate concretions of the Ohio Shale. Ohio J. Sci., 57: 114—129.

Cook, P.J., 1970. Repeated diagenetic calcitization, phosphatization and silification in the Phosphoria Formation. Bull. Geol. Soc. Am., 81: 2107—2116.

Cook, P.J., 1972. Petrology and geochemistry of the phosphate deposits of northwest Queensland, Australia. Econ. Geol., 67: 1193—1213.

D'Anglejan, B.F., 1967. Origin of marine phosphorites off Baja California, Mexico. Mar. Geol., 5: 15—44.

Dietz, R.S. and Emery, K.O., 1950. Submarine phosphorite deposits off California and Mexico. Calif. J. Mines Geol., 46: 7—15.

Draus, F.J., Miklos, F.L. and Leung, S.W., 1962. Carbonic anhydrase in a bovine submaxillary gland extract. J. Dental Res., 41: 497.

Ellis, S.E., 1963. Phosphate minerals in a calculus from the reticulorumen of the sheep. Mineral. Mag., 33: 600—607.

Ennever, J., Vogel, J.J. and Levy, B.M., 1974. Lipid and bone matrix calcification in vitro. Proc. Soc. Exp. Biol. Med., 145: 1386—1388.

Ennever, J., Vogel, J.J. and Streckfuss, J.L., 1971. Synthetic medium for calcification of *Bacterionema matruchotii.* J. Dental Res., 50: 1327—1330.

Fahey, J.J., 1962. Saline minerals in the Green River Formation. U.S. Geol. Surv. Prof. Pap., 405: 1—50.

Fisher, D.J., 1958. Pegmatite phosphates and their problems. Am. Mineral., 43: 181—207, 609—610.

Fisher, D.J., 1966. Cacoxenite from Arkansas. Am. Mineral., 51: 1811—1814.

Fisher, D.J., 1973. Geochemistry of minerals containing phosphorus. In: E.J. Griffith, A. Beeton, J.M. Spencer and D.T. Mitchell (Editors), Environmental Phosphorus Handbook, John Wiley, New York, NY, pp. 141—152.

Fisher, D.J. and McConnell, D., 1969. Aluminum-rich apatite. Science, 164: 551—553.

Fitzgerald, R.J. and McDaniel, E.G., 1960. Dental calculus in the germ-free rat. Arch. Oral Biol., 2: 239—240.

Fleisch, H. and Neuman, W.F., 1961. Mechanism of calcification: role of collagen, polyphosphates, and phosphatase. Am. J. Physiol., 200: 1296—1300.

Francis, M.D. and Webb, N.C., 1971. Hydroxyapatite formation from a hydrated calcium monohydrogen phosphate precursor. Calcif. Tissue Res., 6: 335—342.

Frondel, C., 1943. Mineralogy of the calcium phosphates in insular phosphate rock. Am. Mineral., 28: 215—232.

Gibson, R.J., 1974. Descriptive human pathological mineralogy. Am. Mineral., 59: 1177—1182.

Goldberg, E.D. and Parker, R.H., 1960. Phosphatized wood from the Pacific sea floor. Bull. Geol. Soc. Am., 71: 631—632.

Goldberg, P.S. and Nathan, Y., 1975. The phosphate mineralogy of et-Tabun cave, Mount Carmel, Israel. Mineral. Mag., 40: 253—258.

Gulbrandsen, R.A., 1966. Chemical composition of phosphorites of the Phosphoria Formation. Geochim. Cosmochim. Acta, 30: 769—778.

Hammen, C.S., Hanlon, D.P. and Lum, S.C., 1962. Oxidative metabolism of *Lingula.* Comp. Biochem. Physiol., 5: 185—191.

Haseman, J.F., Brown, E.H. and Whitt, C.D., 1950. Some reactions of phosphate with clays and hydrous oxides of iron and aluminum. Soil Sci., 70: 257—271.

Hass, W.H. and Lindberg, M.L., 1946. Orientation of the crystal units of conodonts. J. Paleontol., 20: 501—504.

Hastings, A.B., 1935. Chemical analysis of otoliths and endolymphatic sac deposits in *Amblystoma tigrinum.* J. Comp. Neurol., 61: 295—296.

Heinen, W. and Lauwers, A.M., 1974. Hypophosphite oxidase from *Bacillus caldolyticus.* Arch. Microbiol., 95: 267—274.

Herman, J.R., Mason, B. and Light, I., 1958. Fluorine in urinary tract calculi. J. Urol., 80: 263—268.

Horowitz, D., 1967. Diagenetic replacement and solution of biogenic phosphate by silica. J. Sediment. Petrol., 37: 1238—1240.

Hutchinson, G.E., 1950. Survey of contemporary knowledge of biogeochemistry. 3. The biogeochemistry of vertebrate excretion. Bull. Am. Mus. Nat. Hist., 96: 554 pp.

Hutton, C.O., 1941. The chemical composition and optical properties of the mineral substance composing an enterolith. N.Z. J. Sci. Technol., 23: 9B—12B.

Hutton, C.O., 1945. The nature of an enterolith. N.Z. J. Sci. Technol., 26: 304—307.
Irving, J.T., 1958a. Sudanophil inclusions in ameloblasts, odontoblasts and cells of the oral epithelium. Nature, 181: 569—570.
Irving, J.T., 1958b. A histological stain for newly calcified tissue. Nature, 181: 704—705.
Jensen, A.T. and Rowles, S.L., 1957. Lattice constants and magnesium contents of some naturally occurring whitlockites. Nature: 179, 912—913.
Johan, Z., 1976. La sénégalite, $Al_2(PO_4)(OH)_3 \cdot H_2O$, un nouveau minéral. Lithos, 9: 165—171.
Kaneshima, K., 1962. Geochemistry of the phosphate rocks in the Ryukyu Islands. Nippon Kagaku Zasshi, 83: 1007—1017, 1074—1078 (Japanese, English summary).
Kasakov, A.V., 1937. The phosphorite facies and the genesis of phosphorites. Trans. Sci. Inst. Fertilizers Insecto-Fungicides, 142: 95—113 (in Russian).
Kato, A., 1965. Redondite. J. Mineral. Soc. Jpn., 7: 298—304 (in Japanese).
Klement, R., 1936. Der Carbonatgehalt der anorganischen Knochensubstanz und ihre Synthese. Ber. Dtsch. Chem. Ges., 69: 2232—2238.
Konta, J., 1956. Crystal orientation of two phosphates in the scales of Permian fish of the group *Palaeoniscidae*. Acta Geol., 4: 113—121.
Krumbein, W.C. and Garrels, R.M., 1952. Origin and classification of chemical sediments in terms of pH and oxidation reduction potentials. J. Geol., 60: 1—33.
Lagergren, C., 1956. Biophysical investigations of urinary calculi. Acta Radiol., Supplement 133: 1—71.
LeGeros, R.Z., Contiguglia, S.R. and Alfrey, A.C., 1973. Pathological calcifications associated with uremia. Calcif. Tissue Res., 13: 173—185.
Lehr, J.R., Brown, E.H., Frazier, A.W., Smith, J.P. and Thrasher, R.D., 1967. Crystallographic Properties of Fertilizer Compounds. Chemical Engineering Bulletin 6, Tennessee Valley Authority, 166 pp.
Little, M.F., 1961. Studies on the inorganic carbon dioxide component of human enamel. II. The effect of acid on enamel CO_2. J. Dental Res., 40: 903—914.
Lonsdale, K., Sutor, D.J. and Wooley, S., 1968. Composition of urinary calculi by X-ray diffraction. Collected data from various localities. Br. J. Urol., 40: 33—36, 402—411.
Love, J.D., 1964. Uraniferous phosphatic lake beds of Eocene age in intermontane basins of Wyoming and Utah. U.S. Geol. Surv., Prof. Pap., 474E: 1—66.
Lowenstam, H.A., 1967. Lepidocrocite, an apatite mineral and magnetite in teeth of chitons (*Polyplacophora*). Science, 156: 1373—1375.
Lowenstam, H.A., 1972. Phosphatic hard tissues of marine invertebrates: their nature and mechanical function, and some fossil implications. Chem. Geol., 9: 153—166.
Lowenstam, H.A. and McConnell, D., 1968. Biologic precipitation of fluorite. Science, 162: 1496—1498.
Łoziński, J., 1973. Rare-earth elements in fossil bones. Ann. Soc. Geol. Pologne, 43: 407—430 (Polish with English summary, pp. 431—435).
Manheim, F., Rowe, G.T. and Jipa, D., 1975. Marine phosphorite formation off Peru. J. Sediment. Petrol., 45: 243—251.
Martini, J., 1978. Sasaite, a new phosphate mineral from West Driefontein Cave, Transvaal, South Africa. Mineral. Mag., 42: 401—404.
Mason, B., 1966. Principles of Geochemistry, 3rd edn, John Wiley, New York, NY, 329 pp.
Matheson, R.S., 1948. The Dandaragan phosphate deposits. Miner. Resour. West. Austr., Bull., 4: 1—64.
McConnell, D., 1935. Spherulitic concretions of dahllite from Ishawooa, Wyoming. Am. Mineral., 20: 693—698.
McConnell, D., 1938. A structural investigation of the isomorphism of the apatite group. Am. Mineral., 23: 1—19.

McConnell, D., 1941. Barrandita, mineral constitutivo de los depósitos de fosfato de la Isla Gran Roque, D.F., Venezuela. Col. Ing. Venezuela, Rev. an., 19, no. 140: 115—119.
McConnell, D., 1943. Phosphatization at Malpelo Island, Colombia. Bull. Geol. Soc. Am., 54: 701—716.
McConnell, D., 1950. The petrography of rock phosphates. J. Geol., 58: 16—23.
McConnell, D., 1952. Viséite, a zeolite with the analcime structure and containing linked SiO_4, PO_4 and H_xO_4 groups. Am. Mineral., 37: 609—617.
McConnell, D., 1960. The crystal chemistry of dahllite. Am. Mineral., 45: 209—216.
McConnell, D., 1962. Dating of fossil bones by the fluorine method. Science, 136: 241—244.
McConnell, D., 1963a. Thermocrystallization of richellite to produce a lazulite structure (calcium lipscombite). Am. Mineral., 48: 300—307.
McConnell, D., 1963b. Inorganic constituents in the shell of the living brachiopod *Lingula*. Bull. Geol. Soc. Am., 74: 363—364.
McConnell, D., 1965. Precipitation of phosphates in sea water. Econ. Geol., 60: 1059—1062.
McConnell, D., 1970a. Arsenic in phosphorite and apatite. Econ. Geol., 65: 64—65.
McConnell, D., 1970b. Crystal chemistry of bone mineral: hydrated carbonate apatites. Am. Mineral., 55: 1659—1669.
McConnell, D., 1971. Crystal chemistry of phosphorite. Econ. Geol., 66: 1085—1086.
McConnell, D., 1973a. Apatite: Its Crystal Chemistry, Mineralogy, Utilization, and Geologic and Biologic Occurrences. Springer, New York/Wien, 111 pp.
McConnell, D., 1973b. Biomineralogy of phosphates and physiological mineralization. In: E.J. Griffith, A. Beeton, J.M. Spencer and D.T. Mitchell (Editors), Environmental Phosphorus Handbook, John Wiley, New York, NY, pp. 425—442.
McConnell, D., 1974. Are vashegyite and kingite hydrous aluminum phyllophosphates with kaolinite-type structures? Mineral. Mag., 39: 802—806.
McConnell, D., 1976a. Hypothetical phyllophosphate structure for taranakite. Am. Mineral., 61: 329—331.
McConnell, D., 1976b. Crystallinity among some hydrous aluminium phosphates. Mineral. Mag., 40: 609—610.
McConnell, D. and Foreman, Jr., D.W., 1971. Texture and composition of bone. Science, 172: 971—972.
McConnell, D. and Foreman, Jr., D.W., 1974. The structure of kehoeite. Can. Mineral., 12: 352—353.
McConnell, D., Frajola, W.J. and Deamer, D.W., 1961. Relation between the inorganic chemistry and biochemistry of bone mineralization. Science, 133: 281—282.
McConnell, D. and Ward, Peter, 1978. Nautiloid uroliths composed of phosphatic hydrogel. Science, 199: 208—209.
McKelvey, V.E., Swanson, R.W. and Sheldon, R.P., 1953. The Permian phosphorite deposits of the western United States. Congres Geologique International, Alger, 1952. Comptes Rendus de la Dix-neuvième Session: Origine des Gisements de Phosphates des Chaux, 11: 45—64.
McKie, D., 1958. Notes on some minerals from Tanganyika. Rec. Geol. Surv. Tanganyika, 5: 81—94 (1955).
Milton, C. and Axelrod, J.M., 1951. Calculi and other stones found in mammals. J. Mammal., 32: 139—154.
Mitchell, R.S. and Sherwood, W.C., 1958. Dahllite pseudomorphs after pyrite concretions from the Big Horn Basin, Wyoming. Am. Mineral., 43: 600—603.
Mrose, M.D., 1971. New mineral data for hydrated phosphates and sulfates. U.S. Geol. Surv., Prof. Pap., 750-A: 115.
Nordin, B.E.C., Bluhm, M. and MacGregor, J., 1962. In vitro and in vivo studies with

bone-seeking isotopes. In: P. Lacroix and A.M. Budy (Editors), Radio-Isotopes and Bone. Blackwell, Oxford, pp. 105—126.

Nriagu, J.O., 1972. Stability of vivianite and ion-pair formation in the system $Fe_3(PO_4)_2$-H_3PO_4-H_2O. Geochim. Cosmochim. Acta, 36: 459—470.

Nriagu, J.O. and Dell, C.I., 1974. Diagenetic formation of iron phosphates in recent lake sediments. Am. Mineral., 59: 934—946.

Orvig, T., 1967. Evolution of some calcified tissues of early vertebrates. In: A.E.W. Miles (Editor), Structural and Chemical Organization of Teeth, Vol. I, Academic, London, 525 pp.

Owens, J.P., Altschuler, Z.S. and Berman, R., 1960. Millisite in phosphorite from Homeland, Florida. Am. Mineral., 45: 547—561.

Pärkö, A., 1975. Dental caries prevalent in the Rapakivi granite and olivine diabase areas of Laitila, Finland. Proceedings of the Finnish Dental Society, 81, supplement I, 55 pp.

Parker, R.J., 1971. The petrography and major element geochemistry of phosphorite nodule deposits on the Agulhas Bank, South Africa. S. Afr. Nat. Committee Oceanogr. Res., Bull., 2: 1—94.

Parker, R.J., 1975. The petrology and origin of some glauconitic and glauco-conglomeratic phosphorites from the South African continental margin. J. Sediment. Petrol., 45: 230—242.

Patrick, Jr., W.H., Gotoh, S. and Williams, B.G., 1973. Strengite dissolution in flooded soils and sediments. Science, 179: 564—565.

Patwardhan, A.M. and Ahluwalia, A.D., 1973. A note on the origin of Mussoorie phosphorite in the Lower Himalaya, India, and its paleogeographic implications. Miner. Deposita, 8: 379—387.

Pecora, W.T., Hearn, Jr., B.C. and Milton, C., 1962. Origin of spherulitic phosphatic nodules in basal Colorado Shale, Bearpaw Mountains, Montana. U.S. Geol. Surv., Prof. Pap., 450: B30—B35.

Pietzner, H., Vahl, J., Werner, H. and Ziegler, W., 1968. Zur chemischen Zusammensetzung und Mikromorphologie der Conodonten. Paleontographica, Abt. A, 128: 115—152. (English summary, 146—147.)

Prien, E.L. and Frondel, C., 1947. Studies in urolithiasis: I. The composition of urinary calculi. J. Urol., 57: 949—991.

Pryce, M.W., 1972. Biphosphammite: a second occurrence. Mineral. Mag., 38: 965.

Retief, D.H., Cleaton-Jones, P.E., Turkstra, J. and de Wet, W.J., 1971. The quantitative analysis of sixteen elements in normal human enamel and dentine by neutron activation analysis and high-resolution gamma spectroscopy. Arch. Oral Biol., 16: 1257—1267.

Rhodes, F.H.T. and Bloxam, T.W., 1971. Phosphatic organisms in the Paleozoic and their evolutionary significance. Phosphate in fossils. Proceedings of the North American Paleontological Convention, Chicago, 1969, Part K, Allen Press, Lawrence, KS, pp. 1485—1513.

Rhodes, F.H.T. and Wingard, P., 1957. The structure, chemical composition and affinities of the neurodontiforms. J. Paleontol., 31: 448—454.

Romankevich, Ye.A. and Baturin, G.N., 1972. Composition of the organic matter in phosphorites from the continental shelf of southwest Africa. Geochem. Int., 9: 464—470.

Rooney, T.P. and Kerr, P.F., 1967. Mineralogic nature and origin of phosphorite, Beauford County, North Carolina. Bull. Geol. Soc. Am., 78: 731—748.

Rosenquist, I.T., 1970. Formation of vivianite in Holocene clay sediments. Lithos, 3: 327—334.

Sakae, T. and Sudo, T., 1975. Taranakite from the Onino-Iwaya limestone cave at Hiroshima Prefecture, Japan: a new occurrence. Am. Mineral., 60: 331—334.

Salvador, P. and Fayos, J., 1972. Some aspects of the structural relationship between "Messbach-type" and "Lucin-type" variscites. Am. Mineral., 57: 36—44.

Scharpf, Jr., L.G., 1973. Transformations of naturally occurring organophosphorus compounds in the environment. In: E.J. Griffith, A. Beeton, J.M. Spencer, and D.T.

Mitchell (Editors), Environmental Phosphorus Handbook, John Wiley, New York, NY, pp. 393—412.

Schroeder, H.E. and Bambauer, H.U., 1966. Stages of calcium phosphate crystallization during calculus formation. Arch. Oral Biol., 11: 1—14.

Seitz, M.G. and Taylor, R.E., 1974. Uranium variations in a dated fossil bone series from Olduvai Gorge, Tanzania. Archaeometry, 16: 129—135.

Shannon, R.D. and Prewitt, C.T., 1969. Effective ionic radii in oxides and fluorides. Acta Crystallogr., B25: 925—946.

Shepard, C.U., 1869. On a mineral phosphate from the island of Redonda, W.I. Am. J. Sci., 47: 428—429.

Simpson, D.R., 1968. Substitutions in apatite: I. Potassium-bearing apatite. Am. Mineral., 53: 432—444.

Simpson, E.S., 1912. Unusual type of petrification from Dandaragan. J. Nat. Hist. Sci. Soc. West. Aust., 4: 33—37.

Skinner, H.C.W., 1973. Phase relations in the CaO-P_2O_5-H_2O system from 300° to 600°C at 2 kb H_2O pressure. Am. J. Sci., 273: 545—600.

Stauffer, C.R., 1938. Conodonts of the Olentangy Shale. J. Paleontol., 12: 411—443.

Stenzel, H.B., 1934. Decapod crustaceans from the middle Eocene of Texas. J. Paleontol., 8: 38—56.

Stow, S.H., 1969. The occurrence of arsenic and the color-causing components in Florida land-pebble phosphate rock. Econ. Geol., 64: 667—671.

Swaine, D.J., 1962. The Trace-Element Content of Fertilizers. Technical Communication 52, Commonwealth Agricultural Bureaux, Farnham Royal, England, 306 pp.

Swanson, V.F. and Legal, C.C., 1967. Mineral contaminants in Florida phosphate rock. Fertilizer Industry Roundtable, Proceedings, 17th, 1967, pp. 67—71.

Tooms, J.S., Summerhayes, C.P. and Cronan, D.S., 1969. Geochemistry of marine phosphates and manganese deposits. Mar. Biol. Ann. Rev., 7: 49—100.

Trautz, O.R., Zapanta-LeGeros, R. and LeGeros, J.P., 1964. Effects of magnesium on various calcium phosphates. II. J. Dental Res., 43: 751. (Abstract).

Trueman, N.A., 1965. The phosphate, volcanic and carbonate rocks of Christmas Island, Indian Ocean. J. Geol. Soc. Austr., 12: 261—283.

Tsubota, G., 1959. Phosphate reduction in the paddy field. Soil Plant Food, 5: 10—15.

Turekian, K.K., 1968. Oceans. Prentice-Hall, Englewood Cliffs, N.J., 120 pp.

Van Tassel, R., 1972. L'acide lithofellique, constituant cristallin majeur de certains bézoards. Soc. Fr. Minéral. Cristallogr., Bull., 95: 106—114.

Van Wambeke, L., 1971. The problem of cation deficiencies in some phosphates due to alteration processes. Am. Mineral., 56: 1366—1384.

Velikanov, V.A., 1975. Distribution controls of phosphorite concretions in Kalyus sediments in the Vendian of Podoliya. Lithol. Miner. Resour. (USSR), 10: 733—738. (Transl. of Litologia Poleznye i Iskopaemye.)

Vinogradov, A.P., 1953. The elementary chemical composition of marine organisms. Sears Foundation for Marine Research, Memoir 2, 647 pp.

Watabe, N., 1956. Dahllite identified as a constituent of prodissoconch 1 of *Pinctada martensii* (Dunker). Science, 124: 630.

Westerden, E.M. and Little, K., 1958. Some observations on the composition of calculus. J. Dental Res., 37: 749—750 (Abstract).

Whitton, B.A., 1967. Phosphate accumulation by colonies of *Nostoc*. Plant Cell Physiol., 8: 293—296.

Willcox, N.R., 1953. The origin of beds of phosphatic chalk with special reference to those at Taplow, England. Congres Geologique International, Alger, 1952. Comptes Rendus de la Dix-neuvième Session: Origine des Gisements de Phosphates de Chaux, 11: 119—133.

Wyckoff, R.W.G., 1971. Trace elements and organic constituents in fossil bones and teeth. Phosphate in fossils. Proceedings of the North American Paleontological Convention, Chicago, 1969, Part K, Allen Press, Lawrence, KS, pp. 1514—1524.

Chapter 3.2

THE PHOSPHORUS CYCLE: QUANTITATIVE ASPECTS AND THE ROLE OF MAN

U. PIERROU *

Valthornsvagen 39, Uppsala, S-572 50 (Sweden)

CONTENTS

INTRODUCTION

Phosphorus is essential for living organisms and is not exchangeable with other elements in biological systems. It is an important constituent of the genetic and information-transfer molecules, deoxyribose- and ribose-nucleic acids, and also of the energy-carrying molecule adenosine triphosphate (ATP) and its di- and monophosphate precursors, ADP and AMP. The special form of AMP called cyclic adenosine monophosphate has a function in controlling different enzymes. Phosphorus is a macro-nutrient but its availability is often in the ng g^{-1} range. The effects of phosphorus in nature are, therefore, very profound. Phosphorus discharged by a single person in one year (about 2 kg P) is sufficient for the growth of 1 Mg of plant material (Vallentyne, 1974), a fact which serves to illustrate the link between urban communities and eutrophication.

Before dealing with the different subcycles, one should perhaps consider a simplified model of the global phosphorus cycle shown in Fig. 3.2.1. The turnover rate of this cycle is regulated by the rate of diagenesis of phosphorus-containing sediments into phosphate rock. This process takes 0.1—1 Gy (Broecker, 1974) which implies that a period of more than 1 Gy is

* Present address: Institute of Limnology, University of Uppsala, Box 557, S-75122 Uppsala, Sweden.

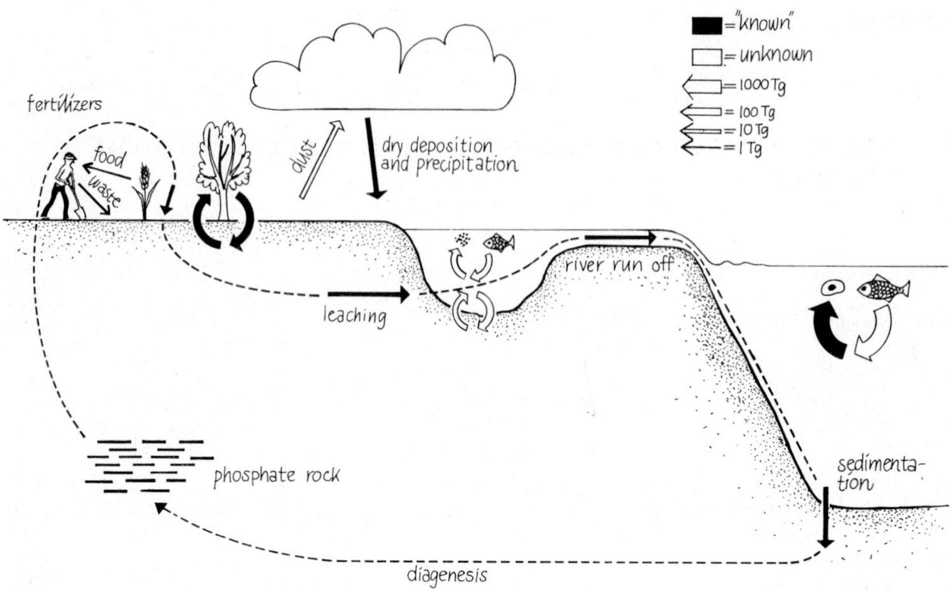

Fig. 3.2.1. Simplified model of phosphorus fluxes within the global phosphorus cycle (from Pierrou, 1976, by permission).

required for one global cycle of phosphorus to be completed. There have been discussions about the formation of phosphorus nodules on the ocean floor ever since these nodules became targets of planned phosphorus mining. Most of these nodules are old, probably more than 100 ky, and are at present being eroded rather than formed. Some phosphorus nodules are forming at the present time under restricted conditions in a few areas of the ocean (Stumm, 1973). Thus, except in terms of the long-term geological record for which limited data are available, the phosphorus "cycle" can be viewed as a unidirectional transport of phosphorus from phosphate rock to marine and, to some extent, freshwater sediments.

PHOSPHORUS IN THE ATMOSPHERE

The role of the atmosphere in the phosphorus cycle seems to be poorly understood. Since it does not exist in the form of stable gaseous compounds, phosphorus in the atmosphere is either adsorbed on particulate matter, e.g. dust (including pollen) and exhaust fumes or dissolved in sea-spray. The fallout of phosphorus, as dry deposition and precipitation, has been estimated to be within the range 3.6—9.2 Tg P y^{-1} for terrestrial ecosystems, 0.054—0.140 Tg P y^{-1} for freshwater ecosystems, and 2.6—3.5 Tg y^{-1} for the marine ecosystem. This gives a total fallout from the atmosphere of 6.3—12.8 Tg P y^{-1} (Pierrou, 1976). It should be noted, however, that Emery et al. (1955)

estimated the fallout over the marine ecosystem to be zero. The influx of phosphorus to the atmosphere due to high-temperature combustion of organic matter has been estimated to be about 0.08 Tg P y^{-1} (Pierrou, 1976) of which 0.07 Tg P y^{-1} is the result of the burning of coal (Bertine and Goldberg, 1971). It follows therefore that dust and sea-spray appear to be the major sources of phosphorus in the atmosphere.

TERRESTRIAL PHOSPHORUS

The transfer of phosphorus from the terrestrial biomass to soil as dead organic matter has been estimated to be 136.4 Tg P y^{-1}: 133.3 Tg P y^{-1} is derived from plants and 3.1 Tg P y^{-1} from animal material (Pierrou, 1976). The uptake of phosphorus by plants from soil was calculated by Bazilevich (1974) to be 178 Tg P y^{-1}, while Stumm (1973) estimated it to be 236.8 Tg P y^{-1} including that of the freshwater ecosystems. The terrestrial biota has been calculated to absorb 0.065 Tg P y^{-1} from aquatic ecosystems and 0.063 Tg P y^{-1} from industrially made foodstuffs and pharmaceutical products (Pierrou, 1976). An important aspect on which quantitative data are not yet available is the bacterial cycling of phosphorus within soils. This "internal" cycle helps in making phosphorus available for plants.

The "natural" influx of phosphorus to soils is hard to assess since no measurements appear to have been made on that proportion of atmosphere fallout of phosphorus (3.6—9.2 Tg P y^{-1}: loc. cit.) which is due to sea spray. According to Hutchinson (1952), the deposition of guano contributes about 0.01 Tg P y^{-1} to terrestrial phosphorus. On 1972 figures, man-made annual contributions to soil phosphorus were 9.93 Tg in the form of superphosphates (FAO, 1975) and 1.1 Tg as human excreta used as a fertilizer (Pierrou, 1976).

Much of the phosphorus in the soil is immobilized in the form of complexes with iron, aluminium and calcium, thereby becoming inaccessible to plants. According to Phillips and Webb (1971) soluble phosphorus rarely migrates more than 2 or 3 cm from a fertilizer granule before being immobilized by reactions with soil components. Some soil components, such as humic acids, increase the solubility of phosphorus compounds. Other processes which diminish the availability of phosphorus to terrestrial plants are the leaching of soluble phosphorus and the erosion of soils containing phosphorus. The leaching rate has been calculated to be within the range 2.5—12.3 g P y^{-1} (Pierrou, 1976). The erosion of soil will be discussed in connection with river transport of phosphorus.

AQUATIC PHOSPHORUS

The most important flux of the freshwater phosphorus cycle is the large amount of phosphorus transported by river runoff. This flux has been esti-

TABLE 3.2.1

Phosphorus inventories (Tg P) (from Pierrou, 1976, by permission)

Biomass:	
Human	<1
Terrestrial	1805
Marine	128
Freshwater	<1
Waters:	
Fresh	90
Marine	120,000—128,000
Soil:	160,000
Rocks:	
Total solid sphere	1.1×10^{13}
Mineable	6500—59,000

mated at between 1.9 and 2 Tg P y^{-1} (Stumm, 1973, Gulbrandsen and Robertson, 1973). Although not specifically defined in these papers, the values appear to refer only to dissolved and particulate phosphorus, and do not include the suspended sediments that are carried by rivers to the ocean. Emery et al. (1955) estimated the total phosphorus transport by rivers, including sediments, to be 14 Tg P y^{-1}. According to Pierrou (1976) the total phosphorus transport by rivers is in the order of 17.4 Tg P y^{-1}: the dissolved and particulate phosphorus is estimated to be 3.7 Tg P y^{-1}, while the sediment loading is 13.7 Tg P y^{-1}. An important feature of the sediment loading is that only 4.6 Tg P y^{-1} appears to be the result of "natural" erosion. The additional 9.1 Tg P y^{-1} is believed to be due to the increase in erosion caused by human activities such as deforestation and extensive agricultural activities.

In freshwater ecosystems, the amounts of phosphorus introduced to sediments are estimated to be of the order of 1 Tg P y^{-1} while the amounts released from sediments annually are estimated to be less than 1 Tg P (Pierrou, 1976). Emery et al. (1955) calculated that the amount of phosphorus deposited in ocean sediments is 13 Tg P y^{-1}. The amount of phosphorus released from ocean sediments is unknown but is probably relatively small, as the reducing conditions (lack of oxygen) required occur relatively rarely in the ocean. The uptake of phosphorus by phytoplankton in the ocean has been variously calculated as 1300 Tg P y^{-1} (Emery et al., 1955) and 990 Tg P y^{-1} (Stumm, 1973). A similar estimate, about 1000 Tg P y^{-1}, can be made for the amount of phosphorus deposited in oceanic detritus (Pierrou, 1976).

The large storage of phosphorus in the ocean (see Table 3.2.1) and the large internal circulation within the ocean can absorb relatively large additions of phosphorus without causing any noticeable effects on the concentration of phosphorus in the water. However, even small increases in phosphorus concentration will increase the fraction of the ocean bottom covered by

Chapter 4

BIOGEOCHEMISTRY OF IRON

D.G. LUNDGREN [1] and W. DEAN [2,*]

Department of Biology [1] *and Geology* [2]*, Syracuse University, Syracuse, NY 13210 (U.S.A.)*

CONTENTS

OVERVIEW OF THE SUBJECT

Iron serves as a vital component of the chemical architecture of the biosphere. Although carbon, hydrogen, oxygen and nitrogen dominate the biosphere, iron is present at about 7.7 g m^{-2} of land mass, but, in biomass, iron ranks only behind the trace elements calcium, potassium, silicon, magnesium, sulfur, aluminium, phosphorus and chlorine (Deevey, 1970).

* Present address: United States Geological Survey, Lakewood, CO 80225, U.S.A.

When the relative composition of the biosphere is compared to that of the lithosphere, a new dimension for iron is noted. Iron now constitutes about 1.3 atoms of every 100 atoms of the lithosphere and, on a weight per cent basis, iron in crustal abundance ranks fourth in number, only behind oxygen, silicon, and aluminium (Table 7.3 in Geochemistry, Wedepohl, 1971, p. 60). On a comparative basis, the lithosphere is a dry metallic aluminium silicate, whereas the biosphere is wet and carbonaceous (Deevey, 1970). Iron is very limited in the hydrosphere and atmosphere.

Although iron is relatively scarce in the biosphere, it is intrinsic and necessary, and not a mere contaminant. All forms of life (with the possible exception of *Lactobacillus*) require iron for such physiological functions as electron transport, oxygen transport or respiration, and fermentation. *Lactobacilli* have apparently learned to live in the absence of a nutritional supply of iron, thereby acquiring a bypass around the iron-containing enzymes involved in physiological reactions supporting life (Neilands, 1974).

In nature, iron exists as common minerals, including those of *carbonates* (siderite, $FeCO_3$); *oxides* (hematite, Fe_2O_3; magnetite, Fe_3O_4; limonite, $2Fe_2O_3 \cdot H_2O$); *sulfides* (troilite, FeS; pyrite, FeS_2); *sulfates* (melanterite, $FeSO_4 \cdot 7H_2O$) and *silicates* (thuringite, $7FeO \cdot 3(Al,Fe)_2O_3 \cdot 5SiO_2 . nH_2O$, chamosite, $4FeO \cdot Al_2O_3 \cdot 3SiO_2 \cdot 3H_2O$). Minerals of these types are mostly of sedimentary origin; however, mineral silicates can also be of igneous and hydrothermal origin. Iron is frequently associated with organic material in nature, and is perhaps the most versatile of the biocatalytic elements. Its many chemical reactivities are due to its two valence states, Fe^{2+} and Fe^{3+} as well as the range of oxidation-reduction potentials associated with the $Fe^{2+} \rightleftharpoons Fe^{3+}$ transition in various iron-containing compounds.

GEOCYCLE OF IRON

Deevey (1970) describes the cycling of minerals and points out some interesting features for those elements which are neither readily soluble nor volatile, and thus uncouple the atmospheric component of a complete mineral cycle, i.e., biosphere, hydrosphere, lithosphere, and atmosphere. A cycle of iron (Fig. 4.1) occurs within the biosphere and, when iron is needed by an organism, it is supplied from available iron reservoirs. There is a constant movement of terrestrial iron towards a hydrospheric sink and within the biosphere, what iron is needed, is channeled to the organism while the remainder is wasted. Since oceans contain roughly 3 μg Fe l^{-1} and fresh-water rivers and lakes contain only slightly more, iron availability for aquatic life is a problem. Muds and deeper waters of various aquatic bodies do contain considerably higher levels of iron. Whenever iron enters the hydrosphere, precipitation and settling occurs.

The concentrations of iron in microorganisms, plants and animals, on a dry weight basis, are low, which agrees with the established physiological

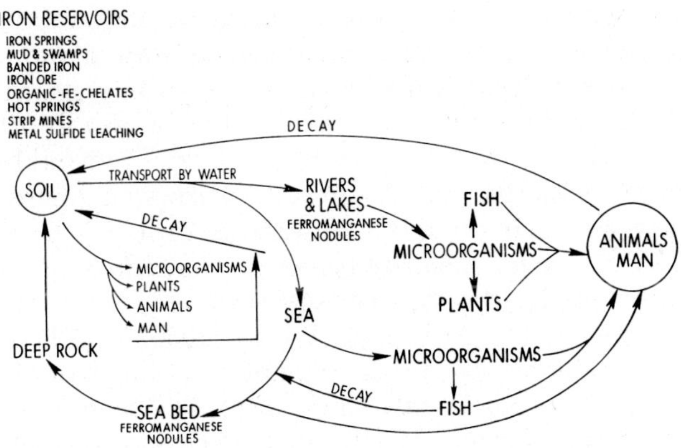

Fig. 4.1. Movement of iron through the biosphere and its availability to microorganisms, plants and animals.

functions of this element, i.e., biocatalyst, structural element, and activator.

Iron occurs in two oxidation states, the divalent (Fe^{2+}) ion or trivalent (Fe^{3+}) ion, and sedimentary rocks contain iron in these various forms with ferric oxides being the most common. When iron is weathered out of the rocks, it is not retained in solution but, depending upon conditions, it is redeposited as oxides or hydroxides. In addition, Fe^{3+} can replace aluminium in some silicate minerals. An important chemical feature of iron (in solution) is its tendency to form complexes with organic materials. Such complexes are considerably more stable and consequently survive in solution or in the soil for longer periods of time. Specific examples of Fe-organic complexes will be discussed in later sections.

There is little doubt that iron transformation within the cycle depends upon the influence of many different microorganisms. However, those organisms known to be directly involved in transforming major deposits of iron are limited and, in this chapter, special importance will be given to the so-called "iron bacteria" group which directly influences the iron reservoirs of the world.

THE IRON BACTERIA GROUP

The organisms belonging to the iron bacteria group have been taken to include: (1) bacteria able to grow at the expense of energy from the oxidation of reduced iron; (2) bacteria producing distinct precipitates of iron generally associated with the cells; (3) bacteria causing reductions of ferric ions.

For a historical perspective of the biological meaning of the term "iron bacteria", the reader is referred to the important review of Pringsheim

(1949). Much of that review is concerned with the problem of "separating facts from an overwhelming mass of rash and controversial statements". Although only bacteria which utilize ferrous iron as a source of energy are properly classified as "iron bacteria", history reveals that a group of filamentous bacteria having the sole power to precipitate ferric hydroxide from iron-bearing solutions was included in the group. European microbiologists dominated the early research on the filamentous type of "iron bacteria", i.e., *Crenothrix*, *Gallionella*, *Spirophyllum*, *Leptothrix* and *Cladothrix*. The research was mostly descriptive and suffered from the absence of reliable cultural methods. Although Winogradsky's (1888) concept restricted the "iron bacteria" to those capable of autotrophic growth at the expense of ferrous iron, *Gallionella* and *Leptothrix* seemed to be legitimate members of the group. Harder (1919), an American geologist, investigated the "iron bacteria" from the standpoint of their geological significance. His monograph is a *classic* and readers are urged to consult this reading for an early account of how "iron bacteria" might be involved in the formation of iron-ore deposits, their distribution in nature, and their activity in natural iron-bearing waters.

Early investigations were meagre but proved the ability of bacteria to grow at the expense of ferrous iron. Before the 1950's, it was apparent that additional information on the physiology of "iron bacteria" was needed (Starkey and Halvorson, 1927; Starkey, 1945). True "iron bacteria" are characterized by the accumulation of an abundance of ferric hydrate. Baas Becking and Parks (1927) have shown that the reaction involving the oxidation of ferrous iron is capable of being utilized by microorganisms for energy. The reaction, however, yields little energy per unit of iron oxidized and since the oxidation product is insoluble, large amounts of ferric hydrate will accumulate. This accumulation of ferric hydrate should be a characteristic of "iron bacteria" and should be present far in excess of cell substance produced at the expense of oxidation. Where cell substance is mostly filamentous material and slimes, there is good reason to believe that growth is at the expense of organic material and not ferrous iron (Starkey, 1945).

More recently, Wolfe (1963) has discussed concepts of the "iron bacteria", paying particular attention to *Gallionella* as an iron bacterium. *Gallionella* is considered to be a gradient organism; that is, an organism capable of competing, within the proper environment, with non-biologically mediated ferrous iron oxidations. Culture studies on *Gallionella* indicated an autotrophic means of growth at the expense of ferrous iron (Kucera and Wolfe, 1957). Drabkova and Stravinskaya (1969) reported *Gallionella* to be the dominant organism influencing the dynamics of iron in Lake Krasnoe, U.S.S.R.: *Metallogenium* and *Siderocapsa* were also inhabitants of the lake environment. Few cultural methods were described in this report and it is difficult to assess the nutritional role of iron for these bacteria. Oxidation of ferrous iron by members of the *Thiobacillus* group has been unequivocally demonstrated by Colmer et al. (1950) and by Temple and Colmer (1951). *Thiobacillus ferrooxidans* is a true bacterium and lives as a chemoautotroph, according to the autotrophic concept of Winogradsky (1888).

The stability of the various types of iron compounds formed in nature will determine the profile of the iron bacteria selected. Aristovskaya and Zavarzin (1971) have adopted an Eh/pH iron stability diagram from the earlier work of Baas Becking et al. (1956, 1960) and Garrels and Christ (1965) to depict the growth regions within the natural environment of the main groups of iron bacteria. Figure 4.2 shows the domains of some iron compounds in terms of Eh and pH and the regions of selection of iron bacteria. Ionic forms of iron can exist where the acidity is low and such soluble forms readily serve the thiobacillus-type of iron-oxidizing organism as an energy source. Specifically, such an environment might be acid mine drainage or an acid hot spring where acidity is associated with the oxidation of pyrite (FeS_2) or other minerals of sulfur. A recent observation was made of growth on Fe^{2+} at low pH values of *Sulfolobus* which was isolated from a sulfur-containing hot spring (Brierely and Murr, 1975). Also, Walsh and Mitchell (1972) have isolated *Metallogenium* from an acidic, iron-bearing stream.

Iron carbonates and sulfides are substrates for *Leptothrix* and *Gallionella* organisms. Relatively low levels of these iron-containing compounds can serve as nutrients under conditions of neutral pH, low oxygen pressure and

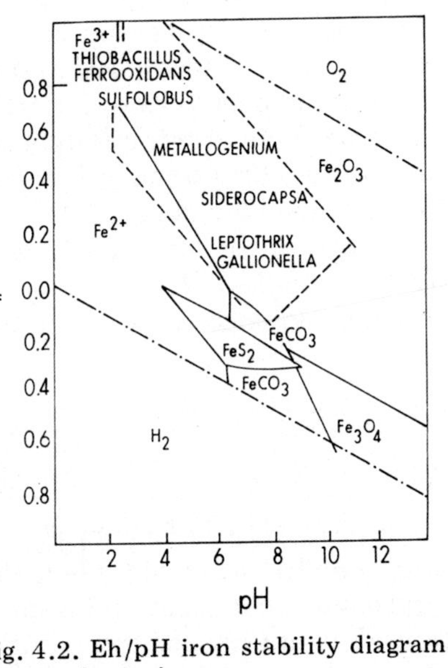

Fig. 4.2. Eh/pH iron stability diagram showing the natural domains of the main groups of the iron bacteria.

low Eh; a continuous iron supply is necessary to support microbial growth. Other organisms that function within a short distance of the Fe^{2+}/Fe^{3+} borderline, as shown in Fig. 4.2, are the *Siderocapsa* which oxidize organic complexes of iron. The complexes exist as stable chelate compounds of reduced iron, resisting oxidation to ferric hydroxides. *Siderocapsa* utilize the organic part of the complex for nutrition, liberating the ferrous iron. A rapid oxidation of the liberated iron occurs, resulting in the formation of ferric hydroxide precipitates which accumulate upon the surface of the cell.

Heterotrophic bacteria (and possibly some autotrophs) can cause the reduction of ferric iron either by producing reducing substances or by directly reducing the Eh of the milieu below that needed for the stability of ferric oxides. Relatively little research has been reported on the biological reduction of ferric iron.

Each major group of iron bacteria will now be considered in somewhat more detail.

Iron-oxidizing thiobacilli

Energy metabolism. One member of the genus *Thiobacillus*, *Thiobacillus ferrooxidans* stands out as a true Fe^{2+}-oxidizing bacterium. Considerable attention has been given this bacterium, motivated by both fundamental interests in the nature of chemoautotrophy and because of the organism's importance to applied biology. Included in the latter are the industrial leaching of metals from sulfide ores and the pollution of waters by acid mine drainage from coal mines (see Chapter 6.3). Recent reviews concerned with *T. ferrooxidans* have dealt with its general biology (Lundgren et al., 1974), its role in acid mine drainage (Lundgren et al., 1972; Lundgren, 1975; Dugan, 1975) and its role in leaching metals from sulfide ores (Tuovinen and Kelly, 1972). The present authors consider the organism *T. ferroxidans* to be synonymous with *Ferrobacillus ferrooxidans*, and have supported this stand (Unz and Lundgren, 1961; Silverman and Ehrlich, 1964; Tuovinen and Kelly, 1972).

In the biosphere, the existence of a high concentration of soluble iron (Fe^{2+}) requires a pH below 4; the optimum pH for growth of *T. ferrooxidans* is in a range of 2—3.5. Generally, the following equations describe the biological oxidation of ferrous iron:

$$4\ FeSO_4 + 2\ H_2SO_4 + O_2 \rightarrow 2\ Fe_2(SO_4)_3 + 2\ H_2O \qquad (1)$$

$$2\ Fe_2(SO_4)_3 + 12\ H_2O \rightarrow 4\ Fe(OH)_3 + 6\ H_2SO_4 \qquad (2)$$

The organism consumes O_2 in the ratio of one mole O_2 per four moles of Fe^{2+} and the oxidized product is Fe^{3+}. Energy (ATP) is derived when the electrons liberated from iron are coupled to cytochrome *c* via cytochrome *c* reductase. The electron(s) is then transported from cytochrome *c* to cytochrome *a* via cytochrome oxidase and is subsequently accepted by oxygen. During electron transfer, at least one phosphorylation site is coupled to iron oxidation. The activation energy for iron oxidation is about 46.9 kJ mol^{-1} (Lundgren, 1975; Guay et al., 1975). Energetically, assuming a value of 29.1 kJ mol^{-1} for iron oxidation at pH 2, insufficient energy would be available from the transport of a single electron for the formation of ATP (Tuovinen and Kelly, 1972). A more likely event is that two ions of Fe^{2+} are oxidized and a pair of electrons is transported yielding approximately 59.0 kJ for the synthesis of ATP; 0.5 or 1 ATP is probably formed per Fe^{3+} produced (Tuovinen and Kelly, 1972).

Ferric sulfate, the product of the oxidation, reacts with water to form ferric hydroxide and sulfuric acid (reaction (2)). This reaction is spontaneous and leads to a net increase in acid formation in the environment. Furthermore, the reaction is important in maintaining an electron-proton balance during the oxidation of iron. The electrons liberated from Fe^{2+} are presumed to be accompanied by protons. However there is no biological source of protons since the substrate is known not to enter the bacterium (Dugan, 1975): these protons are provided by the non-biological reaction (2). Various ferric hydroxide complexes result as reactions occur between ferric hydroxide and sulfates. These ferric hydroxy-sulfate complexes have a buffering capacity which controls the pH of the environment affecting the transformation of iron. Complex formation can also compete with reaction (2) so that less H_2SO_4 is produced (Dugan, 1975). The yellowish and brownish red precipitates observed in acid mine drainage are forms of hydroxide complexes similar to those noted in the rusting of iron.

Some common forms of reduced iron serving as inorganic energy substrates for bacteria are those of sulfide minerals such as pyrite, FeS_2; chalcopyrite, $CuFeS_2$; and arsenopyrite, FeAsS. Bacteria can attack (metabolize) minerals directly, resulting in a direct transformation of the element.

However, the mechanism for direct oxidation is poorly understood (Silverman and Ehrlich, 1964). Iron is also made available for microbial oxidation after dissociation of the sulfide complexes by a chemical oxidation of the sulfide moiety of the mineral. A strong chemical oxidizing agent is the Fe^{3+} ion itself. Singer and Stumm (1970) showed that, under acidic conditions and in the absence of bacteria, Fe^{3+} was a much more effective catalyst of pyrite oxidation than was ferrous iron. However, in the presence of bacteria, the rate of pyrite oxidation in the presence of Fe^{2+} was higher; the reduced iron was biologically oxidized to ferric iron which then oxidized the pyrite:

$$14\ Fe^{3+} + FeS_2 + 8\ H_2O \rightarrow 15\ Fe^{2+} + 2\ SO_4^{2-} + 16\ H^+ \qquad (3)$$

Singer and Stumm (1970) proposed that, in natural acidic environments, a cycling of iron occurs which is dependent upon the iron-oxidizing bacteria to supply Fe^{3+} ion for pyrite oxidation.

Although reduced iron is essentially in the ionic and soluble form when it serves as an energy source for autotrophs, the product of the oxidation (Fe^{3+}) rapidly forms insoluble precipitates. Depending upon environmental conditions and the availability of anions, different froms of metal precipitate occur.

Ecological and cultural studies of iron-oxidizing bacteria have identified a number of unusual abilities of the bacterium. Thiobacilli are able to tolerate high concentrations (up to 10 g l^{-1}) of Zn^{2+}, Ni^{2+}, Cu^{2+}, Co^{2+}, Mn^{2+}, and Al^{3+} (Tuovinen et al., 1971). Metal tolerance depends upon the energy source for bacterial growth and may result from a selection, within a population of cells, of metal-tolerant organisms. *T. ferrooxidans* also grows in the presence of uranium at concentrations exceeding by a factor of 100 that required to inhibit most heterotrophic bacteria (Tuovinen and Kelly, 1972).

There are a number of other important physiological events that must occur if the iron-oxidizing bacteria are to survive and perform in metal transformations. These include CO_2 fixation, generation of reducing power (NADH and NADPH), heterotrophic metabolism and reduced sulfur oxidation.

CO_2 fixation and reducing power. Even though the production of energy from Fe^{2+} oxidation is a separate function from CO_2 fixation, the two processes are closely linked and regulated in iron-oxidizing bacteria. All of the cell's carbon can come from CO_2. The pathway for the endergonic fixation is a reductive pentose phosphate cycle involving two unique enzymic steps: (1) phosphorylation of ribulose 5-phosphate, catalyzed by phosphoribulose kinase to yield ribulose 1,5-diphosphate and; (2) ribulose-1,5-diphosphate carboxylase which catalyzes a reaction between one molecule of CO_2 and one molecule of ribulose-1,5-diphosphate with a concomitant dismutation to yield two molecules of 3-phosphoglyceric acid (Maciag and Lundgren, 1964; Din et al., 1967; Gale and Beck, 1967).

The synthesis of ribulose diphosphate carboxylase is subject to induction and repression. During heterotrophic growth this enzyme is repressed from 0 to 100% of the autotrophic level depending upon what carbon source is being metabolized. Fructose and glutamate are particularly effective in repressing the carboxylase (Tabita and Lundgren, 1971b). In this *T. ferrooxidans*, the phosphoribulose kinase is probably also a regulatory enzyme since adenosine monophosphate is a competitive inhibitor of the enzyme (Gale and Beck, 1966).

An ATP-dependent reduction of pyridine nucleotides (NADH, NADPH) provides the organism with reducing power for CO_2 assimilation. In the iron-oxidizing bacteria, this aspect of metabolism has received little attention; however, in other thiobacilli, reduction of pyridine nucleotides has been investigated (Aleem, 1970). Since the redox potentials of $NAD^+/NADH$ and $NADP^+/NADPH$ are low ($E_0 = -0.320$ and -0.324 V, respectively), and the Fe^{2+}/Fe^{3+} redox potential is high ($E_0 = +0.77$ V), iron cannot under normal redox conditions reduce the pyridine nucleotides. These bacteria carry out reverse electron transport to produce NADH in the presence of ATP, reduced iron and cytochromes (Aleem et al., 1963; Tikhonova et al., 1967).

Heterotrophic metabolism. The importance of the thiobacilli in iron transformations is primarily dependent on their ability to meet energy requirements from the oxidation of reduced iron. Since organic compounds can have a negative relationship with biological iron oxidation, they could have an influence upon metal transformations. Although a number of laboratory studies have shown the adaptability of iron-grown cultures to use organic substrates for energy and carbon, little is known about the organisms' capacity for heterotrophy (or mixotrophy) in nature. It is assumed that in nature, prior to growth on organic substrates, new enzyme synthesis is necessary. Enzyme synthesis would depend upon substrate concentrations, pH, temperature and other environmental parameters. Tuovinen and Kelly (1972) and Lundgren et al. (1974) have reviewed a number of major enzyme systems in iron-grown cells which are repressed or induced, depending upon growth conditions.

Results of laboratory studies have shown that some strains of *T. ferrooxidans* do not adapt to heterotrophy and those that do adapt become obligately heterotrophic after 7—10 transfers on glucose (without Fe) (Shafia and Wilkinson, 1969). Taxonomically this is a problem, for iron oxidation is a constitutive and primary characteristic for the classification of *T. ferrooxidans.* However, at present not enough research on all probable parameters needed for growth of *T. ferrooxidans* on different substrates has been done, so that an explanation of the initial observation of obligate heterotrophy is impossible. Guay and Silver (1975) reported that, after brief exposure to glucose of their culture of *T. ferrooxidans*, a new strain of *Thiobacillus* was isolated which could oxidize glucose or sulfur but not ferrous iron. The organism could be switched back and forth on the two growth substrates. These investigators suggested the name *Thiobacillus acidophilus* for the new isolate. The two organisms, *T. ferrooxidans* and *T. acidophilus*, had different G + C contents but no DNA hybridization studies on the two thiobacilli were reported. *T. acidophilus* was closely related to glucose-grown *T. ferrooxidans*, as reported by other investigators (Tabita and Lundgren, 1971a; Shafia et al., 1972). The recent data of Guay and Silver (1975) support the idea that laboratory cultures of *T. ferrooxidans* were heterogeneous and contained both iron- and sulfur-oxidizing bacteria.

Results of laboratory studies seem to indicate that, among the *Thiobacillus* group, we have a whole range of organisms possessing varying capacities for both inorganic and organic modes of metabolism. Therefore, within the biosphere, a dynamic situation exists and, as environmental changes occur, different organisms with specific metabolic advantages would evolve. Relative to primary iron transformations, the effects on organism change (or evolution) could be either positive and/or negative, depending upon the total ecology of the milieu.

Inorganic sulfur oxidation. Although this chapter is principally concerned with iron transformations and its cycling, one cannot overlook the inorganic sulfur oxidizing capacity of *T. ferrooxidans*. Chapters 6.1—6.4 of this treatise deal with the sulfur cycle but a few comments are relevant within the framework of the present discussion.

Many of the mineral sulfides found in nature exist as iron mineral combinations; consequently, the oxidation of any one reduced mineral can have an effect upon the biological transformation of another. *T. ferrooxidans* plays a dominant role in nature where sulfides are found and appears to have a direct influence over the vital activities of other thiobacilli, such as *Thiobacillus thiooxidans*. *T. ferrooxidans* has been associated with mineral sulfide leaching in Arizona, Utah, California, Pennsylvania and West Virginia in the United States, as well as in Australia, British Columbia, England, Russia, the Congo, and Mexico. Natural leaching of metals involves complex interactions between the organism and solid and aqueous mineral phases. Unfortunately, there is little information about such interactions in nature except that, in all instances, the primary organism is *T. ferrooxidans*. We are forced, therefore, to extrapolate from laboratory studies designed to test concepts and the literature on such studies has been steadily growing over the last few years (Silverman and Ehrlich, 1964; MacDonald and Clark, 1970; Torma, 1971; Tuovinen and Kelly, 1972; Torma and Legualt, 1973; Torma and Tabi, 1973; Silver and Torma, 1974; Torma et al., 1974; Torma and Subramanian, 1974; Guay et al., 1975).

Margalith et al. (1966) demonstrated that *T. ferrooxidans* grown on iron could oxidize sulfur but the rate of oxidation was lower than with iron; however, sulfur-grown cells of *T. ferrooxidans* could also oxidize sulfite, dithionate, thiosulfate, tetrathionate and sulfide but not thiocyanate (Silver, 1970). Details of the specific metabolic reactions for sulfur oxidation by *T. ferrooxidans* have been outlined in an earlier publication (Lundgren et al., 1974).

Iron-oxidizing Sulfolobus

Sulfolobus (or *Sulfolobus*-like organisms) is an acidophilic organism found in acid hot springs and is capable of oxidizing ferrous iron as well as sulfur. This dual metabolic capacity for inorganic oxidations in an acid milieu is comparable to that of *T. ferrooxidans*. The organism lacks the conventional cell wall of procaryotes (Brock et al., 1972) and it has the ability to grow at elevated temperatures (45—75°C), and to tolerate high concentrations of metals. The organism has potential use in commercial metal leaching (Brierley and Murr, 1975).

Iron-oxidizing Metallogenium

Metallogenium is an acid tolerant, iron-oxidizing organism isolated by Walsh and Mitchell (1972). The organism lacks a conventional cell body and its optimum growth is at pH 4.1; it requires ferrous iron for growth and shows encrustations of ferric iron in its viable filaments. The iron-oxidizing bacterium resembles the rose garland form of the *Metallogenium* associated with manganese deposition. The new isolate of Walsh and Mitchell (1972) was unable to grow on manganese. The role of iron in the nutrition of this iron-oxidizing strain is unclear; it may grow autotrophically or it may require an heterotrophic energy-yielding metabolism. It is worthy of note that the ΔF (kJ mol^{-1}) for ferrous iron oxidation will depend upon the pH of the environment. For example at pH 1.5, $\Delta F = -32.7$ and at pH 3.0, $\Delta F = -24.7$. Determinations of ΔF above pH 3.0 are difficult for at this pH the driving force for the oxidation of ferrous iron is not the potential of the Fe^{2+}/Fe^{3+} couple but becomes the removal of the ferric iron through precipitation as $Fe(OH)_3$ (Lees et al., 1969). Walsh and Mitchell (personal communication) suggest that there may be a pH-dependent succession of iron-oxidizing bacteria in the natural environment, initially involving the acid-tolerant, iron-oxidizing *Metallogenium* sp. at pH values of 4.5 to 3.5 where the ionic form of Fe^{2+} can exist. At pH's below 4.0 the chemolithotrophic *Thiobacillus* would then predominate and oxidize iron. In certain environments, *Sulfolobus* may also attack iron and thus influence metal transformations in nature.

With growth of *Metallogenium* at pH 4, one sees the first sign of the need of a complexing agent for stabilizing soluble iron. *Metallogenium* is sensitive to non-complexed ferrous iron; KH phthalate was an effective complexing agent in the growth studies of Walsh and Mitchell (1972).

Leptothrix group

These bacteria are not normal inhabitants of agricultural soils but are found in swamps, muds and freshwater lakes. The organisms also inhabit unnatural environments such as pipelines and filter systems. Their growth depends upon the proper balance of electrons (Eh) and protons (pH) when the proper concentrations of the ionic form of Fe^{2+} can exist to support growth. However, their energy source is probably organic material. The particular domain for growth of the organism is shown in Fig. 4.2 and it is an area where ferrous carbonate and pyrite predominate. The bacterium is also found in the interphase of iron-rich hypolimnetic water and overlying oxygenated water (Jones, 1975). *Leptothrix* are filamentous, difficult to isolate and maintain in pure culture, and generally poorly understood at the molecular and biochemical level.

Filaments can become heavily encrusted with hydrated ferric (and manganous) oxides which has led many investigators to attempt to prove an autotrophic existence for these bacteria at the expense of iron oxidation (Pringsheim, 1949). To date, it seems that the iron oxidation mediated by these bacteria is probably by non-specific reactions at the filaments which are not coupled to the ATP-generating system of the cells enclosed within the sheath. Mulder (1963) has reported on his observations of this group of bacteria.

Gallionella group

Gallionella and its physiology correlates well with that predicted by the conditions which establish the domain of the bacteria shown in the Eh/pH diagram in Fig. 4.2. The organism grows at about pH 6.0 in a medium containing ferrous sulfide and CO_2; it is also able to grow at higher pH's. In laboratory studies, flocculent colonies of *Gallionella* containing ferric hydroxide have been noted attached to the sides of glass vessels containing a mineral medium. Such development has been taken to indicate chemo-autotrophic growth but final proof of autotrophy is still lacking. With *Gallionella*, however, the evidence for autotrophy is stronger than that for *Leptothrix* (Kucera and Wolfe, 1957), although attempts to grow the organism on inorganic media have been unsuccessful. *Gallionella* appears as a tuft of screw-shaped filaments on the end of which can sometimes be seen terminal cells. The stalks are believed to originate from the secretion of ferric hydroxide from the side of the cell. Some investigators feel that the only live elements of *Gallionella* are the filaments (or stalks) which are not too unlike the non-cellular *Metallogenium*. Mechanistically, the specific role iron plays in the metabolism of this bacterium is still uncertain.

Siderocapsa group

These organisms develop in an environment in which free Fe^{2+} iron is unstable and the metal is retained in solution by organic binding. The organic compounds binding iron serve as substrates for the bacteria. The *Siderocapsa* group contains organisms that vary in shape and possess thin or thick capsules; all are able to deposit iron (or manganese) oxides on or in the capsules. There is usually more than one cell embedded in the capsule. Bacteria from lake water have now been grown in pure culture by Dubinina and Zhdanov (1975) who, on the basis of their results, question the present classification of these microbes. *Siderocapsa* are heterotrophic and may be related to the common soil *Arthrobacter*. Dubinina and Zhdanov propose that the present family *Siderocapsa* be abolished and that the naturally occurring bacteria so far described as *Siderocapsa* be considered as forms of *Arthrobacter*.

Heterotrophic bacteria and the reduction of iron

Enzymic reduction of Fe^{3+} by bacteria can occur in the natural environment. Unfortunately, few studies have been reported but these have demonstrated the ease with which the reductive reactions occur with the appropriate heterotrophs. Woolfolk and Whitley (1962) showed that micrococci could oxidize molecular hydrogen at the expense of Fe^{3+}; ferric hydroxide in suspension was converted to ferrous hydroxide. The hydrogenase enzyme involved in the reaction is rather widely distributed among organisms found in iron-bearing environments and it is logical to assume that the reaction is carried out by many indigenous bacteria.

With organisms such as *Bacillus circulans*, *Bacillus megaterium* and *Aerobacter aerogenes*, the reduction of ferric compounds takes place in the presence of a number of hydrogen donors. Ferric hydroxide was reduced when the chelator, 2,2′-dipyridyl, was present with bacteria and the appropriate substrate system (Bromfield, 1954). The reaction appeared to be effected by dehydrogenase enzymes and depended on the presence of the iron-complexing agent. The reduction by bacteria is undoubtedly important in increasing the nutrient availability of iron and in determining its movement in soil profiles. Iron sulfide (FeS, FeS_2) formation can occur in anaerobic sediments in the presence of organic matter and the bacteria, *Desulfovibrio* spp. *Desulfovibrio* produces H_2S which reacts with various reactable iron compounds (e.g. limonite), producing a reduced iron derivative (Berner, 1962; see Chapter 6.2). No specific instances of ferric iron reduction by autotrophic bacteria are known to the present authors but, since autotrophs are found in iron-rich areas and the bacteria do contain hydrogenases, it can be assumed that they too have the potential to reduce ferric iron.

Iron transformations

Those microorganisms discussed in the preceding sections contribute to the transformation of relatively large amounts of mineral matter. How they accomplish this is influenced by their metabolism and the conditions of the environment. The organisms play important roles in pedogenesis, and their effects may be of a direct or indirect nature. The biological cycle of iron transformation develops with great intensity in soils and can have a significant effect upon soil morphology. As an example, consider the redox reactions where direct oxidations occur with ferrous iron yielding ferric-type products. The reactions are important to the organism for energy and support of autotrophic growth. Since the energy yields are low, large amounts of ferrous iron must be oxidized to support growth. Many of those areas supporting the afore-mentioned transformations are man-made and result from coal and mineral mining. Exposure of pyrite to oxygen during coal mining establishes an aerobic enrichment condition needed for biological iron transformation; chemical reactions accounting for the changes were discussed earlier (see p. 217).

Other iron-bearing minerals besides pyrite in soil include biotite, hornblende, augite, limonite, vermiculite, magnetite and hematite. Mobilization of chemical elements from these minerals occurs through weathering brought about by microorganisms (Aristovskaya and Zavarzin, 1971). An important microbial agent in weathering is slime or capsular material. Other contributing agents are metabolic products secreted into soil solutions such as organic acids and phenols. Organic acids are found free and they react with minerals by forming complexes which in turn cause weathering of the elements. Considerable levels of dissolved ferrous iron can be maintained in soils, as well as aquatic systems, in the presence of organic acids such as tannic acid, gallic acid, glutamic acid and tartaric acid, to name only a few. Theis and Singer (1975) described a model for relating the behavior of iron in aquatic systems in the presence of organic matter and oxygen. The model is shown diagrammatically below:

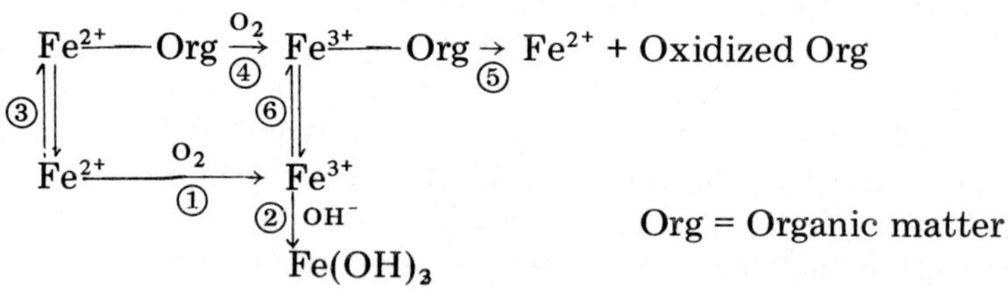

At neutral pH's, Fe^{2+} is oxidized rapidly to Fe^{3+} which precipitates as $Fe(OH)_3$ (Reactions (1), (2)). If significant concentrations of organic matter are present, the complexing reaction (Reaction (3)) will effectively compete with the oxygenation reaction. Environmental factors, including pH and type of organic material, affect the rate of the reaction. Reaction (4) proceeds at a slow rate to form the Fe^{3+}-complex which is unstable and becomes reduced by the organic compound (Reaction (5)). This frees the Fe^{2+} which can be recycled or utilized by any available microorganism. Both organo-iron compounds and ferrous and ferric hydroxide sols migrate in the soil under the peptizing action of humic matter. Humic acid can, depending upon the concentration of interaction compounds, promote iron dissolution or fixation in the soil (Aristovskaya and Zavarzin, 1971).

Many microorganisms form mineral acids (nitric and sulfuric) besides organic acids. Whenever sulfur is present, *Thiobacillus* spp. can oxidize the reduced sulfur compounds and produce sulfuric acid. In areas where intense nitrification is possible and where neutral or alkaline reactions occur, intense weathering by nitrifiers is also possible.

Agents which solubilize iron minerals have been isolated from soil fungi by Arrieta and Grez (1971), who implicated complexing agents in the mechanism of iron solubilization from augite and biotite. The solubilizing agents were probably siderochromes of the ferrichrome-type. Little specific research has been done on the extracting power of siderochromes but it is known that these compounds have very strong binding constants (about 10^{30}) for ferric ions. Siderochromes and related compounds (Bayers, 1974) are produced by many microorganisms to facilitate uptake of iron into the cell. Microbial cells secrete different types of chelating agents, with varying affinities for iron, for this purpose. Two specific categories of biological chelating agents are produced; these are hydroxyamic acids (siderochromes) and phenolic acids (catechols). Such agents are undoubtedly common in nature and may be effective in the solubilization of iron from minerals and inorganic and organic complexes.

The readers are referred to reviews by Silverman and Ehrlich (1964), Zajic (1969) and Aristovskaya and Zavarzin (1971) for a more in-depth discussion of the biotransformation of iron.

PRECAMBRIAN BANDED IRON FORMATIONS

Most of the world's iron ore resources are concentrated in the so-called banded iron formations (BIF) of Precambrian age found in many areas of the world (Table 4.1A, B, C). These deposits are thick (up to hundreds of metres) and extensive (single deposits extending for thousands of km^2). As a group, they form the most extensive resources of iron ore to be found on land. Banded iron formations are not all of Precambrian age, but most are. In fact, the majority of the Lake Superior-type deposits occur within a narrow age bracket in the late Middle Precambrian (Huronian) between 1.8 and 2.5 Gy B.P. (Goldich et al., 1961; Goldich, 1973). This narrow time restriction on a large number of strikingly similar deposits from all over the world has prompted a great many students of the Precambrian to speculate that these deposits represent a period of very fundamental change in the biological and chemical evolution of the oceans and atmosphere. Therefore, the BIF's are important not only because they are a great economic resource, but also because they provide clues to an important period in the evolution of the earth's lithosphere, hydrosphere, atmosphere, and biosphere. They contain a number of direct and indirect fossil links to the early evolution of life on earth, they provide information on the circulation of the Precambrian oceans, they provide evidence on the silica, sulfur, iron, and carbon cycles in the Precambrian, and their complex mineralogy and geochemistry provide valuable information on the behavior of a number of elements in geological and biological systems.

The BIF's characteristically consist of alternating laminations of chert (SiO_2) and an iron mineral, such as siderite ($FeCO_3$), chamosite (a complex

TABLE 4.1A

Occurrences of iron-rich beds of Pre age in North America and South America

Region	Locality	Character	Thickness (m)
Canadian shield: approximately Animikie (or Huronian) age	Iron-River-Crystal Falls district, Michigan	Cherty iron-formation, mostly siderite facies	180
	Amasa Oval and West Marquette, Michigan	Chert-magnetite, chert-hematite, chert-siderite	240
	Belcher-Nastapoka Islands, Canada	Cherty iron-formation oxide, silicate (greenalite), siderite facies	80
	Labrador trough, Labrador-Quebec	Cherty iron-formation; oxide, silicate, siderite facies	180
	Lake Albanel range, Mistassini district, Quebec	Cherty iron-formation; oxide, silicate, siderite facies	210
	Mesabi-Gunflint range, Minnesota-Ontario	Cherty iron-formation; oxide, silicate, siderite facies	240
	Cuyana district, Minnesota	Mostly silicate-siderite iron-formation	Several hundred
	Gogebic, Marquette, and Menominee districts, Wisc.-Michigan	Cherty oxide-formation; oxide, silicate, siderite facies	610 (Marquette district)
Pre-Animikie age	Michipicoten district, Ontario	Helen iron-formation; chert, siderite, pyrite	Siderite beds, to 110; pyrite beds, to 35
		Lower iron-formation; chert-magnetite, chert-hematite	Lenticular typically to 90 thick, 900 long
	Vermilion district, Minnesota	Mostly chert-hematite, chert-magnetite	Few hundred or less
Rocky Mountain U.S.A.	Skunk Creek belt, Michigan	Banded magnetite-quartz-silicate	30
	Southwest Montana	Banded quartz-magnetite, mostly	Beds, 15
	Atlantic City, Wyoming	Banded quartz-magnetite	50

TABLE 4.1A (continued)

Region	Locality	Character	Thickness (m)
	Seminoe Mountains, Wyoming	—	90
	Arizona-New Mexico	Banded jasper-magnetite	10 (lenticular)
South America	Guiana Shield, Venezuela	Banded magnetite-quartz, hematite-quartz	210 (Cerro Bolivar)
	Amapa, Brazil	Banded quartz-hematite	75
	Mato Grosso, Brazil, and Santa Cruz, Bolivia	—	300 (Morro do Urucum)
	Minas Gerais, Brazil (Quadrilatero Ferrifero)	Banded quartz-hematite (itabirite)	To 1000; average about 250
	Minas Gerais, Brazil	Siderite iron-formation	Lenses, to 15

TABLE 4.1B
Occurrences of iron-rich beds of Precambrian age in Europe and Asia

Area	Locality	Character	Thickness (m)
Scandinavia	Sydvaranger and others, Norway; Stripa and others, central Sweden; Porkonen and Jussaari, Finland	Banded quart-hematite, quartz-magnetite	Generally few tens of meters or less
U.S.S.R.	Krivoi Rog, Ukraine	Cherty iron-formation; oxide, silicate, carbonate facies	200
	Kursk (Kursk magnetic anomaly)	Banded quartz-hematite, quartz-magnetite	500
	Belozerkoye area, Zaporozh'ye region	Banded quartz-hematite	500
	Central Kazakhstan	Banded iron-formation	Believed to be thick
	Southern Urals	Siderite layers in shale and limestone	Lenses, 60
	Eastern Siberia	Hematite-jasper	Lenses in zone, 200

TABLE 4.1B (continued)

Area	Locality	Character	Thickness (m)
China	Chahar province, northern China	Oolitic hematite (minette type)	4 or less
	Southern Manchuria and adjacent areas of Korea	Banded hematite-quartz, magnetite-quartz	400 (at Miao Erh Kou)
India	Singhbhum and adjoining areas of Bonai and Keonjar	Banded hematite-jasper	1000 (possible fold repetition)
	Bailadila, Rowgat, and adjacent deposits of Madhya Pradesh	Banded quartz-magnetite, quartz-hematite	460
	Hyderabad	Quartz-magnetite, quartz-hematite, schist	45
	Mysore State	Banded hematite-quartz	60
	Salem-Trichinopoly district, southern India	Banded magnetite-quartz	Beds, to 30

iron silicate), magnetite (Fe_3O_4), or hematite (Fe_2O_3). The mineralogy of deposits or portions of deposits is the main criterion used for subdivision into geochemical facies (James, 1954, 1966; Goodwin, 1956). Individual laminations range in thickness from a fraction of a mm to several cm, but are usually in the order of 1 to 2 cm. Typical ore rock contains between 25 and 35% iron, with iron content being roughly inversely proportional to silica content (James and Sims, 1973).

Because of their economic potential, the geographic and stratigraphic extents of the BIF's, as well as their mineralogy and chemical composition are well known. However, the origin of the deposits is still very speculative. It is generally agreed that the minerals are the result of chemical precipitation, but there is little agreement among investigators regarding such important questions as the source of the iron, depth of water, and importance of diagenetic processes. Of particular importance is the question of the role of organisms in the origin of banded iron formations. Most genetic processes discussed in the literature are basically inorganic in nature. For example, James (1966), in a comprehensive discussion of the chemistry of iron-rich sedimentary rocks, devotes less than one page to the role of organic activity, and concludes that organisms were not of first-order importance in the concentration of iron. And yet most workers who have studied Precambrian banded iron formations in particular have been unable to escape the conclusion that microorganisms were involved in their origin, if only indirectly.

TABLE 4.1C

Occurrences of iron-rich beds of Precambrian age in Australia and South Africa

Locality	Character	Thickness (m)
Australia		
Roper Bar, northern Australia	Oolitic hematite and greenalite; magnetite, siderite	Beds, 12
Constance Range, northern Australia	Oolitic hematite and greenalite (?); siderite	10
Yampi Sound, northwestern Australia	Banded hematite-quartz, magnetite-quartz	40
Western Australia		Many thin beds
Middleback Ranges, South Australia		60

Locality	Stratigraphic position	Character	Thickness
South Africa			
Transvaal System of Transvaal	Pretoria Series, 5 horizons:		
	1. Waaikrall	Magnetite and silicates	4
	2. Daspoort	Oolitic hematite and chamosite	1
	3. Clayband	Magnetite-chamosite-oolite	1
	4. Pisolitic ironstone	Arenaceous hematite chamosite	2
	5. Magnetic quartzite (lowest unit)	Magnetite-hematite-chamosite oolite	8
	Dolomite Series, Thabazimbi	Banded quartz-hematite	60—80
Transvaal System of Cape Province	Pretoria Series, Griquatown Stage: Postmasburg belt (Gamagara Rand)	Banded chert, magnetite, silicates, siderite	760 in south; 300 in north
Witswatersrand System of Transvaal and adjacent areas	Three horizons, main one in Hospital Hill Slate Zone	Banded quartz-hematite, quartz-magnetite	As much as several hundred
Swaziland System of Transvaal, Natal, Zululand, Northern Cape Province; also Southern Rhodesia	Beds at different horizons	Banded quartz-magnetite, quartz-hematite; magnetite-chamosite-siderite in places	Generally 6—30

Harder (1919) in his classic paper on the geologic relations of iron-depositing bacteria, was motivated by "the difficulty of explaining the accumulation of large masses of ferric oxide by simple chemical processes" and by "the almost universal presence in nature of organisms capable of precipitating iron from solution". In the almost 60 years since Harder's paper, geologists and microbiologists have still not been able to agree on *how* organisms were involved in the formation of the Precambrian iron formations, and *which* organisms were responsible.

Most of the discussions relating organic activity to the origin of banded iron formations have centered on three areas: (1) the presence of fossil microorganisms in cherts from banded iron formations; (2) the similarities between laminations in banded iron formations and modern algal and bacterial stromatolites; and (3) role of photosynthesis in the evolution of oxygen in the ocean and atmosphere, and consequently in the origin of banded iron formations.

Fossil microorganisms

The best preserved microfossils in the Precambrian iron formations are found in the cherts. The roles played by these organisms in the formation of the cherts is discussed in greater detail in Chapter 7.1. For our purposes here, we are interested in whether these forms are indeed the fossilized remains of microorganisms, and what role they may have played in the formation of the iron minerals.

Rod-shaped, spheroid, and (rarely) stellate forms have been found in cherts from many banded iron formations (e.g. Glaessner, 1962; Cloud, 1965; Barghoorn and Tyler, 1965; Barghoorn and Schopf, 1966; LaBerge, 1967, 1973; Cloud and Licari, 1968; Schopf, 1970; Barghoorn, 1971; Schopf and Oehler, 1976; Walter et al., 1976). Because these fossils are so ubiquitous in cherts of the banded iron formations, it is reasonable to at least suspect that there is a genetic relationship between the organisms and the chert, and possibly the iron minerals as well. The above authors assumed that the organisms were procaryotic photoautotrophs, and were either bacteria or blue-green algae, or a common heterotrophic ancestor (Olson, 1970). LaBerge (1973) concluded that spherical structures he observed in "astronomical numbers" in cherts from numerous Precambrian BIF's were indeed silica-secreting microorganisms, either bacteria or blue-green algae. However, Oehler (1976) recently demonstrated that quartz microspheres formed by experimental hydrothermal crystallization of silica gel closely resemble the spherical structures described by LaBerge (1973). He suggests that the Precambrian forms may have had a similar origin and are therefore not microfossils at all. Engel et al., (1968) concluded that spheroidal and cup-shaped forms from the Onverwacht Series of South Africa (ca. 3.2 Gy old) probably were organic remains with algal affinities. They occur in sedimentary rocks intimately associated with kerogen-bearing carbonaceous substances which appear to form parts of the walls of the spheres and cups. They are also associated with kerogen-bearing filamentous forms which also appear to be microfossils. However, Engel and colleagues also describe spherical forms of clearly inorganic origin from Onverwacht basalts which appear very similar to the spheroidal forms in the Onverwacht sedimentary rocks. In fact, they report that several microbiologists, not knowing what the enclosing rock was, described the forms as seen in thin sections of the basalts as being definite remains of unicellular life.

Barghoorn and Tyler (1965) observed a number of organic forms in the Gunflint chert from the Lake Superior banded iron formations, and described 8 form genera with 12 species. Most were filamentous, although spherical and stellate forms were also described. Some of the forms suggested taxonomic affinities with modern bacteria, while others appeared to be more closely related to modern filamentous and coccoid blue-green algae. Most of the forms described by Barghoorn and Schopf (1966) and Schopf and Barghoorn (1967) from the Fig Tree chert of South Africa were spheroidal, resembling coccoid blue-green algae. Although most of the variability in form taxa described by Barghoorn and his associates may indeed be a reflection of actual variability in biologic taxa, modern blue-green algal cells are known to exhibit variability of form within species, and even within individuals during their life cycles and post-mortem degradation. This could account for some of the observed variability in form of Precambrian microfossils (Awramik et al., 1972).

In spite of the problems involved in recognizing whether a form in rock is of organic origin, and whether form variability is intra- or inter-taxon variability, there remain many unequivocal microfossils, most notably in cherts from the Gunflint, Fig Tree, Bitter Springs, and Frere Formations. (Cloud, 1965; Barghoorn and Tyler, 1965; Barghoorn and Schopf, 1966; Schopf and Barghoorn, 1967; Schopf, 1968; Schopf and Blacic, 1971; Knoll and Barghoorn, 1975; Schopf and Oehler, 1976; Walter et al., 1976). Most of these microfossils are apparently the remains of un-nucleated (procaryotic) organisms with bacterial and blue-green algal affinities. However, microfossils from the Bitter Springs chert of Australia (ca. 0.85 Gy old) have been interpreted by Schopf and Oehler (1976) as being the remains of nucleated (eucaryotic) organisms. They further conclude that these organelle-bearing "higher" forms may have appeared as early as 1.4 Gy ago.

Additional evidence of the existence of microorganisms in the Precambrian BIF's is provided by the occurrence of stromatolites within the iron formations.

Stromatolites

Stromatolites are variously shaped, laminated masses of sediment or sedimentary rock in which the laminations are due, at least in part, to the trapping and binding of sediment by organisms, most commonly by mats of filamentous blue-green algae or bacteria. The masses may be planar, domal, columnar, hemispherical, or spherical, and may range in size from less than 1 mm to planar laminated sheets covering tens of square kilometres. Most modern and ancient stromatolites occur in carbonate sediments and sedimentary rocks, but they can form by trapping and binding of any sediment particles. Of particular importance to the origin of Precambrian banded iron formations are the siliceous stromatolites commonly associated with hot springs (e.g. Walter, 1972; Doemel and Brock, 1974). Widespread recognition of the importance of organic involvement in the formation of laminated sediments has only come in the past 15—20 years. However, in recent years there has been a tendency toward over-interpretation to the point that many investigators equate laminated sediments with stromatolites (see Dean et al., 1975 for a discussion of this problem). Furthermore, because many modern stromatolites are forming in supratidal environments, there has been a tendency to imply that most ancient stromatolites also must have formed in supratidal environments. The important point is that true stromatolites are the result of direct control by microorganisms on the deposition of sediment. If the organisms controlling sedimentation were photosynthetic (as most organisms forming stromatolites today are), then the trapping and binding mats must have lived in the photic zone, i.e., from supratidal to a depth of about 100 m. The interested reader is referred to a recently published collection of papers edited by M.R. Walter (Walter, 1976) for a comprehensive review of all aspects of stromatolites. The papers of Awramik (1976) on the Gunflint stromatolites, and Mendelsohn (1976) and Trudinger and Mendelsohn (1976) on mineral deposits associated with stromatolites are particularly appropriate.

The masses of laminated chert characteristic of all Precambrian BIF's occur in a wide variety of external forms comprising just about every form type described for modern and ancient stromatolites. Hofmann (1969) devised a complex form taxonomy of stromatolites from the Gunflint Formation of the Lake Superior region, and has tied this classification into those of Logan et al. (1964), Rezak (1957), and Soviet geologists (e.g. Korolyuk, 1960; Krylov, 1963; Komar, 1966) which are based mainly on carbonate stromatolites.

Unfortunately, most so-called stromatolites do not contain micro-structures which can be identified as organic, and consequently an organic interpretation rests mainly on the laminations and external morphology. However, comparison of true stromatolites in Precambrian iron formations (e.g. Barghoorn and Tyler, 1965; Cloud, 1965; Glaessner et al., 1969; Walter et al., 1976) with recognizable unequivocal organic remains leaves little doubt that most of the lamination in the BIF's and other Precambrian strata are indeed true stromatolites.

Direct organic control on the formation of stromatolites, often difficult to prove in rocks, is that of trapping and binding of sediment particles. However, microorganisms may exert indirect biochemical controls which are of even greater importance when considering the precipitation of iron but are even more difficult to prove. One indirect influence would be the autooxidation of iron by oxygen production (Pringsheim, 1949; Silverman and Ehrlich, 1964). In addition, variations in Eh and pH due to variations in rates of photosynthetic CO_2 consumption and O_2 production could affect the precipitation of both silica and iron. For example, Walter (1972) concluded that some very fine laminations in siliceous sinter from the Precambrian Gunflint Formation were not stromatolites but due to diurnal pH-Eh changes caused by photosynthesis.

Filamentous and stalked bacteria of the *Leptothrix* and *Gallionella* groups are commonly encrusted with ferric and manganic oxides and hydroxides, but it is not known whether this is the result of direct enzymic oxidation of iron and manganese, of extracellular (or extra-sheath) excretion of iron and manganese oxides, or of simple adsorption of preformed iron and manganese oxides onto the surface of the bacterium (Pringsheim, 1949; Silverman and Ehrlich, 1964; Wolfe, 1963). According to Wolfe (1963) bacteria belonging to the *Gallionella* group are capable of direct iron oxidation in the absence of an organic carbon source. The fact that extant bacteria from this group can live under near-neutral pH conditions makes them much more plausible contributors to the formation of Precambrian iron ores than true iron-oxidizing bacteria (such as *T. ferrooxidans*) which only exist under very acid conditions (Fig. 4.2). *Gallionella* is apparently involved directly in the formation or recent iron sedimentary deposits forming today in a bay of a volcanic island in the Aegean Sea (Puchelt et al., 1973). Regardless of the specific role of iron in the metabolism of the organism, colonies of these filamentous and/or stalked bacteria would be capable of producing iron-rich stromatolitic laminae. Unfortunately, diagenesis has removed much of the evidence; organic matter has been oxidized and destroyed, opaline silica has been remobilized and recrystallized into chert, and iron mineralogy has been altered by oxidation (e.g. siderite and pyrite to hematite and magnetite). Based on the diversity of microorganisms in modern stromatolites (e.g. Doemel and Brock, 1974; Gebelein, 1969), and the diversity of microfossil remains in the banded iron formations, the Precambrian stromatolites were probably complex communities and not monospecific mats. They are certainly the most abundant fossils of the early and middle Precambrian. The decline in the late Precambrian is probably due to the appearance of burrowing and grazing organisms (metazoans) approximately 0.7 Gy ago (Awramik, 1971; Garrett, 1970). Until then, the microorganisms had the Precambrian seas pretty much to themselves.

Photosynthesis, the evolution of oxygen, and the origin of Precambrian banded iron formations

The origin of the earth's atmosphere is a subject of such enormous complexity, and has so many implications and ramifications that it almost defies synthesis. Fortunately, the subject has been eloquently summarized at several different levels by Preston Cloud and his associates (e.g. Cloud, 1968, 1973, 1974; Cloud and Gibor, 1970). Our main concern here is that the Precambrian BIF's represent the very first evidence of oxygen in the hydrosphere.

Because the atmosphere during earliest Precambrian time (i.e. prior to ca. 3.5 Gy ago) probably consisted mainly of CH_4, NH_3, H_2, and H_2O (Holland, 1964), the oceans at this time must have been reducing (no free oxygen) and alkaline (because of the ammonia). The first hints of life are the fossil forms described by Engel et al. (1968) from the Onverwacht Series (3.2 Gy old) of South Africa. These first organisms were almost certainly anaerobic procaryotic heterotrophs. It is commonly assumed that nucleated (eucaryotic) photoautotrophs did not appear until the late Precambrian, 1.4 to 2 Gy ago (Cloud, 1973; Schopf and Oehler, 1976). However, it is possible that photosynthesis may have begun as far back as 3.35 Gy ago (Cloud, 1974) based mainly on carbon isotope data of Oehler et al. (1972). Photosynthetic fixation of carbon is the main carbon isotopic fractionation process in nature, preferentially utilizing lighter carbon. Oehler and associates reported that organic matter in the Middle and Upper Onverwacht Series was isotopically higher ($\delta^{13}C_{PDB}$ usually less than $-25‰$) than organic matter in the Lower Onverwacht sediments ($\delta^{13}C_{PDB}$ $-16.5‰$), suggesting photosynthetic carbon isotopic fractionation. Cloud (1974) points out that, if the conclusion of Oehler and her associates is correct, then the origin of life must have been even earlier because it is very unlikely that the first organisms were capable of photosynthesis, a process which evolved from heterotrophic ancestors (Olson, 1970; Margulis, 1971).

Although the bulk of evidence supports the origin of O_2 as biologic, Brinkmann (1969) indicates that oxygen was probably also being supplied by photodissociation of water vapour in the upper atmosphere. However, any oxygen which was produced, either by photosynthesis or by photodissociation, would be quickly utilized by a number of oxygen sinks or oxygen depressors including ferrous iron and other non-carbon elements in the ambient waters as well as by ions and molecules such as NH_3 and CH_4 in the upper atmosphere. The latter reactions would lead to the formation of carbon dioxide and nitrogen gas:

$$CH_4 + 2\,O_2 \rightarrow CO_2 + 2\,H_2O \qquad (4)$$

$$4\,NH_3 + 3\,O_2 \rightarrow 2\,N_2 + 6\,H_2O \qquad (5)$$

The evolution of CO_2 and N_2 probably required at least 1 Gy because there is no evidence of free oxygen until the appearance of the Proterozoic record showing terrestrial sediments containing grain coatings and interstitial fillings of ferric oxide (Cloud, 1968). The production of CO_2 by oxidation of CH_4 and the likely conversion of CO to CO_2 utilizing photolytic O_2 coupled with the decrease in ammonia, undoubtedly had an effect on the pH of sea water, which became less alkaline and possibly slightly acid. This would explain the perplexing presence of carbonate rocks older than 3.3 Gy but lack of carbonates associated with the banded iron formations between 3.2 and 2 Gy old (Cloud, 1973).

But what about the banded iron formations themselves? They first appeared in the early Precambrian of south-west Greenland (ca. 3.7 Gy B.P., Appel, 1974) and in the Upper Onverwacht Series of South Africa (ca. 3.2 Gy B.P.). They contain the oldest fossil evidence and appear just after the possible appearance of photosynthesis, and they certainly contain oxidized iron. Cloud (1973) has suggested that the reduced bottom waters of the early and middle Precambrian oceans contained considerably higher concentrations of iron, as Fe^{2+}, than the present oceans. The mixing of the formerly stratified waters would have brought waters containing this Fe^{2+} into the photic zone containing microorganisms including the sediment-binding blue-green algae. The latter were living precariously and were dependent upon the presence of a major oxygen depressing ion such as Fe^{2+} to maintain an anaerobic environment. Cloud (1974) visualizes a mutually dependent relationship with the biological generation of O_2 in the hydrosphere in balance with Fe^{2+} ion:

$$4\ FeO + biol\ O_2 \rightarrow 2\ Fe_2O_3 \qquad (6)$$

The Fe_2O_3 could then react with sedimented carbonaceous matter as:

$$6\ Fe_2O_3 + C \rightarrow 4\ Fe_3O_4 + CO_2 \qquad (7)$$

This reaction accounts for the presence of magnetite and the limited carbon found in BIF's. Such oxygen-releasing photosynthesizers could grow nicely as long as iron and other essential nutrients were available. Whenever a shortage of one or the other minerals occurred, growth died off and a new band or iron-rich lamina was not produced until a second growth cycle occurred. Cloud's idea is that an episodicity of either ferrous iron supply or of microbial growth or activity in the presence of a constant supply of iron accounts for the banding observed. The productivity of the procaryotic organisms in the ocean at the time of the BIF's must have been at least partly self-regulated through chemical stresses occurring in an ever changing environment. Bacteria were then and are today the hardiest of all living species and one would expect that biological competition was very active as it is today. Although Cloud's (1974) ideas relative to the origin of banded iron are attractive, alternative explanations are available and include seasonal upwelling, variation in temperature, and variation in light intensity. Seasonal upwelling as a cause of banded iron ores was suggested by Drever (1974) in his shallow water model; deep anaerobic water containing Fe^{2+} upwells onto a shallow platform and precipitates ferric oxide. This model does require that the surface waters are oxidizing which implies a buildup of excess free O_2 in the atmosphere before 3.2 Gy ago. A shallow water upwelling model was also assumed by Holland (1964).

It is hypothesized further (Cloud, 1974) that during the expansion of procaryotic photosynthesizers a burst of biochemical diversification followed leading to the origin of oxygen-mediated enzymes. Those organisms possessing these enzymes were eventually free from a dependence upon external oxygen-depressors. By about 2 Gy ago, eucaryotic photoautotrophs with oxygen-mediating enzymes had evolved and free oxygen could accumulate in the ocean and atmosphere, reducing the Fe^{2+} concentration to about 1 mg l^{-1} (the concentration of present oxygenated ocean bottom waters). Cloud's model not only explains the appearance of BIF's in the Onverwacht Series soon after the appearance of chemical indications of photosynthesis but it also explains the termination of extensive BIF's about 2 Gy ago, and the onset of conspicuous detrital hematite-rich sediments called red beds.

Although the chemical data suggest that photoautotrophs may have been present at the time of deposition of the BIF's, most of the organisms whose fossil remains are so abundant in the cherts were probably anaerobic heterotrophs or organisms associated with the oxidation of ferrous iron and manganese, such as *Metallogenium* and *Siderococcus.* These bacteria could have assisted in keeping the oxygen level low while deriving their energy from iron or manganese oxidation (Cloud, 1973, 1974).

Any carbon-fixing process, photoautotrophic or chemoautotrophic, would remove CO_2 from the ocean and hence the atmosphere. This would require that the CO_2 concentration of the atmosphere before these autotrophic processes began was considerably higher than at present. For example, Garrels et al., (1973) have estimated that in order to produce all of the ferric iron, sulfate, and free O_2 that has been formed since the evolution of the earth's crust, a total of 5.9×10^{20} mol of CO_2 must have been depleted from the original CO_2 reservoir.

FERROMANGANESE NODULES

Nodules, concretions, or crusts rich in iron and manganese, collectively referred to as "ferromanganese nodules," were first collected on the sea floor during the Challenger Expedition in the 1870's. Since then, they have been the subject of numerous investigations dealing mainly with their mineralogy and geochemistry. With rapid advancement of marine technology, these nodules are considered potential ores not only of iron and manganese, but also of certain trace elements (especially Co, Ni, and Cu) often concentrated in the nodules. As of 1976, no actual mining of ferromanganese nodules has been done, although one organization has filed a mining claim. Although considerable time, effort, and funds have been applied to the study of the distribution of nodules on the sea floor, and the location of the most economically important areas, little effort has been spent on understanding the origin of the nodules. Most of the literature on ferromanganese nodules has appeared in the past ten years as a result of the "rediscovery" of ferromanganese nodules and their economic potential. For a review of ferromanganese nodules, the interested reader is referred to the slightly dated but still excellent book by Mero (1965). More recent collections of papers dealing with all aspects of ferromanganese nodules can be found in Horn et al. (1972), Morgenstein (1973), National Science Foundation (1973), Cronan and Varentsov (1978) and Glasby (1977).

Location of nodules

Any theories which attempt to explain the formation of marine ferromanganese nodules must take into consideration the fact that they also occur in freshwater (lakes, streams, and bogs). Some investigators consider that marine and freshwater deposits are fundamentally different. For example, Ehrlich (1972 and personal communication) feels that freshwater concretions are not true manganese nodules but are "bog iron ores". However, most deposits described as "bog ores" are actually encrustations on rocks in streams, concretions in lakes, or at least formed in streams or lakes and subsequently covered by bog vegetation in the course of lake succession (e.g. Ljunggren, 1953, 1955a; Landergren, 1948; Kuznetsov et al., 1963). For example, Arrhenius (1967) points out that many of the so-called bog ores in Sweden which are now presently located in bogs actually formed in shallow pools which were later covered over by vegetation.

The similarities between bog and lake ores have also been pointed out by Kuznetsov et al. (1963) who summarized investigations by a number of Soviet microbiologists. They concluded that the abundance of filamentous iron bacteria *Leptothrix* and *Crenothrix* in both lacustrine and true bog environments "is probable evidence of their participation in the formation of the bog ore." In addition, work by Perfil'ev and other Soviet microbiologists (in Kuznetsov et al., 1963) has shown that a species of *Metallogenium* is the principal ore-forming agent in lake ores in a number of Soviet lakes. They conclude that microorganisms play a dual role in the formation of "bog ores" in general. First, they create a reducing environment in subsoils (and also in subsurface lake sediments) which facilitates migration of ferrous iron, and secondly they cause the precipitation of ferric hydroxide. The only real differences between lacustrine and true bog deposits, in terms of transport

TABLE 4.2

Elemental composition of freshwater and marine ferromanganese nodules

Element	Oneida Lake [1]	English Lakes [2]	Lake Michigan [3]	Canadian Lakes [4]	Lake Ontario [6]	Pacific Ocean [5]
Mn (%)	24.9	1.4—13	1.3—14	15.7—35.9	17.0	8.2—50.1
Fe (%)	15.1	5.9—39	8—32	11.7—40.2	20.6	2.4—26.6
Ca (%)	0.70				3.1	0.8—4.4
Mg (%)	0.38			0—1.4	2.2	1.0—2.4
Na (%)	0.23	0.18—1.0			0.06	1.5—4.7
K (%)	0.55	0.5—2.9			0.24	0.3—3.1
Sr (%)	0.26	0.003—0.01	0.01			0.024—0.16
Cu ($\mu g\ g^{-1}$)	27	5—20	1—14	6—16	363	280—16,000
Zn ($\mu g\ g^{-1}$)	99	0.03—3000	43—284	250—1940	1996	400—800
Co ($\mu g\ g^{-1}$)	20	20—60	11—86	153—230	643	140—23,000
Ni ($\mu g\ g^{-1}$)	25	10—40		95—373	2385	1600—20,000
Cr ($\mu g\ g^{-1}$)	29	4—60				10—70
Pb ($\mu g\ g^{-1}$)	64	30—8000		10—29		200—3600

Sources: 1, Dean et al. (1973); 2. Gorham and Swaine (1965); 3. Edgington and Callender (1970); 4. Harriss and Troup (1970); 5. Mero (1965); 6. Cronan and Thomas (1970).

and precipitation of iron, are differences in pH and abundance of humic substances which would form organic complexes with iron and facilitate its migration in ground water. A good example of an environment intermediate between a true bog and a clearwater lake is a dystrophic lake with high humic content and low pH. It is not surprising that many freshwater ferromanganese deposits in North America and Scandanavia occur in, but are by no means restricted to, dystrophic lakes.

The main reason for assuming different origins for marine and freshwater (including stream, lake, and "bog") deposits is the fact that freshwater deposits generally contain more iron than manganese, and marine deposits contain more manganese than iron (Table 4.2). However, there are many iron-rich marine deposits and many manganese-rich freshwater deposits. We see no reason to separate genetically ferromanganese deposits formed in various freshwater environments, just as there is no reason to separate genetically ferromanganese deposits formed in various marine environments. In fact, most shallow-water marine ferromanganese nodules have more in common with freshwater nodules than with deep-sea nodules in terms of composition, form, and rate of formation.

A comparison of North American freshwater ferromanganese nodules with shallow-water and deep-sea marine nodules is presented elsewhere (Dean and Ghosh, 1978). The main point here is that although there are differences in these environments, these are mainly differences in sources of iron, manganese, and other elements, and in rates and degrees of processes operating in formation of the deposits. In spite of the fact that freshwater ferromanganese nodules occur in very different chemical and physical environments, the basic mineral and chemical compositions of iron and manganese oxides and hydroxides from both environments are essentially the same. The main difference between freshwater and marine ferromanganese nodules is the fact that freshwater nodules form at rates which are at least three orders of magnitude faster than deep-sea nodules (Dean et al., 1973). As a consequence of faster growth, freshwater nodules do not have time to adsorb high concentrations of trace metals which characterize economically important deep-sea nodules (Table 4.2).

Inorganic mechanisms for nodule formation

Most theories which have been proposed to explain the origins of both marine and freshwater ferromanganese nodules have been based mainly on inorganic physico-chemical mechanisms. The main questions which still need to be answered are where did the iron and manganese come from, and how do iron and manganese become concentrated into nodules and crusts? Bostrom (1967) postulated four main sources of iron and manganese in the oceans: (1) volcanic activity; (2) hydrothermal activity; (3) direct precipitation from sea water; and (4) upward migration of ferrous and manganous ions in sediment pore waters due to reduction below the sediment-water interface. Although the first two sources may help to explain the occurrence of ferromanganese nodules and crusts in certain areas of the ocean (e.g., along mid-oceanic ridges and on volcanic sea mounts), they certainly do not apply to iron and manganese deposition in lakes. Sea water (and lake water) cannot be ruled out as possible direct sources of iron and manganese in nodules, although the nodules commonly contain concentrations of both elements which are at least five orders of magnitude greater than in the overlying waters. The fourth source, diagenetic remobilization of Fe^{2+} and Mn^{2+} under reducing conditions in the sediments, has been recognized as a dominant source of iron and manganese in some areas, particularly in shallow-water marine environments (e.g. Lynn and Bonatti, 1965; Bonatti and Nayudu, 1965; Manheim, 1965; Price, 1967; Glasby, 1972; Ku and Glasby, 1972). It is also potentially the most important source of iron and manganese in lakes.

Assuming that there is a sufficient supply of iron and manganese from one or more of the above sources, the problem still remains as to how they are precipitated and concentrated into nodules and crusts. As a result of a number of experimental studies, it is now known that ferric oxides and hydroxides will precipitate at pH and Eh values much lower than those necessary to precipitate manganese oxides and hydroxides (Hem, 1963, 1972; Collins and Buol, 1970, 1972; Krauskopf, 1956, 1957). Under oxidizing conditions (Eh 100—400 mV), a relatively high pH (>8) would be required to precipitate manganese which normally tends to remain in solution. Iron, on the other hand, can precipitate under conditions of lower pH and/or lower Eh. Burns and Brown (1972) suggested the following mechanism for the formation of marine ferromanganese nodules: At pH values greater than about 7, colloidal ferric hydroxide will precipitate around any negatively charged nucleus. Colloidal ferric hydroxide has a positive zero charge potential and is very active. The positively-charged ferric hydroxide colloids will then adsorb negatively-charged hydrolyzed manganous ions which would be subsequently oxidized to MnO_2 or a hydrated Mn oxide or hydroxide (e.g., birnessite or torodkite) under conditions of sufficiently high pH and Eh. The negatively charged Mn surface would then attract additional ferric hydroxide colloids, and so on. The precipitation would thus be autocatalytic, and would explain the alternation of Fe-rich and Fe-poor bands characteristic of both marine and freshwater ferromanganese nodules.

Organic mechanisms for nodule formation

Inorganic mechanisms for the formation of ferromanganese nodules and crusts imply that it should be theoretically possible to form nodules in marine and freshwater environments without the aid of organisms. However, organic interactions could accelerate nodule formation in a number of direct and indirect ways. The ability of microorganisms to affect the large scale precipitation of both iron and manganese is well known (Silverman and Ehrlich, 1964; Zajic, 1969, pp. 96—120). The most extensive investigations of the role of microbes in the origin of ferromanganese nodules have been carried out by Ehrlich and his colleagues (Ehrlich, 1963, 1966, 1968, 1971, 1972; Trimble and Ehrlich, 1968, 1970). Although Ehrlich concentrated his research mainly on the manganese bacteria, many of his conclusions can also be applied to the iron bacteria. The finding of iron and manganese bacteria associated with ferromanganese nodules, and the ease with which bacteria can cause the precipitation of iron and manganese in the laboratory strongly suggest that microbes play a role in the formation of ferromanganese nodules, even though they may not be necessary for nodule formation. Ehrlich (1972) does not view microbes as the only cause of nodule formation; the main role of microbes in nodule formation is the enzymic catalysis of Mn^{2+} oxidation. The oxidation of Fe^{2+} to $Fe(OH)_3$ may also be aided by microbiological activity but probably not by the true iron-oxidizing bacteria. Instead, this is accomplished by heterotrophic organisms that can actively cause the precipitation of $Fe(OH)_3$, under neutral or alkaline conditions, through metabolic reactions (Zajic, 1969, pp. 96—120 and Fig. 4.2). Alternatively, precipitation may be caused by passive processes which cause adsorption of iron and manganese oxides and hydroxides onto cell surfaces (Zapffe, 1931).

Direct organic contribution to ferromanganese nodule formation was also proposed by Graham and Cooper (1959) and Graham (1959) to explain ferromanganese-rich coatings on Foraminifera. Their results suggested that a protein-rich material coated agglutinated arenaceous forams providing a substrate for other organisms capable of extracting iron, manganese, and other trace metals from sea water. Dugolinsky et al. (1977) also suggested that *Foraminifera* may be important in initiating ferromanganese nodule growth by concentrating trace elements. Greenslate, (1973, 1974a, 1974b) found that cavities in planktonic skeletal debris, especially diatoms, were apparently serving as nuclei for incipient ferromanganese nodule development. Greenslate suggested that bacteria associated with the decay of the planktonic organisms, rather than the organisms themselves, may actually be responsible for the accumulation of the Mn-rich coatings. These findings are also important in that the coatings are low in iron, suggesting that manganese concentration mechanisms may be independent of iron-concentrating mechanisms. Mn-rich coatings have also been reported on zygospores of *Chlamydomonas* isolated from soils (Schulz-Baldez and Lewin, 1975).

In addition to the above direct contributions of microorganisms to the formation of ferromanganese nodules, several indirect contributions may also be possible, and may indeed be essential to the formation of extensive ferromanganese deposits, especially those highly enriched in trace metals other than iron and manganese. For example, the most extensive deposit of freshwater ferromanganese nodules yet reported is to be found in Oneida Lake, New York (Dean et al., 1973; Dean and Ghosh, 1978). Oneida Lake is also very eutrophic, with intense algal blooms during the summer months. Because of its large surface area and shallow depth, wind mixing of the entire water mass prevents summer thermal stratification. Therefore, unlike most eutrophic lakes, Oneida has high planktonic algal productivity without benthonic anaerobiosis. The combination of high rate of algal photosynthesis and wind mixing during the summer months results in high pH and Eh conditions over most of the lake bottom favoring the precipitation of manganese as well as iron. Standing crop algal biomass in Oneida Lake may be as high as 10^5 cells ml^{-1} during the summer (Greeson, 1971). Using Greeson's values for average algal productivity in Oneida Lake, and reported concentrations of trace metals in algae, Dean and Ghosh (1978) have estimated that an average of 8 Mg of iron and 3.2 Mg of manganese are tied up in algae at any one time in the lake. Because of well-oxygenated bottom waters, most of the rain of algal debris is oxidized, releasing iron and manganese to a high pH, well-oxidized bottom environment. Based on Greeson's extensive water chemistry data for Oneida Lake and its tributaries, Dean and Ghosh (1978) estimated that there is an average net transport to the sediments of about 300 kg of iron and 64 kg of manganese per day. Regardless of whether microorganisms are involved, directly or indirectly, in nodule formation, the concentration, transport, and release of iron and manganese by algae in a high pH, oxidized bottom environment are probably the main reasons for the extensive development of ferromanganese deposits in Oneida Lake.

A similar conclusion was reached by Greenslate et al. (1973) for deep sea nodules. Nodules with high concentrations of transition metals (especially copper and nickel) occur in the high productivity equatorial zone of the Pacific. Nodules are not very abundant within this zone because of the high rate of biogenic carbonate sedimentation, but those that are present contain unusually high concentrations of Cu and Ni. The Cu and Ni (as well as Fe, Mn, and other metals) are concentrated in plankton, transported to the bottom in organic debris, and released by decay. The organic decay also causes reducing conditions in the sediments permitting upward diffusion of metals which are then oxidized at the surface and incorporated into nodules. Most interest in deep-sea mining of ferromanganese nodules has been focused on the zone of relatively high Cu-Ni nodules between the Clipperton and Clarion fracture zones just north of the equatorial Pacific high productivity zone (Horn et al., 1972). Within this zone, sedimentation is much slower than in the equatorial zone because surface productivity is lower, and the bottom is below the carbonate compensation depth so that no biogenic carbonate debris accumulates. Nodules high in Cu and Ni are particularly abundant within this zone, although the Cu and Ni concentrations are not as high as in the equatorial zone. Horn et al. (1972) suggested that the reason for the abundance of nodules in this zone was because sediments within this zone are siliceous biogenic oozes (mostly radiolaria) with very high porosities facilitating upward diffusion of metals. If the hypothesis of Greenslate et al. (1973) is correct, then the rain of biogenic debris (radiolaria plus any calcareous plankton which make it to the bottom and are subsequently dissolved) is also the main transport mechanism of metals in the nodules.

In summary, the results of many experimental studies illustrate that a number of indirect contributions of organisms go into the formation of ferromanganese nodules. Plankton can concentrate metals and transport them to the sediments after their death; decay of the organic debris releases these metals and at the same time creates reducing conditions within the sediments. The high porosity of siliceous biogenic sediment facilitates upward diffusion of the reduced transition elements which are then oxidized at the sediment-water interface. Greenslate (1973) has stressed the fact that, in the genesis of ferromanganese nodules in the oceans, there is an intimate association between many transition elements and biological processes. We have little knowledge of these associations at the present time but we must agree with the Soviet microbiologists (e.g. Perfil'ev, 1927; Butkevich, 1928; Kalinenko, 1952; Imshenetskii, 1961; Kuznetsov et al., 1963, pp. 165—177) that microorganisms have played an active role in the formation of both marine and freshwater ferromanganese deposits. As Kuznetsov et al. (1963, p. 174) state "... there are fewer data in support of the physicochemical theory of the formation of iron concretions than there are to support their biological origin."

CONCLUDING REMARKS

In this chapter, we have attempted to relate current ideas on the roles of microorganisms to the genesis of two very different types of iron deposits — one very ancient (Precambrian BIF's) and one modern (ferromanganese nodules and crusts); one comprising the most extensive resource of iron ore on land, and the other comprising the most extensive mineral deposit in the oceans. We can conclude that organisms were almost certainly involved, directly or indirectly, in the formation of both types of deposits, but exactly how and to what extent they are involved remain unanswered questions even for modern deposits.

An historical perspective of the "iron bacteria" has been given and over the years geomicrobiologists have studied a variety of bacteria concerned with iron transformations. The absence of pure culture techniques and modern technical sophistication has prevented the clear elucidation of those organisms attacking iron for use as a major nutrient. It is remarkable that it took from 1888, when the concept of chemoautotrophy was advanced by Winogradsky, until 1950 before the existence of a bacterium which is able to grow at the expense of ferrous iron oxidation was unequally demonstrated (Colmer et al., 1950).

If nothing else, the reader should have some feel for the difficulty in recognizing organic involvement in mineral genesis. It is a tribute to the careful work of researchers such as Barghoorn, Cloud, Oehler, and Schopf, to mention only a few, that organic involvement should even be suspected in the formation of the Precambrian iron ores, and yet they have demonstrated that microorganisms were certainly present in abundance, and were almost certainly involved in some way in iron ore genesis. Direct organic involvement in the formation of both types of iron deposits may have been in providing a substrate, passively trapping and binding sediment particles, actively causing the precipitation of ferric oxides and hydroxides, or concentrating and transporting iron as a micronutrient. Indirect organic involvement may have been in the form of oxidation-reduction changes brought about by O_2 production, CO_2 consumption, or bacterial sulfate reduction. The latter process is extremely important in the formation of iron sulfides and is discussed in Chapters 6.1 and 6.3.

ACKNOWLEDGEMENTS

The authors are appreciative of the assistance given by Mrs. Charlotte Stephenson and Mrs. Linda Campbell in the preparation of the manuscript. This work was supported, in part, by a grant from the National Science Foundation (PCM73-02228), "Biochemical Ecology of Iron-Oxidizing Bacteria."

REFERENCES

Aleem, M.I.H., 1970. Oxidation of inorganic nitrogen compounds. Annu. Rev. Plant Physiol., 21: 67—90.

Aleem, M.I.H., Lees, H. and Nicholas, D.J.D., 1963. Adenosinetriphosphate-dependent reduction of nicotinamide adenine dinucleotide by ferrocytochrome *c* in chemoautotrophic bacteria. Nature, 200: 759—761.

Appel, P.U., 1974. On an unmetamorphosed iron-formation in the Early Precambrian of south-west Greenland. Mineral Deposita, 9: 75—82.

Aristovskaya, T.V. and Zavarzin, G.A., 1971. Biochemistry of iron in soil. In: A.D. McLaren and J.J. Skujins (Editors), Soil Biochemistry, Vol. 2, Marcel Dekker, New York, NY, pp. 385—408.

Arrieta, L. and Grez, R., 1971. Solubilizations of iron-containing minerals by soil microorganisms. Appl. Microbiol., 22: 487—490.

Arrhenius, O., 1967. Ore, iron, artefacts and corrosion. Sver. Geol. Unders. Arsb., Ser. C, 626, 39 pp.

Awramik, S.M., 1971. Precambrian columnar stromatolite diversity: reflection of metazoan appearance. Science, 174: 825—826.

Awramik, S.M., 1976. Gunflint stromatolites: microfossil distribution in relation to stromatolite morphology. In: M.R. Walter (Editor), Stromatolites. Elsevier, Amsterdam, pp. 311—320.

Awramik, S.M., Golubić, S. and Barghoorn, E.S., 1972. Blue-green algal cell degradation and its implication for the fossil record (abs.). Geol. Soc. Am., Abstracts with Programs, 4: 438.
Barghoorn, E.S., 1971. The oldest fossils. Sci. Am., 224: 30—42.
Barghoorn, E.S. and Tyler, S.A., 1965. Microorganisms from the Gunflint chert. Science, 147: 563—577.
Barghoorn, E.S. and Schopf, J.W., 1966. Microorganisms 3 billion years old from the Precambrian of South Africa. Science, 152: 758—763.
Baas Becking, L.G.M. and Parks, G.S., 1927. Energy relationship in the metabolism of autotrophic bacteria. Physiol. Rev., 7: 85—106.
Baas Becking, L.G.M., Ferguson Wood, E.J. and Kaplan, I.R., 1956. Biological processes in the estuarine environment, VIII. Iron bacteria as gradient organisms. K. N. Akad. Wet. Amsterdam. Afd. Natuurk., Proc. Ser. C Biol. Med. Sci., 59: 398—407.
Baas Becking, L.G.M., Kaplan, I.R. and Moore, D., 1960. Limits of the material environment in terms of pH and oxidation-reduction potential. J. Geol., 68: 243—284.
Bayers, B.R., 1974. Iron transport in gram positive and acid fast bacilli. In: J.B. Neilands (Editor), Microbial Iron Metabolism. Academic, New York, NY, pp. 83—105.
Berner, R.A., 1962. Experimental studies on the formation of sedimentary iron sulfides. In: M.L. Jensen (Editor) Biogeochemistry of Sulfur Isotopes. National Science Foundation Conference at Yale University, New Haven, CT, pp. 107—120.
Bonatti, E. and Nayudu, Y.R., 1965. Origin of manganese nodules on ocean floor. Am. J. Sci., 263: 17—39.
Bostrom, K., 1967. The problem of excess manganese in pelagic sediments. In: P. Abelson (Editor), Researches in Geochemistry, Vol. 2. Wiley, New York, NY, pp. 421—452.
Brierley, C.L. and Murr, L.E., 1975. Leaching: use of a thermophilic and chemoautotrophic microbe. Science, 179: 488—490.
Brinkmann, R.T., 1969. Dissociation of water and the evolution of oxygen in the terrestrial atmosphere. J. Geophys. Res., 74: 5355—5366.
Brock, T.D., Brock, K.M., Belly, R.T. and Weiss, R.L., 1972. Sulfolobus — new genus of sulfur oxidizing bacteria living at low pH and high temperature. Arch. Microbiol., 84: 54—56.
Bromfield, S.M., 1954. Reduction of ferric compounds by soil bacteria. J. Gen. Microbiol., 11: 1—6.
Burns, R.G. and Brown, B.A., 1972. Nucleation and mineralogical controls on the composition of manganese nodules. In: D.R. Horn (Editor), Ferromanganese Deposits on the Ocean Floor. Proceedings of a conference held at Arden House in January 1972. NSF-IDOE, pp. 51—62.
Butkevich, V.S., 1928. The formation of marine iron-manganese deposits and the microorganisms taking part in it. Tr. Mosk. Nauch. Inst., 3: 1—81.
Cloud, P.E., 1965. Significance of the Gunflint (Precambrian) microflora. Science, 148: 27—35.
Cloud, P.E., 1968. Atmospheric and hydrospheric evolution on the primitive earth. Science, 160: 729—736.
Cloud, P.E., 1973. Paleoecological significance of the banded iron-formation. Econ. Geol., 68: 1135—1143.
Cloud, P.E., 1974. Evolution of ecosystems. Am. Sci., 62: 54—66.
Cloud, P.E. and Licari, G.R., 1968. Microbiotas of the banded iron formations. Proc. Nat. Acad. Sci., 61: 779—786.
Cloud, P.E. and Gibor, A., 1970. The oxygen cycle. Sci. Am., 223: 110—120.
Collins, J.F. and Buol, S.W., 1970. Effects of fluctuations in Eh-pH environment on iron and/or manganese equilibria. Soil Sci., 110: 111—118.
Collins, J.F. and Buol, S.W., 1972. Patterns of iron and manganese precipitation under specified Eh-pH conditions. Soil Sci., 110: 157—162.

Colmer, A.R., Temple, K.L. and Hinkle, M.E., 1950. An iron-oxidizing bacterium from the acid drainage of some bituminous coal mines. J. Bacteriol., 59: 317—328.
Cronan, D.S. and Thomas, R.L., 1970. Ferromanganese concretions in Lake Ontario. Can. J. Earth Sci., 7: 1346—1349.
Cronan, D.S. and Varentsov, I.M. (Editors), 1978. Monograph on the Geology and Geochemistry of Manganese. 25th International Geological Congress, Sydney, Australia.
Dean, W.E., Ghosh, S.K., Krishnaswami, S. and Moore, W.S., 1973. Geochemistry and accretion rates of freshwater ferromanganese nodules. In: M. Morgenstein (Editor), The Origin and Distribution of Manganese Nodules in the Pacific and Prospects for Exploration. Hawaii Institute of Geophysics, pp. 13—20.
Dean, W.E., Davies, G.R. and Anderson, R.Y., 1975. Sedimentological significance of nodular and laminated anhydrite. Geology, 3: 367—372.
Dean, W.E. and Ghosh, S.K., 1978. Geochemistry of freshwater ferromanganese deposits in North America. In: D.S. Cronan and I.M. Varentsov (Editors), Monograph on the Geology and Geochemistry of Manganese, 25th International Geological Congress, Sydney, Australia.
Dean, W.E. and Ghosh, S.K., 1978. Factors contributing to the formation of ferromanganese nodules in Oneida Lake, New York. U.S. Geol. Surv. J. Res., 6: 231—240.
Deevey, E.S., 1970. Mineral Cycles. In: The Biosphere. A Scientific American Book, pp. 83—92.
Din, G.A., Suzuki, I. and Lees, H., 1967. Carbon dioxide fixation and phosphoenolpyruvate carboxylase in *Ferrobacillus ferrooxidans*. Can. J. Microbiol., 13: 1413—1420.
Doemel, W.N. and Brock, T.D., 1974. Bacterial stromatolites: origin of laminations. Science, 184: 1083—1085.
Drabkova, V.G. and Stravinskaya, E.A., 1969. Role of bacteria in the dynamics of iron in Lake Krasnoe. Microbiology, 38: 306—309.
Drever, J.I., 1974. Geochemical model for the origin of Precambrian banded iron formations. Geol. Soc. Am. Bull., 85: 1099—1106.
Dubinina, G. and Zhdanov, A.V., 1975. Recognition of the iron bacteria "Siderocapsa" as *Arthrobacter* and description of *Arthrobacter siderocapsulatus* sp. nov. J. System. Bacteriol., 25: 340—349.
Dugan, P.R., 1975. Bacterial ecology of strip mine areas and its relationship to the production of acid mine drainage. Ohio J. Sci., 76: 266—279.
Dugolinsky, B.K., Margolis, S.V. and Dudley, W.C., 1977. Biogenic influences on growth of manganese nodules. J. Sediment. Petrol., 47: 428—445.
Edgington, D.N. and Callender, E., 1970. Minor element geochemistry of Lake Michigan ferromanganese nodules. Earth Planet. Sci. Lett., 8: 97—100.
Ehrlich, H.L., 1963. Bacteriology of manganese nodules. 1. Bacterial action on manganese nodule enrichments. Appl. Microbiol., 11: 15—19.
Ehrlich, H.L., 1966. Reaction with manganese by bacteria from marine ferromanganese nodules. Dev. Ind. Microbiol., 7: 43—60.
Ehrlich, H.L., 1968. Bacteriology of manganese nodules. II. Manganese oxidation by cell-free extract from a manganese nodule bacterium. Appl. Microbiol., 16: 197—202.
Ehrlich, H.L., 1971. Bacteriology of manganese nodules. V. Effect of hydrostatic pressure on bacterial oxidation on Mn(II) and reduction of MnO_2. Appl. Microbiol., 21: 306—310.
Ehrlich, H.L., 1972. The role of microbes in manganese nodule genesis and degradation. In: D.R. Horn (Editor), Ferromanganese Deposits on the Ocean Floor. Proceedings of a conference held at Arden House in January 1972. NSF-IDOE, pp. 63—70.
Engel, A.E.J., Nagy, B., Nagy, L.A., Engel, C.G., Kremp, G.O.W. and Drew, C.M., 1968. Alga-like forms in Onverwacht Series, South Africa: oldest recognized lifelike forms on earth. Science, 161: 1005—1008.

Gale, N.L. and Beck, J.V., 1966. Competitive inhibition of phosphoribulokinase by AMP. Biochem. Biophys. Res. Commun., 24: 792—795.

Gale, N.L. and Beck, J.V., 1967. Evidence for Calvin cycle and hexose monophosphate pathway in *Thiobacillus ferrooxidans*. J. Bacteriol., 94: 1052—1060.

Garrels, R.M. and Christ, C.L., 1965. Solution, Minerals, and Equilibria. Harper, New York, NY, 460 pp.

Garrels, R.M., Perry, E.A. and MacKenzie, F.T., 1973. Genesis of Precambrian iron-formations and the development of atmospheric oxygen. Econ. Geol., 68: 1173—1179.

Garrett, P., 1970. Phanerozoic stromatolites: noncompetitive ecologic restriction by grazing and burrowing animals. Science, 169: 171—173.

Gebelein, C.D., 1969. Distribution, morphology and accretion rate of recent subtidal algal stromatolites, Bermuda. J. Sediment. Petrol., 39: 49—69.

Glaessner, M.F., 1962. Precambrian fossils. Biol. Rev., 37: 467—494.

Glaessner, M.F., Preiss, W.V. and Walter, M.R., 1969. Precambrian columnar stromatolites in Australia: morphological and stratigraphic analysis. Science, 164: 1056—1058.

Glasby, G.P., 1972. The geochemistry of manganese nodules from the northwest Indian Ocean. In: D.R. Horn (Editor), Ferromanganese Deposits on the Ocean Floor. Proceedings of a conference held at Arden House in January 1972. NSF-IDOE, pp. 93—104.

Glasby, G.P. (Editor), 1977. Marine Manganese Deposits. Elsevier, Amsterdam, 536 pp.

Goldich, S.S., 1973. Ages of Precambrian banded iron-formations. Econ. Geol., 68: 1126—1134.

Goldich, S.S., Neir, A.O., Baadsgaard, H., Hoffman, J.H. and Kreuger, H.W., 1961. The Precambrian geology and geochronology of Minnesota. Minn. Geol. Surv. Bull., 41: 193.

Goodwin, A.M., 1956. Facies relations in the gunflint iron formation. Econ. Geol., 51: 565—595.

Gorham, E. and Swaine, D.J., 1965. The influence of oxidizing and reducing conditions upon the distribution of some elements in lake sediments. Limnol. Oceanogr., 10: 268—279.

Graham, J.W., 1959. Metabolically induced precipitation of trace elements from sea water. Science, 129: 1428—1429.

Graham, J.W. and Cooper, S.C., 1959. Biological origin of manganese rich deposits of the sea floor. Nature, 183: 1050—1051.

Greenslate, J., 1973. Relation of sediment chemistry to ferromanganese nodules. Trans. Am. Geophys. Union, 54: 342—346.

Greenslate, J., 1974a. Microorganisms participate in construction of manganese nodules. Nature, 249: 181—183.

Greenslate, J., 1974b. Manganese and biotic debris association in some deep-sea sediments. Science, 186: 529—531.

Greenslate, J., Frazer, J.Z. and Arrhenius, G., 1973. Origin and deposition of selected transition elements in the sea bed. In: National Science Foundation, International Decade for Ocean Exploration (NSF-IDOE), Phase I report: Inter-University program of research on ferromanganese deposits of the ocean floor, pp. 13—20.

Greeson, P.E., 1971. Limnology of Oneida Lake with emphasis on factors contributing to algal blooms. U.S. Geological Survey Open-File Report, Washington, DC, 185 pp.

Guay, R. and Silver, M., 1975. *Thiobacillus acidophilus* sp. nov; isolation and some physiological characteristics. Can. J. Microbiol., 21: 281—288.

Guay, R., Torma, A.E. and Silver, M., 1975. Oxydation de l'ion ferreau et mise en solution de l'uranium d'un mineral par *Thiobacillus ferrooxidans*. Ann. Microbiol., 126B: 209—219.

Harder, E.C., 1919. Iron-depositing bacteria and their geologic relations. U.S. Geol. Surv. Prof. Pap., 113: 1—89.

Harriss, R.C. and Troup, A.G., 1970. Chemistry and origin of freshwater ferromanganese concretions. Limnol. Oceanogr., 15: 702—712.

Hem, J.D., 1963. Chemical equilibria and rates of manganese oxidation. U.S. Geological Survey Water-Supply Paper 1667-A, United States Government Printing Office, Washington, DC, 64 pp.

Hem, J.D., 1972. Chemical factors that influence availability of iron and manganese in aqueous systems. Geol. Soc. Am. Bull., 83: 443—450.

Hofmann, H.J., 1969. Stromatolites from the Proterozoic Animikie and Sibley Groups, Ontario. Can. Geol. Surv. Pap., 68—69, 77 pp.

Holland, H.D., 1964. On the chemical evolution of the terrestrial and Cytherean atmosphere. In: P.S. Brancazio and H.S.W. Cameron (Editors), The Origin and Evolution of the Atmosphere and Oceans. John Wiley, New York, NY, pp. 86—101.

Horn, D.R., Horn, B.M. and Delach, M.N., 1972. Distribution of ferromanganese deposits in the world ocean. In: D.R. Horn (Editor), Ferromanganese Deposits on the Ocean Floor. Proceedings of a conference held at Arden House in January 1972. NSF-IDOE, pp. 9—18.

Imshenetskii, A.A. (Editor), 1961. Achievements in Soviet microbiology. Akad. Nauk SSSR, Inst. Mikrobiol. (Trans. from Russian and Consultants Bureau, New York). 35 pp.

James, H.L., 1954. Sedimentary facies of iron formations. Econ. Geol., 49: 235—293.

James, H.L., 1966. Chemistry of the iron-rich sedimentary rocks. U.S. Geol. Surv. Prof. Pap., 440-W, 60 pp.

James, H.L. and Sims, P.K., 1973. Precambrian iron-formations of the world. Econ. Geol., 68: 913—914.

Jones, J.G., 1975. Some observations on the occurrence of the iron bacterium *Leptothrix ochracea* in fresh water, including reference to large experimental enclosures. J. Appl. Bacteriol., 39: 347—350.

Kalinenko, V.O., 1952. The geochemical activity of the bacterial colony. Izv. Akad. Nauk SSSR, Ser. Geol. No. 1, 145 pp.

Knoll, A.H. and Barghoorn, E.S., 1975. Precambrian eukaryotic organisms — Reassessment of evidence. Science, 190: 52—54.

Komar, V.A., 1966. Upper Precambrian stromatolites in the northern part of the Siberian Platform and their stratigraphic significance. Akad. Nauk SSSR, Geol. Inst. Trans., 154: 122 pp.

Korolyuk, I.K., 1960. Stromatolites of the Lower Cambrian and Proterozoic of the Irkutsk Cirque. Akad. Nauk SSSR, Geol. Inst. Trans., 1: 112—161.

Krauskopf, K.B., 1956. Factors controlling the concentrations of thirteen rare metals in seawater. Geochim. Cosmochim. Acta, 9: 1—32.

Krauskopf, K.B., 1957. Separation of manganese from iron in sedimentary processes. Geochim. Cosmochim. Acta, 12: 61—84.

Krylov, I.N., 1963. Columnar branching stromatolites of the southern Urals and their significance for the stratigraphy of the Upper Precambrian. Akad. Nauk SSSR, Geol. Inst. Trans., 69: 133 pp.

Ku, T.L. and Glasby, G.P., 1972. Radiometric evidence for the rapid growth rate of shallow-water, continental margin manganese nodules. Geochim. Cosmochim. Acta, 36: 699—703.

Kuznetsov, S.I., Ivanov, M.V. and Lyalikova, N.N., 1963, Introduction to Geological Microbiology, P.T. Broneer (Translator). McGraw-Hill. New York, NY, 252 pp.

Kucera, S. and Wolfe, R.S., 1957. A selective enrichment method for *Gallionella ferruginea*. J. Bacteriol., 74: 347—350.

LaBerge, G.L., 1967. Microfossils and Precambrian iron-formations. Geol. Soc. Am. Bull., 78: 331—342.
LaBerge, G.L., 1973. Possible biological origin of Precambrian iron-formations. Econ. Geol., 68: 1098—1109.
Landergren, S., 1948. On the geochemistry of Swedish iron ores and associated rocks. Sveriges Geol. Undersokn. Arsbok Ser. C: Avhandl. och Uppsat. 496, 182 pp.
Lees, H., Kwok, C. and Suzuki, I., 1969. The thermodynamics of iron oxidation by the ferrobacilli. Can. J. Microbiol., 15: 43—46.
Ljunggren, P., 1953. Some data concerning the formation of manganiferous and ferrous bog ores. Geol. Foeren. Stockholm Foerh., 75: 277—297.
Ljunggren, P., 1955a. Geochemistry and radioactivity of some Mn and Fe bog ores. Geol. Foeren. Stockholm Foerh., 77: 33—44.
Ljunggren, P., 1955b. Differential thermal analysis and X-ray examination of Fe and Mn bog ores. Geol. Foeren. Stockholm Foerh., 77: 135—147.
Logan, B.W., Rezak, R. and Ginsburg, R.N., 1964. Classification and environmental significance of algal stromatolites. J. Geol., 72: 68—86.
Lundgren, D.G., 1975. Microbiological problems in strip mine areas: relationship to the metabolism of *Thiobacillus ferrooxidans.* Ohio J. Sci., 75: 280—287.
Lundgren, D.G., Vestal, J.R. and Tabita, F.R., 1972. The microbiology of mine drainage pollution. In: R. Mitchell (Editor), Water Pollution Microbiology, John Wiley, New York, NY, pp. 69—88.
Lundgren, D.G., Vestal, J.R. and Tabita, F.R., 1974. The iron-oxidizing bacteria. In: J.B. Neilands (Editor), Microbial Iron Metabolism, Academic, New York, NY, Ch. 18, pp. 457—472.
Lynn, D.C. and Bonatti, E., 1965. Mobility of manganese in diagenesis of deep sea sediments. Mar. Geol., 3: 457—474.
MacDonald, D.G. and Clark, R.H., 1970. The oxidation of aqueous ferrous sulphate by *Thiobacillus ferrooxidans.* Can. J. Chem. Eng., 48: 669—676.
Maciag, J.W. and Lundgren, D.G., 1964. Carbon dioxide fixation in the chemoautotroph *Ferrobacillus ferrooxidans.* Biochem. Biophys. Res. Commun., 17: 603—607.
Manheim, F.T., 1965. Manganese-iron Accumulations in the Shallow Marine Environment. University of Rhode Island Occasional Publications, no. 3. Kingston, RI, pp. 217—276.
Margalith, P., Silver, M. and Lundgren, D.G., 1966. Sulfur oxidation by the iron bacterium *Ferrobacillus ferrooxidans.* J. Bacteriol., 92: 1706—1709.
Margulis, L., 1971. The origin of plant and animal cells. Am. Sci., 59: 230—235.
Mendelsohn, F., 1976. Mineral deposits associated with stromatolites. In: M.R. Walter (Editor), Stromatolites. Elsevier, Amsterdam, p. 645—662.
Mero, J.L., 1965. The Mineral Resources of the Sea. Elsevier, Amsterdam, 312 pp.
Morgenstein, M., 1973. The origin and distribution of manganese nodules in the Pacific and prospects for exploration. Symposium, Honolulu Hawaii, 23—25 July, 1973, Valdevia Manganese Exploration Group, Univ. Hawaii, and IDOE-NSF. 175 pp.
Mulder, E.G., 1963. Some observations of the *Sphaerotilus-Leptothrix* Group. In: H. Heukelekian and N.C. Dondero (Editors), Principles and Applications in Aquatic Microbiology. John Wiley, New York, NY, pp. 98—112.
National Science Foundation, 1973. International Decade for Ocean Exploration. Inter-university program of research on ferromanganese deposits on the ocean floor — Phase 1 Report, 385 pp.
Neilands, J.B., 1974. Iron and its role in microbial physiology. In: J.B. Neilands (Editor), Microbial Iron Metabolism. Academic, New York, NY, pp. 1—31.
Oehler, D.Z., Schopf, J.W. and Kvenvolden, K.A., 1972. Carbon isotopic studies of organic matter in Precambrian rocks. Science, 175: 1246—1248.

Oehler, J.H., 1976. Hydrothermal crystallization of silica gel. Geol. Soc. Am. Bull., 87: 1143—1152.

Olson, J.M., 1970. The evolution of photosynthesis. Science, 168: 438—446.

Perfil'ev, B.V., 1927. New data on the role of microbes in ore formation. Izv. Geol. Kom., Vol. 45.

Puchelt, H., Schock, H.H. and Schroll, E., 1973. Rezente marine Eisenerze auf Santorin, Griechenland, I. Geochemie, Enstehung, Mineralogie, Hanert, H., II. Bakterogenese von Eisenhydroxyidsedimenten. Geol. Rundsch., 62: 786—812.

Price, N.B., 1967. Some geochemical observations on manganese-iron oxide nodules from different depth environments. Mar. Geol., 5: 511—538.

Pringsheim, E.G., 1949. Iron bacteria. Biol. Rev., 24: 200—250.

Rezak, R., 1957. Stromatolites of the belt series in Glacier National Park and vicinity, Montana. U.S. Geol. Surv. Prof. Pap., 294-D: 127—151.

Schopf, J.W., 1968. Microflora of the Bitter Springs Formation, late Precambrian, central Australia. J. Paleontol., 42: 651—688.

Schopf, J.W., 1970. Precambrian microorganisms and evolutionary events prior to the origin of vascular plants. Biol. Rev., 45: 319—352.

Schopf, J.W. and Barghoorn, E.S., 1967. Alga-like fossils from the Early Precambrian of South Africa. Science, 156: 508—512.

Schopf, J.W. and Blacic, J.M., 1971. New microorganisms from the Bitter Springs Formation (Late Precambrian) of the north-central Amadeus Basin, Australia. J. Paleontol., 45: 925—960.

Schopf, J.W. and Oehler, D.Z., 1976. How old are the eukaryotes? Science, 193: 47—49.

Sculz-Baldez, M. and Lewin, R.A., 1975. Manganese encrustation of zygospores of a *Chlamydomonas (Chlorophyta-volvocales)*. Science, 188: 1119—1120.

Shafia, F. and Wilkinson, R.F., 1969. Growth of *Ferrobacillus ferrooxidans* on organic matter. J. Bacteriol., 97: 256—260.

Shafia, F., Brinson, K.R., Heinzman, M.W. and Brady, J.M., 1972. Transition of chemolithotrophic *Ferrobacillus ferrooxidans* to obligate organotrophy and metabolic capabilities of glucose-grown cells. J. Bacteriol., 111: 56—65.

Silver, M., 1970. Oxidation of elemental sulfur and sulfur compounds and CO_2 fixation of *Ferrobacillus ferrooxidans.* Can. J. Microbiol., 16: 845—870.

Silver, M. and Torma, A.E., 1974. Oxidation of metal sulfides by *Thiobacillus ferrooxidans* grown on different substrates. Can. J. Microbiol., 20: 141—147.

Silverman, M.P. and Ehrlich, H.L., 1964. Microbial formation and degradation of minerals. Adv. Appl. Microbiol., 6: 163—206.

Singer, P.C. and Stumm, W., 1970. Acid mine drainage: the rate limiting step. Science, 167: 1121—1124.

Starkey, R.L., 1945. Precipitation of ferric hydrate by iron bacteria. Science, 102: 532—533.

Starkey, R.L. and Halvorson, H.O., 1927. Studies on the transformation of iron in nature. II. Concerning the importance of microorganisms in the solution and precipitation of iron. Soil Sci., 24: 381—402.

Tabita, R. and Lundgren, D.G., 1971a. Heterotrophic metabolism of the chemolithotrophic *Thiobacillus ferrooxidans.* J. Bacteriol., 108: 334—342.

Tabita, R. and Lundgren, D.G., 1971b. Utilization of glucose and the effect of organic compounds on the chemolithotroph *Thiobacillus ferrooxidans.* J. Bacteriol., 108: 328—337.

Temple, K.L. and Colmer, A.R., 1951. The autotrophic oxidation of iron by a new bacterium: *Thiobacillus ferrooxidans.* J. Bacteriol., 62: 605—611.

Theis, T.L. and Singer, P.C., 1975. Complexation of iron (II) by organic matter and its effect on iron (II) oxygenation. Environ. Sci. Technol., 8: 569—573.

Tikhonova, G.V., Lisenkova, L.L., Doman, N.C. and Sjulachev, V.P., 1967. Electron transport pathways in *Thiobacillus ferrooxidans*. Biochemistry, 32: 599—605.

Torma, A.E., 1971. Microbiological oxidation of synthetic cobalt, nickel and sulfides by *Thiobacillus ferrooxidans*. Rev. Can. Biol., 30: 209—216.

Torma, A.E. and Subramanian, K.N., 1974. Selective bacterial leaching of a lead sulphide concentrate. Int. J. Miner. Process., 1: 125—134.

Torma, A.E., Legault, G., Kougioumoutzakis, D. and Ouellet, R., 1974. Kinetics of bio-oxidation of metal sulphides. Can. J. Chem. Eng., 52: 515—517.

Torma, A.E. and Tabi, M., 1973. Mise en solution des metaux de minerals sulfures par voie bacterienne. L'Ingenieur, 294: 1—8.

Torma, A.E. and Legalt, G., 1973. Role de la surface des minerals sulfures lors de leur biodegradation par *Thiobacillus ferrooxidans*. Ann. Microbiol., 124A: 111—121.

Trimble, R.B. and Ehrlich, H.L., 1968. Bacteriology of manganese nodules. III. Reduction of MnO_2 by 2 strains of nodule bacteria. Appl. Microbiol., 16: 695—702.

Trimble, R.B. and Ehrlich, H.L., 1970. Bacteriology of manganese nodules IV. Induction of an MnO_2-reductase system in a marine bacillus. Appl. Microbiol., 19: 966—972.

Trudinger, P.A. and Mendelsohn, F., 1976. Biological processes and mineral deposition. In: M.R. Walter (Editor), Stromatolites. Elsevier, Amsterdam, pp. 663—672.

Tuovinen, O.H. and Kelly, D.P., 1972. Biology of *Thiobacillus ferrooxidans* in the microbiological leaching of sulphide ores. Z. Allg. Mikrobiol., 12: 311—346.

Tuovinen, O.H., Niemela, S.I. and Gyllenberg, H.G., 1971. Tolerance of *Thiobacillus ferrooxidans* to some metals. Antonie van Leeuwenhoek; J. Microbiol. Serol., 37: 489—496.

Unz, R.F. and D.G. Lundgren, 1961. A comparative nutritional study of three chemoautotrophic bacteria: *Ferrobacillus ferrooxidans*, *Thiobacillus ferrooxidans* and *Thiobacillus thiooxidans*. Soil Sci., 92: 302—313.

Walsh, F. and Mitchell, R., 1972. An acid-tolerant iron-oxidizing Metallogenium. J. Gen. Microbiol., 72: 369—376.

Walter, M.R., 1972. A hot spring analogue for the depositional environment of Precambrian iron formations of the Lake Superior region. Econ. Geol., 67: 965—980.

Walter, M.R. (Editor), 1976. Stromatolites. Elsevier, Amsterdam, 790 pp.

Walter, M.R., Goode, A.D.T. and Hall, W.D.M., 1976. Microfossils from a newly discovered Precambrian stromatolitic iron formation in Western Australia. Nature, 261: 221—223.

Wedepohl, K.H., 1971. Geochemistry. Translated from the German by E. Althaus. Holt, Rinehart and Winston, New York, NY, 231 pp.

Winogradsky, S.N., 1888. Uber Eisenbacterin. Bot. Ztg., 46: 262—270.

Wolfe, R.S., 1963. Iron and Manganese Bacteria. In: H. Heukelekian and N.C. Dondero (Editors), Principles and Applications in Aquatic Microbiology. John Wiley, New York, NY, pp. 82—97.

Woolfolk, C.A. and Whiteley, H.R., 1962. Reduction in inorganic compounds with molecular hydrogen by *Micrococcus lactilyticus*. J. Bacteriol., 84: 647—658.

Zajic, J.E., 1969. Microbial Biogeochemistry. Academic, New York, NY, 345 pp.

Zapffe, C., 1931. Deposition of manganese. Econ. Geol., 26: 799—832.

Chapter 5

BIOGEOCHEMISTRY OF MANGANESE MINERALS

K.C. MARSHALL

School of Microbiology, The University of New South Wales, P.O. Box 1, Kensington, N.S.W. 2033 (Australia)

CONTENTS

OCCURRENCE OF MANGANESE IN NATURE

Manganese is found in a limited range of mineral deposits, and in variable concentrations in soils, waters and living organisms. The composition of some common manganese minerals is given in Table 5.1. The major deposits

TABLE 5.1

Some important manganese minerals

Mineral	Formula	Source
Rhodochrosite	$MnCO_3$	Sedimentary
Pyrolusite	β-MnO_2	
Birnessite	δ-MnO_2	
Manganite	γ-$MnOOH$	
Hausmannite	Mn_3O_4	
Pyrochroite	$Mn(OH)_2$	
Manganosite	MnO	
Todorokite	(Mn(II), CaNaK) (Mn(IV), Mn(II), Mg)$_6$O$_{12}$ · 3 H_2O	
Braunite	3 Mn_2O_3 · $MnSiO_2$	Igneous, hydrothermal, sedimentary
Pyromanganite	(Mn, Fe) (SiO_3)	
Rhodonite	(Mn, Fe, Ca) (SiO_3)	
Jacobsite	$MnFeO_4$	
Alabandite	MnS	

are sedimentary in origin and consist of carbonates, oxides and hydroxides of manganese. Since the biogeochemical properties of iron and manganese are similar (Zajic, 1969, p. 20), the simultaneous occurrence of these elements is a regular feature in sedimentary deposits. Sulfides of manganese, unlike those of iron, are rare and their formation by biogeochemical means is doubtful (Silverman and Ehrlich, 1964).

The sedimentary origin of major manganese deposits suggests a large scale conversion of soluble manganese into relatively insoluble forms, particularly

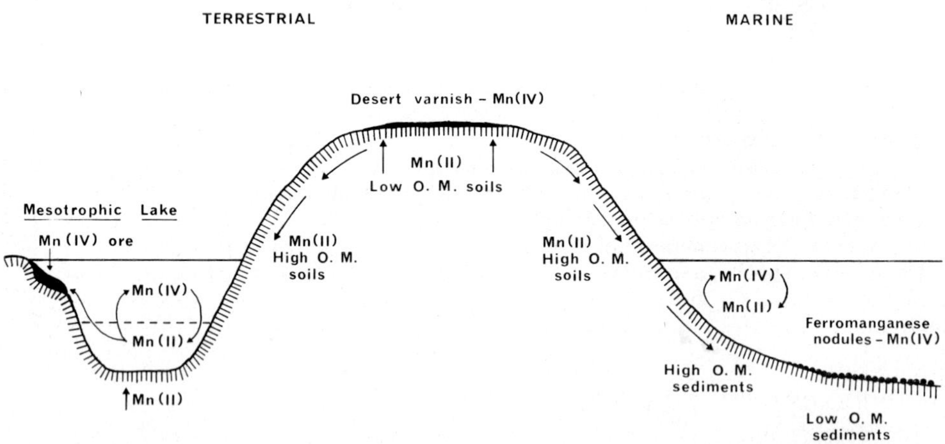

Fig. 5.1. Biogeochemical manganese cycle in marine and terrestrial systems. Arrows indicate the mobilization of manganese as Mn(II) by microbial activity, leading to microbially-induced deposition of manganese as Mn(IV) under suitable conditions.

in aquatic environments. A diagrammatic representation of the biogeochemical manganese cycle is given in Fig. 5.1. In areas containing high levels of organic matter, microorganisms create conditions suitable for the mobilization of manganese as the reduced (Mn)II form (Crerar et al., 1972). Migration of the Mn(II) to sites of low organic matter content and high redox potential provides opportunities for microbial oxidation of Mn(II) to Mn(IV), a process often leading to the formation of substantial manganese ore deposits. The chemistry of manganese in aqueous systems is complex, and the subject is reviewed here to provide a background for the subsequent consideration of the role of biological factors in manganese cycling.

CHEMISTRY OF MANGANESE

Manganese occurs in a number of valency states, the existence of a particular valency state depending, to a large extent, on the pH and redox potential (Eh) of the system. An adequate description of manganese transformations depends on the methods available for distinguishing between manganous manganese and manganese present in the higher oxidation states, Mn(III) and Mn(IV). According to Morgan and Stumm (1965a), analytical methods employed in determining manganese levels in waters should distinguish between higher and lower oxidation states, and between different physical states of manganese. The methods should be sufficiently sensitive to detect the low levels of total manganese found in most waters. The occurrence and distribution of manganese in its different forms in aqueous systems may be determined by the use of *o*-tolidine for Mn in higher oxidation states and formaldoxime for soluble Mn(II), combined with membrane filtration for distinguishing between "soluble" and "insoluble" forms (Morgan and Stumm, 1965a). Bromfield and David (1976) have used atomic absorption techniques to determine soluble manganese and, following dissolution in H_2SO_4 containing H_2O_2, that in the particulate fraction. These authors also described the use of $CuSO_4$ to desorb Mn(II) from manganese oxides. Oxidized forms of manganese are routinely detected in microbial cultures by the production of a blue colour in the presence of benzidine, although the value of this test has been questioned by Ivarson and Heringa (1972). Krumbein and Altmann (1973) have described the use of Berbelin-blue I to determine the higher oxidation states of manganese.

Solubility of manganese species

In the normal pH range (pH 6-9) of natural waters, soluble divalent manganese consists of Mn^{2+} and $MnOH^+$ (Morgan and Stumm, 1965b). As shown in Fig. 5.2, the solubility of the divalent forms in carbonate-containing waters is governed largely by the solubility product of $MnCO_3$ and the

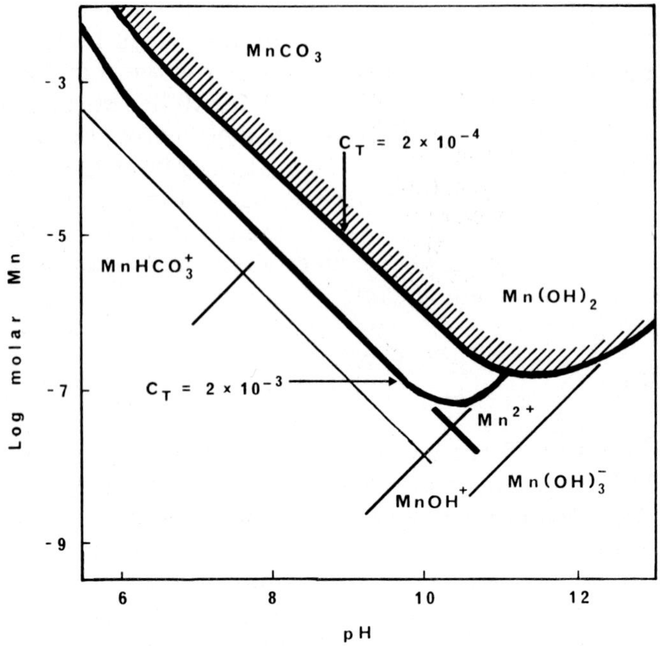

Fig. 5.2. Maximum soluble Mn(II) at two different concentrations of total carbonic species at 25°C (redrawn from Morgan and Stumm, 1965b).

pH of the water. Similarly, manganese solubility in sulfide-containing hypolimnetic waters is controlled by the solubility product of MnS (Morgan and Stumm, 1965b; Delfino and Lee, 1968). The solubility of some manganese species can be modified by complex formation, but complexes of Mn(II) with ligands other than OH^- probably are rare except in water containing high levels of dissolved organic matter. Stable complex formation between Mn(II) and various organic acids found in soils and waters has been reported by Schnitzer (1969) and Crerar et al. (1972).

Mn(III) is thermodynamically unstable and does not occur in soluble form except in the presence of strong complexing agents (Morgan and Stumm, 1965b). MnO_2 is the only higher valency form that is thermodynamically stable in natural waters, but the solubility of MnO_2 is so low that soluble Mn(IV) is undetectable within the pH range 3-10. Morgan and Stumm (1965b) suggest that insoluble non-stoichiometric higher valency oxides (MnO_x, where $1 < x < 2$) may exist as metastable phases in natural waters. Mn(IV) has such a high affinity for OH^- that other organic and inorganic ligands cannot compete successfully with OH^- for coordination to the Mn(IV).

Eh-pH relationships

A comparison of the behaviour of manganese species with that of iron species is essential in any consideration of Eh-pH relationships in aqueous systems. The formation of insoluble, higher valency forms of iron in mixed iron-manganese systems, for instance, results in the disappearance of Mn(II) from solution under Eh-pH conditions where this normally would not be predicted.

Eh-pH stability diagrams derived from thermodynamic data for iron and manganese are shown in Fig. 5.3. These diagrams indicate the dominant, stable species of iron or manganese under the specified conditions. The positions of the stability boundaries vary with the concentration (or activity) of the divalent species in the system under consideration. Such diagrams are given by Morgan and Stumm (1965b), Dubinina (1973), Crerar et al. (1972) and Crerar and Barnes (1974) for aquatic systems and by Collins and Buol (1970) for soil conditions. Inspection of Fig. 5.3 reveals that the predicted pH level necessary for precipitation of either iron or manganese depends on the redox potential of the system. Conversely, the Eh level required for conversion from the divalent form to a higher valency state depends largely on the pH of the environment. Crerar et al. (1972) have pointed out that, at a constant pH, Mn solubility will increase by seven orders of magnitude on reducing the Eh from 0.7 to 0.3 V. It is obvious from Fig. 5.3 that the pH level for the limit of stability of Fe(II) is considerably lower than that for Mn(II) at any particular Eh value and, consequently, Mn solubility exceeds

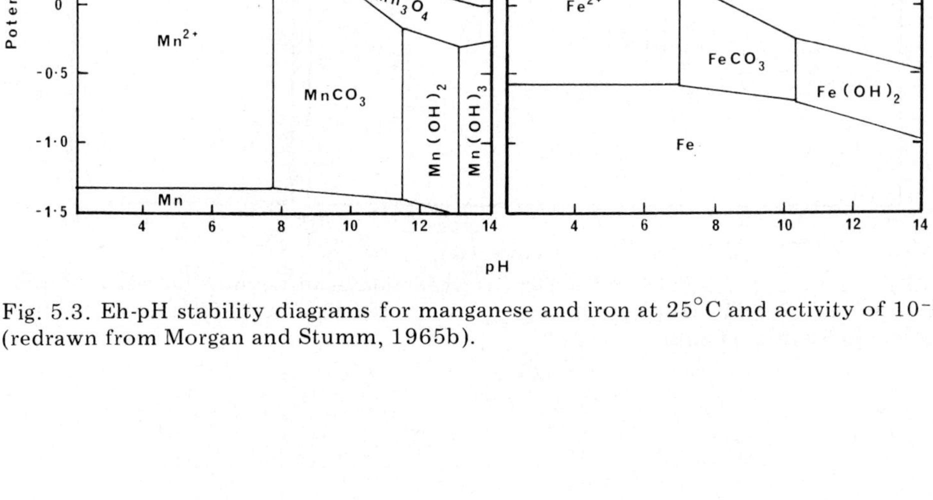

Fig. 5.3. Eh-pH stability diagrams for manganese and iron at 25°C and activity of 10^{-5} M (redrawn from Morgan and Stumm, 1965b).

that of Fe by six or seven orders of magnitude for any given Eh and acid pH (Crerar et al., 1972).

Although the stability diagrams are derived theoretically, Collins and Buol have presented data confirming the slopes of the equilibrium lines between the soluble and solid phases for both elements. Morgan and Stumm (1965b) stress that potentials shown in such Eh-pH stability diagrams are not necessarily identifiable with measurable electrode potentials in natural aqueous systems, as these potentials often represent mixed potentials resulting from distinctly different oxidation and reduction processes occurring at the electrode.

Kinetics of oxidation

Rates of oxygenation of both Mn(II) and Fe(II) are strongly pH-dependent. Measurable oxygenation of Mn(II) within several hours is observed only at pH values in excess of 8.5 (Fig. 5.4), with the manner of Mn(II) disappearance suggesting an autocatalytic reaction. On the other hand, oxygenation of Fe(II) is rapid at pH values above 6.5 (Fig. 5.4) with rates conforming to first-order reaction kinetics. An important consequence of the autocatalytic

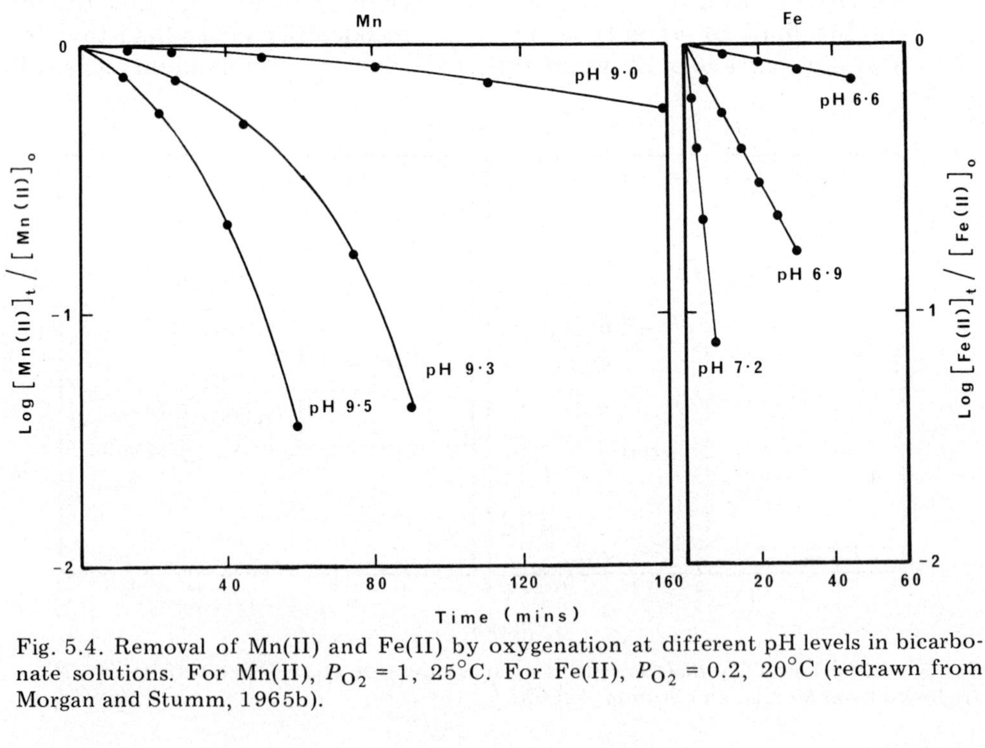

Fig. 5.4. Removal of Mn(II) and Fe(II) by oxygenation at different pH levels in bicarbonate solutions. For Mn(II), P_{O_2} = 1, 25°C. For Fe(II), P_{O_2} = 0.2, 20°C (redrawn from Morgan and Stumm, 1965b).

nature of Mn(II) oxygenation is that rates of reaction must be slow at the low concentration of manganese found in most natural waters.

Morgan and Stumm (1964, 1965b) determined that all of the Mn(II) removed during oxygenation cannot be accounted for as MnO_2. The product is non-stoichiometric, yielding various average degrees of oxidation ranging from $MnO_{1.3}$ to $MnO_{1.9}$. Such products and the autocatalytic nature of the reaction can be explained by the sorption of Mn(II) onto incipiently-formed hydrous MnO_2, with the relative proportions of Mn(II) and Mn(IV) in the solid phase depending on pH and other variables. The overall reaction may be visualized as follows:

$$\mathrm{Mn(II)} + O_2 \xrightarrow{\text{slow}} MnO_2(s) \tag{1}$$

$$\mathrm{Mn(II)} + MnO_2(s) \xrightarrow{\text{fast}} \mathrm{Mn(II)} \cdot MnO_2(s) \tag{2}$$

$$\mathrm{Mn(II)} \cdot MnO_2(s) + O_2 \xrightarrow{\text{slow}} 2\ MnO_2(s) \tag{3}$$

where $MnO_2(s)$ represents an approximation of the solid-phase Mn(IV) state. X-ray diffraction analysis of manganese oxides prepared in the presence of excess base possessed a low degree of crystallinity but resembled manganous manganite or δ-MnO_2 ($MnO_{1.8}$ to $MnO_{1.95}$). Oxidation products in solutions at pH 9.5 having a composition of approximately $MnO_{1.3}$ (or Mn_3O_4), were somewhat amorphous, but had X-ray patterns resembling that of hausmannite (Morgan and Stumm, 1964, 1965b). Hem (1964) has reported a slow, but significant, removal of Mn(II) from solution at pH 8.0 in the presence of fine particulate materials (quartz, orthoclase, plagioclase), yet little or no Mn(II) disappeared from solution under similar conditions in the absence of particulate material.

Metal ions and complexing agents do not markedly influence reaction rates, although hydroxy carboxylic acids do catalyze the oxidation of Mn(II) (Mulder, 1964). On the other hand, hydroxy and hydroxy carboxylic groups can rapidly reduce the solid-phase MnO_2 under appropriate conditions (Morgan and Stumm, 1965b).

Surface chemistry of solid-phase MnO_2

Solid metal hydroxides exhibit an amphoteric behaviour, and Morgan and Stumm (1964) have shown that OH^- and H^+ are potential-determining ions for $MnO_2(s)$. The potential-determining role of OH^- ions at pH values above the zero point of charge (pH 2.8) of $MnO_2(s)$ may be visualized as a binding of OH^- ions or as a dissociation of H^+ ions from surface OH groups. Under such conditions, the solid phase MnO_2 is capable of interacting with cations. Morgan and Stumm reported that sorption of Mn(II) to $MnO_2(s)$ is pH dependent (Fig. 5.5) and may be interpreted as an ion-exchange reaction. Sorption capacities at pH values of 7.5 and 9.0 were found to be 0.5 and 2.0

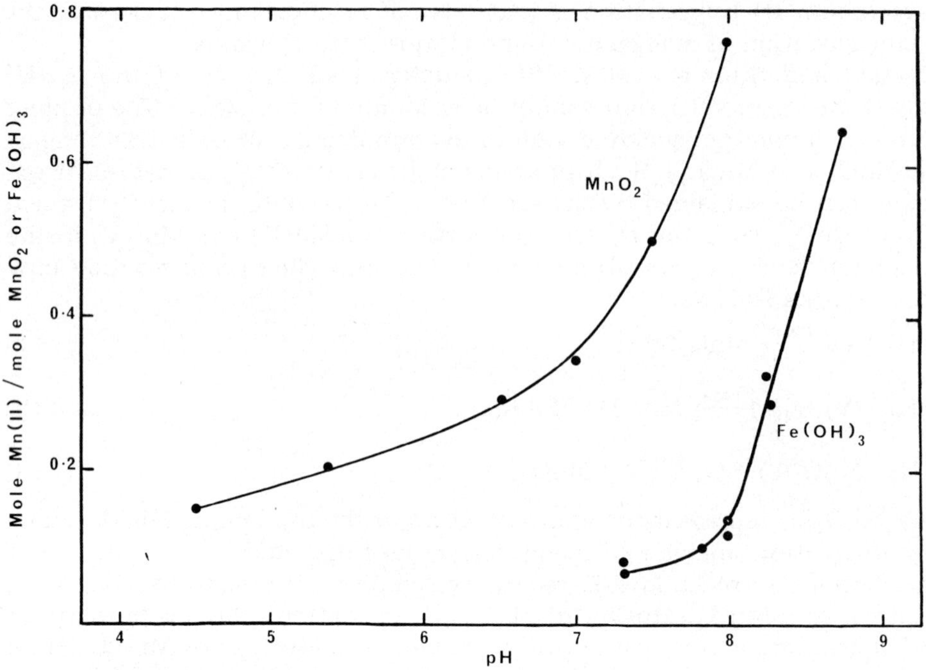

Fig. 5.5. Mn(II) sorption by MnO_2 and $Fe(OH)_3$ as a function of pH at 25°C (redrawn from Morgan and Stumm, 1965b).

moles of Mn(II) per mole of MnO_2(s), respectively. A lack of a simple exchange of H^+ ions in equivalent proportions (2 moles of H^+ per mole of Mn(II)) was explained, in part, by an exchange of other cations (Na^+, K^+) for Mn(II) at the MnO_2(s) surface. The affinity of MnO_2(s) for Zn(II), Ni(II) and Co(II) is slightly less, and for Mg^{2+} and Ca^{2+} ions is significantly less, than for Mn(II).

The magnitude of the cation-sorption capacity of MnO_2(s) can be explained by the large specific surface area of the oxide. Morgan and Stumm (1965b) have compared the specific surface area (300 m^2 g^{-1}) and cation-exchange capacity at pH 8.3 (1.5 meq g^{-1}) for δ-MnO_2 ($MnO_{1.9}$ to MnO_2) with the values for montmorillonitic clay (750 m^2 g^{-1}; 1 meq g^{-1}). However, Lee (1965) has urged caution in making such a comparison because of differences in procedures used in the determination of these parameters in the different materials. Lee also pointed out that the sorption capacity of MnO_2(s) may alter with the age of the precipitate.

The affinity of MnO_2(s) is higher for H^+ and polyvalent cations than for alkali metal ions. Consequently, the charge characteristics and colloidal stability of the MnO_2(s) is dependent on both the H^+ ion concentration and on the concentration of polyvalent metal ions. The addition of polyvalent

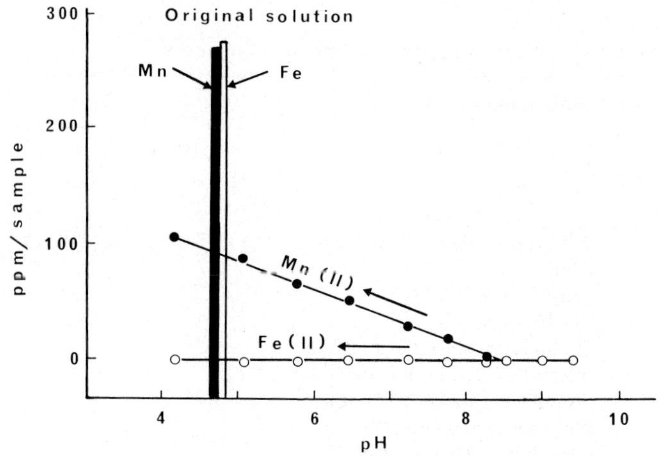

Fig. 5.6. Graph showing original levels of Mn(II) and Fe(II) in solution prior to addition of alkali, and subsequent release of Mn(II), but not of Fe(II), into solution following acidification of the precipitated oxides (redrawn from Collins and Buol, 1970).

cations to a stable suspension of MnO_2(s) displaces H^+ ions from the solid phase, leading to a reduction of charge and to a decrease in colloidal stability.

Mn(II) also sorbs onto solid-phase ($Fe(OH)_3$ (Fig. 5.4), but the affinity of $Fe(OH)_3$(s) for divalent manganese is lower than that of MnO_2(s). Morgan and Stumm (1965b) obtained sorption capacities of 0.3 and 0.1 mol of Mn(II) sorbed per mol of $Fe(OH)_3$(s) and MnO_2(s), respectively, at pH 8.0. Sorption to both solid species is of significance in natural systems containing mixed iron and manganese solutions. In fact, Collins and Buol (1970) have reported that precipitation of ferric products results in Mn(II) removal under Eh-pH conditions well within the field of stability of Mn(II). This was attributed to both occlusion and sorption of Mn(II) by the solid-iron precipitate. On acidification of previously precipitated iron and manganese mixtures, Collins and Buol found release of some Mn(II) into solution, but Fe(II) was not detected (Fig. 5.6). They suggested an exchange of H^+ ions for (Mn(II) on the solid metal hydrates, and supported the view that higher oxides of manganese are mixtures of Mn(II) and Mn(IV), whereas those of iron must be entirely Fe(III).

MICROBIAL TRANSFORMATIONS OF MANGANESE

Microorganisms involved in manganese transformations

Microorganisms have been implicated in transformations of manganese in soils, aquatic environments, and in large-scale manganese deposits. The

TABLE 5.2

A partial list of microorganisms implicated in transformations of manganese

Microorganisms	References
(A) Oxidation	
Bacteria	
Arthrobacter	Ehrlich (1966), Van Veen (1973), Bromfield (1974)
Bacillus manganicus	Beijerinck (1913)
Bacillus spores	Van Veen (1973)
Clonothrix	Wolfe (1960)
Hyphomicrobium	Tyler and Marshall (1967a)
Marine bacteria	Ehrlich (1966), Krumbein (1971)
Metallogenium personatum	Perfil'ev and Gabe (1969)
Nocardia	Schweisfurth (1968)
Neumanniella polymorpha	Ten Khak-Mun (1969)
Pedomicrobium	Aristovskaya (1961)
Pseudomonas	Zavarzin (1962), Van Veen (1973)
Siderocapsa (= *Arthrobacter*)	Dubinina et al., (1974)
Spherotilus discophorus	Johnson and Stokes (1966)
(= *Leptothrix discophora*)	Mulder and Van Veen (1963)
Fungi	
Cephalosporium	Ivarson and Heringa (1972), Timonin et al., (1972)
Cladosporium	Tyler and Marshall (1967a)
Coniothyrium fuckelii	Tyler and Marshall (1967a), Mulder and Van Veen (1968), Timonin et al., (1972)
Periconia spp.	Timonin et al., (1972)
Algae	
Chlorococcum humicola	Bromfield (1976)
Synergistic combinations	
Bacterial-bacterial	Bromfield and Skerman (1950), Zavarzin (1962)
Bacterial-fungal	Zavarzin (1961)
Bacterial-algal	Kutuzova (1975), Bromfield (1976)
(B) Reduction	
Bacillus spp.	Trimble and Ehrlich (1968), Dubinina et al., (1974)
Micrococcus	Bautista and Alexander (1972)
Many bacteria	Ehrlich et al., (1972)
Aspergillus niger	Barea et al., (1971)
Pichia guillermondii	Bautista and Alexander (1972)

microorganisms listed in Table 5.2, while not a complete list, give some impression of the range of different morphological types involved. They include conventional bacteria (*Arthrobacter*, *Pseudomonas*), prosthecate bacteria (*Pedomicrobium*, *Hyphomicrobium*, *Metallogenium*), sheathed bacteria (*Leptothrix discophora*, as given in Bergey, 1974, is used here in preference to the name *Sphaerotilus discophorus*), fungi (*Cephalosporium*, *Cladosporium*), and possible synergistic mixtures (*Corynebacterium-Chromobacterium*; *Coniotherium-Metallogenium*). Many, but not all, of the microorganisms capable of manganese transformations catalyze similar transformations of iron (Silverman and Ehrlich, 1964).

Manganese-oxidizing and -reducing microorganisms may be isolated from the same site within particular ecosystems (Ehrlich et al., 1972), but conditions necessary for oxidation and reduction are different and these processes are unlikely to occur simultaneously. Specific situations where microorganisms are significant in manganese transformations will be considered later.

Manganese-oxidizing microorganisms deposit insoluble, higher valency forms on the outer surface of the organisms. Few investigators have determined the chemical composition of these manganese oxides. Bromfield (1958) subjected manganese oxides produced by *Corynebacterium* sp. strain B (= *Arthrobacter*, Bromfield, 1974) to X-ray and electron diffraction analysis, but failed to find any evidence of crystallinity and concluded that the oxides were amorphous. The oxides were probably hydrated and Bromfield calculated values for x in the formula MnO_x ranging from 1.76 to 1.88. Bromfield and David (1976) clearly demonstrated that cells of this *Arthrobacter*, and the manganese oxide they form, rapidly absorb Mn(II) from aqueous solutions, but found no evidence for rapid oxidation of the adsorbed Mn(II) by non-biological reactions. Plants grown under sterile conditions were able to obtain manganese from these bacterial manganese oxides, and Bromfield (1958) demonstrated that root exudates contained substances which dissolved the oxides. The solubilizing activity of these exudates increased with increasing acidity. Ivarson and Heringa (1972) have shown that manganese oxides produced by *Cephalosporium* sp. on media at pH 4.5 were of variable chemical composition. X-ray diffraction patterns were identical with either hausmannite (Mn_3O_4) or birnessite (δ-MnO_2). The fungus was isolated from manganese pans in soils underlying acid peat deposits in Newfoundland. Natural manganese oxides from these pans were found to be amorphous. The obvious crystallinity in the oxides associated with *Cephalosporium* and its absence in the natural oxides and the oxides associated with *Corynebacterium* may be related to organic matter levels at the sites of manganese oxide deposition. Organic matter appears to inhibit crystallization of these metal hydrates (Schwertmann, 1966).

Mechanisms of microbial transformations of manganese

Several different mechanisms have been proposed for the transformation of manganese from one valency state to another by microorganisms. These mechanisms involve indirect effects resulting from changes to the microenvironment around the microbial surface, or direct effects of enzymic oxidation or reduction.

Alterations to microenvironments

Eh modification. Many microorganisms alter the Eh of their own microenvironment as a result of O_2 production or consumption, or by the excretion of reducing compounds. Such activities may cause a shift in the stability of a particular manganese species, as can be seen in Fig. 5.3. For example,

Fig. 5.7. "Microaerophilic" deposition of manganese in an agar medium by *Hyphomicrobium* T37 (photograph by P.A. Tyler).

microbial utilization of an available organic energy source in certain soils and lacustrine sediments results in a lowering of the Eh and a rapid, non-specific solubilization of MnO_2(s) (Crerar et al., 1972). Some algae are involved in manganese and iron deposition (Kindle, 1932; Pringsheim, 1949; Bromfield, 1976), and it has been suggested that these organisms generate an oxidizing environment through O_2 production or CO_2 consumption. However, Mulder (1964) and other investigators have observed maximum manganese precipitation by microorganisms at apparently reduced O_2 levels or microaerophilic conditions (Fig. 5.7). This "microaerophilic" oxidation may not be indicative of a requirement for lowered O_2 levels but, rather, of a requirement for higher CO_2 levels (N. Uren, personal communication, 1969) or of a decrease in pH in a CO_2-enriched atmosphere (Bromfield, 1974).

pH modification. Excretion of acidic or alkaline metabolic by-products by microorganisms may alter the pH of the microenvironment and result in a non-specific reduction or oxidation of manganese (see Fig. 5.3). Such pH modification could alter the sorption capacity for Mn(II) of solid-phase manganese or manganese-iron mixtures. Thus, larger quantities of Mn(II) would sorb to preformed MnO_2(s) or $Fe(OH)_3$(s) if the pH increased (see Fig. 5.5), or Mn(II) would be released from the solid phase if the pH decreased (see Fig. 5.6).

It is necessary to view such apparently simple relationships with some caution. Not all microorganisms that excrete alkaline by-products are able to precipitate manganese oxides, and Bromfield (1974) has shown that manganese oxidation by *Arthrobacter* is inhibited if the medium pH becomes alkaline too rapidly. The optimum for Mn(II) oxidation by resting cell preparations of this organism is pH 6.5, with no oxidation occurring at pH 5.4 or 7.9 (Bromfield and David, 1976). These authors have stressed the fact that oxidation of Mn(II) in a mildly alkaline reaction mixture containing the *Arthrobacter* can be initiated by *acidification.* In addition, the Mn/Fe ratios of many natural deposits in aquatic systems are higher than the ratios in the aqueous phase (Gorham and Swaine, 1965) suggesting a selective enrichment of manganese. A microbially-induced increase in pH should result in earlier and more extensive deposition of iron in preference to manganese (see Fig. 5.3) rather than the reverse.

Sorption of preformed oxides to cell surfaces. Various authors have suggested that preformed MnO_2(s) present in colloidal suspension is concentrated by selective sorption to the surface of certain microorganisms (see Silverman and Ehrlich, 1964). Factual evidence for this mechanism of concentration is lacking. Colloidal MnO_2(s) is unlikely to form chemically under many of the Eh-pH conditions where extensive microbial deposition of manganese is observed, except following alteration of the microenvironment by the microorganisms.

Microbial utilization of organic complexes. The presence of significant levels of soluble organic matter in some natural habitats (Geering et al., 1969; Rashid, 1971) provides opportunities for complex formation with soluble manganese species, resulting in a degree of stabilization that would not be predicted by reference to an Eh-pH field diagram such as Fig. 5.3 (Crerar et al., 1972). Microbial utilization of the organic moiety of the complex, either as an energy source or as a source of other essential nutrients, may release soluble manganese. If the microenvironmental conditions around the cell are suitable, then the manganese may be precipitated by purely chemical means (Starkey, 1945; Graham, 1959). This mechanism would be of particular significance when the organic-Mn complex is sorbed at solid-liquid interfaces and microorganisms are attracted to, and take advantage of the concentrated organic nutrient. Möse and Brantner (1966) and Brantner (1970) have shown that some bacteria utilize the organic portion of a citrate-manganese complex in media and deposit the manganese around the bacteria. However, Bromfield and Skerman (1950) have advised against the use of hydroxyacids, such as citrate, in media for isolation of manganese-oxidizing organisms, because some isolates capable of manganese deposition on these media fail to oxidize manganese in soils.

Enzymic transformations

Chemolithotrophic and/or mixotrophic oxidation. The oxidation of an inorganic species, such as Mn(II), invites speculation on the possibility of some microorganisms obtaining energy from the process. The oxidation of Mn(II) may serve as a sole energy source for growth and CO_2 fixation in certain microorganisms (*chemolithotrophic*) or, in other microorganisms, provide all or part of the energy required to assimilate organic carbon (*mixotrophic*). According to Ehrlich (1976), the standard free energy change (ΔF^0) for the following equation:

$$Mn^{2+} + \frac{1}{2} O_2 + 2\,OH^- \rightarrow \delta\text{-}MnO_2 + H_2O \qquad (4)$$

is -148 kJ mol^{-1}. Even lower energy yields have been calculated for the effective concentration of soluble manganese in the aqueous phase of marine systems. Ehrlich (1976) has stated, however, that such calculations would be in error since most of the Mn(II) transformed by microorganisms is sorbed to solid MnO_2 or other iron-manganese oxides and, therefore, the effective concentration available for microbial oxidation must be significantly higher than in the bulk aqueous phase.

Earlier evidence suggesting the possibility of chemolithotrophic growth of microorganisms on Mn(II) has been reviewed by Mulder (1964) and Johnson and Stokes (1966). One of the most likely candidates for successful chemolithotrophic growth on Mn(II) is *Leptothrix discophora.* Mulder (1964) presented evidence favouring the production by this organism of hydroxy-

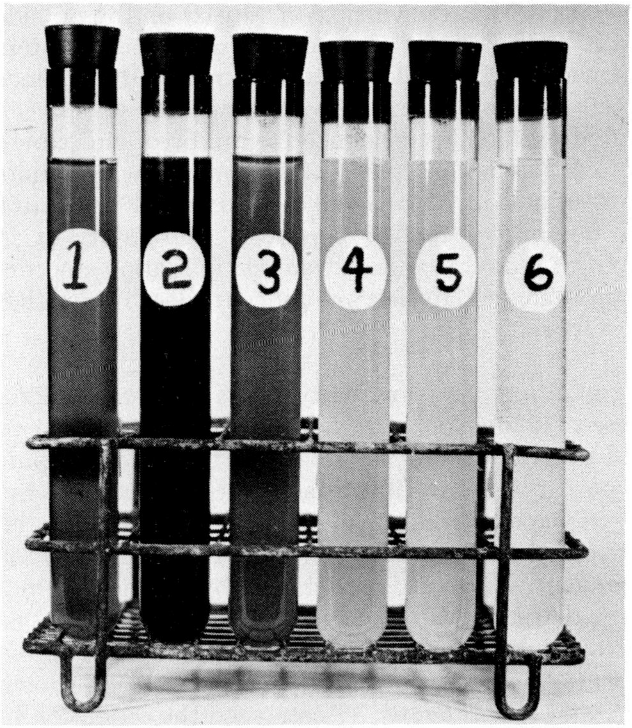

Fig. 5.8. Oxidation of Mn(II) by *Leptothrix discophora*. Treatments as follows: 1, suspension of manganese-grown cells; 2, as in 1 plus 8 μmol of $MnSO_4$ ml^{-1}; 3, as in 2 except cell suspension heated at 93°C for 5 min; 4, suspension of cells grown without $MnSO_4$; 5, as in 4 plus 8 μmol of $MnSO_4$ ml^{-1}; 6, control with phosphate buffer plus $MnSO_4$ but without bacterium (from Johnson and Stokes, 1966).

carboxylic acids which could catalyze the chemical oxidation of Mn(II), but he did not exclude the possibility of a direct enzymic oxidation. Evidence for the induction by Mn(II) in *L. discophora* of a heat-labile, Mn(II)-oxidizing enzyme was presented by Johnson and Stokes (1966). The effects of manganese induction and heating on this process is illustrated in Fig. 5.8. More recently, Ali and Stokes (1971) have reported chemolithotrophic growth of *L. discophora* on Mn(II), provided trace quantities of biotin, thiamine and cyanocobalamin were provided. These authors also reported that the organism grew mixotrophically in a $MnSO_4$-casamino acid medium supplemented with biotin and thiamine. Chemolithotrophic oxidation by a similar organism was claimed by Hogan (quoted by Ehrlich, 1976) to be coupled to an electron transport sequence which included cytochrome oxidase. Mulder (1972) and van Veen (1972) have questioned the above claims for chemolithotrophic growth of *L. discophora* on manganese. These authors have presented evidence indicating that conversion of Mn(II) on the outside of the sheath results from the formation of an insoluble manganese-protein complex that is not lost on washing the cells. Bacteria grown in the absence of Mn(II) produce an extracellular protein that is lost on washing.

Ehrlich (1976) has provided preliminary evidence for mixotrophic growth in strain BIII 45, a bacterium isolated from a marine ferromanganese nodule. Manganese stimulated the assimilation of tritiated L-leucine and the oxidation of Mn(II) appeared to be coupled to ATP synthesis. The ability of an organism to grow mixotrophically could be an advantage in marine environments where both Mn(II) and organic materials are concentrated by sorption to MnO_2(s) and other oxide surfaces.

Other forms of enzymic oxidation. The possibility exists that some microorganisms possess manganese-oxidizing enzymes which are not coupled to energy-generating systems. The oxidation of Mn(II) by normal heterotrophic bacteria may be of this type. Bromfield (1956) has demonstrated that the oxidation of Mn(II) by *Corynebacterium* strain B is enzymic, and indicated that a catalase system is involved in the oxidation. *Arthrobacter* strain 37, isolated from a marine manganese nodule, was shown by Ehrlich (1966, 1968) to oxidize Mn(II) provided that MnO_2(s) is present in the culture system. Ehrlich suggested that only the Mn(II) sorbed to the solid phase is oxidized by the bacterium; that is, the bacterium catalyzes the reaction shown in eqn (5.3). Cell-free extracts of *Arthrobacter* 37 catalyzed the oxidation of Mn(II) in the presence of synthetic Mn-Fe oxide or crushed manganese nodule (Ehrlich, 1968). The manganese-oxidizing activity of the extracts was heat-labile and inhibited by $HgCl_2$ and *p*-chloromercuribenzoate, suggesting that the activity resides in an enzyme protein.

Enzymic reduction. Several authors have reported reduction of Mn(IV) to Mn(II) by bacteria using either endogenous or externally supplied substrates as electron donors (Hochster and Quastel, 1952; Ehrlich, 1966; Trimble and Ehrlich, 1968, 1970). Bromfield and David (1976) have shown that a normally manganese-oxidizing *Arthrobacter* sp. was capable of reducing bacterial manganese oxide in deep, static cultures, and that the rate of reduction was increased greatly by the addition of methylene blue. The reduction of Mn(IV) by *Bacillus* strain 29 and a coccus, strain 32, both isolated from ferromanganese nodules, is enzymic, with part of the Mn(IV)-reducing system inducible in the presence of MnO_2(s) or Mn(II) (Trimble and Ehrlich, 1970). Resting-cell suspensions of "unadapted" cultures, grown in the absence of MnO_2(s), were unable to reduce Mn(IV) without the addition of an electron carrier (ferricyanide), but could be "adapted" by growing the culture in contact with MnO_2(s) (Ehrlich, 1966). Oxygen is needed for culture adaptation to Mn(IV) reduction, and oxygen does not interfere with the process of Mn(IV) reduction by these organisms (Trimble and Ehrlich,

1968). The absence of any significant increase in the rate of Mn(IV) reduction under anaerobic conditions is contrary to the generally accepted view that such conditions are essential (see Alexander, 1961). It seems necessary to define the precise Eh-pH conditions at the bacterium- and MnO_2(s)-water interfaces, since O_2 availability is not necessarily reflected in the Eh level.

Trimble and Ehrlich (1968) have suggested that adapted cultures may use Mn(IV) in preference to oxygen as the terminal electron acceptor, and that Mn(IV) and O_2 do not compete with each other in such cultures. They have summarized the process of microbial solubilization of MnO_2(s) as follows:

$$\text{glucose} \xrightarrow{\text{bacteria}} n\ c^- + n\ H' + \text{end products} \tag{5}$$

$$n\ MnO_2 + 2n\ e^- + 2n\ H^+ \xrightarrow[\text{or unadapted cells} + Fe(CN_6)^{3-}]{\text{adapted cells}} n\ Mn(OH)_2 \tag{6}$$

$$n\ Mn(OH)_2 + 2n\ H^+ \rightarrow n\ Mn^{2+} + 2n\ H_2O \tag{7}$$

Trimble and Ehrlich indicate that the $Mn(OH)_2$ would be brought into solution by lactic and pyruvic acids produced by *Bacillus* 29 and pyruvic acid produced by coccus 32. It was shown that pyruvic acid did not promote dismutation between MnO_2(s) and sorbed Mn(II) by complexing the Mn(II).

TABLE 5.3

Effect of sucrose and yeast extract levels on manganese oxidation by *Arthrobacter* sp. and on the medium pH after 8 d incubation (Bromfield, 1974)

Sucrose content of medium (%)	Yeast extract content of medium (%)						
	0	0.1	0.2	0.3	0.4	0.5	1.0
0	6.3	8.3	8.75	8.7	8.9	8.8	8.9
0.1	5.5	6.8	8.6	8.6	8.8	8.9	9.0
0.2	5.4	5.5	7.8	8.7	8.8	8.9	8.9
0.3	5.4	5.6	7.3	8.4	8.8	8.8	8.8
0.4	5.4	5.4	5.5	8.2	8.6	8.8	8.8
0.5	5.3	5.4	5.5	7.5	8.4	8.5	8.6
1.0	5.4	5.4	5.4	5.5	6.3	7.0	8.0

Numbers represent pH after 8 d; the boxed area indicates those treatments showing manganese oxidation after 8 d.

Effect of medium composition on microbial manganese transformations

Manganese oxidation generally is thought to be inhibited on media rich in organic nutrients. For instance, manganese was not oxidized by a soil bacterium when yeast extract in the medium exceeded 0.01% (Bromfield, 1956), by *Leptothrix discophora* when both glucose and peptone exceeded 0.2% Mulder et al., 1969), and by fungi when added biomalt exceeded about 0.6% (Schweisfurth, 1972). Stokes and Powers (1965) have described the induction of smooth colony forms of *Sphaerotilus discophorus* and their failure to oxidize Mn(II) on a 0.5% peptone-basal medium supplemented with certain carbohydrates. Smooth colonies also were induced by 2% peptone or 0.5% tryptose, but Mn(II) oxidation was not impaired.

Bromfield (1974) has made an extensive study of the effects of sucrose and yeast extract on the oxidation by manganese by a soil *Arthrobacter*. Using a basal salts-$MnSO_4$ medium of initial pH 6.3, manganese oxidation occurred in the absence of added substrate and at levels of 1% of both sucrose and yeast extract (Table 5.3). Vigorous growth occurred over a wide range of substrate concentrations, but manganese oxidation did not occur on sucrose media where the pH fell below 5.5, or on yeast extract media that became alkaline too rapidly. Various patterns of MnO_2(s) deposition were observed at different substrate concentrations. Low levels of yeast extract ($<$0.025%) in media gave densely packed particles with uniformly dark colonies (Fig. 5.9a). The size of the MnO_2(s) particles increased with increasing substrate concentration and colony diameter (Fig. 5.9b). On media containing sucrose alone or in combination with relatively high levels of yeast extract, MnO_2(s) was deposited as discrete and often large specks within and beyond the margin of the colony (Figs. 5.9c and d). Bromfield also reported that non-oxidizing colonies, on media that had become alkaline, rapidly oxidized Mn(II) when exposed to CO_2-enriched air (up to 5% CO_2). The pH of the medium was reduced to below 8.15 in the presence of the CO_2, and under these conditions manganese oxidation commenced.

MANGANESE TRANSFORMATIONS AT SOLID-LIQUID INTERFACES

In natural habitats, and particularly in aquatic systems, the levels of available organic and inorganic nutrients often are insufficient to support substantial microbial growth. Nutrient cations, however, are attracted to and concentrated at negatively-charged solid surfaces. In addition, macromolecules and other organic materials sorb to the surfaces forming monolayers, which alter the physical properties of the surface (Baier et al., 1968) and provide a relatively concentrated source of organic nutrients (ZoBell and Anderson, 1936; Stark et al., 1938). This may include a concentration of organic-manganese complexes. Microbial growth in nutrient-deficient aqueous systems generally is stimulated when the solid-liquid interfacial area

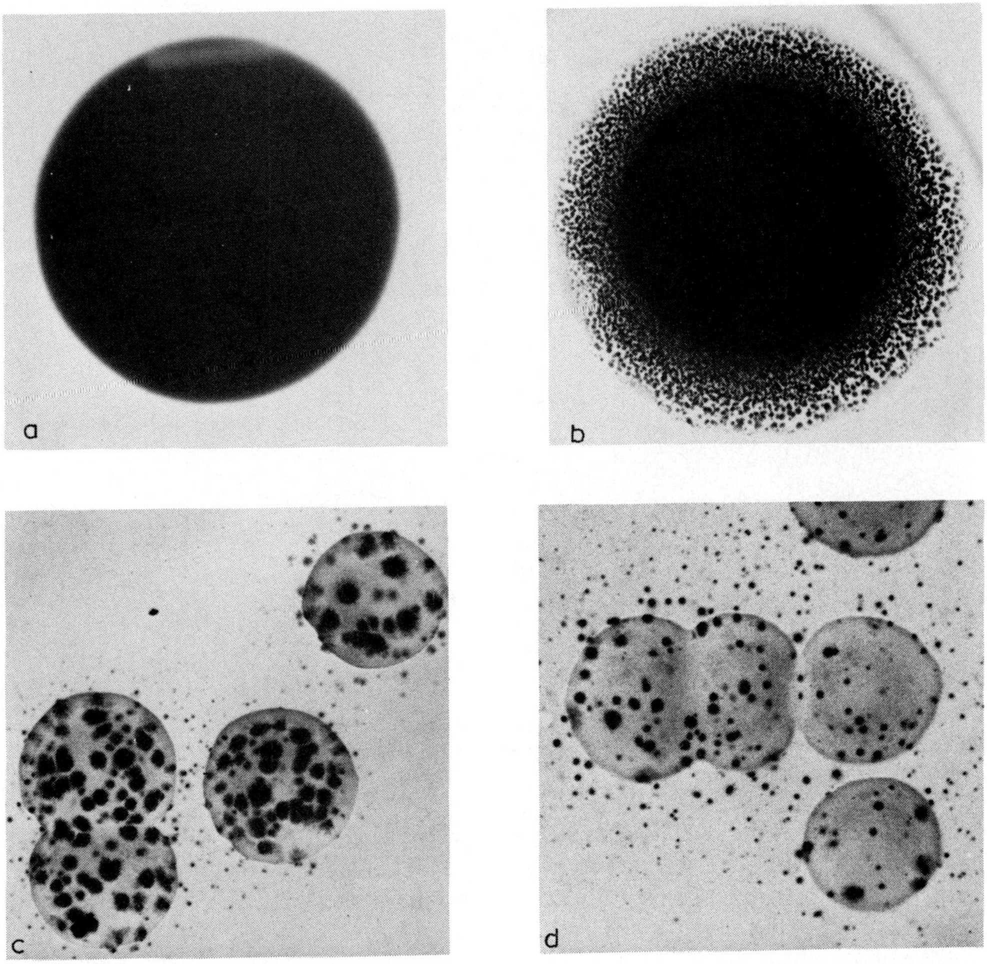

Fig. 5.9. Variation in form of manganese deposition by *Arthrobacter* sp. on different media. Yeast extract (Y) and sucrose (S) levels as follows: (a) Y, 0.0016%, S, 0.001%, 25×; (b) Y, 0.05%, S. 0.037%, 6×; (c) S, 0.018%, 25×; (d) S, 0.018%, 25× (from Bromfield, 1974).

is increased (Heukelekian and Heller, 1940; ZoBell, 1943; Jannasch, 1958). In dynamic aqueous systems, it is advantageous for microorganisms to firmly adhere to solid surfaces and take advantage of continual nutrient replenishment at the solid-liquid interface. Details of the mechanisms of attraction of bacteria to solid surfaces and of selective adhesion of some species to surfaces have been presented in considerable detail by Marshall (1975, 1976).

The accumulation of manganese-rich deposits on the inner surfaces of

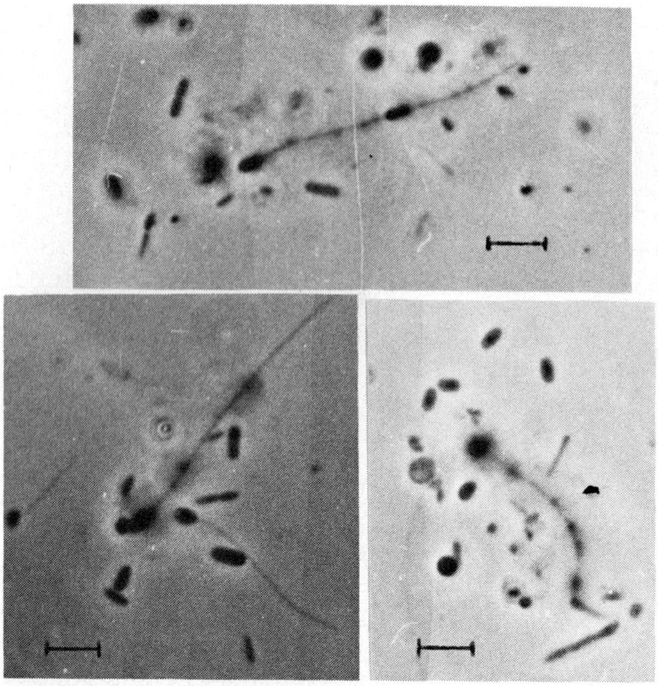

Fig. 5.10. Different bacterial types colonizing glass slides immersed in Lake King William, Tasmania,water for 48 h. Note commencement of manganese deposition by stalked, budding bacteria. Bar = 2 μm.

pipes in large-scale water-distribution systems (Zapffe, 1931; Schweisfurth and Mertes, 1962; Tyler and Marshall, 1967a) demonstrates the advantage afforded to manganese-oxidizing bacteria in aquatic systems at solid-liquid interfaces. Because of high flow-rates and of low nutrient and Mn(II) levels, microbial oxidation of Mn(II) within the aqueous phase may be negligible. Manganese-oxidizing bacteria adsorb at the surface, utilizing accumulated nutrients and depositing manganese at the pipeline surface. The Mn-oxidizing bacteria are not necessarily primary colonizers and, in the case of hyphomicrobium-types (Tyler and Marshall, 1967a), may be dependent on some prior modification to the surface by the pioneer species (Marshall, unpublished results). An early stage in the successional sequence showing the selective deposition of MnO_2(s) by sorbed hyphomicrobia is found in Fig. 5.10.

Another example of the role of solid-liquid interfaces in microbial transformations of manganese is the importance of Mn(IV), in the presence or absence of Fe(III), in the oxidation of Mn(II) by the marine bacterium BIII 45 isolated from a ferromanganese nodule (Ehrlich, 1976). This bacterium is mixotrophic, obtaining maximum energy from the oxidation of Mn(II) at a concentration of approximately 4 mM Mn(II). The concentration of Mn(II) in seawater ranges from 0.2 to 8.0 $\mu g\ l^{-1}$ (Crerar and Barnes, 1974) and, consequently, is inadequate to serve as a useful energy source for the bacterium. Concentration of Mn(II) by sorption to $MnO_2(s)$ or to ferromanganese nodule surfaces is sufficient to provide an adequate energy source for such bacteria. In fact, various marine bacteria studied by Ehrlich (1966, 1968, 1976) only oxidize Mn(II) where it is sorbed to $MnO_2(s)$ or to solid ferromanganese material. Numerous bacteria were found on surfaces of ferromanganese nodules examined under the scanning electron microscope (LaRock and Ehrlich, 1975), and Ehrlich et al. (1972) have shown that many of the bacteria adhering to marine manganese nodules are capable of Mn(II) oxidation or Mn(IV) reduction.

The sheathed, manganese-oxidizing *Lepothrix discophora* is found in nature attached to surfaces. Ehrlich (1976) has pointed out that the optimum level of Mn(II) for mixotrophic growth of *L. discophora* is 163 $mg\ l^{-1}$ (Ali and Stokes, 1971), yet many freshwater systems contain only 0.1 to 1 $mg\ l^{-1}$ of Mn(II). Obviously, *L. discophora* takes advantages of the higher concentrations of Mn(II) at solid-liquid interfaces.

FORMATION OF MANGANESE MINERALS

Since most major manganese deposits are of sedimentary origin, it is relevant to consider aspects of microbial transformations of manganese in freshwater and marine systems, including the conditions necessary for microbial involvement in the formation of extensive sedimentary manganese deposits.

Manganese transformations in stratified waters

Thermal stratification occurs in many moderately deep freshwater lakes in summer when the surface water is significantly warmer than the bottom water. The surface water (*epilimnion*) is regularly circulated as a result of wind action, whereas the more dense bottom water (*hypolimnion*) does not circulate. These waters are separated by a zone (the *thermocline*) characterized by a steep-temperature gradient with depth (Fig. 5.11). If the thermocline is below the level of effective light penetration (compensation level), then the oxygen supply in the hypolimnion may be rapidly depleted. The extent of this depletion depends on the overall nutrient status of the lake, as this controls the extent of primary productivity in the epilimnion and, hence, the supply of decomposable organic matter ultimately entering the hypolimnion. In cooler weather, the temperature of the epilimnion approaches that of the hypolimnion and the entire lake begins circulating (turnover) and oxygen enters the bottom waters. Lakes which mix periodically

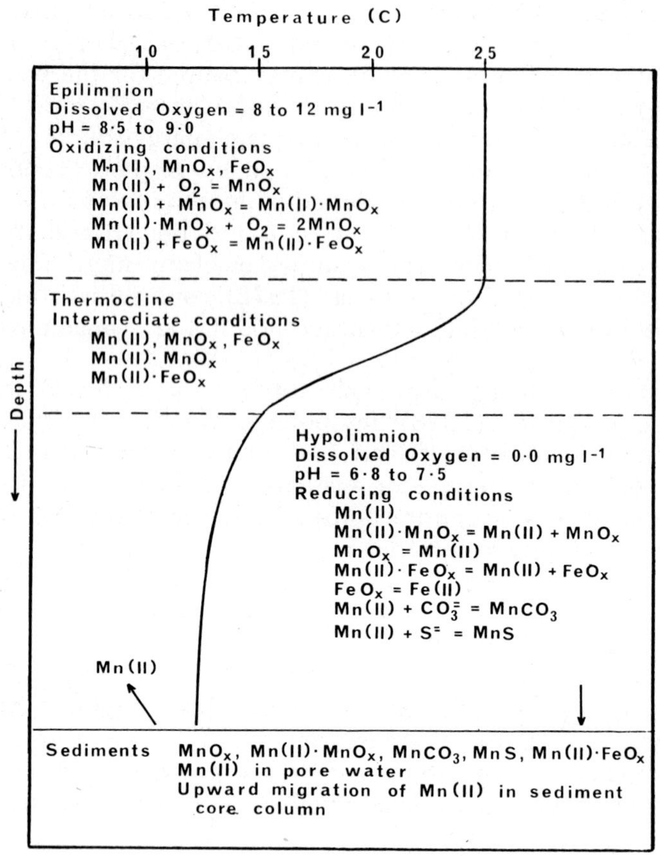

Fig. 5.11. Vertical distribution of manganese species in the stratified Lake Mendota, Wisconsin (from Delfino and Lee, 1968).

are termed *holomictic*. *Meromictic* lakes are permanently stratified as a result of an intrusion of soluble salts in the bottom waters giving a permanent density difference between the upper and lower layers, which are separated by a transition zone or *chemocline*. Free oxygen is absent from the hypolimnion of meromictic lakes.

Conditions in the epilimnion are oxidizing and the pH can exceed 8.5 because of photosynthetic activity (Delfino and Lee, 1968). Manganese is present in this zone mainly as MnO_2(s) or as Mn(II) sorbed to iron and manganese oxides. In this state, the Mn(II) is probably oxidized rapidly to Mn(IV) by microorganisms associated with the particulate oxides (Ehrlich, 1966, 1968). The development of manganese-oxidizing bacteria in lakes is governed by the presence of oxygen and the occurrence of Mn(II). Dubinina et al. (1974) have studied the occurrence of such bacteria with depth in the meromictic Lake Gek-Gel′ (U.S.S.R.) where, they reason, the presence of a relatively constant chemocline zone should favour the development of manganese-oxidizing bacteria. The maximum populations of the manganese- and iron-oxidizing species *Metallogenium* and *Siderocapsa* were found at a depth approximating that where oxygen was almost depleted and where adequate Mn(II) was available (Fig. 5.12). *Siderocapsa* was obtained in pure culture for the first time, but the authors believe it may be a form of *Arthrobacter*. Few of the manganese-oxidizing bacteria were detected in the hypolimnion. In a study of the distribution of particulate iron and manganese oxides with with depth in Lake Kizaki-Ko (Japan), Tanaka (1965) reported a distinct geochemical separation of iron and manganese particles in the zone of near

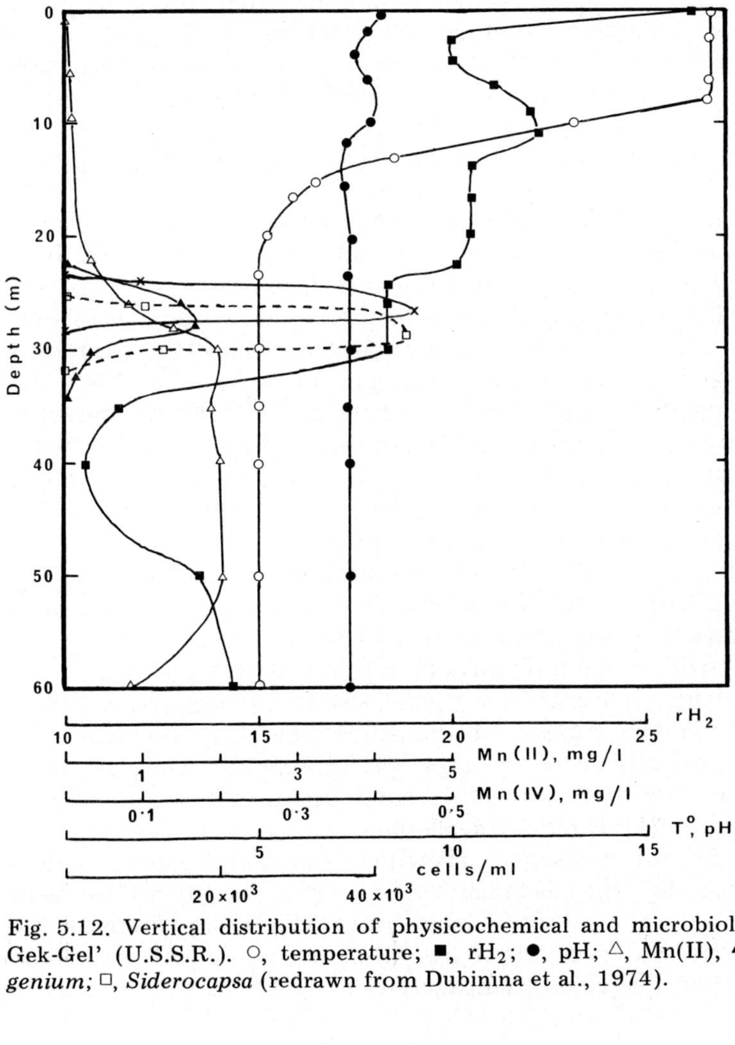

Fig. 5.12. Vertical distribution of physicochemical and microbiological factors in Lake Gek-Gel' (U.S.S.R.). ○, temperature; ■, rH_2; ●, pH; △, Mn(II), ▲; Mn(IV); × *Metallogenium;* □, *Siderocapsa* (redrawn from Dubinina et al., 1974).

oxygen depletion. Tanaka suggested that this resulted from an earlier oxidation then sedimentation of the iron oxide particles, followed later by the manganese oxide particles.

The upper portion of the thermocline in a nutrient rich (eutrophic) lake contains some oxygen, but the lower portion of the zone may contain little or no oxygen. Thus, the thermocline can represent a transition zone in which the manganese would be in a relatively unstable state. $MnO_2(s)$, and Mn(II) in the free and sorbed state, are found in this zone (Fig. 5.11).

During thermal stratification of a eutrophic lake, the hypolimnion would be essentially anoxic and manganese would occur primarily as Mn(II)), as found in Lake Mendota (U.S.A.) by Delfino and Lee (1968) and Nordbytjernet (Norway) by Hongve (1974). Microbial activity certainly is responsible for reduction of Mn(IV) to Mn(II) under such conditions (Dubinina et al., 1974). The presence of dissolved sulphide and carbonate in the hypolimnion may result in the precipitation of MnS and $MnCO_3$ if conditions are suitable (Morgan and Stumm, 1965b; Delfino and Lee, 1968; Hongve, 1974).

Manganese concretions and crusts in freshwater lakes and streams

Gorham and Swaine (1965) have described a variety of ferromanganese deposits in freshwater lakes and streams. "Oxidate crusts" apparently form above basal glacial deposits in Lake Windermere (England) where Mn(II) and Fe(II) are continually oxidized and deposited, provided organic lake mud deposits are thin or absent. Manganese crusts also form on stony or gravelly shores near the water surface of Nova Scotian and Fennoscandian lakes, but not on the soft, organic bottom muds. Crerar et al. (1972) have shown in diagrammatic form how the position of manganese deposits in lacustrine situations varies with the trophic state of the individual lake. It has been suggested that oxidate crust formation near lake margins may involve the induction of high pH and oxygen levels due to photosynthetic algal activity (Kindle, 1935). Using the capillary peloscope method of Perfil'ev and Gabe (1969), Kutuzova (1975) studied the development of iron- and manganese-oxidizing microorganisms in the surface layer of bottom deposits and in the water above an ore field in Lake Krasnyi (U.S.S.R.). It was found that iron and manganese-oxidizing *Metallogenium* bacteria were involved, but that these bacteria developed on the surface of the euglenoid alga *Trachelomonas volvocina.* Oxygen production by the alga was thought to stimulate manganese oxidation by the *Metallogenium* under conditions where oxygen might be limiting and Mn(II) is readily available.

Using capillary and slit peloscopes (Perfil'ev and Gabe, 1969) in mud samples from lakes in the Karelian ASSR and Karelian Peninsula (U.S.S.R.), Perfil'ev and Gabe (1965) have provided a most extensive and revealing study of the development of iron and manganese microzones in submerged muds. They described manganese-rich, black-brown deposits in upper mud layers resulting from the activities of the iron- and manganese-oxidizing organism *Metallogenium personatum*. The only organism found in lower layers of iron-rich, orange deposits was the iron-oxidizing bacterium *Siderococcus limoniticus*. Gabe and Rabinovich (1965) have emphasized the importance of Eh relationships in the development of the stratified manganese and iron microlayers. In the lower reduced mud layers, Fe(II) and Mn(II) are available, and diffuse into the higher, more oxidized layers. The separation between the iron-rich and manganese-rich layers results from the greater mobility of manganese, the oxidation of Fe(II) at lower Eh values than Mn(II), and the fact that the predominantly manganese-oxidizing *M. personatum* is less microaerophilic than the iron-oxidizing *S. limoniticus*. The *Metallogenium* was detected in peloscopes at the lower boundary of the visually distinguishable oxidized region.

Metallogenium personatum has been detected in most peloscopic studies of freshwater manganese deposits. Perfil'ev and Gabe (1965, 1969) describe a cycle of growth that is characterized by extreme polymorphism. Motile buds are produced and attach to solid surfaces, where they germinate and produce a single or verticillate trichospherical microcolony consisting of

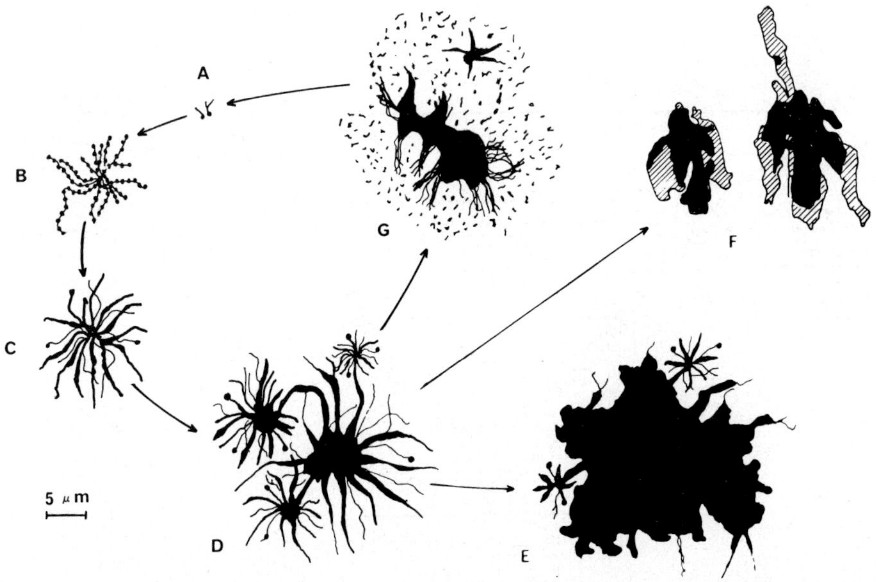

Fig. 5.13. A simplified growth cycle of *Metallogenium personatum*. A, germination of round, reproductive cells; B, young trichospherical microcolony showing chains of coccoidal cells; C, early stage of manganese deposition on trichospherical microcolony; D, development of secondary trichospherical microcolonies; E, heavy mineralization of trichospherical microcolony; F, clavate microcolony with heavy manganese deposition; G, zoogloeal-form with clusters of small coccoidal cells (drawn from photographs in Perfil'ev and Gabe, 1965).

radiating, filamentous, black-brown processes, tapering towards the tips (Fig. 5.13). The radial processes consist of chains of coccoidal cells, which become encrusted with manganese oxide. A later zoogloeal stage may yield radial-lobate or clavate forms heavily masked by manganese deposits. All stages can form motile solitary cells.

Dubinina (1973) has presented evidence for the role of *Metallogenium* in the deposition of manganese in the littoral zone (shallow area near the shore) of Lake Punnus-Järvi (U.S.S.R.). This is a holomictic lake, and the suggested cycle of manganese transformations and final deposition is shown in Fig. 5.14. Manganese enters the lake as Mn(II) and as suspended MnO_2(s). Microbial oxidation of the Mn(II) leads to a sedimentation of the larger solid manganese particles. The Mn(IV) is reduced and remobilized in the sediments and hypolimnion, and migrates along the upper layers of sediment towards the littoral, where stable oxidizing conditions exist and manganese concretions are actively increasing in size. Manganese-oxidizing *Metallogenium* organisms have been isolated from the epilimnion and from littoral zone concretions by selective culture techniques. In addition, electron microscopy has revealed the existence of these organisms on the surface of the manganese concretions. Oxidation of Mn(II) by *Metallogenium* in this lake

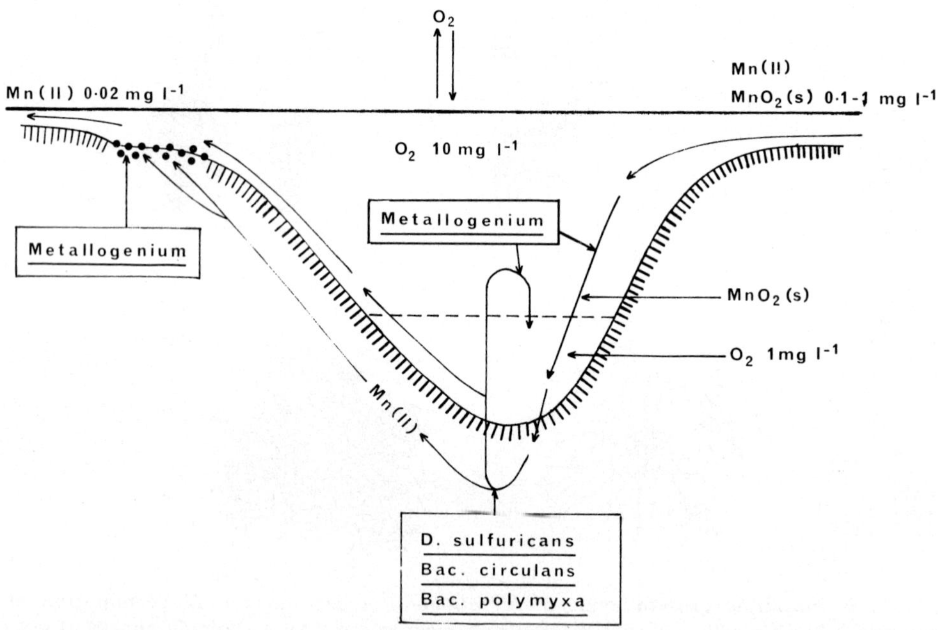

Fig. 5.14. Participation of specific groups of microorganisms * in the formation of lacustrine manganese ore in Lake Punnus-Järvi (U.S.S.R.) (redrawn from Dubinina, 1973 and Kuznetsov, 1975).

* Editors' note: *D. sulfuricans* = *Desulfovibrio desulfuricans.*

takes place at Eh values of 250-400 mV under weakly acidic conditions where autoxidation is not possible (see Fig. 5.3).

Gorham and Swaine (1965) found a wide range of Fe and Mn levels in ferromanganese concretions from different sources *. Concretions from Fennoscandian freshwaters contain more Fe than those from English lakes. An inverse relationship between Fe and Mn was found in the freshwater and marine concretions examined, with Fe predominant in freshwater crusts and Mn in the marine nodules. If microorganisms are involved in deposition of these concretions, then the wide range in Fe : Mn ratios may reflect site differences in terms of supply of Fe(II) and Mn(II), pH and Eh conditions, trophic state of the waters (Crerar et al., 1972) and the types of microorganisms involved. Very detailed comparisons of the microflora associated with the zones of deposition in these areas are warranted. Few estimates have been made of the time taken for significant build-up of ferromanganese deposits. Gorham and Swaine (1965) quote examples of the development of visible coatings on rocks within 25 years in Nova Scotian lakes, and of the possibility of ore-scraping for economic use every 50 years in lakes of southern Sweden. A rapid deposition of manganese has been reported in hydroelectric pipelines (see later section).

Marine manganese crusts and nodules

The presence of manganese nodules in ocean sediments depends on the Eh and organic matter status of the particular sediment. Volkov et al. (1975) classified the sediments in a transect from the Japanese coast to the open northern Pacific Ocean on their organic matter content and oxidation state as follows: (1) reduced sediments of the continental slope; (2) transitional sediments (weakly reduced) of the marginal zone of the northwest basin; (3) oxidized pelagic sediments of the open ocean (red oozes). Mn(IV) was not detected in the reduced coastal sediments, whereas it was present in the transitional and pelagic zones. Manganese nodules appeared in early samples of the transitional zone, and ranged in size from tiny grains to large boulders with a layered structure. The form of the manganese in nodules and crusts remains controversial, but Crerar and Barnes (1974) suggest that the major forms are birnessite and todorokite. Although both minerals may occur in the same nodule, birnessite is dominant in nodules on ridges and sea-mounts whereas todorokite is dominant in near-shore and abyssal floor nodules.

Manganese-oxidizing bacteria and fungi have been detected in 70% of sediment samples from the Bay of Biscay and most samples from the North Sea (Krumbein, 1971). Ehrlich et al. (1972) found that on nodules from the central and eastern Pacific Ocean, 13 to 18% of the total bacteria were capable of oxidizing Mn(II), and over 50% were able to reduce Mn(IV) to Mn(II). Sorokin (1971) also found bacteria developing on manganese con-

---

* Elemental compositions of ferromanganese nodules are given on p. 238.

cretions from the central Pacific Ocean, but failed to detect organisms resembling *Metallogenium.* More recently, LaRock and Ehrlich (1975) have examined freshly-collected nodules from Blake Plateau in the Atlantic Ocean by scanning electron microscopy. They observed numerous microcolonies of rod- and coccus-shaped bacteria firmly adhering to the nodule surfaces by means of obvious bridging polymers. The frequency of bacterial occurrence on nodules based on microcolony distribution was estimated to be about 6×10^5 bacteria cm^{-2}.

Ehrlich has made an extensive study of the physiology and biochemistry of manganese-oxidizing and -reducing bacteria isolated from marine manganese nodules. Bacteria from the nodules oxidized Mn(II) in laboratory culture only in the presence of MnO_2(s) or a synthetic iron-manganese oxide Ehrlich, 1963, 1966, 1968). Cell-free extracts of *Arthrobacter* 37 exhibited enzymic activity which accelerated manganese accretion to the solid oxides (Ehrlich, 1968). The marine bacterium strain BIII 45 may grow mixotrophically, deriving energy from the oxidation of Mn(II) (Ehrlich, 1976). As shown in Table 5.4, bacteria isolated from manganese nodules are capable of oxidizing Mn(II) at hydrostatic pressures equivalent to those experienced at depths where the nodules are located (Ehrlich, 1971, 1974). Culture BIII 39 oxidized Mn(II) at 48 MPa and 4°C, despite the fact that it was unable to grow at pressures beyond 41.2 MPa at 15°C.

Ehrlich (1974) also demonstrated active manganese reduction by bacteria at pressures up to 41.2 MPa. The MnO_2-reductase activity of *Bacillus* 29 was stimulated by Na^+, K^+, Mg^{2+} and Ca^{2+}. The optimum temperature for MnO_2-reductase activity varied with the system tested; i.e., 40°C for induced whole cells, 25°C for induced glucose-linked activity in cell extracts, 30°C for uninduced glucose-linked, ferricyanide activity, and 40°C for uninduced

TABLE 5.4

Mn(II) oxidation by culture BIII 39 at various hydrostatic pressures. Incubated for 17 h at 4°C (adapted from Ehrlich, 1974)

Pressure (MPa)	Cell No. ($\times 10^{-9}$)	Mn(II) oxidized (nmol ml^{-1})
0.1	6.1	50
	4.8	40
34.3	4.8	20
	4.3	40
41.2	3.8	50
	4.4	50
48.0	4.4	30
	4.1	30
54.9	4.1	trace
	4.3	trace

ferrocyanide-linked activity (Ghiorse and Ehrlich, 1974). Acceleration of MnO_2(s) reduction in ferromanganese nodules by *Bacillus* GJ33 also released copper, nickel and cobalt. Negligible amounts of iron were solubilized under these conditions. The close correlation between manganese solubilization and nickel and cobalt release led Ehrlich et al. (1973) to suggest that nickel and cobalt are more tightly bound to the Mn(IV) oxide matrix than is copper. A similar relationship between manganese and both nickel and cobalt in marine and terrestrial manganese nodules was reported by McKenzie (1975) as a result of electron microprobe studies.

Ehrlich (1975) has recently reviewed aspects of microbial participation in the formation of ferromanganese concretions in the deep sea. Other aspects of the biogeochemistry of ferromanganese nodules are given by Lundgren and Dean in Chapter 4.

Deposition of manganese in pipelines

Microbial oxidation of Mn(II) and the deposition of MnO_2(s) on surfaces of water reticulation and hydroelectric pipelines is of frequent occurrence throughout the world (Zapffe, 1931; Wolfe, 1960; Schweisfurth and Mertes, 1962; Tyler and Marshall, 1967a). These deposits create problems due to blockage of reticulation pipes, staining of waters, and increased friction effects leading to a reduced head and, hence, a reduced power output in hydroelectric systems (Schweisfurth and Mertes, 1962).

Various microorganisms have been implicated in the deposition of manganese in pipeline systems. These include sheathed bacteria (Wolfe, 1960; Mulder, 1964), Gram-negative bacteria (Schweisfurth and Mertes, 1962), *Nocardia* sp. (Schweisfurth, 1968) and some fungi (Tyler and Marshall, 1967a; Schweisfurth, 1969). In Tasmanian pipeline deposits, Tyler and Marshall (1967a, b, c) found that the dominant organism was a stalked, budding bacterium, which resembled *Hyphomicrobium* (Hirsch and Conti, 1964) under certain growth conditions and *Pedomicrobium* (Aristovskaya, 1961) under other conditions. The pleomorphy of this organism has been considered in more detail by Tyler and Marshall (1967d), Bauld et al. (1971) and Bauld and Tyler (1971). Because the hyphomicrobia are an unusual group of bacteria and because their stalk diameter is near the resolution limit of the light microscope, Tyler and Marshall (1967a) suggested that these bacteria may be present, but unrecognized, in other pipeline manganese deposits. An investigation of deposits from other parts of the world (Tyler and Marshall, 1967c; Tyler, 1970), revealed that hyphomicrobia did, in fact, dominate most, but not all, deposits.

The concentration of Mn(II) in waters entering these pipelines is very low, 0.01 to 0.07 $\mu g\ l^{-1}$ in the system studied by Tyler and Marshall (1967a), but bacteria attached to the pipeline walls are able to concentrate significant quantities of manganese. Tyler and Marshall (1967b) have compared a pipeline to an elongated continuous culture vessel into which lake water is fed at a relatively constant rate. Initially, various bacteria adhere to the pipeline surfaces (Fig. 5.10), but manganese-oxidizing forms eventually dominate the surface. This provides a selective advantage as in the "take-over" phenomenon described for continuous culture systems by Munson and Bridges (1964). The mechanism whereby hyphomicrobia adhere so effectively to surfaces has been studied by Marshall and Cruickshank (1973), who concluded that the large end of the pear-shaped *Hyphomicrobium* cells is relatively hydrophobic and is attracted to any non-aqueous phase, such as found at solid-water and oil-water interfaces. Firm adhesion at surfaces was shown to depend on bridging between the cell and the solid surface at extracellular polymers.

Aristovskaya (1963) and Tyler and Marshall (1967b) have related the production of branched stalks and buds by these organisms to the efficient development of a three-dimensional manganese deposit. Older cells, isolated from the nutrient medium by an insoluble manganese deposit, maintain cytoplasmic continuity with actively metabolizing young buds that are free of manganese. Repetition of the budding process, following the initiation of manganese deposition around maturing cells, ensures an efficient and active development of the colonial manganese-oxidizing organism. The efficiency of this microbial process can be judged by the fact that the deposits of about 1 cm thick can develop on surfaces of 1 m radius pipelines within 6 months of cleaning. For a pipeline 1 km long, this represents approximately 62 m^3 of wet deposit containing from 30 to 50% manganese (on a dry weight basis).

Although most of the waters involved contain more iron than manganese, low levels of iron generally are found in pipeline manganese deposits (Tyler and Marshall, 1967a; Tyler, 1970). This suggests that a specific mechanism for manganese deposition exists. Tyler and Marshall (1967a) have provided evidence that microbial involvement is required for initiation of manganese deposition. Deposition of manganese was prevented by autoclaving a large sample of natural lake water (from Lake King William, Tasmania), but commenced shortly after inoculation of the autoclaved water with a small sample of unsterilized water. On the other hand, treatment of natural water with sodium azide inhibited deposition of manganese and the inhibition was still evident following inoculation as above. Tyler and Marshall also demonstrated that the lack of manganese deposition in pipelines associated with a different lake system (Great Lake) was the result of negligible amounts of Mn(II) in the water, and not to the absence of specific manganese-oxidizing bacteria.

Manganese deposition in soils

Manganese is present in relatively large quantities in many soils, and even occurs as nodules (Taylor et al., 1964; McKenzie, 1975) and crusts or "desert varnish" (Krumbein, 1969; Crerar et al., 1972). Because of the vast array of soil types with very different physical, chemical and biological properties, it is difficult to generalize on the forms of manganese present. Aspects of the chemical state of manganese in soil solutions have been considered by Leeper (1947), Jones and Leeper (1951), Taylor et al. (1964), Geering et al. (1969) and Collins and Buol (1970). In general, manganese in soils exists as water-soluble manganese, exchangeable Mn^{2+} on soil colloids, organically bound Mn, and as various Mn oxides.

Mn(II) is readily oxidized when added to neutral soils (Mann and Quastel, 1946) and numerous investigators have isolated manganese-oxidizing microorganisms from soils (Gerretsen, 1937; Leeper and Swaby, 1940; Bromfield and Skerman, 1950; Timonin, 1950; Aristovskaya, 1961, 1963; Mulder and van Veen, 1968; Ivarson and Heringa, 1972; Timonin et al., 1972; Bromfield, 1976). Some authors have used hydroxyacids (malate, citrate) in media for isolation of manganese-oxidizing microorganisms, but Bromfield and Skerman (1950) found that organisms oxidizing manganese in the presence of citrate may not oxidize manganese in soils. They recommend that hydroxyacids should not be used in isolation media. Van Veen (1973) has concluded that the oxidation of Mn(II) in soils with pH values between 5.5 and 8 cannot be explained by the catalytic effects of hydroxyacids, but must be catalyzed by specific manganese-oxidizing microorganisms (Bromfield and David, 1976). As mentioned earlier, Bromfield (1956) has provided sound evidence for enzymic oxidation of Mn(II) by resting-cell preparations of a soil *Corynebacterium* (= *Arthrobacter*, Bromfield, 1974).

Biological oxidation of Mn(II) is slow or absent in soils below pH 5.0 (Leeper and Swaby, 1940; Van Veen, 1973), although Ivarson and Heringa (1972) have reported the oxidation of Mn(II) below pH 5.0 by several species of *Cephalosporium* isolated from manganese pans underlying peat deposits in Newfoundland. Bromfield (1976) also found a slow, but significant, oxidation of Mn(II) by the alga *Chlorococcum humicola* at pH levels below 5.0. This alga was isolated from manganese deposited at the surface of a cultivated acid soil. Bromfield noted that deposition of MnO_2(s) on the soil surface by the alga may actually take place at a higher pH level, since bacteria associated with the alga readily raise the pH of soil agar to near 6.0. Krumbein (1969) has found manganese-oxidizing lichens, algae, fungi and bacteria associated with ferromanganese crusts on desert soils in the Negev, Israel.

More direct observations of microorganisms in podzolic soils by Aristovskaya (1961, 1963), using the capillary pedoscope methods of Perfil'ev and Gabe (1969), revealed the existence of manganese-oxidizing *Pedomicrobium* and *Metallogenium* species. Three species of the branching, budding *Pedomicrobium* were described by Aristovskaya; *P. ferrugineum* oxidizing Fe(II), *P. manganicum* oxidizing Mn(II), and *P. podzolicum* oxidizing both Fe(II) and Mn(II). Aristovskaya has presented evidence suggesting that iron- and manganese-oxidizing microorganisms play an important role in the genesis of podzolic soils and in the formation of ferromanganese concretions in such soils. Ten Khak-Mun (1967, 1969, 1973) has described the activities of *Pedomicrobium*, *Metallogenium* and *Neumanniella* in the oxidation of Mn(II) in soils in South Sakhalin (U.S.S.R.) and in the formation of ferromanganese concretions in brown forest soils.

Zavarzin (1961, 1963) has claimed that the deposition of manganese in discrete areas near certain fungal hyphae results from the activity of a bacterium which he named *Metallogenium symbioticum*. This controversial bacterium was described by Zavarzin (1963) as possessing thin (10-20 nm diameter) filaments, or araia, surrounded by MnO_2(s). Examination by electron microscopy revealed that round motile cells are formed in the filaments by a process of budding (Zavarzin, 1964). Dubinina (1969, 1970) has suggested that *M. symbioticum* resembled the mycoplasmas and may parasitize the fungal host. Other workers have observed similar structures around various species of manganese-oxidizing fungal colonies (Tyler and Marshall, 1967a; Schweisfurth, 1969; Mirchink et al., 1970; Schweisfurth and Hehn, 1972; Timonin et al., 1972), but all attempts to isolate the proposed bacterium have failed. Walsh and Mitchell (1972) have described a similar, but free-living organism involved in the oxidation of Fe(II) in acid-mine wastes, and have detected increases in protein and nucleic acids with growth of the organism. Generally, there is some scepticism regarding the validity of *M. symbioticum* (Timonin et al., 1972; Bromfield, 1974). Bromfield (1974) observed that manganese deposition at a distance from *Arthrobacter* colonies is substrate-dependent (see Figs. 5.8c and d), and rejected the possibility of an association with *M. symbioticum.* He suggested that manganese deposition away from bacterial or fungal colonies may involve precipitation of manganese by a diffusible microbial by-product, such as the manganese-complexing protein reported by Van Veen (1972).

The mobilization of manganese is greatest in soils of high organic matter content (Crerar et al., 1972), particularly in very moist soils where microbial activity induces reducing conditions.

Micropalaeontology and stromatolites

The ubiquitous occurrence of manganese-oxidizing microorganisms in sites where deposition of manganese is actively proceeding has led to a search for possible evidence of microbial activity in fossil manganese deposits. Shternberg (1967) has presented photographic evidence for structures resembling *Metallogenium personatum* in Tetri-Tskaro (U.S.S.R.) sedimentary manganese deposits (Palaeozoic, containing 29.4% Mn and 5.0% Fe) and in manganite and pyrolusite ores from the Chiatura (U.S.S.R.) sedimentary deposit (Oligocene). Earlier reports by Barghoorn of pleomorphic, budding, bacterial-like structures in the Lower Gunflint Iron Formation, Canada (2.0

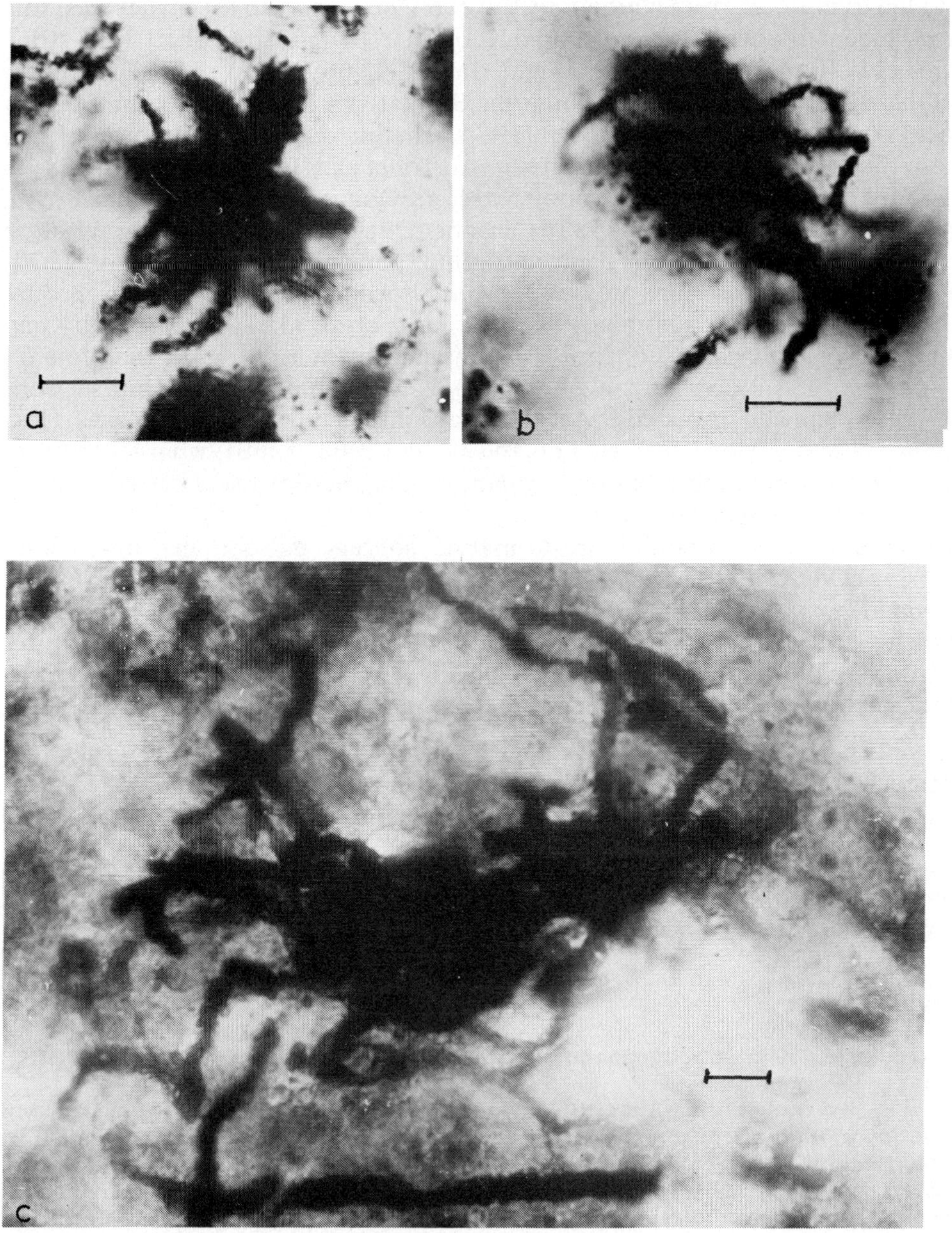

Fig. 5.15. Structures resembling *Metallogenium personatum* in fossil deposits. (a), manganese-containing structures in the Cooley Dolomite of the McArthur Group, N.T., Australia (from Muir et al., 1974); (b) as in (a) (unpublished photograph by M.D. Muir); (c) structures from minor iron formations in the McMinn Formation of the Roper Group, N.T., Australia (unpublished photograph by C.J. Peat and M.D. Muir). Bar = 10 μm.

Gy) led to a tentative assignment of the name *Eoastrion* to the structures. In a more recent report, Barghoorn and Knoll (1975) indicated that these structures closely resembled Perfil'ev and Gabe's (1965, 1969) published photographs of *Metallogenium personatum.* Similar structures have subsequently been described by Muir et al. (1974) in the Cooley Dolomite of the McArthur Group, Northern Territory, Australia (1.5 Gy), by Kline (1975) in the Paradise Creek Formation, northwestern Queensland, Australia (1.6 Gy), by Knoll and Barghoorn (1975) in the Duck Creek Dolomite, Western Australia (1.8 Gy), and by Peat and Muir (personal communication, 1976) in minor iron formations in the McMinn Formation of the Roper Group, Northern Territory, Australia (1.3 Gy). Muir et al. (1974) have used X-ray microprobe analysis to confirm the presence of manganese in the *Metallogenium*-like structures of the Cooley Deposit. Examination of thin sections of these deposits revealed typical trichospherical-like microcolonies (Fig. 5.15a, b) as described by Perfil'ev and Gabe (1965, 1969), whereas sections of the iron-containing McMinn Formation (Fig. 5.15c) resemble the clavate forms of *Metallogenium.*

Ferromanganese nodules from marine sources exhibit an "onionskin" layering (Crerar and Barnes, 1974) and Monty (1973) has suggested that the ferromanganese nodules from Blake Plateau (Atlantic Ocean) resemble a form of bacterial stromatolite. Monty has observed bacterial-like structures in thin sections of these nodules and a general stromatolite appearance to the whole structure.

ACKNOWLEDGEMENTS

The author gratefully acknowledges the provision of unpublished data by Dr. Marjorie D. Muir *, Department of Geology, Royal School of Mines, London, and by Dr. Henry L. Ehrlich, Department of Biology, Rensselaer Polytechnic Institute, Troy, NY.

REFERENCES

Alexander, M., 1961. Introduction to Soil Microbiology. John Wiley, New York, NY, 472 pp.

Ali, S.H. and Stokes, J.L., 1971. Stimulation of heterotrophic and autotrophic growth of *Sphaerotilus discophorus* by manganous ions. Antonie van Leeuwenhoek; J. Microbiol. Serol., 37: 519—528.

Aristovskaya, T.V., 1961. Accumulation of iron in breakdown of organomineral humus complexes by microorganisms. Dokl. Akad. Nauk USSR, 136: 954—957 (English Translation).

Aristovskaya, T.V., 1963. Natural forms of existence of soil bacteria. Microbiology, 32: 564—568.

* Present address: Bureau of Mineral Resources, Canberra, Australia.

Baier, R.E., Shafrin, E.G. and Zisman, W.A., 1968. Adhesion: Mechanisms that assist or impede it. Science, 162: 1360—1368.

Barea, J.M., Olivares, J., Aguilar, A. and Callao, V., 1971. Microbial reduction of MnO_2 and its applications to agricultural soils deficient in assimilable Mn. I. Microbiological and analytical studies of a mould from soil capable of Mn solubilization. Agrochimica, 15: 472—479.

Barghoorn, E.S. and Knoll, A.H., 1975. *Eoastrion* and the *Metallogenium* problem. In: Chemical Evolution of the Precambrian. College Park Colloquia on Chemical Evolution, Abstracts, University of Maryland, p. 24.

Bauld, J. and Tyler, P.A., 1971. Taxonomic implications of reproductive mechanisms of *Hyphomicrobium*-facies and *Pedomicrobium*-facies of a pleomorphic budding bacterium. Antonie van Leeuwenhoek; J. Microbiol. Serol., 37: 417—424.

Bauld, J., Tyler, P.A. and Marshall, K.C., 1971. Pleomorphy of a budding bacterium on various carbon sources. Antonie van Leeuwenhoek; J. Microbiol. Serol., 37: 409—416.

Bautista, E.M. and Alexander, M., 1972. Reduction of inorganic compounds by soil microorganisms. Soil Sci. Soc. Am. Proc., 36: 918—920.

Beijerinck, M.W., 1913. Oxydation des Mangancarbonates durch Bakterien und Schimmelpilze. Folia Microbiol. (Delft), 2: 123—134.

Bergey's Manual of Determinative Bacteriology., 1974. R.E. Buchanan and N.E. Gibbons (Editors), Williams and Wilkins, Baltimore, 8th Ed.

Brantner, H., 1970. Investigations on biological oxidation of iron and manganese. Zentralbl. Bakteriol. Parasitenkd., Infektionskr. Hyg., Abt. II, 124: 412—426 (from Microbiology Abstracts, 1970, No. B7755).

Bromfield, S.M., 1956. Oxidation of manganese by soil microorganisms. Aust. J. Biol. Sci., 9: 238—252.

Bromfield, S.M., 1958. The properties of a biologically formed manganese oxide, its availability to oats and its solution by root washings. Plant Soil, 9: 325—337.

Bromfield, S.M., 1974. Bacterial oxidation of manganous ions as affected by organic substrate concentration and compostion. Soil Biol. Biochem., 6: 383—392.

Bromfield, S.M., 1976. The deposition of manganese oxide by an alga on acid soil. Aust. J. Soil Res., 14: 95—102.

Bromfield, S.M. and David, D.J., 1976. Sorption and oxidation of manganous ions and reduction of manganese oxide by cell suspensions of a manganese oxidising bacterium. Soil Biol. Biochem., 8: 37—43.

Bromfield, S.M. and Skerman, V.B.D., 1950. Biological oxidation of manganese in soils. Soil Sci., 69: 337—347.

Collins, J.F. and Buol, S.W., 1970. Effects of fluctuations in the Eh-pH environment on iron and/or manganese equilibria. Soil Sci., 110: 111—118.

Crerar, D.A. and Barnes, H.L., 1974. Deposition of deep-sea manganese nodules. Geochim. Cosmochim. Acta, 38: 279—300.

Crerar, D.A., Cormick, R.K. and Barnes, H.L., 1972. Organic controls on the sedimentary geochemistry of manganese. Acta Mineral. Petrogr., Szeged, 20: 217—226.

Delfino, J.J. and Lee, G.F., 1968. Chemistry of manganese in Lake Mendota, Wisconsin. Environ. Sci. Technol., 2: 1094—1100.

Dubinina, G.A., 1969. Inclusion of *Metallogenium* among the Mycoplasmatales. Dokl. Akad. Nauk USSR (Biological Sciences Section), 184: 87—90 (English Translation).

Dubinina, G.A., 1970. Untersuchungen über die Morphologie von *Metallogenium* und die Beziehungen zu *Mycoplasma*. Z. Allg. Mikrobiol., 10: 309—320.

Dubinina, G.A., 1973. The significance of microbiological processes in lacustrine manganese ore formation. Int. Ver. Theor. angew. Limnol. Verh., 18: 1261—1276.

Dubinina, G.A. Gorlenko, V.M. and Suleimanov, Ya.I., 1974. A study of microorganisms involved in the circulation of manganese, iron, and sulfur in meromictic Lake Gek-Gel'. Microbiology, 42: 817—822.

Ehrlich, H.L., 1963. Bacteriology of manganese nodules. I. Bacterial action on manganese in nodule enrichments. Appl. Microbiol., 11: 15—19.

Ehrlich, H.L., 1966. Reactions with manganese by bacteria from marine ferromanganese nodules. Dev. Ind. Microbiol., 7: 279—286.

Ehrlich, H.L., 1968. Bacteriology of manganese nodules. II. Manganese oxidation by cell-free extract from a manganese nodule bacterium. Appl. Microbiol, 16: 197—202.

Ehrlich, H.L., 1971. Bacteriology of manganese nodules. V. Effect of hydrostatic pressure on bacterial oxidation of Mn^{II} and reduction of MnO_2. Appl. Microbiol. 21: 306—310.

Ehrlich, H.L., 1974. Response of some activities of ferromanganese nodule bacteria to hydrostatic pressure. In: R.R. Colwell and R.Y. Morita (Editors), Effect of the Ocean Environment on Microbiol Activities. University Park Press, Baltimore, pp. 208—221.

Ehrlich, H.L., 1975. The formation of ores in the sedimentary environment of the deep sea with microbial participation: The case for ferromanganese concretions. Soil Sci., 119: 36—41.

Ehrlich, H.L., 1976. Manganese as an energy source for bacteria. In: J.O. Nriagu (Editor), Environmental Biochemistry. Ann Arbor Science Publishers, Ann Arbor, MI, pp. 633—644.

Ehrlich, H.L., Ghiorse, W.C. and Johnson, G.L., 1972. Distribution of microbes in manganese nodules from the Atlantic and Pacific oceans. Dev. Ind. Microbiol., 13: 57—65.

Ehrlich, H.L., Yang, S.H. and Mainwaring, J.D., Jr., 1973. Bacteriology of manganese nodules. VI. Fate of copper, nickel, cobalt, and iron during bacterial and chemical reduction of the manganese (IV). Z. Allg. Mikrobiol., 13: 39—48.

Gabe, D.R. and Rabinovich, V.A., 1965. Physiochemical conditions of development of *Metallogenium* Perf. in mud. In: B.V. Perfil'ev, D.R. Gabe, A.M. Gal'perina, V.A. Rabinovich, A.A. Sapotnitskii, É.É. Sherman and É.P. Troshanov, Applied Capillary Microscopy. The Role of Microorganisms in the Formation of Iron-Manganese Deposits. Consultants Bureau, New York, pp. 55—68.

Geering, H.R., Hodgson, J.F. and Sdano, C., 1969. Micronutrient cation complexes in soil solution: IV. The chemical state of manganese in soil solution. Soil Sci. Soc. Am. Proc., 33: 81—85.

Gerretsen, F.C., 1937. Manganese deficiency of oats and its relation to soil bacteria. Ann. Bot., New Series, 1: 207—230.

Ghiorse, W.C. and Ehrlich, H.L., 1974. Effects of seawater cations and temperature on manganese dioxide-reductase activity in a marine *Bacillus*. Appl. Microbiol. 28: 785—792.

Gorham, E. and Swaine, D.J., 1965. The influence of oxidizing and reducing conditions upon the distribution of some elements in lake sediments. Limnol. Oceanogr., 10: 268—279.

Graham, J.W., 1959. Metabolically induced precipitation of trace elements from sea water. Science, 129: 1428--1429.

Hem. J.D., 1964. Deposition and solution of manganese oxides. United States Geological Survey Water Supply Paper, 1667-B, 42 pp.

Heukelekian, H. and Heller, A., 1940. Relation between food concentration and surface for bacterial growth. J. Bacteriol., 40: 547—558.

Hirsch, P. and Conti, S.F., 1964. Biology of buffing bacteria. I. Enrichment, isolation and morphology of *Hyphomicrobium* spp. Arch. Mikrobiol., 48: 339—357.

Hochster, R.M. and Quastel, J.H., 1952. Manganese dioxide as terminal hydrogen acceptor in the study of respiratory systems. Arch. Biochem. Biophys., 36: 132—146.

Hongve, D., 1974. Hydrographical features of Nordbytjernet, a manganese-rich meromictic lake in southeast Norway. Arch. Hydrobiol., 74: 227—246.

Ivarson, K.C. and Heringa, P.K., 1972. Oxidation of manganese by microorganisms in manganese deposits of Newfoundland soil. Can. J. Soil Sci., 52: 401—416.

Jannasch, H.W., 1958. Studies on planktonic bacteria by means of a direct membrane filter method. J. Gen. Microbiol., 18: 609—620.

Johnson, A.H. and Stokes, J.L., 1966. Manganese oxidation by *Sphaerotilus discophorus.* J. Bacteriol. 91: 1543—1547.

Jones, L.H.P. and Leeper, G.W., 1951. The availability of various manganese oxides to plants. Plant Soil, 3: 141—153.

Kindle, E.M., 1932. Lacustrine concretions of manganese. Am. J. Sci., 224: 496—504.

Kindle, E.M., 1935. Manganese concretions in Nova Scotia lakes. Trans. R. Soc. Can. Section IV, 29: 163—180.

Kline, G.L., 1975. *Metallogenium*-like microorganisms from the Paradise Creek formation, Australia. The Geological Society of America, Cordilleran Section, 71st Annual Meeting, Los Angeles, California, p. 336.

Knoll, A.H. and Barghoorn, E.S., 1975. A Gunflint-type flora from the Duck Creek Dolomite, Western Australia. In: Chemical Evolution of the Precambrian. College Park Colloquia on Chemical Evolution, Abstracts, University of Maryland, p. 61.

Krumbein, W.E., 1969. Über den Einfluss der Mikroflora auf die exogene Dynamik (Verwitterung und Krustenbildung). Geol. Rundsch., 58: 332—363.

Krumbein, W.E., 1971. Manganese-oxidizing fungi and bacteria in recent shelf sediments of the Bay of Biscay and the North Sea. Naturwissenschaften, 58: 56—57.

Krumbein, W.E. and Altmann, H.J., 1973. A new method for the detection and enumeration of manganese oxidizing and reducing microorganisms. Helgol. Wiss. Meeresunters., 25: 347—356.

Kutuzova, R.S., 1975. Study of iron-manganese microorganisms by the overgrowth method. Microbiology, 44: 129—133.

Kuznetsov, S.I., 1975. The role of microorganisms in the formation of lake bottom deposits and their diagenesis. Soil Sci. 119: 81—88.

LaRock, P.A. and Ehrlich, H.L., 1975. Observations of bacterial microcolonies on the surface of ferromanganese nodules from Blake Plateau by scanning electron microscopy. Microb. Ecol. 2: 84—96.

Lee, G.F., 1965. Discussion in Morgan and Stumm (1965b), pp. 124—126.

Leeper, G.W., 1947. The forms and reactions of manganese in the soil. Soil Sci., 63: 79—94.

Leeper, G.W. and Swaby, R.J., 1940. Oxidation of manganous compounds by microorganisms in soil. Soil. Sci., 49: 163—169.

Mann, P.J.G. and Quastel, J.H., 1946. Manganese metabolism in soils. Nature, 158: 154—156.

Marshall, K.C., 1975. Clay mineralogy in relation to survival of soil bacteria. Annu. Rev. Phytopathol., 13: 357—373.

Marshall, K.C., 1976. Interfaces in Microbial Ecology. Harvard University Press, Cambridge, MA, 156 pp.

Marshall, K.C. and Cruickshank, R.H., 1973. Cell surface hydrophobicity and the orientation of certain bacteria at interfaces. Arch. Mikrobiol., 91: 29--40.

McKenzie, R.M., 1975. An electron microprobe study of the relationships between heavy metals and manganese and iron in soils and ocean floor nodules. Aust. J. Soil Res., 13: 177—188.

Mirchink, T.G., Zaprometova, K.M. and Zvyagin'tsev, D.G., 1970. Satellite fungi of manganese-oxidizing bacteria. Microbiology, 39: 327—330.

Monty, C., 1973, Les nodules de manganese sont des stromatolithes océaniques. C. R. Acad. Sci., 276: 3285—3288.

Morgan, J.J. and Stumm, W., 1964. Colloid-chemical properties of manganese dioxide. J. Colloid Sci., 19: 347—359.

Morgan, J.J. and Stumm, W., 1965a. Analytical chemistry of aqueous manganese. J. Am. Water Works Assoc., 57: 107—119.

Morgan, J.J. and Stumm, W., 1965b. The role of multivalent metal oxides in limnological transformations as exemplified by iron and manganese. In: O. Jaag (Editor), Proceedings Second Water Pollution Research Conference, Pergamon, New York, NY, 1: 103—131.

Möse, J.R. and Brantner, H., 1966. Mikrobiologische Studien an manganoxidierenden Bakterien. Zentralbl. Bakteriol., Parasitenkd., Infektionskr. Hyg., Abt. II, 120: 480—495.

Muir, M.D., Spicer, R.A., Grant, P.R. and Giddens, R., 1974. X-ray microanalysis in the SEM for the determination of elements in modern and fossil micro-organisms. In: J.V. Sanders and D.J. Goodchild (Editors), Electron Microscopy 1974, Vol. II Biological, 8th International Congress on Electron Microscopy, Canberra, The Australian Academy of Science, Canberra, pp. 104—105.

Mulder, E.G., 1964. Iron bacteria, particularly those of the *Sphaerotilus-Leptothrix* group, and industrial problems. J. Appl. Bacteriol., 27: 151—173.

Mulder, E.G., 1972. Le cycle biologique tellurique et aquatique du fer et du manganese. Rev. Ecol. Biol. Sol, 9: 321—348.

Mulder, E.G. and van Veen, W.L., 1963. Investigations on the *Sphaerotilus-Leptothrix* group. Antonie van Leeuwenhoek; J. Microbiol. Serol., 29: 121—153.

Mulder, E.G. and van Veen, W.L., 1968. Effect of microorganisms on the transformation of mineral fractions in soil. Trans. 9th Int. Cong. Soil Sci., Adelaide, 4: 651—660.

Mulder, E.G., Lie, T.A. and Woldendorp, J.W., 1969. Biology and Soil Fertility in Soil Biology. Reviews of Research, UNESCO Natural Resources Research IX, Paris.

Munson, R.J. and Bridges, B.A., 1964. "Take-over" — an unusual selection process in steady-state cultures of *Escherichia coli.* J. Gen Microbiol., 37: 411—418.

Perfil'ev, B.V. and Gabe, D.R., 1965. The use of the microbial landscape method to investigate bacteria which concentrate manganese and iron in bottom sediments. In: B.V. Perfil'ev, D.R. Gabe, A.M. Gal'perina, V.A. Rabinovich, A.A. Sapotnitskii, É.É. Sherman and É.P. Troshanov, Applied Capillary Microscopy. The Role of Microorganisms in the Formation of Iron-Manganese Deposits. Consultants Bureau, New York, pp. 9—54.

Perfil'ev, B.V. and Gabe, D.R., 1969. Capillary Methods of Investigating Micro-organisms. Oliver and Boyd, Edinburgh, 627 pp.

Pringsheim, E.G., 1949. The filamentous bacteria *Sphaerotilus*, *Leptothrix*, *Cladothrix* and their relation to iron and manganese. Trans. R. Soc., Ser. B, 233: 453—482.

Rashid, M.A., 1971. Role of humic acids of marine origin and their different molecular weight fractions in complexing di- and tri-valent metals. Soil Sci., 111: 298—306.

Schnitzer, M., 1969. Reactions between fulvic acid, a soil humic compound, and inorganic soil constituents. Soil Sci. Soc. Am. Proc., 33: 75—81.

Schternberg, L.E., 1967. Biogenic structures in manganese ores. Microbiology, 36: 710—712.

Schweisfurth, R., 1968. Untersuchungen über manganoxydierende und-reduzierende Mikroorganismen. Mitt. Int. Ver. Theor. Angew. Limnol., 14: 179—186.

Schweisfurth, R., 1969. Manganoxidierende Pilze. Zentralbl. Bakteriol., Parasitenk., Infektionskr. Hyg., Abt. I Original, 212: 486—491.

Schweisfurth, R., 1972. Manganoxidierende Pilze II. Untersuchungen an Laboratoriumskulturen. Z. Allg. Mikrobiol., 12: 667—671.

Schweisfurth, R. and Hehn, G.V., 1972. Licht- und electronen-mikroskopische Untersuchungen sowie Kulturversuche zum Metallogenium-Problem. Zentralbl. Bakteriol., Parasitenkd., Infektionskr. Hyg., Abt. I, Original A, 220:357—361.

Schweisfurth, R. and Mertes, R., 1962. Microbiologische und chemische Untersuchungen über Bildung und Bekämpfung von Manganschlamm-Ablagerungen in einer Druckleitung für Talsperrenwasser. Arch. Hyg. Bakteriol., 146: 401—417.

Schwertmann, U., 1966. Inhibitory effect of soil organic matter on the crystallization of amorphous ferric hydroxide. Nature, 212: 645—646.

Silverman, M.P. and Ehrlich, H.L., 1964. Microbial formation and degradation of minerals. In: W.W. Umbreit (Editor), Advances in Applied Microbiology, Academic, New York, NY, 6: 153—206.

Sorokin, Yu. I., 1971. Microflora of iron-manganese concretions from the ocean floor. Microbiology, 40: 493—495.

Stark, W.H., Stadler, J. and McCoy, E., 1938. Some factors affecting the bacterial population of freshwater lakes. J. Bacteriol., 36: 653—654.

Starkey, R.L., 1945. Transformation of iron by bacteria in water. J. Am. Water Works Assoc., 37: 963—984.

Stokes, J.L. and Powers, M.T., 1965. Formation of rough and smooth strains of *Sphaerotilus discophorus.* Antonie van Leeuwenhoek; J. Microbiol. Serol., 31: 157—164.

Tanaka, M., 1965. Discussion in Morgan and Stumm (1965b), pp. 118—123.

Taylor, R.M., McKenzie, R.M. and Norrish, K., 1964. The mineralogy and chemistry of manganese in some Australian soils. Aust. J. Soil Res., 2: 235—248.

Ten Khak-Mun., 1967. Iron- and manganese-oxidizing microorganisms in soils of South Sakhalin. Microbiology, 36: 276—281.

Ten Khak-Mun., 1969. A new soil microorganism which oxidizes manganese. Dokl. Akad. Nauk USSR, 188: 697—699 (in Russian).

Ten Khak-Mun., 1973. The role of microorganisms in the formation of Mn-Fe concretions in brown forest soils of the Far East. Dokl. Akad. Nauk USSR, 209: 1203—1205 (in Russian).

Timonin, M.I., 1950. Soil microflora and manganese deficiency. Trans. 4th Int. Congr. Soil Sci., Amsterdam, 3: 97—99.

Timonin, M.I., Illman, W.I. and Hartgerink, T., 1972. Oxidation of manganous salts of manganese by soil fungi. Can. J. Microbiol., 18: 793—799.

Trimble, R.B. and Ehrlich, H.L., 1968. Bacteriology of manganese nodules. III. Reduction of MnO_2 by two strains of nodule bacteria. Appl. Microbiol., 16: 695—702.

Trimble, R.B. and Ehrlich, H.L., 1970. Bacteriology of manganese nodules. IV. Induction of an MnO_2-reductase system in a marine bacillus. Appl. Microbiol., 19: 969—972.

Tyler, P.A., 1970. Hyphomicrobia and the oxidation of manganese in aquatic systems. Antonie van Leeuwenhoek; J. Microbiol. Serol., 36: 567—578.

Tyler, P.A. and Marshall, K.C., 1967a. Microbial oxidation of manganese in hydro-electric pipelines. Antonie van Leeuwenhoek, J. Microbiol. Serol., 33: 171—183.

Tyler, P.A. and Marshall, K.C. 1967b. Form and function in manganese-oxidizing bacteria. Arch. Mikrobiol. 56: 344—353.

Tyler, P.A. and Marshall, K.C., 1967c. Hyphomicrobia — a significant factor in manganese problems. J. Am. Water Works Assoc., 59: 1043—1048.

Tyler, P.A. and Marshall, K.C., 1967d. Pleomorphy in stalked, budding bacteria. J. Bacteriol., 93: 1132—1136.

van Veen, W.L., 1972. Factors affecting the oxidation of manganese by *Sphaerotilus discophorus.* Antonie van Leeuwenhoek; J. Microbiol. Serol., 38: 623—626.

van Veen, W.L., 1973. Biological oxidation of manganese in soils. Antonie van Leeuwenhoek; J. Microbiol. Serol., 39: 657—662.

Volkov, I.I., Rozanov, A.G. and Sokolov, V.S., 1975. Redox processes in diagenesis of sediments in the nortwest Pacific Ocean. Soil Sci., 119: 28—35.

Walsh, F. and Mitchell, R., 1972. An acid-tolerant iron-oxidizing *Metallogenium.* J. Gen. Microbiol., 72: 369—374.

Wolfe, R.S., 1960. Microbial concentration of iron and manganese in water with low concentrations of these elements. J. Am. Water Works Assoc., 52: 1335—1337.

Zajic, J.E., 1969. Microbial Biogeochemistry. Academic, New York, NY, p. 20.

Zapffe, C., 1931. Deposition of manganese. Econ. Geol., 26: 799—832.
Zavarzin, G.A., 1961. Symbiotic culture of a new microorganism oxidizing manganese. Microbiology, 30: 343—345.
Zavarzin, G.A., 1962. Symbiotic oxidation of manganese by two species of *Pseudomonas*. Microbiology, 31: 481—482.
Zavarzin, G.A., 1963. Structure of *Metallogenium*. Microbiology, 32: 864—867.
Zavarzin, G.A., 1964. *Metallogenium symbioticum*. Z. Allg. Mikrobiol., 4: 390—395.
ZoBell, C.E., 1943. The effect of solid surfaces upon bacterial activity. J. Bacteriol., 46: 39—56.
ZoBell, C.E. and Anderson, D.Q., 1936. Observations on the multiplication of bacteria in different volumes of sea water and the influence of oxygen tension and solid surfaces. Biol. Bull. (Woods Hole), 71: 324—342.

Chapter 6.1

THE BIOLOGICAL SULFUR CYCLE

P.A. TRUDINGER

Baas Becking Geobiological Laboratory, P.O. Box 378, Canberra City, A.C.T. 2601 (Australia)

Contents

INTRODUCTION

Sulfur can exist in a number of valence states between +6 and −2. In the natural environment, the most abundant forms of sulfur have valencies of +6 (sulfates, sulfate esters), 0 (elemental sulfur) and −2 (sulfides, reduced organic sulfur) but sulfur dioxide and sulfites (+4) and polysulfur compounds with mixed valence states (e.g. thiosulfate, polythionates) are produced transiently.

About 90% of crustal sulfur occurs in the forms of minerals in deep oceanic and sedimentary rocks, the remainder being largely accounted for as oceanic sulfate (Table 6.1.1). The element undergoes continuous cycling between reservoirs (Holser and Kaplan, 1966; Chapter 6.4) and, within this broad geochemical frame-work, a well-defined biological oxidation-reduction cycle has become established (Goldhaber and Kaplan, 1974; Fig. 6.1.1), which may have had its origins during the early stages of biological evolution (Peck, 1966).

Since sulfur, in the form of cysteine, methionine and other organic molecules (Freney, 1967), is an essential component of living matter, all organisms play a role in the biological sulfur cycle, but certain classes of bacteria transform large amounts of sulfur in energy-yielding reactions. The major types of sulfur metabolism, and the associated organisms, are shown in Table

6.1.2 and have been the subjects for several reviews (e.g. Trudinger, 1969; Kelly, 1972) and a major Symposium (Van Egeraat and Huntjens, 1975).

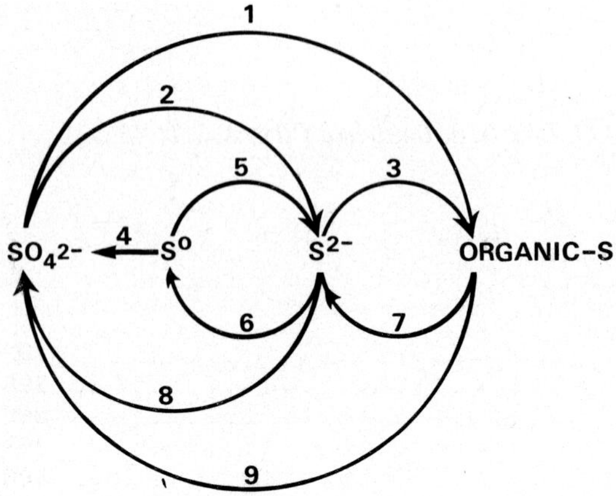

Fig. 6.1.1. Biological cycling of major sulfur pools. (Number explained in Table 6.1.2.)

TABLE 6.1.1

Estimated sulfur reservoirs (After Holser and Kaplan, 1966, Ref. 1, and Ericksson, 1963, Ref. 2.)

Ref.	Reservoir	Tg × 10^{-6} S		
1	Deep oceanic rocks			
	Sediments	75	±	20
	Mafic rocks	2300	±	800
	Sedimentary rocks			
	Sandstone	250	±	60
	Shale	2000	±	580
	Limestone	380	±	110
	Evaporites	5100	±	1600
	Volcanics	50	±	18
	Connate water	27	±	5
	Total sediments	7800	±	1700
	Freshwater	0.003	±	0.002
	Ice	0.006	±	0.002
	Atmosphere	3.6	±	
	Sea	1280	±	55
2	Organic reservoir			
	Land plants	0.6	×	10^{-3}
	Marine plants	0.024	×	10^{-3}
	Dead organic matter	5.0	×	10^{-3}
	Total organic	5.62	×	10^{-3}

TABLE 6.1.2

Major classes of organisms involved in cycling of sulfur

Organism	Reactions [a]	Habitat	Remarks
Dissimilatory reducers [b]			
Desulfovibrio, *Desulfotomaculum*	2	Water-logged soils; anoxic waters and sediments (marine or fresh water).	Strict anaerobes; sulfur metabolism linked to energy production; *Dm. nigrificans* thermophilic.
Desulfuromonas	5	Anaerobic, H_2S-containing waters and muds of marine or estuarine environments	Strict anaerobes; sulfur metabolism linked to energy production.
Assimilatory reducers			
Bacteria, fungi, algae, plants	1, 3	Ubiquitous.	Aerobic or anaerobic; sulfur metabolised for synthesis of cellular constituents.
Chemolithotrophs [b]			
Thiobacillus, *Beggiatoa*	4, 6, 8	Wide variety of soils and H_2S-containing marine and fresh water environments.	Mainly strict aerobes — some utilize NO_3^{2-} anaerobically; autotrophic, facultatively autotrophic or mixotrophic deriving energy from sulfur oxidation; autotrophic status of *Beggiatoa* uncertain.
Photolithotrophs [b]			
Chlorobium, *Chromatium*	4, 6, 8	H_2S-containing muds and stagnant waters exposed to light; sulfur springs.	Photosynthetic, strict anaerobes; sulfur metabolism linked to energy fixation.
Heterotrophic microorganisms	7, 9 (some involved in 4, 6, 8).	Ubiquitous.	Reaction 7 may take place anaerobically; overall conversion to SO_4^{2-} aerobic.

[a] Numbers refer to stages in Fig. 6.1.1.

[b] Representative organisms only — for comprehensive listing of specialized sulfur microorganisms see Buchanan and Gibbons (1974).

ASSIMILATION OF SULFUR

The most important metabolic reaction is the assimilation of sulfur into organic forms which ultimately require the reduction of oxidized sulfur to the oxidation level of sulfide. This reduction is effected by the majority of microorganisms (bacteria, algae, fungi) and plants and, because of its abundance, sulfate is the dominant precursor of reduced sulfur. Pathways of assimilatory sulfate reduction are discussed briefly in Chapter 6.2 and depicted in Fig. 6.2.1 (p. 317).

Animals represent an extension of the assimilatory chain in that, although unable to reduce sulfate, they utilize preformed sulfur amino-acids synthesised by plants and other organisms.

The consequences of assimilation of sulfur in terms of mineral cycling are hard to assess. Although only a small fraction of crustal sulfur is at any time present in organic form (Table 6.1.1) its turnover, in a geological sense, is extremely rapid. Fresh organic matter commonly has a C : S ratio of about 50 : 1. Current estimates of global primary organic matter production are in the order of 150 PgC y^{-1} (Riley, 1944) which would require the annual assimilation of 1—2 Pg of sulfur. In terms of elemental sulfur, this is equivalent to the annual deposition of an 0.2 mm layer over the entire land and ocean surface of the earth. There is experimental evidence that organic sulfur may be a potential source of sulfur for sulfide mineralization (Weil, 1955, 1958; Weil et al., 1954; Lambert, 1973), but it is probable that most assimilated sulfur is rapidly reoxidized to sulfate.

DISSIMILATORY SULFATE REDUCTION

Of more immediate geochemical relevance is the *dissimilatory* reduction of sulfate to hydrogen sulfide which is a major topic for discussion in Chapter 6.2. Sometimes called "sulfate respiration" the process involves oxidation of organic matter (or hydrogen) with the transfer of electrons to sulfate instead of oxygen as in the majority of respiratory systems. The process is accompanied by a net release of free energy which is utilized by the organism for growth (e.g. eqn (1), Wake et al., 1977).

$$2\ \text{lactate} + SO_4^{2-} \rightarrow 2\ \text{acetate} + 2\ H_2O + 2\ CO_2 + S^{2-} (\Delta G' = 88\ \text{kJ}) \qquad (1)$$

Dissimilatory sulfate reduction is a rare metabolic process which is carried out by a few bacterial species belonging to the genera *Desulfovibrio*, *Desulfotomaculum* and the newly-described *Desulfomonas*. The bacteria, however, are widely distributed in the natural environment and they are probably responsible for most of the H_2S formation on earth at temperatures below about 100°C. Their activities have been detected under environmental conditions of Eh, +350 to −500 mV; pH, 4.2 to 10.4; pressure, 0.1 to 100

MPa; temperature, 0 t0 104°C and salinity <1% to saturated NaCl (ZoBell, 1958), although conditions for optimal activity are considerably more restricted (Trudinger et al., 1972).

A major constraint is an absolute requirement for anoxic conditions and the organisms are particularly active in sediments, aqueous basins and waterlogged soils where restricted water movement and oxidation of organic matter by aerobic organisms has led to a depletion of oxygen. Nevertheless, bacterial sulfate reduction may occur in oxidized sediments due to the presence of reduced microniches (Jørgensen, 1977a).

Laboratory studies on dissimilatory sulfate-reducing bacteria indicate that only lactate, pyruvate and a few other simple organic molecules are capable of serving as energy-yielding substrates (Postgate, 1959; Skyring et al., 1977). Cappenberg (1974) reported that β-fluorolactate inhibited H_2S production in bottom deposits of Lake Vechten and caused the accumulation of lactate. This suggests that lactate was the main organic substrate during sulfate reduction in the muds. These results imply that complex organic matter produced by living organisms must first be degraded by fermentative and other processes before being available for the reduction of sulfate. This notion is supported by experiments of Sorokin (1962) indicating that the absence of sulfate reduction in the sub-surface layers of Black Sea sediments was due to the absence or "depression" of saprophytic organisms. Nedwell and Floodgate (1972a) noted a correlation between increased sulfate reduction activity and increased numbers of heterotrophic bacteria during summer in the 19—20 cm sub-surface layer of sediments near Menai Bridge, Anglesey.

The possibility of hydrocarbon utilization by sulfate-reducing bacteria is of interest in view of the association of H_2S and sulfate-reduction with petroliferous environments (see Chapters 6.2 and 6.4). Such utilization has been reported on a number of occasions (Tausson and Alishima, 1932; Tausson and Veselov, 1934; Novelli and ZoBell, 1944; Rosenfeld, 1947; ZoBell, 1950) but many of the results have been criticized by Postgate (1959) on the grounds that authentic pure cultures were not used and the purity of reagents was not specified. These criticisms may not apply to the experiments of Davis and Yarbrough (1966) who demonstrated the formation of radioactive carbon dioxide from ^{14}C-labelled methane, ethane and n-octadecane in cultures of *Desulfovibrio desulfuricans*. The oxidations, however, took place to only a small extent (less than 0.002% in 22 days) and required the presence of additional organic matter (lactate). Recently, Wade et al. (1977) concluded, on thermodynamic grounds, that only short-chain alkynes are potential hydrocarbon substrates for sulfate-reducing bacteria. On present evidence, therefore, crude oils appear unlikely to be primary organic sources for bacterial sulfate reduction, but may possibly be utilized after partial degradation by other hydrocarbon-degrading organisms.

Most dissimilatory sulfate-reducing bacteria appear incapable of oxidizing organic matter beyond the level of acetate. Recently, however, Widdel and Pfennig (1977) isolated an organism, *Desulfotomaculum acetoxidans*, from anaerobic fresh and seawater muds which couples sulfate reduction to the oxidation of acetate to CO_2 and water. In some situations, therefore, sulfate may be the terminal electron acceptor for the complete oxidation of organic matter.

OXIDATION OF REDUCED SULFUR

While some reduced sulfur, either of biogenic or non-biogenic origin, may accumulate in the environment as metal sulfides and elemental sulfur, or be incorporated into fossil organic matter, most is eventually oxidized to sulfate, a process in which microbial activities play a major role.

The mechanism of oxidation of organic sulfur in nature is somewhat problematical. Cysteine, the principal organic sulfur component of living systems, is readily hydrolysed to H_2S by a wide range of heterotrophic microorganisms and it is often assumed that this reaction, followed by sulfide oxidation, is a major route for mineralization of organic sulfur. Nedwell and Floodgate (1972b) reported that rates of H_2S production from organic matter in an intertidal mud flat were greater than (at 5—10°C) or comparable with (at 20–30°C) those of sulfate reduction although the extent of organic-sulfur mineralization by this route was not determined.

There is evidence, however, that in soils at least some organic sulfur is oxidized while remaining attached to the organic moiety (Freney, 1960, 1967). Volatile methylated sulfides are known products of organic sulfur degradation (Challenger, 1951; Freney, 1967) and have been proposed as components of the global sulfur cycle (Lovelock et al., 1972). Organisms utilizing methyl sulfides as sole sources of energy have recently been described (Sivelä and Sundman, 1975). The quantitative significance of the latter processes, however, has yet to be evaluated and at the present time most of our information relates to the oxidative transformations of *inorganic* sulfur compounds.

Two classes of bacteria are specially adapted for sulfur oxidation (Table C.1.2): *Chemolithotrophic sulfur bacteria* which utilize the energy released during oxidation by oxygen for fixation of CO_2 into organic matter (e.g. eqn (2))

$$CO_2 + H_2S + O_2 + H_2O \rightarrow [CH_2O] + SO_4^{2-} + 2\ H^+ \qquad (2)$$

and *Photolithotrophic sulfur bacteria* which carry out photosynthetic carbon fixation using sulfide (and other sulfur compounds) as an "oxidant sink" (e.g. eqns (3) and (4)).

$$CO_2 + 2\ H_2S \xrightarrow{\text{light}} [CH_2O] + H_2O + S^0 \qquad (3)$$

$$2\ CO_2 + H_2S + 2\ H_2O \xrightarrow{\text{light}} 2[CH_2O] + SO_4^{2-} + 2\ H^+ \qquad (4)$$

Typical chemolithotrophic bacteria are the thiobacilli (Vishniac and Santer, 1957; Trudinger, 1967). Many of these are strict autotrophs but facultative autotrophs and mixotrophic species have been described (e.g. Trudinger, 1969; Rittenberg, 1969).

As a genus, the thiobacilli occupy a range of ecological niches. *Thiobacillus thioparus* and *Thiobacillus neapolitanus*, for example, grow best under neutral to mildly alkaline conditions while *Thiobacillus thiooxidans* is acidophilic with an optimum pH close to 2. Some thiobacilli (e.g. *Thiobacillus denitrificans*) can utilize nitrate and an alternative electron acceptor and can, therefore, function under anaerobic conditions. In general, the thiobacilli oxidize most reduced forms of inorganic sulfur including sulfides, elemental sulfur, thiosulfate and polythionates: *Thiobacillus ferroxidans* can also grow at the expense of the oxidation of ferrous iron under acidic conditions. These properties equip the organisms for activities of considerable geochemical and economic importance which are discussed in detail in Chapter 6.3.

The thiobacilli are generally restricted to temperatures below about 50°C. The geothermal habitats where temperatures may reach 95°C a new genus of sulfur-oxidizing bacteria, *Sulfolobus*, has recently been reported (Brock et al., 1972; Fliermans and Brock, 1972; de Rosa et al., 1975; Bohlool, 1975) members of which have an optimum growth temperature of 70–75°C and may remain active at 85°C. They are acidophilic (pH optimum 2–3) and according to Mosser et al. (1973) may be responsible for the generation of sulfuric acid in solfataras. In the Yellowstone National Park solfataras, Moose Pool and Sulfur Caldron, Mosser et al. recorded oxidation rates of 67 and 190 g $S^0\ m^{-2}\ d^{-1}$, respectively, which appeared to be entirely under biological control.

Two principal groups of photolithotrophic sulfur organisms are the green and the purple bacteria (Pfennig, 1977) exemplified by *Chlorobium* and *Chromatium*, respectively. They are obligate anaerobes and, therefore, occupy a relatively restricted niche in the natural environment. In meromictic lakes, for example, were sulfate is reduced in the bottom waters and sediments, photosynthetic bacteria may be sharply stratified at the H_2S-O_2 boundary the depth of which is determined by the light gradient (Kuznetsov, 1959; Sorokin, 1970a).

A geochemically significant property of the photolithotrophic sulfur bacteria is the production of sulfur from sulfide when the latter is in excess (van Neil, 1932). Sulfur may accumulate extracellulary, as is the case with *Chlorobium*, or as intracellular globules (in *Chromatium*) and subsequently is oxidized to sulfate when the supply of sulfide becomes limited.

Although the association of lithotrophic bacteria and sulfur oxidation is clear-cut there is growing evidence that a good deal of the oxidative transformations of inorganic sulfur compounds in the natural environment may be brought about by heterotrophic microorganisms for which the physiological function of sulfur metabolism has yet to be defined. Tuttle and Jannasch (1972) reported that "typical" thiobacilli are rare in the marine environment while Caldwell et al. (1975) described enrichment of fluorescent pseudomonds in thiosulfate-media inoculated with decomposing plant material from a sub-tropical sulfur spring. Vitolins and Swaby (1969), in an analysis of Australian soils, detected relatively few thiobacilli but large numbers of sulfur- and thiosulfate-oxidizing heterotrophs which, they concluded, were mainly responsible for sulfur oxidation in such soils. Recently, Jørgensen (1977) reported high population densities (av. 5—20 g m^{-2} over a period of 1 year) of the colourless sulfur bacteria, *Beggiatoa* spp., in the sediments of the brackish fjord, Limfjorden, Denmark. He suggested that these heterotrophs may play a significant role in sulfur cycling in the sediments. The roles of the various classes of colourless and photosynthetic sulfur bacteria in the sulfur cycle have recently been discussed by Kuenen (1975) and Pfennig (1975).

In laboratory systems, sulfur-oxidizing bacteria often produce thiosulfate and polythionates from sulfide and elemental sulfur (Trudinger, 1967; Kelly, 1968) but the extent to which these reactions take place in the natural environment is uncertain. Although significant concentrations of thiosulfate have been detected in natural waters such as the Black Sea (Sorokin, 1962; Tuttle and Jannasch, 1973) and meromictic freshwater lakes (Sorokin, 1970a), much of this may arise by chemical oxidation of H_2S (Cline and Richards, 1969; Chen and Morris, 1972; Avrahami and Golding, 1968) and it has been suggested (Sorokin, 1970b, c; 1964a) that the main function of bacteria in these environments is to oxidize the products of chemical sulfide oxidation to sulfate.Whatever its origin, thiosulfate may reach concentrations in natural waters comparable with those of sulfide (Sorokin, 1970a, b). Little attention has been given, however, to possible geochemical reactions involving thiosulfate although a possible role in pyrite formation has been suggested (Volkov and Ostroumov, 1957).

SULFURETA

Baas Becking (1925) introduced the term "sulfuretum" to describe an ecological community of sulfate-reducing and sulfur-oxidizing organisms. Simple sulfureta may be constructed by mixing pure cultures of *Desulfovibrio* spp. with photosynthetic sulfur bacteria (*Chromatium* spp. or *Chlorobium* spp.) and providing a suitable source of carbon and electrons for the sulfate reducer (Butlin and Postgate, 1954; Gemerden, 1967). Reoxidation of sulfide produced by the reduction of sulfate supports the growth of the photoautotroph and organic matter synthesized by the latter organism may provide additional oxidizable carbon for the sulfate-reducer. Matheron and Baulaigue (1976) recently described a stable mixed culture of *D. desulfuricans*, *Chlorobium* sp. and the non-sulfur heterotroph, *Escherichia coli*. The mixture utilized glucose and the biochemical sequence may be formulated as follows:

E. coli

2 glucose + H_2O → 2 lactate + acetate + ethanol + 2 CO_2 + 2 H_2

D. desulfuricans

2 lactate + sulfate → 2 acetate + sulfide + 2 CO_2 + 2 H_2O

ethanol + 0.5 sulfate → acetate + 0.5 sulfide + H_2O

2 H_2 + 0.5 sulfate → 0.5 sulfide + 2 H_2O

Chlorobium

2 sulfide + 4 CO_2 + 4 H_2O + n acetate → 4 + n[CH_2O] + 2 sulfate

Since glucose and other fermentable compounds are products of photosynthesis, this system probably represents an approximation to the reality of the natural environment.

A "natural' mixed population, or consortium, "*Chloropseudomonas ethylica*", was isolated from H_2S-containing stagnant waters (Shaposhnikov et al., 1960) and originally classified as a novel photosynthetic sulfur autotroph. It later proved to be a syntrophic mixture of *Chlorobium limicola* and an heterotrophic H_2S-producing bacterium (Gray et al., 1973). A similar situation may apply to two other "natural" mixed populations, *Chlorochromatium* and *Pelochromatium* (Pfennig, 1975). Both represent an association of photosynthetic organisms (*Chlorobium*) with, as yet uncharacterized, colourless bacteria. Mixed populations of the type described often grow at faster rates than the individual organisms in isolation and their viability may be maintained for extended periods of time without subculturing (Pfennig, 1975). This suggests that each organism provides factors, nutritional or otherwise, which favour the development and survival of its companion, a fact which could have important ecological implications.

Further examination of "*Chloropseudomonas*" (Pfennig and Biebl, 1976) revealed an interesting variation of the sulfuretum. A new organism, designated *Desulfuromonas acetoxidans*, which reduces sulfur at the expense of acetate, ethanol or propanol, was isolated from "*Chloropseudomonas*" cultures and from an anaerobic, sulfide-containing marine mud: more oxidized forms of sulfur were not metabolized. Together with photosynthetic green sulfur bacteria, *Desulfuromonas* forms a syntrophic mixture in which sulfur is continually recycled between sulfide and its

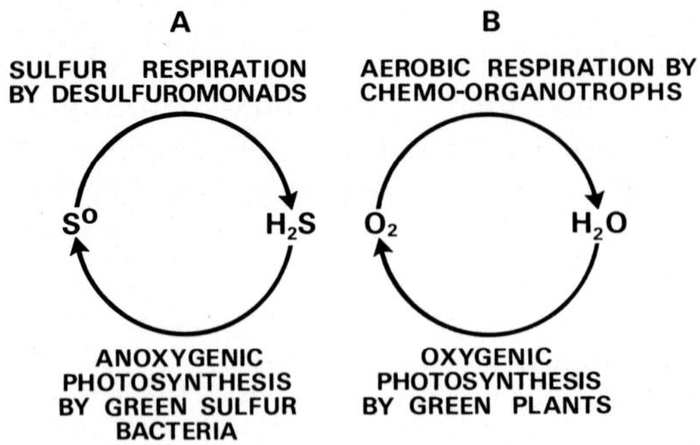

Fig. 6.1.2. Comparison of elemental sulfur-linked and oxygen-linked respiratory-photosynthetic cycles.

elemental form (Fig. 6.1.2A). This property is also shared by the newly described spirillum 5175 which reduces sulfur, sulfite and thiosulfate, but not sulfate, at the expense of oxidation of formate or hydrogen (Wolfe and Pfennig, 1977). It may also apply to strains of sulfate-reducing bacteria which have the ability to reduce elemental sulfur (Biebl and Pfennig, 1977). Pfennig and Biebl (1976) recognized the analogy between the cycle shown in Fig. 6.1.2A and that of planetary oxygen cycle (Fig. 6.1.2B) and one may speculate whether the former not only plays a role in the cycling of sulfur in present-day anaerobic environments but may be representative of a primitive stage in the evolution of the sulfur cycle and photosynthesis during the early anoxic period of earth's history. This speculation perhaps takes on added significance in view of the current concept that photosynthetic bacteria were the evolutionary precursors of oxygen-evolving cyanophytes (blue-green algae *) (Klein and Cronquist, 1967; Margulis, 1970) and of the recent discovery that the cyanophyte, *Oscillatoria limnetica*, can carry out both oxygenic photosynthesis and anaerobic bacterial-type photosynthesis linked to H_2S oxidation (Cohen et al., 1975a, b).

In the natural environment, large scale, and more complex sulfureta develop wherever sulfate reduction has become established (Fenchel and Riedl, 1970; Caldwell and Tiedje, 1975; Fenchel and Jørgensen, 1977). Numerous terrestrial, freshwater and marine examples have been reported in the literature but rarely has an attempt been made to analyse a single environment in terms of the biota and the chemical and biochemical processes associated with both the reductive and oxidative aspects of the sulfur cycle. One notable exception is the Black Sea where the work of Sorokin (1962, 1964b, 1970b) and Jannasch and his colleagues (Tuttle and Jannasch, 1973;

---

* See footnote on p. 12.

Sen Gupta and Jannasch, 1973; Jannasch et al., 1974) has revealed a semi-quantitative picture of the interrelationships of the sulfur, nitrogen and carbon (both photosynthetic and chemosynthetic) cycles and provided a preliminary description of the types or organisms involved.

Inevitably, however, analyses of complex natural environments are subject to considerable fluctuations and uncertainties and a large mass of data must be assembled over a long period of time before a statistically realistic model can be advanced. Jørgensen and Fenchel (1974) attempted to circumvent some of the problems associated with the natural environment by studying the kinetics of the sulfur cycle in an experimental system designed to resemble a natural, reduced, shallow sea-water sediment. The "sediment", consisting of sand mixed with chopped *Zostera marina* (eel grass) leaves, was deposited under circulating sea-water at 22°C and illuminated for 8 h d^{-1} for seven months. Rates of sulfate reduction obtained by direct measurements on core material with ^{35}S-sulfate were comparable with those reported for marine sediments (see p. 304) indicating that the system was reasonably analogous to a natural environment. In the early stages of the experiment, there was a successive development of chemolithotrophic sulfur bacteria (*Beggiatoa*) and purple sulfur bacteria (*Chromatium*) as had earlier been observed by Fenchel (1969) in artificial sulfureta. A significant finding of Jørgensen and Fenchel (1974) and Jørgensen (1977c) was that sulfate reduction can account for the oxidation of more than half the organic matter degraded in marine sediments. This indicates the potential importance of the sulfur cycle in the overall biochemical ecology of sediments.

Another experimental approach to the study of sulfureta in sediments was devised by Hallberg et al. (1976). Closed plexiglass boxes (Schippel et al., 1973) equipped with sampling ports and electrodes for continuous measurement of pH, Eh and sulfide-ion activity are anchored in soft bottom sediments, and physical, chemical and biological characteristics of the sediment-water system monitored over a period of time. In a 9-month experiment on Baltic Sea sediments, sulfate reduction proceeded rapidly in the early stages followed by sulfide reoxidation due to the development of the photosynthetic sulfur bacteria *Chromatium* and *Chlorobium*.

KINETICS OF BIOLOGICAL SULFUR CYCLING

Table 6.1.3 lists some rates of hydrogen sulfide production determined on ground waters and surface muds of sediments. The values range over about five orders of magnitude but, in general, the highest rates are found in saline sediments which may reflect, in part, the greater availability of nutrients in these environments compared with ground waters and freshwater sediments. Indeed, vigorous sulfate reduction is often a consequence of pollution of natural waters (e.g. Koyama et al., 1965).

TABLE 6.1.3

Rates of sulfate reduction in ground waters and surface muds of sediments [a]

	SO_4^{2-} content (mmol l^{-1})	Rate (μmol H_2S l^{-1} d^{-1})	Ref.
Ground waters (U.S.S.R.)			
Shor-su S deposit	0.7— 7.3	0.3— 5.3	1
Gaurdak S deposit	9.6—12.2	1.1— 14.3	1
Cis-Carpathian S deposits	1.2—10.2	0.3—112	1
Kosha-Naur oil fields	Flooded with sea water	2.9—58.8	1
Freshwater sediments (U.S.S.R.)			
Lake Beloe	0.5—2.1	$6—18 \times 10^{-3}$	1
Rybinski reservoir	0.4—6.2	2×10^{-2}—2.7	1
Gor'kii reservoir	0.3—1.4	1.2—86.5	1
Kiubyshev reservoir	1.1—5.3	3.3—89.4	1
Lake Belovod	4.5—6.5	2.0— 3.6	1
Lake Belovod	—	ca. 4.4	2
Lake Gel Gel	0.2—5.6	0.9—134	2
Lake Sakavo	7.7	22—140	3
Lake Chernyi Kichiyer	1.1	416	3
Lake Bol'shoy Kichiyer	0.45	193	3
Lake Konon'yer	0.4	123—163	3
Lake Kuznechikha	0.4	13—114	3
Lake Pomaretskoe	3.3	900	4
Saline sediments			
Lake Solenoe	up to 43.8	11—600	1
Lake Mogil'noe	27.1	29	1
Barents Sea littoral zone	up to 28.1	267—714	1
Krasnovodsk Bay	up to 22.9	114—247	1
Black Sea	up to 22.9	0.8— 72	5
Lake Veisovoe	31	86—600	6
Repnoe Lake		488	7
Lizard Island (Coral island)	30.1	40—100	8
Aarhus Bay	—	439—1013	9
Anglesey — intertidal and flat	—	178—2000	10
Experimental sedimentary systems			
Sea water sediment from Branford Bay, Connecticut	b	16 [c]	11
Baltic Sea sediment	b	32 [c]	12
Sand-eelgrass	b	82—95	13

References: 1 = Ivanov, 1968; 2 = Sorokin, 1970a; 3 = Chebotarev et al., 1975; Matrosov et al., 1975; 4 = Gorlenko et al., 1974b; 5 = Sorokin, 1962; 6 = Chebotarev et al., 1974; 7 = Chebotarev et al., 1973; 8 = Skyring and Chambers, 1976; 9 = Jørgensen, 1978); 10 = Nedwell and Floodgate, 1972a; 11 = Nakai and Jensen, 1964; 12 = Hallberg et al., 1976; 13 = Jørgensen and Fenchel, 1974.

[a] Where necessary, 1 kg of sediment has been taken as 1 litre.

[b] Sulfate concentration of sea water decreasing during experiment.

[c] Initial apparently zero-order rate.

Note: The values in this Table were determined by direct measurements of sulfate reduction (usually of $^{35}SO_4^{2-}$). Other rates, calculated on the basis of sedimentation and diffusion models, may be found in Berner (1972) and Goldhaber and Kaplan (1975).

A major parameter leading to variability of sulfate reduction is the supply of utilizable organic matter (Sorokin, 1962; Ivanov, 1968; Berner, 1970; Ramm and Bella, 1974), and deposition of particulate organic matter at the sediment-water interface probably accounts to a large extent for the fact that, even in anoxic basins, the rates of reduction in surface muds may be several orders of magnitude greater than in the overlying waters (Chebotarev et al., 1974; Matrosov et al., 1975; Sorokin, 1962). Sweeney (1972) noted a general correlation between the organic carbon and pyrite contents of recent marine sediments while Goldhaber and Kaplan (1975) demonstrated a positive relationship between rates of sulfate reduction and sedimentation which accords with the trend towards higher organic carbon contents of rapidly depositing sediments (Berner, 1972).

Bacterial sulfate reduction appears to proceed to considerable depths in marine sediments but rates computed from changes in interstitial water sulfate concentrations, with suitable corrections for diffusion and sedimentation, are generally orders of magnitude below those in surface muds (Goldhaber and Kaplan, 1975). Again, this probably reflects a depletion of utilizable organic matter in the deeper layers by microbial utilization and conversion to more intractable humates and kerogens.

The variability of reduction rates reported in Table 6.1.3 is probably not due solely to differences in organic matter availability, but the nature of other controlling factors cannot be readily assessed. While Postgate (1951) reported that the rate of sulfate reduction by *D. desulfuricans* was independent of sulfate concentration above 1 mM, the results of Nakai and Jensen (1964) and Hallberg et al. (1976) suggest deviation from zero-order kinetics below about 2—3 mM sulfate in experimental sedimentary systems. Nevertheless, it is unlikely that sulfate limitation is a major kinetic barrier in the natural sulfur cycle except perhaps in some ground waters and freshwater sediments. Further speculation on the reasons for variable sulfate reduction rates in these complex systems is, however, not warranted at this stage.

It should be emphasised that most of the data in Table 6.1.3 are point measurements which are not necessarily representative of the overall rates of sulfate reduction in a particular environment. Nevertheless, Trudinger et al. (1972) and Rickard (1973) considered that the average rates in the Black Sea and other sediments may be "typical" of euxinic environments and concluded that they were of sufficient magnitude to account for synsedimentary sulfide ore deposition (see also, Temple, 1964).

Since, at the present time, there is no general large-scale accumulation of reduced sulfur in sediments and soils, the combined rates of biological and chemical oxidation of sulfide can be assumed to be in the same order as those of sulfate reduction. Unfortunately, the few recorded rates of sulfide oxidation in the environment are not directly comparable with those of sulfate reduction. Aside from difficulties posed by the experimental methods, in most cases considerable time intervals elapsed between analyses of the two reactions. Sorokin (1970a), reported rates of up to 10 μmol l^{-1} d^{-1} for ^{35}S-labelled sulfide oxidation in the freshwater, meromictic Lake Belovod: in the most active zone, chemical, chemosynthetic and photosynthetic oxidations each accounted for about one third of the overall rate. Similar overall oxidation rates were observed in Lake Gel Gel. In the Black Sea rates of up to 16 μmol S^{2-} m^{-2} d^{-1} have been recorded in the oxic-anoxic transition zone of the water column (Sorokin, 1970b). Extremely high rates (about 15 mmol l^{-1} d^{-1}) for photosynthetic sulfide oxidation were recorded by Blackburn et al. (1975) in the upper 3 mm of sediment from an organic-rich, sandy marine area in Denmark under laboratory conditions and artificial light.

An indirect assessment of photosynthetic sulfide oxidation in Lake Repnoe, U.S.S.R., was made by Chebotarev et al. (1975). Using data from Gorlenko et al. (1974a) on CO_2 fixation by phototrophic bacteria, they calculated H_2S oxidation according to eqn (3), on the assumption that sulfur, not sulfate, was the oxidation product. The calculated rate of 22 mmol m^{-2} d^{-1} was considerably higher than that found in the Black Sea (see above) but agreed well with the rate of sulfate reduction (25 mmol m^{-2} d^{-1}) measured in the sediment plus water column in the same area of the lake.

Jørgensen and his colleagues have made assessments of sulfur fluxes in sediments based on analyses of rates of sulfate reduction and the concentrations of reduced and oxidized sulfur compounds. Jørgensen and Cohen (1977) examined in the littoral sediments of Solar Lake on the Sinai coast of the Gulf of Elat. The sediments consist of stromatolitic cyanophytic mats up to 1 m in depth (Cohen et al., 1977a, b, c; Krumbein et al., 1977) which appear to supply the organic substrate for sulfate reduction. The most active zone was the upper 10 cm, corresponding to the first 10—20 years of the life of the mat. Rates of sulfate reduction in this zone were in the order of 67 mmol S m^{-2} d^{-1} and accounted for well over 90% of the total reduction of sulfate by the sediment: only 0.15% of H_2S produced was fixed in the sediments the remainder, presumably, being reoxidized at the sediment surface. Below 20 cm, rates of sulfate reduction fell to 10—50 μmol S m^{-2} d^{-1} with most of the H_2S being trapped within the sediment as iron sulfides.

A similar study to the one just described was undertaken by Jørgensen (1977c) on the coastal marine sediments of Limfjorden, Denmark. Sulfate reduction (on average 9.5 mmol m^{-2} d^{-1}) was largely confined to the upper 20 cm of sediment and over 90% of the H_2S produced appeared to be reoxidized at the sediment-water interface: only about 7% of H_2S was converted to iron sulfides. The development of the sulfur cycle in sediments has been described by Jørgensen and Fenchel (1974) using the experimental system described earlier (see p. 303). The results, illustrated in Fig. 6.1.3, showed that the rate of sulfate reduction remained almost constant over the period studied and was accompanied by increasing pools of H_2S, FeS and S^0

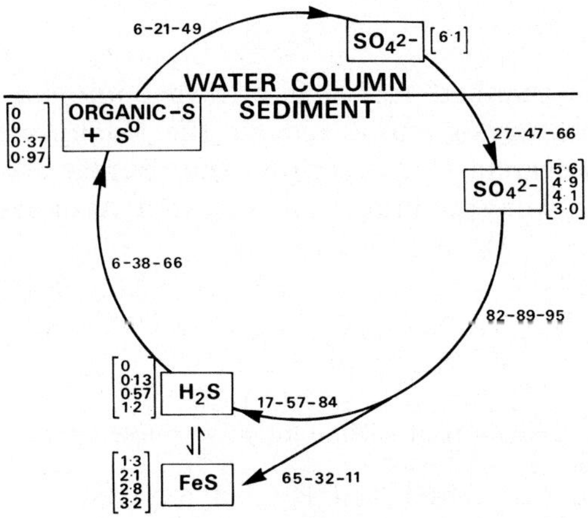

Fig. 6.1.3. Evolution of the sulfuretum in a model sediment (modified and redrawn from Jørgensen and Fenchel, 1974). Figures in brackets are the mean pool sizes in μmol cm^{-3} on days 7, 19, 41 and 76 of the experiment. Other figures are the mean fluxes in μmol cm^{-3} d^{-1} between these days.

(+ organic-S) and rates of oxidative processes. Although it is a reasonable assumption that the deficit between H_2S produced by sulfate reduction and sulfide trapped in sediments represents reoxidation, this may not necessarily take place in situ: a variable amount of H_2S may escape to the atmosphere to be oxidised elsewhere. Hansen et al. (1978) demonstrated H_2S-emission from two shallow coastal areas in Denmark averaging 1.5 and 38 mmol m^{-2} d^{-1}. The emissions took place largely at night when photosynthetic sulfide oxidation was suppressed, and represented a significant proportion of H_2S produced by sulfate reduction: possible contributions from organic sulfur and reduction of elemental sulfur were, however, not determined. Nriagu and Coker (1976) estimated that the emission of sulfur from Lake Ontario sediments (1.3 mmol m^{-2} y^{-1}) was only about 1% of the annual input of sulfur. The loss was, however, sufficient to cause marked enrichment in ^{34}S in the sulfur which remained fixed in sediments. (For a discussion of isotope effects, see Chapter 6.2.)

Realistic fluxes of sulfur (or indeed any element) in the natural environment are difficult to obtain. The environment itself is inhomogeneous and the biochemical systems complex, and there are major problems associated with the extrapolation of point measurements of limited duration to a regional, long-term scale. Nevertheless, the recent studies in the field are giving at least a semiquantitative picture of some of the biogeochemical reactions of sulfur which have important implications in the mineral formation and dissolution (see Chapters 6.2—6.4).

ACKNOWLEDGEMENTS

L.A. Plumb and G.W. Skyring are thanked for their critical comments and advice. The Baas Becking Laboratory is supported by the Commonwealth Scientific and Industrial Research Organization, the Bureau of Mineral Resources and the Australian Mineral Industrial Research Association Ltd.

REFERENCES

Avrahami, M. and Golding, R.M. 1968. The oxidation of sulfide ion at very low concentrations in aqueous solutions. J. Chem. Soc., A: 647—651.

Baas Becking, L.G.M., 1925. Studies on the sulfur bacteria. Ann. Bot., 39: 613—650.

Berner, R.A., 1970. Sedimentary pyrite formation. Am. J. Sci., 268: 1—23.

Berner, R.A., 1972. Sulfate reduction, pyrite formation and the oceanic sulfur budget. In: D. Dyrssen and D. Jagner (Editors), The Changing Chemistry of the Oceans. Nobel Symposium 20. Almquist and Wiksell, Stockholm, pp. 347—361.

Biebl, H. and Pfennig, N., 1977. Growth of sulfate-reducing bacteria with sulfur as electron donor. Arch. Microbiol., 112: 115—117.

Blackburn, T.H., Kleiber, P. and Fenchel, T., 1975. Photosynthetic sulfide oxidation in marine sediments. Oikos, 26: 103—108.

Bohlool, B.B., 1975. Occurrence of *Sulfolobus acidocaldarius*, an extremely thermophilic acidophilic bacterium, in New Zealand hot springs. Arch. Mikrobiol., 106: 171—174.

Brock, T.D., Brock, K.M., Belly, R.T. and Weiss, R.L., 1972. Sulfolobus: a new genus of sulfur-oxidizing bacteria living at low pH and high temperature. Arch. Mikrobiol., 84: 54—88.

Buchanan, R.E. and Gilbons, N.E. (Editors), 1974. Bergey's Manual of Determinative Bacteriology, 8th edn. The Williams and Wilkins Co., Baltimore, 1268 pp.

Butlin, K.R. and Postgate, J.R., 1954. The microbiological formation of sulfur in Cyrenaican lakes. In: J.L. Cloudsley-Thompson (Editor), Biology of Deserts. Institute of Biology, London, pp. 112—122.

Caldwell, D.E. and Tiedje, J.M., 1975. The structure of anaerobic bacterial communities in the hypolimnia of several Michigan lakes. Can. J. Microbiol., 21: 377—385.

Caldwell, D.E., Caldwell, S.J. and Tiedje, J.M., 1975. An ecological study of the sulfur-oxidizing bacteria from the littoral zone of a Michigan lake and a sulfur spring in Florida. Plant Soil, 43: 101—114.

Cappenberg, T.E., 1974. The interrelations between sulfate-reducing and methane-producing bacteria in bottom sediments of a freshwater lake. II. Inhibition experiments. Antonie van Leeuwenhoek; J. Microbiol. Serol., 40: 297—306.

Challenger, F., 1951. Biological methylation. Adv. Enzymol., 12: 429—491.

Chebotarev, E.N., 1975. Microbiological hydrogen sulfide formation in the freshwater karst lakes Bol'shoi Kichier and Chernyi Kichier. Microbiology, 43: 939—943.

Chebotarev, E.N., Gorlenko, V.M. and Kachalkin, V.I., 1973. Microbiological hydrogen sulfide formation in Lake Repnoe (Slavyansk lakes). Microbiology, 42: 465—479.

Chebotarev, E.N., Gorlenko, V.M. and Kachalkin, V.I., 1974. Microbial production of hydrogen sulfide in Lake Veisovoe (Slavyansk lakes). Microbiology, 43: 271—274.

Chebotarev, E.N., Matrosov, A.G., Kudryavtseva, A.I., Zyakun, A.M. and Ivanov, M.V., 1975. Fractionation of stable isotopes during microbiological processes in Slavonic lakes. Microbiology, 44: 268—271.

Chen, K.Y. and Morris, J.C., 1972. Kinetics of oxidation of aqueous sulfide by O_2. Environ. Sci. Technol., 6: 529—537.

Cline, J.D. and Richards, F.A., 1969. Oxygenation of hydrogen sulfide in sea water at constant salinity, temperature and pH. Environ. Sci. Technol., 3: 838—843.

Cohen, Y., Padan, E. and Shilo, M., 1975a. Facultative anoxygenic photosynthesis in the cyanobacterium *Oscillatoria limnetica*. J. Bacteriol., 123: 855—861.

Cohen, Y., Jørgensen, B.B., Padan, E. and Shilo, M., 1975b. Sulphide-dependent anoxygenic photosynthesis in the cyanobacterium *Oscillatoria limnetica*. Nature, 257: 489—492.

Cohen, Y., Krumbein, W.E. and Shilo, M., 1977a. Solar Lake (Sinai). 2. Distribution of photosynthetic microorganisms and primary production. Limnol. Oceanogr., 22: 609—620.

Cohen, Y., Krumbein, W.E. and Shilo, M., 1977b. Solar Lake (Sinai). 3. Bacterial distribution and production. Limnol. Oceanogr., 22: 621—634.

Cohen, Y., Krumbein, W.E., Goldberg, M. and Shilo, M., 1977c. Solar Lake (Sinai). I. Physical and chemical limnology. Limnol. Oceanogr., 22: 597—608.

Davis, J.B. and Yarbrough, H.F., 1966. Anaerobic oxidations of hydrocarbons by *Desulfovibrio desulfuricans*. Chem. Geol., 1: 137—144.

de Rosa, M., Gambacorta, A. and Bu'lock, J.D., 1975. Extremely thermophilic acidophilic bacteria convergent with *Sulfolobus acidocaldarius*. J. Gen. Microbiol., 86: 156—184.

Ericksson, E., 1963. The yearly circulation of sulfur in nature. J. Geophys. Res., 68: 4001—4008.

Fenchel, T., 1969. The ecology of marine microbenthos. IV. Structure and function of the benthic ecosystem, its chemical and physical factors and the microfauna communities with special reference to the ciliated protozoa. Ophelia, 6: 1—182.

Fenchel, T.M. and Jørgensen, B.B., 1977. Detritus food chains of aquatic ecosystems. The role of bacteria. Adv. Microb. Ecol., 1: 1—58.

Fenchel, T.M. and Riedl, R.J., 1970. The sulfide system — a new biotic community underneath the oxidized layer of marine sand bottoms. Mar. Biol., 7: 255—268.

Fliermans, C.B. and Brock, T.D., 1972. Ecology of sulfur-oxidizing bacteria in hot acid soils. J. Bacteriol., 111: 343—350.

Freney, J.R., 1960. The oxidation of cysteine to sulphate in soil. Aus. J. Biol. Sci., 13: 387—392.

Freney, J.R., 1967. Sulfur-containing organics. In: A.D. McLaren and G.H. Peterson (Editors). Soil Biochemistry. Marcel Dekker, New York, pp. 229—259.

Gemerden, H. van., 1967. On the Bacterial Sulfur Cycle in Inland Waters. Thesis Leiden — quoted by Pfennig (1975).

Goldhaber, M.B. and Kaplan, I.R., 1974. The sulphur cycle. In: E.D. Goldberg (Editor), The Sea. Ideas and Observations on Progress in the Study of the Seas. John Wiley, New York, NY, pp. 569—655.

Goldhaber, M.B. and Kaplan, I.R., 1975. Controls and consequences of sulfate reduction rates in recent marine sediments. Soil Sci., 119: 42—55.

Gorlenko, V.M., Chebotarev, E.N. and Kachalkin, V.I., 1974a. Microbiological processes of oxidation of hydrogen sulfide in the Repnoe Lake (Slavonic lakes). Microbiology, 42: 643—646.

Gorlenko, V.M., Chebotarev, E.N. and Kachalkin, V.I., 1974b. Participation of microorganisms in the circulation of sulfur in Pomyaretskoe Lake. Microbiology, 43: 772—776.

Gray, B.H., Fowler, C.F., Nugent, N.A., Rigopoulos, N. and Fuller, R.C., 1973. Reevaluation of *Chloropseudomonas ethylica* strain 2-K. International Journal of Systematic Bacteriology, 23: 256—264.

Hallberg, R.O., Bågander, L.E. and Engvall, A.G., 1976. Dynamics of phosphorus, sulphur and nitrogen at the sediment-water interface. In: J.O. Nriagu (Editor), Environmental Biogeochemistry, Vol. 1. Ann Arbor Science, Ann Arbor, MI, pp. 295—380.

Hansen, M.H., Ingvorsen, K. and Jørgensen, B.B., 1978. Mechanisms of hydrogen sulfide release from coastal marine sediments to the atmosphere. Limnol. Oceanogr., 23: 68—76.

Holser, W.T. and Kaplan, I.R., 1966. Isotope geochemistry of sedimentary sulfates. Chem. Geol., 1: 93—135.

Ivanov, M.V., 1968. Microbiological Processes in the Formation of Sulfur Deposits. Israel Program for Scientific Translations, Jerusalem, 298 pp.

Jannasch, H.W., Trüper, H.G. and Tuttle, J.H., 1974. The microbial sulfur cycle of the Black Sea. In: E.T. Degens and D.A. Ross (Editors), The Black Sea: its Geology, Chemistry and Biology. American Association of Petroleum Geologists, Memoir 20: 419—425.

Jones, H.E., Trudinger, P.A., Chambers, L.A. and Pyliotis, N.A., 1976. Metal accululation by bacteria with particular reference to dissimilatory sulphate-reducing bacteria, Z. Allg. Mikrobiol., 16: 425—435.

Jørgensen, B.B., 1977a. Bacterial sulfate reduction within reduced microniches of oxidized marine sediments. Mar. Biol., 41: 7—17.

Jørgensen, B.B., 1977b. Distribution of colourless sulfur bacteria (*Beggiatoa* spp.) in a coastal marine sediment. Mar. Biol., 41: 19—28.

Jørgensen, B.B., 1977c. The sulfur cycle of a coastal marine sediment (Limfjorden, Denmark). Limnol. Oceanogr., 22: 814—832.

Jørgensen, B.B., 1978. A comparison of the methods for the quantification of bacteriological sulfate reduction in coastal marine sediments. III. Estimation from chemical and bacteriological field data. Geomicrobiol. J., 1:49—64.

Jørgensen, B.B. and Cohen, Y., 1977. Solar Lake (Sinai). 5. The sulfur cycle in cyanobacterial mats. Limnol. Oceanogr., 22: 657—666.

Jørgensen, B.B. and Fenchel, T., 1974. The sulfur cycle of a marine sediment model system. Mar. Biol., 24: 189—201.

Kaplan, I.R., Emery, K.O. and Rittenberg, S.C., 1963. The distribution and isotopic abundance of sulfur in recent marine sediments off Southern California. Geochim. Cosmochim. Acta, 27: 297—331.

Kelly, D.P., 1968. Biochemistry of oxidation of inorganic sulphur compounds by microorganisms. Aust. J. Sci., 31: 165—173.

Kelly, D.P., 1972. Transformations of Sulphur and its Compounds in Soils. International Symposium on Sulphur in Agriculture, Versailles, 1970. Institute Nationale Recherches Agronomique, pp. 217—232.

Klein, R.M. and Cronquist, A., 1967. A consideration of the evolutionary and taxonomic significance of some biochemical. micromorphological and physiological characters of the thallophytes. Q. Rev. Biol. 42: 105—296.

Koyama, T., Nakai, N. and Kamata, E., 1965. Possible discharge rate of hydrogen sulfide from polluted coastal belts in Japan. J. Earth Sci., 13: 1—11.

Krumbein, W.E., Cohen, Y. and Shilo, M., 1977. Solar Lake (Sinai). 4. Stromatolytic cyanobacterial mats. Limnol. Oceanogr., 22: 635—656.

Kuenen, J.G., 1975. Colourless sulfur bacteria and their role in the sulfur cycle. Plant Soil, 43: 49—76.

Kuznetsov, S.I., 1959. Die Rölle der Mikroorganismen im stoffkreislauf der Seen. VEB Deutscher Verlag der Wissenschaften, Berlin, 301 pp.

Lambert, I.B., 1973. Post-depositional availability of sulphur and metals and formation of secondary textures and structures in stratiform sedimentary sulphide deposits. J. Geol. Soc. Aust., 20: 205—215.

Lovelock, J.E., Maggs, R.J. and Rasmussen, R.A. 1972. Atmospheric dimethyl sulphide and the natural sulphur cycle. Nature, 237: 452—453.
Margulis, L., 1970. Origin of Eukaryotic Cells. Yale University Press, New Haven, 349 pp.
Matheron, R. and Baulaigue, R., 1976. Bactéries fermentatives, sulfato-réductrices et phototrophes en cultures mixtes. Arch. Mikriobiol., 109: 319—320.
Matrosov, A.G., Chebotarev, Ye.N., Kudryavtsera, A.J., Zyukun, A.M. and Ivanov, M.V., 1975. Sulfur isotope composition in freshwater lakes containing H_2S. Geochem. Int., 12: 217—221.
Mosser, J.L., Mosser, A.G. and Brock, T.D., 1973. Bacterial origin of sulfuric acid in geothermal habitats. Science, 179: 1323—1324.
Nakai, N. and Jensen, M.L., 1964. The kinetic isotope effect in the bacterial reduction and oxidation of sulphur. Geochim. Cosmochim. Acta, 28: 1893—1912.
Nedwell, D.B. and Floodgate, G.D., 1972a. The effect of microbial activity upon the sedimentary sulphur cycle. Mar. Biol., 16: 192—200.
Nedwell, D.B. and Floodgate, G.D., 1972b. Temperature-induced changes in the formation of sulphide in a marine sediment. Mar. Biol., 14: 18—24.
Novelli, G.D. and ZoBell, C.E., 1944. Assimilation of petroleum hydrocarbons by sulfate-reducing bacteria. J. Bacteriol., 47: 447—448.
Nriagu, J.O. and Coker, R.D., 1976. Emission of sulfur from Lake Ontario sediments. Limnol. Oceanogr., 21: 485—489.
Peck, H.D., 1966. Some Evolutionary Aspects of Inorganic Sulfur Metabolism. Lectures on Theoretical and Applied Aspects of Microbiology. University of Maryland, 22 pp.
Pfennig, N., 1975. The phototrophic bacteria and their role in the sulfur cycle. Plant Soil, 43: 1—16.
Pfennig, N., 1977. Phototrophic green and purple bacteria: a comparative, systematic survey. Annu. Rev. Microbiol., 31: 275—290.
Pfennig, N. and Biebl, H., 1976. *Desulfuromonas acetoxidans* gen. nov. and sp. nov., a new anaerobic, sulfur-reducing, acetate-oxidizing bacterium. Arch. Mikrobiol., 110: 3—12.
Postgate, J.R., 1951. The reduction of sulphur compounds by *Desulphovibrio desulphuricans*. J. Gen. Microbiol., 5: 725—738.
Postgate, J.R., 1959. Sulphate reduction by bacteria. Annu. Rev. Microbiol., 13: 505—520.
Ramm, A.E. and Bella, D.A., 1974. Sulfide production in anaerobic microcosms. Limnol. Oceanogr., 19: 110—118.
Rickard, D.T., 1973. Limiting conditions for synsedimentary sulfide ore formation. Econ. Geol., 68: 605—617.
Riley, G.A., 1944. Carbon metabolism and photosynthetic efficiency of the eart as a whole. II. Carbon metabolism and photosynthetic efficiency. Am. Sci., 32: 132—134.
Rittenberg, S.G., 1969. The roles of exogenous organic matter in the physiology of chemolithotophic bacteria. Adv. Microb. Physiol., 3: 159—196.
Rosenfeld, W.D., 1947. Anaerobic oxidation of hydrocarbons by sulfate-reducing bacteria. J. Bacteriol., 54: 664—665.
Schippel, F.A., Bågander, L.E., and Hallberg, R.O., 1973. An apparatus for subaqueatic in situ measurements of sediment dynamics. Stockholm Contrib. Geol., 24: 105—108.
Sen Gupta, R. and Jannasch, H.W., 1973. Photosynthetic production and dark assimilation of carbon dioxide in the Black Sea. Int. Rev. Ges. Hydrobiol. Hydrogr., 58: 625—632.
Shaposhnikov, V.V., Kondrat'eva, E.N. and Federov, V.D., 1960. A new species of green sulphur bacteria. Nature, 187: 167—168.
Sivelä, S. and Sundman, V., 1975. Demonstration of *Thiobacillus*-type bacteria which utilize methyl sulphides. Arch. Mikrobiol., 103: 303—304.
Skyring, G.W. and Chambers, L.A., 1976. Biological sulphate reduction of carbonate sediments of a coral reef. Aust. J. Mar. Freshwater Res., 27: 595—602.

Skyring, G.W., Jones, H.E. and Goodchild, D., 1977. The taxonomy of some new isolates of dissimilatory sulfate-reducing bacteria. Can. J. Microbiol., 23: 1415—1425.
Sorokin, Yu.I., 1962. Experimental investigation of bacterial sulfate reduction in the Black Sea using S^{35}. Microbiology, 31: 329—335.
Sorokin, Yu.I., 1964a. On the trophic role of chemosynthesis in water bodies. Int. Rev. Ges. Hydrobiol. Hydrogr., 49: 307—324.
Sorokin, Yu.I., 1964b. On the primary production and bacterial activities of the Black Sea. J. Cons. Perm. Int. Explor. Mer, 29: 41—60.
Sorokin, Yu.I., 1970a. Interrelations between sulphur and carbon turnover in meromictic lakes. Arch. Hydrobiol., 66: 391—446.
Sorokin, Y. I., 1970b. Experimental investigation of the rate and mechanism of oxidation of hydrogen sulfide in the Black Sea using ^{35}S. Oceanology, 10: 37—46.
Sorokin, Yu.I., 1970c. Mechanism of the chemical and biological oxidation of sodium, calcium and iron sulfides. Microbiology, 39: 220—224.
Sweeney, R.E., 1972. Pyritization during Diagenesis of Marine Sediments. Thesis University of California, Los Angeles, CA, 184 pp.
Tausson, W.O and Alishima, W.A., 1932. The reduction of sulphates by bacteria in the presence of hydrocarbons. Mikrobiologiya, 1: 229—261.
Tausson, W.O. and Veselov, I.J., 1934. The bacteriology of decomposition of cyclical compounds in the reduction of sulphates. Mikrobiologiya, 3: 360-369.
Temple, K.L., 1964. Syngenesis of sulfide ores: An evaluation of biochemical aspects. Econ. Geol., 59: 1473—1491.
Trudinger, P.A., 1967. The metabolism of inorganic sulphur compounds by thiobacilli. Rev. Pure Appl. Chem., 17: 1—23.
Trudinger, P.A., 1969. Assimilatory and dissimilatory metabolism of inorganic sulfur compounds by microorganisms. Adv. Microb. Physiol., 3: 111—158.
Trudinger, P.A., Lambert, I.B. and Skyring, G.W., 1972. Biogenic sulfide ores — a feasibility study. Econ. Geol., 67: 1114—1127.
Tuttle, J.H. and Jannasch, H.W., 1972. Occurrence and types of *Thiobacillus*-like bacteria in the sea. Limnol. Oceanogr., 17: 532—543.
Tuttle, J.H. and Jannasch, H.W., 1973. Sulfide and thiosulfate-oxidizing bacteria in anoxic marine basins. Mar. Biol., 20: 64—70.
van Egeraat, A.W.S.M. and Huntjens, J.L.M. (Editors), 1975. The Sulphur Cycle. Proceedings of the meeting on the sulphur cycle, Wageningen, 1974. Plant Soil, 43: 228 pp.
van Neil, C.B., 1932. On the morphology and physiology of the purple and green sulphur bacteria. Arch. Mikrobiol., 3: 1—112.
Vishniac, W. and Santer, M., 1957. The *Thiobacilli*. Bacteriol. Rev., 29: 195—213.
Vitolins, M.I. and Swaby, R.J., 1969. Activity of sulphur-oxidizing microorganisms in some Australian soils. Aust. J. Soil. Res., 7: 171—183.
Volkov, I.I. and Ostroumov, E.A., 1957. The form of sulphur compounds in the interstitial waters of Black Sea bottom sediments. Geochemistry, 4: 397—406.
Wake, L.V., Christopher, R.K., Rickard, A.D., Andersen, J.E. and Ralph, B.J., 1977. A thermodynamic assessment of possible substrates for sulfate-reducing bacteria. Aust. J. Biol. Sci., 30: 155—172.
Weil, R., 1955. Réproduction expérimentale des sulfures metalliques des sédiments biogènes. Compte Rendu del Société des Savantes, Section Sciences. Gauthier-Villars, Paris, pp. 117—125.
Weil, R., 1958. Recherches expérimentales sur quelques aspects de la géochimie de la biosphère. Geochim. Cosmochim. Acta, 14: 166.
Weil, R., Hocart, H. and Monier, J.C., 1954. Synthèses minérales et milieux organigues. Bull. Soc. Fr. Mineral. Cristallogr., 77: 1084—1101.

Widdel, F. and Pfennig, N., 1977. A new anaerobic, sporing, acetate-oxidizing, sulfate-reducing bacterium. *Desulfotomaculum* (emend.) *acetoxidans*. Arch. Microbiol., 112: 119--122.
Wolfe, R.S. and Pfennig, N., 1977. Reduction of sulfur by spirillum 5175 and syntrophism with *Chlorobium*. Appl. Environ. Microbiol., 33: 427—433.
ZoBell, C.E., 1950. Assimilation of hydrocarbons by bacteria. Adv. Enzymol., 10: 443—486.
ZoBell, C.E., 1958. Ecology of sulfate-reducing bacteria. Prod. Mon. Pa. Oil Prod. Assoc., 22: 12—29.

Note added in proof

Since this chapter was prepared, six new genera of strictly anaerobic sulfate-reducing bacteria have been identified. They include coccoidal, rod-shaped and filamentous organisms, chemolithotrophs which grow on CO_2, H_2 and sulfate, and organisms capable of completely oxidizing higher fatty acids (N. Pfennig, 1979, personal communication).

Chapter 6.2

REDUCTIVE REACTIONS IN THE SULFUR CYCLE

H.R. KROUSE

Department of Physics, The University of Calgary, 2920, 24th Avenue, N.W., Calgary, Alberta T2N 1N4 (Canada)

R.G.L. McCREADY

Lethbridge Research Station, Agriculture Canada, Lethbridge, Alberta T1J 4B1 (Canada)

CONTENTS

INTRODUCTION

The reduction of sulfate generates the major part of hydrogen sulfide in the natural environment and is believed to be a significant factor in the formation of minerals of metal sulfides and elemental sulfur. Anaerobic reduction of sulfate by organic matter is thermodynamically favourable (Thorstenson, 1970), but at terrestrial near-surface temperatures and pressures the reaction appears not to take place without biological participation. Although all microorganisms and higher plants assimilate and reduce sulfate for the synthesis of cell constituents, only a few specialized species of bacteria produce large amounts of H_2S from sulfate. It is to these specialized microorganisms, which use sulfate primarily as an electron acceptor in a respiratory pathway, that the term "sulfate reducer" is usually applied. Electrons for the reaction may be supplied by organic compounds or molecular hydrogen.

The classification of sulfate-reducing bacteria has recently been summarized by LeGall and Postgate (1973), Campbell (1975) and by Postgate (1975). The classical example of a sulfate-reducing bacterium is *Desulfovibrio desulfuricans* Beijerinck (Beijerinck, 1895). Presently five species of *Desulfovibrio* are recognised: *D. desulfuricans*, *D. vulgaris*, *D. salexigens*, *D. africanus* and *D. gigas:* all are non-sporulating. Spore-forming sulfate-reducing organisms belong to the genus *Desulfotomaculum* and there are three recognised species: *Dm. nigrificans* (formerly classified as *Clostridium nigrificans*), *Dm. orientis* and *Dm. ruminus.*

REDUCTION OF SULFUR COMPOUNDS BY ASSIMILATORY ORGANISMS

Sulfate reduction

The metabolism of sulfur has been the topic of detailed reviews by Peck (1962), Trudinger (1969), Roy and Trudinger (1970), Schiff and Hodson (1973), Goldhaber and Kaplan (1974), Kaplan (1975) and Siegel (1975). Most microorganisms assimilate sulfate-sulfur which they reduce to the level of sulfide and incorporate into the sulfur-containing metabolites such as cysteine, methionine and glutathione. Figure 6.2.1 describes current ideas on various aspects of the assimilatory pathway in the yeast *Saccharomyces cerevisiae.* The initial step of sulfate activation by ATP (adenosine triphosphate) to form APS (adenylyl sulfate) is believed to be the same in all organisms. The pathway presented in Fig. 6.2.1 is common to the enterobacteria and fungi whereas the algae * reduce APS to sulfide via two enzymic steps catalyzed by APS: thiolsulfotransferase and thiosulfonate reductase. Variations in the assimilatory pathway which are found in different species of orga-

---

* and plants (Editors' note.)

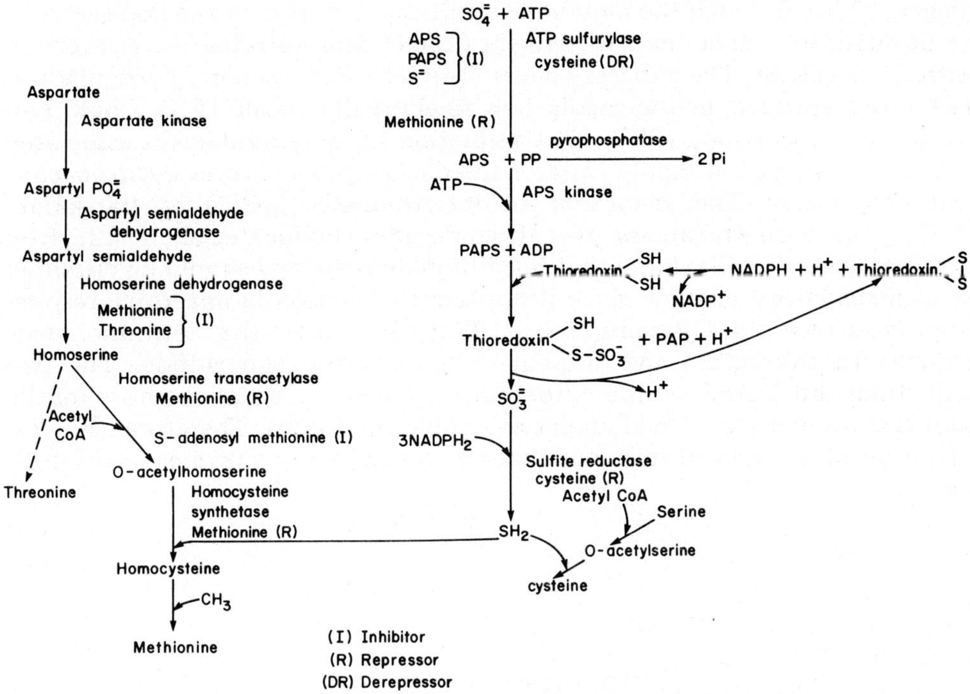

Fig. 6.2.1. Regulation of assimilatory sulfate metabolism in *Saccharomyces cerevisiae*. (Cherest et al., 1969; Siegel, 1975; de Vito and Dreyfuss, 1964). Abbreviations as follows: ADP: adenosine 5′-diphosphate; AMP: adenosine 5′-phosphate (adenylic acid); APS: adenylyl sulphate; ATP: adenosine 5′-triphosphate; $NADP^+$, NADPH: nicotinamide-adenine dinucleotide phosphate (oxidised and reduced); PAP: adenosine 3′,5′-diphosphate; PAPS: 3′-phosphoadenylyl sulphate; P_i: orthophosphate ion; PP_i: pyrophosphate ion.

nisms have been recently described by Siegel (1975). The assimilatory reduction of sulfate contributes very little sulfide to the environment, although after death and decay of the organisms, the sulfur is recycled.

Reduction of less oxidized compounds

While sulfate reduction has claimed the most attention by earth scientists, assimilatory organisms also reduce less oxidized forms of sulfur and may thus play a wider role in the geochemical cycling of sulfur than at present recognized.

In laboratory studies, many assimilatory microorganisms release H_2S during growth on sodium sulfite as their sole sulfur source (Kaplan and Rittenberg, 1964; Krouse et al., 1967; McCready et al., 1974, 1975). For over a century, numerous studies have attested to the reduction of elemental sulfur to sulfide by many microorganisms, plants and animal tissue (Roy and Trudinger, 1970) although the metabolic significance of this conversion seems to be in question. Kaplan and Rittenberg (1964) demonstrated the conversion with *S. cerevisiae*. The process occurs also with *Bacillus* spp., particularly if sulfur is suspended in the media in a finely divided state (F.D. Cook, personal communication, 1975). The reduction of polythionates is associated with members of the family Enterobacteriaciae, genera *Proteus*, *Citrobacter*, and *Salmonella*. The reduction of tetrathionate ($S_4O_6^{2-}$) to thiosulfate ($S_2O_3^{2-}$) has been known for over three decades (Pollock et al., 1942). Roy and Trudinger (1970) stated that tetrathionate reductase should be classified as a dissimilatory enzyme since it appears to function in anaerobic respiration. Most recently, Oltmann et al. (1975) have shown that tetrathionate is reduced to thiosulfate and subsequently to sulfite and sulfide. The two reductions are linked via the cytochrome system to oxidative phosphorylation and production of ATP under anaerobic conditions. Therefore, the classification of the responsible reductases as dissimilatory enzymes seems justified.

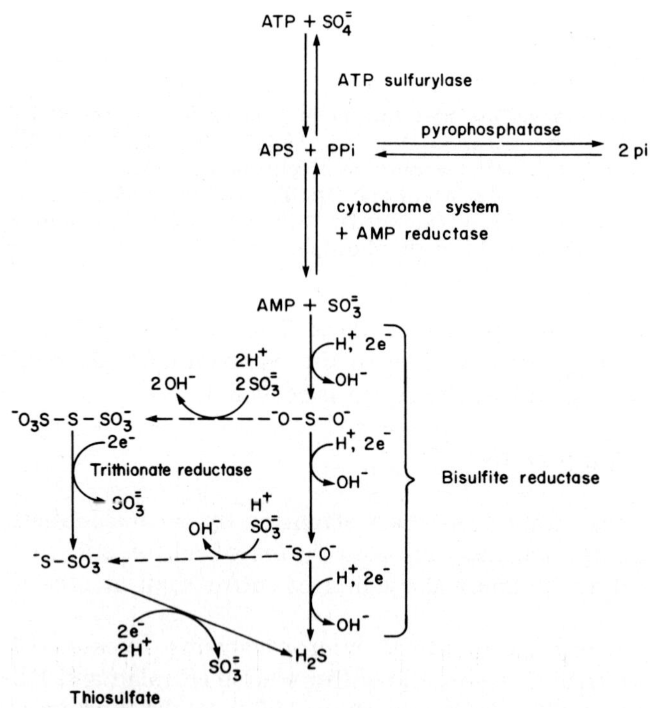

Fig. 6.2.2. Pathways of dissimilatory sulfate metabolism (Peck, 1962; Kobayashi et al., 1974; Lee and Peck, 1971). See Fig. 6.2.1 for abbreviations.

TABLE 6.2.1.

Reduction of inorganic sulfur compounds by microorganisms

Metabolic reduction	Genera involved	References
	Salmonella	Kline and Schoenhard (1969)
$SO_4^{2-} \rightarrow SO_3^{2-}$	*Bacillus*	Hunt (1974); Šmejkal et al. (1971a)
$SO_3^{2-} \rightarrow H_2S$	*Saccharomyces*	Kaplan and Rittenberg (1964); McCready et al. (1974)
	Salmonella	Krouse et al. (1967)
	Escherichia	McCready (unpublished results)
	Clostridium	McCready et al. (1975)
$SO_4 \rightarrow H_2S$	*Bacillus*	Bromfield (1953).
	Saccharomyces	Wainwright (1970); McCready et al. (1974)
$S_4O_6^{2-} \rightarrow S_2O_3^{2-} \rightarrow SO_3^{2-} + H_2S$	*Proteus*	Oltmann et al. (1975)
	Citrobacter	
	Salmonella	
$S_2O_3^{2-} \rightarrow H_2S$	*Proteus*	Tarr (1933)
$S^0 \rightarrow H_2S$	*Bacillus*	F.D. Cook (unpublished results)
	Saccharomyces	Kaplan and Rittenberg (1964)

Synergistic reductions of sulfate

Although the major source of biologically-produced sulfide in nature is believed to be due to the activity of the dissimilatory, anaerobic sulfate-reducing microorganisms, there is evidence that stepwise reductions of sulfate are carried out in the environment (Fig. 6.2.2). Krouse et al. (1970) and Šmejkal et al. (1971a) have isolated pairs of organisms from thermal springs in western Canada which, together, catalysed sulfide formation from sulfate. The first organism, an unidentified *Bacillus*, appears to reduce sulfate to sulfite; the second organism, a *Clostridium*, in turn reduced the sulfite to hydrogen sulfide. Many such synergistic pairs have been isolated from other environments (F.D. Cook, personal communication, 1975). When assessing terrestrial sulfate reduction, biogeochemists have tended to emphasise the role of *Desulfovibrio* and ignore those of other organisms. However, the concept of synergistic pairs could involve many combinations of organisms (Table 6.2.1) and their contribution to the formation of sulfide in the environment may not be insignificant in comparison to that of the classical sulfate reducers.

DISSIMILATORY SULFATE REDUCTION

As already mentioned the anaerobes, *Desulfovibrio* spp. and *Desulfotomaculum* spp., are believed to be the main contributors of biologically-produced sulfide in the environment. Cultures of *Dm. nigrificans* reducing sulfate produce from 191 to 244 mg H_2S l^{-1} of medium in 48 h at 55°C, (ZoBell, 1958). *D. desulfuricans* will tolerate hydrogen sulfide concentrations of 1—2.5 g l^{-1} (Miller, 1949, 1950a).

The enzymes involved in the initial stages of dissimilatory sulfate reduction are fairly well known. Energy is expended (46 kJ mol^{-1}) to activate sulfate to APS which is then reduced to sulfite. There is a net gain of energy, however, due to the coupling of this reaction to oxidation of organic matter. While there is general agreement on the pathway to the stage of sulfite formation, controversy exists regarding the details of the pathway for sulfite reduction (Kobayashi et al., 1969, 1972, 1974; Lee and Peck, 1971; Akagi et al., 1974; Chambers and Trudinger, 1975; Jones and Skyring, 1975; Siegel, 1975). This is partly due to different results found in experiments with cell extracts which lack the integrity of the whole cell, and those with whole cells which forbid the isolation of intracellular intermediates of the sulfite reduction pathway. Figure 6.2.3 presents a current concept of the dissimilatory pathway. An interesting feature of the pathway is that sulfur may follow two parallel routes, one of which involves the recycling of sulfite and the formation of the intermediates, trithionate and thiosulfate. Using highly purified enzyme preparations, Saks and Siegel (Siegel, 1975, p. 252) have shown that, at low sulfite concentrations (10 μM), bisulfite reductase catalyses the stoichiometric reduction of sulfite to H_2S. As the sulfite concentration is increased, first thiosulfate and then trithionate are formed in increasing proportions. At sulfite concentrations of 1 mM or greater, trithionate becomes the predominant product.

Chambers and Trudinger (1975) presented evidence that thiosulfate and trithionate are not normal intermediates during sulfate reduction. They were also unable to demonstrate any significant pools of sulfur compounds other than sulfide during sulfate reduction. The latter observations conflict with

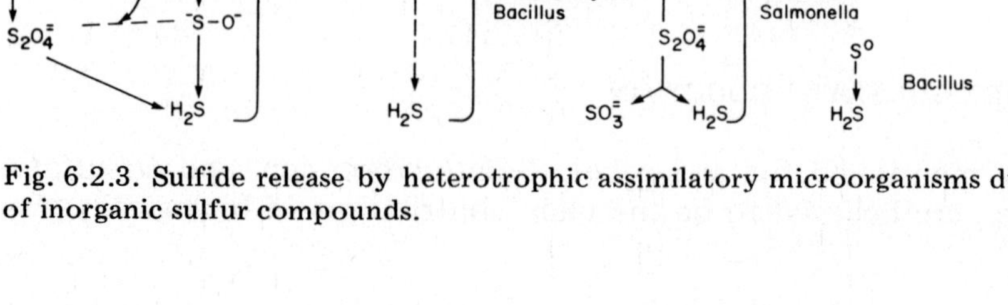

Fig. 6.2.3. Sulfide release by heterotrophic assimilatory microorganisms during reduction of inorganic sulfur compounds.

those of Jobson (1975) who examined the products of sulfate reduction in the effluent of continuous cultures of an isolate from crude oil provisionally classified as *D. vulgaris* var. *oxamicus*. Sulfite and thiosulfate were detected in amounts ranging respectively from 0—1.7% and 0—5.8% of the influent sulfate. The possibility that thiosulfate and trithionate arise as artificial side-products of inorganic chemical reactions has been discussed by Kobayashi et al. (1974) and Siegel (1975).

It is possible that parts of the dissimilatory pathway shown in Fig. 6.2.2 are induced in microorganisms not normally associated with dissimilatory reduction, in the presence of high concentrations of oxidized sulfur compounds. The particular case of *Clostridium pasteurianum* (McCready et al. 1975) is considered later (see p. 328).

ENVIRONMENTAL LIMITS OF BIOLOGICAL REDUCTION OF SULFUR COMPOUNDS

In contrast to the broad range of physical-chemical conditions under which chemical reductions of sulfur compounds occur, biological reductions are restricted to rather narrow ranges of pH, Eh, and temperature as well as requiring specific inorganic and organic compounds. Most organisms listed in Tables 6.2.1 and 6.2.2 are mesophilic, i.e., growth is limited to the temperature range 10 to 40°C. One exception is the thermophilic *Dm. nigrificans* which grows best around 55°C. Barophilic strains of sulfate reducers isolated from deep oil wells have reproduced and generated sulfide by sulfate reduction at 104°C but only under compressions of several hundred atmospheres (ZoBell, 1958). Sulfate reducers can function near the freezing point of water since they are commonly found in abundance on the ocean floor in locations where the temperature is below 5°C (ZoBell, 1961). However, the optimum growth temperature for *Desulfovibrio* spp. is generally 20 to 30°C (Temple, 1964).

There are certain metabolic requirements which must be fulfilled to obtain maximal growth and sulfate reduction by *Desulfovibrio* spp. in the laboratory. Butlin et al. (1949) showed that *Desulfovibrio* spp. have an absolute requirement for inorganic iron. Later, Postgate (1956) determined the requirement to be 10 to 15 μg atoms Fe l^{-1}. In addition K^+, Mg^{2+}, CO_3^{2-}, and $H_2PO_4^-$ (but not Ca^{2+} or Br^-) ions have been reported as essential for the growth of marine *Desulfovibrio* spp. (Hata, 1960). Some organisms require unusually high salt concentrations for growth. Marine organisms are generally moderate halophiles and grow best between 1.5 and 5% sodium chloride which are lethal concentrations for most freshwater and soil organisms. On the other hand, obligate halophiles are capable of reducing sulfate in nutrient solutions containing 30% NaCl (ZoBell, 1958).

All the microorganisms listed in Table 6.2.1 are facultative aerobes with the exception of the strictly anaerobic *Clostridium*, and their growth is limited to approximately the same pH and temperature ranges as with *Desulfovibrio*. It should be noted that sulfide production by these assimilatory organisms occurs only during anaerobic growth.

A wide variety of organic and inorganic compounds inhibit microbial activity by interfering with the uptake of metabolites and enzymic reactions (Davis and Feingold, 1962; Pelczar and Reid, 1958). In particular, the heavy metals are extremely toxic to the heterotrophs but are tolerated at higher levels by *Desulfovibrio* spp. due to their precipitation as metal sulphides (see p. 343). Other factors affecting the activity of sulfate-reducers are discussed in Chapter 6.1.

The ideal environment for bacterial sulfate reduction and sulfide mineral formation is anoxic, non-toxic, and near neutral pH. It is also characterized by adequate organic matter and mesophilic temperatures. In the absence of such an ideal environment, it must be remembered that microorganisms are diversified and adaptable. For example, bacteria may function in a microenvironment that differs distinctly from the surrounding macroenvironment. Trudinger et al. (1972) concluded that few geochemical factors by themselves can prevent sulfate reduction and sulfide mineralization.

ANTIQUITY OF BIOLOGICAL SULFATE REDUCTION

Bacterial sulfate reduction has been invoked as a factor in the formation of ancient mineral deposits and it is relevant to enquire how long sulfate-reducing bacteria have functioned on earth. Because of the structural delicacy of microorganisms, they seldom leave recognisable traces of their existence, and positive identification of fossilized bacteria is extremely difficult.

Estimates of biological antiquity can sometimes be made based on comparisons of amino-acid sequences of specific proteins which reflect changes in the hereditary DNA in organisms over geological time (Dickerson, 1971). Trudinger et al. (1972) outlined a number of such molecular evolution studies and concluded that some proteins of sulfate-reducers, and hence the bacteria, had been evolving for a period greater than 1 Gy. Another approach to the problem involves consideration of fossilized organic matter since the bacteria could not have functioned in the absence of a suitable carbon source. It has been argued that sulfate reducers were probably among the more primitive life forms (Peck, 1968; Klein and Cronquist, 1967; Postgate, 1968), so that they should pre-date occurrences of more complex fossilized organisms. Many reports in recent years have established the presence of organic material in very old sediments. These include electron microscopic findings of fragments of fossilized algae in dolomites of the Bulawayo Formation, South Africa (~2.6 Gy B.P.; Oberlies and Prashnowsky, 1968); pristane and phytane, judged to be biogenic hydrocarbons associated with

3 Gy old sedimentary material in the iron ore Soudan Formation of Minnesota, U.S.A. (Belsky et al., 1965); and the remains of bacteria-like organisms and pristane and phytane in 3 Gy old black siliceous rocks of the Fig-Tree Formation, Swaziland, South Africa (Barghoorn and Schopf, 1965, 1966). A detailed discussion on Precambrian palaeobiology can be found in Schopf (1975). A carbon isotope discontinuity between the lower Onverwacht cherts (Theespruit Formation) and overlying groups in the earlier Precambrian of South Africa may signify the origin of specific biochemical processes (Oehler et al., 1972).

A controversial approach to determining the antiquity of biological sulfate reduction involves the interpretation of natural sulfur isotope abundance variations (see p. 349). During bacterial reduction, $^{32}SO_4^{2-}$ and $^{34}SO_4^{2-}$ are converted at different rates leading to wide variations in the $^{34}S/^{32}S$ abundance ratios in sulfides. However, wide isotope abundance variations may arise from other processes so that this approach must be exercised with caution. There are also problems in assuring that the rocks containing sulfur are dated accurately. For example, the lead isotopic composition of galena is generally not helpful. The lead isotopes may fit a multistage model, but the history of the sulfur may be entirely unrelated. Thode et al. (1953) first estimated the occurrence of sulfur isotope fractionation at 0.7—0.8 Gy B.P. A number of workers since have found large sulfur isotope abundance variations in much older rocks. Of particular interest are the data of Chukhrov et al. (1970) on the lazurite deposit of Malaya Bystraya. Here, the Precambrian country rocks are the famed "stinking dolomite" (so called because of H_2S release upon striking) with an average thickness of 450 m. Both the sulfide and sulfate in these rocks are unusually enriched in the heavier isotope, ^{34}S, and Chukhrov et al. (1970) concluded that these data provided ample evidence for the existence of sulfate-reducing bacteria 3 Gy ago or earlier.

To summarize then, several lines of evidence suggest that biological sulfate reduction was an early evolutionary development and we may, with some confidence, conclude that it has been a significant geochemical factor throughout most of the geological record.

KINETICS AND THE SELECTIVITY OF SULFUR ISOTOPES DURING MICROBIAL REDUCTIONS UNDER LABORATORY CONDITIONS

As foreshadowed in the previous section, isotopic data are frequently used in attempts to determine the geochemical history of sulfide mineralisations. Isotopes of an element differ in their masses. In many processes, which are mass dependent, the isotopic species participate at different rates. If the conversion is complete, the isotopic composition of the product is the same as the reactant. However, if the process is examined under the conditions of partial reaction then the isotopic abundances in the product, remaining reactant, and intermediates will generally differ depending upon the isotopic selectivity in individual reaction steps, the extent of reaction, and the reaction kinetics. Numerous isotopic selectivity experiments have been conducted to elucidate the microbiological reduction mechanisms and the principles of sulfur isotope fractionation in nature. One motive for these investigations is to devise a diagnostic test for differentiating biological and chemical processes in sulfide ore genesis (see p. 349).

There are understandably some difficulties in describing reaction kinetics during microbiological conversions. Unlike chemical reactions, transport of components into and out of cells is involved. The cell population itself is in a dynamic state, and processes occurring within the cells may be accompanied by chemical reactions in the surrounding medium.

Rees (1973; see p. 330) demonstrated that the overall isotope fractionation during bacterial sulfate reduction can be influenced by a change in the kinetics of sulfate uptake from first- to second-order due to saturation of enzymic activation, or permeation sites. The sulfate concentration at which the change in kinetics occurs is uncertain and indeed may not be the same for all organisms and all physiological conditions. Postgate (1951) found that the specific rate of H_2-linked sulfate reduction by washed cells of *D. desulfuricans* was independent of sulfate concentration between 1 and 100 mM. On the other hand, Harrison (1957) reported that with lactate as the electron donor, reduction was first-order with respect to sulfate below 10 mM sulfate but became independent of sulfate concentration above 10 mM.

Theoretical sulfur isotope fractionation during a one-step first-order conversion

Since first-order kinetics with respect to sulfate ion is indicated under some conditions for bacterial sulfate reduction, it is useful to consider the isotopic behaviour of a simple one-step first-order conversion (Fig. 6.2.4a). The term *kinetic isotope effect* describes the competing reactions (1) and (2),

$$A_{32} + B \xrightarrow{k\,32} X_{32} + Y \qquad (1)$$

$$A_{34} + B \xrightarrow{k\,34} X_{34} + Y \qquad (2)$$

where A and B are reactants and X and Y are products. The subscripts 32 and 34 signify that the molecule contains ^{32}S and ^{34}S, respectively, while k_{32} and k_{34} are the isotopic rate constants. Assuming the reaction is first-order with respect to A, we may write for each of the isotopic reactions

$$\frac{dX}{dt} = K(A_0 - X)(B) \qquad (3)$$

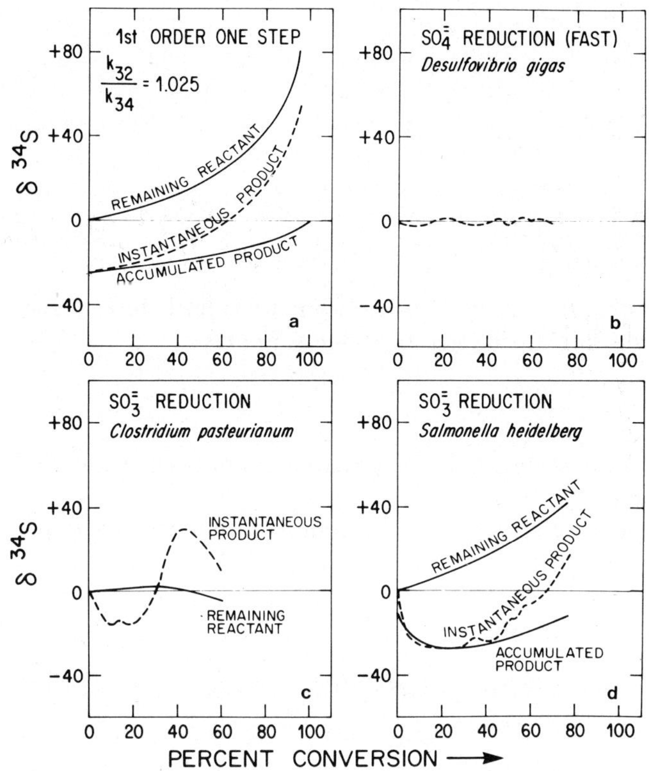

Fig. 6.2.4. Isotopic behaviour of product H_2S and other components during reduction of sulfate and sulfite by a variety of microorganisms.

where A_0 is the initial concentration of reactant, so that $A_0 - X$ is the remaining reactant at any time "t".

At time $t = 0$, X_{32} and $X_{34} = 0$.

Integration of the equations for the two isotopic reactions and division yields:

$$\frac{k_{32}}{k_{34}} = \frac{\ln\{(A_{32})_0/((A_{32})_0 - X_{32})\}}{\ln\{(A_{34})_0/((A_{34})_0 - X_{34})\}} \tag{4}$$

The fraction, F, of molecules that have reacted is given by

$$F = \frac{X}{A_0} = \frac{X_{32} + X_{34}}{(A_{32})_0 + (A_{34})_0} = \frac{X_{32}}{(A_{32})_0} \cdot \frac{1 + X_{34}/X_{32}}{1 + (A_{34}/A_{32})_0} \tag{5}$$

If the ratio of isotopic abundance ratios for the product and initial reactant, a quantity that is usually measured in the laboratory, is designated

'r' where

$$r = \frac{X_{34}/X_{32}}{(A_{34}/A_{32})_0} \tag{6}$$

$$\text{then } \frac{k_{32}}{k_{34}} = \frac{\ln\left[1 - F \cdot \dfrac{1 + (A_{34}/A_{32})_0}{1 + X_{34}/X_{32}}\right]}{\ln\left[1 - rF \cdot \dfrac{1 - (A_{34}/A_{32})_0}{1 + X_{34}/X_{32}}\right]} \tag{7}$$

Since the natural abundances of ^{34}S and ^{32}S are about 4% and 95%, respectively, the error encountered in simplifying expression (7) to

$$\frac{k_{32}}{k_{34}} = \frac{\ln(1-F)}{\ln(1-rF)} \tag{8}$$

is less serious than other uncertainties in laboratory experimentation or in evaluating natural conversions.

Nakai and Jensen (1964) developed an expression which relates k_{32}/k_{34} to "R" where

$$R = \frac{X_{34}}{X_{32}} \frac{A_{34}}{A_{32}} \tag{9}$$

If "F" is the ratio of the concentration of residual reactant $(A_0 - X)$ to the initial reactant A_0, then

$$R = \frac{F^{(k_{34}/k_{32})^{-1}} - F}{1 - F} \tag{10}$$

Figure 6.2.4a depicts the behaviour of components for a first-order one-step conversion with k_{32}/k_{34} equal to 1.025. The initial product is depleted in ^{34}S by 25‰ *. As the reaction proceeds, the remaining reactant becomes progressively enriched in ^{34}S. One must distinguish between "instantaneous" and "accumulated" product when defining "isotope fractionation factor". Whereas the product formed at any instant is depleted in ^{34}S with respect to the remaining reactant by $(k_{32}/k_{34} - 1) \times 1000$‰, the isotopic difference between the remaining reactant and accumulated product ($\Delta\delta$-value) increases to large $\delta^{34}S$ values as the reaction proceeds consistent with expres-

* Sulfur isotope abundances are measured on a δ-scale which represents the deviation in parts per 1000 (‰) of the $^{34}S : {}^{32}S$ abundance ratio in a sample in comparison to that of a reference thus:

$$\delta^{34}S(‰) = \left[\frac{(^{34}S/^{32}S)_{sample}}{(^{34}S/^{32}S)_{reference}} - 1\right] \times 1000$$

For laboratory studies, the reference is usually the initial reactant, while meteoritic troilite is taken as the reference for terrestrial abundances.

sion (10). Hoefs (1973) presents the isotopic changes in components by plotting $\delta^{34}S$ values of the remaining reactant or instantaneous product against the logarithm of F. Straight line plots arise for a first order, one-step reaction. It may be noted that some expressions derived for a first-order reaction are equally valid for zero-order kinetics (Rees, 1973).

Experimental findings on sulfate reduction

Harrison and Thode (1957) found that, during inorganic reduction of sulfate at temperatures below 100°C, the ratio of the isotopic rate constants was 1.022 assuming first-order kinetics. On theoretical and chemical grounds they identified this fractionation with the initial S—O bond rupture. Whereas it has been known for some time that different chemical reducing agents cause different nitrogen isotope effects during nitrate and nitrite reduction (Brown and Drury, 1967, 1969; Wellman, 1969), only recently have comparable observations been made with sulfate reduction (C.J. Downes and T.H. Donnelly, unpublished results, 1976). Using chromous ions in the presence of halide catalysts, they found that the fractionation increased (k_{32}/k_{34} = 1.012—1.022) with increasing activity of the catalyst, and they suggested that fractionation was determined by the nature of the activated complex.

Jones and Starkey (1957) reported that the isotopic fractionation during sulfate reduction by *D. desulfuricans* displayed a large temperature dependence. However, Harrison and Thode (1958) and Kaplan and Rittenberg (1964) presented evidence that it was the rate of reduction which influenced the isotopic fractionation and that other parameters were effective only in so far as they altered the reduction rate. Harrison and Thode (1958) explained instantaneous depletions of ^{34}S in the evolved H_2S ranging from 0 to —25‰ on the basis of a two-step mechanism. The values near zero resulted when one step with negligible isotopic discrimination competed for rate control while values near —25‰ corresponded to S—O bond rupture controlling the reaction rate. Kaplan and Rittenberg (1964) found with *D. desulfuricans* that the isotopic fractionation depended inversely upon the reduction rate with the added finding that the evolved H_2S could be depleted in ^{34}S by as much as —46‰ when ethanol was used as the carbon source, thus exceeding the value identified with S—O bond rupture by Harrison and Thode (1957).

Nakai and Jensen (1964) studied sulfate reduction in a natural mud culture over a period of 65 days. They assumed the reaction to be first-order and found a rate constant of around 0.02 d^{-1} and a k_{32}/k_{34} value of 1.02. However, Sakai (cited in Goldhaber and Kaplan, 1974) suggested that this system may have exhibited zero-order kinetics in which case the rate was about 0.5 mg S d^{-1}. The question of reaction kinetics is probably unresolvable since it is difficult to reproduce data in laboratory experiments with pure cultures and defined media, let alone evaluate the unknowns arising when a natural mud culture is transferred to the laboratory.

The first intensive study of isotope fractionation using continuous cultures was reported by Chambers et al. (1975). Using *D. desulfuricans*, they found a general similarity between sulfur isotope fractionation by continuous cultures and data previously reported for resting cells and batch cultures. The approximate inverse relationship between the isotopic fractionation and the reduction rate was also found, the maximum depletion of ^{34}S in the product H_2S being —35‰ as compared to the —46‰ found by Kaplan and Rittenberg (1964) with ethanol as substrate. Jobson (1975) also used a chemostat to study sulfur isotope fractionation during sulfate reduction by an isolate from a crude oil, provisionally identified as *D. vulgaris* var. *oxamicus*. Hydrogen sulfide exhibited $\delta^{34}S$ values of —9 to —10‰ with respect to the influent sulfate.

Sulfur isotope fractionation during sulfite reduction and synergistic sulfate reduction

Although instantaneous isotope fractionation factors of 1.025 or less have been observed during microbiological sulfite reduction to H_2S (Kemp and Thode, 1968; Krouse et al., 1967) factors as high as 1.05 have been achieved during sulfite reduction by *S. cerevisiae* (Kaplan and Rittenberg, 1964; McCready et al., 1974) and *Salmonella* spp. (Krouse et al., 1967; Krouse and Sasaki, 1968). With the latter organism the instantaneous fractionation factor was 1.02 below concentrations of 0.01% (w/v) Na_2SO_3 but reached values in excess of 1.04 at higher concentrations. These organisms are not directly involved in mineral cycling (the latter is enteric) but they are, nevertheless, participants in the overall sulfur cycle.

Unusual isotopic selectivity patterns are noted during sulfite reduction by *Clostridium* spp. and are illustrated in Fig. 6.2.4c. During early stages of the conversion, the H_2S becomes increasingly depleted in ^{34}S; i.e., a normal kinetic isotope effect prevails. Then a minimum $\delta^{34}S$ value is reached (as low as —20‰), after which the instantaneously produced H_2S becomes isotopically heavier and more enriched in ^{34}S than the unreacted SO_3^{2-} and the intermediates in the culture vessel; i.e. an "inverse" kinetic isotope effect is observed. After reaching a maximum $\delta^{34}S$ value (up to +45‰), the ^{34}S content of the H_2S decreases again. This behaviour was first observed (although not so pronounced) with an unidentified clostridium from a thermal spring in western Canada (Šmejkal et al., 1971b). It has been confirmed in over 200 experiments with the soil organism *Clostridium pasteurianum* grown under a range of carbon sources and temperatures, as well as mixtures of sulfite and sulfate and various organic sulfur compounds (McCready et al., 1975, 1976; Laishley et al., 1976). The isotopic trends observed with *Clostridium* spp. reinforce the desirable practice of carrying out reductions to the greatest extent possible in laboratory studies. Many data in the literature correspond to small percentage conversions. Had this been practiced with *Clostridium*, the interesting inverse isotope effects would have been missed.

The step-wise reduction of sulfate to sulfide by pairs of organisms provides opportunities for selective isotopic utilization from at least two pools external to the cells — the initial sulfate and products of the first organism's metabolism. Hunt (1974) measured changes in sulfur and oxygen isotopic compositions during sulfate metabolism by such a pair of organisms. Initially, both the $\delta^{34}S$ and $\delta^{18}O$ values of the unreacted sulfate increased. Sulfate then became isotopically lighter and then heavier again in contrast to the progressive relative enrichment in ^{34}S and ^{18}O predicted from normal kinetic isotope effects. Trudinger and Chambers (1973) presented evidence based on experiments with $H_2{}^{35}S$ that all stages of the pathway of sulfate reduction in *D. desulfuricans* are reversible. Should this reversibility apply to other organisms then a possible explanation can be advanced for the isotopic results found with step-wise sulfate reduction. Since inverse isotope effects can occur during sulfite reduction by the second organism, a clostridium, back reactions could cause the isotopic inversions observed in the unreacted sulfate.

Multi-step models of biological sulfate and sulfite reduction

As discussed in the previous sections, the isotopic fractionation patterns during sulfate and sulfite reduction may deviate considerably from the simple one-step reaction with first-order kinetics. The effects are summarized in Fig. 6.2.4. Figure 6.2.4b illustrates negligible isotopic selectivity during fast SO_4^{2-} reduction by *D. gigas* isolated from a spring in western Canada. Figure 6.2.4c shows the inversion in the isotopic behaviour of released H_2S during sulfite reduction by *C. pasteurianum*. Figure 6.2.4d shows ^{34}S depletion in the product H_2S during sulfite reduction by *Salmonella heidelberg* which are greater than the fractionation which Harrison and Thode (1957) associated with S—O bond rupture. Figure 6.2.4 illustrates another feature which is sometimes found in growth experiments, namely that isotopic selectivity is lower during the initial stages or logarithmic growth phase. This phenomenon was also observed with nitrogen isotopes during nitrate and nitrite reduction (Wellman et al., 1968; Cook et al., 1973) and has not been adequately explained. One possibility is that substrate uptake mechanisms are induced during the initial growth stages.

It is clear that many steps are involved in a microbiological reduction. The overall isotopic fractionation is a complex function of the isotopic selectivities of the individual steps and may be further modified due to the formation of intermediates. Tentative explanations for isotopic fractionations greater than those predicted on the basis of chemical systems, include additive effects of successive S—O bond ruptures and sulfur isotope exchange among intermediates (Kaplan and Rittenberg, 1964; Kemp and Thode, 1968; Krouse et al., 1968; Trudinger and Chambers, 1973).

As previously discussed, Harrison and Thode (1958) invoked a two-step model to account for the range of isotopic fractionation encountered during sulfate reduction by *D. desulfuricans.* Rees (1973) developed a steady-state multi-step model for isotope fractionation during bacterial reduction. His approach differed from previous attempts in that he included the possibility of zero-order kinetics for describing the uptake of sulfate. His reaction scheme is basically of the form

$$A \underset{K_{ba}}{\overset{N_{ab}}{\rightleftharpoons}} B \overset{K_{bc}}{\longrightarrow} C$$

where A represents free sulfate, B an intermediate state and C the final product, N_{ab}, K_{ba} and K_{bc} are rates. At higher sulfate concentrations, N_{ab} is considered constant; i.e. zero-order kinetics prevail. The reverse reaction from B and A, as well as the forward reaction from B to C, are considered to be first-order. With the condition that B is zero at time $t = 0$, the expression for the concentration of the intermediate B as a function of time and reaction rates is

$$B = \frac{N_{ab}}{K_{ba} + K_{bc}} \{1 - e^{-(K_{ba}+K_{bc})t}\} \tag{11}$$

In time, B approaches the value $N_{ab}/(K_{ab} + K_{bc})$; i.e., a steady-state value. If the sulfate concentration is lowered sufficiently, the flow from A to B ceases to become independent of A and the reaction scheme changes to a first-order kinetic system. In the specific case of sulfate reduction by *D. desulfuricans*, Rees (1973) expanded the reaction scheme to include: A, external SO_4^{2-}; B, internal sulfate; C, APS; D, sulfite; E, end-product (H_2S). He assigned an inverse isotope effect of +3‰ for sulfate uptake, zero isotope effect for K_{bc}, and −25‰ for K_{cd} and K_{de}. This model can account for the data of laboratory sulfate-reduction experiments including the large isotopic selectivities observed occasionally.

The inverse isotope effects observed during sulfite reduction by *C. pasteurianum* are not so readily explained. While the model of Rees (1973) has all intermediates in series, a model with additional parallel loops would provide latitude to explain almost any conceivable isotopic behaviour. A simple model with two parallel pathways which subsequently converged was suggested by Laishley et al. (1976) to explain the pattern shown in Fig. 6.2.4c. This speculation is interesting in view of the branched sulfite-reduction pathway illustrated in Fig. 6.2.3. These considerations suggest that a search for biochemical evidence of branched pathways in *Clostridium* spp. (and other organisms) might be rewarding. Recently, Chambers and Trudinger (1976, unpublished results) have observed thiosulphate (but not polythionate) formation in culture of *C. pasteurianum* growing on sulfite and producing sulfide. They suggested that this could be the result of the well-established chemical reaction between sulfide and sulfite. Therefore, branching of pathways may involve chemical reactions external to the cell as well as the biochemical mechanisms.

Summary of sulfur isotope fractionation in laboratory reductions

Microbial growth in natural environments may approximate continuous culture, batch culture, or resting cell states depending upon factors such as temperature and nutrient supply (Brock, 1966). Isotopic data are now available from laboratory experiments for sulfate-reducing bacteria growing in all three states, but the identification of specific microbiological conversion in nature on the basis of laboratory data alone is far from quantitative. Very large $\Delta\delta^{34}S$ values between sulfate and sulfide are frequently encountered in nature. Although some of these can be explained on the basis of extensive sulfate reduction in enclosed or semiclosed systems (Fig. 6.2.4a), in many cases the instantaneous isotopic fractionation factor is as high as 1.05 (see p. 328). Such large fractionations have been the exception in the laboratory (Ford, 1960; Kaplan and Rittenberg, 1964). Kemp and Thode (1968) concluded that the maximum instantaneous fractionation factor to be expected was 1.025. Consequently, these authors suggested the possibilities of a multistage process of sulfate-reduction, sulfide-oxidation and isotopic exchange between sulfate and sulfide to account for the $\Delta\delta^{34}S$ value of 50‰ between sulfate and sulfide found, for example, in the Black Sea. However, the steady-state model of bacterial sulfate reduction proposed by Rees (1973) permits these large fractionations to arise from the internal mechanisms of the bacteria.

On the basis of current knowledge, one might conclude that instantaneous fractionation factors less than 1.012 or greater than 1.022 are due mainly to bacterial sulfate reduction. Values between 1.012 and 1.022 could represent either biological or chemical sulfate reduction.

BIOLOGICAL SULFATE REDUCTION IN NATURE

Biological sulfate reduction requires organic matter and anoxic conditions. Locations which best fit these requirements include polluted river waters, lagoons, fjords undergoing stagnation, pore waters in some marine sediments, ground waters, sewage, industrial wastes and pipe lines (Postgate, 1959, 1965). Recently, sulfate reduction has been identified in the anoxic zone of pyritic uranium mine wastes (McCready, 1976, unpublished data).

The question of rates of bacterial sulfide production is important to the problem of sulfide ore deposition. Measurements of H_2S in air or sulfide ions in water do not account for that which has been oxidised or precipitated. Determinations in sediments exclude sulfide which escaped into the overlying water. Incubations of mud or water in the laboratory are not entirely satisfactory, but they may give an assessment of the sulfide-generating capacity of a system. A more direct in situ method of assessing sulfide production involves the use of radioactive sulfate. Sorokin (1962) used $^{35}SO_4^{2-}$ in the Black Sea (natural SO_4^{2-} concentration ~2 g l^{-1}) and measured activities in the silt corresponding to sulfide productions of 0.3 to 2.3 mg l^{-1} d^{-1}. Comparable reduction rates were found, using ^{35}S techniques, in Linsley Pond, Connecticut, U.S.A. where the sulfate concentration is two orders of magnitude lower (Stuiver, 1967). This comparison emphasizes that sulfate concentration is only one parameter governing sulfide production: organic matter and environmental variables such as temperature are very important in regulating sulfide production. Ivanov (1964), Trudinger et al. (1972) and Rickard (1973) summarised rates of sulfate reduction reported for a number of sediments and ground waters (see also Chapter 6.1). They concluded that these were comparable with the rates of sulfide deposition for several stratiform ore bodies.

With abundant evidence for sulfate reduction in the hydrosphere, the question arises as to the actual site of the sulfate-reducing activity. This depends upon oxygen input to the system, organic matter concentration, and other factors. Suitable conditions are often encountered at, or just below, the water-sediment interface. Here, the population of sulfate-reducers is highest because of the availability of sulfate and organic matter. In some non-eutrophic lakes a secondary population maximum may arise at depths around 3 m in the sediments (Kuznetsov et al., 1963). Sulfate reduction occurs both within the water column and the sediments of the Black Sea and a number of the lakes examined by Ivanov (1964) and Sorokin (1970).

Many studies have been made of sulfate reduction in natural environments, but in this section only a few examples are considered. Others are to be found in review articles by Silverman and Ehrlich (1964), Ivanov (1964) and Trudinger et al. (1972).

Groundwater

Water within the ground occurs in two distinct zones. The water table marks the upper level of the saturated lower zone where the spaces between the mineral grains are filled with groundwater. In the upper zone of aeration, the moisture adhering to the grains is known as vadose water. Erosion by ground water consists essentially of solution of chemical precipitates such as limestone and gypsum. The solution process may create small cavities in the order of 1 cm in size, called vugs, or larger caverns in soluble rock. Groundwater may also deposit minerals on the interior walls of a vug to form a geode. These minerals may include galena, sphalerite and pyrite. Deposits may also occur as local mineral concentrations called concretions in isolated places on insoluble rock beds. Fossils found at the centre of concretions may have served as nuclei for the mineral depositions. Organic matter in wood, etc., may be replaced by petrifying minerals including pyrite. Fractures in rock can fill with minerals precipitated from groundwater to form veins.

Sulfate reduction to sulfide may occur over a significant depth range in groundwater. Sulfide formation may be small scale as in vugs or sulfide species may enter streams to precipitate in a quiet bay or lagoon. An example of the latter appears to be the copper sulfides in Pernatty lagoon in South Australia (Lambert et al., 1971; Donnelly et al., 1972). Wallhaüser and Puchelt (1970) identified *D. desulfuricans* in 42 samples of connate water from the Carboniferous of the Ruhr Basin.

The occurrence of native sedimentary sulfur deposits in sulfate-carbonate rocks of oil-gas-bearing strata provides strong evidence that bacterial sulfate reduction played a role in their formation (see p. 357, 421), and there is clear evidence for sulfate reduction taking place in ground waters of these deposits at the present time. Data from Ivanov (1964) on bacterial activity in ground waters etc., are summarized in Table 6.2.2. The data are from many locations in a given deposit and are intended to emphasise the overall ranges for each parameter. Since the variability with location is considerable, the reader should be cautious in drawing conclusions from data in any given row in Table 6.2.2.

Although much of the bacterial sulfate reduction in hydrology can be discussed under the general heading of groundwater, it will be treated in separate sections. For example, springs are considered in the next section while the role of freshwater recharge in sulfate reduction in oil-formation waters is examined on p. 421. It is sufficient at this point to recognize that groundwater dissolves sulfate minerals, provides nutrients for subsequent reduction by microorganisms to sulfide, and bears the necessary cations to promote metallic sulfide formation.

Springs

In many thermal and cold water springs, sulfate ions have been dissolved from evaporite strata by circulating waters and then reduced biologically to sulfide. In some instances at least, reduction appears to involve the dissimilatory sulfate-reducing bacteria. Wallhaüsser and Puchelt (1970), for example, found *D. desulfuricans* in 25 diverse German and Austrian sulfur springs. The extent of conversion varies with factors such as temperature, available organic matter, and near surface aeration which tends to reoxidise the sulfide. In the case of the organically rich spring, Smrdáky, in Czechoslovakia, the conversion of sulfate to sulfide was calculated to be over 97% (Šmejkal et al., 1971b). In instances where the in situ reduction to sulfide is low, the

TABLE 6.2.2

Bacterial sulfate reduction and sulfide oxidation in groundwaters of sulfur deposits (After Ivanov, 1964)

Location	Sulfate (mg SO_4 l^{-1})	Sulfide (mg H_2S l^{-1})	Sulfide production (mg H_2S l^{-1} d^{-1})	Sulfate-reducers (organisms $\times 10^{-3}$ l^{-1})	Sulfide-oxidisers (organisms $\times 10^{-3}$ l^{-1})	S^0 formation (mg S l^{-1} d^{-1})
Shor-su	70—4956	4—1627	0.009—0.179	10—100	0—1000	63—70
Cis-Carpathian	117—1700	0—515	0.009—3.82	0—10	0—100	$\leqslant$500
Baurdak	992—1170	142—222	0.037—0.485		0.10—10^4	
Kosha-Naur oil fields	Flooding with seawater	$\leqslant$272	0.10—2.00			
Kara Kum	2767—5104	35—38		0.10—3.2 $\times 10^4$		

biological potential for significant reduction may exist but it is curbed upon emergence of the water. For example, one water sample extracted from Cold Sulphur Spring (near Jasper, Alberta, Canada) at 8°C, continued to generate sulfide for several months at ambient laboratory temperature without an added carbon source. After some nine months, a carbon source was added which maintained sulfide production for over three years. At the end of one year, the sulfate was depleted and the organisms were found to be producing sulfide from intermediate sulfur compounds (A. Sasaki and Krouse, 1968, unpublished results).

On the basis of the experimental trends outlined on p. 327, sulfur isotope data have been used in unravelling the sulfur biogeochemistry of springs. The relationships between finely divided iron sulfides in the beds of springs and biogenic sulfide have been determined on isotopic evidence.

Instantaneous fractionation factors calculated from data on springs are summarized in Fig. 6.2.5. The most frequently encountered $\Delta\delta^{34}$S-value ($SO_4^{2-} - H_2S$) is around 30‰ : some of the higher $\Delta\delta^{34}$S values are encountered in situations where there is evidence of synergistic sulfate reduction (see p. 328). Despite errors which arise in calculating the extent of conversion because of lack of data on inputs and outputs of sulfur compounds, the field data are generally consistent with laboratory experiments. For example, in the Smrdáky spring, the fast sulfate conversion is accompanied by a low instantaneous isotopic fractionation ($\Delta\delta^{34}S = +0.1$‰) in agreement with laboratory data (see p. 325; Fig. 6.2.4b). Sulfur isotope data for a number of springs around the world have been summarized by Krouse (1976).

Ponds, lakes and fjords

It is appropriate to discuss sulfate reduction in ponds, lakes and fjords together because each exhibits similar variations of either a seasonal or epi-

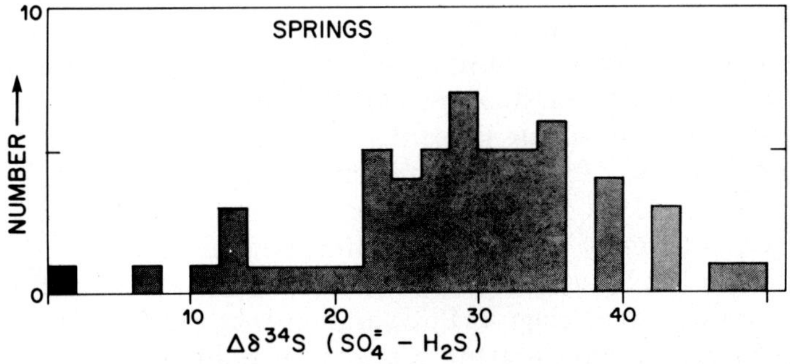

Fig. 6.2.5. Isotopic differences between sulfate and sulfide in natural springs where bacterial sulfate reduction is occurring. The $\Delta\delta^{34}$S values calculated for reduction at zero time.

sodic nature. Seasonal factors which affect sulfate reduction include temperature and biological factors such as the accumulation of falling leaves, either on the surface or bottom, leading to enrichment in organic matter and depletion of oxygen. Episodic events include water "overturn" which mixes normally stratified waters and aerates normally anaerobic zones. On a geological time-scale, the release of phosphates and other anthropogenic wastes can be considered episodic and has made certain lakes more eutrophic. The examples cited in this section will illustrate the above points. Extensive descriptions of sulfate reduction in lakes have been given by Hutchinson (1957) and Ivanov (1964). Table 6.2.3 is a compilation of sulfate reduction data in bodies of water cited in the latter reference. This table shows that measured bacterial reduction rates range over five orders of magnitude and must depend in a complex way on factors such as sulfate concentration, temperature, and nutrient supply (see also Chapter 6.1).

In tracer studies (with ^{35}S-sulfate) on Linsley Pond, Connecticut, U.S.A., Stuiver (1967) showed that large quantities of sulfate disappeared from the hypolimnion (relatively undisturbed cooler bottom region, $\sim 8°C$) and the metalimnion (intermediate region with a large temperature gradient, ~ 20 to $8°C$). Normally, the sulfate concentration decreased with depth (7 to 2 mg S l^{-1} from 0 to −12 m), but during an overturn event in November 1964, this trend temporarily disappeared and the concentrations were scattered around 6 mg S l^{-1} over the depth range. Sulfate reduction in the more aerobic epilimnion was about 10% of the rate found in the lower regions, while the dominant sulfate loss was associated with the mud-water interface. Sulfur isotope data were consistent with biological sulfate reduction, the produced sulfide being depleted in ^{34}S by about 11‰, and residual dissolved sulfate enriched in ^{34}S by about 1‰ compared with the source sulfate (Nakai and Jensen, 1964). While vertical diffusion of sulfate ion was found to be very slow, horizontal movement of sulfate to the sediments with subsequent bacterial reduction was the dominant transport process. This phenomenon was not totally explained, but the basin geometry was one contributing factor. Sulfur transport in the form of organic matter was found to be negligible but the organic material served to support bacterial metabolism. Sulfate and sulfide in the water, and poorly-formed iron sulfides in the sediments, accounted for essentially all of the sulfur present. Very little hydrogen sulfide gas was believed to find its way from the hypolimnion to the surface.

Lake Mendota, Wisconsin, U.S.A., is a possible case where human activity caused a sufficient oxygen deficiency to promote sulfate reduction to sulfide (Nriagu, 1968). However, on the basis of organic matter content, total sulfur and sulfide-sulfur in the sediments, Nriagu (1968) argued that the formation of sulfide from decomposing organic matter was highly competitive with sulfate reduction.

Genovese (1963) described a situation in Lake Faro in which the conditions for anaerobic sulfate reduction were destroyed and subsequently re-established. The normal average H_2S concentration in the lake of 70 mg l^{-1} dropped to zero in about one month after opening of a canal to the Tyrrhenian Sea. When the canal was subsequently closed, the H_2S concentration increased to about one quarter of its former value over a three-month period.

Emerson (1976) carried out physical-chemical analyses of interstitial waters in cores of Greifensee in north-central Switzerland. The lake has an area of 8.6 km^2 and a mean depth of 19 m. The lake has become increasingly eutrophic, the hypolimnion being anoxic for about six months of the year. Whereas the sulfate concentration in the water is 18 mg l^{-1}, it is below 2 mg l^{-1} (the limit of detection) in the interstitial waters. The pS^{2-} ranges from 12 near the bottom surface to 15 at a depth of 16 m (log a_{HS^-} from −6 to −9 at pH 7.05) and identifies a zone of active sulfate reduction near the water-sediment interface. This also causes a large gradient of ferrous iron with depth because of diagenesis of iron sulfide minerals.

An example of sulfate reduction in a shallow fjord estuary is the upper basin of the Pettaquamscutt River (Rhode Island, U.S.A.) studied by Orr and Gaines (1973). It is normally well-stratified water but, every few years, there is "overturn" which mixes near-surface oxygenated water with the bottom water containing biogenic sulfide. In a time span of just over one month after an overturn in November 1971, 80% of the sulfide was oxidised. After stratification was re-established, the sulfide remained below 4 m depth and accumulated at a calculated rate of 48 g S m^{-3} y^{-1}.

Bacterial sulfate reduction has been studied in several stratified lakes of the Arctic and Antarctic. Nakai et al. (1975) found that $\delta^{34}S$ values of dissolved sulfate in the Antarctic lake, Vanda, increased with depth, especially below 60 m where dissolved sulfide accumulated. Whereas the near surface sulfate had a $\delta^{34}S$ value near +15‰, a maximum of +49‰ was reached near the bottom. Krouse (unpublished data, 1969) found similar trends for a lake on Ellesmere Island in the Canadian Arctic. This lake is believed to have been a fjord until uplift removed its connection to the Arctic Ocean a few thousand years ago (Hattersley-Smith et al., 1970). The surface waters originate from glacier melt but the bottom waters are oceanic. In Lake Vanda (Nakai et al., 1975) the dissolved sulfide had $\delta^{34}S$ values ranging from +8 to +11‰ so that the $\Delta\delta^{34}S$ value (sulfate − sulfide) was near 35‰. Since the surface sulfate was depleted in $\delta^{34}S$ in comparison to its postulated source, it was reasoned that some biogenic sulfide migrating upwards was oxidised by bacteria, thus contributing isotopically lighter sulfate to the near surface waters.

Generally in volcanic lakes, sulfide-oxidising bacteria dominate (see p. 358). However, Ivanov (1964) reported the presence of sulfate reducers in the waters of the Kipyaschchee and Goryachee lakes * in the Golovnin Caldera, Kunashir Island (up to 100 organisms g^{-1} in the case of Kipyaschchee) and populations of about 10^4 organisms g^{-1} in the muds of both lakes. These

* These lakes are called Itibisinai and Honto, respectively, in the Japanese literature.

results are of particular interest since the lake waters have a pH of near 3. According to Ivanov (1964), sulfur-reducing bacteria are also present.

Seas and oceans

Table 6.2.3 shows that a lake or sea may not necessarily be distinguished as such on the basis of bacterial sulfate reduction. While there tends to be a greater amount of reduction in the saline sediments, rates are often unrelated to sulfate concentration (see also Chapter 6.1). In Lake Belovod, with a relatively high sulfate content, sulfate reduction is limited by the paucity of organic matter. In Lake Solenoe, sulfate reduction is slower along the eastern banks where the sewage has received more treatment. In contrast, extremely high rates of reduction (3 mg S^{2-} kg^{-1} d^{-1}) occur in certain bays of the Kuibyshev reservoir because of the presence of dead phytoplankton. In the Barents Sea littoral, the highest rates of sulfate reduction (24 mg S^{2-} kg^{-1} d^{-1}) occur in thickets of rockweed and under dead mussels (Ivanov and Ryzhova, 1960). In Krasnovodsk Bay, sulfate reduction in silt exceeds daily rates of 8 mg S^{2-} kg^{-1} only in places where gales carry decomposed organic matter. In mineralised areas or those devoid or organic refuse, sulfide production is minimal.

The presence of hydrogen sulfide in deeper waters of the Black Sea was reported as early as 1890 (Andrusov, 1890; Lebedintsev, 1893). Throughout its history, the Black Sea has been sporadically connected through the Bosporus to the Aegean sea, the last incursion of sea water beginning about 8000 y B.P. The current salinity is over 20‰ (compared to about 35‰ for open oceans). The sea water tends to sink to the bottom upon entry while the fresher waters flow back through the Bosporus on the surface. Estimates by various scientists in the U.S.S.R. suggest that the time for a complete change-over is of the order of 200 y or more (Vinogradov et al., 1962). Consequently, there are alterations of sulfide concentrations in the sediments depending upon the sulfate content of the water during deposition. Roughly 90% of the water now contains dissolved sulfide. There is an oxygen-hydrogen sulfide interface at about 150 m depth. Sorokin (1962) used radioactive methods to measure sulfate reduction in the sediments of the Black Sea. Rates in the surface silts range from 0.03 to 2.3 mg S^{2-} l^{-1} d^{-1} and decreased markedly at depth below 5 cm.

There has been considerable controversy over the formation of sulfide in the water column because of an apparent scarcity of suitable organic matter. Kriss and Rukina (1949) ascribed a large percentage of the soluble sulfide to the breakdown of organic matter whereas other authors believed that up to 99% of the sulfide was due to bacterial sulfate reduction (Danilchenko and Chigirin, 1926; Redfield et al., 1963; Sokolova and Karavaiko, 1964). Sulfur isotope studies by Vinogradov et al. (1962) tend to favour sulfate reduction as the dominant source of sulfide. Carbon isotope measurements by Deuser (1970) together with sulfur-carbon ratio determinations on marine organic

TABLE 6.2.3

Bacterial sulfate reduction in lakes and seas of the U.S.S.R. (After Ivanov, 1964)

Location	Sulfate ($mg SO_4$)		Sulfide ($mg H_2S$)		Reduction rate ($mg H_2S d^{-1}$)		Sulfate reducers ($organisms \cdot 10^{-3}$)		Organic matter
	Water l^{-1}	Mud kg^{-1}	Water l^{-1}	Mud kg^{-1}	Water l^{-1}	Mud kg^{-1}	Water l^{-1}	Mud kg^{-1}	
Gorkii Reservoir		32—132		6—288		0.04—2.94		1—1860	
Lake Beloe (Vologda Region)		48—205		3.5—56.1		2×10^{-4}— 6×10^{-4}			
(Kosino)	3		1.25	45			0	0	High
Lake Kipyashchee			0—32				10	10	
Lake Goryachee			0—6.8				0	10	
Lake Bol'shoe Kucheer	50		7.61				1.0	470	High
Lake Chernoe Kucheer	324		57.0				3.88	2	High
Rybinsk Reservoir		34—595		63—935		6×10^{-4}— 0.092		6—48	
Kuibyshev Reservoir		108—508	2—6	⩽2000	0.002—0.23	0.133—3.04		12—120	
Lake Belovod	834	428—635	40	⩽270	0	0.067—0.127	0.13	10—50	Low
Lake Mogil'noe		2600				1.00			
Black Sea		⩽2000				0.03—2.31		1—12	
Lake Solenoe	4050	⩽4200	37—150	⩽1600	0.0036—0.020	0.388— 18.97	10	10—100	High
Littoral of Krasnovodsk Bay		⩽2200		⩽550		3.87—8.41		1—50	
Littoral of Barents Sea		⩽2700				9.09—24.26			

matter, suggested that a maximum of 3—5% of sulfide could be derived from organic matter.

The Dead Sea is depleted in sulfate with respect to its main feeder, the Jordan River (Bentor, 1961). One reason for sulfate depletion is gypsum crystallisation. Although gypsum precipitation has been identified at certain times in the upper water column (Neev and Emery, 1967), it is not found in significant amounts in the sediments of deeper water. Recent studies have shown the sulfide ion concentration to be 0.3 mg l^{-1} or one order of magnitude lower than that in the Black Sea. Using sulfur and carbon isotope abundances, Nissenbaum and Kaplan (1976) examined the production of hydrogen sulfide in the Dead Sea and concluded it was due to biological reduction of sulfate. Previously, Nissenbaum (1975) had reported the presence of an anaerobic clostridium-like sulfate reducer in the Dead Sea sediment at depths greater than 300 m.

The Red Sea is an example of a basin which exhibits a considerable temperature range because of geothermal activity. The deeper Atlantis brine is at 56°C compared with the "normal" Red Sea water at 22°C, and appears to be devoid of sulfate reducers. Sulfate-reducing bacteria were identified in the intermediate zone between the brine and cooler water (Trüper, 1969; Watson and Waterbury, 1969), but Kaplan et al. (1969) tended to view this transition region as one of net sulfide oxidation. The latter authors questioned the presence of in situ bacterial sulfate reduction on the bases of increased sulfate concentration, depletion of ^{34}S in sulfate, and the availability of oxygen. Kaplan et al. (1969) suggested that the sulfide in Atlantis II arose from a thermochemical reduction of sulfate by organic-rich shales. On the other hand, in interstitial waters of cooler basins such as Discovery Deep, the sulfate is relatively enriched in ^{34}S and the sulfides significantly depleted in ^{34}S, a trend which is characteristic of bacterial sulfate reduction. This is consistent with the finding of *D. desulfuricans* in sediments of Discovery Deep (Watson and Waterbury, 1969). However, it is generally agreed that the numbers of sulfate-reducing bacteria even outside the hot brine area are very low.

An example of bacterial sulfate reduction in an hypersaline, stratified lagoon environment is described by Spiro (1977). In bituminous shales of the Ghareb Formation of Maastrichtian age at Nebi Muso (near Jericho, Jordan), he found single crystal calcite emplacement of gypsum infilling of foraminifera and a predominance of even over odd carbon numbers among C_{28}—C_{30} n-alkanes. He interpreted these as evidence of sulfate-reduction activity during deposition of these shales in a hypersaline lagoon. In this environment, an upper oxygenated layer is depicted which was rich in planktonic foraminifera, mainly near the margins. Organic matter accumulated on the anoxic bottom. Empty foraminiferal tests initially filled with gypsum, but as bacterial sulfate reduction progressed, the gypsum was dissolved and replaced by carbonate.

In open oceans, biogenic sulfide production is particularly related to the availability of suitable organic matter and the degree of aeration. On the basis of data from the northwestern part of the Pacific Ocean, Strakhov (1972) estimates the minimum organic carbon content supporting sulfate reduction to be 0.3% of the sediment. He cites an apparent anomaly when profiles near Japan are compared to those off New Guinea. In the former, sulfides are present while no sulfate reduction is evident in the latter. Organic carbon in the Japanese profiles ranged from 0.80 to 1.74% whereas it ranged from 0.38 to 0.88% off New Guinea. The difference in organic carbon content is not the explanation since abundant sulfides are found in the Kurile Kamchatka area and in the Marianas trench where organic carbon contents range from 0.40 to 0.90%. The explanation advanced by Strakhov (1972) relates to the forms of the organic matter present. Cores from the New Guinea area contained foraminifera tests in contrast to the terrigenous deposits in cores from the other areas. The presence of biogenic, particularly foraminiferal, calcium carbonates in the sediments increases the total organic carbon demand since organic matter is required for cementation in the animals walls.

With suitable anoxic conditions and abundant organic matter, the pore waters in sediments may be totally depleted of sulfate within a depth of 1 to 2 m. In pore waters of Saanich Inlet, British Columbia, where organic carbon contents reached 5%, Nissenbaum et al. (1972) found essentially total sulfate depletion at depths of less than 0.5 m. This depth corresponds to a time span of hundreds of years. In contrast, some off-shore cores from the JOIDES * Deep Sea Drilling Programme show that, in more slowly accumulating sediments, there is negligible sulfate depletion despite apparent continuation of sulfate reduction for millions of years, and to depths of several hundred meters. A number of these extreme, as well as intermediate cases are summarised by Goldhaber and Kaplan (1974).

If metals, particularly iron, are not available to precipitate the biogenic sulfide, then dissolved sulfide builds up in the pore waters and may even reach toxic levels. When iron is present the dissolved sulfide is significantly lower in concentration. The concentration profiles for dissolved sulfide in sediments often show a depletion in the upper layers and a maximum at a depth of a meter or less. The depletion is interpreted by Goldhaber and Kaplan (1974) to reflect reactions between iron oxide and dissolved sulfide to yield iron sulfides. As a consequence of different reactivities of iron oxides to aqueous sulfide, a depth may be reached where the sulfide production rate exceeds removal as iron sulfide. Volkov et al. (1972) reported that, in sediments off the Japan Depression, the free hydrogen sulfide concentration reaches as high as 150 mg l^{-1} which is roughly 50% higher than that found in the Black Sea and is comparable to the maximum concentration observed at Saanich Inlet, British Columbia, by Nissenbaum et al. (1972). The

* See Glossary.

high sulfide concentration of the former sediments is surprising in view of its greater aeration in comparison to the latter two sites. For comparison, the maximum dissolved sulfide content noted in a core from the Gulf of California is 0.16 mg l^{-1} (Goldhaber and Kaplan, 1974).

Mathematical solutions of the differential equations relating the concentrations of materials being consumed or generated in interstitial waters to rates of biological conversions and sedimentations have been treated by Berner (1971), Graf and Anderson (1975), and Lasaga and Holland (1975).

FORMATION OF METAL SULFIDES

An important consequence of biological sulfate reduction is the precipitation of metals as sulfides. This reaction has probably had significant effects on the geochemistry of sulfur and the formation of metal sulfide deposits.

Laboratory syntheses

Several laboratory investigations have been undertaken to determine the extent to which metal sulfide formation may take place in biological milieu.

In aqueous solutions, H_2S ionizes in the following manner:

$$H_2S \rightleftharpoons H^+ + HS^- \rightleftharpoons 2\ H^+ + S^{2-} \tag{12}$$

The sulfide ion may be oxidized by ferric ions to form free sulfur:

$$2\ Fe^{3+} + S^{2-} \rightarrow S^0 + 2\ Fe^{2+} \tag{13}$$

Ferrous ions react with sulfide ions to yield amorphous ferrous sulfide:

$$Fe^{2+} + S^{2-} \rightarrow FeS \tag{14}$$

The term "hydrotroilite" has been historically applied to the hydrated form of ferrous sulfide and it has been identified for some time as a common constituent of reducing sediments (Galliher, 1933). It is now believed that the initial iron sulfide precipitate is poorly crystallized mackinawite, $FeS_{0.94}$ (Ward, 1970).

The genesis of pyrite has claimed the attention of many workers. Pyrite has been synthesized chemically in the laboratory under a variety of temperatures and pressures (Berner, 1964a; Roberts et al., 1969; Rickard, 1969). Pyrite framboids, so named because of their raspberry-like texture when viewed under a microscope, are found in clay sediments and silts, or as infillings of foram, diatom, or radiolarian tests. Biotic and abiotic mechanisms have been proposed for their formation (Schneiderhöhn, 1923; Schouten, 1946; Love, 1965; Rickard, 1970). More recently, structures have been synthesized in the laboratory which resemble the pyrite framboids found in marine sediments (Berner, 1969; Farrand, 1970; Sunagawa et al., 1971; Sweeney and Kaplan, 1973). On the basis of experiments with stable isotope labelling techniques and scanning electron microscopy (SEM), Sweeney and Kaplan (1973) concluded that the initial amorphous iron sulfide or mackinawite was transformed by sulfurisation to pyrite via the following sequence:

$$FeS_{0.9} \xrightarrow{\text{limited oxygen}} FeS_{1.1} \longrightarrow Fe_3S_4 \longrightarrow FeS_2$$

mackinawite (tetragonal)	pyrrhotite (hexagonal)	greigite (melnikovite) (cubic, spheres)	pyrite (cubic, spheres, framboids)

Berner (1969) demonstrated that pyrite framboids could be synthesized during reaction of excess elemental sulfur with fresh amorphous iron sulfide. Sweeney and Kaplan (1973) found that, at temperatures of 60—85°C, freshly precipitated iron sulfide in aqueous suspensions, or as undried filtrates, reacted within a few days with sulfur to produce pyrite spheres and framboids. At higher temperatures and under drier conditions, there was a greater tendency to form greigite and cubic pyrite crystals. Under strictly anaerobic conditions, neither greigite spheres nor pyrite framboids were produced. Hexagonal platelets of pyrrhotite tended to be converted to framboidal aggregates of pyrite.

The question arises as to whether pyrite can be formed without the assistance of chemical sulfurisation. Reports suggest that at low oxygen tensions, or in reducing environments, ferrous sulfide may be converted to pyrite intracellularly by some organisms. Issatschenko (1929) identified crystalline inclusions in an unidentified bacterium as pyrite and speculated that they resulted from reaction (15):

$$FeS + H_2S + \tfrac{1}{2}\,O_2 \rightarrow FeS_2 + H_2O \qquad (15)$$

Upon death and cell lysis, the small pyrite crystals were liberated. Intracellular pyrite has also been identified in anaerobic cultures of sulfate-reducing bacteria and has been attributed to the following coupled reactions (Zobell, 1961):

$$FeS + H_2S \rightarrow FeS_2 + H_2 \qquad (16)$$

$$4\,H_2 + H_2SO_4 \rightarrow H_2S + 4\,H_2O \qquad (17)$$

Further, ZoBell (1961) proposed another possible reaction for the production of pyrite within bacteria or sediments, which depends upon sulfite being released during sulfate reduction:

$$2\,FeS + 2\,H_2S + H_2SO_3 \rightarrow 2\,FeS_2 + 3\,H_2O + S^0 \qquad (18)$$

Numerous laboratory experiments attest to the feasibility of biogenic formation of sulfides from metals other than iron. Miller (1950b), using sulfate-reducing bacteria, demonstrated the formation of sulfides of antimony, bismuth, cobalt, cadmium, nickel, lead, zinc and iron. Baas Becking and Moore

(1961) obtained covellite (CuS), digenite (Cu_2S_n), acanthite (Ag_2S), galena (PbS) and sphalerite (ZnS) by incubating various carbonate or oxide minerals in artificial sea water with sulfate-reducing bacteria; positive mineralogical identification of the reaction products was made by X-ray diffraction. Recently, continuous cultures of *D. desulfuricans* have been used to simulate processes occurring in nature. Using this technique Freke and Tate (1961) and Hallberg (1970) have studied the biogenic formation of iron sulfides. Subsequently, Vosjan and Van der Hoek (1972) showed that $HgCl_2$ or $CuCl_2$, at a concentration of 0.1 g l^{-1}, were non-toxic to active cultures of *D. desulfuricans* and that these metals were precipitated as the respective sulfides.

Booth and Mercer (1963) reported that *D. desulfuricans*, *Dm. orientis* and *Dm. nigrificans* could not tolerate ionic copper concentrations greater than 50 mg l^{-1} and therefore rejected the role of bacteria in copper-sulfide ore deposition. In contrast, Temple and LeRoux (1964a) showed that cultures of *Desulfovibrio* spp., which were actively producing sulfide, were protected from copper toxicity provided the copper concentration was stoichiometrically less than the concentration of sulfide being produced by the organism. One can envisage a protective sulfide-ion zone around the bacterium reacting with the metal ions, thereby preventing uptake of metal ions into the cell. Further, Kramarenko (1962) suggested that the sulfate-reducing bacteria play a direct role in metal sulfide precipitation and are not merely hydrogen-sulfide generators. Bacterial cultures containing 70—100 mg H_2S l^{-1} caused, on average, 94% precipitation of added soluble molybdenum compared with about 43% by bacteria-free solutions containing 300—1000 mg H_2S l^{-1}. Thus the microorganisms appear able, by some unknown mechanism, to alter the solubility product of the metal sulfides to accomplish precipitation. Jones et al. (1976) have demonstrated a superior ability of sulfate reducers over non-sulfate reducing bacteria to concentrate metal ions.

Predictably, the formation of metal sulfides may occur at a site removed from that of bacterial sulfate reduction. Leleu and Goni (1974) used sulfate-reducing bacteria to provide hydrogen sulfide at a reasonably rapid rate to a lead chloride solution in an adjacent flask. Galena precipitated rapidly and produced dendritic and other incomplete forms such as found in some lead-zinc deposits. Similar experiments by Leleu et al. (1975) produced wurtzite.

In an interesting laboratory model system, Temple and LeRoux (1964b) separated a slurry of sorbed metal ions (Fe, Pb, Cu, Zn) from a culture of *Desulfovibrio* by an agar partition. Banded metal-sulfide formation in the agar attested to desorption and diffusion of the metals as well as diffusion of the sulfide. Lambert and Bubela (1970) carried out related experiments with finely-ground sediments in test tubes. A nutrient medium inoculated with *Dm. nigrificans* was introduced into the lowest sediment layer. Metals were introduced at the top of the test tubes as aqueous solutions containing equal molarities of zinc, copper, and lead acetates. After incubation at 50°C for one month, metal sulfides had precipitated as distinct sharp bands. Experiments using sodium sulfide solution at ambient temperatures gave identical results. Monomineralic sulfide bands also formed in related experiments carried out with metal-enriched clays and slowly settling suspensions. These finds may help in elucidating the mechanism of formation of those sulfide deposits which are characterized by large numbers of closely spaced metal-sulfide bands and which are conformable within their sedimentary host rocks.

Metal sulfides in nature

Authigenically formed iron sulfides are found in a variety of sedimentary environments including the beds of thermal springs, high nutrient lakes and streams, swamps, lagoons, land-locked sea basins (e.g., Black Sea) and oceanic sediments. In rapidly deposited sediments, the sulfides in greatest abundance are the acid-labile minerals, mackinawite and greigite. Such appears to be the case in the upper 20 cm of sediments in Greifensee, Switzerland (Emerson, 1976) and Lake Mendota in Wisconsin, U.S.A. (Nriagu, 1968). Slower rates of sediment deposition permit more complete transformations to pyrite. The amount of free sulfur in the sediments strongly influences the formation of pyrite. In the Dead Sea, only small amounts of poorly crystalline iron sulfides are formed in deep sediments because of a paucity of elemental sulfur (Nissenbaum and Kaplan, 1976). In the case of the Black Sea, Volkov (1961) concluded that if the ratio of free sulfur to hydrotroilite was 1.5 to 2.0, up to 90% of the hydrotroilite could be converted to pyrite. In sediments east of Japan, Volkov et al. (1972) noted that a ratio of $S^0 : S^{2-} \simeq 1$ usually favoured pyrite formation. At one horizon where $S^0 : S^{2-} \simeq 0.009$, only 40% of the sulfide had been converted to pyrite as compared to 95% or better for most other samples.

Goldhaber and Kaplan (1974) summarised data on a number of pyrite occurrences in various seas as well as sediments recovered by the JOIDES Deep Sea Drilling Project. More recently, Krouse et al. (1977) studied pyrites found at depths of 300 m in sediments recovered from the mid-Atlantic ridge during Leg 37 of the D.S.D.P. These pyrites were unusually depleted in ^{34}S ($\delta^{34}S = -40$ to -50‰) and photomicrographs showed that many grains, typically 40 to 50 μm in diameter, had incorporated single or multiple circular features resembling the tests of foraminifera (Fig. 6.2.6).

In marine sediments, the amounts of reduced sulfur are often far greater than the sulfate originally enclosed by the sediment in porewater. This arises because of diffusion of sulfate into the sediment. In sediments from the Gulf of California, Berner (1964b) found values of reduced sulfur exceeding the expected concentrations by upwards of a factor of 10. Strakhov (1972) depicts the phenomenon in terms of a volume index which gives

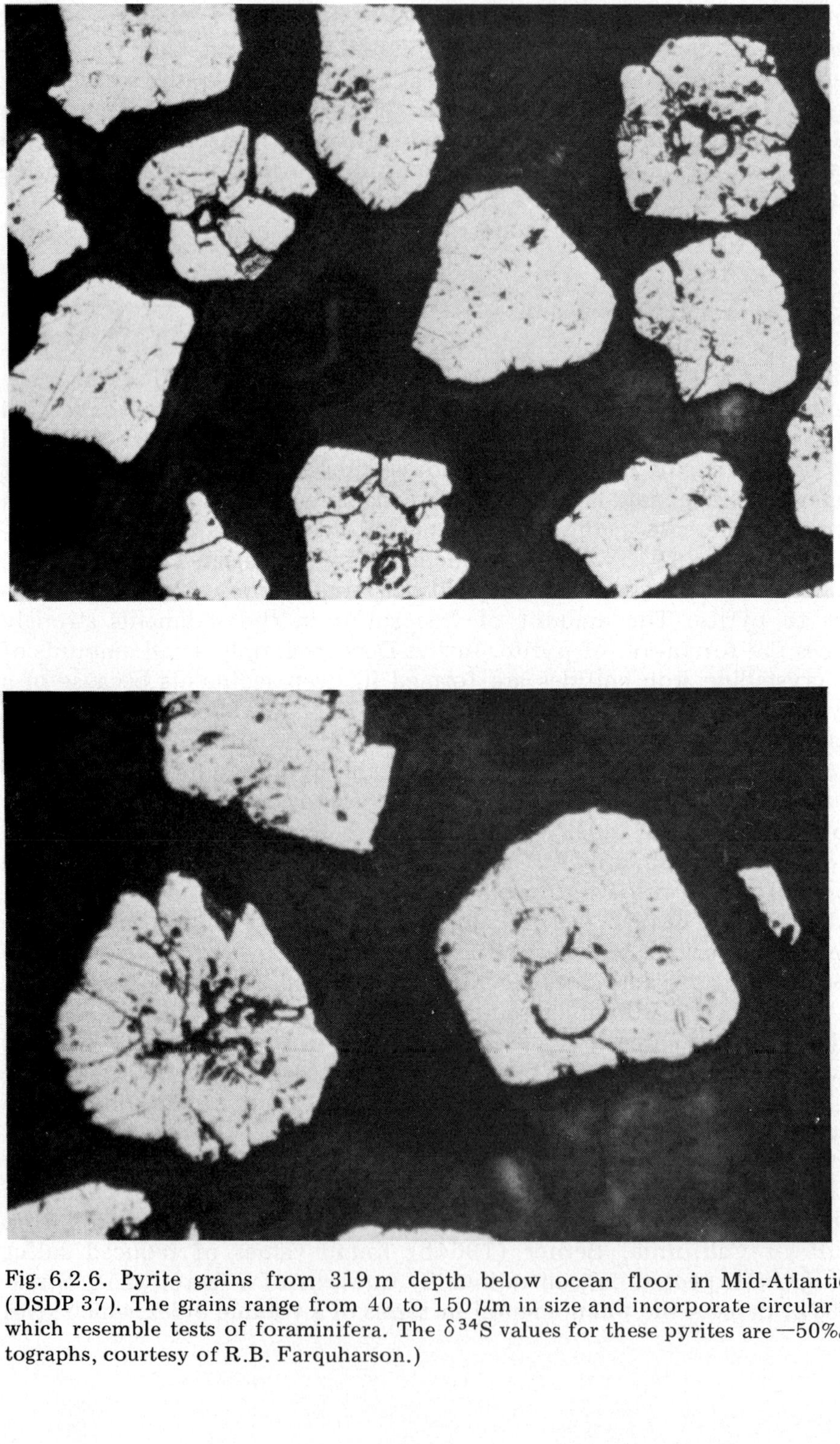

Fig. 6.2.6. Pyrite grains from 319 m depth below ocean floor in Mid-Atlantic Ridge (DSDP 37). The grains range from 40 to 150 μm in size and incorporate circular features which resemble tests of foraminifera. The $\delta^{34}S$ values for these pyrites are $-50‰$. (Photographs, courtesy of R.B. Farquharson.)

the number of litres of seawater which would have to diffuse into 1 kg of sediment ooze and yield all of its sulfate. The calculated volumes range from 0.2 to >10 l of seawater per kg of sediment.

Since the pyrite content of sediments may be of the order of 1%, the question arises as to the source of iron. Iron can enter sediments by many modes as discussed by Goldhaber and Kaplan (1974). These authors favoured transport in the form of iron oxide coatings on clay particles in view of the abundances of such particles (Carroll, 1958) and the potential chemical reactivity of these surfaces with sulfide. The occurrence of metal sulfides other than iron appears to be rare in recent sediments. One report (Arrhenius et al., 1957) describes relatively high metal concentrations (0.6 to 1.5% Zn; 0.1 to 0.5% Cu; 0.05 to 0.15% Sn; 0.03 to 0.1% Pb) in skeletal fish debris from pelagic sediments, i.e. about 100 times the concentrations found in living fish in the vicinity. In that the specific forms of the metals in the debris were not identified, one cannot evaluate the possibility that H_2S released during putrefaction, caused the enrichment of metals by sulfide formation. In the absence of reports on their occurrence in recent sediments, we can only surmise that since a variety of metal sulfides can form in association with sulfate-reducing bacteria in laboratory studies, it may be that natural formation occurs but identification is difficult because of small concentrations and/or complexing with organic matter. Nissenbaum and Swaine (1976) examined the roles that humic substances played in metal concentration and partitioning between humates and sulfides in recent sediments of Saanich Inlet, British Columbia, Canada and the Dead Sea. Whereas plankton in Saanich Inlet contained (mg l^{-1}) 3 Cu, 9 Ni and 165 Zn (Presley et al., 1972), the humates contained ($\mu g\ g^{-1}$) 80—4000 Cu, 100—1000 Ni and 550—3000 Zn. In terms of partitioning, Cu, most of the Zn, and Mo were associated with humic acids while Co, Ni and Fe were associated with the sulfide phase. Although it is desirable to concentrate metals to promote formation of their sulfides, it is difficult to evaluate whether the preferential concentration in certain organic fractions assists or inhibits in this regard. Kovalev and Generalova (1974) have carried out laboratory experiments with iron humate and fulvate complexes which suggest that the organic acids inhibit pyrite formation.

Despite the apparent low abundances of sulfides of other than iron in modern sediments, sulfate-reducers appear capable of participating in secondary sulfide formation Lyalikova and Sokolova (1965) reported that *D. desulfuricans* was generally present in those parts of the copper deposits of central Kazakstan which were not highly acidic as the result of pyrite oxidation. The rocks enclosing the secondary ore had sufficient organic matter to support the reduction of sulfate necessary to form these commercial deposits.

There remains the problem of explaining many non-magmatic metal sulfide ore bodies of the world where direct microbial participation seems unlikely since temperatures appear to be too high. Of these, the most economically important are the base metal-sulfide deposits wherein iron sulfides (FeS_2 and/or $Fe_{1-x}S$), sphalerite (ZnS), galena (PbS), chalcopyrite ($CuFeS_2$), and bornite (Cu_5FeS_4) are found conformably layered (stratiform), within bedding planes of sediments (stratabound). They are frequently associated with carbonates. A number of these deposits have been discussed in detail by Stanton (1972). The term "Mississippi Valley type" has been used as a loose categorisation for deposits similar to those found in the Pb-Zn districts in the states of Illinois, Wisconsin, and Missouri, U.S.A. *. In order to form these base metal sulfides, Zn, Pb and Cu must be concentrated by two of three orders of magnitude over the abundances normally associated with continental shelf sediments. It seems unlikely that biological sources can provide these metal concentrations. Since sediment-volcanic intercalation and/or volcanic overlay or underlay of the sedimentary facies are common, volcanism is the likely source of the metals. In the formation processes, the solubilities of ions in solution had to be high enough to transport sufficient quantities for ore deposition. Beales (1975) points out that if the host rock pore space were merely a migration route and space for sulfide precipitation, then Mississippi Valley-type ore deposits should be equally distributed between sandstone and limestone hosts. This argument is based on the finding of petroleum reservoirs in approximately equal quantities in sandstone and limestone. Since the base metal sulfides are specific to carbonate host rocks (and associated anhydrite), any mechanism which works equally well for both host rock types would not seem applicable. This argument favours localised sulfate reduction as the source of sulfide ion. Anderson (1973, 1975) examined solubility data for sphalerite and galena under a variety of conditions. Although saline brines can, theoretically, transport lead and sulfide ions simultaneously while depositing galena in ore quantities, the conditions required are too acidic for typical basin environments. Thus Anderson (1975) concluded that the most likely precipitation mechanism was the interaction of two solutions, one bearing metal ions and the other bearing sulfide ions.

In addition to biological sulfate reduction, other sources for the sulfide ions have been proposed for Mississippi Valley type deposits. Skinner (1967) suggested a mechanism by which rising hot brines, carrying soluble metal-chloride complexes, thermally degraded organo-sulfur compounds to release sulfide slowly and precipitate the metals. He also proposed that porphyrin degradation might contribute other metals such as nickel and cobalt found in these deposits. Barton (1967) suggested that chemical reduction of sulfate by methane or other organic molecules would provide the sulfide ions. Other fluids containing sulfide include volcanic exhalations under ocean bottom

* A symposium devoted to the Mississippi Valley type ore deposits is the subject of Economic Geology, Monograph No. 3 (J.S. Brown, Ed., 1967).

conditions, simple magmatic-hydrothermal sources, and leaching from rocks by descending meteoric waters. Roedder (1967) argued that fluid inclusion data did not support the latter three mechanisms in the case of the Mississippi Valley deposits. Replacement of pre-existing sulfides as discussed by Lovering (1963) is an alternate mechanism for PbS and ZnS ore formation.

Although metal sulfides are currently depositing in locations in the bottom of the Red Sea (Degens and Ross, 1969), most data come from records of past geological events. The reader can comprehend that the delineation of ore-forming mechanisms is not easy. The interaction of sulfide-bearing and metal-bearing solutions mentioned above is a possibility but other proposals have merit. In the case of the Red Sea, it is suggested that brine rises in fractures of the Atlantis II Deep. On occasions, sulfide is ejected in this brine possibly as the result of hydrothermal organic reduction of sulfate (Kaplan et al., 1969). Metals in the basin are usually precipitated by this sulfide but excess metal-rich brines may escape to react with biogenic sulfide in other basins, e.g., Discovery Deep (see p. 340). Thus it is important not to confuse the deposition mechanisms with the question of source. Sulfur isotopes are informative in this regard as shown in the study of the Red Sea by Kaplan et al. (1969). Another problem discussed in the next section is that biogenic sulfide may undergo a number of physical-chemical alterations before it is manifested as an ore deposit.

Hallberg (1972) provided an interesting analysis in which sedimentary sulfide formation was described in energy symbols combined in a circuit system as well as in geochemical and biological terms. He visualised two different lines of formation dependent upon whether or not the metals had reacted with oxygen prior to metal-sulfide formation.

SULFUR ISOTOPES AND THE ORIGIN OF SULFIDE ORE BODIES

Early work by Thode et al. (1949) established that there are large variations in the isotopic composition of sulfur compounds in nature. Since then sulfur isotope abundance data have been frequently used to elucidate many terrestrial processes including the genesis of sulfide ore bodies. The mechanisms of isotopic fractionation (alteration of relative isotopic abundances) can be broadly categorised under exchange processes or the kinetic isotope effects discussed on pp. 324ff. Isotopic exchange may be represented by the reaction:

$$aA_{32} + bB_{34} \rightleftharpoons aA_{34} + bB_{32} \qquad (19)$$

where the subscripts refer to ^{32}S and ^{34}S, and "a" and "b" are the number of moles of A and B, respectively. A and B are two different sulfur-containing molecules in the case of chemical exchange, or two states of one compound during physical exchange processes. At equilibrium, the following relationship holds

$$K = \left[\frac{A_{34}}{A_{32}}\right]^a \Big/ \left[\frac{B_{34}}{B_{32}}\right]^b = \left[\frac{Q_{A_{34}}}{Q_{A_{32}}}\right]^a \Big/ \left[\frac{Q_{B_{34}}}{Q_{B_{32}}}\right]^b \tag{20}$$

where K is the equilibrium constant and Q is the thermodynamic partition function. Theoretical calculations of the partition function ratios of a number of sulfur compounds have been carried out (Tudge and Thode, 1950; Sakai, 1957). In general, compounds at higher oxidation states tend to become enriched in the heavy isotope. The isotopic fractionation factor, α, is defined by the expression:

$$\alpha = \left[\frac{A_{34}}{A_{32}}\right] \Big/ \left[\frac{B_{34}}{B_{32}}\right] \tag{21}$$

If there are "n" exchangeable atoms in a molecule then $\alpha = K^{1/n}$. The equilibrium constant K has a temperature dependence given by the expression ln K = constant x T $(K)^{-2}$. Experimental determinations of K for exchange among sulfide minerals can prove effective in elucidating the thermal history of ore deposits (Kajiwara and Krouse, 1971; Czamanske and Rye, 1974; Smith et al., 1977).

Figure 6.2.7 summarises data from many investigations. Many occurrences are omitted, including sulfur compounds in the atmosphere and freshwaters, since the intention of Fig. 6.2.7 is to emphasise features which are relevant to sulfide ore deposition. Meteoritic troilite is chosen as the reference for the $\delta^{34}S$ scale because its isotopic composition is remarkably uniform and close to the arithmetic mean of those of terrestrial sulfur. Deep-seated or magmatic primary sulfides tend to fall near $\delta^{34}S = 0$, with a bias towards slight enrichments in ^{34}S, as shown by data from magmatic deposits, basic sills and carbonatites (Thode et al., 1962; Shima et al., 1963; Ryznar et al., 1967; Grinenko et al., 1970; Schwarcz, 1973; Mitchell and Krouse, 1975). Volcanic sulfur compounds tend to have larger isotopic variations either as a result of isotope-exchange processes or the presence of other than primary sulfur components in the systems.

In earlier sections, natural environments were described in which the difference in $\delta^{34}S$ values between unconverted sulfate and bacterially produced sulfide ranged up to 60‰. Therefore, it is to be expected that biogenic sedimentary sulfides should display large isotopic variations as shown in 6.2.7. In oceanic sediments, there is a general tendency for the $\delta^{34}S$ values for dissolved sulfate and sulfide to increase with depth suggesting that the sediments approximate the closed system illustrated in Fig. 6.2.4a. In many cases, the isotopic difference between sulfate and instantaneously produced sulfide is near 30‰, which is, interestingly, the situation most often encountered in springs (Fig. 6.2.5). In contrast, however, pyrites tend not to show such large depth trends in their isotopic composition. This has been interpreted in terms of sulfate reduction, with pyrite formation occurring pre-

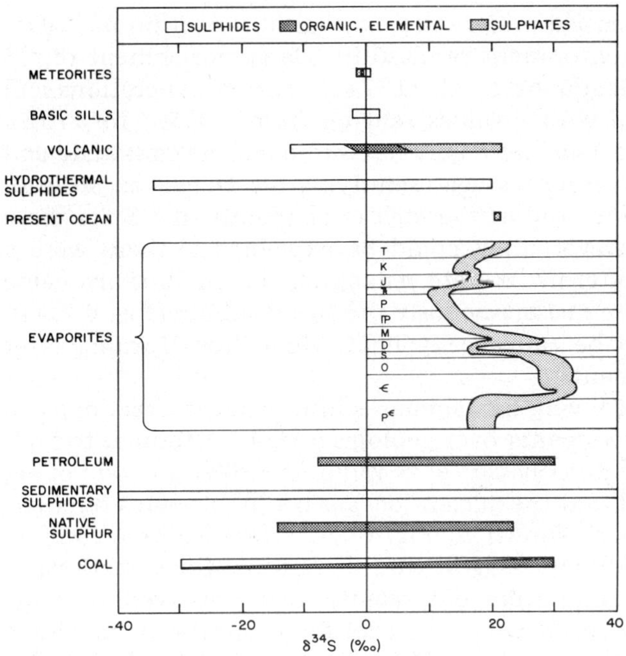

Fig. 6.2.7. Sulfur isotope abundance variations in nature (evaporitic curve after Kaplan, 1975).

dominantly near the sediment surface where the supply of sulfate is unlimited (Kaplan et al., 1963; Hartmann and Nielsen, 1969). While sulfate reduction persists in deeper zones, the amount of pyrite formed is small in comparison to that formed prior to deep burial. Goldhaber and Kaplan (1974) summarised a number of studies of sulfur isotope fractionation during bacterial sulfate reduction in recent sediments. They claimed that the data indicated a two-stage fractionation process with larger fractionations (corresponding to instantaneous fraction factors of the order of 1.04) in the upper 15—40 cm of the sediment. They also suggested that the reduction might be zero-order with respect to sulfate in nutrient-rich zones near the sediment surface, and might change to first-order after burial or depletion of nutrients. It should be noted, however, that in at least one of the sediments (Santa Barbara Basin) examined by Goldhaber and Kaplan (1974), which exhibited this trend, the concentration of pore-water sulfate never fell below 20 mM. This is considerably higher than the sulfate concentration at which a change from zero-to first-order kinetics has been observed in the laboratory (see p. 324).

Sulfur-isotope variations in concretions are interesting in that their δ-values may fall outside the range in Fig. 6.2.7. Sakai (1971) reported barite concretions in the Japan Sea with δ-values as high as +87‰. A plausible explanation is that extensive preferential biological reduction of $^{32}SO_4^{2-}$ in a somewhat restricted environment resulted in a large enrichment of ^{34}S in the unreacted sulfate. Bogdanov et al. (1971) reported concretions and individual crystals of pyrite with δ-values ranging from —47‰ to +70‰ over a depth range of 310 to 1459 m in gray nearshore-marine carbonate and clastic rocks of lower Carboniferous age underlying the Dzhezkazgan lead-copper ore beds. The range as well as the high enrichments of ^{34}S for these sulfides are surprising. However, they could be explained if there were a closed environment with extensive sulfate reduction, the concretions being formed at various stages from instantaneously produced sulfide (Fig. 6.2.4a). Those forming early would be isotopically light while those forming later would have high ^{34}S enrichments.

The question arises as to whether similar sulfur isotope fractionation occurred in fresh oceanic sediments over geological time. Although today's oceanic sulfate has a markedly constant $\delta^{34}S$ value of +20‰ (Thode et al., 1961), evaporite data reveal that the ancient oceans had $\delta^{34}S$ values fluctuating between +6 and +40‰ as shown in Fig. 6.27. These isotopic fluctuations are discussed on p. 409 and suggest that, if isotopic fractionations in sediments over geological time were equivalent, then preserved biogenic metal sulfides should show shifts which parallel the evaporite data. Thode and Monster (1965) noted that organic sulfur in petroleum was depleted in ^{34}S by at least 15‰ compared with contemporaneous evaporites, and Sangster (1968) reported a similar depletion in the average isotopic compositions of strata-bound metal sulfide deposits of equivalent ages. These apparent correlations, however, must be treated with caution. For example, Sasaki and Krouse (1969) found that Pb-Zn sulfide deposits of Pine Point, N.W.T., Canada have a narrow isotopic distribution and a mean $\delta^{34}S$ value close to those of the associated marine evaporites. In contrast, the Cambrian Nairne Pyrite deposit of South Australia was found to have $\delta^{34}S$ values ranging from —13 to —21‰ with the most frequent value near —20‰ (Jensen and Whittles, 1969). Since oceanic sulfate from the Cambrian up to the Silurian was generally more enriched in ^{34}S ($\sim$ + 30‰), this represents a ^{34}S depletion of —50‰ if the sulfides arose from marine sulfate. In the Mount Gunson copper deposits, Pernatty Lagoon, South Australia, $\delta^{34}S$ values of lagoon sulfides range from —16 to +26‰ while those of the off-lagoon sulfides range from —2 to —9‰ (Donnelly et al., 1972). By comparison, groundwater sulfates and gypsum samples have $\delta^{34}S$ values ranging from +13 to +19‰. In metal sulfide deposits of the Walton-Cheverie area of Nova Scotia, Canada, Boyle et al. (1976) found the $\delta^{34}S$ values of +34 to +6‰ for dissolved sulfate and sulfate minerals while sulfides and sulfosalts had $\delta^{34}S$ values of —40 to —1‰. They concluded that the sulfides and sulfosalts were derived from biological sulfate reduction of deep circulating brines. In summary, sedimentary ore deposits may exhibit narrow isotopic distributions with mean δ-values close to contemporaneous sea water (Pine Point), narrow distributions with the mean δ-values far removed from contemporaneous sea water (Nairne), and wide patterns in isotopic composition (Pernatty Lagoon, Walton-Cheverie).

Schwarcz and Burnie (1973) reviewed sulfur isotope abundances in strata-bound sulfide deposits in clastic sediments not associated with volcanic rocks and concluded that two patterns were evident. One was a broad distribution ranging from around the $\delta^{34}S$ of sea water to values 25‰ lower while the other was a narrow distribution around a $\delta^{34}S$ of −50‰ with respect to oceanic sulfate. The former pattern was identified with shallow marine or brackish-water environments while the latter occurred in deep, euxinic basins. They explained the first distribution on the basis of Fig. 6.2.4a and assumed a closed system with an average k_{32}/k_{34} ratio of about 1.025. The second distribution pattern applied to deep basins and Schwarcz and Burnie (1973) concluded that the systems were fully open, in which case the isotopic selectivity was much larger as is the case in a number of modern euxinic basins (e.g. Black Sea; see pp. 338, 412).

Although sulfur isotope values in fresh sediments can be positively equated with bacterial sulfate reduction, care must be taken in interpreting isotopic data on ore bodies. Sulfur isotope compositions far removed from $\delta^{34}S = 0$, and/or varying widely within an ore deposit, have often been interpreted as evidence of biogenic origin since bacterial sulfate reduction is an effective means of achieving large isotope fractionations. However, Ohmoto (1972) has now demonstrated that, theoretically, such isotopic patterns may arise as the result of isotopic exchange between sulfur species in hydrothermal solutions. Using available isotopic fractionation factors and thermodynamic data on minerals and ions in solutions, he evaluated how temperature, pH, oxygen fugacity (fO_2) and sulfur concentration (fS_2) affected the distribution of sulfur species and their isotopic compositions. Although his calculations assume ideal conditions such as equilibrium isotopic exchange among the species, his conclusions may be generally applied; namely that $\delta^{34}S$ values of minerals forming from hydrothermal solutions may be changed by several units ‰ by small changes in fO_2 and pH. Isotopic fractionation at 250°C is comparable to that achieved microbiologically at much lower temperatures. Therefore, wide variability of $\delta^{34}S$ values in an ore deposit is not a sufficient criterion for identifying biogenesis. Rye and Ohmoto (1974) evaluated the mean sulfur isotopic compositions of the sulfur sources of a number of hydrothermal ore deposits, on the assumption that exchange processes were responsible for the observed isotopic patterns. They identified three main sources of sulfur; deep-seated sources, adjacent country rocks, and marine evaporites.

The obvious question is "Which interpretation is more plausible in the case of a given ore deposit?" Exchange among co-existing sulfides can be readily achieved in the laboratory even using dry reagents (Kajiwara and Krouse, 1971; Smith et al., 1977) whereas exchange in solution between sulfide and sulfate ions is favoured by low pH values (Igumnov, 1976; Robinson, 1978). Evidence of non-isotopic equilibrium among the sulfides means either that part of the system did not equilibrate originally or subsequent processes have altered the isotopic distribution. A sulfide could be biogenic but its $^{34}S/^{32}S$ distribution might reflect subsequent exchange phenomena. In interpreting isotope distributions, attention must be paid to all relevant geological, geochemical and mineralogical data which may provide clues as to the environment prevailing at the time of deposition and the post-depositional history of the deposit.

THE FORMATION OF ELEMENTAL SULFUR

There is considerable evidence that, in nature, bacterial sulfate reduction plays an important role in the formation of some deposits of elemental sulfur. Free sulfur is not, however, produced by sulfate-reducers per se and its formation depends, therefore, on chemical or biological oxidation of sulfide. Microorganisms capable of effecting the latter reaction are discussed in Chapters 6.1 and 6.3 while isotopic selectivities associated with this conversion are summarised in Table 6.4.2 (see p. 406). As discussed in Chapters 6.1 and 6.3 colourless sulfide-oxidising bacteria, e.g. *Beggiatoa*, and *Thiobacillus*, inhabit aerobic zones of ponds, etc. while in the underlying anaerobic zones, where light can penetrate, photosynthetic oxidisers, such as *Chromatium* and *Chlorobium*, are active.

Isotopic data have proved useful in delineating situations in which both reductive and oxidative processes operate. Krouse et al. (1970) described such a situation in thermal springs of the Canadian Rockies. In one case, the primary sulfate had a $\delta^{34}S$ value near +25‰ while sulfate salts and elemental sulfur precipitated in the algal mat had negative $\delta^{34}S$ values close to those associated with sulfide dissolved in the water. *Beggiatoa* were intermixed with the algae. It was concluded that sulfide produced by sulfate reduction was oxidized to sulfur and sulfate in a localized environment in which free interchange of sulfate with the surrounding macroenvironment was restricted.

The origin of sulfur in Lake Eyre, South Australia was studied by Baas Becking and Kaplan (1956). They reported the presence of very active sulfate-reducing bacteria and proposed that the carbon source for these organisms was derived from photosynthetic organisms followed by decomposition in the silt by other organisms. Sulfur isotope studies supported the idea of bacterial sulfate reduction in that elemental sulfur was depleted in ^{34}S by 20‰ in comparison to gypsum from nearby cliffs which were considered to be the sulfate source (Kaplan et al., 1960). In a mud sample under a salt crust, dissolved sulfate was enriched in ^{34}S by 8‰ compared to the gypsum.

This is consistent with the isotopic pattern expected of a closed system (see p. 353). In contrast to the relatively small scale sulfur production in Lake Eyre, elemental sulfur in the lakes of Cyrenaica in the northern part of the Libyan desert may comprise one half of the silt (Butlin and Postgate, 1954). The shallow lake waters (~1 m depth) smell strongly of hydrogen sulfide, which varies in concentration from 15—20 mg l^{-1} at the surface to over 100 mg l^{-1} at the bottom. The dissolved sulfide : sulfate ratio in the bottom waters is about 0.15. The lakes are supplied by warm sulfuretted springs of the sodium chloride type enriched in sulfate ion (1848 mg l^{-1}). Massive algal mats of several cm thickness cover the littorals of three of the lakes. These contain large amounts of native sulfur and the presence of *Chromatium* and *Chlorobium*, cellulose-decomposing bacteria, sulfate-reducers and other types of bacteria, was demonstrated. Butlin and Postgate (1954) concluded that, in these lakes, sulfur is formed by oxidation of the product of sulfate reduction (H_2S) by photosynthetic bacteria and that the sulfate-reducers utilised organic matter produced by coloured sulfur bacteria. Again, sulfur isotope abundances are consistent with bacterial sulfate reduction (Macnamara and Thode, 1951; Harrison and Thode, 1958; Kaplan et al., 1960). Annual sulfur recovery from these lakes has ranged upwards to 180 Mg.

Sulfur formation along the coast of eastern India near the village of Kona (Masulipatam, Madras) was described by Iya and Sreenivasayi (1944, 1945). Clays in certain coastal areas of the Bay of Bengal may be flooded during monsoons for several months at a time so that they become logged with sea water; halophilic sulfate-reducers develop abundantly in the lower black clays. Diffusing sulfide ions are oxidised near the surface and produce colloidal and crystalline sulfur to a depth of some 20 cm.

Elemental sulfur also forms in many of the larger water bodies in which sulfate reduction occurs. In the Black Sea, for example, elemental sulfur is found in both the water and the surface silt and is probably formed largely at an oxygen-hydrogen sulfide interface which exists around 100 to 200 m depth. Sorokin (1964) identified this interface as a zone of chemosynthesis and showed that maximum chemosynthesis corresponded to a maximum in a number of chemoautothrophs which were assumed to be thiobacilli. Jannasch et al. (1973) found aerobic sulfide- and thiosulfate-oxidising bacteria to be most active within the oxygen-hydrogen sulfide interface while anaerobic photosynthetic sulfide oxidisers were not found in offshore waters. In the Black Sea, and presumably other environments, the net formation of elemental sulfur is a function of the balance between biological sulfide production and oxidation on the one hand, and depletion by further oxidation of sulfur to sulfate, or incorporation of reduced sulfur into pyrite, on the other.

Nissenbaum and Kaplan (1966) favour a lagoonal environment for the origin of a sulfur deposit from Upper Pleistocene sandstone in Beeri, Israel. There, elemental sulfur is closely associated with organic matter throughout the sediment. The data are consistent with intermittent flooding of a shallow coastal basin. The $^{34}S/^{32}S$ abundances imply that bacterial sulfate reduction was almost complete when the lagoon was separated from the sea. The shallow environment would also be conducive to sulfide oxidation and algal growth which would contribute organic matter to the sediment. Ivanov (1964) considers that the Krasnovodsk sulfur deposits in U.S.S.R. have been formed syngenetically in sediments of a lagoonal basin of the Akchagyl stage of the Neocene. They are characterised by nodular features and the absence of circulating groundwaters in the enclosing rock. Numbers of thiobacilli in the sulfur nodules are as high as 10^5 organisms g^{-1} of rock.

Biogenic sulfur deposits may also occur in deep groundwaters. In the Carpathian sulfur deposits, downward percolating surface waters apparently bear organic matter as nutrients for sulfate-reducing bacteria, as well as providing dissolved oxygen for elemental sulfur formation. The sulfide production is seasonally dependent. Bacteria participate in the oxidation forming about 0.5 g $S^0 l^{-1}$ d^{-1} (Ivanov, 1960; Sokolova, 1960; Invanov and Kostruva, 1961; Ivanov and Ryzhova, 1961). Lein et al. (1976) measured $^{34}S/^{32}S$ abundances in sulfur-calcite ores of the Gaurdak deposit in the eastern part of Turkmen S.S.R. Celestite exhibited $\delta^{34}S$ values between +22 and +27‰ whereas the subsurface waters contained sulfate and sulfide ions with $\delta^{34}S$ values ranging from 35 to 48‰, and 3 to 13‰, respectively. Native sulfur was consistently depleted in ^{34}S as compared to adjacent sulfate minerals. The data are consistent with solution of the celestite followed by extensive microbiological sulfate reduction to sulfide which is then oxidized to native sulfur. Karavaiko et al. (1963) suggested that the Quaternary sulfur deposits of Kara Kum, north of Askkhabad arose from bacterial sulfate reduction in oil formation waters in Cretaceous strata. Ivanov (1964) classifies together the Kara Kum deposits, sectors of the Shor-Su deposits in Uzbekistan, and smaller deposits in western Turkmenia on the basis that sulfate reduction and sulfur formation were separated in space. Hence sulfur is deposited in rocks of various lithological and chemical composition and not accompanied by secondary calcite. Table 6.2.3, based on Ivanov (1964), summarises data from these deposits which describe the activity of sulfide oxidisers.

D. desulfuricans has been implicated by Al-Sawaf (1977) in the formation of economic sulfur deposits in the middle Miocene, Lower Fars Formation of Northwestern Iraq. In the richest structure at Mishraq, sulfur is deposited in the lower part of the formation to a thickness of some 70 m. Most of the groundwater is of the sulfate type ($CaSO_4$ up to 2 g l^{-1}), but even in areas of low sulfate concentration there is significant biological sulfate reduction, particularly where the temperatures are higher (up to 50°C) and the supply of hydrocarbons ample. Circulation of fresh meteoric waters in the groundwater system was identified as the agent for oxidation of H_2S to S^0 but it also served to leach sulfur out of some strata.

Salt dome deposits of the world have been described as dried up sulfureta (see p. 300) although current biological sulfate reduction is often evident. The best known occurrences are in Texas and Louisiana where elemental sulfur is associated with calcite, anhydrite, gypsum, petroleum and waters within the "cap rock" of anhydrite and calcite situated above salt beds. Hundreds of domes have pushed up through the sediments of the Gulf Coast geosyncline. Elucidation of the role played by bacteria in these environments was among the earlier triumphs of stable isotope studies which clearly implicated microbial sulfate reduction (Thode et al., 1954; Feely and Kulp, 1957). Similar studies of calcite-sulfur ores were carried out by Dessau et al. (1962) on the Sicilian sulfur deposits, and by Vinogradov et al. (1964) on the Shor-su deposit. Furthermore, carbon isotope abundances provided evidence that the calcite in the cap rock originated from oxidised petroleum. The domes are pictured as originating at a stress in a salt bed at a depth of 10 km or more. As the salt plus rose to within 3 km of the surface, anhydrite capping commenced. Near 1.5 km depth, the conditions were suitable for bacterial action provided petroleum seeped into the cap rock *. The carbon dioxide from the biodegradation of petroleum caused calcite formation in the cap rock. Soon thereafter, relative stability was achieved within a few hundred metres of the surface. Some H_2S escaped while some was oxidised to elemental sulfur by dissolved oxygen in the descending surface waters. In this regard, it is difficult to assess the degree to which the oxidation occurred by purely chemical means or with biological assistance (see also Chapters 6.1 and 6.3). In some individual salt domes of the Gulf Coast, about 100 Tg of petroleum were oxidised while 10 Tg of sulfur were produced. In some cases anhydrite reduction reached 1 Pg. Often the conversion of sulfide to elemental sulfur was less than 1%. Feely and Kulp (1957) proposed that a chemical reaction of sulfate with hydrogen sulfide might be one mechanism for the formation of elemental sulfur in these domes. Davis et al. (1970), however, argued that the reaction was thermodynamically unfavourable at neutral and alkaline pH and that an alternative suggestion offered by Feely and Kulp (1957), namely oxidation by oxygenated groundwater, was more tenable. Some cap rocks of the world were not as productive as those of the Gulf Coast simply because the available anhydrite or carbon source was low. One example described by Ivanov et al. (1971) is the Romny salt cupola within the Dneiper-Donets Basin. The anhydrite occurs mainly as seams comprising about 5% of the total rock. Sulfur and carbon isotope data coupled with an assessment of the environment, led these authors to conclude that sulfate-reducing bacteria functioned under rather unfavourable conditions in comparison to the salt domes of the Gulf Coast.

Davis and Kirkland (1970) described the deep-seated (down to 400 m) widespread economic native sulfur deposits of western Texas as having fea-

* Editors' footnote: The question of hydrocarbon utilization by sulfate reducers is discussed in Chapter 6.1.

tures which were comparable with salt domes. In particular, the sulfur isotope composition ($\delta^{34}S$: +6.7‰ for S^0 and +26.6‰ for associated anhydrite) and carbon isotope values for the epigenetic calcite ($\delta^{13}C$ from −24.1 to −38.0‰) were interpreted as evidence of microbiological sulfate reduction. Although bacterial sulfate reduction in salt domes is well established, an evaluation of the microbiological participation in the oxidation has not been as thoroughly researched. In the Romny salt cupola, Ivanov et al. (1971) found single cells of *Thiobacillus thioparus* associated with one specimen of gypsum from the sub-salt breccia but, in general, a microbiologically reducing environment prevailed.

Occurrences of elemental sulfur in peat, coal, and petroleum are described in Chapter 6.4. The role of sulfate reducers in these environments is suggested by the fact that fossil fuels formed in marine environments, where sulfate is in abundant supply, have significantly more sulfide and native sulfur than those formed under freshwater conditions. In fact, a general geological feature of native sedimentary sulfur deposits is their location in sulfate-carbonate rocks and proximity to oil-gas-bearing strata and hydrologic zones where sulfate waters mix with chloride brines (Ivanov, 1964).

It may be noted that sulfides of volcanic origin are frequently oxidised to elemental sulfur apparently with bacterial participation. Lake Sernoe in the Kuibyshev region of the U.S.S.R. is fed by fumarolic springs, and elemental sulfur is deposited at a rate of over 100 kg d^{-1}. The presence of *Thiobacillus thioparus* suggested that microbial factors are involved in this oxidation (Ivanov, 1964; Sokolova, 1962). Ljunggren (1960) described Lake Ixpaco in Guatamala where the mud contained up to 60% elemental sulfur. This was thought to be a consequence of oxidation of volcanic sulfide by Beggiatoaceae. In hot acid springs in Yellowstone Park, Brock et al. (1976) reported significant populations of *Sulfolobus acidocaldaris*, an autotrophic organism capable of oxidising sulfur compounds at low pH and high temperature. On the other hand, Ivanov (1964) found that very few oxidisers were present in the Golovnin volcanic lakes, Kumashir Island, despite the presence of sulfate and sulfate-reducers. It is also noted that Sato (1960) did not find elemental sulfur crystals formed during inorganic oxidation of pyrite but rather, very reactive S_2 molecules. This supports the concept that the transformation of pyrites to sulfur in concretions or other environments is mainly by microbiological oxidation (Sass et al., 1965; Nissenbaum and Rafter, 1967).

REFERENCES

Akagi, J.M., Chan, M. and Adams, V., 1974. Observations on the bisulfite reductase (P582) isolated from *Desulfotomaculum nigrificans*. J. Bacteriol., 120: 240—244.

Al-Sawaf, F.D.S., 1977. Sulfate reduction and sulfur deposition in the lower Fars Formation, Northern Iraq. Econ. Geol., 72: 608—618.

Anderson, G.M., 1973. Hydrothermal transport and deposition of galena and sphalerite near 100°C. Econ. Geol., 68: 480—482.

Anderson, G.M., 1975. Precipitation of Mississippi Valley-type ores. Econ. Geol., 70: 937—942.

Andrusov, N.I., 1890. Preliminary report on participation in the Black Sea Deep expedition in 1890. Russk. Geogr. Obshch. Izv., 26: 398—409.

Arrhenius, G.O.S., Bramlette, M.N. and Picciotto, E.E., 1957. Localization of radioactive and stable heavy nuclides in ocean sediments. Nature, 180: 85—86.

Baas Becking, L.G.M. and Kaplan, I.R., 1956. The microbiological origin of the sulfur nodules of Lake Eyre. Trans. R. Soc. South Australia, 79: 52—65.

Baas Becking, L.G.M. and Moore, D., 1961. Biogenic sulphides. Econ. Geol., 56: 259—272.

Barghoorn, E.S. and Schopf, J.W., 1965. Microorganisms from the late Precambrian of Central Australia. Science, 150: 337—339.

Barghoorn, E.S. and Schopf, J.W., 1966. Microorganisms 3 billion years old from the Precambrian of South Africa. Science, 152: 758—763.

Barton, P.B., Jr., 1967. Possible role of organic matter in the precipitation of the Mississippi Valley ores. In: J.S. Brown (Editor), Genesis of Statiform Lead-Zinc-Barite-Fluorite Deposits (Mississippi Valley Type Deposits). Econ. Geol., Monograph 3, pp. 363—369.

Beales, F.W., 1975. Precipitation mechanisms of Mississippi Valley-type ore deposits. Econ. Geol., 70: 943—948.

Beijerinck, W.M., 1895. Über Spirillum desulfuricans als Ursache von Sulfat-reduction. Zentralbl. Bakteriol. Parasitenkd., 11: 1—9; 49—59; 104—114.

Belsky, T., Johns, R.B., McCarthy, E.D., Burlingame, A.L., Richter, W. and Calvin, M., 1965. Evidence of life in a sediment two and a half billion years old. Nature, 206: 446—447.

Bentor, Y.K., 1961. Some geochemical aspects of the Dead Sea. Geochim. Cosmochim. Acta, 25: 239—260.

Berner, R.A., 1964a. Iron sulfides formed from aqueous solution at low temperatures and atmospheric pressure. J. Geol., 72: 293—306.

Berner, R.A., 1964b. Distribution and diagenesis of sulfur in some sediments from the Gulf of California. Mar. Geol., 1: 117—140.

Berner, R.A., 1969. The synthesis of framboidal pyrite. Econ. Geol., 64: 383—384.

Berner, R.A., 1971. Principles of Chemical Sedimentology. Internat. Series in Earth and Planet. Science. McGraw-Hill, New York, NY, 240 pp.

Bogdanov, Yu.V., Golubchina, M.N., Prilutsky, R.E., Toksubayev, A.I. and Feoktistov, V.P., 1971. Some characteristics of sulfur isotopic composition of iron sulfides in Paleozoic Sedimentary Rocks at Dzhezkazgan (in Russian), Geokhimiya, 11: 1376—1378; in English: Geochem. Int., 1971, 856—858.

Booth, G.H. and Mercer, S.J., 1963. Resistance to copper of some oxidizing and reducing bacteria. Nature, 199: 622.

Boyle, R.W., Wanless, R.K. and Stevens, R.D., 1976. Sulfur isotope investigations of the barite, manganese and lead-zinc-copper-silver deposits of the Walton-Cheverie area, Nova Scotia, Canada. Econ. Geol., 71: 749—762.

Brock, T.D., 1966. Principles of Microbial Ecology. Prentice-Hall, New Jersey, 306 pp.

Brock, T.D., Cook, S., Petersen, S. and Mosser, J.L., 1976. Biogeochemistry and bacteriology of ferrous iron oxidation in geothermal habitats. Geochim. Cosmochim. Acta, 40: 493—500.

Bromfield, S.M., 1953. Sulfate reduction in partially sterilized soils exposed to air. J. Gen. Microbiol., 8: 378—390.

Brown, L.L. and Drury, J.S., 1967. Nitrogen isotope effects in the reduction of nitrate, nitrite and hydroxylamine to ammonia. I. In sodium hydroxide solution with Fe(II). J. Chem. Phys., 46: 2833—2837.

Brown, L.L. and Drury, J.S., 1969. Nitrogen isotope effects in the reduction of nitrate, nitrite, and hydroxylamine to ammonia. II. The MgO and $CuSO_4$ systems. J. Chem. Phys., 51: 3771—3775.

Butlin, K.R. and Postgate, J.R., 1954. The microbiological formation of sulfur in Cyrenaican lakes. In: J. Cloudsley-Thompson (Editor), Biology of Deserts. Symposium Inst. Biol., London, pp. 112—122.

Butlin, K.R., Adams, M.E. and Thomas, M., 1949. The isolation and cultivation of sulfate-reducing bacteria. J. Gen. Microbiol., 3: 46—59.

Campbell, L.L., 1975. Genus *Desulfotomaculum*. In: R.E. Buchanan and N.E. Gibbons (Editors) Bergey's Manual of Determinative Bacteriology, 8th ed. Williams and Wilkin Co., Baltimore, pp. 572—573.

Carroll, D., 1958. Role of clay minerals in the transportation of iron. Geochim. Cosmochim. Acta, 14: 1—27.

Chambers, L.A. and Trudinger, P.A., 1975. Are thiosulfate and trithionate intermediates in dissimilatory sulfate-reduction? J. Bacteriol., 123: 36—40.

Chambers, L.A., Trudinger, P.A., Smith, J.W. and Burns, M.S., 1975. Fractionation of sulfur isotopes by continuous cultures of *Desulfovibrio desulfuricans*. Can. J. Microbiol., 21: 1602—1607.

Cherest, H., Eichler, F. and deRobichon-Szulmajster, H., 1969. Genetic and regulatory aspects of methionine biosynthesis in *Saccharomyces cerevisiae*. J. Bacteriol., 97: 328—336.

Chukhrov, F.V., Vinogradov, V.I. and Ermilova, L.P., 1970. On the isotopic composition of some Precambrian strata. Miner. Deposita, 5: 209—222.

Cook, F.D., Wellman, R.P. and Krouse, H.R., 1973. Nitrogen isotope fractionation in the nitrogen cycle. Proceedings of the Symposium on Hydrochemistry and Biogeochemistry, Tokyo, Japan, Biogeochemistry, Vol. 2. Clarke Co., Washington, DC, pp. 49—64.

Czamanske, G.K. and Rye, R.O., 1974. Experimentally determined sulfur isotope fractionations between sphalerite and galena in the temperature range 600° to 275°C. Econ. Geol., 69: 17—25.

Danilchenko, P.T. and Chigirin, N.I., 1926. Origin of H_2S in the Black Sea. Acad. Sci. URSS, Trav. Lab. Zool. Sta. Biol., Sebastopol, Ser. 2.10: 141—191.

Davis, J.B. and Kirkland, D.W., 1970. Native sulfur deposition in the Castile Formation, Culberson County, Texas. Econ. Geol., 65: 107—121.

Davis, J.B., Stanley, J.P. and Custard, H.C., 1970. Evidence against oxidation of hydrogen sulfide by sulfate ions to produce elemental sulfur in salt domes. Am. Assoc. Pet. Geol. Bull., 54: 2444—2447.

Davis, P.D. and Feingold, D.S., 1962. Antimicrobial Agents: Mechanism of action and use in metabolic studies. In: E.C. Gunsalas and R.Y. Stanier (Editors), The Bacteria, Vol. 4, Academic, New York, NY, pp. 343—397.

Degens, E.T. and Ross, D.A., 1969. Hot Brines and Recent Heavy Metal Deposits in the Red Sea. Springer, New York, NY, 600 pp.

Dessau, G., Jensen, M.L. and Nakai, N., 1962. Geology and isotopic studies of Sicilian sulfur deposits. Econ. Geol., 57: 410—438.

Deuser, W.G., 1970. Carbon-13 in Black Sea waters and implications for the origin of hydrogen sulfide. Science, 168: 1575—1577.

deVito, P.C. and Dreyfuss, J., 1964. Metabolic regulation of adenosine triphosphate sulfurylase in yeast. J. Bacteriol., 88: 1341—1348.

Dickerson, R.E., 1971. The structure of cytochrome *c* and the rates of molecular evolution. J. Mol. Evol., 1: 26—45.

Donnelly, T.H., Lambert, I.B. and Dale, D.H., 1972. Sulphur isotope studies of the Mount Gunson Copper Deposits, Pernatty Lagoon, South Australia. Miner. Deposita, 7: 314—322.

Emerson, S., 1976. Early diagenesis in anaerobic lake sediments: chemical equilibria in interstitial waters. Geochim. Cosmochim. Acta, 40: 925—934.

Farrand, M., 1970. Framboidal sulphides precipitated synthetically. Miner. Deposita, 5: 237—247.

Feely, H.W. and Kulp, J.L., 1957. The origin of the Gulf Coast salt-dome sulfur deposits. Bull. Am. Assoc. Pet. Geol., 41: 1802—1853.

Ford, R.W., 1957. Sulphur Isotope Effects in Chemical and Biological Processes. Ph.D. Thesis, McMaster University, Hamilton, Ontario, Canada, 96 pp.

Freke, A.M. and Tate, D., 1961. The formation of magnetic iron sulfide by bacterial reduction of iron solutions. J. Biochem. Microbiol. Technol. Eng., 3: 29—39.

Galliher, E.W., 1933. The sulphur cycle in sediments. J. Sediment. Petrol., 3: 51—63.

Genovese, S., 1963. The distribution of the H_2S in the Lake of Faro (Messina) with particular regard to the presence of "red water". In: C.H. Oppenheimer (Editor), Marine Microbiology, C.C. Thomas, Springfield, IL, pp. 194—204.

Goldhaber, M.B. and Kaplan, I.R., 1974. The sulfur cycle. In: E.D. Goldberg (Editor), The Sea, Vol. 5. John Wiley, New York, NY, pp. 569—655.

Graf, D.L. and Anderson, D.E., 1975. Properties of one-dimensional diagenetic models. J. Geol., 83: 331—348.

Grinenko, L.N., Kononova, V.A. and Grinenko, V.A., 1970. Isotope composition of sulphide sulphur in carbonatites. Geochem. Int., 6: 45—53.

Hallberg, R.O., 1970. An apparatus for the continuous cultivation of sulfate-reducing bacteria and its application to geomicrobiological purposes. Antonie van Leeuwenhoek; J. Microbiol. Serol., 36: 241—254.

Hallberg, R.O., 1972. Sedimentary sulfide mineral formation: An energy circuit system approach. Miner. Deposita, 7: 189—201.

Harrison, A.G., 1957. Isotope Effects in the Reduction of Sulphur Compounds. Ph.D. Thesis, McMaster University, Hamilton, Ontario, Canada.

Harrison, A.G. and Thode, H.G., 1957. The kinetic isotope effect in the chemical reduction of sulphate. Trans. Faraday Soc., 53: 1648—1651.

Harrison, A.G. and Thode, H.G., 1958. Mechanism of bacterial reduction of sulphate from isotope fractionation studies. Trans. Faraday Soc., 54: 84—92.

Hartmann, U.M. and Nielsen, H., 1969. $\delta^{34}S$-Werte in rezenten Meeressedimenten und ihre Deutung am Beispiel einiger Sedimentprofile aus der westlichen Ostsee. Geol. Rundsch., 58: 621—655.

Hata, Y., 1960. Inorganic nutrition of marine sulfate-reducing bacteria. J. Shimonoseki Coll. Fisheries, 9: 39 (cited in Trüper, 1969).

Hattersley-Smith, G., Keys, J.E., Serson, H. and Mielke, J.E., 1970. Density stratified lakes in northern Ellesmere Island. Nature, 225: 55—56.

Hoefs, J., 1973. Stable Isotope Geochemistry. Springer, New York, NY, 140 pp.

Hunt, R.N., 1974. Oxygen Isotope Studies on Sulfates. Ph.D. Thesis, University of Alberta, Edmonton, Canada, 221 pp.

Hutchinson, G.E., 1957. A Treatise on Limnology. John Wiley, New York, NY, 1014 pp.

Igumnov, S.A., 1976. Experimental study of isotope exchange between sulphide and sulphate sulphur in hydrothermal solution (in Russian). Geokhimiya, 4: 497—503.

Issatschenko, B.L., 1929. Zur frage der biogenischen bildung des pyrits. Int. Rev. Gesamten Hydrobiol. Hydrogr., 22: 99—101.

Ivanov, M.V., 1960. Microbiological investigation of Carpathian sulfur deposits. I: Investigations of the Nemirov and Luben deposits. Mikrobiologiya, 29: 109—113.

Ivanov, M.V., 1964. Microbiological Processes in the Formation of Sulfur Deposits. Translated by Israel Program for Scientific Translations, 1968, 298 pp.

Ivanov, M.V. and Kostruva, M.F., 1961. Microbiological studies of Carpathian sulfur deposits. III. Hydrogen sulfide formation in Yazov sulfur field. Mikrobiologiya, 30: 130—134.

Ivanov, M.V. and Ryzhova, V.N., 1960. The rate of hydrogen sulfide production in some soils of the Barents Sea littoral. Dokl. Akad. Nauk. S.S.S.R., 130: 187—188.

Ivanov, M.V. and Ryzhova, V.N., 1961. Microbiological studies of Carpathian sulfur deposits. IV: Study of conditions for activity of sulfate-reducing bacteria in underground waters of Rozdol. Mikrobiologiya, 30: 280—285.

Ivanov, M.V., Lein, A.Yu., Grinenko, V.A., Galimov, E.M. and Vetrov, Yu.I., 1971. The origin of native sulfur in the caprock or Romny salt cupola. Lithology Miner. Resour. (U.S.S.R.), 6: 569—579.

Iya, K.K. and Sreenivasayi, M., 1944. A preliminary study of the bacterial flora associated with sulfur deposits on the East Coast (Masulipatam). Curr. Sci., 13: 316—317.

Iya, K.K. and Sreenivasayi, M., 1945. Studies in the formation of sulfur at Kona (Masulipatam). Curr. Sci., 14: 243—244.

Jannasch, H.W., Trüper, H.G. and Tuttle, J.H. (1973). Microbial cycle in Black Sea. In: E.T. Degens and D.A. Ross (Editors), The Black Sea — Geology, Chemistry, Biology. A.A.P.G. Mem. 20, pp. 419—425.

Jensen, M.L. and Whittles, A.W.G., 1969. Sulfur isotope of the Nairne pyrite deposit, South Australia. Miner. Deposita, 4: 241--247.

Jobson, A.M., 1975. Physiological Characterisation of a *Desulphovibrio sp.* Isolated from Crude Oil. Ph.D. Thesis, University of Alberta, 194 pp.

Jones, G.E. and Starkey, R.L., 1957. Fractionation of the stable isotopes of sulphur by microorganisms and their role in deposition of native sulphur. Appl. Microbiol., 5: 111—118.

Jones, H.E. and Skyring, G.W., 1975. Effect of enzymic assay conditions on sulfite reduction catalyzed by desulfoviridin from *Desulfovibrio gigas*. Biochim. Biophys. Acta, 377: 52—60.

Jones, H.E., Trudinger, P.A., Chambers, L.A. and Pyliotis, N.A., 1976. Metal accumulation by bacteria with particular reference to dissimilatory sulfate-reducing bacteria. Z. Allg. Mikrobiol., 16: 425—435.

Kajiwara, Y. and Krouse, H.R., 1971. Sulfur isotope partitioning in metallic sulfide systems. Can. J. Earth Sci., 8: 1397—1408.

Kaplan, I.R., 1975. Stable isotopes as a guide to biogeochemical processes. Proc. R. Soc., B-189: 183—211.

Kaplan, I.R. and Rittenberg, S.C., 1964. Microbiological fractionation of sulfur isotopes. J. Gen. Microbiol., 34: 195—212.

Kaplan, I.R., Emery, K.O. and Rittenberg, S.C., 1963. The distribution and isotopic abundance of sulfur in recent marine sediments off Southern California. Geochim. Cosmochim. Acta, 27: 297—331.

Kaplan, I.R., Rafter, T.A. and Hulston, J.R., 1960. Sulphur isotopic variations in nature. Part 8: Application to some biogeochemical problems. N.Z. J. Sci., 3: 338—361.

Kaplan, I.R., Sweeney, R.E. and Nissenbaum, A., 1969. Sulfur isotope studies on Red Sea geothermal brines and sediments. In: E.T. Degens and D.A. Ross (Editors), Hot Brines and Recent Heavy Metal Deposits in the Red Sea. Springer, New York, NY, pp. 263—271.

Karavaiko, G.I., Ivanov, M.V. and Pomerants, L.B., 1963. Microbiological investigation of a Kara Kum sulphur deposit. Izv. Akad. Nauk, SSSR, Ser. Biol., 2: 249—260.

Kemp, A.L.W. and Thode, H.G., 1968. The mechanism of the bacterial reduction of sulphate and of sulphite from isotope fractionation studies. Geochim. Cosmochim. Acta, 32: 71—91.

Klein, R.M. and Cronquist, A., 1967. A consideration of the evolutionary and taxonomic significance of some biochemical, micromorphological and physiological characters in the thallophytes. Q. Rev. Biol., 42: 105—296.

Kline, B.C. and Schoenhard, D.E., 1969. Accumulation of sulfite by a sulfate-using revertant of *Salmonella pullorum*. J. Bacteriol., 100: 365—369.

Kobayashi, K., Seki, Y. and Ishimoto, M., 1974. Biochemical studies on sulfate-reducing bacteria. XIII: Sulfite reductase from *Desulfovibrio vulgaris* — mechanism of trithionate, thiosulfate and sulfide formation and enzymatic properties. J. Biochem., Tokyo, 75: 519—529.

Kobayashi, K., Tachibana, S. and Ishimoto, M., 1969. Intermediary formation of trithionate in sulfite reduction by sulfate-reducing bacterium. J. Biochem., Tokyo, 65: 155—157.

Kobayashi, K., Takahashi, E. and Ishimoto, M., 1972. Biochemical studies on sulfate-reducing bacteria. XI: Purification and some properties of sulfite reductase, desulfoviridin. J. Biochem., 72: 879—887.

Kovalev, V.A. and Generalova, V.A., 1974. Modeling of sulphide formation from iron humate complexes (in Russian). Geokhimiya, 3: 425—533.

Kramarenko, L.E., 1962. Bacterial biogenesis in underground waters of some mineral fields and their geological importance. Mikrobiologiya, 31: 694—701.

Kriss, A.E. and Rukina, E.A., 1949. Reductive and oxidative processes in the sulfurated region of the Black Sea (in Russian). Mikrobiologiya, 18: 402—415.

Krouse, H.R., 1976. Sulphur isotope variations in thermal and mineral waters. Proc. Int. Symp. on Water-Rock Interaction (IAGC), Prague, 1974: 340—347.

Krouse, H.R. and Sasaki, A., 1968. Sulphur and carbon isotope fractionation by *Salmonella heidelberg* during anaerobic sulphite reduction in trypticase soy broth medium. Can. J. Microbiol., 14: 417—422.

Krouse, H.R., Brown, H.M. and Farquharson, R.B., 1977. Sulphur isotope compositions of sulphides and sulphates. D.S.D.P. Leg 37. Can. J. Earth Sci., 14: 787—793.

Krouse, H.R., Cook, F.D., Sasaki, A. and Smejkal, V., 1970. Microbiological fractionation in springs of Western Canada. In: Recent Developments in Mass Spectroscopy, Proc. Int. Conf. on Mass Spect., Kyoto, Japan, University of Tokyo Press, Tokyo, pp. 629—639.

Krouse, H.R., McCready, R.G.L., Husain, S.A. and Campbell, J.N., 1967. Sulphur isotope fractionation by *Salmonella sp.* Can. J. Microbiol., 13: 21—25.

Krouse, H.R., McCready, R.G.L., Husain, S.A. and Campbell, J.N., 1968. Sulphur isotope fractionation kinetic studies of sulfite reduction in growing cells of *Salmonella heidelberg*. Biophys. J., 8: 109—124.

Krouse, H.R. and Sasaki, A., 1968. Sulphur and carbon isotope fractionation by *Salmonella heidelberg* during anaerobic sulphite reduction in trypticase soy broth medium. Can. J. Microbiol., 14: 417—422.

Kuznetsov, S.I., Ivanov, M.V. and Lyalikova, N.N., 1963. Introduction to Geological Microbiology. McGraw-Hill, New York, NY, 252 pp.

Laishley, E.J., McCready, R.G.L., Bryant, R. and Krouse, H.R., 1976. Stable isotope fractionation by *Clostridium pasteurianum*. In: J.O. Nriagu (Editor), Environmental Biogeochemistry. Ann Arbor Science Publ., Ann Arbor, MI, pp. 327—349.

Lambert, I.B. and Bubela, B., 1970. Banded sulphide ores: the experimental production of monomineralic sulphide bands in sediments. Miner. Deposita, 5: 97—102.

Lambert, I.B., McAndrew, J. and Jones, H.E., 1971. Geochemical and bacteriological studies of the cupriferous environment at Pernatty Lagoon, South Australia. Proc. Aust. Inst. Mining Met., 240: 15—23.

Lasaga, A.C. and Holland, H.D., 1975. Mathematical aspects of nonsteady-state diagenesis. Geochim. Cosmochim. Acta, 40: 257—266.

Lebedintsev, A.A., 1893. Preliminary report on chemical studies of the Black and Azov Seas in the autumn of 1891. Zap. Novoross. Obshch. Estetv., 16: 149.

Lee, J.P. and Peck, H.D., 1971. Purification of the enzyme reducing bisulfite to trithionate from *Desulfovibrio gigas* and its identification as desulfoviridin. Biochem. Biophys. Res. Commun., 45: 583—589.

LeGall, J. and Postgate, J.R., 1973. The physiology of sulfate-reducing bacteria. Adv. Microbiol. Physiol., 10: 81—133.

Lein, A.Yu., Ivanov, M.V., Matrosov, A.G. and Zyakun, A.M., 1976. Genesis of sulphur-calcite ores of the Gaurdak deposit. Geokhimiya, 3: 422—432.

Leleu, M. and Goni, J., 1974. Sur la formation biogéochimique de stalactites de galène. Miner. Deposita, 9: 27—32.

Leleu, M., Gugalski, T. and Goni, J., 1975. Synthèse de wurtzite par voie bacteriénne. Miner. Deposita, 10: 323—329.

Ljunggren, P., 1960. A sulfur mud deposit formed through bacterial transformation of fumarolic hydrogen sulfide. Econ. Geol., 55: 531—538.

Love, L.G., 1965. Micro-organic material with diagenetic pyrite from the Lower Proterozoic Mount Isa shale and a carboniferous shale. Proc. Yorkshire Geol. Soc., 35: 187—202.

Lovering, T.S., 1963. Deposition of sulfides from sulfur-deficient solutions. Econ. Geol., 56: 68—97.

Lyalikova, N.N. and Sokolova, G.A., 1965. A microbiological characterization of some ore deposits in central Kazakstan. Mikrobiologiya, 34: 335—343.

MacNamara, J. and Thode, H.G., 1951. The distribution of S^{34} in nature and the origin of native sulfur deposits. Research, 4: 582—584.

McCready, R.G.L., 1975. Sulfur isotope fractionation by *Desulfovibrio* and *Desulfotomaculum* species. Geochim. Cosmochim. Acta, 39: 1395—1401.

McCready, R.G.L., Kaplan, I.R. and Din, G.A., 1974. Fractionation of sulfur isotopes by the yeast *Saccharomyces cerevisiae*. Geochim. Cosmochim. Acta, 38: 1239—1253.

McCready, R.G.L., Laishley, E.J. and Krouse, H.R., 1975. Stable isotope fractionation by *Clostridium pasteurianum*. I: $^{34}S/^{32}S$: Inverse isotope effects during SO_4^{2-} and SO_3^{2-} reduction. Can. J. Microbiol., 21: 235—244.

McCready, R.G.L., Laishley, E.J. and Krouse, H.R., 1976. Biogeochemical implications of inverse sulfur isotope effects during reduction of sulfur compounds by *Clostridium pasteurianum*. Geochim. Cosmochim. Acta, 40: 979—981.

Miller, L.P., 1949. Rapid formation of high concentrations of hydrogen sulfide by sulfate reducing bacteria. Contrib. Boyce Thompson Inst., 15: 437—465.

Miller, L.P., 1950a. Tolerance of sulfate-reducing bacteria to hydrogen sulfide. Contrib. Boyce Thompson Inst., 16: 73—83.

Miller, L.P., 1950b. Formation of metal sulfides through the activities of sulfate-reducing bacteria. Contrib. Boyce Thompson Inst., 16: 85—89.

Mitchell, R.H. and Krouse, H.R., 1975. Sulphur isotope geochemistry of carbonatites. Geochim. Cosmochim. Acta, 39: 1505—1513.

Nakai, N. and Jensen, M.L., 1964. The kinetic isotope effect in the bacterial reduction and oxidation of sulfur. Geochim. Cosmochim. Acta, 28: 1893—1912.

Nakai, N., Wada, H., Kiyosu, Y. and Takimoto, N., 1975. Stable isotope studies on the origin and geological history of water and salts in the Lake Vanda area, Antarctica. Geochem. J., 9: 7—24.

Neev, D. and Emery, K.O., 1967. The Dead Sea. Geol. Surv. Israel Bull., 41: 147 pp.

Nissenbaum, A., 1975. The microbiology and biogeochemistry of the Dead Sea. Microbiol. Ecol. 2: 139—161.

Nissenbaum, A. and Kaplan, I.R., 1966. Origin of the Berri (Israel) sulfur deposit. Chem. Geol., 1: 295—316.

Nissenbaum, A. and Kaplan, I.R., 1976. Sulfur and carbon isotopic evidence for biogeochemical processes in the Dead Sea ecosystem. In: J.O. Nriagu (Editor), Environmental Biogeochemistry, Ann Arbor Sci. Publ., Ann Arbor, MI, pp. 309—325.

Nissenbaum, A. and Rafter, T.A., 1967. Sulfur isotopes in altered pyrite concretions from Israel. J. Sediment. Petrol., 37: 961—962.

Nissenbaum, A. and Swaine, D.J., 1976. Organic matter-metal interactions in recent sediments: the role of humic substances. Geochim. Cosmochim. Acta, 40: 809—816.

Nissenbaum, A., Presley, B.J. and Kaplan, I.R., 1972. Early diagenesis in a reducing fjord, Saanich Inlet, British Columbia. I. Chemical and isotopic changes in major components of interstitial water. Geochim. Cosmochim. Acta, 36: 1007—1027.

Nriagu, J., 1968. Sulfur metabolism and sedimentary environment: Lake Mendota, Wisconsin. Limnol. Oceanogr., 13: 430—439.

Oberlies, F. and Prashnowsky, A.A., 1968. Biogeochemische und elektronen-mikroskopische Untersuchung prakambrischer Gesteine. Naturwissenschaften, 55: 25—28.

Oehler, D.Z., Schopf, J.W. and Kvendolden, K.A., 1972. Carbon isotopic studies of organic matter in Precambrian rocks. Science, 175: 1246—1248.

Ohmoto, H., 1972. Systematics of sulfur and carbon isotopes in hydrothermal ore deposits. Econ. Geol., 67: 551—578.

Oltmann, L.F., Van der Beek, E.G. and Stouthamer, A.H., 1975. Reduction of inorganic sulfur compounds by facultatively aerobic bacteria. Plant Soil, 43: 153—169.

Orr, W.L. and Gaines, A.G., Jr., 1973. Observations on rate of sulfate reduction and organic matter oxidation in the bottom waters of an estuarine basin: the upper basin of the Pettaquamscutt River (Rhode Island). In: Advances in Organic Geochemistry. Editions Technip, Paris, France, pp. 791—812.

Peck, H.D., 1962. Symposium on metabolism of inorganic compounds V. Comparative metabolism of inorganic sulfur compounds in microorganisms. Bacteriol. Rev., 26: 67—94.

Peck, H.D., 1968. Energy-coupling mechanisms in chemolithotrophic bacteria. Annu. Rev. Microbiol., 22: 489—518.

Pelczar, M.J. and Reid, R.D., 1958. Microbiology. McGraw-Hill, New York, NY. 229 pp.

Pollock, M.R., Knox, R. and Gell, P.G.H., 1942. Bacterial reduction of tetrathionate. Nature, 150: 94.

Postgate, J.R., 1951. The reduction of sulfur compounds by *Desulphovibrio desulphuricans*. J. Gen. Microbiol., 5: 725—738.

Postgate, J.R., 1956. Iron and the synthesis of cytochrome C_3. J. Gen. Microbiol., 15: 186—193.

Postgate, J.R., 1959. Sulfate reduction by bacteria. Annu. Rev. Microbiol., 13: 505—520.

Postgate, J.R., 1965. Recent advances in the study of sulfate-reducing bacteria. Bacteriol. Rev., 29: 425—441.

Postgate, J.R., 1968. The sulphur cycle. In: G. Nickless (Editor), Inorganic Sulphur Chemistry. Elsevier, Amsterdam, pp. 259—279.

Postgate, J.R., 1975. Genus *Desulfovibrio*. In: R.E. Buchanan and N.E. Gibbons (Editors), Bergey's Manual of Determinative Bacteriology, 8th edn. Williams and Wilkins Co., Baltimore, pp. 418—420.

Presley, B.J., Kolodny, Y., Nissenbaum, A. and Kaplan, I.R., 1972. Early diagenesis in a reducing fjord, Saanich Inlet, British Columbia. II: Trace element distribution in interstitial water and sediment. Geochim. Cosmochim Acta, 36: 1073—1090.

Redfield, A.C., Ketchum, B.H. and Richards, F.H., 1963. The influence of organisms on the composition of seawater. In: M.H. Hill (Editor), The Sea, Vol. 2. Interscience, New York, NY, pp. 26—27.

Rees, C.E., 1973. A steady-state model for sulphur isotope fractionation in bacterial reduction processes. Geochim. Cosmochim. Acta, 37: 1141—1162.

Rickard, D.T., 1969. The chemistry of iron sulphide formation at low temperatures. Stockholm Contr. Geol., 20: 67—95.

Rickard, D.T., 1970. The origin of framboids. Lithos, 3: 269—293.

Rickard, D.T., 1970. Covellite formation in low temperature aqueous solutions. Miner. Deposita, 7: 180—188.

Rickard, D.T., 1973. Limiting conditions for synsedimentary sulfide ore formation. Econ. Geol., 68; 605—617.

Robbins, P.W. and Lipmann, F., 1958. Enzymatic synthesis of adenosine-5′-phosphosulfate. J. Biol. Chem., 233: 686—690.

Roberts, W.M.B., Walker, A.L. and Buchanan, A.S., 1969. The chemistry of pyrite formation in aqueous solution and its relation to the depositional environment. Miner. Deposita, 4: 18—29.

Robinson, B.W., 1978. Sulphur isotope equilibria between sulphur solute species at high temperature. Proc. Int. Conf. on Stable Isotopes, Lower Hutt, N.Z., Aug. 1976, D.S.I.R. Bull., 220: 203—206.

Roedder, E., 1967. Environment of deposition of stratiform (Mississippi Valley-type) ore deposits, from studies of fluid inclusions. In: J.B. Brown (Editor), Genesis of Stratiform Lead-Zinc-Barite-Fluorite Deposits (Mississippi Valley-type Deposits). Econ. Geol. Monograph, 3: 349—361.

Roy, A.B. and Trudinger, P.A., 1970. The Biochemistry of Inorganic Sulfur Compounds. Cambridge University Press, London, 400 pp.

Rye, R.O. and Ohmoto, H., 1974. Sulfur and carbon isotopes and ore genesis: A review. Econ. Geol., 69: 826—842.

Ryznar, G., Campbell, F.A. and Krouse, H.R., 1967. Sulfur isotopes and the origin of the Quemont ore body. Econ. Geol., 62: 664—678.

Sakai, H., 1957. Fractionation of sulfur isotopes in nature. Geochim. Cosmochim. Acta, 12: 150—169.

Sakai, H., 1971. Sulfur and oxygen isotopic study of barite concretions from banks in the Japan Sea off the Northeast Honshu, Japan. Geochem. J., 5: 79—93.

Sangster, D.F., 1968. Relative sulphur isotope abundances of ancient seas and strata-bound sulphide deposits. Proc. Geol. Assoc. Can., 19: 79—91.

Sasaki, A. and Krouse, H.R., 1969. Sulfur isotopes and the Pine Point lead-zinc mineralization. Econ. Geol., 64: 718—730.

Sass, E., Nathan, Y. and Nissenbaum, A., 1965. Mineralogy of certain pyrite concretions from Israel and their alteration products. Mineral. Mag., 35: 84—87.

Sato, M., 1960. Oxidation of sulfide ore bodies. II: Oxidation mechanisms of sulfide minerals at 25°C. Econ. Geol., 55: 1202—1231.

Schiff, J.A. and Hodson, R.C., 1973. The metabolism of sulfate. Annu. Rev. Plant Physiol., 24: 381—414.

Schneiderhöhn, H., 1923. Chalkoeraphishe Untersuchung des Mansfelder Kupferschiefers. Neues Jahrb. Mineral. Geol. Palaeontol., Beilage-B, 47: 1—38.

Schopf, J.W., 1975. Precambrian paleobiology: Problems and perspectives. Annu. Rev. Earth Planet. Sci., 3: 213—249.

Schouten, C., 1946. The role of sulfur bacteria in the formation of the so-called sedimentary copper ores and pyritic ore bodies. Econ. Geol., 41: 517—538.

Schwarcz, H.P., 1973. Sulfur isotope analyses of some Sudbury, Ontario ores. Can. J. Earth Sci., 10: 1444—1459.

Schwarcz, H.P. and Burnie, S.W., 1973. Influence of sedimentary environments on sulfur isotope ratios in clastic rocks: a review. Miner. Deposita, 8: 264—277.

Shima, M., Gross, W.H. and Thode, H.G., 1963. Sulphur isotope abundances in basic sills, differentiated granites and meteorites. J. Geophys. Res., 68: 2835—2847.

Siegel, L.M., 1975. Biochemistry of the sulfur cycle. In: D.M. Greenberg (Editor), Metabolic Pathways, Vol. 7, Metabolism of Sulfur Compounds. Academic, New York, NY, pp. 217—286.

Silverman, M.P. and Ehrlich, H.L., 1964. Microbial formation and degradation of minerals. Adv. Appl. Microbiol., 6: 153—206.

Skinner, B.J., 1967. Precipitation of Mississippi Valley-type ores. A possible mechanism. In: J.S. Brown (Editor), Genesis of Stratiform Lead-Zinc-Barite-Fluorite Deposits (Mississippi Valley-type Deposits). Econ. Geol. Monograph, 3: 363—369.

Šmejkal, V., Cook, F.D. and Krouse, H.R., 1971a. Studies of sulphur and carbon isotope fractionation with microorganisms isolated from springs of Western Canada. Geochim. Cosmochim. Acta, 35: 787—800.

Šmejkal, V., Michalicek, M. and Krouse, H.R., 1971b. Sulphur isotope fractionation in some springs of the Carpathian mountain system in Czechoslovakia. Cazs. Mineral. Geol., 16: 275—283.

Smith, J.W., Doolan, S. and McFarlane, E.F., 1977. A sulfur isotope geothermometer. Chem. Geol., 19: 83—90.

Sokolova, G.A., 1960. Microbiological production of sulfur from sulfide seam waters. Mikrobiologiya, 29: 888—893.

Sokolova, G.A., 1962. Microbiological sulfur formation in Sulfur Lake. Mikrobiologiya, 31: 324—327.

Sokolova, G.A. and Karavaiko, G.I., 1964. Physiology and Geochemical Activity of Thiobacilli. Israel Program for Sci. Transl., Jerusalem, 1968, 283 pp.

Sorokin, Y.I., 1962. Experimental investigation of bacterial sulphate reduction in the Black Sea using ^{35}S. Mikrobiologiya, 31: 329—335.

Sorokin, Y.I., 1964. On the primary production and bacterial activities in the Black Sea. J. Cons. Int. Explor. Mer., 29: 41—60.

Sorokin, Y.I., 1970. Interrelations between sulphur and carbon turnover in meromictic lakes. Arch. Hydrobiol., 66: 391—446.

Spiro, B., 1977. Bacterial sulphate reduction and calcite precipitation in hypersaline deposition of bituminous shales. Nature, 269: 235—237.

Stanton, R.L., 1972. Ore Petrology. McGraw-Hill, New York, NY, 713 pp.

Strakhov, N.M., 1972. The balance of reducing processes in the sediments of the Pacific Ocean. Lithol. Miner. Resourc. (U.S.S.R.), 7: 455—479.

Stuiver, M., 1967. The sulfur cycle in lake waters during thermal stratification. Geochim. Cosmochim. Acta, 31: 2151—2167.

Sunagawa, I., Endo, Y. and Nakai, N., 1971. Hydrothermal synthesis of framboidal pyrite. Soc. Min. Geol. Japan, Spec. Issue, 2: 10—14.

Sweeney, R.E. and Kaplan, I.R., 1973. Pyrite framboid formation: laboratory synthesis and marine sediments. Econ. Geol., 68: 618—634.

Tarr, H.L.A., 1933. The enzymic formation of hydrogen sulphide by certain heterotrophic bacteria. Biochem. J., 27: 1869—1874.

Temple, K.L., 1964. Syngenesis of sulfide ores: An evaluation of biochemical aspects. Econ. Geol., 59: 1473—1491.

Temple, K.L. and LeRoux, N.W., 1964a. Syngenesis of sulfide ores: Sulfate-reducing bacteria and copper toxicity. Econ. Geol., 59: 271—278.

Temple, K.L. and LeRoux, N.W., 1964b. Syngenesis of sulfide ores: Desorption of adsorbed metal ions and their precipitation as sulfides. Econ. Geol., 59: 647—655.

Thode, H.G. and Monster, J., 1965. Sulfur isotope geochemistry of petroleum evaporites and ancient seas. Am. Assoc. Petrol. Geol. Mem., 4: 367—377.

Thode, H.G., Dunford, H.B. and Shima, M., 1962. Sulfur isotope abundances in rocks of the Sudbury District and their geological significance. Econ. Geol., 57: 565—578.

Thode, H.G., Macnamara, J. and Collins, C.B., 1949. Natural variations in the isotopic composition of sulfur and their significance. Can. J. Res., B27: 361—373.

Thode, H.G., Macnamara, J. and Fleming, W.H., 1953. Sulfur isotope fractionation and geological and biological time scales. Geochim. Cosmochim. Acta, 3: 235—243.

Thode, H.G., Monster, J. and Dunford, H.B., 1961. Sulfur isotope geochemistry. Geochim. Cosmochim. Acta, 25: 159—174.

Thode, H.G., Wanless, R.K. and Wallouch, R., 1954. The origin of native sulphur deposits from isotope fractionation studies. Geochim. Cosmochim. Acta, 5: 286—298.

Thorstenson, D.C., 1970. Equilibrium distribution of small organic molecules in natural waters. Geochim. Cosmochim. Acta, 34: 745—770.

Trudinger, P.A., 1969. Assimilatory and dissimilatory metabolism of inorganic sulfur compounds by microorganisms. Adv. Microb. Physiol., 3: 111—158.

Trudinger, P.A. and Chambers, L.A., 1973. Reversibility of bacterial sulfate reduction and its relevance to isotope fractionation. Geochim. Cosmochim. Acta, 37: 1775—1778.

Trudinger, P.A., Lambert, I.B. and Skyring, G.W., 1972. Biogenic sulfide ores: a feasibility study. Econ. Geol., 67: 1114—1127.

Trüper, H.G., 1969. Bacterial sulfate reduction in the Red Sea hot brines. In: E.T. Degens and D.A. Ross (Editors), Hot Brines and Recent Heavy Metal Deposits in the Red Sea. Springer, New York, NY, pp. 263—271.

Tudge, A.P. and Thode, H.G., 1950. Thermodynamic properties of isotopic compounds of sulfur. Can. J. Res., 28: 567—578.

Vinogradov, A.P., Grinenko, V.A. and Ustinov, V.I., 1962. Isotopic composition of sulfur compounds in the Black Sea. Geochem. Int., 5: 975—997.

Vinogradov, A.P., Grinenko, V.A. and Ustinov, V.I., 1964. Isotopic composition of the sulphur and carbon in ore of the Shor-su deposit of Uzbekistan (in Russian). Geokhimiya, 11: 1075—1085.

Volkov, I.I., 1961. Iron sulfides, their interdependence and transformation in the Black Sea bottom sediments. Tr. Inst. Okeanol. Akad. nauk SSSR, 50: 68—92.

Volkov, I.I., Rozznov, A.G., Zhabina, N.N. and Yagodinskaya, T.A., 1972. Sulfur in sediments of the Pacific Ocean east of Japan. Litol. Polezn. Iskop., 4: 50—64.

Vosjan, J.H. and Van der Hoek, G.J., 1972. A continuous culture of *Desulfovibrio* on a medium containing mercury and copper ions. Neth. J. Sea Res., 5: 440—444.

Wainwright, T., 1970. Hydrogen sulfide production by yeast under conditions of methionine, pantothenate, or vitamin B_6 deficiency. J. Gen. Microbiol., 61: 107—119.

Wallhaüsser, K.H. and Puchelt, H., 1970. Sulfate-reducing bacteria in sulfur springs and underground water of Germany and Austria. Contr. Miner. Petrol., 13: 12—30.

Ward, J.C., 1970. The structure and properties of some iron sulfides. Rev. Pure Appl. Chem., 20: 175—206.

Watson, S.W. and Waterbury, J.B., 1969. The sterile hot brines of the Red Sea. In: E.T. Degens and D.A. Ross (Editors), Hot Brines and Recent Heavy Metal Deposits in the Red Sea. Springer, New York, NY, pp. 272—281.

Wellman, R.P., 1969. Nitrogen Isotope Fractionation in Chemical and Microbiological Systems. Ph.D. Thesis, University of Alberta, Edmonton, Canada, 159 pp.

Wellman, R.P., Cook, F.D. and Krouse, H.R., 1968. Microbiological alteration of the ^{15}N abundance. Science, 161: 269—270.

ZoBell, C.E., 1958. Ecology of sulfate-reducing bacteria. Prod. Mon., 22: 12—29.

ZoBell, C.E., 1961. Importance of microorganisms in the sea. Proc. Low Temp. Microbiol. Symp., Camden, NJ, pp. 107—132.

Chapter 6.3

OXIDATIVE REACTIONS IN THE SULFUR CYCLE

B.J. RALPH

School of Biological Technology, The University of New South Wales, P.O. Box 1, Kensington, N.S.W. 2033 (Australia)

CONTENTS

INTRODUCTION

Of the numerous sulfide and polysulfide compounds which the majority of the chemical elements will form with sulfur under appropriate conditions (Jellinek, 1968), only about two hundred and fifty possess physical and chemical characteristics which enable them to accumulate and persist as stable, crystalline phases in the geological environment. The mineral sulfides are comparatively stable compounds but, even in the protected situations of deposits at depth, they may undergo phase changes and electrochemical transformations (Nickel et al., 1974). In more accessible regions of natural ore deposits, or following the extraction of ores in mining operations, interactions with water, oxygen, carbon dioxide and other chemical substances occur with sulfide minerals, and degradative reactions proceed at relatively high rates. Such transformations of sulfide minerals are generally manifested by the generation of acidity, the conversion of the sulfur moiety to sulfate and the solubilization of metallic components. The involvement of biological agencies in the accelerated degradation of sulfide minerals has been unequivocally demonstrated by a number of investigators (Rudolfs and Helbronner, 1922; Carpenter and Herndon, 1933; Colmer and Hinkle, 1947).

Their conclusions have relied essentially upon comparisons of the kinetics of degradation under sterile and non-sterile conditions, and the consistent presence in non-sterile laboratory experiments and in active field situations of a limited number of microorganism species whose biochemical and physiological characteristics are compatible with the environmental situations which apply. The organisms most commonly associated with sulfide mineral degradation are *Thiobacillus* spp. which derive energy from the oxidation of reduced sulfur compounds and iron-oxidizing organisms such as *Thiobacillus ferrooxidans* and *Metallogenium* spp.

It is highly characteristic of sulfide mineral degradation in the field that a gradual but accelerating increase of acidity in the milieu occurs, eventually stabilizing at levels determined by a complex array of factors, and persisting at these levels for long periods of time. Frequently the increase in acidity is roughly paralleled by solubilization of metals. Less obviously, but of the greatest importance with respect to the understanding of the mechanisms involved, a succession of microbial populations is established, each component of which prepares the way for the development of the next step in the overall process. In general terms, the extent to which the sulfide mineral degradation proceeds depends very largely on the sequential development of the microbial succession.

Field situations are normally extremely complex. Mineral deposits differ in composition: the number and relative abundance of individual sulfide minerals vary widely, as do also the associated gangue minerals (see, for example, Roberts, 1960, Lawrence and Savage, 1975). The development of the microbial succession is influenced by the physical and chemical characteristics of the mineral components, both sulfidic and non-sulfidic, by climatic circumstances such as temperature range and rainfall, and by the availability at reaction sites of water, oxygen, carbon dioxide and other materials, both in their roles as chemical reactants and as microbial nutrients. Situations can arise in which further development of the microbial succession is inhibited and in such cases, extensive degradation of the sulfide minerals is unlikely to occur. If the development of the succession proceeds, two general situations can arise, each of which can lead to greatly accelerated rates of breakdown. Firstly, at acidity levels below about pH 4.5, microbial populations can develop which can derive energy for growth from the oxidation of ferrous iron. The catalytic ability of these organisms for the production of ferric ions outstrips that of any other agents and provides a continuing supply of a most potent reagent for the oxidative degradation of a wide range of sulfide minerals. Secondly, the continuing development of microbial populations in successions up to and including the vigorous iron-oxidizing types generates considerable heat energy and, within the confines of a poorly-heat-conducting rock mass can raise the temperature to a level at which a diversity of abiotic chemical reactions involving sulfides can occur at significant rates.

Comprehensive and quantitative models of the sequence of events which occur in a sulfidic ore mass during degradation of the sulfide minerals have not yet been evolved but a great deal of information has accumulated over the past thirty years on particular facets of the overall processes. Apart from the intrinsic interest and the fundamental importance of processes which are involved in the geochemical cycling of sulfur and various metallic elements, the mechanisms of sulfide mineral degradation and the factors which influence them are of increasing practical significance in respect of the recovery of metals from low-grade ores and in the limitation and control of environmental damage by acid and heavy metal pollution.

MICROBIAL POPULATIONS ASSOCIATED WITH SULFIDE MINERAL DEGRADATION

The exposure of sulfide-bearing material to water and the atmosphere provides habitats for the build-up of microbial populations which resemble those in pyritic soils (Bloomfield, 1972) but differ in significant respects from those in the more common soil types (Vitolins and Swaby, 1969). The complex associations which develop are dominated by organisms with sulfur- and iron-oxidizing abilities, and with tolerance to high levels of acidity and metal ion concentration (Rao and Berger, 1971; Tuovinen et al., 1971a, b; Tuovinen and Kelly, 1974a, 1974b, 1974c; Imai et al., 1973), together with a diverse array of metal and acid-tolerant heterotrophic bacteria, fungi, yeasts, algae and protozoa (Joseph, 1953; Marchlewitz and Schwartz, 1961; Kuznetsov et al., 1963; Ehrlich, 1963; Moss and Andersen, 1968; Tuttle et al., 1968; Arrieta and Grez, 1971; Lundgren et al., 1972; Updegraff and Duoros, 1972; Bhurat et al., 1973; Belly and Brock, 1974; Dugan, 1975; Groudev et al., 1978; Madgwick and Ralph, 1977). Typically, the microbial populations vary qualitatively and quantitatively with time until a steady-state situation is reached, this persisting until the sulfide minerals are depleted or some other controlling factor imposes a long-term stability. Considerable variations in the microbial populations due to climatic patterns, can be superimposed upon the general trends (Khalid and Ralph, 1976). Concomitantly, acidity is generated by the activities of the sulfur-oxidizing organisms and the pH of percolating waters and effluent solutions gradually falls. Associated non-sulfidic minerals, usually present in overwhelming excess, exert a controlling influence and may stabilize the pH at a number of different levels. If this influence is not paramount and other factors such as oxygen availability do not become limiting, the pH can fall to very low levels (pH 2 or lower), provided that the absolute amount of sulfide mineral is adequate. Chemical modifications of sulfide minerals occur along with the oxidation of sulfur moieties and the generation of acidity, but substantial release of metals in soluble forms in effluent waters is characteristically associated with pH levels below 4.

The principal microbial populations whose activities generate acidity are members of the *Thiobacillus* group (Table 6.3.1) but, under some conditions, other types of organisms may contribute (Table 6.3.2). The temperature within sulfide-bearing rock masses may rise considerably above ambient (Ehrlich and Fox, 1967; Beck, 1967) and elemental sulfur is a product of some of the reactions occurring during sulfide degradation. Under such circumstances, some formation of sulfuric acid could arise from the activities of *Sulfolobus acidocaldarius* (Brock et al., 1972; Fliermans and Brock, 1972; Shivvers and Brock, 1973; Bohlool, 1975) or *Sulfolobus*-like organisms (De Rosa et al., 1975). It is of interest to note that the chemoautotrophic and thermophilic microorganism isolated by Brierley and Brierley (1973) can utilize both elemental sulfur and ferrous iron as an energy substrate and degrade molybdenite and chalcopyrite (Brierley and Murr, 1973), and that

TABLE 6.3.1

Various taxonomic groupings of the *Thiobacilli*

Species	DNA base composition		Multivariate analysis group No. [b]	FAME * profile group No. [c]
	Moles %	Group [a]		
T. rubellus [d]	65 ± 3			
T. delicatus [d]	67 ± 3			
T. trautweinii	66	1	0	
T. novellus	66—68	1	1	II
T. denitrificans	64	1	2	II
T. thioparus **	62—66	1	3	II
T. intermedius			7	II
T. acidophilus [e]	63			
Thiobacillus sp. ***,[f]	66			
T. neapolitanus	56	2	4	I
T. ferrooxidans	57	2	6	I
T. thiooxidans †	51—52	3	5	III

* FAME = Fatty Acid Methyl Esters.
** Includes *T. thiocyanoxidans.*
*** As yet uncharacterised.
† Includes *T. concretivorus.*

[a] Jackson et al., 1968;
[b] Hutchinson et al., 1966.
[c] Agate and Vishniac, 1973; Dunn et al., 1977.
[d] Mizoguchi et al., 1976.
[e] Guay and Silver, 1975.
[f] Williams and Hoare, 1972.

TABLE 6.3.2

Principal sulfur-oxidizing and iron-oxidizing bacteria involved in sulfide mineral degradation

Species	Inorganic substrates	Terminal electron acceptor	Facultative heterotrophy [a]	Lowest pH tolerated
Metallogenium spp.	Fe^{2+}	O_2	+	3.5
Sulfolobus spp. [b]	S^0, S^{2-}, Fe^{2+}	O_2	—	<2.0
Thiobacillus denitrificans	RCS	O_2, NO_3^-	—	5.0
T. ferrooxidans	Fe^{2+}, RSC	O_2	+	<2.0
T. intermedius	RSC	O_2	+	2.0—2.8
T. neapolitanus	RSC	O_2	—	2.8
T. novellus	RSC	O_2	+	5.0
T. perometabolis [c]	RSC	O_2	+	3.0
T. thiooxidans	RSC	O_2	+	<2.0
T. thioparus	RSC	O_2	—	3.5

RSC = Reduced sulfur compounds (S^{2-}, S^0, SO_3^{2-}, $S_2O_3^{2-}$, polythionates).

[a] There appears to be a spectrum of abilities with respect to facultative heterotrophy, even with well-defined species.

[b] Strains have been described with a temperature range of 45—90°C.

[c] This organism requires RSC and organic matter simultaneously for optimum growth.

another facultative, thermophilic *Thiobacillus*-like bacterium (Brierley and Le Roux, 1977) can oxidise pyrite, pentlandite and chalcopyrite at 50°C. The oxidation of elemental sulfur is not limited to chemolithotrophic organisms and a limited role in the overall processes of sulfide degradation may be played by the oxidation of liberated sulfur by heterotrophic populations such as *Streptomyces* spp. (Yagi et al., 1971) and *Arthrobacter* spp. (Ehrlich, 1962).

The dynamic flow experiments of Baker and Wilshire (1970) on the microbial degradation of pyrite suggest that the low pH levels (2—3) typically associated with extensive metal solubilization from sulfide minerals and favourable for the development of the most acidophilic of the *Thiobacilli*, are not achieved solely by the activities of sulfur-oxidizing microbial populations. There is considerable evidence that a decrease in pH below 4.5 and an accompanying increase in acidity are frequently the result of ferric iron degradation of pyrite, and that a rate-limiting step is the regeneration of ferric ions from the ferrous ions produced in sulfide degradation reactions (Singer and Stumm, 1970). *T. ferrooxidans* can catalyse the ferrous-ferric oxidation at rates 10^5—10^6 greater than abiotic rates (Keenan, 1969; Lacey and Lawson, 1970) and is most active in the range of pH 2.0—4.5. The discrepancy between the low abiotic rate of ferrous iron oxidation in the range pH 2.5—4.5 and the observable generation of acidity over this range in many field situations has led to the postulation of a pH-dependent succession of iron-oxidizing bacteria (Walsh and Mitchell, 1972a). An acid-tolerant, filmentous iron-oxidizing bacterium of the genus *Metallogenium* has been isolated and characterized (Walsh and Mitchell, 1972b). The organism accelerates ferrous iron oxidation and resultant acidity production by a factor greater than 200 in the range pH 3.5—4.5, with an optimum at pH 4.1. An interesting method of biological control of acid mine pollution is based on breakage of the ecological succession of iron bacteria by taking advantage of the inhibition of *Metallogenium* populations by increased ferrous iron concentrations (Walsh and Mitchell, 1972c, 1975).

Above pH 4.5, the chemical oxidation rate of ferrous iron is relatively fast and this diminishes its effectiveness as an energy source for microbial growth. Nevertheless, stalked iron bacteria such as *Sphaerotilus natans* and *Gallionella ferruginea* are widely distributed in ferruginous natural waters in the pH range 6—9. *G. ferruginea* is probably a chemolithotroph (Buchanan and Gibbons, 1974) and is able to use ferrous iron as a primary energy source (Alexander, 1971), but whether this organism is to be regarded as an early member of a pH-dependent succession of iron bacteria with analogous roles to the pH-dependent sulfur bacteria is not clear and requires further investigation.

As noted above, the microbial populations isolated from natural leaching situations contain a considerable array of heterotrophs some of which display metal and acid tolerances comparable to those of the sulfur and iron bacteria. Detailed information on the nature and characteristics of these heterotrophic species is sparse; some authors have commented on the difficulties of purifying cultures of organisms such as *T. ferrooxidans* from contaminating, heterotrophic satellites (Zavarsin, 1972). Manning (1975) has described a new, readily-prepared, solid medium, based on purified agar and containing no added phosphate. Its use has permitted the differentiation of colony types of acidophilic iron-oxidizing bacteria and the detection of what appears to be a new group of acidophilic, heterotrophic organisms. Recently, evidence for a mutualism between *T. ferrooxidans* and nitrogen-fixing *Beijerinckia lacticogenes* strains has been described by Tsuchiya et al. (1974) and for an enhancement of sulfide mineral degradation by such an organism pair (Trivedi and Tsuchiya, 1975). These observations are of considerable interest in respect of the provision of nitrogen sources for autotrophic and other microbial populations. The nitrogen-fixing abilities of certain *T. ferrooxidans* strains were suggested by the experiments of Mackintosh (1971) and recently unequivocally demonstrated by the same author (Mackintosh, 1976).

BIOCHEMISTRY OF THE THIOBACILLI

The biochemistry and physiology of the thiobacilli have been intensively investigated in recent years and the accumulating information has been

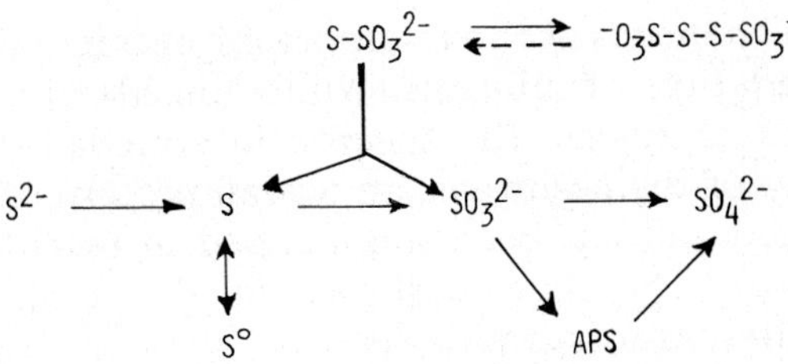

Fig. 6.3.1. Patterns of inorganic sulfur oxidation by thiobacilli.

progressively summarized in a number of excellent reviews (Vishniac and Santer, 1957; Peck, 1962, 1968; Trudinger, 1967, 1969; Kelly, 1968, 1971; Roy and Trudinger, 1970; Suzuki, 1974; Aleem, 1975). The bulk of research on the biochemistry of sulfur oxidation by autotrophic bacteria has been carried out with elemental sulfur and with soluble substrates such as hydrogen sulfide and soluble sulfides, polysulfides, sulfite, thiosulfate and polythionates. Information on the metabolism of these compounds is critical

TABLE 6.3.3

Reactions of significance in the biological oxidation of sulfide minerals

	F' (kJ mol^{-1})
Reactions yielding energy for growth	
$H_2S + 2\ O_2 \rightarrow H_2SO_4$	669
$S^0 + 1.5\ O_2 + H_2O \rightarrow H_2SO_4$	494
$S_2O_3^{2-} + 2\ O_2 + H_2O \rightarrow 2\ SO_4^{2-} + 2\ H^+$	883
$5\ S_2O_3^{2-} + 8\ NO_3^- + H_2O \rightarrow 10\ SO_4^{2-} + 4\ N_2 + 2\ H^+$	3736
$SO_3^{2-} + 0.5\ O_2 \rightarrow SO_4^{2-}$	251
$4\ FeSO_4 + 2\ H_2SO_4 + O_2 \rightarrow 2\ Fe_2(SO_4)_3 + 2\ H_2O$	117
Reactions involving iron	
$FeS_2 + 3.5\ O_2 + H_2O \rightarrow Fe^{2+} + 2\ SO_4^{2-} + 2\ H^+$	
$Fe^{2+} + 0.25\ O_2 + H^+ \rightarrow Fe^{3+} + 0.5\ H_2O$	
$Fe^{3+} + 3\ H_2O \rightarrow Fe(OH)_3 + 3\ H^+$	
$FeS_2 + 14\ Fe^{3+} + 8\ H_2O \rightarrow 15\ Fe^{2+} + 2\ SO_4^{2-} + 16\ H^+$	
$FeS_2 + Fe_2(SO_4)_3 \rightarrow 3\ FeSO_4 + 2\ S^0$	
General ferric sulfate leaching reaction	
$2\ RS + 2\ Fe_2(SO_4)_3 + 2\ H_2O + 3\ O_2 \rightarrow 2\ RSO_4 + 4\ FeSO_4 + 2\ H_2SO_4$	

to the elucidation of the intimate mechanisms at the microorganism-mineral sulfide interface during the biodegradation of minerals. While considerable gaps exist in respect of particular mechanisms, the specific intermediates involved and the detailed enzymology of the pathways, an overall pattern of inorganic sulfur oxidation has emerged and has been summarized in recent reviews (Suzuki, 1974; Aleem, 1975). The scheme shown in Fig. 6.3.1 accommodates most of the experimental information.

The mechanisms by which the solid, insoluble mineral sulfides are biologically oxidized have been less intensively studied and the details of the interactions are incompletely understood. As will be discussed in greater detail in later sections, not only are mechanisms facilitating the oxidation of sulfur operative but an important role is played by iron oxidation reactions. The reactions of greatest significance in the biological oxidation of sulfide minerals are summarized in Table 6.3.3.

GENERAL ASPECTS OF THE BIODEGRADATION OF SULFIDE MINERALS

The interaction of microorganisms with mineral sulfides is a particular case of the general interaction of organisms with solid, insoluble substrates. Some aspects of the general rationale of such transformations developed by Nickerson (1969) in his studies of the breakdown of natural organic polymers might be applied to the mineral field with profit. In the case of mineral sulfides, two general groups of mechanisms can be distinguished which have been commonly described by a number of investigators as "*indirect*" and "*direct*" mechanisms. The distinction in particular cases is often blurred. *Indirect* mechanisms include modes of degradation in which an attacking chemical species is produced or regenerated by the microorganism. In some cases, an attacking species such as ferric iron may have its origin in the mineral (e.g., chalcopyrite) and be initially released by a direct attack mechanism but, subsequently, may play a role in an indirect process. *Direct* mechanisms are those in which components of the sulfide mineral are catalytically modified to chemical forms which can escape from the mineral structure into solution or into new combinations within a solid phase. They probably also embrace those degradation mechanisms in which the organism accelerates a rate-limiting step in chemical or electrochemical reactions to which the mineral is prone.

The practical significance of metal release from sulfide minerals has focussed attention on the mechanisms which operate under the conditions of the highest metal concentrations in the aqueous phase of field situations, and on the associated organisms. The most studied substrates have been iron and copper sulfide minerals and the unique position of *T. ferrooxidans* as the principal terminal member of the pH-dependent successions of both sulfur-oxidizing and iron-oxidizing bacteria has led to intensive investigation of this organism. The factors which govern the activities of this organism have consequently determined the physico-chemical parameters of such mechanism studies. The results of these investigations have thrown some light on the relative roles of various postulated mechanisms and have yielded a very large amount of information on the kinetics of the breakdown of sulfide minerals by *T. ferrooxidans* and the controlling factors (Duncan and Trussell, 1964; Beck, 1969; Torma et al., 1970, 1972, 1974; Tuovinen and Kelly, 1972; Brock, 1975; Torma, 1977). There is no comparable body of information on the mechanistic roles of the *Metallogenium* species which occupy the penultimate niche in the succession of iron-oxidizing organisms in a pH range in which metal release and solubilization may be considerable. At higher pH levels, conditions are not favourable for the solubilization of some metals and there is a paucity of information on the details of mechanisms which lead to sulfate formation, the generation of acidity and modifications to the mineral structure, without substantial metal release and solubilization (Silver and Torma, 1974). The preoccupation with the very important final role of *T. ferrooxidans* has to some extent obscured the significance of organisms earlier in the succession and, to date, there does not appear to have been a systematic study of the mechanisms by which these organisms degrade or modify sulfide minerals under the conditions appropriate to their optimum activity. There are also indications in the literature (Ehrlich, 1962) that some heterotrophs significantly degrade sulfide minerals under non-acidic conditions; these mechanisms require further investigation.

A specific organism will exert maximum influence upon the rate of degradation of a particular sulfide mineral within a limited range of values of various parameters. Provided that bacterial nutrients per se are not limiting, the most important factors will be those controlling the energy flow through the mineral-organism system. The status of the system in this respect can be defined in terms of the redox potential and the hydrogen ion concentration. The nutritional needs of particular microorganisms include requirements for levels of Eh and pH within specific ranges in the immediate milieu, but their metabolic activities will bring about changes in these parameters. The field situation is dynamic and microbial generation of changes in Eh-pH and other conditions contributes largely to the commonly-observed succession of microbial populations, as well as influencing the chemical behaviour of the inorganic components of the system. The empirical relationship (Baas Becking et al., 1960) between the occurrence of particular types of microbial populations and the Eh-pH of the specific location arises from the interplay of these processes.

In the same context, the metal sulfide substrate is a component of an Me-S-H_2O system which may be far from equilibrium and which may contain a number of metastable molecular and ionic species. While the availability of the relevant thermodynamic data, especially for metastable systems, is far from complete, Eh-pH diagrams have been constructed in connection with the development of hydrometallurgical processes for most of the Me-S-H_2O ternary systems of the metals of greatest practical interest (Cu, Pb, Zn, Fe, Sn, Ag, As, Sb, Cd, Mo, Co, Ni, Mn and Hg) (Majima and Peters, 1968). Methods for the construction of Eh-pH diagrams are described by Garrels and Christ (1965) and Pourbaix (1966). These diagrams provide a direct graphical representation of chemical equilibria and are of the greatest significance in the understanding of geochemical phenomena, as well as providing a firm basis for the elucidation of possible mechanisms by which mineral-microorganism interactions may occur.

The pH-dependence of the solubility of sulfide minerals is of considerable interest in relation to their interaction with successions of microorganism populations which are themselves pH-dependent. In pure water, the solubility of the majority of sulfide minerals is extremely low (pK_{sp}: ZnS, 24; PbS, 27; HgS, 53) (Sillen and Martel, 1964), but the potential remains of reaction with water to yield hydrated metal cations and sulfide ions. The latter are strong bases and further hydrolysis can yield HS^- ions or even H_2S. The equilibrium concentration of S^{2-} ions in solution and consequently the solubility of the metal sulfide in water is highly pH-dependent.

Some of the metal sulfides which occur as minerals have interesting electrical and magnetic properties (Jellinek, 1968) and these open up the possibility of envisaging mechanisms of biodegradation in electrochemical terms where rate-limiting steps in anodic or cathodic reactions are amenable to intervention by organisms (Corrans et al., 1972; Torma and Legault, 1973). In this context, many of the experimental techniques used in the study of corrosion phenomena of metals and alloys, are applicable to the investigation of biological mineral sulfide oxidation.

Finally, some other characteristics of the mineral sulfide substrates need to be borne in mind in the evaluation of mechanisms of breakdown. Sulfide minerals frequently do not display a simple stoichiometry but exhibit a range of compositions, and their strict chemical definition as substrates may be further complicated by the presence of other metallic ions in solid solution, as imperfections in the crystal lattices or by lattice substitution. Even museum-grade specimens may contain inclusions of other chemicals of submicron dimensions, the detection of which may be difficult and the effects of which upon degradation phenomena may be significant. Many recent investigations on mineral sulfide biodegradation have employed synthetic minerals to ensure strict definition of chemical composition but it is open to question as to whether the criteria employed to establish identity with naturally-occurring forms (chemical analysis and X-ray diffraction) are always adequate (Jessup, 1976). The abandonment of the use of synthetic speciments of pyrrhotites for crystallographic and stability studies is relevant because of the difficulty of avoiding formation of metastable phases (Morimoto et al., 1975). Variation in composition may be accompanied by changes in crystalline form (e.g. pyrrhotites) and some sulfide minerals of identical chemical composition exhibit more than one crystalline form (e.g., iron and zinc sulfides). The limited amount of published information suggests significant variations in the behaviour of different crystalline forms under microbial attack. Leathen et al. (1953) noted a high degree of resistance to oxidation of museum-grade pyrite in the presence of acidophilic, iron-oxidizing bacteria, but a ready attack upon high-grade specimens of marcasite. In a more detailed investigation of these two iron sulfide minerals, Silverman et al. (1961) confirmed the resistance of coarsely crystalline pyrite to oxidative attack by *T. ferrooxidans* and *T. thiooxidans* but observed that both these organisms degraded marcasite (FeS_2). A diversity of finely divided pyritic materials were oxidized at enhanced rates in the presence of *T. ferrooxidans*. It was suggested that some unknown factors such as lattice imperfections or the presence of impurities in the minerals affected the ability of the organisms to accelerate oxidation. Similar results have been reported by Khalid and Ralph (1977) in studies of the relative recalcitrance to oxidative breakdown of a range of zinc sulfide minerals (sphalerite, ZnS, wurtzite, ZnS, marmatite, (Zn, Fe)S, and a synthetic zinc sulfide) in the presence of *T. ferrooxidans*, *T. thiooxidans* and *T. thioparus*. The samples were finely divided and of equivalent surface area. The hexagonal structure of zinc sulfide in wurtzite proved less amenable to microbial attack than the cubical form of sphalerite and the substitution of iron for some of the zinc in the marmatitic form further facilitated biodegradation.

MECHANISMS OF OXIDATIVE ATTACK

Various ways in which microorganisms might facilitate the rate of oxidation of sulfide minerals have been summarized and discussed by Vanselow (1976). The principal modes suggested by a number of investigators are:

(i) Excretion of surface-active material such as phospholipids by *Thiobacillus* spp. (Starkey et al., 1956; Schaeffer and Umbreit, 1963; Shively and Benson, 1967; Barridge and Shively, 1968; Beebe and Umbreit, 1971) may affect the degree of aggregation in slurries and facilitate the mass transfer rates at phase boundaries of some reactants such as oxygen. These observations have been extended to synthetic surfactants (Duncan et al., 1964) and it has been noted that the influence of wetting agents may vary with different sulfide minerals (Trussell et al., 1964).

(ii) The influence of microorganisms upon the pH of the microenvironment may impose a different stoichiometry upon overall reactions and lead to different end-products. For example, numerous authors have noted the formation of elemental sulfur during the non-biological leaching of sulfide minerals (see Wadsworth, 1972) whereas in the presence of sulfur-oxidizing organisms, sulfuric acid is an end-product. The resulting acidity may enhance attack by oxygen and affect the concentration of metal and sulfide ions in solution. On the other hand, the sulfate ion may form insoluble precipitates with some metal ions and inhibit oxygen attack. Other influences upon the pH of the system can result from the presence of iron and the iron-oxidizing bacterium, *T. ferrooxidans.* The oxidation of ferrous iron consumes hydrogen ions but subsequent hydrolysis of ferric ions is acid-generating (eqns (1) and (2));

$$Fe^{2+} + 0.25\ O_2 + H^+ \rightarrow Fe^{3+} + 0.5\ H_2O \qquad (1)$$

$$Fe^{3+} + 3\ H_2O \rightarrow Fe(OH)_3 + 3\ H^+ \qquad (2)$$

Re-solution of precipitated ferric hydroxide consumes hydrogen ions. These reactions provide a buffering effect against rapid pH fluctuations by maintaining the pH in the range 2.0—2.2 due to precipitation and resolubilization of ferric iron (Duncan and Walden, 1972). Other acid-generating reactions which involve ferric and other ions are the formation of jarosites (eqn (3)):

$$3\ Fe^{3+} + A^+ + 2\ SO_4^{2-} + 6\ H_2O \rightarrow A \cdot Fe_3(SO_4)_2 \cdot (OH)_6 + 6\ H^+ \qquad (3)$$

The usual jarosites formed in these situations are hydronium, potassium or ammonium forms in which A is hydrogen, potassium or ammonium ion. In leaching situations, jarosite formation can strip out essential microbial nutrients such as potassium and ammonium ions (Duncan and Walden, 1972; L.A.V. Sulligoi, 1972, personal communication).

(iii) Sulfide minerals are generally prone to electrochemical processes which can rapidly lead to the coating of surfaces by a sulfur or sulfur-rich passivation layer, formed by removal of metal ions from the sulfide lattice. The formation of this layer imposes a rate-limiting step in the dissolution of the mineral which can be overcome by the removal of the sulfur passivation film by sulfur-oxidizing bacteria (Corrans et al., 1972; Vanselow, 1976).

(iv) There is a good deal of evidence for direct, enzymically-mediated attack by microorganisms on either or both of sulfide and metallic ions in mineral sulfide lattices (Tuovinen and Kelly, 1972). The diverse abilities of different species of organisms, the particular range of physico-chemical parameters within which their capabilities can be exerted, and the specific characteristics of the sulfide mineral substrates suggest that a number of direct action mechanisms may be possible.

(v) The so-called "indirect" mechanisms of degradation of sulfide minerals involve attack by a reactant which is cyclically regenerated by the organism. The role of ferric ions in this process has been intensively studied; other chemical species such as polysulfide ions may be involved in some circumstances.

ELECTROCHEMICAL DEGRADATION MECHANISMS

Since sulfides display electrical conductivity, their oxidation can be regarded as analogous to the oxidative corrosion of metals, and the course of some of the reactions involved might be anticipated to be affected by the presence of microorganisms. This approach to the microbial degradation of sulfide minerals is well illustrated by the studies of Corrans (1970) on the bacterial leaching of copper sulfide minerals by *T. ferrooxidans* strains. In this investigation (Corrans et al., 1972), massive samples of synthetic chalcocite (Cu_2S) and covellite (CuS) were allowed to interact with the organism under a variety of conditions. The potentials of the sulfide minerals were monitored during the course of the experiments and the final state of the minerals examined by electron-microprobe analysis of cross-sections and by optical and scanning electron microscopy. The following sequence of anodic reactions occurred (eqns (4)-(6)):

$$5\ Cu_2S \rightarrow Cu_9S_5 + Cu^{2+} + 2\ e^- \quad (4)$$

$$Cu_9S_5 \rightarrow 5\ CuS + 4\ Cu^{2+} + 8\ e^- \quad (5)$$

$$CuS \rightarrow S + Cu^{2+} + 2\ e^- \quad (6)$$

Each sequential reaction took place at a slightly higher equilibrium oxidation potential and resulted in a shrinking of the solid lattice (densities: chalcocite, 5.6; covellite, 4.6; rhombic sulfur, 2.07 g cm^{-3}) (Weast, 1974). The cathodic reactions were:

$$O_2 + 4\ H^+ + 4\ e^- \rightarrow 2\ H_2O \quad (7)$$

$$Fe^{3+} + e^- \rightarrow Fe^{2+} \quad (8)$$

The data indicated that the cathodic reaction, the reduction of oxygen on the CuS surface, is the rate-limiting step in the oxidation of both chalcocite and covellite, and that this impediment arises from passivation of the entire mineral surface by a sulfur or sulfur-rich film. This film is removed by organisms with sulfur-oxidizing ability and degradation proceeds. The subsequent investigations of Vanselow (1976) on the reduction of oxygen by covellite (CuS), using a sensitive modified oxygraph technique for the measurement of oxygen uptake, confirmed the existence of the sulfur passivation film on the mineral surface and provided an estimate of its thickness (1—1.5 atoms thick (Golding et al., 1977)). A *T. thioparus* strain, without ability to oxidize CuS itself but able to oxidize elemental sulphur, oxidized the passivation film to sulfuric acid. A *T. ferrooxidans* strain in the absence of iron, gave similar results.

The concept of microbially-mediated mineral sulfide oxidation as a particular case of electrochemical corrosion phenomena warrants considerable extension, and the techniques developed by Corrans, Vanselow and their collaborators could be usefully applied to a variety of other sulfide mineral systems.

INDIRECT MECHANISMS OF OXIDATIVE ATTACK

Very early in the study of sulfide mineral biodegradation, it was discerned that microbially-regenerated ferric ions constituted a very potent reagent for indirect attack (for literature, see Kuznetsov et al., 1963; Silverman and Ehrlich, 1964; Zajic, 1969). The ubiquitous occurrence of iron in natural situations and the wide range of reactions with sulfide minerals that can occur, combine to make the ferric ion, either as such or in complex form, the most important chemical species involved in indirect attack mechanisms. The general reaction, known to be applicable to a number of sulfide minerals under aerobic conditions (Bryner et al., 1954; Ivanov et al., 1961), may be expressed as:

$$MeS + 2\ Fe^{3+} + H_2O + 1.5\ O_2 \rightarrow Me^{2+} + 2\ Fe^{2+} + SO_4^{2-} + 2\ H^+ \quad (9)$$

Under anaerobic conditions, and in the absence of microorganisms, reactions such as the following may occur (Silverman, 1967):

$$Fe_2(SO_4)_3 + FeS_2 \rightarrow 3\ FeSO_4 + 2\ S \quad (10)$$

In the presence of an organism with iron-oxidizing ability such as *T. ferrooxidans*, ferrous ions produced by the oxidation of a metallic sulfide can be re-oxidized to ferric ions and a cyclic mechanism established (eqn (11)):

$$Fe^{2+} + 0.25\ O_2 + H^+ \rightarrow Fe^{3+} + 0.5\ H_2O \quad (11)$$

The overall reaction is expressed by eqn (12):

$$MeS + 2\ O_2 \rightarrow Me^{2+} + SO_4^{2-} \quad (12)$$

This mechanism of oxidative attack has two facets, namely the regeneration of ferric ions by the organism and the chemical interaction of ferric ions with the sulfide mineral. Singer and Stumm (1970) have shown that the rate of oxidation of ferrous iron by oxygen in abiotic systems is a function of pH. At pH values greater than 4.5, the kinetic relationship is described by eqn. (13):

$$-\frac{d[Fe^{2+}]}{dt} = K[Fe^{2+}][O_2][OH^-]^2 \quad (13)$$

where $K = 8.0 \times 10^{15}\ l^2\ mol^{-2}\ Pa^{-1}\ min^{-1}$ at 25°C. At pH values below 3.5, the reaction proceeds at a rate independent of pH:

$$-\frac{d[Fe^{2+}]}{dt} = K'[Fe^{2+}]\ [O_2] \quad (14)$$

where $K' = 1.0 \times 10^{-5}\ Pa^{-1}\ min^{-1}$ at 25°C. The half-life of the chemical oxidation reaction is short at higher pH (a few minutes at pH 6) but rises to approximately 1000 days under strongly acidic conditions (pH 2.5). In the presence of microorganisms capable of deriving metabolic energy from the oxidation of ferrous ions, the very slow rate of abiotic ferrous ion oxidation at low pH is greatly accelerated.

There now appears to be a substantial consensus that most of the numerous isolations of acidophilic, chemoautotrophic and facultatively heterotrophic, iron- and sulfur-oxidizing bacteria from acidic, sulfidic mineral habitats are all biochemical and physiological variants of the species *T. ferrooxidans* (Colmer and Hinkle, 1947; Colmer et al., 1950). The variations in nutritional and metabolic characteristics already reported are considerable in range (Unz and Lundgren, 1961; Ivanov and Lyalikova, 1962; Razzell and Trussell, 1963a; Hutchinson et al., 1966; Groudev and Genchev, 1978), and it is probable that the full extent is not yet delineated. The various species of *Sulfolobus* and *Thermoplasma* which have been described (Brock et al., 1972; Brierley and Brierley, 1973; Brierley and Murr, 1973; de Rosa et al., 1975; Millonig et al., 1975, Langworthy, 1977) resemble *T. ferrooxidans* in their acidophilic mode and in the ability to oxidize iron and sulfur (and in some cases, organic compounds) but are clearly distinguished by the extreme thermophilic habit (up to 90°C), their internal ultrastructure and other morphological characteristics (de Rosa et al., 1974) and by the unusual nature of some chemical components (de Rosa et al., 1971; 1976).

The rates at which *T. ferrooxidans* strains oxidize soluble iron are remarkably high under appropriate conditions (Silverman and Lundgren, 1959a, b; Beck, 1960) and Landesman et al. (1966a) have drawn attention to the enhancement which occurs with low cell densities and young cells. The use of improved assay techniques has thrown further light on the kinetics of the iron oxidation reaction and the effects of a number of physico-chemical factors (Steiner and Lazaroff, 1974; Vanselow, 1976). The pH optimum for the iron oxidation as recorded by various investigators varies considerably from 1.75 (Landesman et al., 1966a), to 2.0—2.5 (Bodo and Lundgren, 1974) and 2.5—3.8 (Schnaitman et al., 1969).

The ability to grow on ferrous iron is retained by *T. ferrooxidans* which has been previously grown on sulfur (Margalith et al., 1966) and the constitutive nature of the iron-oxidizing ability in *T. ferrooxidans* is clearly shown by Colmer's (1962) experiments in which growth resumed on ferrous iron after 85 transfers over 14 months on a thiosulfate medium lacking oxidizable iron. *T. ferrooxidans* exhibits a rather unusual anion requirement in that sulfate ion is required in substantial quantities for both growth and iron oxidation (Lazaroff, 1963; Beck, 1969; Tuovinen et al., 1971b, 1975). The sulfate requirement is not absolutely obligatory (Bodo and Lundgren, 1971), as it can be replaced to a limited extent by other divalent anions (HPO_4^{2-} and $HAsO_4^{2-}$) but not by monovalent anions (BO_3^-, NO_3^- or Cl^-). Ferrous iron oxidation can be depressed if sulfate and sulfuric acid are replaced by chloride or nitrate salts and hydrochloric acid (Schnaitman et al., 1969). The natural habitats of *T. ferrooxidans* frequently have a high sulfate content, this anion arising from the oxidation of the sulfide moiety during the degradation of sulfide minerals. It has been suggested that sulfate is involved as a complexing agent in ferrous iron oxidation (Dugan and Lundgren, 1965), and the inhibitory effects on bacterial iron oxidation noted at sulfate concentrations above 0.216 M (Steiner and Lazaroff, 1974) could result from the formation of different complexes with either the ferrous iron substrate or the ferric iron product. It has been observed that inhibition of the oxidation process can also occur with ferrous ion (Landesman et al., 1966a), ferric ion (Wong et al., 1974), phosphate ion (Beck and Shafia, 1964; Beck, 1969) and molybdate ion (Schnaitman et al., 1969) at some concentration levels.

Ferrous iron oxidation by whole cells of *T. ferrooxidans* is remarkably resistant to inhibition by other metals and increased tolerance to specific metals can be rapidly developed (Sadler and Trudinger, 1967; Tuovinen and Kelly, 1972). This increased tolerance appears to be due to selection of resistant mutants. This characteristic of *T. ferrooxidans* populations is ecologically significant and of considerable importance in the recovery of metals by bacterial leaching. The iron-oxidizing ability of *T. ferrooxidans* is affected by a number of organic materials. The inhibitory effects are probably related to interference with the autotrophic metabolism, either by directly affecting enzymes of carbon dioxide fixation and iron oxidation, or by the inhibitory compound (or a product of its metabolism) acting as an over-effective end-product inhibitor of enzymes in a branched biosynthetic pathway (Tuovinen and Kelly, 1972).

As mentioned earlier (see p. 374), organisms other than *T. ferrooxidans* have been demonstrated to catalyse the oxidation of ferrous iron at low pH. The *Sulfolobus*-like organism isolated by Brierley and Brierley (1973) had a temperature optimum of 70°C and oxidized both sulfur and ferrous iron; the rates of oxidation are considerably lower than those recorded for *T. ferrooxidans.* The isolate, however, was able to oxidize molybdenite, MoS_2, at 60°C and the rate was increased by addition of ferrous sulfate. The organism showed a unique tolerance to molybdenum (2 g l^{-1}) (Brierley and Murr, 1973). Whether organisms of this type play a significant role in the oxidation of sulfide minerals under the conditions of elevated temperature known to exist in leaching heaps, remains to be demonstrated.

The acid-tolerant *Metallogenium* isolated by Walsh and Mitchell (1972b) also possesses marked iron-oxidizing ability which is greatest in a pH range of 2.5—5.0. The isolation and purification of this filamentous bacterium has presented some difficulties and, in spite of its important role in natural situations, it does not appear to have been intensively studied. An interesting feature of its biochemistry is its sensitivity to inhibition by quite low concentration levels of ferrous ion and this characteristic has shown some promise in laboratory tests as a means for the reduction of mine drainage pollution (Walsh and Mitchell, 1975).

While the overwhelming mass of evidence to date suggests a unique role for ferric iron as an agent in indirect mechanisms of oxidative attack on sulfide minerals, the possibility cannot be excluded that complexed cuprous ions might, under some circumstances, play a similar role. In abiotic systems, a solution of copper sulfate in acidified aqueous acetonitrile, in which the cuprous ion is stabilized, is a more potent oxidant of copper sulfide minerals than an acidified ferric sulfate solution (Muir et al., 1976). Other organic compounds can stabilize cuprous ions and there is some evidence that *T. ferrooxidans* can derive growth energy from the $Cu^{1+} \rightarrow Cu^{2+}$ oxidation (Nielsen and Beck, 1972; Golding et al., 1974). A cyclic cuprous-cupric system can be envisaged which might have a role in an indirect attack mechanism of sulfide minerals.

DIRECT MECHANISMS OF OXIDATIVE ATTACK

The oxidative degradation of sulfide minerals can be accelerated by micro-organisms in the absence of ferric ions and at pH and Eh levels which preclude any significant contribution by a ferrous-ferric cyclic mechanism (Mizoguchi et al., 1976). Even at low pH, in the presence of iron, there is evidence that mechanisms are operative other than indirect attack by ferric or hydrogen ions (Bryner and Jameson, 1958; Razzell and Trussell, 1963b; Duncan and Trussell, 1964). Further examples of non-iron-mediated attack on sulfide minerals are cited by Silverman and Ehrlich (1964). In a study on the bacterial oxidation of iron pyrites by *T. ferrooxidans*, Silverman (1967) advanced the view that the organism played at least two roles in the degradation of the mineral, one involved with the maintenance of a ferrous-ferric cycle and the second a "direct contact" mechanism, independent of the action of ferric ions, requiring intimate physical contact between the bacteria and the mineral under aerobic conditions. These investigations established the view that, by some relatively direct intervention, the organism could modify the chemical state of metal and sulfide ions within the lattice structure of the mineral.

The investigations of Landesman et al. (1966a, b) clarify the effects of the various conditions controlling the optimum oxidation rates of ferrous iron, sulfur and reduced sulfur compounds by *T. ferrooxidans*. Experiments on soluble iron, sulfur and iron-containing sulfide minerals (chalcopyrite, $CuFeS_2$, bornite, Cu_5FeS_4, and pyrite) established that iron and sulfur can be oxidized simultaneously. With a mixed iron-sulfur substrate a rate of oxidation, equal to that of the sum of the maximum rates of oxidation of the two substrates individually was observed with both S-adapted and Fe-adapted cells. Subsequently, Duncan et al. (1967) established the differential susceptibility of the bacterial oxidation of ferrous iron and sulfur to N-ethyl maleimide and sodium azide, and determined the effect of these inhibitors on pyrite and chalcopyrite oxidation. Decreased rates of oxidation in the presence of each inhibitor suggested that both iron and sulfide were simultaneously oxidized, and that enzymic mechanisms were involved. The relative rates of oxidation of iron and sulfur depended on the previous history of the cells. When chalcopyrite-grown cells were used to oxidize chalcopyrite, 68—74% of the oxygen uptake was the result of sulfide oxidation, but with iron-grown cells, two rates resulted. During the initial rapid rate, 80—90% of the oxygen uptake was due to iron oxidation but during the second phase, the rates were similar to those with chalcopyrite-grown cells. When chalcopyrite-grown cells were used to oxidize pyrite, all the oxygen uptake was due to sulfide oxidation; iron-grown cells oxidized pyrite at a constant and more rapid rate and only 20—30% of the total oxygen uptake was due to sulfide oxidation. Further evidence for the direct oxidation of sulfide minerals by *T. ferrooxidans* was obtained by Beck and Brown (1968) who compared the efficiencies of carbon dioxide fixation on the substrates ferrous iron, sulfur, pyrite and chalcopyrite. The results suggested that both iron and sulfide ions were attacked but that relatively more iron than sulfur was being oxidized. Further information on the influence of different growth substrates on the subsequent oxidative abilities of *T. ferrooxidans* against various metal sulfides has been reported by Silver and Torma (1974).

The effects of ferrous and/or ferric sulfate addition on the leaching of chalcopyrite, chalcocite (Cu_2S) and marmatite by *T. ferrooxidans* have been studied by Duncan and Walden (1972). The addition of ferrous sulfate had no influence on the rate or extent of copper release from chalcopyrite, whereas the addition of ferric iron depressed the solution of copper by about 50%. It was concluded from these studies that, under the conditions present in the leaching system (i.e., pH 1.8—2.0), the presence of ferric iron plays a minor role in the release of copper and zinc from their metallic sulfides.

The possibility of an indirect mechanism involving a cyclic cuprous-cupric ion system has been referred to earlier, and there is some evidence, in spite of the reservations expressed by Zajic (1969), that *T. ferrooxidans* can catalyse an oxidation of cuprous to cupric ions in some copper sulfide minerals and other compounds containing cuprous copper, and that energy can be obtained from this oxidation. Nielsen and Beck (1972) reported that the oxidation of chalcocite is coupled to carbon dioxide fixation by *T. ferrooxidans.* They noted that the only detectable oxidized products were cupric ions and more oxidized forms of the sulfide minerals, namely, digenite and covellite, and that the apparent source of energy for the carbon dioxide fixation was provided by oxidation of the cuprous copper of the chalcocite. This conclusion is not unequivocable since Silver and Torma (1974) have detected digenite ($Cu_{9-x}S_5$), antlerite ($Cu_3SO_4(OH)_4$), covellite (CuS) and metallic copper in the residues from the microbial oxidation of chalcocite by *T. ferrooxidans* suggesting that at least some of the sulfide ion of chalcocite is oxidized and may contribute energy for CO_2 fixation. Imai et al. (1973) have presented strong evidence, based on the inhibition of iron oxidation by quinacrine, for a direct mechanism of chalcocite oxidation by *T. ferrooxidans* involving the enzymic oxidation of Cu^{+} to Cu^{2+} with ferric iron as a cofactor. The observation of Sakaguchi et al. (1976) that the optimum ferric iron concentration for the microbiological oxidation of synthetic chalcocite by *T. ferrooxidans* is 4—10 mM is not in conflict with the findings of Imai et al. (1973).

It is clear from the foregoing evidence that some microorganisms can mediate direct oxidative attacks on sulfide materials. It also underlines the scantiness of the understanding of the intimate details of the mechanisms involved. The only organisms studied in any detail are various strains of *T. ferrooxidans* and the range of sulfide minerals studied is extremely limited. The evidence suggests that enzymic systems are involved and the recent delineation of the enzymic profiles of *T. ferrooxidans* by Tuovinen et al. (1976) provides a background for further investigation. Intermediates are probably involved in both metal and sulfide oxidation and a role for polythionate ions, based on spectroscopic evidence, has been suggested by Vanselow (1976). The findings of Gaidarjiev et al. (1975) and of Groudev and Genchev (1978) represent important additions to the understanding of direct attack mechanisms. These authors have studied the leaching activity of 120 strains of *T. ferrooxidans* towards chalcopyrite, ferrous iron and elemental sulfur, and delineated patterns of enzymic activity. The strains which possessed the highest oxidative activity towards chalcopyrite were characterised by strong sulfur-oxidizing ability and high activity levels of those enzymes which participate in the transformations of inorganic sulfur compounds. A high level of rhodanese activity appeared particularly significant. There is considerable need for more information on the enzymology of direct attack mechanisms, including the characterization of the enzymes and cofactors involved, possible excretion patterns, adsorption to the mineral surface (cf. Nickerson, 1969), and related problems. The state of the mineral residues, after microbial attack, also requires close attention particularly under those circumstances in which no substantial solution of metal occurs.

The concept of direct attack mechanisms as "direct contact" phenomena requiring "intimate physical contact" between the mineral surface and the organism requires some precise definition. Vanselow (1976) has thrown some light on the possible physical interactions between cells and minerals in studies on the effects of dilution on the rates of oxygen uptake by *T. ferrooxidans* strains in the presence of synthetic covellite. The dilution of a slurry will lower the rate of copper sulfide oxidation per unit volume, and the relationship between the dilution factor, and the factor by which the oxidation rate is lowered, will depend upon the nature of the physical interaction between the cells and the mineral particles. Three principal situations were postulated, namely, that in which the oxidation was carried out by cells permanently adsorbed to the mineral surface, that in which the cells were reversibly adsorbed, and that in which the oxidation was effected by suspended cells. The theoretical responses to dilution experiments in each case were deduced. The data indicated that some strains of thiobacilli oxidized the mineral while physically separate from it. These experiments and others concerned with the nature of the intermediates involved in the microbial oxidation of covellite lead to the general conclusion that the behaviour of the microbial populations did not reflect a mechanistic necessity for physical contact. It was noted that an evolutionary adaptation which favoured the adsorption of cells on surfaces would be advantageous when the concentration of intermediates was so low that their diffusion through the liquid phase was rate-limiting.

The characteristically rapid adaptation of some *Thiobacillus* spp. to different substrates, to which selection mechanisms probably contribute, suggests that the isolates employed by some investigators may have been quite heterogeneous in physiological and biochemical abilities and may perhaps have been mixed cultures. There is some evidence that mixed cultures of closely related organisms may collectively greatly enhance the rate of sulfide mineral degradation (Ralph and L.A.V. Sulligoi, unpublished results); this phenomenon requires further investigation.

THE MICROBIAL OXIDATION OF SULFIDE MINERALS IN FIELD SITUATIONS

The conditions under which sulfide minerals exist in the field differ considerably from the laboratory situations which have been described in the preceding sections. Mineral deposits are characteristically heterogeneous with respect to both the sulfide mineral content and the associated gangue minerals, and there is some evidence of interactive effects in the leaching behaviour of mixtures of sulfide minerals as opposed to their performance in isolation (Razzell and Trussell, 1963b). The enhancement of chemical leaching rates of mixtures of sulfide minerals by galvanic effects has been reported by a number of investigators (see Majima and Peters, 1968), but such effects do not appear to have been investigated under conditions in which microbial oxidation is occurring. The gangue minerals may also contribute to such effects since relatively large electrical potential differences may develop between different rocks under some circumstances (Taylor and Fischer, 1973).

The physical form of sulfide mineral components may vary from a state of fine sub-division to massive form with consequent variation in available surface area, and this may be further affected by the mode of association of the gangue material. The effect of surface area on rates of mineral sulfide oxidation is well established (Torma et al., 1970, 1972; Torma and Legault, 1973).

The nature of the gangue minerals plays a major role in the determination of the micro-environment of the sulfide mineral surfaces. Physical screening effects may limit access of water, dissolved nutrients, gases and the microbial populations; the rate-limiting factors in oxidative degradation may in fact lie in the diffusion characteristics of the gangue material (Auck and Wadsworth, 1973) rather than in the chemical characteristics of the mineral or the biochemical abilities of the organisms. Further limitations may be imposed by the chemical nature of the gangue in respect of its influence on pH and redox potential and the adsorption of essential reactants. It will also play a major role in its contribution to the matrix component of the water potential of the micro-environment, a parameter which is known to affect the rate of microbial leaching (Brock, 1975). The temperature profiles of the ore mass are likely to affect the rates of both chemical and microbial reactions and to have considerable influence upon the nature of the microbial populations. The thermal properties of the gangue materials may be such as to facilitate rises in temperature to a point where chemical reactions are facilitated (Bryner et al., 1967). It is interesting to speculate that heat energy might accumulate to a level at which anaerobic, endothermic chemical reactions could occur; for example, the interaction of water with chalcopyrite (Roberts and Williams, 1976) (eqns (15) and (16)):

$$6\ CuFeS_2 + 8\ H_2O \rightarrow 3\ Cu_2S + 2\ Fe_3O_4 + 8\ H_2S + S^0 \qquad (15)$$

$$15\ CuFeS_2 + 16\ H_2O \rightarrow 3\ Cu_5FeS_4 + 4\ Fe_3O_4 + 16\ H_2S + 2\ S^0 \qquad (16)$$

The reactions involving ferrous hydroxide which yield molecular hydrogen and, under some conditions, reduced nitrogen compounds (Schrauzer and Guth, 1976) could be operative in some field situations.

The heterogeneity of the mineral sulfide substrate is likely to be matched by that of the microbial populations and the diversity of the latter could be extended by the availability of additional substrates such as hydrogen sulfide, sulfur, molecular hydrogen and reduced nitrogen compounds resulting from the chemical reactions postulated above. There is unlikely to be a complete absence of reduced carbon compounds in the field situation, even if only that arising from the detritus of a series of autotrophic populations, and this circumstances will further increase the range of the microbial types. Attention has been drawn earlier to the complexity of the microbial associations in field situations and to the enhancement of sulfide mineral oxidation rates by mixed microbial populations (L.A.V. Sulligoi, 1973, personal communication; Tsuchiya et al., 1974). Interactive effects of this kind must be common in natural situations but very little is known of the intimate mechanisms involved.

The hydrology of field situations, whether arising from climatic patterns or of a contrived kind, has been much studied in relation to the pollutional aspects of metal and acid release by sulfide mineral oxidation (Andersen and

TABLE 6.3.4

Summary of factors affecting the biological oxidation of sulfide minerals in field situations

1. Physical characteristics of body of sulfidic material
   - Geometry of the body of sulfidic material
   - Particle-size distribution
   - Permeability to water and gases
   - Thermal conductivity and heat capacity
   - Location in relation to influent water, drainage and water table.

2. Characteristics of sulfidic material
   (a) Sulfide mineral components
   - Present or absence of iron pyrite
   - Nature and diversity of other metallic sulfides
   - Particle size of individual sulfides
   - Distribution of individual sulfides

   (b) Gangue mineral components
   - Permeability to water and gases
   - Capacity for acid neutralization
   - Stability to acid or heat degradation and nature of degradation products
   - Sorptive capacity for water, H^+, metal ions, other chemical compounds (including organic matter)
   - Ion-exchange properties
   - Redox characteristics

3. Composition of influent water
   - Solutes present (oxygen, carbon dioxide, carbon compounds, nitrogen compounds, other microbial nutrients)
   - Redox status (largely controlled by oxygen level, ferrous/ferric ion ratio)
   - Microbial populations (diversity and magnitude)
   - Naturally-occurring toxic materials (e.g. mercury and molybdenum compounds)

4. Climatic influences
   - Rainfall and run-off patterns, influencing the availability of water, water-logging, erosion of body of material
   - Ambient temperature variations (seasonal and diurnal)

Allman, 1968; Davy, 1975). Transport by water movement of end-products of sulfide oxidation at one site to other locations can influence the development of microbial populations and may facilitate chemical interactions leading to the modification of as yet unoxidized components. For example, a metal interchange reaction of the kind indicated below can lead to a stabilization of the concentration of solubilized copper and a conversion of chalcopyrite to the more readily oxidizable covellite (E. Peters, 1976, personal communication) (eqn (17)):

$$Cu^{2+} + CuFeS_2 \rightarrow Fe^{2+} + 2\ CuS \quad (17)$$

There is some evidence that the nickel-iron exchange in pyrrhotite, which has been studied by Ewers (1971) at higher temperatures (⩾200°C), may also occur during the bacterial leaching of nickel-iron sulfide minerals. The principal factors affecting the oxidation of sulfides in field situations are summarized in Table 6.3.4.

While some models of sulfide mineral leaching in natural or heap situations have considerable utility for the management of metal recovery operations (Harris, 1969), the complexity of such situations has so far defied efforts to construct models which accommodate all the observed phenomena, and there is a need for intensification of multi-disciplinary approaches to the overall problem. Considerably greater understanding of the interactions between microbial populations and mineral arrays in the dynamic patterns of the field context is needed before this task can be accomplished.

THE BIOLOGICAL OXIDATION OF ELEMENTAL SULFUR

Sulfur is one of the relatively few elements which occurs comparatively abundantly in free form in nature. It does not react readily with oxygen under abiotic conditions within the normal climatic temperature range and, like sulfide mineral ore bodies, massive, undisturbed accumulations of elemental sulfur may remain chemically unchanged over geological time. Indeed, the inertness of sulfur under appropriate conditions has commended it as a suitable form for the disposal of sulfur dioxide in gas streams (Johnson, 1972). Sulfur arises in nature by both chemical and biological processes, both frequently being associated with volcanic and hydrothermal activities. The roles of microorganisms in the biogenesis of elemental sulfur deposits (see Chapter 6.2) have been reviewed by Silverman and Ehrlich (1964) and Postgate (1965).

The mining of massive sulfur deposits and the exposure of the element to air and water permits the development of populations of sulfur-oxidizing bacteria, with concomitant formation of acidity and sulfate ions. The occurrence of *T. thiooxidans* and *T. thioparus* in the Rozdol deposit in Russia has been described by Karavaiko (1959; 1961). Similar events can occur during the industrial usage of elemental sulfur and the amenability of sulfur to bacterial oxidation has been widely exploited agriculturally for modification of soil acidity, supply of sulfate ion and for in situ solubilisation of rock phosphate (Starkey, 1950; Gleen and Quastel, 1953; Vitolins and Swaby, 1969). While most attention has been focussed on chemolithotrophic thiobacilli, such as *T. thiooxidans* and *T. thioparus*, an ability to oxidise elemental sulfur has been shown to be possessed by a number of heterotrophs such as the 35 species of *Streptomyces* examined by Yagi et al. (1971) and *Arthrobacter* (Ehrlich, 1962).

The most interesting examples of microbial oxidation of elemental sulfur are those which occur naturally in geothermal habitats, such as hot springs and solfatara soils. In both situations, the temperature is raised by steam from depth and elemental sulfur is present in large quantities from the spontaneous oxidation of hydrogen sulfide present in the steam (Schoen and Rye, 1970). Such hot springs and soil areas are highly acid due to sulfuric acid produced by the biological oxidation of the elemental sulfur. In moderately hot habitats of this kind (40—50°C), thermophilic forms of *Thiobacillus* spp. have been shown to commonly occur (Kaplan, 1956; Schoen and Ehrlich, 1968; Schwartz and Schwartz, 1965) and are probably the principal biological agents responsible for the oxidation of sulfur. At higher temperatures (60—90°C), the dominant and very widely distributed sulfur-oxidizing organisms belong to the genus *Sulfolobus* of which the type species is *S. acidocaldarius* (Brock et al., 1972). The contribution of other workers to the characterisation of these and closely-related organisms has been referred to earlier. These habitats probably define the limits of temperature and pH within which microbial activity is possible (Brock, 1969; Brock and Darland, 1970; Mosser et al., 1973).

The observations of Brock and Gustafson (1976) and Brock (1977) on the ability of acidophilic bacteria of the genera *Thiobacillus* and *Sulfolobus* to reduce ferric iron when growing on elemental sulfur as an energy source have interesting implications, both in respect of the bio-oxidation of sulfur per se and in the oxidative degradation of metallic sulfides. Elemental sulfur is an intermediate in some oxidation mechanisms of metallic sulfides and the possibility of its further oxidation under anaerobic conditions with ferric iron as the electron acceptor adds a further component to the spectrum of intermeshing chemical and biological reactions that could be operative in sulfidic mineral degradation in the field.

Various aspects of the mechanisms of microbial oxidation of sulfur have been referred to earlier, but it is clear that further investigation of such aspects as the nature of the frequently-observed close attachment of the cells to the sulfur surface and the penetration or otherwise of cell membranes by elemental sulfur (Kaplan and Rittenberg, 1962) would be of interest. Elemental sulfur occurs in a number of solid allotropic forms, its chemical activity is profoundly affected by a number of impurities, and it is photosensitive (Meyer, 1968). There is a paucity of information on the effects of variation of these factors on the amenability or otherwise of elemental sulfur to microbial attack.

REFERENCES

Agate, A.D. and Vishniac, W., 1973. Characterisation of *Thiobacillus* species by gas-liquid chromatography of cellular fatty acids. Arch. Mikrobiol., 89: 257—267.

Aleem, M.I.H., 1975. Biochemical reaction mechanisms in sulfur oxidation by chemosynthetic bacteria. Plant Soil, 43: 587—607.

Alexander, M., 1971. Microbial Ecology. John Wiley, New York, NY, 511 pp.

Andersen, J.E. and Allman, M.B., 1968. Some operational aspects of heap leaching at Rum Jungle. Proc. Aust. Inst. Min. Metal., 225: 27—31.

Arrieta, L. and Grez, R., 1971. Solubilisation of iron-containing minerals by soil microorganisms. Appl. Microbiol., 22: 487—490.

Auck, Y.T. and Wadsworth, M.E., 1973. Physical and chemical factors in copper dump leaching. In: D.J. Evans and R.S. Shoemaker (Editors), International Symposium on Hydrometallurgy, Chicago. American Institute of Mining Engineering, New York, pp. 645—700.

Baas Becking, L.G.M., Kaplan, I.R. and Moore, D., 1960. Limits of the natural environment in terms of pH and oxidation—reduction potentials. J. Geol., 68: 243—284.

Baker, R.A. and Wilshire, A.G., 1970. Microbiological factors in acid mine drainage formation: a pilot plant study. Environ. Sci. Technol., 4: 401—407.

Barridge, J.K. and Shively, J.M., 1968. Phospholipids of the *Thiobacilli.* J. Bacteriol., 95: 2182—2185.

Beck, J.V., 1960. A ferrous-ion-oxidizing bacterium. I. Isolation and some general physiological characteristics. J. Bacteriol., 79: 502—509.

Beck, J.V., 1967. The role of bacteria in copper mining operations. Biotechnol. Bioeng., 9: 487—497.

Beck, J.V., 1969. *Thiobacillus ferrooxidans* in its relation to the solubilisation of ores of copper and iron. In: D. Perlman (Editor), Fermentation Advances, Academic, New York and London, pp. 747—771.

Beck, J.V., and Brown, D.G., 1968. Direct sulfide oxidation in the solubilisation of sulfide ores by *Thiobacillus ferrooxidans.* J. Bacteriol., 96: 1433—1434.

Beck, J.V. and Shafia, F.M., 1964. Effect of phosphate ion and 2,4-dinitrophenol on the activity of intact cells of *Thiobacillus ferrooxidans.* J. Bacteriol., 88: 850—857.

Beebe, J.L. and Umbreit, W., 1971. Extracellular lipid of *Thiobacillus thiooxidans.* J. Bacteriol., 108: 612—614.

Belly, R.T. and Brock, T.D., 1974. Ecology of iron-oxidizing bacteria in pyritic materials associated with coal. J. Bacteriol., 117: 726—732.

Bhurat, M.C., Dwivedy, K.K., Jayaram, K.M.V. and Dar, K.K., 1973. Some results of microbial leaching of uranium ore samples from Narwapahar, Bhatin and Keruadungri, Singhbhum District, Bihar. Nat. Metal. Lab. Tech. J., 15: 47—51.

Bloomfield, C., 1972. The oxidation of iron sulphides in soils in relation to the formation of acid sulfate soils, and of ochre deposits in field drains. J. Soil Sci., 23: 1—16.

Bodo, C. and Lundgren, D.G., 1971. Uptake and oxidation of iron by *Thiobacillus ferrooxidans.* Abstrats of the 71st Annual Meeting of the American Society for Microbiology, p. 64.

Bodo, C. and Lundgren, D.G., 1974. Iron oxidation by cell envelopes of *Thiobacillus ferrooxidans.* Can. J. Microbiol., 20: 1647—1652.

Bohlool, B.B., 1975. Occurrence of *Sulfolobus acidocaldarius,* an extremely thermophilic acidophilic bacterium in New Zealand hot springs. Arch. Mikrobiol., 106: 171—174.

Brierley, C.L. and Brierley, J.A., 1973. A chemoautotrophic and thermophilic microorganism isolated from an acid hot spring. Can. J. Microbiol., 19: 183—188.

Brierley, C.L. and Murr, L.E., 1973. Leaching: Use of a thermophilic and chemoautotrophic microbe. Science, 179: 488—490.

Brierley, J.A. and Le Roux, N.W., 1977. A facultative thermophilic *Thiobacillus*-like bacterium: oxidation of iron and pyrite. Conference Bacterial Leaching 1977. GBF Monograph Series, No. 4 (August, 1977), pp. 55—66.

Brock, T.D., 1969. Microbial growth under extreme conditions. Symposia of the Society for General Microbiology, No. 19. Microbial Growth, pp. 15—41.

Brock, T.D., 1975. Effect of water potential on growth and iron oxidation by *Thiobacillus ferrooxidans.* Appl. Microbiol., 29: 495—501.

Brock, T.D., 1977. Ferric iron reduction by sulfur- and iron-oxidizing bacteria. Conference Bacterial Leaching. GBF Monograph Series, No. 4 (August 1977), p. 47.

Brock, T.D. and Darland, G.K., 1970. Limits of microbial existence: temperature and pH. Science, 169: 1316—1318.

Brock, T.D. and Gustafson, J., 1976. Ferric iron reduction by sulfur- and iron-oxidising bacteria. Appl. Environ. Microbiol., 32: 567—571.

Brock, T.D., Brock, K.M., Belly, R.T. and Weiss, L.R., 1972. *Sulfolobus*: A new genus of sulfur-oxidizing bacteria living at low pH and high temperature. Arch. Mikrobiol., 84: 54—68.

Bryner, L.C., Beck, J.F., Davis, D.B. and Wilson, D.G., 1954. Microorganisms in leaching sulfide minerals. Ind. Eng. Chem., 46: 2578—2592.

Bryner, L.C. and Jameson, A.K., 1958. Microorganisms in leaching sulfide minerals. Appl. Microbiol., 6: 281—287.

Bryner, L.C., Walker, R.B. and Palmer, R., 1967. Some factors influencing the biological and non-biological oxidation of sulfide minerals. Trans. Soc. Min. Eng., 238: 56—61.

Buchanan, R.E. and Gibbons, N.E. (Editors), 1974. Bergey's Manual of Determinative Bacteriology, Williams and Wilkins, Baltimore, MD, 8th edn, p. 160.

Carpenter, L. and Herndon, L., 1933. Acid mine drainage from bituminous coal mines. University of West Virginia Engineering Experimental Station Research Bulletin No. 10, 37 pp.

Colmer, A.R., 1962. Relation of the iron-oxidizer, *Thiobacillus ferrooxidans* to thiosulphate. J. Bacteriol., 83: 761—765.

Colmer, A.R. and Hinkle, M.E., 1947. The role of microorganisms in acid mine drainage. Science, 106: 253—256.

Colmer, A.R., Temple, K.L. and Hinkle, M.R., 1950. An iron-oxidizing bacterium from the acid drainage of some bituminous coal mines. J. Bacteriol., 59: 317—328.

Corrans, I.J., 1970. Studies on the Bacterial Leaching of Natural and Synthetic Iron Copper Sulfide Minerals. M.Sc. Thesis, University of New South Wales, 117 pp.

Corrans, I.J., Harris, B. and Ralph, B.J., 1972. Bacterial leaching: An introduction to its application and theory and a study of mechanism of operation. J. South African Inst. Min. Metal., 72: 221—230.

Davy, D.R., 1975. Rum Jungle Environmental Studies. Australian Atomic Energy Commission Research Establishment, Lucas Heights, N.S.W., 302 pp.

de Rosa, M., Gambacorta, A., Minale, L. and Bu'Lock, J.D., 1971. Bacterial triterpenes. Chem. Commun., pp. 619—620.

de Rosa, M., Gambacorta, A., Millonig, G. and Bu'Lock, J.D., 1974. Convergent characters of extremely thermophilic acidophilic bacteria. Experientia, 30: 866—868.

de Rosa, M., Gambacorta, A. and Bu'Lock, J.D., 1975. Extremely thermophilic acidophilic bacteria convergent with *Sulfolobus acidocaldarius*. J. Gen. Microbiol., 86: 156—164.

de Rosa, M., de Rosa, S., Gambacorta, A. and Bu'Lock, J.D., 1976. Isoprenoid triether lipids from Caldariella. Phytochemistry, 15: 1995—1996.

Dugan, P.R., 1975. Bacterial ecology of strip mine areas and its relationship to the production of acidic mine drainage. Ohio J. Sci., 75: 266—279.

Dugan, P.R. and Lundgren, D.G., 1965. Energy supply for the chemoautotroph *Ferrobacillus ferrooxidans*. J. Bacteriol., 89: 825—834.

Duncan, D.W., and Trussell, P.C., 1964. Adavances in microbiological leaching of sulfide ores. Can. Metall. Q., 3: 43—55.

Duncan, D.W. and Walden, C.C., 1972. Microbiological leaching in the presence of ferric iron. Dev. Ind. Microbiol., 13: 66—75.

Duncan, D.W., Landesman, J. and Walden, C.C., 1967. Role of *Thiobacillus ferrooxidans* in the oxidation of sulfide minerals. Can. J. Microbiol., 13: 397—403.

Duncan, D.W., Trussell, P.C. and Walden, C.C., 1964. Leaching of chalcopyrite with *Thiobacillus ferrooxidans*: Effect of surfactants and shaking. Appl. Microbiol., 12: 122—126.

Dunn, N.W., Ralph, B.J. and Rhodes, S.H., 1977. Further observations on the FAME profiles of *Thiobacillus* species. GIAM V Symposium, Bangkok, November, 1977, to be published.

Ehrlich, H.K., 1962. Microbial association with some mineral sulfides. In: M.L. Jensen (Editor), Biogeochemistry of Sulfur Isotopes. Yale University, New Haven, CT, pp. 153—168.

Ehrlich, H.L., 1963. Microorganisms in acid drainage from a copper mine. J. Bacteriol. 86: 350—352.

Ehrlich, H.L. and Fox, S.I., 1967. Environmental effects on bacterial copper extraction from low-grade copper sulfide ores. Biotechnol. Bioeng., 9: 471—485.

Ewers, W.E., 1971. Nickel-iron exchange in pyrrhotite. Annual Conference of Australian Institute of Mining and Metallurgy, Paper No. 14, 14-1—14-23.

Fliermans, C.B. and Brock, T.D., 1972. Ecology of sulfur-oxidizing bacteria in hot acid soils. J. Bacteriol., 111: 343—350.

Gaidarjiev, S., Groudev, S. and Genchev, F., 1975. Direct mechanism of bacterial oxidation of sulphide minerals. Paper 35 presented at 11th International Mineral Processing Congress, Cagliari, Italy (April, 1975), under auspices of "Regione Autonoma Della Sardegna".

Garrels, R.S.M. and Christ, C.L., 1965. Solutions, Minerals and Equilibria. Harper and Row, New York, NY, Evanston and London, 437 pp.

Gleen, H. and Quastel, J.H., 1953. Sulfur metabolism in soil. Appl. Microbiol., 1: 70—77.

Golding, R.M., Harris, B., Ralph, B.J., Rickard, P.A.D. and Vanselow, D.G., 1977. The nature of the passivation film on covellite exposed to oxygen. Conference Bacterial Leaching. GBF Monograph Series, No. 4, pp. 191—200.

Golding, R.M., Rae, A.D., Ralph, B.J. and Sulligoi, L., 1974. A new series of polynuclear copper dithiocarbamate — copper halide polymers. Inorg. Chem., 13: 2499—2504.

Groudev, S. and Genchev, F., 1978. Mechanisms of bacterial oxidation of chalcopyrite. Private communication of ms, in press.

Groudev, S.N., Genchev, F.N. and Gaidarjiev, S.S., 1978. Observations on the microflora in an industrial copper dump leaching operation. In: L. Murr, J.A. Brierley and A.E. Torma (Editors) Metallurgical Applications of Bacterial Leaching and Related Microbiological Phenomena. Academic, New York, NY, pp. 253—274.

Guay, R. and Silver, M., 1975. *Thiobacillus acidophilus* sp. nov.; isolation and some physiological characteristics. Can. J. Microbiol., 21: 281—288.

Harris, J.A., 1969. Development of a theoretical approach to the heap leaching of copper sulfide ores. Proc. Aust. Inst. Min. Metal., 230: 81—92.

Hutchinson, M., Johnstone, K.I. and White, D., 1966. Taxonomy of the acidophilic Thiobacilli. J. Gen. Microbiol., 44: 373—381.

Imai, K., Sakaguchi, H., Sugio, T. and Tano, T., 1973. On the mechanism of chalcocite oxidation by *Thiobacillus ferrooxidans.* J. Ferment. Technol., 51: 856—870.

Ivanov, V.I. and Lyalikova, N.N., 1962. On the taxonomy of the iron-oxidizing *Thiobacilli.* Mikrobiologiya, 31: 468—469.

Ivanov, V.I., Nagirnyak, F.I. and Stepanov, B.A., 1961. Bacterial oxidation of sulfide ores. I. The Role of *Thiobacillus ferrooxidans* in the oxidation of chalcopyrite and sphalerite. Mikrobiologiya, 30: 688—692.

Jackson, J.F., Moriarty, D.J.W. and Nicholas, D.J.D., 1968. Deoxyribonucleic acid base composition and taxonomy of *Thiobacilli* and some nitrifying bacteria. J. Gen. Microbiol., 53: 53—60.

Jellinek, F., 1968. Sulfides. In: G. Nickless (Editor), Inorganic Sulfur Chemistry. Elsevier, Amsterdam, pp. 669—747.

Jessup, K.L., 1976. Physico-chemical and Theoretical Studies of Some Transition Metal-sulfur Systems. Ph.D. Thesis, University of New South Wales, 238 pp.

Johnson, C.F., 1972. Paper presented at AIME meeting, San Francisco, February. Referred to in Reports on the Progress of Applied Chemistry, Vol. 58, 1975, p. 348.

Joseph, J.N., 1953. Microbiological study of acid mine waters: Preliminary report. Ohio J. Sc., 53: 125—127.

Kaplan, I.R., 1956. Evidence of microbiological activity in some of the geothermal regions of New Zealand. N.Z. J. Sci. Technol., 37: 639—663.

Kaplan, I.R. and Rittenberg, S.C., 1962. Fractionation of isotopes in relation to the problem of elemental sulfur transport by microorganisms. Nature, 194: 1098—1099.

Karavaiko, G.I., 1959. Znacheniye Biologicheskogo Faktora v Okislenii Soyedineniy Sery Rozdol'skogo Mestorozhdeniya (The significance of the biological factor in the oxidation of sulfur compounds in the Rozdol'skoye Deposit). Mikrobiologiya, 28: 846—850.

Karavaiko, G.I., 1961. O Mikrozonal'nom Rasprostranenii Okislitel'nykh Protsessov v Sernoy Rude Rozdol'skogo Mestorozhdeniya (On the microzonal distribution of oxidizing processes in the sulfur ore of the Rozdol'skoye Deposit. Mikrobiologiya, 30: 286—288.

Keenan, E.A., 1969. Bacterial Beneficiation of Uranium Minerals. Ph.D. Thesis. University of New South Wales, 156 pp.

Kelly, D.P., 1968. Biochemistry of oxidation of inorganic sulfur compounds by microorganisms. Aust. J. Sci., 31: 165—173.

Kelly, D.P., 1971. Autotrophy: Concepts of lithotrophic bacteria and their organic metabolism. Ann. Rev. Microbiol., 25: 177—210.

Khalid, A.M. and Ralph, B.J., 1976. Unpublished information on the microbial ecology of the Rum Jungle Mine area, Northern Territory, Australia. (Australian Atomic Energy Commission Research Contract 74/F/40.)

Khalid, A.M. and Ralph, B.J., 1977. The Leaching behaviour of various zinc sulphide minerals with three *Thiobacillus* species. Conference Bacterial Leaching. GBF Monograph Series, No. 4; pp. 165—173.

Kuznetsov, S.I., Ivanov, M.V. and Lyalikova, N.N., 1963. Introduction to Geological Microbiology. McGraw-Hill, New York, NY, 252 pp.

Lacey, D.T. and Lawson, F., 1970. Kinetics of the liquid-phase oxidation of acid ferrous sulfate by the bacterium *Thiobacillus ferrooxidans.* Biotechnol. Bioeng., 12: 29—50.

Landesman, J., Duncan, D.W. and Walden, C.C., 1966a. Iron oxidation by washed cell suspensions of the chemoautotroph, *Thiobacillus ferrooxidans.* Can. J. Microbiol., 12: 25—33.

Landesman, J., Duncan, D.W. and Walden, C.C., 1966b. Oxidation of inorganic sulfur compounds by washed cell suspensions of *Thiobacillus ferrooxidans.* Can. J. Microbiol., 12: 957—964.

Langworthy, T.A., 1977. Comparative lipid composition of heterotrophically and autotrophically grown *Sulfolobus acidocaldarius.* J. Bacteriol., 130: 1326—1332.

Lawrence, L.J. and Savage, E.N., 1975. Mineralogy of the titaniferous porphyry copper deposits of Melanesia. Aust. Inst. Min. Metal. Proc., 256: 1—14.

Lazaroff, N., 1963. Sulfate requirement for iron oxidation by *Thiobacillus ferrooxidans.* J. Bacteriol., 85: 78—83.

Leathen, W.W., Braley, S.A. and McIntyre, L.D., 1953. The role of bacteria in the formation of acid from certain sulfuric constituents associated with bituminous coal. II. Ferrous iron oxidizing bacteria. Appl. Microbiol., 1: 65—68.

Lundgren, D.G., Vestall, J.R., and Tabita, F.R., 1972. The microbiology of mine drainage. In: R. Mitchell (Editor), Water Pollution Microbiology, Wiley-Interscience, New York, pp. 69—88.

Mackintosh, M.E., 1971. Nitrogen fixation by *Thiobacillus ferrooxidans* species. J. Gen. Microbiol., 66: i—ii.
Mackintosh, M.E., 1976. Fixation of $^{15}N_2$ by *Thiobacillus ferrooxidans.* Proc. Soc. Gen. Microbiol. 4: 23.
Madgwick, J.C. and Ralph, B.J., 1977. The metal-tolerant alga *Hormidium fluitans* (Gay) Heering from acid mine drainage waters in Northern Australia and Papua-New Guinea. Conference Bacterial Leaching. GBF Monograph Series, No. 4, pp. 85—91.
Majima, H. and Peters, E., 1968. Electrochemistry of sulfide dissolution in Hydrometallurgical systems. VIII International Mineral Processing Congress, Leningrad.
Manning, H.L. 1975. New medium for isolating iron-oxidizing and heterotrophic acidophilic bacteria from acid mine drainage. Appl. Microbiol., 30: 1010—1016.
Marchlewitz, B. and Schwartz, W., 1961. Investigation of microbial associations in acid mine waters. Z. Allg. Mikrobiol., 1: 100—114.
Margalith, P., Silver, M. and Lundgren, D.G., 1966. Sulfur oxidation by the iron bacterium *Ferrobacillus ferrooxidans.* J. Bacteriol., 92: 1706—1709.
Meyer, B., 1968. Elemental sulfur. In: G. Nickless (Editor), Inorganic Sulfur Chemistry. Elsevier, Amsterdam, pp. 241—258.
Millonig, G., de Rosa, M., Gambacorta, A. and Bu'lock, J.D., 1975. Ultrastructure of an extremely thermophilic acidophilic microorganism. J. Gen. Microbiol., 86: 165—173.
Mizoguchi, T., Sato, T. and Okabe, T., 1976. New sulfur-oxidizing bacteria capable of growing heterotrophically, *Thiobacillus rubellus* nov. sp. and *Thiobacillus delicatus* nov. sp. J. Ferment. Technol., 54: 181—191.
Morimoto, N., Gyobu, A., Mukaiyama, H. and Izawa, E., 1975. Crystallography and stability of pyrrhotites. Econ. Geol. 70: 824—833.
Moss, F.J. and Andersen, J.R., 1968. The effects of environment on bacterial leaching rates. Aust. Inst. Min. Metal. Proc., 225: 15—25.
Mosser, J.L., Mosser, A.G. and Brock, T.D., 1973. Bacterial origin of sulfuric acid in geothermal habitats. Science, 179: 1323—1324.
Muir, D.M., Parker, A.J. and Giles, D.W., 1976. Cuprous Hydrometallurgy Part IV. Rates and equilibria in the reaction of copper sulfides with copper (II) sulfate in aqueous acetonitrile. Hydrometallurgy, 2: 127—140.
Nickel, E.H., Ross, J.R. and Thornber, M.R., 1974. The supergene alteration of pyrrhotite — pentlandite ore at Kambalda, Western Australia. Econ. Geol., 69: 93—107.
Nickerson, W.J., 1969. Microbial transformations of naturally occurring polymers. In: D. Perlman (Editor), Fermentation Advances, Academic, New York and London, pp. 631—634.
Nielsen, A.M. and Beck, J.V., 1972. Chalcocite oxidation and coupled carbon dioxide fixation by *Thiobacillus ferrooxidans.* Science, 175: 1124—1126.
Peck, H.D. Jr., 1962. Comparative metabolism of inorganic sulfur compounds in microorganisms. Bacteriol. Rev., 26: 67—94.
Peck, H.D. Jr., 1968. Energy-coupling mechanisms in chemolithotrophic bacteria. Ann. Rev. Microbiol., 22: 489—518.
Postgate, J.R., 1965. Recent advances in the study of the sulfate-reducing bacteria. Bacteriol. Rev., 29: 425—441.
Pourbaix, M.J.N. (Editor), 1966. Atlas of Electrochemical Equilibria in Aqueous Solutions, Pergamon, Oxford, 644 pp.
Rao, G.S. and Berger, L.R., 1971. The requirement of low pH for growth of *Thiobacillus thiooxidans.* Arch. Mikrobiol., 79: 338—344.
Razzell, W.E. and Trussell, P.C., 1963a. Isolation and properties of an iron-oxidizing *Thiobacillus.* J. Bacteriol., 85: 595—603.
Razzell, W.E. and Trussell, P.C., 1963b. Microbiological leaching of metallic sulfides, Appl. Microbiol., 11: 105—110.

Roberts, W.M.B., 1960. Mineralogy and genesis of White's Orebody, Rum Jungle Uranium Field, Australia. Neues Jahrb. Mineral., Abh., 94: 868—889.
Roberts, D.B. and Williams, D.J., 1976. The reaction of steam with chalcopyrite. Trans. Inst. Min. Metal. (London), Sect. C, 85: 133—138.
Roy, A.B. and Trudinger, P.A., 1970. The Biochemistry of Inorganic Compounds of Sulfur. Cambridge University Press, Cambridge, 400 pp.
Rudolfs, W. and Helbronner, A., 1922. Oxidation of zinc sulfide by microorganisms. Soil Sci., 14: 459—464.
Sadler, W.R. and Trudinger, P.A., 1967. The inhibition of microorganisms by heavy metals. Miner. Deposita, 2: 158—168.
Sakaguchi, H., Torma, A.E. and Silver, M., 1976. Microbiological oxidation of synthetic chalcocite and covellite by *Thiobacillus ferrooxidans*. Appl. Environ. Microbiol., 31: 7—10.
Schaeffer, W.I. and Umbreit, W.W., 1963. Phosphatidyl-inositol as a wetting agent in sulfur oxidation by *Thiobacillus thiooxidans*. J. Bacteriol., 85: 492—493.
Schnaitman, C.A., Korczynski, M.S. and Lundgren, D.G., 1969. Kinetic studies of iron oxidation by whole cells of *Ferrobacillus ferrooxidans*. J. Bacteriol., 99: 552—557.
Schoen, R. and Ehrlich, G.G., 1968. Bacterial origin of sulfuric acid in sulfurous hot springs. Twenty-third International Geological Congress, Vol. 17. Prague, pp. 171—178.
Schoen, R. and Rye, R.O., 1970. Sulfur isotope distribution in solfataras, Yellowstone National Park. Science, 170: 1082—1084.
Schrauzer, G.N. and Guth, T.D., 1976. Hydrogen evolving systems. I. The formation of H_2 from aqueous suspensions of $Fe(OH)_2$ and reactions with reducible substrates, including molecular nitrogen. J. Am. Chem. Soc., 98: 3508—3513.
Schwartz, A. and Schwartz, W., 1965. Geomikrobiologische Untersuchungen. VII. Uber das Vorkommen von Mikroorganismen in Solfataren und Heiben Quellen. Z. Allg. Mikrobiol., 5: 395—405.
Shively, J.M. and Benson, A.A., 1967. Phospholipids of *Thiobacillus thiooxidans*. J. Bacteriol., 94: 1679—1683.
Shivvers, D.W. and Brock, T.D., 1973. Oxidation of elemental sulfur by *Sulfolobus acidocaldarius*. J. Bacteriol., 114: 706—710.
Sillen, L.G. and Martell, A.E., 1964. Stability Constants of Metal-ion Complexes. The Chemical Society, London, 754 pp.
Silver, M. and Torma, A.E., 1974. Oxidation of metal sulfides by *Thiobacillus ferrooxidans* grown on different substrates. Can. J. Microbiol., 20: 141—147.
Silverman, M.P., 1967. Mechanism of bacterial pyrite oxidation. J. Bacteriol., 94: 1046—1051.
Silverman, M.P. and Ehrlich, H.L., 1964. Microbial formation and degradation of minerals. Adv. Appl. Microbiol., 6: 153—206.
Silverman, M.P. and Lundgren, D.G., 1959a. Studies on the chemoautotrophic bacterium *Ferrobacillus ferrooxidans*. I. An improved medium and a harvesting procedure for securing high cell yields. J. Bacteriol., 77: 642—647.
Silverman, M.P. and Lundgren, D.G., 1959b. Studies on the chemoautotrophic bacterium *Ferrobacillus ferrooxidans*. II. Manometric studies. J. Bacteriol., 78: 326—331.
Silverman, M.P., Rogoff, M.H. and Wender, I., 1961. Bacterial oxidation of pyritic materials in coal. Appl. Microbiol., 9: 491—496.
Singer, P.C. and Stumm, W., 1970. Acidic mine drainage: the rate-determining step. Science, 167: 1121—1123.
Starkey, R.L., 1950. Relations of microorganisms to transformations of sulfur in soils. Soil Sci., 70: 55—65.

Starkey, R.L., Jones, G.E. and Frederick, L.R., 1956. Effects of medium agitation and wetting agents on oxidation of sulfur by *Thiobacillus thiooxidans.* J. Gen. Microbiol., 15: 329—334.

Steiner, M. and Lazaroff, N., 1974. Direct method for continuous determination of iron oxidation by autotrophic bacteria. Appl. Microbiol., 28: 872—880.

Suzuki, I., 1974. Mechanisms of inorganic oxidation and energy coupling. Ann. Rev. Microbiol., 28: 85—101.

Taylor, R.M. and Fischer, W.R., 1973. Electric potentials between different rocks. Naturwissenschaften, 63: 298.

Torma, A.E., 1977. The role of *Thiobacillus ferrooxidans* in hydrometallurgical processes. Adv. Biochem. Eng., 6: 1—37.

Torma, A.E. and Legault, G., 1973. Metal sulfides biodegradation by *Thiobacillus ferrooxidans*: Effect of their surfaces. Ann. Microbiol. (Inst. Pasteur), 124A: 111—121.

Torma, A.E., Walden, C.C. and Branion, R.M.R., 1970. Microbiological leaching of a zinc concentrate. Biotechnol. Bioeng., 12: 501—507.

Torma, A.E., Legault, G., Kougiomoutzakis, D. and Ouellet, R. 1974. Kinetics of biooxidation of metal sulfides. Can. J. Chem. Eng., 52: 515—517.

Torma, A.E., Walden, C.C., Duncan, D.W. and Branion, R.M.R., 1972. The effect of carbon dioxide and particle surface area on the microbiological leaching of a zinc sulfide concentrate. Biotechnol. Bioeng., 14: 777—786.

Trivedi, N.C. and Tsuchiya, H.M., 1975. Microbial mutualism in leaching of Cu — Ni sulfide concentrate. Int. J. Miner. Process., 2: 2—14.

Trudinger, P.A., 1967. The metabolism of inorganic sulfur compounds by *Thiobacilli.* Rev. Pure Appl. Chem., 17: 1—24.

Trudinger, P.A., 1969. Assimilatory and dissimilatory metabolism of inorganic sulfur compounds by microorganisms. Adv. Microb. Physiol., 3: 111—157.

Trussell, P.C., Duncan, D.W. and Walden, C.C., 1964. Biological mining. Can. Min. J., 83: 46—49.

Tsuchiya, H.M., Trivedi, N.C. and Schuler, M.L., 1974. Microbial mutualism. Biotechnol. Bioeng., 16: 991—995.

Tuovinen, O.H. and Kelly, D.P., 1972. Biology of *Thiobacillus ferrooxidans* in relation to the microbiological leaching of sulfide ores. Z. Allg. Mikrobiol., 12: 311—346.

Tuovinen, O.H. and Kelly, D.P., (1974 (a) (b) (c). Studies on the growth of *Thiobacillus ferrooxidans.* II. Toxicity of uranium to growing cultures and tolerance conferred by mutation, other metal cations and EDTA. Arch. Mikrobiol., 95: 153—164. III. Influence of uranium, other metal ions and 2,4-dinitrophenol on ferrous iron oxidation and carbon dioxide fixation by cell suspensions. ibid, 95: 165—180. IV. Influence of monovalent metal cations on ferrous iron oxidation and uranium toxicity in growing cultures. ibid, 98: 167—174.

Tuovinen, O.H., Kelly, B.C. and Nicholas, D.J.D., 1975. The uptake and assimilation of sulfate by *Thiobacillus ferrooxidans.* Arch. Mikrobiol., 105: 123—127.

Tuovinen, O.H., Kelly, B.C. and Nicholas, D.J.D., 1976. Enzymic comparison of the inorganic sulfur metabolism in autotrophic and heterotrophic *Thiobacillus ferrooxidans.* Can. J. Microbiol., 22: 109—113.

Tuovinen, O.H., Niemelä, S.I. and Gyllenberg, H.G., 1971a. Tolerance of *Thiobacillus ferrooxidans* to some metals. Antonie van Leeuwenhoek; J. Microbiol. Serol., 37: 489—496.

Tuovinen, O.H., Niemelä, S.I. and Gyllenberg, H.G., 1971b. Effect of mineral nutrients and organic substances on the development of *Thiobacillus ferrooxidans.* Biotechnol. Bioeng., 13: 517—527.

Tuttle, J.H., Randles, C.I. and Dugan, P.R., 1968. Activity of microorganisms in acid mine water. I. Influence of acid water on aerobic heterotrophs of a normal stream. J. Bacteriol., 95: 1495—1503.

Unz, R.F. and Lundgren, D.G., 1961. A comparative nutritional study of three chemo-autotrophic bacteria: *Ferrobacillus ferrooxidans, Thiobacillus ferrooxidans,* and *Thiobacillus thiooxidans.* Soil Sci., 92: 302—313.
Updegraff, D.M. and Duoros, J.D., 1972. The relationship of microorganisms to uranium deposits. In: Developments in Industrial Microbiology, Vol. 13 (Ch. 8, Symposium: Minerals Microbiology), pp. 76—90.
Vanselow, D.G., 1976. Mechanisms of Bacterial Oxidation of the Copper Sulfide Mineral Covellite. Ph.D. Thesis, University of New South Wales, 198 pp.
Vishniac, W. and Santer, M., 1957. The *Thiobacilli.* Bacteriol. Rev., 21: 195—213.
Vitolins, M.I. and Swaby, R.J., 1969. Activity of sulfur-oxidizing microorganisms in some Australian soils. Aust. J. Soil Res., 7: 171—183.
Wadsworth, M.E., 1972. Second Tutorial Symposium on Extractive Metallurgy, Leaching of Low and High Grade Ores, Dump Deposits and Concentrates, Part III, Advances in the leaching of sulfide and oxide minerals. University of Utah.
Walsh, F. and Mitchell, R., 1972a. A pH-dependent succession of iron bacteria. Environ. Sci. Technol., 6: 809—812.
Walsh, F. and Mitchell, R., 1972b. An acid-tolerant iron-oxidizing Metallogenium. J. Gen. Microbiol., 72: 369—376.
Walsh, F. and Mitchell, R., 1972c. Biological control of acid mine pollution. J. Water Pollut. Contr., 44: 763—768.
Walsh, F. and Mitchell, R., 1975. Mine drainage pollution reduction by inhibition of iron bacteria. Water Res., 9: 525—528.
Weast, R.C. (Editor), 1974. Handbook of Chemistry and Physics. Chemical Rubber Company, Cleveland, OH, 1918 pp.
Williams, R.A.D. and Hoare, D.W., 1972. Physiology of a new facultatively autotrophic thermophilic *Thiobacillus.* J. Gen. Microbiol., 72: 555—566.
Wong, C.W., Scharer, J.M. and Reilly, P.M., 1974. A discrimination among microbial iron oxidation mechanisms. Can. J. Chem. Eng., 52: 645—653.
Yagi, S., Kitai, S. and Kimura, T., 1971. Oxidation of elemental sulfur to thiosulfate by *Streptomyces.* Appl. Microbiol., 22: 157—159.
Zajic, J.E., 1969. Microbial Biogeochemistry. Academic, New York and London, 345 pp.
Zavarsin, G.A., 1972. A heterotrophic satellite of *Thiobacillus ferrooxidans.* Mikrobiologiya, 41: 369—370.

Chapter 6.4

BIOGEOCHEMICAL CYCLING OF SULFUR

H.R. KROUSE

Department of Physics, The University of Calgary, 2920, 24th Avenue, N.W., Calgary, Alberta T2N 1N4 (Canada)

R.G.L. McCREADY

Lethbridge Research Station, Agriculture Canada, Lethbridge, Alberta T1J 4B1 (Canada)

CONTENTS

INTRODUCTION

The biological cycling of sulfur has been outlined in Chapter 6.1. In nature, this cycle forms part of an overall global geochemical sulfur cycle which embraces chemical as well as biological transformations and includes anthropogenic sources of sulfur compounds. Reservoirs and fluxes for the global cycle are usually considered in terms of the atmosphere, hydrosphere, biosphere, lithosphere, and pedosphere. Difficulties arise in rigidly defining the spheres. The biosphere requires water and the pedosphere may be considered as made up from the lithosphere, biosphere and hydrosphere.

A version of the global sulfur cycle is shown in Fig. 6.4.1 but this must, in many ways, be regarded as much oversimplified. For example, it does not show the sulfur of plants undergoing transition to fossil fuels with subsequent combustion and release of SO_2 to the atmosphere. It does not show oceanic volcanism or deep circulation of waters to produce thermal springs and fumaroles. However, it depicts transfers of sulfur compounds among the

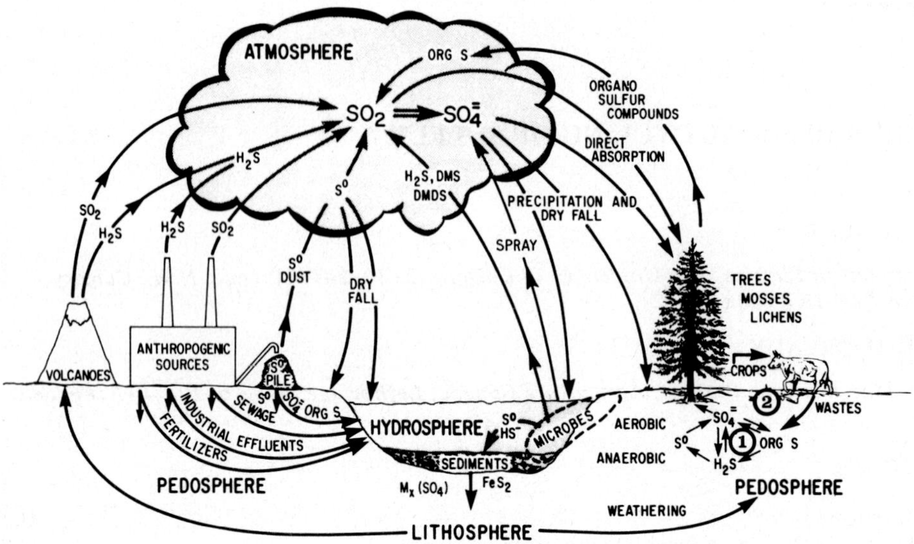

Fig. 6.4.1. Simplified version of the overall sulfur cycle in nature. The encircled 1 refers to the microbiological cycle while the encircled 2 depicts a food cycle with higher animals.

various spheres and provides some examples of perturbations by man. It also shows that sulfur sub-cycles operate within the overall cycle. The microbial sub-cycle, for example, is designated "1" and a food sub-cycle is designated "2".

Biological and abiological processes are frequently coupled in nature and a thorough understanding of biological factors in mineral cycling must encompass the global sulfur cycle and its changing fluxes over geological time.

In this chapter, we do not attempt to derive a quantitative global sulfur cycle. Versions have been presented recently be several authors (Eriksson, 1963; Junge, 1963; Holser and Kaplan, 1966; Robinson and Robbins, 1970; Kellogg et al., 1972; Friend, 1973; Granat et al., 1976).

Although it is relatively easy to find flaws in these recent versions, a major reconstruction will not be meaningful until many more reliable data become available. Whereas some reservoirs and fluxes are reasonably well known, e.g., dissolved sulfate in the oceans and emissions from carefully monitored industries, other estimates may have uncertainties of up to two orders of magnitude. In particular, biogenic fluxes are difficult to assess. Even if steady-state concentrations were established more precisely with a global sampling, it is another matter to determine proper fluxes. Although numbers from the above references will be used to give the reader an appreciation of the quantities of sulfur involved, our prime intention is to evaluate the overall biological participation in the sulfur cycle. Most of the authors cited

TABLE 6.4.1

Sulphur content of terrestrial spheres (Tg) (after Friend, 1973)

Lithosphere	$(12 \pm 6) \times 10^9$
Hydrosphere	$(1.3 \pm 0.1) \times 10^9$
Biosphere (dead)	$5 \leftarrow\!\!\longrightarrow \times 10^3$
Biosphere (living)	$8 \leftarrow\!\!\longrightarrow \times 10^2$
Atmosphere	1.8

above advise against accepting their models in entirety and warn against predictions of concentrations and fluxes resulting from projected increased emissions.

Table 6.4.1 gives estimates of the amounts of sulfur in different "spheres" of our earth. Fluxes between these spheres will be considered on pp. 408ff. It is interesting to note that the atmosphere, with the smallest amount of sulfur, has the largest fluxes associated with it while the converse is true for the lithosphere. This is desirable for life as we know it. Larger amounts of sulfur in the atmosphere could be intolerable, and the hydrosphere in contact with the lithosphere would possess excessive sulfur if fluxes from the lithosphere were larger.

PROBLEMS IN EVALUATING MICROBIAL ACTIVITY IN NATURE

Difficulties in evaluating mineral biogenesis are discussed in Chapter 6.2. Assessment of current natural microbial activities is also fraught with problems. Microorganisms survive in microenvironments of their own creation within soil and sediments. Consequently the evaluation, isolation, or identification of the various species present in any ecological sample is extremely difficult. Upon selection and removal of a sample, changes in chemical and physical properties alter these microenvironments. Further, the microenvironment may differ widely in nutrient quality and quantity, pH, and osmotic pressure from the macroenvironment which can be analysed with ordinary instruments and chemical procedures (Brock, 1971). The physicochemical conditions of the natural environment and the competition among organisms in nature are difficult, if not impossible, to simulate in the laboratory.

As a myriad of different and distinct species of microorganisms may exist in any given ecological sample, it is difficult to determine the total number of different species present using standard microbiological techniques. A multitude of selective media may be required and dormant spores or bacterial cysts present in the sample may develop on the laboratory medium giving the investigator a false impression of which species were most active.

There is also the problem of determining whether a given species is merely surviving or is actively engaged in transforming a sulfur compound. Growth and activity on laboratory media may not reflect the performance of the species in its natural environment. Interactions and competition among microorganisms are suppressed during laboratory culturing procedures so that the observations may bear little resemblance to the actual ecological activity. Similarly, viruses or bacteriophages are usually ignored during ecological analyses. Therefore, normal antagonistic interactions which occur in nature are overlooked during microbial assays.

In summary, it is extremely difficult to evaluate processes such as biological conversions of minerals using current microbiological techniques. Hence biological contributions to the cycling of sulfur have been estimated by the bulk behaviour of systems. The flaw in this approach is that biotic and abiotic processes are not delineated to the extent desirable.

INDIRECT BIOLOGICAL FACTORS WHICH AFFECT THE SULFUR CYCLE

When sulfate minerals are mobilised by sulfate-reducing bacteria, the question arises as to the extent of direct and indirect processes. If an aqueous sulfate ion is reduced, there is a shift in the equilibrium between dissolved and precipitated sulfate, favouring dissolution. On the other hand, the bacteria may be adsorbed to the mineral surface to carry out sulfate reduction in a microenvironment differing from its bulk surroundings. This has not been demonstrated with sulfate reducers but has been recorded during oxidation of sulfur compounds (see p. 419). Finally, organic acids released during metabolism may assist in chemically solubilizing sulfate minerals. Bolze et al. (1974), for example, described the solubilization of barite in the presence of sulfate-reducers. The reaction was inhibited by the presence of soluble sulfate indicating that the organisms attacked barite only when more readily available substrates were not available.

On a macroscopic scale, Goldhaber (1974) suggested that the burrows of benthonic organisms, in the upper half metre of a sediment column, facilitate the supply of sulfate to sediments. This mechanism may account for the relatively constant sulfate concentration in pore-water profiles where the presence of iron sulfide attests to significant sulfate reduction. Roots, in addition to functioning as conduits for sulfate uptake by plants, secrete organic exudates and with branching of their roots through the soil, slough off dead cells that provide carbon sources for sulfur bacteria.

Joshi and Hollis (1977) reported a mutual interdependence between *Beggiatoa* and rice seedlings exposed to hydrogen sulfide. *Beggiatoa* caused a drop in hydrogen sulfide levels in soils and an increase in oxygen release from rice plants while rice plants in turn increased *Beggiatoa* survival in flooded soils. The authors suggest that similar roles may be played by *Beggiatoa* and other plants in swamps, coastal areas, etc.

It can be seen from the few examples cited above that indirect biological factors in sulfur cycling are numerous and difficult to assess quantitatively.

ELUCIDATION OF THE SULFUR CYCLE WITH STABLE ISOTOPES

The potential use of sulfur isotope abundances in identifying reductive processes and the origin of sulfide minerals is discussed in Chapter 6.2. Although the majority of laboratory experiments have been carried out with reducing bacteria, there have been a sufficient number of studies with other bacteria to indicate that stable isotopes can play a role in elucidating the whole sulfur cycle. Indeed attempts have been made to use isotopic and mass balances to estimate global sulfur shifts with time (Holser and Kaplan, 1966).

In contrast to reduction, oxidation processes are characterised by a relatively small sulfur isotopic selectivity (Table 6.4.2). Many examples exist where small isotopic differences between sulfate and sulfide species have helped identify oxidation processes. Field and Lombardi (1972) used the negligible fractionation factors for sulfate-sulfide mineral pairs (0.995 to 1.004) to argue that alunite and barite of the Tolfa district, Italy, were of supergene origin, i.e., derived from surficial oxidation of sulfides. In such studies, it would be convenient if isotope abundances provided a means of distinguishing chemical and biological oxidation. Schoen and Rye (1970) attempted to assess the relative contributions of biological and purely chemical oxidation of sulfide in thermal springs at Yellowstone. Unfortunately, however, it is doubtful whether the characteristics of isotope discrimination during microbial sulfide oxidation under different conditions, and by different organisms, are sufficiently established to carry out such assessments with confidence.

The data in Table 6.4.2 show that isotope selectivity during oxidation of sulfide and elemental sulfur varies with the organism, the form of reactant and, presumably, with other properties of the medium. The limited information on *Beggiatoa* suggests that this organism does not influence isotope fractionation since the results were the same in both sterile and inoculated media. The small isotope selectivities obtained with solid reactants (S_8, FeS, FeS_2) is consistent with the oxidation proceeding layer by layer. Sulfate, produced during the oxidation of soluble sulfide in the presence of *Thiobacillus* or *Chromatium*, may be depleted to a greater or less extent in ^{34}S than that formed by aeration of soluble sulfide in a sterilized medium. With *Rhodopseudomonas*, the isotopic effect is the reverse of that formed for simple aeration of soluble sulfide. In biological systems, intermediate polythionates were found to be enriched in ^{34}S by as much as 11‰ with respect to initial sulfate. On the present evidence, it can only be suggested that some effects may be purely kinetic, thereby enriching sulfate in ^{32}S, while

TABLE 6.4.2

Isotope selectivity during oxidation of sulfur compounds

[a]	Substrate oxidant	T (°C)	$\delta^{34}S$ w.r.t. initial reactant S^0	$S_xO_6^{2-}$	SO_4^{2-}	References [b]
S^{2-}	*Thiobacillus*	—	−3.0 to −6.0	—	−6.0 to −9.0	1
	Thiobacillus	—	−2.5 to −1.2	+3.2 to +10.6	−10.5 to −18.0	2
S_8	*Thiobacillus*	—	—	—	−0.1 to +1.4	2
S^{2-}	*Chromatium*	30	−3.6 to −10.0	+4.9 to +11.2	−0.9 to −2.9	2
S_8	*Chromatium*		—	—	−0.3 to +0.4	2
S^{2-}	*Rhodopseudomonas*	30	−0.2 to −1.0	—	+2.05	3
S_8	Mixed culture	30	—	—	+0.3 to −1.4	4
FeS_2	Mixed culture	30	—	—	−1.0 to −1.7	4
S^{2-}	*Beggiatoa*	24	—	—	−6.0	5
S^{2-}	Air	24	—	—	−6.0	5
FeS	H_2O_2	0—100	+0.5 to +1.8	—	−0.8 to −2.0	6

[a] S^{2-} is used to signify all aqueous sulfide ions.

[b] References: 1, Kaplan and Rafter (1958); 2, Kaplan and Rittenberg (1964); 3, Mekhtieva and Kondrat'eva (1966); 4, Nakai and Jensen (1964); 5, H.R. Krouse, R. Lewin and A. Sasaki (unpublished data; 1967); 6, Lewis and Krouse (1969).

isotopic exchange among intermediates and/or products should favour ^{34}S in sulfur and compounds with higher valence states as predicted by statistical thermodynamics (Tudge and Thode, 1950).

Sulfur isotopes are more fruitful in resolving the relative contributions of reduction or oxidation processes. For example, in the Peace River area of Alberta, Canada, sulfur and sulfate are deposited on the tops of peat occurrences. At depths of 80 cm and lower, pyrites are found with $\delta^{34}S$ values typically around −24‰. Above 80 cm depth, elemental sulfur begins to dominate and has $\delta^{34}S$ values in the range −28 to −30‰. The sulfate deposits very close to the surface have $\delta^{34}S$ values near −33‰. The isotopic data are consistent with the preferential reaction of ^{32}S during sulfide oxidation. This, coupled with the occurrence of the more oxidised species near the surface, leads to the conclusion that the elemental sulfur originated from deeper sulfide sources (Krouse, 1979).

Sulfate encrustations around some thermal springs in the Northwest Territories of Canada have $\delta^{34}S$ values as low as −30‰ whereas those of sulfate in the nearby spring waters are nearer +20‰. With the use of high volume atmospheric sampling equipment, it was found that H_2S and SO_2 in the air are also significantly depleted in ^{34}S. Clearly, the mineralised sulfate was derived from the atmospherically-oxidised biogenic H_2S which evolved from the thermal springs (van Everdingen and Krouse, 1977).

An interesting case where sulfur isotope data appear inconsistent with proposed biogenesis concerns the gypsum deposits (10^3 km^2) in the vicinity of Swakopmund and Walvis Bay, South West Africa. During the summer months, changes in ocean currents destroy coastal plankton and create a lethal environment for fish (Martin, 1963). Decomposition of the plankton and dead fish by facultative microorganisms on the sea bottom releases sulfide from organic sulfhydryl compounds. In addition, organic acids serve as a carbon source for anaerobic sulfate reducers which produce more sulfide. The quantities of H_2S are no doubt sizeable as vividly described by Butlin (1953) and Martin (1963). The latter author suggested that H_2S precipitated with mist and dew had been sufficient to react with the Tertiary and Pleistocene calcretes to produce the gypsum deposits. However, sulfur isotope analyses of 22 gypsum and anhydrite samples extending from the coast inland to over 100 km from Walvis Bay (H.R. Krouse and F. Cagle, 1967, unpublished data) gave $\delta^{34}S$ values of +16.4‰ (S.D. 1.4). This narrow range in isotopic composition over such a large area seems inconsistent with erratic sulfide production from rotting fish and sulfate reduction, so that alternate sources of the sulfate such as the nearby Dwyka shales seem more appropriate.

In addition to sulfate reduction and sulfide oxidation, isotopic fractionation data have been obtained for other conversions in the sulfur cycle. With the exception of sulfate and sulfite reduction by *Saccharomyces cerevisiae* under varied concentrations of added panthothenate (McCready et al.,

1974) assimilation by microorganisms as well as plants and animals results in small isotopic selectivity (Ishii, 1953; Kaplan and Rittenberg, 1964; Mekhtieva, 1971). Reduction of elemental sulfur by *S. cerevisiae* to H_2S was accompanied by negligible isotopic fractionation (Kaplan and Rittenberg, 1964). It was suggested that each S_8 molecule was metabolized by the cell in such a way that there was no interchange of sulfur atoms with the external sulfur pool. Kaplan and Rittenberg (1964) also reported that H_2S released during hydrolysis of the sulfhydryl bond (-C-SH) by *Proteus vulgaris* metabolising cysteine, was slightly depleted in ^{34}S (−5‰ with respect to cysteine-sulfur). More isotopic data from degradation experiments are desirable for interpreting natural situations where organic matter is undergoing decay. Sulfide produced by both organic degradation and sulfate reduction should have an isotopic composition which reflects both processes.

In addition to the problems in distinguishing individual biological conversions as discussed above, there are more general difficulties in using sulfur isotopes to elucidate the natural sulfur cycle. The identification of sources of sulfur compounds is hampered if extensive biological isotopic fractionation has occurred. Another problem arises when product molecules are converted back to reactant molecules and mix with the reactant. For example, H_2S from biological sulfate reduction in springs leaves the aqueous environment to undergo oxidation in the air and may form sulfate minerals, relatively depleted in ^{34}S, some distance away (van Everdingen and Krouse, 1977). When precipitation later dissolves these sulfates and bears them back to the initial stream, the isotopic data reflect this recycling and cannot be readily interpreted.

PROCESSES, FLUXES AND RESERVOIRS IN THE SULFUR CYCLE

Hydrosphere

The hydrosphere plays many roles in the sulfur cycle. Foremost, the biosphere requires water to function. The hydrosphere is a transport medium for ions, gases, particulates and life. It is also an accumulation of assorted chemical solutions. These many facets are exemplified by supergene enrichment of metal sulfides in the vicinity of the water table. The overall mechanism operates as follows. Iron sulfides are oxidised chemically and/or biologically to sulfuric acid and ferric sulfate, which subsequently react with other metal sulfides (see Chapters 4 and 6.3). The soluble products move downward in solution. Upon encountering the hypogene zone where O_2 is absent, reaction with dissolved sulfide or the primary metal sulfides forms supergene enrichments of the sulfide minerals. Depending upon tolerance to metal ions, sulfate reducers might provide sulfide for the processes in the lower zone. The biological regimes will shift vertically, dependent upon the behaviour of the water table. Many other examples can be cited where the biosphere and the physical-chemical properties of the hydrosphere effectively combine to cause mineral solution or deposition (Perel'man, 1967; Lelong et al., 1976). However this section will stress water as a medium while the next section will discuss the associated biology.

Approximately 72% of the earth's surface area is covered by water and 98% of the total volume of water is found in the oceans. The oceans of today are well mixed in terms of sulfate ion concentration and consistency of its $\delta^{34}S$ values near +20‰ (Thode et al., 1961). The total sulfate content of the oceans is one of the more readily estimated quantities in the sulfur cycle with a value of 1.2 to 1.4×10^9 Tg S.

Although the modern ocean is isotopically uniform, sulfur isotope data from evaporites show that ancient oceanic sulfate underwent large excursions in $\delta^{34}S$ values (Fig. 6.2.7, p. 351). Minimum values of less than +10‰ existed in the Permian while maximum values of near +35‰ were attained during the Precambrian and Ordovician. The Devonian and Triassic are characterized by excursions of about 10% . From the Tertiary to present, the δ value has remained close to +20‰. There were probably localised minor variations of shorter time duration superimposed on the general trends but gross features of the behaviour appear to have been worldwide (Nielsen and Ricke, 1964; Thode and Monster, 1965; Holser and Kaplan, 1966; Smirnov et al., 1972; Grinenko and Grinenko, 1974; Davies and Krouse, 1975). It is generally agreed that inputs and removal processes caused the isotopic variations and presumably altered the sulfate content of the oceans. A model of Rees (1970) considered an aspect of evaporite formation overlooked in prior considerations. Crystallisation is associated with small isotopic fractionation while sulfate reduction has larger isotopic selectivity. These two processes compete in sulfate removal. Consequently the isotopic composition may not be unequivocally related to changes in the total sulfur content of the oceans. Holland (1973) examined these isotopic trends and derived relationships which imply that operation of the sulfur cycle caused the lower- and mid-Palaeozoic to be periods of net gain of atmospheric oxygen while the Permian was a period of oxygen loss. Holland (1972) also concluded that $[Ca^{2+}]$, $[HCO_3^-]$ and $[SO_4^{2-}]$ in ocean water could have varied independently by a factor of up to three without changing the mineral sequences in Phanerozoic marine evaporites,

Estimates of the fluxes in the hydrosphere are given in Table 6.4.3. River runoff is one of the larger inputs and includes rock weathering, pedospheric runoff, volcanism and acid mine waters. Berner (1971) assumed comparable weathering rates for pyrite and gypsum. His choice of Ca^{2+} content to estimate weathered sulfate overcomes problems such as the contribution of biological SO_4^{2-} reduction. Using relative abundances of other sulfur components in sedimentary rocks, the total for rock weathering is estimated at 42 Tg S y^{-1}. Pedospheric runoff tends to be evaluated from the difference

TABLE 6.4.3
Sulfur fluxes (Tg y^{-1}) in the hydrosphere

				References [a]
Inputs:	River runoff	89	(Pedospheric, including fertiliser)	1
		42	(Rock weathering)	2
		5	(Volcanism)	1
		4	(Acid mine waters)	3
	Volcanism	5		1
	Wet and dry deposition	71		1
	Solution from sediments	116	(An additional 48 escapes to atmosphere as biogenic sulfide)	1
	SO_2 adsorption	25		1
Removal:	Marine plants	200		4
	Sedimentation	64	(SO_4^{2-}) } A portion of this routed through marine plants	5
		36	(S^{2-}) }	5
	Sea spray sulfate	44		6

[a] References: 1, Friend (1973); 2, Berner (1971); 3, Lundgren (1971); 4, Eriksson (1963); 5, Holser and Kaplan (1966); 6, Eriksson (1960).

between total runoff and other estimated fluxes to rivers. Friend (1973) increased this component above estimates of previous workers because of increased use of fertilizers (see p. 417). Acid mine water runoff tends not to be considered as a separate entity in sulfur cycling but Lundgren (1971) estimated that 4 Tg S as sulfate are produced per annum from the weathering of pyrite mine wastes in the U.S.A. alone. Two volcanic flux entries in Table 6.4.3 signify that volcanic sulfur enters the hydrosphere directly in the oceans, and the portion going to land is largely dissolved in geothermal waters.

In measurements of dissolved sulfate in rivers, the fact that flowing water may not readily mix laterally or vertically has often been ignored. For example, the Mackenzie and Liard rivers of northern Canada were found to require a distance of the order of 400 km below their confluence before lateral mixing was achieved (Krouse and Mackay, 1971). In another study of this same river system, Hitchon and Krouse (1972) used chemical data and factor analyses to relate qualitatively large variations in the sulfur isotope compositions of dissolved sulfate to a number of geological sources, despite the fact that biological activity had probably altered both the concentrations and $^{34}S/^{32}S$ ratios.

Eriksson (1963) estimated that 200 Tg y^{-1} of sulfur was removed as SO_4^{2-} from the hydrosphere by marine plants. This process is complex since plants upon death support microbial sulfate reduction and some of the product sulfide (or a derivative thereof, e.g. S°) is incorporated into organic matter. On the other hand, a large part of the sulfur in the sediments is redissolved, and biologically-produced sulfide ions are oxidized back to sulfate. Therefore, marine plants contribute a high sulfur flux by virtue of their growth and decay. Further, during decay some gaseous sulfur compounds may escape to the atmosphere, thereby causing a loss to the hydrosphere.

Almgren and Hagström (1974) used an Ag/Ag_2S membrane electrode to study the oxidation of HS^- by dissolved oxygen in sea water at pH 8.0 and 8.5. In the presence of excess oxygen, the reaction tended to obey pseudo first-order kinetics. The reaction rate depended upon pH, temperature, and the oxygen to sulfide ratio. With O_2-saturated sea water (0.2 mM) at 23—24°C, and a sulfide concentration of 1 μM, the half-life of the reaction was 38 min at pH 8.0 and 53 min at pH 8.5. An increase of the sulfide concentration to 200 μM increased the half-life to 280 min while a drop in temperature to 9.8°C approximately doubled the half-life. Surprisingly, these workers found that the presence of sulfur-oxidizing bacteria had little influence on the oxidation rate. Since bacteria definitely participate, it was concluded that the rate-determining step occurred during the purely chemical oxidations of sulfide ion species to sulfite, with the bacterial effects occurring later in the reaction. This is consistent with the idea of Sorokin (1971) that, in natural waters like the Black Sea, thiobacilli and other chemo-autotrophs oxidize thiosulfate produced by chemical oxidation of sulfide. It may be noted, however, that Almgren and Hagström (1974), using Sorokin's (1971) data, estimated the half-life of sulfide oxidation in pynocline water of the Black Sea to be in the order of 10^4 min, and they concluded that this high value was caused by some inhibitory factor. Almgren and Hagström also examined data from freshwater experiments of Chen and Morris (1972) from which half-life values of 3000 min were obtained. They suggested that this high value might reflect the absence of catalysts present in sea water or might be due to differences in analytical techniques.

Although 44 Tg of sulfur leave the ocean each year in the form of sulfate spray (Eriksson, 1960), nearly twice this amount is returned in the form of wet and dry deposition. Considerable debate exists as to whether the oceans are net absorbers or emitters of SO_2. Another mechanism of sulfate removal from sea water which is difficult to quantify is direct chemical interaction with the lithosphere. For example, Bjoernsson et al. (1972) reported that sea water was depleted in sulfate after passage through volcanic rocks of the Reykjanes Peninsula, Iceland.

The foregoing discussion illustrates the variety of processes involved in hydrospheric sulfur movement. Examination of Table 6.4.3 suggests that physical processes tend to dominate in input fluxes while biological mechanisms, such as uptake by marine plants and escape of biogenic sulfide, are dominant in removal processes.

Biosphere

In aquatic settings, sulfate reducers are intimately associated with various heterotrophs and autotrophs. These include purple and green sulfur bacteria and thiobacilli which affect the availability of organic matter and alter the distribution of sulfur compounds in their various valence states. In the majority of the sites where elemental sulfur is formed (see Chapter 6.2), oxidation to sulfate also occurs and other intermediate oxidation states may also be found. For example, Volkov et al. (1972) reported the presence of thiosulfate in waters of some sediments from the Pacific Ocean east of Japan. These are interpreted as oxidation products of sulfide rather than intermediates of sulfate reduction since their concentration increased with increasing free sulfide ion content.

Recycling of sulfur in the Black Sea has been the subject of considerable study by a number of workers; aspects of sulfate reduction and formation of elemental sulfur have already been discussed in Chapter 6.2. Above 150—200 m depth, the oxygen content ($0.1\ ml\ l^{-1}$) is sufficient to oxidize all sulfide, although the relative contributions of chemical and microbiological oxidation are uncertain (Sorokin, 1970; see also Chapter 6.1). Kriss (1963) suggested that perhaps half of the dissolved sulfide originated from putrifying organic matter. However, Deuser (1971) calculated that most of the organic matter was recycled in the upper 200 cm and he concluded (Deuser, 1970) that the contribution of organic sulfur to hydrogen sulfide ranged from 3—5%. A further complication was introduced by Sorokin (1972) who compared the rates of sulfide oxidation, sulfate reduction and organic matter production in the Black Sea. He noted that the amounts of sulfide oxidized were considerably in excess of the amounts of sulfide which could be produced, theoretically, by organic-linked sulfate reduction. Clearly then, either the analytical techniques are inadequate or other factors contribute to sulfur cycling in this basin. The sulfur isotope data of Vinogradov et al. (1962) are very good indicators of sulfur cycling in various parts of the Black Sea. Sulfate in the bulk of the water has $\delta^{34}S$ values near +20‰ consistent with its sea-water origin: lighter $\delta^{34}S$ values near the top reflect freshwater inputs. In contrast, the $\delta^{34}S$ values of the dissolved sulfide range from −23.7 to −36.1‰ reflecting preferential bacterial reduction of $^{32}SO_4^{2-}$. The fact that the sulfate is isotopically uniform near +20‰ means that the fraction reduced is very small. In the sediments, however, where sulfide ions abound, the sulfate is enriched in ^{34}S ($\delta^{34}S$, up to +25‰) in upper ooze layers attesting to a limited sulfate supply and enrichment of ^{34}S in the unreacted sulfate (see Chapter 6.2). The δ-values for reduced species range from −16 to −34‰. In more oxygenated surface sediments, the reduced sulfur species make up less than 10% of the concentrations found in the sulfide-rich regions, and the δ-values of the sulfate are lower than elsewhere, ranging down to +11‰: some of this sulfate probably represents oxidised sulfide. In the Neo-Euxinian layers underlying more reduced surface sediments, sulfate is also depleted in ^{34}S in comparison to sea water due to oxidation of reduced sulfur in these strata, when they were surface sediments, by the oxygenated freshwaters of the Neo-Euxinian. This oxidation mechanism was also coupled with inputs of isotopically lighter sulfates from the rivers. In summary, the sulfur isotope data of Vinogradov et al. (1962) provide an appreciation of the variations in the sulfate reduction and oxidation processes at different locations in the Black Sea today as well as how these processes functioned historically.

In open flow systems, it is possible that re-oxidation of sulfide may occur far removed from the site of the sulfate reduction. This appears to be the case in Flysch waters in Czechoslovakia. Sulfur isotope data obtained by Šmejkal et al. (1971) provided evidence of sequential oxidations and reductions of the form:

$$\text{sulfate} \xrightarrow{\text{reduction}} \text{metal sulfides} \xrightarrow{\text{oxidation}} \text{sulfate} \xrightarrow{\text{reduction}} \text{sulfide}$$

Since transport and a considerable time span were involved, the isotopic composition of the sulfide is consistent with a summation of the kinetic isotope effects in both reductions. The initial sulfates, intermediate sulfates and product-sulfide ions had $\delta^{34}S$ values typically near +20, 0 to −13, and −25 to −40‰, respectively. Ŝitborite spring had an unusually low $\delta^{34}S$ value of −62‰ for dissolved sulfide.

A number of botanical factors affect interactions among the various spheres in the sulfur cycle. Mosses and lichens differ from plant root systems in that their rhizoids and rhizines serve for attachment rather than food channels. Hence they interact with atmospheric sulfur compounds rather more directly, by reactions either with gaseous compounds or solutions (rained-out sulfate or solution of dry fallout on bark). They derive little sulfur from their host plants or the pedosphere. In contrast, plants absorb sulfate via their root systems. Sulfate is metabolised within their chloroplasts and the enzymes ATP-sulfurylase. PAPS-synthetase *, sulfite reductase, serine acetyltransferase and O-acetylserine sulfhydrylase have all been found in plant tissue (Siegel, 1975) indicating a close similarity to the bacterial assimilatory pathway (see Chapter 6.2). However, in the past two decades, investigators have found that up to 40% of a plant's sulfur may be sequestered from aerial SO_2 (Severson, 1975). Recent sulfur isotope studies reveal that in the vicinity of sour gas (H_2S rich) processing plants pine and spruce needles can acquire up to 50% of their sulfur from ambient SO_2 (Krouse, 1977a). Therefore, needles and leaves derive sulfur from two sources, the pedosphere and the atmosphere. Penetration of compounds from the atmosphere into the subsoil depends upon the ground water and the biological cover. In a recharge area, penetration is negligible while in an area of net

* It is now thought that PAPS is not involved in sulfate assimilation by plants. (Editors' note.)

drainage, sulfur compounds descend in solution. On the other hand, moss cover is capable of retaining sulfur compounds thereby blocking their passage from the atmosphere to the pedosphere. Sulfur isotope abundances have proved valuable in following these movements of sulfur compounds (Krouse, 1977a, 1979).

The extent of H_2S release by plants as the result of sulfate reduction is another unknown flux in the sulfur cycle. Relatively rapid reduction of sulfate and thiosulfate to H_2S by a thermophilic blue-green alga *Synechococcus lividus* isolated from a thermal spring in Yellowstone has been reported (Sheridan, 1966; Sheridan and Castenholz, 1968). Wilson et al. (1977) described light-dependent emission of H_2S from leaves of a variety of plants at a maximum rate of 8 n mol min^{-1} g^{-1} (fresh-weight) which they judged to be comparable to the activity associated with sulfate uptake. However, the emission was not a steady phenomenon and increased markedly with stresses of root injury, increases in light intensity and increased bisulfite or sulfate ion concentrations. Emissions with bisulfite solutions were higher than with sulfate solutions.

Organic sulfur

The biological conversion of organic sulfur compounds is a very complex subject and its relation to the cycling of inorganic sulfur compounds is far from being understood quantitatively. Sulfur in organic compounds ranges in valency from +6 in sulfate esters to −2 in the sulfhydryls of amino acids, co-factors and vitamins. Sulfate esters such as choline sulfate in fungi (Spencer and Harada, 1960) are associated mainly with eucaryotic organisms. An extensive review of the metabolism of organic sulfate esters was given by Roy and Trudinger (1970), while a number of authors examined in detail the synthesis and dissimilation of selected complex sulfur-containing compounds in a book edited by Greenberg (1975). Houghton and Rose (1976) discuss the liberation of sulfate from sulfate esters by soils while Fitzgerald (1976) has suggested that sulfate ester formation and hydrolysis is a potentially important aspect of the sulfur cycle of aerobic soils.

During the microbial degradation of organic material, large quantities of methyl- or hydrogen sulfide may be evolved. Tarr (1934) found that a variety of aerobic, facultative, and strictly anaerobic heterotrophs were capable of releasing the sulfhydryl groups of cysteine and cystine as H_2S during growth on the amino acids alone or on proteinaceous compounds. *Escherichia coli*, *Proteus vulgaris*, *Pseudomonas aeruginosa*, *Clostridium tetani* and *Clostridium perfringens* (Mitsuhashi, 1949) as well as other bacteria, fungi and actinomycetes (Segal and Starkey, 1969) produce methyl sulfide (CH_3SH), but not H_2S, from methionine and H_2S (no CH_3SH) from cystine or homocystine. Also dimethyl sulfide (CH_3-S-CH_3) is released during the microbial degradation of plant material containing methyl methionine sulfonium salts (Greene and Davis, 1960; Mazelis et al., 1965; Miers, 1966;

Wong and Carson, 1966; Bills and Keenan, 1968; Keenan and Lindsay, 1968; Segal and Starkey, 1969) and is believed to be an oxidation product of methanethiol (Segal and Starkey, 1969). Higher alkyl disulfides have been produced from higher alkyl cysteines in laboratory experiments. For example, ethyl and ethylmethyl sulfide, dipropyl disulfide and diallyl disulfide were evolved when S-ethylcysteine, S-propylcysteine and S-allylcysteine were degraded by various microorganisms (Challenger and Charlton, 1947; Challenger and Liu, 1950; King and Coley-Smith, 1969; Cherest et al., 1970).

Desulfation of organic sulfate esters with the release of sulfate ion is identified with enzymes such as glycosulfatase in the red algae, *Porphyra umbilicus* (Rees, 1961). However organic esters may be reduced to liberate methanethiol or dimethyl sulfide. Young and Maw (1958) reported that the fungus, *Schizophyllum commune*, synthesises dimethyl sulfide and methyl sulfide by the following successive reduction and methylation reactions:

$$SO_4^{2-} \rightarrow SO_3^{2-} \rightarrow CH_3SO_3^- \rightarrow$$

$$\rightarrow CH_3SO_2^- \rightarrow (CH_3)_2SO_2 \rightarrow (CH_3)_2SO \rightarrow CH_3SH \rightarrow (CH_3)_2S$$

Release of alkyl sulfides from bacteria, fungi, algae and marine dinoflagellates has been discussed in reviews by Challenger (1959) and Kadota and Ishida (1972). Examples of microorganisms and their degradation of organosulfur compounds are given in Table 6.4.4. Mechanisms of degradation of organic sulfur compounds (based on Segal and Starkey, 1969) are shown in Fig. 6.4.2.

There has been a tendency in the literature to conclude that biogenic sulfide gas is exclusively H_2S. However, in terms of the overall sulfur cycle, the biological production of compounds such as methyl sulfide, dimethyl sulfide, and dimethyl disulfide cannot be ignored, and evidence suggests that H_2S may not be the pre-eminent biospheric gas released to the atmosphere. The surface waters of oceans are saturated with respect to dimethyl sulfide (Lovelock et al., 1972). Rasmussen (1974) has evaluated the production of methyl sulfides in a series of in vitro experiments and field studies. The production is related to oxygen supply. Aerobic cultures of *Pseudomonas fluorescens* (soil habitat) and *Pseudomonas aeruginosa* (sewage habitat) reducing $MgSO_4$, evolved mostly dimethyl disulfide: dimethylsulfide, dimethyldisulfide and minor amounts of H_2S were evolved when organic sulfur compounds were added to the media. Tidal zone seaweeds, *Entromorpha compressa* and *Fucus furcatus*, gave similar results. Analyses of gaseous sulfur compounds evolved from an anaerobic manure lagoon into an aerobic atmosphere revealed that the concentrations of organic sulfides were two orders of magnitude higher than those of H_2S. In a small pond supporting large populations of green and blue-green algae, the dimethylsulfide concentration in water during summer months reached 4 mg m^{-3} or roughly 40 times the H_2S content. During earlier cooler months, the dimethyl-

TABLE 6.4.4

Sulfide release from organic matter by microorganisms

Metabolic process	Genera involved	References
Cysteine, cystine and homocysteine → H_2S	*Proteus*	Tarr (1934); Kallio and Porter (1950)
	Escherichia	Desnuelle and Fromageot (1939)
	Bacillus	Fromageot (1947)
	Propionibacterium	Desnuelle et al. (1940)
	Pseudomonas	Kadota and Ishida (1972)
S-methylcysteine → CH_3SH	*Neurospora*	
	Pseudomonas	Kadota and Ishida (1972)
	Saccharomyces	
	Scopulariopsis	
Methionine → CH_3SH	*Clostridium*	Mitsuhashi (1949); Wiesendanger and Nisman (1953)
	Pseudomonas	Mitsuhashi (1949)
	Proteus	
	Sarcina	
	Streptomyces	Segal and Starkey (1969)
	Micromonospora	
	Aspergillus	
	Fusarium	
	Candida	
	Escherichia	Mitsuhashi (1949)
	Achromobacter	Ruiz-Herrera and Starkey (1970)
	Propionibacterium	Dystra (1971)
	Schizophylum	Challenger and Charlton (1947)
	Scopulariopsis	Stahl et al., (1949)
Methionine → $(CH_3)_2S$	*Pseudomonas*	Kadota and Ishida (1972)
	Schizophylum	
S-methyl-methionine → $(CH_3)_2S$	*Scopulariopsis*	Challenger and Charlton (1947)
SO_4 + glucose → CH_3SH	*Schizophylum*	Challenger (1959)

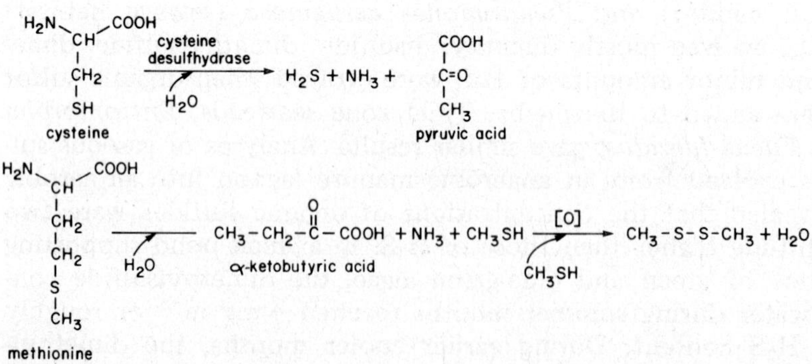

Fig. 6.4.2. Sulfide release during microbiological degradation of organic matter (after Segal and Starkey, 1969).

sulfide and H_2S contents were comparable. Interestingly, ice cover favoured production of dimethyldisulfide bringing it up to concentrations comparable with dimethylsulfide. Rasmussen (1974) concluded that H_2S emissions from waterlogged soils of rice paddies and estuarine muds were related to anaerobic conditions coupled with mass transfer of H_2S to the atmosphere through non-oxidative waters or by a turbulent transfer mechanism. Aerobic environments such as soil, foliage and freshwater favour organic sulfide evolution and negligible H_2S production.

Pedosphere

Sulfur cycling in relation to agriculture is shown by the encircled in 2 in Fig. 6.4.1. This subject is discussed in McLachlan (1975) and some quantitative aspects are covered in Chapter 9.

In the pedosphere, the main outputs of sulfur are represented by river run-off and biogenic H_2S, and inputs by dead organic matter, precipitation, dry deposition and fertiliser application. Friend (1973) assumes that the pedosphere is in dynamic equilibrium and retains a constant sulfur concentration. This is partially supported by experiments cited by Eriksson (1963) where, under SO_2 exposure, soils rich in sulfate evolved H_2S while those poor in sulfate gained sulfate. Removal of sulfur from the pedosphere by river run-off was estimated by Friend (1973) to be 89 Tg S y^{-1} based on differences between total run-off and the volcanic and rock weathering inputs to rivers. Friend also estimated that biogenic H_2S released from the pedosphere was in the order of 58 Tg S y^{-1}, assuming that sulfur inputs and removals from the pedosphere are balanced. The plant-soil cycle probably represents a net transfer of atmospheric sulfur (uptake by live plants) to the pedosphere (plant decay) although some H_2S is released back to the atmosphere in the latter process.

Eriksson's (1963) estimation for fertiliser application in 1960 (10 Tg S y^{-1}) was increased by Friend (1973) to 26 Tg S y^{-1} on the basis of increased use of fertiliser nutrients. This estimation assumed that the sulfate content of fertilisers remained constant. However, until recently there was a trend towards greater use of high-nitrogen fertilisers such as urea and ammonium nitrate rather than $(NH_4)_2SO_4$. In fact, the use of these fertilisers has created sulfur deficiencies in some soils.

Lithosphere

The biological influence on sulfur in the lithosphere may be more fully appreciated by briefly reviewing geological evolution with reference to sulfide mineralisations as summarised by various authors (Watson, 1973; Lacy, 1975). The postulated early events in the sequence are accretion and separation into core, mantle and crust, accompanied by a decline in the thermal gradient with loss of radioactive heating sources. The early crust was very thin and mantle sulfide readily surfaced to be preserved as the volcanogenic metal sulfides in greenstone belts of the Archaean (~3 Gy B.P.). At about this time, the hydrosphere and oxygenated atmosphere were developing and evidence (discussed in Chapter 6.2) suggests that sulfate-reducing bacteria were also present. These events radically altered the earth's geochemistry. Iron-rich seas precipitated banded iron formations all over the world, such as found at Lake Superior, North America. During the Proterozoic, toughening of the crust occurred but despite this, rifting, shearing, etc. tended to produce massive volcanogenic and sedimentary/volcanic concentrations of sulfide as represented by Phalaborwa (S. Africa), Broken Hill and Mount Isa (Australia), and Sudbury (Canada). The question arises as to whether the activity of sulfur bacteria by this time was sufficient to influence these sulfide formations. $\delta^{34}S$ values tend to cluster near 0‰ for some deposits such as Broken Hill (Stanton and Rafter, 1966; Both and Smith, 1975) or range up to quite positive values (~+30‰) in other deposits such as Mount Isa (Solomon and Jensen, 1965). Whereas the former values are consistent with a deep-seated origin, the heavier values may indicate some biological influence although the interpretation of sulfur isotope data must be exercised with care as discussed in Chapter 6.2. If biological activity was a factor in the formation of the Mount Isa deposits (age, ca. 1.7 Gy), and the conclusions of Chukrov et al. (1970) regarding the Malaya Bystraya deposits (age, ca. 3 Gy; see Chapter 6.2), are accepted, it would mean that about 1 Gy may have elapsed between the earliest occurrence of sulfur microorganisms and their meaningful participation in ore genesis.

While the foregoing events were happening, one supercontinent, Pangea, existed and its subsequent breakup appears to have coincided with the formation of lower temperature Mississippi Valley type Pb-Zn deposits (see Chapter 6.2) from about 600 to 200 My B.P. near the area where the Americas, Africa, Europe, and Greenland rifted apart. Biogenic sulfide seems to be more evident in many of these deposits although one might expect a greater magmatic sulfur component closer to the actual rift. It is also noteworthy that whereas sulfide deposits in some shales have persisted for over 3 Gy, evaporites older than 1 Gy are rarely found. This probably implies that sulfate deposits have a shorter life time than sulfides rather than that evaporites did not form earlier in the geological record. Although bacterial oxidative-reductive reactions are difficult to quantify in the above sequence, they no doubt contributed considerably to the current distribution of sulfur in rocks. These can be crudely estimated as 0.03% for all forms in igneous and metamorphic rocks and 0.17 and 0.35% for sulfides and sulfates, respectively, in sedimentary rocks.

The lithosphere contains of the order of 10^{10} Tg of sulfur. Inputs (Friend, 1973) include mantle degassing of 2 Tg S y^{-1} (which may pass through the upper lithosphere rather rapidly), sulfide formation by sulfate reduction in the biosphere (36 Tg S y^{-1}) and evaporite formation (64 Tg S y^{-1}). The amount of sulfur going into fossil fuels tends to be ignored although estimates

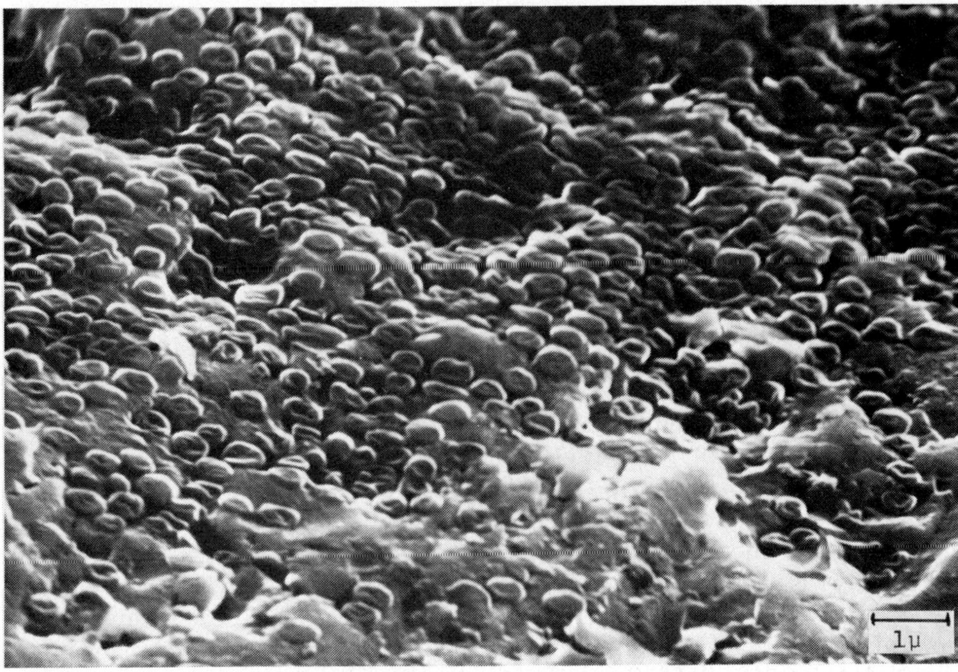

Fig. 6.4.3. Scanning electron micrograph showing selective attachment of *Caldariella* to a pyrite surface. Note that the microorganisms did not attach themselves to the matrix or gangue minerals in the medium. (Courtesy of L.E. Murr and V.K. Berry.)

have been made of the accumulation of dead organic matter on the ocean floor (see p. 412). Petroleum, however, incorporates some sulfur derived from sulfate through dissimilation.

Removal processes include rock weathering estimated at 42 Tg S y^{-1} (see p. 410) and volcanism. Although solution chemistry plays an important part in weathering solid surfaces (see p. 409), there is evidence that microorganisms can be much more directly involved. The oxidisers *Thiobacillus ferrooxidans* (Gormely and Duncan, 1974) and *Caldariella* (Murr and Berry, 1976) have been shown to attach themselves preferentially to pyrite surfaces. Figure 6.4.3 demonstrates the phenomenon observed by the latter authors: the microorganisms did not attach themselves to gangue material in the medium. Removal of sulfur by volcanism is estimated at 10 Tg S y^{-1} and 2 Tg S y^{-1} for crustal and mantle degassing, respectively. The total release can be divided into 2 Tg S y^{-1} to the atmosphere and 5 Tg S y^{-1} each to the land and oceans (Friend, 1973). Mining and drilling activities are removing sulfur from the lithosphere in the form of fossil fuels and metal sulfides. With coal, some of the sulfur is retained in the ash so it might be argued that it did not leave the lithosphere. Amounts of coal, lignite, fuel oils and natural gas consumed yearly are each of the order of 1 Pg (Bertine and Goldberg, 1971). Whereas average values for the sulfur content of coal and oil are about 2 and 0.3% respectively, there is considerable variation throughout the world. In Australia, for example, crude oils from the Bass Strait field have 0.1% or less, and coals are normally in the range 0.5 to 1%. The sulfur extracted yearly as Frasch sulfur, and from metal sulfides and fossil fuels, is not negligible in terms of the lithospheric fluxes. The total world production of sulfur in all forms was slightly under 50 Tg in 1973 (British Sulphur Corporation Ltd., Statistical Supplement, 1974; see also p. 535).

Sulfur in fossil fuels

Sulfur occurs in fossil fuels as organosulfur compounds, sulfide and sulfate minerals, and elemental sulfur (Thiessen, 1945; Kemezys and Taylor, 1964). Combustion of these fuels contributes large quantities of sulfur compounds to the atmosphere (see p. 423).

Coal is believed to be fossilised terrestrial plant remains which initially undergo biochemical changes to peat and humic substances. Peat deposits are usually associated with bogs, swamps, everglades, deltas and heavily vegetated depressions in clay, silt, or sandy areas. The peat-forming layer constitutes less than the upper metre in which decomposition of plant material by fungi and aerobic bacteria occurs. The lignin component of plants (aromatic structures, average formula $C_{12}H_{18}O_9$) is more resistant to bacterial attack than the aliphatic cellulose (average formula $C_6H_{10}O_5$). Biological aspects of early stages of coal formation have been discussed by Given (1972). The biochemical conversion is followed by metamorphic changes which produce the successive ranks, lignite, bituminous coal and anthracite, with increasing temperature and pressure.

Casagrande et al. (1977) found pyrite in peat formed in a marine environment to constitute over 15% of the sulfur present while the pyrite level in freshwater peat was an order of magnitude lower. Carbon-bonded sulfur accounted for 50% of the total sulfur in marine peat, but 70% in freshwater peat. Ester-sulfate constituted 25% of the sulfur in both cases. These authors concluded that the total sulfur found in coal can be incorporated in the peat-forming stage.

Coal and petroleum share a common feature in that their sulfur contents are highly variable and often higher by an order of magnitude than the sulfur content of living plants and animals. Furthermore, as the result of alteration processes, the organosulfur compounds do not resemble those commonly found in biological materials. The only known source which could supply the additional sulfur is sulfide generated from biological sulfate reduction. The incorporation of sulfur into organic matter during alteration in the sediments at relatively low temperatures is not well understood since relevant laboratory experiments have been conducted at much higher temperatures.

The parent material of petroleum is considered to be deposited usually in marine environments, although a non-marine history is not excluded. Detrital organic matter, not destroyed by microorganisms on the ocean bottom, undergoes a number of changes in the sediments. Further burial generates petroleum which tends to migrate from source rocks to reservoir rocks in which it is currently found. The behaviour of sulfur in petroleum depends upon the rate and depth of burial. With increasing pressure and temperature, there is general desulfurisation, producing H_2S and a mature low sulfur crude. Deep burial in carbonate strata causes large amounts of CH_4 and H_2S to form. To explain this observation, Orr (1974, 1975) proposed that sulfate reacted thermochemically with H_2S to form elemental sulfur, which, in turn reacted with organic matter to generate H_2S (eqns (1) and (2)).

$$SO_4^{2-} + 3\,H_2S \rightleftharpoons 4\,S^0 + 2\,OH^- + 2\,H_2O \tag{1}$$

$$4\,S^0 + 1.33(CH_2) + 2.66\,H_2O \rightleftharpoons 4\,H_2S + 1.33\,CO_2 \tag{2}$$

The net result of the combined reactions is the generation of H_2S from sulfate (eqn 3):

$$SO_4^{2-} + 1.33(CH_2) + 0.66\,H_2O \rightarrow H_2S + 1.33\,CO_2 + 2\,OH^- \tag{3}$$

In contrast, in shallow reservoirs bacterial sulfate reduction may occur with degradation of the crude oil (Bailey et al., 1973). The sulphate reducers may not be able to use petroleum per se but only with the participation of petroleum-oxidizing aerobes, transported by freshwater recharge, to provide suitable carbon sources for the sulfate-reducers (see Chapter 6.1).

The existence of sulfate-reducers in oil-field waters has been accepted since the early work of Bastin (1926). ZoBell (1958) provided evidence that sulfate-reducers were indigeneous rather than introduced during drilling. With the cooperation of Shell Oil Company, he sampled a well, drilled "aseptically" at Ventura, California. Core samples of oil-bearing rock around 1.4 km depth yielded sulfate reducers but not heterotrophic aerobes which would be indicators of surface contamination. On the basis of salt tolerance, Rozanova and Khudyakova (1973) claim to have distinguished between bacteria indigenous to a Tertiary oil bed and those which had been introduced by the injected lake water. ZoBell (1958) demonstrated that sulfate-reducers could migrate through oil-bearing sands at rates from 2 mm to 4 cm d^{-1}. Myers and McCready (1966) subsequently demonstrated that bacteria could migrate through Bera sandstone, Mississippian and Early Devonian limestone, and Late Mesozoic sandstone. These findings suggest that migration of bacteria to petroleum reservoirs with surface-water recharge is probably a common occurrence.

Sulfur isotope abundance studies have elucidated a number of problems in the evolution, migration, and alteration of petroleum (Krouse, 1977b). In contrast, only recently have sufficient isotopic data been obtained with coal to attempt meaningful interpretations (Smith and Batts, 1974). Organic sulfur in some Australian coals with less than 1% total sulfur exhibited a relatively narrow range in $\delta^{34}S$ values (+4.6 to +7.3‰) while the $\delta^{34}S$ values for organic matter in more sulfur-rich coals ranged from +2.9 to +24.4‰. The narrow range was interpreted as reflecting a relative constancy in the isotopic composition of freshwater SO_4^{2-} since the Permian, while the wide variations were identified with processes arising from sea water, and/or the products of biogenic sea water SO_4^{2-} reduction, penetrating underlying coal beds.

Biological factors play many roles in the cycling of sulfur minerals associated with fossil fuels. Sulfate reduction may lead to the solution of evaporite sulfates while production of sulfide ions at various stages in the formation and degradation of fossil fuels may lead to deposition of sulfide minerals or elemental sulfur (see Chapter 6.2).

Atmosphere

In comparison to other spheres, the sulfur content of the atmosphere is small, about 1.8 Tg compared with 1.3×10^9 Tg for hydrosphere (Table 6.4.1). However, in terms of the global cycle of sulfur, the atmosphere plays a complex and critical role (Fig. 6.4.1). The residence time for sulfur in the atmosphere is considered to be a few days with wide variations dependent upon meteorological and other factors (Kellogg et al., 1972). This contrasts with the case of the lithosphere, for example, which although by far the largest sulfur reservoir, has a turnover time in the order of millions of years (Holser and Kaplan, 1966). The atmosphere is also the recipient of the majority of anthropogenic sulfur.

The oxidation of sulfur gases in the atmosphere is complex and depends upon variables such as sunlight intensity, temperature, humidity and the nature of particulates. The bulk of the anthropogenic input is as SO_2 although occasionally broken gas pipelines release H_2S. Biogenic gases are usually in the form of organic sulfides and H_2S. H_2S is very rapidly oxidised with possible participation from ozone, O_2, and atomic oxygen. Oxidation of SO_2 to sulfate tends to be slower as shown by data from the plumes of power stations. In coal-fired plumes, studied by the Brookhaven National Laboratory, oxidation of SO_2 seldom exceeded 5% over distances up to 50 km, whereas in plumes from oil-fired power plants, conversions of up to 13% were measured over shorter distances (Newman et al., 1975 a,b). Vanadium is believed to have acted as a catalyst for SO_2 oxidation in the case of the latter. Ammonia is also involved in SO_2 oxidation processes and serves as a neutralizing agent. In its absence, H_2SO_4 forms under humid conditions. In its presence, droplets of ammonium sulfate solution form at higher humidities while particulates form under drier conditions. Since much of the atmospheric ammonia is biogenic this represents one of many ways in which the biological nitrogen cycle influences the sulfur cycle *.

Because of their low abundances, it is difficult to assess concentrations of certain sulfur compounds in the atmosphere, let alone attribute their presence to specific sources. In most industrial areas of the world, emissions are carefully monitored, but the biological fluxes are so variable with location and season that precise evaluation would be very costly and time-consuming. In a few instances, studies of $^{34}S/^{32}S$ abundance variations have provided some appreciation of the relative contributions from industrial and biological sources (Krouse, 1979).

The concentrations of H_2S and other organic sulfides are particularly difficult to ascertain. Robinson and Robbins (1970) estimated a global average for H_2S of 0.2 $\mu g\ m^{-3}$. However, local concentrations of dimethyl sulfide immediately above algal ponds must be a few orders of magnitude higher (see p. 415). The average SO_2 concentrations for the Northern and Southern hemispheres were estimated by Kellogg et al. (1972) to be 0.2 and 0.1 $\mu g\ m^{-3}$, respectively. Sulfate concentrations range from typically 0.5 $\mu g\ m^{-3}$ over non-urban land areas (Junge, 1963), through 0.7 to 5 $\mu g\ m^{-3}$ over oceans (Gillette and Blifford, 1971; Georgii, 1970), to over 20 $\mu g\ m^{-3}$ in industrialised regions. Altshuller (1973) assessed atmospheric SO_2 and SO_4^{2-} in urban and non-urban sites of the U.S.A., and found annual average SO_2 concentrations as high as 448 $\mu g\ m^{-3}$. The sulfate concentration increased linearly with $[SO_2]$ in the 6 to 80 $\mu g\ m^{-3}$ range according to the relation:

$$[SO_4^{2-}] = 0.144\ (\pm\ 0.013)[SO_2] + 4.9\ (\pm\ 0.5)$$

At SO_2 concentrations above 80 $\mu g\ m^{-3}$, the sulfate concentration levelled off and seldom exceeded 20 $\mu g\ m^{-3}$. Friend (1973) selected 1.5 $\mu g\ m^{-3}$ as representative of SO_4^{2-} concentration in continental surface air.

Table 6.4.5 presents fluxes of sulfur compounds in the atmosphere as summarised by Friend (1973). Since the yearly inputs from anthropogenic and natural sources are known to be about two orders of magnitude larger than the atmospheric concentration, removal processes must delicately balance the injections.

Volcanoes inject a variety of sulfur compounds into the atmosphere which correspond to only about 1% of the tltal atmospheric input, or 1.5 to 2 Tg S y^{-1}, (Kellogg et al., 1972; Friend, 1973). The next lowest flux is SO_4^{2-} in sea spray, estimated by Eriksson (1960) to be 44 Tg S y^{-1}. The industrial SO_2 input of 65 Tg S y^{-1} can be broken down into contributions from coal-burning, oil-burning, smelting and petroleum-refining and transportation of 45.4, 5.5, 13.6 and 0.4 Tg S y^{-1}, respectively. The absolute and relative magnitudes of these numbers vary considerably throughout the world. The extensive use

* *Editors' footnote:* For a discussion on the interaction of biogeochemical cycles see Chapter 1.

TABLE 6.4.5

Atmospheric fluxes (Tg y^{-1}) of sulfur compounds (after Friend, 1973)

Inputs:	Volcanoes		2
	Biological sulfides		48 (oceans)
			58 (land)
	Sea spray sulfate		44
	Industrial SO_2		65
		Total	217
Removal:	Precipitation sulfate		86 (wet, land)
			20 (dry, land)
			71 (oceans)
	SO_2 absorption		15 (vegetation)
			25 (oceans)
		Total	217

of coal in north-eastern U.S.A. and Europe contrasts to Alberta, Canada, where natural gas is the preferred fuel, and the dominant anthropogenic emissions come from the sour gas-processing industry. In other areas such as Sudbury, Canada, and Mount Isa, Australia, smelting operations contribute the bulk of sulfur compounds to the atmosphere. Biological sulfide fluxes are usually measured by balance techniques in the overall sulfur cycle. With the exception of rice paddy fields and tidal flats, probably little H_2S escapes to the atmosphere from the hydrosphere because of oxidation in the surface-water layers. Since it is unlikely that all oceanic sulfide production is restricted to near-shore environments, organic sulfide gases are suspected to form a significant source in the oceans.

Removal from the atmosphere is usually in the form of sulfate and to a lesser extent SO_2 (Table 6.4.5). Deposition of sulfate may be in the form of precipitation or dry fallout. By analogy with nuclear fallout, Robinson and Robbins (1970) evaluated dry fallout as 20% of the total deposition. The component of sea-water SO_4^{2-} in precipitation can be estimated by measuring the Cl^- content and assuming that the Cl^-/SO_4^{2-} ratio of sea water holds for sea spray. SO_2 absorption by the ocean is a controversial subject and there is debate as to whether the oceans represent a sink or source of SO_2 (from oxidation of biogenic gaseous sulfides). Absorption of SO_2 by vegetation is difficult to evaluate and there is the possibility that vegetation might release sulfur compounds back into the atmosphere. Other compounds emitted by vegetation, such as terpenes from evergreens (Boyle, 1976), may react with SO_2.

The role of transportation by the atmosphere becomes evident when attempts are made to balance the fluxes of the oceans and land. Eriksson (1963) estimated that 10% of the sea-water salt spray was deposited on land.

From this, Friend (1973) concludes that, while 4 Tg S y^{-1} in the form of spray moves from sea to land, a flow of 8 Tg S y^{-1}, in the form of SO_2 and sulfate, is required in the reverse direction to balance the atmospheric budget.

An interesting phenomenon occurs in the lower stratosphere (16—20 km) in the form of a layer of sulfate particles with concentration ranges from 0.01 to 20 μg m^{-3}. A clear correlation was found between sulfate concentration and well-known volcanic eruptions (Castleman et al., 1973). In the Junge layer, which is located south of the equator, the sulfate concentration increased by two orders of magnitude within a year after the notable eruptions of Mt. Agung (8.5°S) in 1963. Similarly, eruptions of Trident (1963), Surtsey (1963), Sheveluch (1964) and Mt. Redoubt (1966) caused increases in sulfate concentration in samples collected over the Arctic. Sulfur-isotope abundances coupled with the concentration data support the concept of volcanic contributions. During quiescent period, the source may be biogenic (Castleman et al., 1974). The residence time in this layer is estimated to be 1.5 y.

ACKNOWLEDGEMENTS

The authors thank the National Research Council of Canada for financial support of their research over the years. In particular, a negotiated development grant to The University of Calgary for an interdisciplinary sulfur research group (Acronym UNISUL) proved valuable in elucidating processes in the sulfur cycle. Mrs. B. Delay and Miss L. Boreiko of the Department of Physics, The University of Calgary, and Mrs. W. McCready assisted greatly in the mechanical chores.

REFERENCES

Almgren, T. and Hagström, I., 1974. The oxidation rate of sulphide in sea water. Water Res., 8: 395—400.

Altshuller, P., 1973. Atmospheric sulfur dioxide and sulfate distribution and concentration at urban and nonurban sites in United States. Environ. Sci. Technol., 7: 709—712.

Bailey, N.J.L., Krouse, H.R., Evans, C.R. and Rogers, M.A., 1973. Alteration of crude oil by waters and bacteria: evidence from geochemical and isotope studies. Am. Assoc. Petrol. Geol. Bull., 57: 1276—1290.

Bastin, E.S., 1926. The problem of the natural reduction of sulfates. Am. Assoc. Petrol. Geol. Bull., 10: 1270—1299.

Berner, R.A., 1971. Worldwide sulfur pollution of rivers. J. Geophys. Res., 76: 6597—6600.

Bertine, K.K. and Goldberg, E.D., 1971. Fossil fuel combustion and the major sedimentary cycle. Science, 173: 233.

Bills, D.D. and Keenan, T.W., 1968. Dimethyl sulfide and its precursor in sweet corn. J. Agric. Food Chem., 16: 643—645.

Bjoernsson, S., Arnorsson, S. and Tomasson, J., 1972. Economic evaluation of Reykjanes thermal brine area, Iceland. Am. Assoc. Petrol. Geol. Bull., 56: 2380—2391.

Bolze, C.E., Malone, P.G. and Smith, M.J., 1974. Microbial mobilization of barite. Chem. Geol., 13: 141—143.

Both, R.A. and Smith, J.W., 1975. A sulfur isotope study of base-metal mineralization in the Willyama complex, western New South Wales, Australia. Econ. Geol., 70: 308—318.

Boyle, J.M., 1976. Atmospheric Reactions of Sulfur Dioxide and Monoterpenes. M.Sc. Thesis, The University of Calgary, Calgary, Canada, 97 pp.

Brock, T.D., 1971. Microbial growth rates in nature. Bacteriol. Rev., 35: 39—58.

Butlin, K.R., 1953. The bacterial sulfur cycle. Research, 6: 184—191.

Casagrande, D.J., Siefert, K., Berschinski, C. and Sutton, N., 1977. Sulfur in peat-forming systems of the Okefenokee Swamp and Florida Everglades: origins of coal. Geochim. Cosmochim. Acta, 41: 161—167.

Castleman, A.W., Jr., Munkelwitz, H.R. and Manowitz, B., 1973. Contribution of volcanic sulfur compounds to the stratospheric aerosol layer. Nature, 244: 345—346.

Castleman, A.W., Jr., Munkelwitz, H.R. and Manowitz, B., 1974. Isotopic studies of the sulfur component of the stratospheric aerosol layer. Tellus, 26: 223—254.

Challenger, F., 1959. Aspects of the Organic Chemistry of Sulphur. Academic, New York, NY, 253 pp.

Challenger, F. and Charlton, P.T., 1947. Studies on biological methylation. X. The fission of the mono- and disulfide links by molds. J. Chem. Soc., pp. 424—429.

Challenger, F. and Lui, Y.C., 1950. Elimination of methanethiol and methyl sulfide from methyl-mercapto and dimethylsulfonium compounds by molds. Rec. Trav. Chim. Pays-Bas, 69: 334—342.

Chen, K.Y. and Morris, J.C., 1972. Kinetics of oxidation of aqueous sulfide by O_2. Environ. Sci. Technol., 6: 529—537.

Cherest, H., Talbot, G. and deRobichon-Szulmajster, H., 1970. Role of homocysteine synthetase in an alternate route for methionine biosynthesis in *Saccharomyces cerevisiae*. J. Bacteriol., 102, 448—461.

Chukhrov, F.V., Vinogradov, V.I. and Ermilova, L.P., 1970. On the isotopic composition of some Precambrian strata. Miner. Deposita, 5: 209—222.

Davies, G.R. and Krouse, H.R., 1975. Carbon and oxygen isotopic composition of late Paleozoic calcite cements, Canadian Arctic Archipelago — preliminary results and interpretation. Geol. Surv. Can. Paper, 75: 1, Part B, 215—220.

Desnuelle, P. and Fromageot, C., 1939. Anaerobic decomposition of cysteine by Bacterium coli. I. Cysteinase. Enzymologia, 6: 80—87.

Desnuelle, P., Wookey, E. and Fromageot, C., 1940. Anaerobic degradation of cysteine and of cystine by *Propionibacterium pentosaceum*. Enzymologia, 8: 225—240.

Deuser, W.G., 1970. Carbon-13 in Black Sea waters and implications for the origin of hydrogen sulfide. Science, 168: 1575—1577.

Deuser, W.G., 1971. Organic-carbon budget of the Black Sea. Deep Sea Res., 18: 995—1004.

Dystra, G.J., 1971. Formation of dimethyl sulfide by *Propionibacterium shermanii* ATCC 9617. J. Dairy Sci., 54: 168—172.

Eriksson, E., 1960. The yearly circulation of chloride and sulfur in nature; meteorological, geochemical and pedological implications. Part II. Tellus, 12: 63—109.

Eriksson, E., 1963. The yearly circulation of sulfur in nature. J. Geophys. Res., 68: 4001—4008.

Field, C. and Lombardi, G., 1972. Sulfur isotopic evidence for the supergene origin of alunite deposits, Tolfa District, Italy. Miner. Deposita, 7: 113—125.

Fitzgerald, V.W., 1976. Sulfate ester formation and hydrolysis: a potentially important yet often ignored aspect of the sulfur cycle of aerobic soils. Bacteriol. Rev., 40: 698—721.

Friend, J.T., 1973. The global sulfur cycle. In: S.I. Rasool (Editor), Chemistry of the Lower Atmosphere, Plenum, New York, NY, pp. 177—201.

Fromageot, C., 1947. Oxidation of organic sulphur in animals. Adv. Enzymol., 7: 369—407.

Georgii, H.W., 1970. Contribution to the atmospheric sulfur budget. J. Geophys. Res., 75: 2365—2371.

Gillette, D.A. and Blifford, Jr., I.H., 1971. Composition of tropospheric aerosols as a function of altitude. J. Atmos. Sci., 28: 1199.

Given, P.H., 1972. Biological aspects of the geochemistry of coal. In: Advances in Organic Geochemistry, 1971. Pergamon, Oxford, pp. 69—72.

Goldhaber, M.B., 1974. Equilibrium and Dynamic Aspects of the Marine Geochemistry of Sulfur. Ph.D. Thesis, University of California at Los Angeles, 399 pp.

Gormely, L.S. and Duncan, D.W., 1974. Estimation of *Thiobacillus ferrooxidans* concentrations. Can. J. Microbiol., 20: 1453—1455.

Granat, L., Hallberg, R.O. and Rodhe, H., 1976. The global sulphur cycle. In B.H. Svensson and R. Soderlund (Editors), Nitrogen, Phosphorus and Sulphur — Global Cycles. SCOPE Report 7. Ecol. Bull. (Stockholm), 22: 89—134.

Greenberg, D.M., 1975 (Editor), Metabolic Pathways, Vol. 7. Metabolism of Sulfur Compounds. Academic, New York, NY, 614 pp.

Greene, R.C. and Davis, N.B., 1960. Biosynthesis of S-methylmethionine in the jack bean. Biochim. Biophys. Acta, 43: 360—362.

Grinenko, V.A. and Grinenko, L.N., 1974. Geochemistry of Sulfur Isotopes. Nauk, Moscow, 275 pp.

Hitchon, B. and Krouse, H.R., 1972. Hydrogeochemistry of the surface waters of the Mackenzie River drainage basin, Canada — III: Stable isotopes of oxygen, carbon and sulphur. Geochim. Cosmochim. Acta, 36: 1337—1357.

Holland, H.D., 1972. The geologic story of sea water — an attempt to solve the problem. Geochim. Cosmochim. Acta, 36: 637—651.

Holland, H.D., 1973. Systematics of the isotopic composition of sulfur in the oceans during the Phanerozoic and its implications for atmospheric oxygen. Geochim. Cosmochim. Acta, 37: 2605—2616.

Holser, W.T. and Kaplan, I.R., 1966. Isotope geochemistry of sedimentary sulfates. Chem. Geol., 1: 93—135.

Houghton, C. and Rose, F.A., 1976. Liberation of sulfate from sulfate esters by soils. Appl. Environ. Microbiol., 31: 969—976.

Ishii, M.M., 1953. The Fractionation of Sulphur Isotopes in the Plant Metabolism of Sulphates. M.Sc. Thesis, McMaster University, Hamilton, Canada, 40 pp.

Joshi, M.M. and Hollis, J.P., 1977. Interaction of *Beggiatoa* and rice plant: detoxification of hydrogen sulphide in rice rhizosphere. Science, 195: 179—180.

Junge, C.E., 1963. Sulfur in the atmosphere. J. Geophys. Res., 68: 3975—3976.

Kadota, H. and Ishida, Y., 1972. Production of volatile sulfur compounds by microorganisms. Annu. Rev. Microbiol., 26: 127—138.

Kallio, R.E. and Porter, J.R., 1950. The metabolism of cystine and cysteine by *Proteus vulgaris* and *Proteus morganii.* J. Bacteriol., 60: 607—615.

Kaplan, I.R. and Rafter, T.A., 1958. Fractionation of stable isotopes of sulfur by *Thiobacilli.* Science, 127: 517.

Kaplan, I.R. and Rittenberg, S.C., 1964. Microbiological fractionation of sulfur isotopes. J. Gen. Microbiol., 34: 195—212.

Keenan, T.W. and Lindsay, R.C., 1968. Evidence for a dimethyl sulfide precursor in milk. J. Dairy Sci., 51: 112—114.

Kellogg, W.W., Cadle, R.D., Allen, E.R., Lazrus, A.L. and Martell, A.E., 1972. The sulfur cycle. Science, 175: 587—596.

Kemezys, M. and Taylor, G.H., 1964. Occurrence and distribution of minerals in some Australian coals. J. Inst. Fuel, 37: 389—397.

King, J.E. and Coley-Smith, J.R., 1969. Production of volatile alkyl sulphides by microbial degradation of synthetic allün and allün-like compounds in relation to germination of scherotia of *Sclerotium cepivorum* Berk. Ann. Appl. Biol., 64: 303—314.

Kriss, A.E., 1963. Marine Microbiology. Oliver and Boyd, London, Ch. 4, p. 242. (Cited in Deuser, 1970.)

Krouse, H.R., 1977a. Sulphur isotope abundances elucidate uptake of atmospheric sulphur emissions by vegetation. Nature, 265: 45—46.

Krouse, H.R., 1977b. Sulphur isotope studies and their role in petroleum exploration. In: B. Hitchon (Editor), Application of Geochemistry to the Search for Crude Oil and Natural Gas. J. Biochem. Explor., 7: 189—211.

Krouse, H.R., 1979. Sulphur isotopes in our environment. In: P. Fritz and J.Ch. Fontes (Editors), Natural isotopes in Environmental Studies. Elsevier, Amsterdam, in press.

Krouse, H.R. and Mackay, J.R., 1971. Application of H_2O^{18}/H_2O^{16} abundances to the problem of lateral mixing in the Liard-Mackenzie River system. Can. J. Earth Sci., 8: 1107—1109.

Lacy, W.C., 1975. Geological evolution and mineralisation. Aust. Soc. Explor. Geophys. Bull., 7: 2—8.

Lelong, F., Tardy, Y., Grandin, G., Trescases, J.J. and Boulange, B., 1976. Pedogenesis, chemical weathering and processes of formation of some supergene ore deposits. In: K.H. Wolf (Editor) Handbook of Strata-bound and Stratiform Ore Deposits. Elsevier, Amsterdam, pp. 93—173.

Lewis, J.S. and Krouse, H.R., 1969. Isotopic composition of sulfur and sulfate produced by oxidation of FeS. Earth Planet. Sci. Lett., 5: 425—428.

Lovelock, J.E., Maggs, R.J. and Rasmussen, R.A., 1972. Atmospheric dimethyl sulfide and the natural sulfur cycle. Nature, 237: 452—453.

Lundgren, D.G., 1971. Inorganic Sulfur Oxidation by Iron Oxidizing Bacteria. Water pollution control research series, Contract No. 14010 DAY, Environmental Protection Agency, Water Quality Office, Washington, DC, 149 pp.

Martin, H., 1963. A suggested theory for the origin and a brief description of some gypsym deposits of South West Africa. Trans. Geol. Soc. South Africa, 66: 345—351.

Mazelis, M., Levin, B. and Mallison, B., 1965. Decomposition of methyl methionine sulfonium salts by a bacterial enzyme. Biochim. Biophys. Acta, 105: 106—114.

McCready, R.G.L., Kaplan, I.R. and Din, G.A., 1974. Fractionation of sulfur isotopes by the yeast *Saccharomyces cerevisiae.* Geochim. Cosmochim. Acta, 38: 1239—1253.

McLachlan, K.D., 1975 (Editor), Sulphur in Australasian Agriculture. Sydney University Press, Sydney, 261 pp.

Mckhtieva, V.L., 1971. Isotope composition of sulfur of plants and animals from reservoirs of different salinity (in Russian). Geokhimiya, 6: 725—730.

Mekhtieva, V.L. and Kondrat'eva, E.N., 1966. Fractionation of stable isotopes of sulfur by photosynthesizing purple bacteria *Rhodopseudomonas* sp. Dokl. Akad. Nauk SSSR, 166: 465—468.

Miers, J.C., 1966. Formation of volatile sulfur compounds in processed tomato products. J. Agric. Food Chem., 14: 419—423.

Mitsuhashi, S., 1949. Decomposition of thioether derivatives by bacteria. I. Methylmercaptan formation and the properties of the responsible enzyme. Jpn. J. Exp. Med., 20: 211—222.

Murr, L.E. and Berry, V.C., 1976. Direct observations of selective attachment of bacteria on low-grade sulfide ore and other mineral surfaces. Hydrometallurgy, 2: 11—24.

Myers, G.E. and McCready, R.G.L., 1966. Bacteria can penetrate rock. Can. J. Microbiol., 12: 477—484.

Nakai, N. and Jensen, M.L., 1964. The kinetic isotope effect in the bacterial reduction and oxidation of sulfur. Geochim. Cosmochim. Acta, 28: 1893—1912.

Newman, L., Forrest, J. and Manowitz, B., 1975a. Determining the extent of oxidation of sulphur dioxide in power plant plumes from isotopic ratio measurements. In: Isotope Ratios as Pollutant Source and Behaviour Indicators. International Atomic Energy Agency, Vienna, 486 pp.

Newman, L., Forrest, J. and Manowitz, B., 1975b. The application of an isotopic ratio technique to a study of the atmospheric oxidation of sulfur dioxide in the plume from an oil-fired power plant. Atmos. Environ., 9: 969—977.

Nielsen, H. and Ricke, W., 1964. Schwefel-Isotopenverhältnisse von Evaporiten aus Deutschland: Ein Betrag zur Kenntnis von $\delta^{34}S$ im Meerwasser-Sulfat. Geochim. Cosmochim. Acta, 28: 577—591.

Orr, W.L., 1974. Changes in sulfur content and isotopic ratios of sulfur during petroleum maturation — study of Big Horn Basin Paleozoic oils. Am. Assoc. Petrol. Geol. Bull., 50: 2295—2318.

Orr, W L., 1975. Geologic and geochemical controls on the distribution of hydrogen sulfide in natural gas. 7th Int. Meet. on Organic Geochemistry, Sept. 16—19, Madrid. Preprint, 20 pp.

Perel'man, A.I., 1967. Geochemistry of Epigenesis. Translation by N.N. Kohamowski, Plenum, New York, NY, 266 pp.

Postgate, J.R., 1968. The sulphur cycle. In: G. Nickless (Editor), Inorganic Sulphur Chemistry. Elsevier, Amsterdam, pp. 259—279.

Rasmussen, R.A., 1974. Emission of biogenic hydrogen sulfide. Tellus, 26: 254—260.

Rees, C.E., 1970. The sulphur isotope balance of the ocean, an improved model. Earth Planet. Sci. Lett., 7: 366—370.

Rees, D.A., 1961. Enzymic desulfation of porphyran. Biochem. J., 80: 449—453.

Robinson, E. and Robbins, R.C., 1970. Gaseous sulfur pollutants from urban and natural sources. J. Air Pollut. Control Assoc., 20: 233—235.

Roy, A.B. and Trudinger, P.A., 1970. The Biochemistry of Inorganic Compounds of Sulphur. Cambridge University Press, London, 400 pp.

Rozanova, E.P. and Khudyakova, A.I., 1973. Sulfate-reducing bacteria from oil strata of Apsheron. Mikrobiologiya, 42: 136—140.

Ruiz-Herrera, J. and Starkey, R.L., 1970. Dissimilation of methionine by *Achromobacter starkeyi*. J. Bacteriol., 104: 1286—1293.

Schoen, R. and Rye, R.O., 1970. Sulfur isotope distribution in solfatara, Yellowstone National Park. Science, 170: 1082—1084.

Segal, W. and Starkey, R.L., 1969. Microbial decomposition of methionine and identity of the resulting sulfur compounds. J. Bacteriol., 98: 908—913.

Severson, J.G. Jr., 1975. Sulfur and higher plants. In: Sulfur in the Environment. Published by the Missouri Botanical Garden in cooperation with Union Electric Co., St. Louis, MO, pp. 92—111.

Sheridan, R.P., 1966. Photochemical and dark reduction of sulfate and thiosulfate to hydrogen sulfide in *Synechococcus lividus*. Ph.D. Thesis, University of Oregon, U.S.A., 104 pp.

Sheridan, R.P. and Castenholz, R.W., 1968. Production of hydrogen sulfide by a thermophilic blue-green alga. Nature, 217: 1063—1064.

Siegel, L.M., 1975. Biochemistry of the sulfur cycle. In: D.M. Greenberg (Editor), Metabolic Pathways, Vol. 7, Metabolism of Sulfur Compounds. Academic, New York, NY, 217—286.

Šmejkal, V., Cook, F.E. and Krouse, H.R., 1971. Studies of sulphur and carbon isotope

fractionation with microorganisms isolated from springs of Western Canada. Geochim. Cosmochim. Acta, 35: 787—800.
Smirnov, V.I., Petrov, V.P., Vinogradov, F.I., Znamensky, V.C. and Sokolov, A.C., 1972. Geochemistry and Mineralogy of Sulfur. Nauk, Moscow, 296 pp.
Smith, J.W. and Batts, B.D., 1974. The distribution of isotopic composition of sulfur in coal. Geochim. Cosmochim. Acta, 38: 121—133.
Solomon, P.J. and Jensen, M.L., 1965. Sulphur isotopic fractionation in nature with particular reference to Mt. Isa, Queensland. Proc. Eighth Commonwealth Min. Metall. Congr., 6: 1275—1286.
Sorokin, Y.I., 1970. Experimental investigation of the rate and mechanism of oxidation of hydrogen sulfide in the Black Sea using S-35. Okeanologiya, 10: 37—46.
Sorokin, Y.I., 1971. Experimental data on the oxidation rate of hydrogen sulfide in the Black Sea. Okeanologiya, 11, 423—431.
Sorokin, Y.I., 1972. The bacterial population and the processes of hydrogen sulfide oxidation in the Black Sea. J. Cons. Int. Explor. Mer., 34: 423—454.
Spencer, B. and Harada, T., 1960. The role of choline sulphate in the sulphur metabolism of fungi. Biochem. J., 77: 305—315.
Stahl, W.M., McQue, B., Mandels, G.R. and Siu, R.G.H., 1949. Studies on the microbiological degradation of wool. 1. Sulfur metabolism. Arch. Biochem. Biophys., 20: 422—432.
Stanton, R.L. and Rafter, T.A., 1966. The isotopic constitution of sulphur in some stratiform lead-zinc sulphide ores. Miner. Deposita, 1: 16—29.
Tarr, H.L.A., 1934. The enzymatic formation of hydrogen sulphide by certain heterotrophic bacteria, II. Biochem. J., 28: 192—198.
Thiessen, G., 1945. Forms of sulfur in coal. In: H.H. Lowry (Editor) Chemistry of Coal Utilization, Vol. 1. Wiley Interscience, New York, NY, pp. 425—449.
Thode, H.G. and Monster, J., 1965. Sulfur isotope geochemistry of petroleum evaporites and ancient seas. Am. Assoc. Petrol. Geol. Mem., 4: 367—377.
Thode, H.G., Monster, J. and Dunford, H.B., 1961. Sulfur isotope geochemistry. Geochim. Cosmochim. Acta, 25: 159—174.
Tudge, A.P. and Thode, H.G., 1950. Thermodynamic properties of isotopic compounds of sulfur. Can. J. Res., B28: 567—578.
van Everdingen, R.O. and Krouse, H.R., 1977. Sulfur isotope geochemistry of springs in the Franklin Mountains, N.W.T., Canada. In: J. Parquet and Y. Tasdy (Editors). Proc. IAGC Second Water Rock Interaction Symp., Strasbourg, France, Aug. 17—25, 1977. Sciences Geologiques, I 134—I 145.
Vinogradov, A.P., Grinenko, V.A. and Ustinov, V.I., 1962. Isotopic composition of sulfur compounds in the Black Sea. Geokhimiya, 5: 973—997.
Volkov, I.I., Rozanov, A.G., Zhabina, N.N. and Yagodinskaya, T.A., 1972. Sulfur in sediments of the Pacific Ocean east of Japan. Litol. Polezn. Iskop., 4: 50—64.
Watson, J., 1973. Influence of crustal evolution on ore deposition. Inst. Min. Metal. Trans., 82: B107—B113.
Weisendanger, S. and Nisman, B., 1953. La L-methione démercapto-désaminase: un nouvel enzyme à pyridoxalphosphate. Comp. Rend., 237: 764—765.
Wilson, L.G., Bressan, R.A. and Filner, P., 1978. Light-dependent emission of hydrogen sulfide from plants. Plant Physiol., 61: 184—189.
Wong, F.F. and Carson, J.F., 1966. Isolation of S-methyl methionine sulfonium salt from fresh tomatoes. J. Agric. Food Chem., 14: 247—249.
Young, L. and Maw, G.A., 1958. The Metabolism of Sulphur Compounds. Methuen, London, 180 pp.
ZoBell, C.E., 1958. Ecology of sulfate-reducing bacteria. Prod. Mon., 22: 12—29.

Chapter 7.1

EVOLUTIONARY ASPECTS OF BIOLOGICAL INVOLVEMENT IN THE CYCLING OF SILICA

W. HEINEN

Department of Exobiology, University of Nijmegen, Toernooiveld, Nijmegen (The Netherlands)

J.H. OEHLER

Research and Development Department, Continental Oil Company, P.O. Box 1267, Ponca City, OK 74601 (U.S.A.)

CONTENTS

INTRODUCTION

Silicon constitutes 28% of the elemental composition of the earth's crust and is the most abundant crustal element after oxygen (47%). Silicates, compounds which contain SiO_4 tetrahedra in the crystal lattice, account for about one-third of all known mineral species and about 95% of the earth's crust. In view of this high abundance of silicate minerals in the near-surface crustal regime, it is perhaps not surprising that organisms (which also inhabit the near-surface environments of the earth) have found numerous ways of interacting with siliceous materials.

These modes of interaction, which fit into the overall silica cycle as illustrated in Fig. 7.1.1, can be conveniently classified into three major categories: (1) dissolution and decomposition of crustal silicates, (2) uptake and utilization of silica, and (3) release and redeposition of silica in sedimentary environments. The aim of the present chapter is to outline general aspects of biological interactions with siliceous materials and to discuss plausible evolutionary developments in the ability of organisms to participate in the cycling of silica.

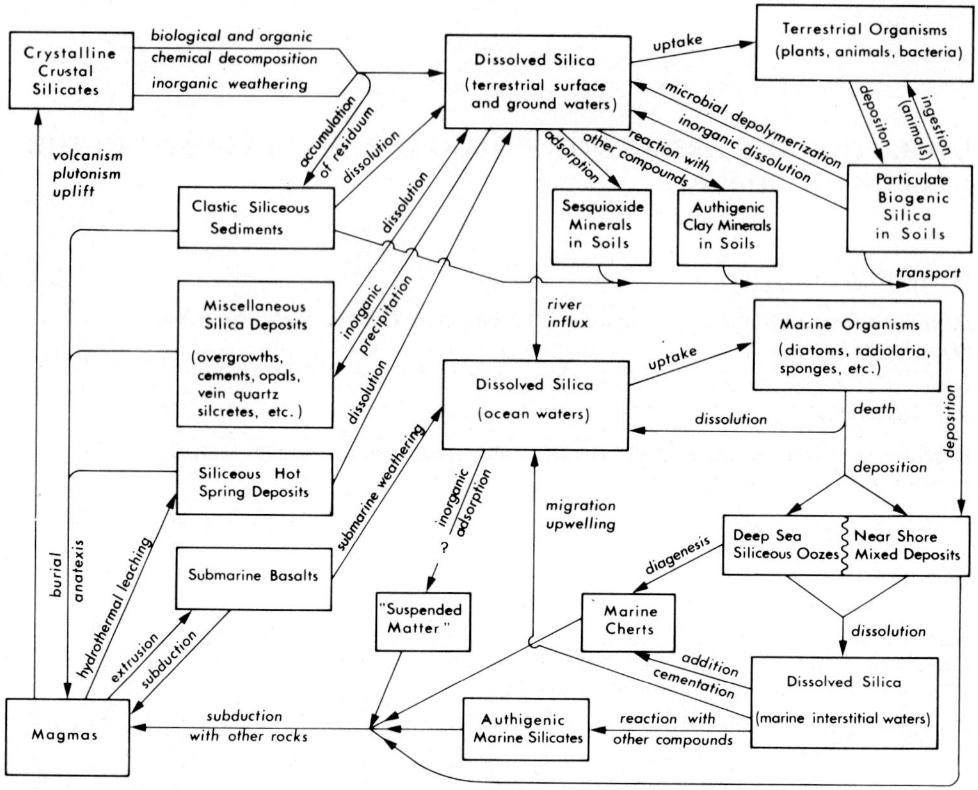

Fig. 7.1.1. Diagrammatic representation of the present day silica cycle.

MODES OF INTERACTION BETWEEN ORGANISMS AND SILICEOUS MATERIALS

Modern organisms interact with siliceous substances in three major ways. They contribute to the decomposition and solubilization of silicate minerals. They utilize silica in a dissimilatory fashion by incorporating it into their cells or bodies in the dissolved form, keeping it there for a period of time in an altered or unaltered condition, and then releasing it again as free silicic acid. They also assimilate silica by taking it up in the dissolved form and polymerizing it within the cells, such that the silica becomes an integral part of the organism itself and is not released prior to death.

Biological interactions with *crystalline* silicates occur only through the first process described above, and the mode of interaction is irreversible in the sense that organisms participate in the destruction of silicate minerals but not in the creation of new crystalline silicates (a possible exception being the reported, but as yet unconfirmed, ability of certain trees to synthesize quartz and cristobalite in their leaves (Wilding and Drees, 1974). Biological interactions with *non-crystalline* forms of silica may be either reversible (as in the dissimilatory process, in which dissolved silica is taken up and dissolved silica is released) or irreversible (as in the assimilatory process, in which dissolved silica is taken up, but no siliceous products are released before death).

At the chemical or biochemical levels, modes of interaction between organisms and siliceous materials can be characterized by a variety of reactions, of which the following are examples.

Equations (1) and (2) describe the solubilization of polymeric (crystalline or non-crystalline) silica.

$$SiO_2 \underset{-H_2O}{\overset{+H_2O}{\rightleftharpoons}} O{=}Si(OH)_2 \underset{-H_2O}{\overset{+H_2O}{\rightleftharpoons}} HO{-}Si(OH)_2{-}OH \qquad (1)$$

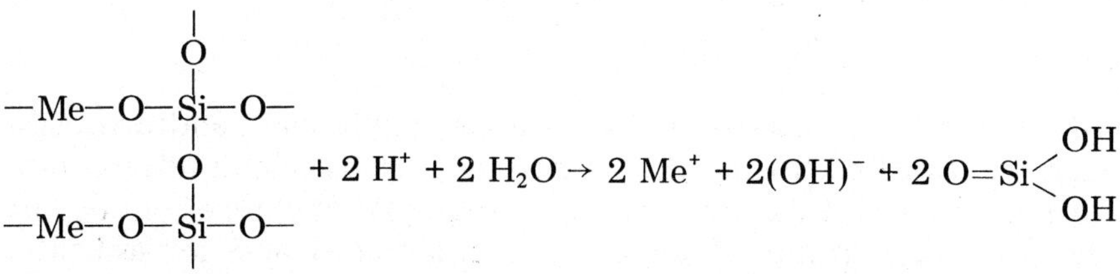

(2)

In the first reaction, soluble silica is the sole product, and it may in turn be repolymerized (but not recrystallized, at least not directly by the organisms involved). The second reaction pertains to a mineral that contains metal ions (Me), such as Ca^{2+} or Mg^{2+}, which are brought into solution, along with the silica, through biological degradation involving extrusion of organic acids: comparable reactions involving minerals with metal ions in other valence states are, of course, also possible.

These types of biologically mediated reactions are rather common and well-documented (e.g., Oberlies and Pohlmann, 1958; Duff and Webley, 1959; Webley et al., 1960, 1963; Silverman and Munoz, 1970, Lauwers and Heinen, 1974). One biological advantage of such reactions is that they make available soluble forms of silica (as well as other inorganic compounds; e.g., eqn. (2)) that can be taken up either by the solubilizing organisms (such as soil bacteria) or by other organisms not directly involved in the solubilization process (such as silica-depositing vascular plants). Another advantage to some types of organisms is that such reactions can provide a means of becoming attached to surfaces of mineral substrates (Oberlies and Pohlmann, 1958; Pohlmann and Oberlies 1960; Meadows and Anderson, 1968). The reverse reaction in eqn. (1) is used by all organisms that deposit silica within their cells and tissues. Generally, however, the mechanism of the polymerization reaction in biological systems is more complex than in indicated in eqn.

(1) and involves the intervention of organic templates or organosilicate intermediates (e.g., Fig. 7.1.2).

This leads to a consideration of reactions in which organosilicates are synthesized and degraded. Equations (3)—(5) describe theoretical reactions by which organosilicon esters or ethers may be formed in biological systems.

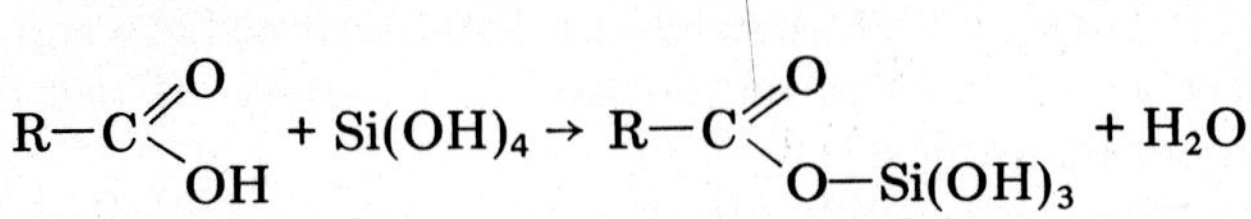
(3)

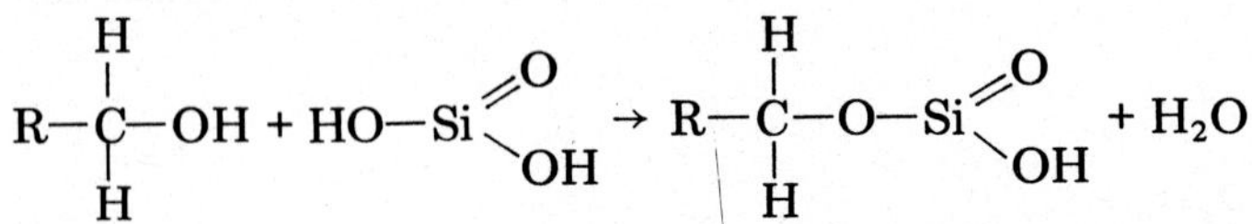
(4)

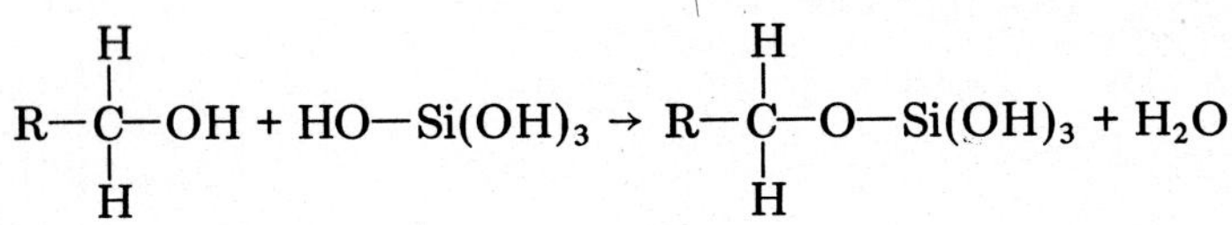
(5)

Equation (3) illustrates a reaction between a carboxylic acid and orthosilicate. Through reactions of this type, compounds such as silicon tetraacetate may be formed. In the following example (eqn. 4), the reaction occurs between the hydroxyl group of an organic compound and metasilicate, which acts as an inorganic acid. The products of such reactions may be analogous to organic phosphate and sulfate esters. Equation (5) illustrates a reaction between the hydroxyl group of an organic compound and orthosilicate. In products of such reactions, the C-O-Si linkage established may be defined as an ether linkage. Further elaborations on these general reactions may produce compounds with a greater number of C-O-Si ester or ether linkages or molecules in which silicon forms bonds between two organic compounds, comparable to the function of phosphorus in phospholipids or nucleotides (e.g., eqn. (6)).

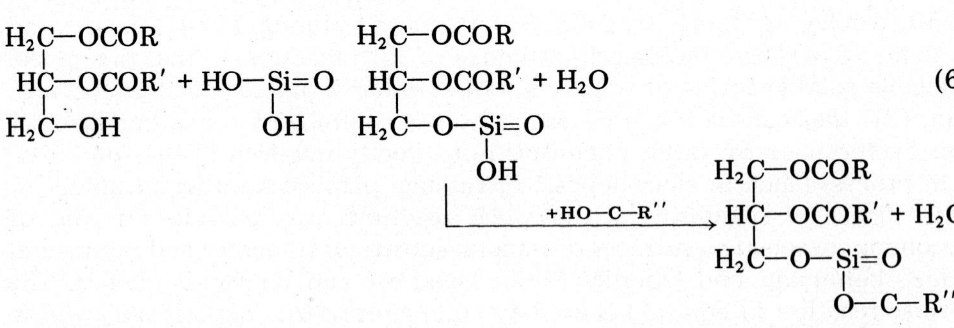
(6)

Equation (7) describes the breakdown of organic compounds with direct

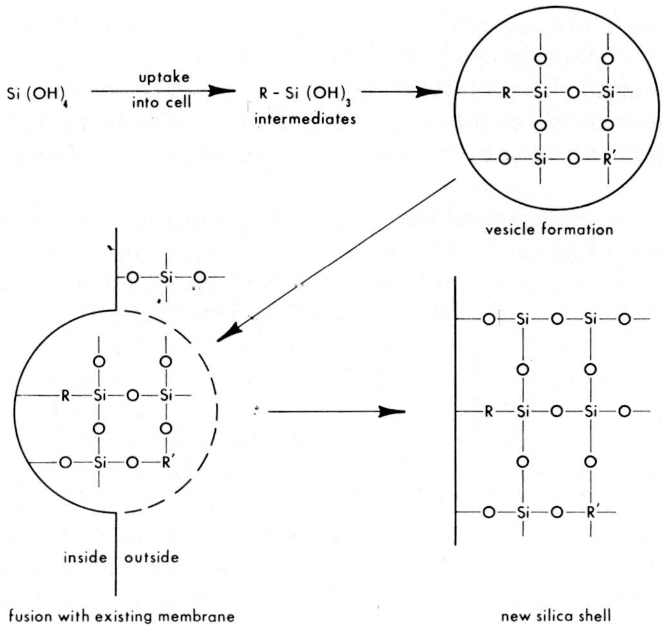

Fig. 7.1.2. Diagrammatic representation of processes involved in the formation of siliceous tests.

bonds between carbon and silicon or molecules containing both C-O-Si and C-Si linkages.

$$R{-}CH_2{-}O{-}\underset{\displaystyle CH_2{-}R'}{\underset{|}{Si}}{=}O + H_2O \rightarrow R{-}CH_2{-}OH + O{=}\underset{\displaystyle OH}{\underset{|}{Si}}{-}CH_2{-}R' \qquad (7)$$

$$\xrightarrow{+H_2O} O{=}Si(OH)_2 + R'{-}CH_3 \qquad (a)$$

$$\xrightarrow{+H_2} O{=}\underset{\displaystyle H}{\underset{|}{Si}}{-}OH + R'{-}CH_3 \qquad (b)$$

$$\xrightarrow{+H_2O} O{=}\underset{\displaystyle H}{\underset{|}{Si}}{-}OH + R'{-}CH_2OH \qquad (c)$$

As is shown, degradation of compounds with these kinds of linkages may lead to a variety of products. In the first case (eqn. (7a)), the products are metasilicate and an organic molecule with a fully reduced terminal carbon atom. The second case (eqn. (7b)) illustrates a hydrogenation/cleavage reaction that leaves both products in a reduced state. In the third case (eqn. (7c)), hydrolysis produces a siliceous analogue of formic acid, as was the case in eqn. (7b), plus an alcohol. The occurrence of various silica compounds will, of course, depend on their stabilities — relative to those of other compounds, For comparison the strengths (kJ mol^{-1}) of various silica bonds compared with their carbon counterparts are as follows: Si-O (25.8) > C-O (20.4), Si-C (18.6) < C-C (19.5) > Si-Si (19.2), Si—H (10.2) < C-H (23.6) (Sidgwick, 1962).

There are several lines of evidence which suggest that organisms synthesize, degrade, and utilize organosilicon compounds. For example, compounds with C-O-Si, C-N-Si, and C-Si linkages have been detected in soils (Gentili and Deuel, 1975; Deuel et al., 1960; Hess et al., 1960; Scheffer and Kroll, 1960). Silicon has been found in the lipid fraction of animal tissues (Holzapfel, 1949), in bone (Carlisle, 1972), and in moulds grown on silicon-bearing media (Holzapfel and Engel, 1954). C-O-Si and Si-H linkages have been found in the protein, non-protein, and cell-wall fractions of bacteria (Heinen, 1965a, b), and carbohydrate-silicate esters have been found in straw and in bacteria (Engel, 1953; Heinen, 1965c). Evidence of a different kind has come from comparisons of certain bactericidal and medicinal carbon compounds with their silicon analogues in terms of their specific effects on biological systems. Examples of such analogous compounds are the bactericide bis(hydroxyphenyl)silane and its carbon counterpart, the allergen 2,2bis(*p*-hydroxyphenyl)propane and its silicon counterpart (p-HO-$C_6H_4)_2Si$ $(CH_3)_2$, and carbamates with the general formula R-Z-$(CH_3)_2CH_2$ $OCONH_2$ (where Z can be either C or Si, and R represents either CH_3, C_2H_5, C_3H_7, or C_4H_9). In some cases, the two analogues produce essentially similar effects, while in other cases, the effects of the two analogues differ markedly. For example, for carbamates such as meprobamate and related compounds, Fressenden and Fressenden (1967) have shown that the carbon and silicon analogues are equally inhibitory toward NADH-oxidase, while the degradation of these compounds in animal tissue follows different pathways. Finally, a third line of evidence is illustrated also by the work of Fressenden and Fressenden (1967) who showed that a strain of *Pseudomonas* bacteria can use either toluene ($C_6H_5CH_3$) or phenylsilane ($C_6H_5SiH_3$) as the sole organic nutrient. Similarly, silicon tetraacetate, tetraethoxysilane, dimethyldiethylsilane and related compounds can be used as the only carbon source by several soil bacteria (Heinen, 1978).

The synthesis and breakdown of organosilicon compounds is fundamental to assimilatory uptake and utilization of siliceous materials and is involved also in certain dissimilatory processes. Studies of the kind described above provide insights into the mechanisms by which such processes operate, and they offer a basis for inferring the evolutionary history of certain modes of biosphere-silicate interaction.

THE BIOGEOCHEMICAL SILICA CYCLE THROUGH GEOLOGIC TIME

The present-day silica cycle is strongly influenced by the activity of organisms (Fig. 7.1.1). In the terrestrial environment, organisms contribute to the decomposition of crustal silicates and to the concentration and redistribution of terrigenous silica. But the effects of biological activity are most pronounced in the marine environment, where cycling of silica is largely controlled by a single class of organisms, the Bacillariophyceae or diatoms. These microscopic algae, together with the silicoflagellates, radiolaria, and siliceous sponges, extract large quantities of dissolved silica from sea water (upwards of 10 Eg y^{-1}; e.g., Heath, 1974) and polymerize it within and around their cells in the form of opaline tests and spicules. This process maintains the oceans in a markedly undersaturated condition with respect to $Si(OH)_4$ and, consequently, inhibits the inorganic chemical precipitation of siliceous deposits on the sea floor. Instead, the formation of siliceous marine deposits occurs principally through deposition and diagenetic transformation of the tests and spicules of dead organisms. Furthermore, these deposits accumulate mainly in those regions of the sea floor which underlie surface waters of high biological productivity, such that their geographic distribution is primarily determined by the distribution (and, hence, the ecological and nutritional requirements) of the organisms from which they are derived.

Palaeontological evidence suggests that, in the marine environment, similar patterns of biological influence have existed throughout the Phanerozoic. Diatoms (and, to a lesser extent, silicoflagellates) have been important constituents of the marine biosphere since the late Mesozoic (Tappan and Loeblich, 1973), and it seems likely that, since that time, they have exerted a degree of control over silica cycling comparable to that observed today. Prior to the advent of diatoms, the dominant silica-utilizers probably were the radiolaria (Tappan and Loeblich, 1973) which, together with the siliceous sponges, apparently originated in the Cambrian. Although radiolaria and sponges undoubtedly played important roles in controlling the abundance and distribution of siliceous phases in pre-Cenozoic oceans, it is uncertain whether their influence was quantitatively comparable to that of modern diatoms. That it may have been, is suggested by the widespread occurrence of biogenic cherts in Palaeozoic and Mesozoic terrains and by the fact that tests of pre-Cenozoic polycystine radiolaria (which did not have to compete with diatoms for silica; e.g., Harper and Knoll, 1975) were thicker and more massive than those of their younger counterparts.

Biological degradation of silicate minerals in the marine environment apparently has not been documented in the fossil record. However, in modern estuarine environments, colonization of hornblende and biotite grains by microorganisms (including bacteria, blue-green algae, and diatoms) has been demonstrated by Frankel (1977) who concluded that the organisms enhance weathering of these minerals by contributing to their physical disintegration. That some chemical degradation also is occurring is suggested by the presence of the diatoms, which presumably are there because of an abundance of dissolved silica being released during the weathering process. It would not be surprising if some of this dissolved silica were being liberated through the biochemical activity of the other colonizers (e.g., the bacteria and blue-green algae); however, this was not demonstrated. If silicate dissolution by bacteria and blue-green algae is taking place, then it would be reasonable to suppose that such activities were carried out by similar organisms in the geologic past, perhaps even as far back as the Precambrian when the biosphere was dominated by bacteria and blue-green algae.

On land, uptake and polymerization of silica by vascular plants has occurred at least since the Eocene, when grasses apparently originated, and probably since the Devonian, when the Equisetales arose. The inference concerning silica-depositing capabilities of ancient equisetaleans, which were dominant members of the late Palaeozoic land flora, is based on studies of modern *Equisetum* (the single extant genus of the order Equisetales), which on a dry weight basis may contain up to 16% SiO_2 (Lewin and Reimann, 1969).

Significant degradation of terrestrial silicate rocks by vascular plants probably began in the Devonian, when tracheophytes first became abundant. Degradation by "lower" organisms, such as bryophytes, fungi, algae, and bacteria, probably dates from a much earlier period. The known fossil record of pre-Devonian terrestrial microorganisms is extremely sparse, but judging from the abundance and evolutionary status of late Precambrian aquatic microorganisms (e.g., Schopf, 1975), it seems reasonable to infer that terrestrial microbes existed prior to the Devonian and perhaps as early as the late Precambrian. It is noteworthy that soils in the strict sense (which contain both mineral and organic matter) cannot have developed before organisms invaded the land; prior to that time, there was only a regolith.

From the foregoing, we can infer that, since the early Palaeozoic, organisms have influenced the cycling of silica in substantially the same ways as they do today, although the magnitudes of the various biologically mediated fluxes in the silica cycle probably have varied. Thus, the *major* elements of the Phanerozoic silica cycle can be summarized as in Fig. 7.1.3A.

Before the Cambrian, biological influences on the silica cycle probably were quantitatively negligible (with the possible exception of microbial degradation of colonized mineral grains). Two decades of intensive study of the Precambrian fossil record have failed to yield any clear evidence of siliceous organisms in deposits of this age. This is despite the fact that the search for Precambrian microfossils has been directed mainly at unmetamorphosed cherts, precisely the lithofacies where one would expect to find evidence of silica-depositing organisms, had they existed during that time. Many unmetamorphosed Precambrain cherts (including those of the banded iron formations) are known to contain abundant quartz micro-

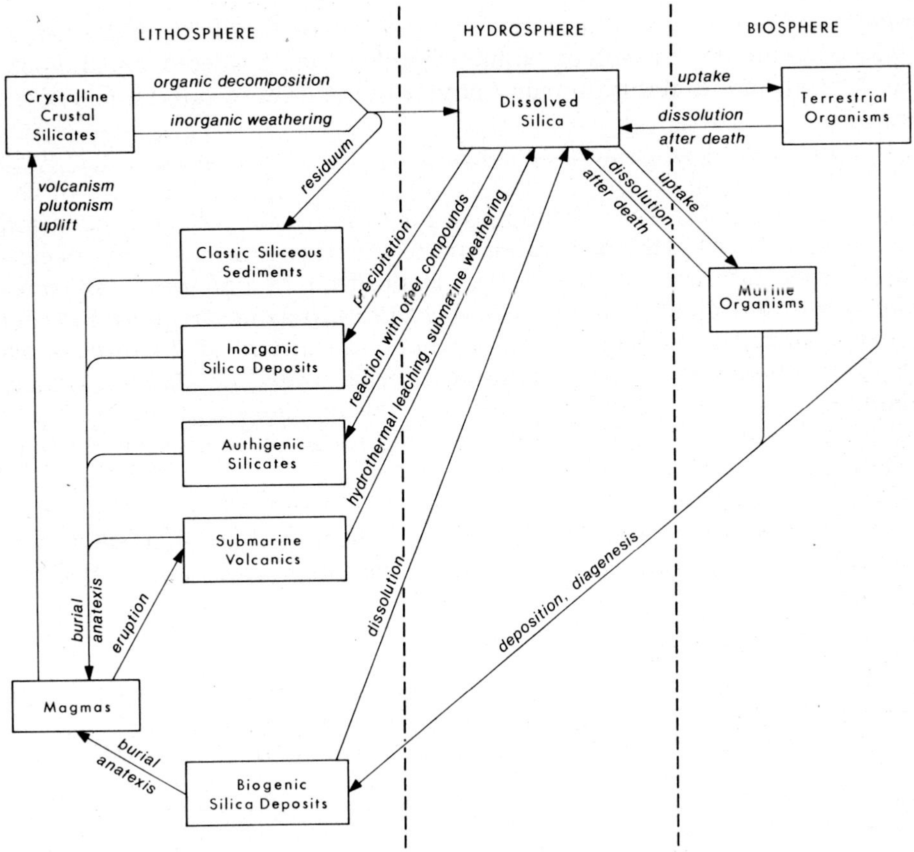

Fig. 7.1.3A. Diagrammatic representation of major elements of the Phanerozoic silica cycle.

spheres (about 5 to 40 μm in diameter), and it has been suggested that these microspheres could be the fossilized remains of siliceous microorganisms (LaBerge, 1967, 1973). However, data from recent studies indicate that the microspheres probably are non-biological structures, formed during initial crystallization of inorganically precipitated siliceous colloids (Oehler, 1976). The apparent absence of siliceous organisms during the Precambrian and their relatively high abundance during the early Palaeozoic is evidence that the ability to extract silica from solution and deposit it in the form of tests and spicules was a Phanerozoic innovation. An intriguing and evidently unsolved question is why this ability was not developed by marine "plants" (algae) until some 400 to 500 My after it had been developed by marine "animals" (protozoans and sponges). A related question is why the time of origin of organisms with siliceous hardparts (the Cambrian) was also the time of origin of organisms with calcareous and phosphatic hardparts and whether this was coincidental or reflects some stimulus that triggered an independently derived, latent ability in many phylogenetic lines to build mineralized frameworks.

In the absence of silica-utilizing organisms, the Precambrian silica cycle (Fig. 7.1.3B) would have been substantially less complex than that of the Phanerozoic (Fig. 7.1.3A). Decomposition of crustal silicates occurred mainly, if not entirely, through inorganic weathering processes. The oceans probably were saturated with $Si(OH)_4$. Deposition of non-clastic siliceous sediments resulted chiefly, if not strictly, from inorganic precipitation and probably occurred principally in areas where the influx of dissolved silica was high or where evaporation produced supersaturated conditions. Hence, Precambrian siliceous sediments probably were deposited in different environments (and possibly in a less restricted range of environments) than were similar sediments of Phanerozoic age.

Despite the fact that there is no known record of silica-depositing organisms before the Cambrian, it seems almost certain that silica was utilized by the Precambrian ancestors of the radiolaria and sponges, because

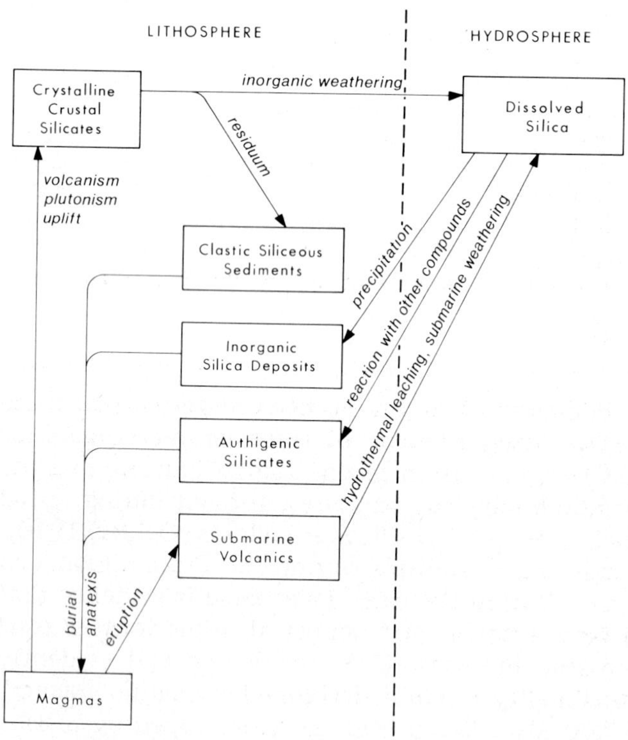

Fig. 7.1.3B. Comparative diagrammatic representation of major elements of the Precambrian silica cycle.

the ability to construct siliceous hardparts is a fairly advanced evolutionary trait which must have been based on a substantial prior history of biochemical experimentation with siliceous materials. An approximate limit on the antiquity of this early period of biochemical experimentation with silica can be estimated by considering a combination of physiological and geological data. Physiological data indicate that uptake of silica by modern organisms is linked with aerobic respiration (Lewin, 1955; Heinen, 1967). Aerobic respiration requires the presence of free oxygen (either gaseous or dissolved in water). Geological data suggest that free oxygen did not become abundant on the primitive earth until about 1.8-2.0 Gy ago, during the middle Precambrian (e.g., Cloud, 1974; Schopf, 1975). Thus, these dates can be taken as a maximum limit for the antiquity of aerobic organisms, and, if we assume that silica uptake by ancient organisms was linked to aerobic respiration (as it is in modern organisms), then these dates also set a maximum limit on the antiquity of biological abilities to actively transport silica through the cell wall and membrane. This type of active transport is fundamental to both assimilatory and dissimilatory utilization of silica.

One can conceive of earlier organisms in which silica diffused passively into and out of the cells, but these would have had to be very primitive organisms indeed to tolerate such a lack of control over the chemical composition of the cytoplasmic fluid. Nevertheless, this kind of situation could have led to an ability in some organisms to expel unwanted silica from the cell (either as inorganic $Si(OH)_4$ or as an organosilicate complex), and the biochemical machinery for doing this could have formed the basis for (and preadapted organisms toward) a later ability to selectively transport silica into the cell. It is possible that development of biosynthetic pathways for the uptake and utilization of silica was initiated and subsequently abandoned by numerous biological groups at various times in the distant geologic past. Indeed, one may speculate that some early organisms may have utilized silica for the formation of organosilicate esters in much the same way that modern organisms use phosphate and sulfate for the synthesis of esterified metabolic intermediates and, further, that the organosilicate esters found in some modern organisms could represent remnants of ancient metabolic pathways that have since been abandoned as a means of energy transfer and storage.

In any event, it would appear that the history of biosphere-silicate interactions was one of initial tolerance of dissolved silica by primitive organisms. This was followed in the middle or late Precambrian by development of biochemical mechanisms for limited silica utilization, and perhaps the development of terrestrial and benthic marine lifestyles which resulted in the breakdown of silicate minerals. Then, in the early Palaeozoic, advanced mechanisms developed for construction of siliceous hardparts by certain marine organisms and for solubilization and uptake of silica by certain terrestrial organisms. During subsequent stages of Earth's history, these various abilities were acquired by other biological groups, but, except for the recent development and widespread use of carbon-silicon plastics ("silicones") by man, the basic patterns of biosphere-silicate interaction seem not to have changed significantly since Palaeozoic times.

The Phanerozoic emergence of silica-depositing organisms resulted in an order-of-magnitude decrease in the dissolved silica content of the oceans (from a saturation value of about 100 μg ml^{-1} to less than 10 μg ml^{-1}). This was a fundamental and permanent alteration in the surface chemistry of the earth, and it provides an excellent example of the profound effects that biological systems are capable of exerting on geochemical processes and mineral cycling.

ACKNOWLEDGMENTS

Preparation of John H. Oehler's contribution to this paper commenced while he was employed with the Commonwealth Scientific and Industrial Research Organization, Division of Mineralogy, in the Baas Becking Geobiological Laboratory, Canberra, Australia. The Baas Becking Geobiological Laboratory is supported by the Commonwealth Scientific and Industrial Research Organization, the Bureau of Mineral Resources, and the Australian Mineral Industries Research Association Limited.

REFERENCES

Carlisle, E.M., 1972. Silicon: An essential element for the chick. Science, 178: 619—621.

Cloud, P., 1974. Evolution of ecosystems. Am. Sci., 62: 54—66.

Deuel, H., Dubach, P., Mehta, N.C. and Bach, R., 1960. Zur Chemie der organischen Substanz des Bodens. Schweiz. Z. Hydrol., 22: 111—121.

Duff, R.B., and Webley, D.M., 1959. 2-ketogluconic acid as a natural chelator produced by soil bacteria. Chem. Ind., 1376—1377.

Engel, W., 1953. Untersuchunge über die Kieselsäureverbindungen im Roggenhalm. Planta, 41: 358—390.

Fressenden, R.J. and Fressenden, J.S., 1967. The biological properties of silicon compounds. Adv. Drug Res., 4: 95—132.

Frankel, L., 1977. Mircoorganism induced weathering of biotite and hornblende grains in estuarine sands. J. Sed. Petrol., 47: 849—854.

Frieden, E., 1972. The chemical elements of life. Sci. Am., 227: 52—60.

R. Gentili and H. Deuel, 1957. Organische Derivative von Tonmineralien. 5. Mitt. Abbau von Phenylmontmorillonite. Helv. Chim. Acta, 40: 106—113.

Harper, H.E., Jr. and Knoll, A.H., 1975. Silica, diatoms, and Cenozoic radiolarian evolution. Geology, 4: 175—177.

Heath, G.R., 1974. Dissolved silica and deep-sea sediments. In: W.H. Hay (Editor), Studies in Paleo-oceanography. Society of Economic Paleontologists and Mineralogists Special Publication No. 20, pp. 77—93.

Heinen, W., 1965a. Siliciumstoffwechsel bei Mikroorganismen. VI Mitt. Enzymatische Veränderungen des Stoffwechsels bei der Umstellung von Phosphat auf Silikat bei *Proteus mirabilis.* Arch. Mikrobiol., 52: 49—68.

Heinen, W., 1965b. Time-dependent distribution of silicon in intact cells and cell-free extracts of *Proteus mirabilis* as a model of bacterial silicon transport. Arch. Biochem. Biophys., 110: 137—149.

Heinen, W., 1965c. Siliciumstoffwechsel bei Mikroorganismen. VII. Verteilung der Kieselsäure in Zell-Fraktionen von *Proteus mirabilis* und der Nachweis von Kohlenhydrat-Kieselsäure-Estern. Arch. Mikrobiol., 52: 69—79.

Heinen, W., 1967. Ion-accumulation in bacterial system. III. Respiration-dependent accumulation of silicate by a particulate fraction from *Proteus mirabilis* cell-free extracts. Arch. Biochem. Biophys., 120: 101—107.

Heinen, W., 1978. Biodegradation of silicon-oxygen-carbon- and silicon-carbon-bonds by bacteria. In: G. Bendz and J. Lindqvist (Editors), Biochemistry of Silicon and Related Problems. Nobel Symposium, Stockholm, Plenum, New York and London, pp. 129—147.

Hess, R., Bach, R. and Deuel, H., 1960. Modelle für Reaktionen zwischen organischen und mineralischen Substanzen im Boden. Experientia, 16: 38—45.

Holzapfel, L., 1949. Silicate in tierischen Geweben in Kombination mit Lipoiden und Cholesterol. Kolloid-Z., 115: 137—141.

Holzapfel, L. and Engel, W., 1954. Der Einfluss organischer Kieselsäureverbindungen auf das Wachstum von *Aspergillus niger* und *Triticum*. Z. Naturforsch., 93: 602—606.

LaBerge, G.L., 1967. Microfossils and Precambrian iron-formations. Geol. Soc. Am. Bull., 78: 331—347.

LaBerge, G.L., 1973. Possible biological origin of Precambrian iron-formations. Econ. Geol., 68: 1098—1109.

Lauwers, A.M. and Heinen, W., 1974. Bio-degradation and utilization of silica and quartz. Arch. Microbiol. 95: 67—78.

Lewin, J.C., 1955. Silicon metabolism in diatoms. III. Respiration and silicon uptake *Navicula pelliculosa*. Can. J. Microbiol., 3: 427—433.

Lewin, J.C. and Reimann, B.E.F., 1969. Silicon and plant growth. Annu. Rev. Plant. Physiol., 20: 289—304.

Meadows, P.S. and Anderson, J.G., 1968. Microorganisms attached to marine sand grains. J. Mar. Biol. Assoc. U.K., 48: 161—175.

Oberlies, F. and Pohlmann, G., 1958. Einwirkung von Mikroorganismen auf Glas. Naturwissenschaften, 45: 487.

Oehler, J.H., 1976. Hydrothermal crystallization of silica gel. Geol. Soc. Am. Bull., 87: 1143—1152.

Pohlmann, G. and Oberlies, F., 1960. Angriff von Glasoberflächen durch tiersches Gewebe. Naturwissenschaften, 47: 58.

Scheffer, F. und Kroll, W., 1960. Die Bedeutung nichtmetallischer Oxyde im organischen Stoffkreislauf des Bodens unter besonderer Berüchsichtigung des katalytischen Einflusses des Kieselsäure auf Huminsäureauf- und -abbaureaktionen. Agrochimica, 4: 97—109.

Schopf, J.W., 1975. Precambrian paleobiology: Problems and perspectives. Annu. Rev. Earth Planet. Sci., 3: 213—249.

Sidgwick, N.V., 1962. Chemical Elements and Their Compounds, Vol. I, Clarendon Press, Oxford, 551 pp.

Silverman, M.P., and Munoz, E.F., 1970. Fungal attack on rocks: Solubilization and altered infrared spectra. Science, 169: 985—987.

Tappan, H., and Loeblich, A.R., Jr., 1973. Evolution of the ocean plankton. Earth-Sci. Rev., 9: 207—240.

Webley, D.M., Duff, R.B., and Mitchell, W.A., 1960. A plate method for studying the breakdown of synthetic and natural silicates by soil bacteria. Nature, 188: 766—767.

Webley, D.M., Henderson, M.E.K., and Taylor, I.F., 1963. The microbiology of rocks and weathered stones. J. Soil Sci., 14: 102—112.

Wilding, L.P., and Drees, L.R., 1974. Contributions of forest opal and associated crystalline phases to fine silt and clay fractions of soils. Clays Clay Miner., 22: 295—306.

Chapter 7.2

BIOLOGICAL AND ORGANIC CHEMICAL DECOMPOSITION OF SILICATES

M.P. SILVERMAN

National Aeronautics and Space Administration, Ames Research Centre, Moffett Field, CA 94035 (U.S.A.)

CONTENTS

INTRODUCTION

The weathering of silicate rocks and minerals, an important concern of geologists and geochemists for many years, traditionally has been approached from strictly physical and chemical points of view. Biological effects were either unrecognized, ignored, or were mentioned in passing to account for such phenomena as the accumulation of organic matter in sediments or the generation of reducing environments. A major exception occurred in soil science where agricultural scientists, studying the factors important in the development of soils and their ability to nourish and sustain various crops, laid the foundation for much of what is known of the biological breakdown of silicate rocks and minerals. The advent of the space age accelerated the realization that many environmental problems and geochemical processes on Earth can only be understood in terms of ecosystems. This in turn, spurred renewed interest and activity among modern biologists, geologists and soil scientists attempting to unravel the intimate relations between biology and the weathering of silicate rocks and minerals of the earth's surface. Some of these efforts are documented in a number of reviews (Jacks, 1953; Ilyaletdinov, 1969; Ivashov, 1971; Krumbein, 1969, 1971, 1972; Aristovskaya, 1973; Syers and Iskandar, 1973).

BIOLOGICAL WEATHERING

Biological weathering of silicate rocks and minerals encompasses both biogeophysical and biogeochemical weathering processes. We can define the former as those processes by which life forms cause mechanical fracturing and disruption of rocks and minerals to produce particles smaller than the original material. Biogeochemical weathering refers to all other processes, direct or indirect, by which living organisms and their metabolic processes and products affect the chemical stability and composition of silicate rocks and minerals.

Initial colonization of silicate rocks

Life on Earth has evolved to the point where almost every conceivable ecological niche has been filled by some form of macroscopic or microscopic life, so that it is virtually impossible to find any region on Earth that is sterile. The ubiquity and diversity of life ensures that, once fresh rocks are exposed at the surface of the earth, some form of life will ultimately colonize the rock and begin to grow.

The pioneering inhabitants of freshly exposed silicate rocks and minerals devoid of organic matter are usually photoautotropic plants and microorganisms (i.e., organisms capable of synthesizing all their required organic carbon photosynthetically as opposed to heterotrophs which depend on exogenous sources of organic carbon). Brock (1973) found that the dominant primary colonizers of the volcanic island of Surtsey were mosses and lichens growing on hard substrata, and vascular plants growing on ash; green algal mats were also present on hard substrata but were less frequent. Liverworts were the dominant plants found on the volcanic ash of Katmai, Alaska (Griggs, 1933), whereas blue-green algae were the primary colonizers of Krakatau following the devastating eruption of 1883 (Treub, 1888). Relatively recent lava flows on the Island of Hawaii were colonized by lichens, mosses and blue-green algae, with lichens being the most abundant and widespread pioneers (Jackson, 1971). Krasil'nikov (1949a) studied the weathering surfaces of basalt, tufa and granite rocks from Armenia. Many were covered with lichens beneath which were abundant heterotrophic microorganisms. Other weathered zones, although devoid of lichens, harbored heterotrophic microorganisms in numbers up to 1.5×10^5 g^{-1} of rock and one must assume that there was an exogenous source of organic nutrients. The surface of weathering nepheline syenites and granites from northern regions of Russia contained diverse populations of lichens, algae, bacteria, actinomycetes, and fungi (Gromov, 1959). The surfaces of rocks taken from quarries and used by man for buildings, monuments, gravestones, etc. offer additional unique opportunities to observe the course of biological colonization and weathering within a precisely known time frame (Krasil'nikov, 1949b; Krumbein and Pochon, 1964; Jaton et al., 1966; Pochon and Jaton, 1967; Krumbein, 1968).

Biogeophysical weathering

Microorganisms in nature tend to accumulate at interfaces. In the oceans, they are more concentrated at the air-sea and sediment-sea interfaces than in the main body of water. In soils and on rock surfaces, they are not uniformly distributed but accumulate in discrete microcolonies attached to mineral surfaces or organic particulates to form a system composed of more or less discontinuous microcolonies, each in its own distinct microhabitat (Stotsky, 1972). Microorganisms can easily and rapidly penetrate cracks, joints and microscopic fissures in rock (Webley et al., 1963; Myers and McCready, 1966). Initial attachment of microorganisms to mineral surfaces is thought to be a sorptive process which depends upon the nature of the mineral and microbial surfaces and the physical and chemical characteristics of the aqueous phase (Marshall, 1971).

Following sorption, many organisms produce mucilaginous materials which bind them more firmly to mineral surfaces. Expansion and contraction of the mucilaginous thallus of many lichens upon wetting and drying, coupled with penetration of rock by rhizines (bundles of fungal hyphae of lichens) results in the tearing loose of rock fragments and thin films of the substratum (Jacks, 1953; Syers and Iskandar, 1973). The roots of higher plants can penetrate underlying rocks and split them into smaller fragments (Jacks, 1953; Carroll, 1970). Addition examples of the role of biological agents in the mechanical disintegration or rocks and minerals are reviewed by Ivashov (1971).

Biogeochemical weathering

In what follows, we shall examine how biological processes and biogenic organic chemicals affect the chemical weathering of silicate rocks and minerals. For convenience, and to accomodate a relatively large list of silicate rocks and minerals, much of the data on biogeochemical weathering agents is assembled in Table 7.2.1. These agents are defined arbitrarily as biological (B) in the case of living organisms, organic acids (O), fulvic acids (FA) and humic acids (HA) based on whether they were the sole or principal biological weathering agent studied in the field or employed by an investigator in his experimental system.

TABLE 7.2.1

Biogeochemical weathering agents of silicate rocks and minerals
B = biological; O = organic acids; FA = fulvic acids; HA = humic acids.

Silicate	Agent	Reference
Rocks		
Andesite	B	Schoen et al. (1974)
	B	Silverman and Munoz (1970)
Basalt	O	Iskandar and Syers (1972)
	B	Jackson and Keller (1970a,b)
	B,O	Silverman and Munoz (1970)
	B	Silverman and Munoz (1971)
Diabase	B	Ribeiro et al. (1973)
Dunite	B	Silverman and Munoz (1970)
Granite	O	Iskandar and Syers (1972)
	B	Silverman and Munoz (1970)
	B	Silverman and Munoz (1971)
	B	Tesic and Todorovic (1958)
	B	Wagner (1966)
	B	Wagner and Schwartz (1967a,b)
Granodiorite	B	Arrieta and Grez (1971)
	B	Schoen et al. (1974)
	B	Silverman and Munoz (1970)
	B	Silverman and Munoz (1971)
Mica schist	B	Tesic and Todorovic (1958)
Peridotite	B	Silverman and Munoz (1970)
Quartzite	B	Silverman and Munoz (1970)
	B	Silverman and Munoz (1971)
Rhyolite	B	Silverman and Munoz (1970)
Sandstone	B	Krumbein (1968)
	B	Krumbein and Pochon (1964)
	B	Pochon and Jaton (1967)
	B,O	Williams and Rudolph (1974)
Silicified limestone	B	Krumbein (1969)
Minerals		
Amorphous		
Genthite	B	Duff et al. (1963)
	B	Henderson and Duff (1963)
Glass	B	Aristovskaya and Kutusova (1968)
	B	Lauwers and Heinen (1974)
	B	Oberlies and Pohlmann (1958a)
Phytoliths	B	Aristovskaya and Kutusova (1968)
Nesosilicates		
Datolite	B	Duff et al. (1963)
Garnet	O	Maksimov et al. (1972)
Olivine	B	Duff et al. (1963)
	B	Henderson and Duff (1963)
	B	Goni et al. (1973a)
	O	Huang and Keller (1970)
Willemite	B	Agbim and Doxtader (1975)

TABLE 7.2.1.(continued)

Silicate	Agent	Reference
Sorosilicates		
Epidote	HA	Baker (1973)
Hemimorphite	B	Agbim and Doxtader (1975)
Inosilicates		
Actinolite	HA	Baker (1973)
Augite	B	Arrieta and Grez (1971)
	B	Duff et al. (1963)
	O	Huang and Keller (1970)
Bustamite	B	Duff et al. (1963)
Diopside	B	Duff et al. (1963)
Enstatite	HA	Baker (1973)
Hornblende	B	Arrieta and Grez (1971)
	B	Duff et al. (1963)
	O	Maksimov et al. (1972)
Hypersthene	B	Duff et al. (1963)
Pectolite	B	Duff et al. (1963)
Rhodonite	B	Duff et al. (1963)
Wollastonite	B	Duff et al. (1963)
	B	Henderson and Duff (1963)
	B	Webley et al. (1963)
Phyllosilicates		
Apophyllite	B	Henderson and Duff (1963)
Biotite	B	Aleksandrov et al. (1967)
	B	Arrieta and Grez (1971)
	HA	Baker (1973)
	B,O	Boyle et al. (1967)
	B	Eno and Reuszer (1955)
	B	Henderson and Duff (1963)
	O	Iskandar and Syers (1972)
	O	Maksimov et al. (1972)
	B	Mortland et al. (1956)
	O,FE,HA	Ponomareva and Ragim-Zade (1969)
	FA,HA	Sokolova (1969)
	B	Tesic and Todorovic (1958)
	B	Wagner (1966)
	B	Wagner and Schwartz (1967a,b)
	B	Weed et al. (1969)
Chlorite	O	Maksimov et al. (1972)
	B	Pryor (1975)
Illite	O	Huang and Keller (1971)
	B	Pryor (1975)
Kaolinite	O	Huang and Keller (1971)
	O,FA,HA	Kononova et al. (1964)
	B	Pryor (1975)
Lepidomelane	O,FA,HA	Kononova et al. (1964)
Leuchtenbergite	FA	Kodama and Schnitzer (1973)
Mixed-layer clay	B	Pryor (1975)
Montmorillonite	B	Anderson et al. (1958)
	O	Huang and Keller (1971)
	O,FA,HA	Ponomareva and Ragim-Zade (1969)

TABLE 7.2.1 (continued)

Silicate	Agent	Reference
Muscovite	B	Aleksandrov et al. (1967)
	B	Antipov-Karatayev et al. (1966)
	B	Babak and Pressman (1969)
	B	Duff et al. (1963)
	B	Eno and Reuszer (1951)
	B	Eno and Reuszer (1955)
	B	Goni et al. (1973a,b)
	B	Henderson and Duff (1963)
	O	Huang and Keller (1970)
	O	Mazkimov et al. (1972)
	O,FA,HA	Ponomareva and Ragim-Zade (1969)
	FA,HA	Sokolova (1969)
	B	Tesic and Todorovic (1958)
	B	Weed et al. (1969)
Phlogopite	B	Duff et al. (1963)
	B	Henderson and Duff (1963)
	B	Tesic and Todorovic (1958)
	B	Weed et al. (1969)
Saponite	B	Duff et al. (1963)
	B	Henderson and Duff (1963)
Serpentine	B	Duff et al. (1963)
Talc	B	Duff et al. (1963)
	O	Maksimov et al. (1972)
Thuringite	FA	Kodama and Schnitzer (1973)
Vermiculite	B	Duff et al. (1963)
	B	Henderson and Duff (1963)
	O,FA,HA	Ponomareva and Ragim-Zade (1963)
	B,O	Sawhney and Voigt (1969)
Tectosilicates		
Albite	B	Antipov-Karatayev et al. (1966)
	B,O	Goni et al. (1973a)
	O	Huang and Kiang (1972)
	B,O	Leleu et al. (1973)
	FA,HA	Sokolova (1969)
Bytownite	B	Huang and Kiang (1972)
Chabazite	B	Duff et al. (1963)
Feldspar	HA	Baker (1973)
	B	Goni et al. (1973a)
	B	Oberlies and Pohlmann (1958b)
	B	Wagner and Schwartz (1967b)
Harmotome	B	Henderson and Duff (1963)
Heulandite	B	Henderson and Duff (1963)
Labradorite	B	Duff et al. (1963)
	O	Huang and Keller (1970)
	O	Huang and Kiang (1972)
	FA,HA	Sokolova (1969)
Leucite	B	Henderson and Duff (1963)
	B	Wagner (1966)
	B	Wagner and Schwartz (1967b)
Microcline	B	Eno and Reuszer (1951)
	B	Eno and Reuszer (1955)
	B,O	Goni et al. (1973a)
	O	Huang and Keller (1970)

TABLE 7.2.1.(continued)

Silicate	Agent	Reference
	O	Maksimov et al. (1972)
	O,FA,HA	Ponomareva and Ragim-Zade (1969)
Natrolite	B	Duff et al. (1963)
	B	Henderson and Duff (1963)
Nepheline	B	Aristovskaya and Kutusova (1968)
	B	Aristovskaya et al. (1969)
	B	Duff et al. (1963)
	B	Henderson and Duff (1963)
	O,FA,HA	Kononova et al. (1964)
	O,FA,HA	Ponomareva and Ragim-Zade (1969)
	FA,HA	Sokolova (1969)
	B	Wagner (1966)
	B	Wagner and Schwartz (1967b)
Oligoclase	O	Huang and Kiang (1972)
	B	Müller and Förster (1963)
Orthoclase	B	Babak and Pressman (1969)
	B	Duff et al. (1963)
	B,O	Goni et al. (1973a)
	B,O	Leleu et al. (1973)
	O	Maksimov et al. (1972)
	B	Müller and Förster (1961)
	B	Müller and Förster (1963)
	B	Wagner (1966)
	B	Wagner and Schwartz (1967a)
Plagioclase	B	Aristovskaya and Kutusova (1968)
	B,O	Goni et al. (1973a)
	O	Huang and Kiang (1972)
	O,FA,HA	Kononova et al. (1964)
Quartz	B	Aristovskaya and Kutusova (1968)
	B	Bertrand (1973)
	B	Lauwers and Heinen (1974)
	B	Oppenheimer and Master (1965)
Stilbite	B	Henderson and Duff (1963)
Synthetic silicates		
$CaSiO_3$	B	Duff et al. (1963)
	B	Jackson and Voigt (1971)
	B	Webley et al. (1963)
$MgSiO_3$	B	Webley et al. (1963)
$SrSiO_3$	B	Duff et al. (1963)
$ZnSiO_3$	B,O	Agbim and Doxtader (1975)
	B	Duff et al. (1963)
	B	Webley et al. (1963)
Miscellaneous		
Aluminosilicates	B	Aleksandrov and Zak (1950)
Granitic sand	B	Berthelin (1971)
	B	Berthelin and Dommergues (1972)
Greensand	B	Eno and Reuszer (1951)
	B	Eno and Reuszer (1955)
Soil	B	Berthelin and Kogblevi (1974)
	B	Berthelin et al. (1974)
	B	Duff et al. (1963)

It is evident from Table 7.2.1 that most of the biogeochemical weathering studies have been concerned with the effects of living organisms (B). Most studies were carried out with microorganisms such as bacteria, fungi, actinomycetes, algae and lichens, either singly or in mixtures of species as they occur naturally. A few studies have been carried out with higher plants such as wheat (Mortland et al., 1956; Bertrand, 1973) and *Equisetum* (Lauwers and Heinen, 1974) and with marine animals such as oysters, clams and mullet (Anderson et al., 1958), the ghost shrimp *Callianassa major* and with *Onuphis microcephala*, a polychaete annelid (Pryor, 1975).

Table 7.2.1 also indicates the wide variety of silicate rocks and minerals that are susceptible to biogeochemical weathering. The minerals are classified on the basis of the internal arrangement of silica tetrahedra, the fundamental structural unit of silicate minerals. In order of increasing complexity, they may be classed as *nesosilicates* (independent silica tetrahedra linked by other metal cations); *sorosilicates* (two to six tetrahedra linked together); *inosilicates* (single or double chains of tetrahedra); *phyllosilicates* (two-dimensional sheets of linked tetrahedra); and *tectosilicates* (a three-dimensional framework of tetrahedra). More detailed descriptions of these mineral classes are given by Keller (1955), Krauskopf (1967) and Loughnan (1969).

Before proceeding further with a discussion of biological and organic chemical breakdown of silicate rocks and minerals, it seems appropriate to summarize first the current concepts of abiological (chemical) weathering. Several authors appear to agree that water is the single most important agent in the chemical weathering of silicates (Degens, 1965; Krauskopf, 1967; Loughnan, 1969; Berner, 1971; Carroll, 1970). All silicate minerals are soluble to some extent in pure water. More important is the primary role played by hydrogen ions. In the system pure water-silicate mineral, hydrogen ions arise from both the normal dissociation of water molecules and by hydrolysis. The latter occurs at the mineral-water interface through the hydrolytic action of charged mineral surface atoms possessing unsatisfied valencies. The hydrogen ions may then exchange with other cations at the mineral surface. Their small size also permits relatively easy diffusion into crystal lattices where the high ratio of charge to radius of the hydrogen ion disrupts the internal lattice charge balance. Subsequent rearrangement of the crystal lattice to a more stable configuration often results in diffusion of other cations out of the crystal lattice, a new lattice arrangement, or complete breakdown of the crystal depending on the silicate mineral in question. Loughnan (1969) has summarized these concepts succinctly in the generalized eqn (1):

$$M^+[\text{mineral}]^- + H^+OH^- \rightleftharpoons H^+[\text{mineral}]^- + M^+OH^- \qquad (1)$$

where $M^+[\text{mineral}]^-$ represents the initial unweathered mineral, H^+-$[\text{mineral}]^-$ is the residual weathered mineral, and M^+ represents a cation. The equation, represented as an equilibrium reaction, indicates that any process that changes the concentration of any of the reactants or products will affect the extent of the reaction. It is the major thrust of much of what follows to examine how biological processes affect this equilibrium and, therefore, cause biogeochemical weathering of silicate rocks and minerals.

Enzymes. Soils contain extracellular enzymes that catalyse the degradation of organic macromolecules to form lower-molecular-weight compounds which serve as carbon and energy sources for the soil biota. The extracellular enzymes are synthesized and secreted by the animals, plants and micro-organisms present in soil or are released from dead and dying cells (Skujins, 1967). There is no unequivocal evidence for the existence of extracellular enzymes that directly catalyse the degradation of silicate minerals to release lower-molecular-weight inorganic silicates, although there are some species of bacteria that elaborate voluminous extracellular slime capsules which attack silicate minerals when in intimate physical contact with them (Alexandrov and Zak, 1950; Tesic and Todorovic, 1958; Aristovskaya et al., 1969; Goni et al., 1973b). How capsular material attacks silicate minerals remains unknown. Enzymes are involved, however, indirectly in silicate mineral degradation through their participation in other biochemical reactions leading to the formation of products which are capable of attacking silicate rocks and minerals (Tables 7.2.1 and 2).

Biogenic hydrogen ions. It is evident from eqn (1) that an increase in the concentration of H^+ will increase the breakdown of silicate minerals, and there are many reports of silicate rock and mineral degradation associated with biogenic lowering of pH (Henderson and Duff, 1963; Müller and Förster, 1963; Antipov-Karatayev et al., 1966; Aristovskaya and Kutusova, 1968; Aristovskaya et al., 1969; Arrieta and Grez, 1971; Jackson, 1971; Silverman and Munoz, 1971). Living organisms can increase the concentration of H^+ by the formation of organic and inorganic acids.

Carbon dioxide dissolved in water leads to the formation of carbonic acid and a consequent increase in H^+. Ponnamperuma (1967) has calculated that water at 25°C, in equilibrium with the normal concentration of CO_2 in the earths's atmosphere (0.03% by volume), will attain a pH of 5.63. The weathering action of this weak acid over geologic time is well known to geologists (Krauskopf, 1967). Ponnamperuma's calculations also indicate that increased atmospheric CO_2 concentrations will result in further decreases in pH, down to pH 3.97 with one amosphere of CO_2. Respiratory CO_2 concentrations in soil atmospheres can be 10 to 100 times greater than the normal 0.03% in the earth's atmosphere (Stotsky, 1972). Thus, pH values considerably lower than 5.63 can be achieved through respiration. Similarly, respiratory activity in shallow waters and tidal flats, especially at night when photosynthetic CO_2 assimilation is halted, can cause a marked decrease in pH (Oppenheimer and Master, 1965).

TABLE 7.2.2

Biogenic organic acids active in the breakdown of silicate rocks and minerals

Organic acid	Reference
acetic	Agbim and Doxtader (1975), Berthelin (1971), Berthelin and Dommergues (1972), Berthelin and Kogblevi (1974), Henderson and Duff (1963), Huang and Keller (1970), Huang and Kiang (1972), Schalscha et al. (1967)
aspartic	Huang and Keller (1970, 1971), Huang and Kiang (1972)
butyric	Berthelin (1971), Berthelin and Dommergues (1972), Berthelin and Kogblevi (1974)
citric	Agbim and Doxtader (1975), Berthelin (1971), Berthelin and Dommergues (1972), Berthelin et al. (1974), Boyle et al. (1967), Goni et al. (1973a), Henderson and Duff (1963), Huang and Keller (1971), Huang and Kiang (1972), Iskandar and Syers (1972), Leleu et al. (1973), Maksimov et al. (1972), Müller and Förster (1961), Ponomareva and Ragim-Zade (1969), Sawhney and Voigt (1969), Schalscha et al. (1967), Silverman and Munoz (1970), Webley et al. (1963), Williams and Rudolph (1974)
formic	Berthelin and Dommergues (1972), Berthelin et al. (1974), Henderson and Duff (1963)
gluconic	Wagner and Schwartz (1967a)
p-hydroxybenzoic	Iskandar and Syers (1972)
2-ketogluconic	Duff et al. (1963), Webley and Duff (1965), Webley et al. (1963)
lactic	Agbim and Doxtader (1975), Berthelin (1971), Berthelin and Dommergues (1972), Berthelin and Kogblevi (1974), Berthelin et al. (1974), Wagner and Schwartz (1967a)
lichen acids	Iskandar and Syers (1972), Schatz (1962), Schatz et al. (1956), Schatz et al. (1957), Syers and Iskandar (1973), Williams and Rudolph (1974)
malonic	Agbim and Doxtader (1975), Schatz et al. (1954)
oxalic	Berthelin and Dommergues (1972), Berthelin and Kogblevi (1974), Bertelin et al. (1974), Boyle et al. (1967), Dormaar (1968), Goni et al. (1973a), Henderson and Duff (1963), Leleu et al. (1973), Matvayeva (1969), Müller and Förster (1961), Sawhney and Voight (1969), Silverman and Munoz (1970), Wagner and Schwartz 1967a), Webley et al. (1963)
propionic	Berthelin and Dommergues (1972)
salicylic	Huang and Keller (1970, 1971), Huang and Kiang (1972), Iskandar and Syers (1972), Schalscha et al. (1967)
succinic	Agbim and Doxtader (1975), Berthelin (1971), Berthelin and Dommergues (1972), Berthelin et al. (1974), Müller and Förster (1961)
tannic	Huang and Keller (1971)
tartaric	Huang and Keller (1970, 1971), Kononova et al. (1964), Müller and Förster (1961), Schalscha (1967)

The second principle source of biogenic hydrogen ions is organic acids. Stevenson (1967) has reviewed the distribution and pedogenic activity of organic acids in soil. Organic acids are synthesized by soil microorganisms, excreted by plant roots, or leached into the soil from surface litter, especially in forested soils. Table 7.2.2 lists only those organic acids that various investigators have employed or have identified as biogenic in studies specific to the degradation of silicate rocks and minerals. Many other soil organic acids not yet identified as being associated with silicate rock and mineral degradation are discussed in the review by Stevenson (1967). Fulvic and humic acids are also associated with silicate rock and mineral breakdown; they are excluded from Table 7.2.2 because they have already been referenced in Table 7.2.1.

In addition to organic acids, strong biogenic inorganic acids can also play a role in silicate rock and mineral degradation (Krumbein, 1968). Extensive weathering of granite by sulfuric acid produced by the sulfur-oxidizing bacterium *Thiobacillus thiooxidans* was demonstrated in laboratory studies (Wagner and Schwartz, 1967a). Intensive weathering in the area of a hot spring was attributed to the activity of sulfur-oxidizing bacteria which produced sulfuric acid from the hydrogen sulfide exsolved from deep thermal water (Schoen et al., 1974). Additional examples of the role of biogenic inorganic acids may be found in the review by Krumbein (1972).

Metal-organic complexes. Some organic matter of biological origin can chelate, or otherwise bind a variety of cations and metals in the geochemical cycle (Mortenson, 1963; Saxby, 1969), and solubilize relatively insoluble inorganic compounds including silicates (Mandl et al., 1952, 1953; Mandl and Neuberg, 1956; Neuberg et al., 1961). In addition to forming water-soluble salts with inorganic cations, many biogenic organic acids active in the breakdown of silicate rocks and minerals (Table 7.2.2) are chemically polyfunctional, i.e., they contain more than one carboxyl group (oxalate, succinate, etc.) or one or more hydroxyl group (citrate, lactate, gluconate, etc.), and can form chelates with inorganic ions. Higher-molecular-weight organic polymers, such as fulvic and humic acids, although less completely characterized, possess carboxyl, hydroxyl and amide groups and also form complexes with metals (Mortenson, 1963). Kononova (1961) and Carroll (1970) have discussed the various roles of organic matter in pedogenesis.

Equation (1) indicates that removal of the weathering products of silicate minerals (M^+ and $H^+[\text{mineral}]^-$) will allow the basic weathering process to proceed even further. There is abundant evidence that biogenic acids and other organic compounds accomplish this by metal-organic complexing. Bacterial weathering of albite and muscovite was accompanied by the production of metal-organic complexes (Antipov-Karatayev et al., 1966). Organic complexes with Fe, Al, Si, Ca and Mg were formed as a result of bacterial attack on muscovite (Tsyurupa, 1964). Arrieta and Grez (1971) showed that unidentified substances in acidic fungal culture fluids chelated the iron released biologically from iron-containing silicate minerals. Metals such as Ca, Mg, Mn, Zn, Sr and Ni in the divalent state are strongly chelated by 2-ketogluconic acid of bacterial origin (Duff et al., 1963). Chelation by oxalic acid and three other non-volatile acids of microbial origin caused a significant increase in soluble Si, Fe, Al, Mn, Ca in a brown forest soil (Berthelin et al., 1974).

Natural and artificial chelating agents (Schatz et al., 1956), lichens (Schatz et al., 1956, 1957) and lichen acids (Schatz, 1962) remove metals from rocks by chelation. Schalscha et al., (1967) concluded that chelation is implicated in rock weathering by keeping sparingly-soluble metals in solution. Six lichen compounds released greater amounts of Ca than of Mg, Fe, and Al from silicate rocks and minerals; this was attributed more to metal-complex formation than to reactions directly involving hydrogen ions donated by the lichen compounds (Iskandar and Syers, 1972). Iskandar and Syers also found that the more soluble citric, salicylic and *p*-hydroxybenzoic acids released much greater amounts of these cations than did the more sparingly-soluble lichen acids. Similarly, Williams and Rudolph (1974) showed that fungi produced extracellular acid products that were 2 to 5 times more active in chelating Fe than the lichen acid, squamatic acid.

Boyle et al. (1967) observed that the greater the chelating ability of a biogenic acid the more Fe and Al it removed from biotite. Huang and Kiang (1972) found citric acid to be more effective than other acids in extracting Al and Ca from Ca-rich plagioclase, presumably because of its greater complexing ability. Strongly complexing organic acids increase the total weight of clay minerals dissolved by distilled water by factors of 5 to 75 (Huang and Keller, 1971), and may alter the ratio of Si to other metals, notably Al and Fe, dissolved from some silicate minerals (Huang and Keller, 1970). Indeed, the relative distribution of pH-dependent Al ion species in aqueous solution is markedly altered by Al-salicylate complexes within the pH range 1.5 to 7.5 (Huang and Keller, 1972).

Infrared spectra of humic acids and fulvic acids after reaction with clay minerals gave evidence of strong complexing with Si and Al (Tan, 1975). Perdue et al. (1976) found a strong positive correlation between the amount of dissolved organic matter and the sum of the concentration of Fe and Al in natural waters, and they suggested that competition between Fe and Al for available organic complexing sites may determine their relative abundances. A large proportion of organic matter in some tropical soils occurs in the form of iron and aluminium complexes with fulvic acid (Griffith and Schnitzer, 1975). Fulvic and humic acids may differ in their ability to keep certain metals in solution as complexes. Ponomareva and Ragim-Zade (1969) reported that fulvic acid from a sandy podzol, when complexed with Al, tended to form a gel at certain Al concentrations, whereas humic acid from a chernozem did not. Bloomfield et al. (1976) found that aerobically decomposed plant matter mobilized Cu, Mn, Co, Ni, Pb, Zn and Cd from the oxides, partly in association with colloidal organic matter and partly in true solution as metal-organic complexes.

In addition to a role in removing metals already released by hydrogen ion attack on silicate minerals, organic compounds capable of forming complexes with metals may also be involved in direct removal of metals from silicate minerals without hydrogen ion participation. One can visualize direct metal-organic complex formation at unweathered mineral surfaces. Subsequent rearrangements of the residual mineral crystal surface structure could result in an insoluble solid, in which case further organic complexing of interior metals could not occur. However, if the residual mineral surface dissolved in the aqueous environment to expose fresh unweathered mineral surfaces, weathering by metal-organic complexing could continue. Iskander and Syers (1972) reported that lichen compounds released more Fe, Al, Ca and Mg from biotite, granite and basalt than could be accounted for by hydrogen ions alone. Lichen acids removed Fe from granite and muscovite in solutions buffered at pH 7.4 (Schatz, 1962). The sodium salts of salicylic and citric acids removed appreciable quantities of Fe and Al from augite, epidote, biotite and granodiorite in the pH range 6.82 to 7.70 (Schalscha et al., 1967).

Organisms as a sink for weathering products. Living organisms may also enhance weathering processes by acting as a sink for the soluble products of weathering. This is not surprising because the ultimate source of all inorganic nutrients required for life must be the rocks and minerals of the earth. Lovering (1959) reviewed the literature on the ash content of higher plants and noted the relatively large number of species that accumulate Si and Al in their tissues. He estimated that a forest of tropical plants averaging 2.5% silica content and new growth (dry wt) of 10 tons y^{-1} could extract all the silica to a depth of one foot in an acre of basalt in only 5,000 y. Lauwers and Heinen (1974) reported that monomeric silica, released by mineralization of polymerized silica or quartz, was taken up by the higher plant *Equisetum* and the bacterium *Proteus mirabilis*. Tesic and Todorovic (1958) reported an absolute nutritional requirement for Si by silicate bacteria and the accumulation of Si in their slime layers. Wheat plants grown from seeds on sterile soil accumulated appreciable Si, although growth in unsterilized soil resulted in nearly twice as much Si in the plant (Bertrand, 1973). The organic geochemistry of silica was reviewed by Siever and Scott (1963).

The ability of living cells to accumulate relatively large concentrations of K against a concentration gradient is well known. Eno and Reuszer (1951, 1955) reported that the fungus *Aspergillus niger* accumulated K in its mycelium during growth in the presence of biotite, muscovite, greensand and microcline. They attributed the release of K from these minerals in part to a shift in K equilibrium as a result of K removal by *A. niger*. Müller and

Förster (1961, 1963) reported a similar release and uptake of K by the mycelium of *A. niger* and a variety of other soil fungi when incubated with orthoclase and oligoclase. Weed et al. (1969) demonstrated that fungi weathered biotite, muscovite and phlogopite to vermiculite by acting as a sink for the K released from these minerals. Wheat plants apparently function in this manner during the alteration of biotite to vermiculite (Mortland et al., 1956).

Jackson and Voigt (1971) found a higher percentage of bacteria that dissolved calcium silicates in the rhizosphere of Eastern redcedar (*Juniperus virginiana*) than in the rhizosphere of Eastern white pine (*Pinus strobus*). These observations, and the fact that redcedar and white pine both accumulate Ca in their tissue, prompted these authors to suggest an interesting symbiotic relationship between the bacteria and the trees in which the roots of the latter supply organic nutrients for the silicate-dissolving bacteria and the bacteria supply dissolved Ca for the trees.

pH and Eh. As indicated in eqn (1), consumption of H^+ and diffusion of M^+ out of silicate minerals to yield $M^+—OH^-$ will tend to make the external aqueous phase alkaline, especially when M^+ represents alkalin, or alkaline earth, metals such as Na, K, Ca and Mg. Thus, many silicate minerals pulverized under pure water give rise to pH's on the alkaline side of neutrality (summarized in Krauskopf, 1967; Loughnan, 1969).

Loughnan (1969) discussed the solubility in relation to pH of some of the common products of chemical weathering of silicate minerals, In general, the hydroxides of Na, K and Ca are soluble at all pH's, and $Mg(OH)_2$ is soluble at $pH < 10$. Aluminium oxide is soluble at pH's < 4 and >10, whereas SiO_2 is slightly soluble at $pH < 9$ and increasingly soluble at higher pH values. Titanium hydroxide is soluble at $pH < 5$, but TiO_2 is soluble only at $pH < 2$. The hydroxide of trivalent iron is soluble only below pH 2.5, but $Fe(OH)_2$ is soluble below about pH 8.5.

Since abiological weathering normally produces alkaline solutions, the major influence of biological activity in relation to pH and the solubility of weathering products will occur through the generation of acid or particularly strong alkaline solutions. Biogenic acid production (Table 7.2.2) has already been discussed and is widespread among the organisms studied in relation to silicate breakdown (Table 7.2.1). It undoubtedly influences the quantity of Al, Fe or Ti that can be maintained in solution. Biogenic alkali production, although less widely documented with respect to dissolving or maintaining weathering products in solution, is exemplified by the work of Aristovskaya and Kutusova (1968) who showed that the quantity of SiO_2 dissolved from quartz or amorphous phytoliths was directly related to the high pH that developed in microbial culture solutions.

The solubility and mobilization of multivalent cations of metals such as Fe, Mn and Ti is related to their valence states and, thus, to the Eh of aqueous solutions. Some microorganisms, by removing dissolved oxygen or synthesizing reducing agents, can create reducing solutions. Others may accelerate the rate of oxidation of reduced ions. Excellent reviews on these and other mechanisms involved in the biochemistry and biological oxidation and reduction of Fe (Aristovskaya and Zavarzin, 1971) and other multivalent metals such as Mn, Mo, Cu, V (Ehrlich, 1971) in soil have been published recently.

Rate and extent of biological and organic chemical weathering

The rates of biological and organic chemical weathering of silicates are difficult to assess in the field due to the problem of distinguishing between biological and abiological processes. Consequently, most investigators have turned to the laboratory where the variables can be selected and appropriate abiological controls employed. The most common method employed involves a closed system in which both silicates and a single biological species or organic chemical are placed in a vessel with an appropriate solution. The rock- or mineral-forming inorganic elements that appear in solution after a suitable incubation period are then taken as a measure of weathering.

Among the variables to be considered is the relative susceptibility of different silicate mineral structures to weathering (see review by Loughnan, 1969). Goldich (1938) concluded that the relative resistance to weathering of some common rock-forming silicates followed the order: olivine < augite < horneblende < biotite < K-feldspar < muscovite < quartz. Table 7.2.3 lists the quantity of some of the mineral-forming elements released to solution from these minerals by pure cultures of microorganisms in closed systems. Only some of the soluble elements from a given silicate mineral were determined by the various investigators. In addition, there are large differences in the quantities of a single element released from the same mineral (e.g., biotite and Si; muscovite and K, Al, Si) and in the incubation times employed. These facts make comparisons of the data in Table 7.2.3 difficult. Nevertheless, one can conclude, at least, that olivine is very susceptible to microbial attack and that significant microbial breakdown of the silicate minerals in Goldich's sequence can occur in a matter of days or weeks. Other laboratory studies with living organisms, organic acids, fulvic and humic acids (Table 7.2.1) also reveal significant weathering of silicate rocks and minerals within a comparable time frame.

The weathering of silicate rocks and minerals in nature is usually envisaged as a relatively slow process, taking place over geologic time. But biological and organic chemical weathering can be remarkably rapid in the laboratory where significant breakdown of silicates within days or weeks appears to be the rule rather than the exception. However, direct extrapolation of laboratory findings to natural events may be premature because of the many variables that cannot be controlled under natural conditions. Such

TABLE 7.2.3

Microbial solubilization of elements from silicate minerals in the Goldich stability sequence (Goldich, 1938)

Mineral	Element solubilized	Per cent	Time (days)	Reference
Olivine	Mg	54.0	8	Duff et al. (1963)
	Mg	52.0	7	Henderson and Duff (1963)
	Fe + Mg	60.0	15	Goni et al. (1973a)
Augite	Mg	4.6	8	Duff et al. (1963)
	Ca	10.7	8	Duff et al. (1963)
	Fe	0.74	21	Arrieta and Grez (1971)
Hornblende	Mg	4.0	8	Duff et al. (1963)
Biotite	K	9.45	5	Aleksandrov et al. (1967)
	Fe	2.2	21	Arrieta and Grez (1971)
	Al	4.54	5	Aleksandrov et al. (1967)
	Al	3.0	7	Henderson and Duff (1963)
	Si	3.05	5	Aleksandrov et al. (1967)
	Si	14.0	7	Henderson and Duff (1963)
K-feldspar	K	1.0	8	Duff et al. (1963)
Muscovite	K	7.14	5	Aleksandrov et al. (1967)
	K	2.86	49	Antipov-Karatayev et al. (1966)
	K	10.0	8	Duff et al. (1963)
	K	26.8	18	Goni et al. (1973b)
	Al	2.84	5	Aleksandrov et al. (1967)
	Al	0.25	49	Antipov-Karatayev et al. (1966)
	Al	7.6	8	Duff et al. (1963)
	Al	4.0	7	Henderson and Duff (1963)
	Si	2.39	5	Aleksandrov et al. (1967)
	Si	0.09	49	Antipov-Karatayev et al. (1966)
	Si	9.0	7	Henderson and Duff (1963)
Quartz	Si	0.4	70	Aristovskaya and Kutusova (1968)

variables as the number and different kinds of living organisms present, their interactions with one another, and with the kinds and amounts of organic matter present initially or as metabolic products, the availability of water and fluctuations in temperature, pH and Eh, etc., all acting singly or in different combinations, make predictions of natural events uncertain. One can only acknowledge the great potential that exists for rapid and extensive biological and organic chemical weathering of silicate rocks and minerals in the natural environment.

Earth is a planet with an atmosphere, abundant liquid water, and an extensive biota that appears to be ubiquitous. The fossil record shows the presence of multicellular life forms from the late Precambrian to the present, and there is little doubt that microscopic forms of life have been present on Earth for several billion years. It may well be impossible to study chemical weathering on Earth with complete assurance that the observed results were not influenced to some extent by biological activity. By contrast, Earth's moon is an example of a planetary body with no atmosphere, no water and no life, and Mars has a thin atmosphere, shows evidence for the presence of liquid water at some time in its history, and may or may not have harbored living systems. Thus, the ultimate solution to the problem may lie in comparative studies of silicate weathering on Earth and other planetary bodies in our solar system before the relative contributions of abiological and biological weathering on Earth can be assessed.

REFERENCES

Agbim, N.N. and Doxtader, K.G., 1975. Microbial degradation of zinc silicates. Soil Biol. Biochem. 7: 275—280.

Aleksandrov, V.G. and Zak, G.A., 1950. Bacteria that destroy aluminosilicates (silicate bacteria) (in Russian). Mikrobiologiya, 19: 97—104.

Aleksandrov, V.G., Ternovskaya, M.I. and Blagodyr, R.N., 1967. Decomposition of aluminosilicates by silicate bacteria (in Russian). Vestn. S'kh. Nauki, 12: 39—43.

Anderson, A.E., Jonas, E.C. and Odum, H.T., 1958. Alteration of clay minerals by digestive processes of marine organisms. Science, 127: 190—191.

Antipov-Karatayev, I.N., Tsyurupa, I.G. and Alferova, V., 1966. Regularities in the biochemical decomposition of albite and muscotive (in Russian). Kora Vyvetrivaniya Akad. Nauk SSSR, Inst. Geol. Rudn. Mestorozhdenii, Petrogr. Mineral. Geokhim., 7: 53—88.

Aristovskaya, T.V., 1973. Geochemical activity of soil microorganisms as an integral part of biogeocenosis, (in Russian). In: Ye.M. Lavrenko and T.A. Rabotnov (Editors), Problemy Biogeotsenologii. Nauka Press, Moscow, pp. 11—23.

Aristovskaya, T.V., Daragon, A.Yu., Zykina, L.V. and Kutusova, R.S., 1969. Microbiological factors in the migration of certain mineral elements in soil (in Russian). Pochvovedeniye, 9: 95—104.

Aristovskaya, T.V. and Kutusova, R.S., 1968. Microbiological factors in the extraction of silicon from slightly-soluble natural compounds. Sov. Soil Sci., 12: 1653—1659.

Aristovskaya, T.V. and Zavarzin, G.A., 1971. Biochemistry of iron in soil. In: A.D. McLaren and J. Skujins (Editors), Soil Biochem., 2: 385—408.

Arrieta, L. and Grez, R., 1971. Solubilization of iron-containing minerals by soil microorganisms. Appl. Microbiol., 22: 487—490.

Babak, N.M. and Pressman, L.M., 1969. Silicate bacteria in Moldavian soils and their role in the breakdown of some aluminosilicate minerals (in Russian). Tr. Mold. Nauchno Issled. Inst. Oroshaemogo Zemled. Ovoshchevod., 10: 115—122 (Chem. Abstr., 1971, 75: 31864n).

Baker, W.E., 1973. The role of humic acids from Tasmanian podzolic soils in mineral degradation and metal mobilization. Geochim. Cosmochim. Acta, 37: 269—281.

Berner, R.A., 1971. Principles of Chemical Sedimentology. McGraw-Hill, New York, NY, 240 pp.

Berthelin, J., 1971. Altération microbienne d'une arène granitique. Note préliminaire. Sci. Sol, 1: 11—29.

Berthelin, J. and Dommergues, Y., 1972. Rôle de produits du métabolisme microbienne dans la solubilisation des minéraux d'une arène granitique. Rev. Ecol. Biol. Sol, 9: 397—406.

Berthelin, J. and Kogblevi, A., 1974. Influence de l'engorgement sur l'altération microbienne des minéraux dans les sols. Rev. Écol. Biol. Sol, 11: 499—509.

Berthelin, J., Kogblevi, A. and Dommergues, Y., 1974. Microbial weathering of a brown forest soil: influence of partial sterilization. Soil Biol. Biochem., 6: 393—399.

Bertrand, D., 1973. A propos de l'origine de la silice des graminées. C. R. Acad. Sci. (Paris), Ser. D, 277: 857—859.

Bloomfield, C., Kelso, W.I. and Pruden, G., 1976. Reactions between metals and humified organic matter. J. Soil Sci., 27: 16—31.

Boyle, J.R., Voigt, G.K. and Sawhney, B.L., 1967. Biotite flakes: alteration by chemical and biological treatment. Science, 155: 193—195.

Brock, T.D., 1973. Primary colonization of Surtsey, with special reference to the blue-green algae. Oikos, 24: 239—243.

Carroll, D., 1970. Rock Weathering. Plenum, New York, NY, 203 pp.

Degens, E.T., 1965. Geochemistry of Sediments. Prentice-Hall, Englewood Cliffs, NJ, 342 pp.

Dormaar, J.F., 1968. Infrared absorption spectra of mineral matter in saxicolous lichens and associated mosses. Can. J. Earth Sci., 5: 223—230.

Duff, R.B., Webley, D.M. and Scott, R.O., 1963. Solubilization of minerals and related materials by 2-ketogluconic acid-producing bacteria. Soil Sci., 95: 105—114.

Ehrlich, H.L., 1971. Biogeochemistry of the minor elements in soil. In: A.D. McLaren and J. Skujins (Editors), Soil Biochem., 2: 361—384.

Eno, C.F. and Reuszer, H.W., 1951. The availability of potassium in certain minerals to *Aspergillus niger*. Soil Sci. Soc. Am. Proc., 15: 155—159.

Eno, C.F. and Reuszer, H.W., 1955. Potassium availability from biotite, muscovite, greensand and microcline as determined by growth of *Aspergillus niger*. Soil Sci., 80: 199—209.

Goldich, S.S., 1938. A study of rock weathering. J. Geol., 46: 17—58.

Goni, J., Greffard, J., Gugalski, T. and Leleu, M., 1973a. La géomicrobiologie et la biominéralurgie. Bull. Soc. Francaise Mineral. Cristallogr., 96: 252—266.

Goni, J., Gugalski, T. and Sima, M., 1973b. Solubilisation du potassium de la muscovite par voie microbienne. Bulletin du Bureau de Recherches Géologiques et Minières, Deuxième série, Section IV, No. 1, 31—47.

Griffith, S.M. and Schnitzer, M., 1975. The isolation and characterization of stable metal-organic complexes from tropical volcanic soils. Soil Sci., 120: 126—131.

Griggs, R.F., 1933. The colonization of the Katmai ash, a new and inorganic 'soil'. Am. J. Bot., 20: 92—113 (cited by Brock, 1973).

Gromov, B.V., 1959. The microflora of rock layers and primitive soils of some northern districts of the USSR (in Russian). Microbiologiya, 26: 52—59.

Henderson, M.E.K. and Duff, R.B., 1963. The release of metallic and silicate ions from minerals, rocks, and soils by fungal activity. J. Soil Sci., 14: 236—246.

Huang, W.H. and Keller, W.D., 1970. Dissolution of rock-forming silicate minerals in organic acids: simulated first-stage weathering of fresh mineral surfaces. Am. Mineral., 55: 2076—2094.

Huang, W.H. and Keller, W.D., 1971. Dissolution of clay minerals in dilute organic acids at room temperature. Am. Mineral., 56: 1082—1095.

Huang, W.H. and Keller, W.D., 1972. Geochemical mechanics for the dissolution, transport, and deposition of aluminium in the zone of weathering. Clays Clay Miner., 20: 69—74.

Huang, W.H. and Kiang, W.C., 1972. Laboratory dissolution of plagioclase feldspars in water and organic acids at room temperature. Am. Mineral., 57: 1849—1859.

Ilyaletdinov, A.N., 1969. Participation of microorganisms in rock weathering, (in Russian). Izv. Akad. Nauk SSSR, Ser. Biol., 3: 420—427.

Iskander, I.K. and Syers, J.K., 1972. Metal-complex formation by lichen compounds. J. Soil Sci., 23: 255—265.
Ivashov, P.V., 1971. The significance of biological factors in the weathering of rocks and minerals (in Russian). In: A.S. Khomentovskiy (Editor), Biogeokhimiya Zony Gipergeneza. Nauka Press, Moscow, pp. 30—50.
Jacks, G.V., 1953. Organic weathering . Sci. Prog., 41: 301—305.
Jackson, T.A., 1971. A study of the ecology of pioneer lichens, mosses, and algae on recent Hawaiian lava flows. Pac. Sci., 25: 22—32.
Jackson, T.A. and Keller, W.D. 1970a. Evidence for biogenic synthesis of an unusual ferric oxide mineral during alteration of basalt by a tropical lichen. Nature, 227: 522—523.
Jackson, T.A. and Keller, W.D., 1970b. A comparative study of the role of lichens and 'inorganic' processes in the chemical weathering of recent Hawaiian lava flows. Am. J. Sci., 269: 446—466.
Jackson, T.A. and Voigt, G.K., 1971. Biochemical weathering of calcium-bearing minerals by rhizosphere micro-organisms, and its influence on calcium accumulation in trees. Plant Soil, 35: 655—658.
Jaton, C., Pochon, J., Delvert, J. and Bredillet, M., 1966. Étude du mond-milch des grottes du Cambodge. Ann. Inst. Pasteur, 110: 912—919.
Keller, W.D., 1955. The Principles of Chemical Weathering. Lucas Brothers, Columbia, MO, 88 pp.
Kodama, H. and Schnitzer, M., 1973. Dissolution of chlorite minerals by fulvic acid. Can. J. Soil Sci., 53: 240—243.
Kononova, M.M., 1961. Soil Organic Matter. Pergamon, New York, NY, 450 pp.
Kononova, M.M., Aleksandrova, I.V. and Titova, N.A., 1964. Decomposition of silicates by organic substances in the soil. Sov. Soil Sci., No. 10, 1005—1014.
Krasil'nikov, N.A., 1949a. The role of microorganisms in the weathering of rocks. I. Microflora of the surface layer of rocks (in Russian). Mikrobiologiya, 18: 318—323.
Krasil'nikov, N.A., 1949b. The role of microorganisms in the weathering of rocks. II. Focal distribution of microorganisms on the surface of rocks (in Russian). Mikrobiologiya, 18: 492—497.
Krauskopf, K.B., 1967. Introduction to Geochemistry. McGraw-Hill, New York, NY, 721 pp.
Krumbein, W.E., 1968. Zur Frage der biologischen Verwitterung: Einfluss der Mikroflora auf die Bausteinverwitterung und ihre Abhängigkeit von edaphischen Faktoren. Z. Allg. Mikrobiol., 8: 107—117.
Krumbein, W.E., 1969. Über den Einfluss der Mikroflora auf die exogene Dynamik (Verwitterung und Krustenbildung). Geol. Rundsch., 58: 333—363.
Krumbein, W.E., 1971. Sedimentmikrobiologie und ihre geologischen Aspekte. Geol. Rundsch., 60: 438—471.
Krumbein, W.E., 1972. Rôle des microorganismes dans la genèse la diagenèse et la dégradation des roches en place. Revue Ecol. Biol. Sol, 9: 283—319.
Krumbein, W.E. and Pochon, J., 1964. Écologie bacterienne des pierres alterées des monuments. Ann. Inst. Pasteur, 107: 724—732.
Lauwers, A.M. and Heinen, W., 1974. Bio-degradation and utilization of silica and quartz. Arch. Microbiol. 95: 67—78.
Leleu, M., Sarcia, C. and Goni, J., 1973. Altération experimentale de deux feldspaths naturels par voie microbiologiques directe et simulée. Actes du 6e Congrès International de Géochimie Organique, Reuil-Malmaison, France, pp. 905—924.
Loughnan, F.C., 1969. Chemical Weathering of the Silicate Minerals. Elsevier, New York, 154 pp.
Lovering, T.S., 1959. Significance of accumulator plants in rock weathering. Geol. Soc. Am. Bull. 70: 781—800.

Maksimov, O.B., Prischepo, R.S. and Shvets, T.V., 1972. The geochemical role of humic acid oxidation products. Geochem. Int., 9: 135—141.

Mandl, I., Grauer, A. and Neuberg, C., 1952. Solubilization of insoluble matter in nature. I. The part played by salts of adenosinetriphosphate. Biochim. Biophys. Acta, 8: 654—663.

Mandl, I., Grauer, A. and Neuberg, C., 1953. Solubilization of insoluble matter in nature. II. The part played by salts of organic and inorganic acids occurring in nature. Biochim. Biophys. Acta, 10: 540—569.

Mandl, I. and Neuberg, C., 1956. Solubilization, migration, and utilization of insoluble matter in nature. Adv. Enzymol., 17: 135—157.

Marshall, K.C., 1971. Sorptive interactions between soil particles and microorganisms. In: A.D. McLaren and J. Skujins (Editors), Soil Biochem., 2: 409—445.

Matveyeva, L.A., 1969. Role of the temperature factor during the weathering of minerals (in Russian). In: K.I. Lukashev (Editor), Materialy Seminarii: Geokhimiya Gipergenza Kory Vyvetrivaniya. Belorussian SSR, Academy of Sciences Press, pp. 133—138.

Meyers, G.E. and McCready, R.G.L., 1966. Bacteria can penetrate rock. Can. J. Microbiol., 12: 477—484.

Mortenson, J.L., 1963. Complexing of metals by soil organic matter. Soil Sci. Soc. Am. Proc., 27: 179—186.

Mortland, M.M., Lawton, K. and Uehara, G., 1956. Alteration of biotite to vermiculite by plant growth. Soil Sci. 82: 477—481.

Müller, G. and Förster, I., 1961. Einige methodische Versuche zum Problem der Nährstoff-freisetzung aus Mineralien durch Bodenpilze. Zentralbl. Bakteriol., Parasitenkd., Infeksionskr. Hyg., II Abt., 114: 1—10.

Müller, G. and Förster, I., 1963. Der Einfluss mikroskopischer Bodenpilze auf die Nährstoff-freizetzung aus primaren Mineralien, als Beitrag zur biologischen Verwitterung. Zentralbl. Bakteriol., Parasitenkd., Infeksionskr. Hyg., II Abt., 116: 372—409.

Neuberg, C., Salvesen, R.H. and Oster, G., 1961. Role of phosphoglyceric acid salts in the solubilization of inorganic substances in nature. Arch. Biochem. Biophys., 95: 533—539.

Oberlies, F. and Pohlmann, G., 1958a. Einwirkung von Mikroorganismen auf Glas. Naturwissenschaften, 45: 487.

Oberlies, F. and Pohlmann, G., 1958b. Veränderung von Felspatoberflächen durch Mikroorganismen. Naturwissenschaften, 45: 513—514.

Oppenheimer, C.H. and Master, M., 1965. On the solution of quartz and precipitation of dolomite in sea water during photosynthesis and respiration. Z. Allg. Mikrobiol., 5: 48—51.

Perdue, E.M., Beck, K.C. and Reuter, J.H., 1976. Organic complexes of iron and aluminiium in natural waters. Nature, 260: 418—420.

Pochon, J. and Jaton, C., 1967. Causes of the deterioration of building materials. 2. The role of microbiological agencies in the deterioration of stone. Chem. Ind., 1587—1589.

Ponnamperuma, F.N., 1967. A theoretical study of aqueous carbonate equilibria. Soil Sci., 103: 90—100.

Ponomareva, V.V. and Ragim-Zade, A.I., 1969. Comparative study of fulvic and humic acids as agents of silicate mineral decomposition. Sov. Soil Sci., No. 3, 157—166.

Pryor, W.A., 1975. Biogenic sedimentation and alteration of argillaceous sediments in shallow marine environments. Geol. Soc. Am. Bull., 86: 1244—1254.

Ribeiro, R.M., Moureaux, C. and Mussi Santos, A., 1973. Action des microorganismes sur l'alteration d'une roche basique. Cah. ORSTOM, Ser. Pedol., 11: 57—64.

Sawhney, B.L. and Voigt, G.K., 1969. Chemical and biological weathering in vermiculite from Transvaal. Soil Sci. Soc. Am. Proc., 33: 625—629.

Saxby, J.D., 1969. Metal-organic chemistry of the geochemical cycle. Rev. Pure Appl. Chem., 19: 131—150.

Schalscha, E.B., Appelt, H. and Schatz, A., 1967. Chelation as a weathering mechanism- I. Effect of complexing agents on the solubilization of iron from minerals and granodiorite. Geochim. Cosmochim. Acta, 31: 587—596.
Schatz, A., 1962. Pedogenic (soil-forming) activity of lichen acids. Naturwissenschaften, 49: 518—519.
Schatz, A., Cheronis, N.D., Schatz, V. and Trelawney, G.S., 1954. Chelation (sequestration) as a biological weathering factor in pedogenesis. Pa. Acad. Sci. Proc., 28: 44—51.
Schatz, A., Schatz, V. and Martin, J.J. 1957. Chelation as a biochemical weathering factor. Geol. Soc. Am. Bull., 68: 1792—1793.
Schatz, V., Schatz, A., Trelawney, G.S. and Barth, K., 1956. Significance of lichens as pedogenic (soil-forming) agents. Pa. Acad. Sci. Proc., 30: 62—69.
Schoen, R., White, D.E. and Hemley, J.J., 1974. Argillization by descending acid at Steamboat Springs, Nevada. Clays Clay Mineral., 22: 1—22.
Siever, R. and Scott, R.A., 1963. Organic geochemistry of silica. In: I.A. Breger (Editor), Organic Geochemistry, Pergamon, Oxford, pp. 579—595.
Silverman, M.P. and Munoz, E.F., 1970. Fungal attack on rocks: solubilization and altered infrared spectra. Science, 169: 985—987.
Silverman, M.P. and Munoz, E.F., 1971. Fungal leaching of titanium from rock. Appl. Microbiol., 22: 923—924.
Skujins, J.J., 1967. Enzymes in soil. In: A.D. McLaren and G.H. Peterson (Editors), Soil Biochem., 1: 371—414.
Sokolova, Ye.I., 1969. Role of fulvic acids during the weathering of silicate minerals, (in Russian). In: K.I. Lukashev (Editor)., Materialy Seminarii: Geokhimiya Gipergeneza Kory Vyvetrivaniya. Belorussian SSR, Academy of Sciences Press, pp. 162—171.
Stevenson, F.J., 1967. Organic acids in soil. In: A.D. McLaren and G.H. Peterson (Editors), Soil Biochem., 1: 119—146.
Stotsky, G., 1972. Activity, ecology, and population dynamics of microorganisms in soil. Crit. Rev. Microbiol., 2: 59—137.
Syers, J.K. and Iskandar, I.K., 1973. Pedogenetic significance of lichens. In: V. Ahmadjian and M.E. Hale (Editors), The Lichens. Academic, New York, NY, pp. 225—248.
Tan, K.H., 1975. The catalytic decomposition of clay minerals by complex reaction with humic and fulvic acid. Soil Sci., 120: 188—194.
Tesic, Z.P. and Todorovic, M.S., 1958. Contribution to knowledge of the specific properties of silicate bacteria. Zemljiste Biljka, 8: 233—240.
Treub, M., 1888. Notice sur la nouvelle flore de Krakatau. Ann. Jardin Bot. Buitenzorg, 7: 213—223 (cited by Brock, 1973).
Tsyurupa, I.G., 1964. Some data on complex products of microbial activity and autolysis with soil minerals. Sov. Soil Sci., No. 3, 261—265.
Wagner, M., 1966. Vorkommen und Rolle Oxalat-verwertender Mikroorganismes bei Verwitterungsprozessen. Z. Allg. Mikrobiol., 6: 197—209.
Wagner, M. and Schwartz, W., 1967a. Geomikrobiologische Untersuchungen. VIII. Über das Verhalten von Bakterien auf der Oberfläche von Gesteinen und Mineralien und ihre Rolle bei der Verwitterung. Z. Allg. Mikrobiol., 7: 33—52.
Wagner, M. and Schwartz, W., 1967b. Geomikrobiologische Untersuchungen. IX. Verwertung von Gesteins- und Mineralpulveren als Mineralsalzquelle für Bakterien. Z. Allg. Mikrobiol., 7: 129—141.
Webley, D.M. and Duff, R.B., 1965. The incidence, in soils and other habitats, of microorganisms producing 2-ketogluconic acid. Plant Soil, 22: 307—313.
Webley, D.M., Henderson, M.E.K. and Taylor, I.F., 1963. The microbiology of rocks and weathered stones. J. Soil Sci., 14: 102—112.
Weed, S.B., Davey, C.B. and Cook, M.G., 1969. Weathering of mica by fungi. Soil Sci. Soc. Am. Proc., 33: 702—706.
Williams, M.E. and Rudolph, E.D., 1974. The role of lichens and associated fungi in the chemical weathering of rock. Mycologia, 66: 648—660.

Chapter 7.3

DEPOSITION AND DIAGENESIS OF BIOGENIC SILICA

J.H. OEHLER

Research and Development Department, Continental Oil Company, P.O. Box 1267, Ponca City, OK 74601 (U.S.A.)

CONTENTS

INTRODUCTION

When silica-depositing organisms die, the organic constituents of their cells decompose, and the polymeric silica originally deposited within and around these cells is released, usually in a particulate form. The sources, nature, and ultimate fate of this biogenic silica are the subjects of this chapter. The first section deals with biogenic siliceous deposits on land and the second with such deposits in the sea.

DEPOSITION OF BIOGENIC SILICA IN TERRESTRIAL ENVIRONMENTS

Sources

The main sources of biogenic silica in terrestrial environments are vascular plants, diatoms, and sponges, generally in that order. The principal contributors among vascular plants are the monocotyledons (monocots), particularly members of the family Gramineae, which includes grasses, bamboos, rice, wheat, barley, oats, and maize. On a dry-weight basis, the silica content of such plants commonly is in the range of 3—5%, although values exceeding 20% have been reported for some grasses (Norgren, 1973). In general, dicotyledonous plants (dicots) contain about an order of magnitude less silica than monocots. However, in some environments, such as deciduous forests, dicots are the main sources of biogenic silica. Conifer woods are typically low in silica, although conifer needles may have as much as 7.9% SiO_2 on a dry-weight basis (Norgren, 1973).

In all of these plants, silica is taken up through the roots as dissolved $Si(OH)_4$ and is precipitated within and around the cells as hydrated opaline deposits which often replicate the shapes of the associated cellular structures. Upon death and decomposition of a plant, these siliceous deposits are released to the soil as discrete and generally microscopic structures known as "phytoliths" or "plant opal". Most phytoliths in soils are derived from the aerial parts of plants. However, silicification can be extensive also in the roots and rhizomes of certain grasses, so that soils developed under this kind of vegetation receive significant quantities of biogenic silica from the underground portions of the plants. A rarer type of siliceous deposit in some vascular plants is tabashir (Jones et al., 1966). This material apparently is restricted to the bamboos, where it occurs within the hollow stems as solid, transluscent, opaline masses up to several cm thick.

Diatom frustules and sponge spicules in terrestrial environments are derived chiefly from fresh and brackish water species and from unconsolidated fossiliferous sediments exposed to surface winds and water.

Grazing mammals are another, albeit mainly indirect, source of biogenic silica in terrestrial environments. Such animals as sheep, cattle, and horses consume large quantities of grasses and other silica-depositing plants. The majority of the silica in the consumed plant material passes through the alimentary tract and is redeposited with the faeces and urine, sometimes at a great distance from the original source area. Silica in the faeces is mainly in the particulate form (phytoliths from plants and rarer diatom frustules and sponge spicules ingested with soil). Silica in the urine is in the dissolved form, $Si(OH)_4$, and may be excreted in concentrations as high as $1\ g\ l^{-1}$ (Jones and Handreck, 1967).

Although probably of minor significance, birds also are an indirect source of biogenic silica in soils. They may ingest it with plant material or may pick it up inadvertently with soil that adheres to their feet. This silica can then be transported and redeposited in localities quite distant from the original source area. Finally, man, through the use of herbivore faeces as fertilizer, can be instrumental in contributing substantial quantities of biogenic silica to soils that otherwise might be relatively impoverished in this material.

Characteristics

With the exception of dissolved silica in herbivore urine, biogenic silica in terrestrial environments is deposited in the particulate form (phytoliths, diatom frustules, etc.). These particles are generally microscopic in size, although some may exceed 1 mm in longest dimension. In soils, they tend to be concentrated in the 2-200 μm size fraction. The morphologies of particulate silica in soils are highly variable and can often be used to determine the biological, ecological, and geographical sources of the silica. When dealing with redeposited diatom frustules, particle morphology may also give an indication of the geologic age (and, hence, geographic locality) of the original source deposits.

As a rule, biologically precipitated silica is non-crystalline and optically isotropic. As Lewin and Reimann (1969) have pointed out, reports of crystalline silica in plants (quartz, cristobalite, etc.) should be viewed with caution. Such reports are often based on materials that have been "dry ashed" by heating to temperatures in the region of 700—1000°C or "wet ashed" by treament with dehydrating acids such as H_2SO_4. These kinds of preparatory techniques can alter the physical properties of non-crystalline silica and lead to spurious results. However, Wilding and Drees (1974) have recently reported X-ray diffraction data showing quartz and cristobalite reflections in samples of silica isolated from deciduous tree leaves by a low-temperature (60-65°C) dry-ashing technique. If subsequently confirmed, these results would indicate that some plants are capable of synthesizing crystalline silica within a period of less than one year, a remarkable finding in view of the extremely slow crystallization rates of silica generally observed under earth-surface conditions (e.g., Mackenzie and Gees, 1971).

Phytoliths, diatom frustules, and sponge spicules typically contain about 85% SiO_2 and variable small percentages of other major elements (Ca, Na, K, Al, etc.). In addition, they contain varying amounts of bound water, usually between 5 and 14%. Refractive indices of the particles are generally in the range of 1.41 to 1.48; specific gravities range from about 1.5 to 2.3, but tend to be around 2.1 (Wilding and Drees, 1974).

Distribution in sediments

In most soils, particulate, biogenic silica is concentrated in the organic-rich A horizon, where phytoliths, diatom frustules, and sponge spicules often account for 1-2% of the total soil mass. Wilding et al., (1977) note that soils developed under long periods of grass vegetation typically contain 5-10 times as much biogenic silica (about 0.8-30 g m^{-2}) as those developed in forest environments (ca. 0.5-8 g m^{-2}). The highest concentration of soil phytoliths so far recorded is on the Isle of Réunion (east of Madagascar) where Riquier (1960) found a soil horizon 5-30 cm thick consisting almost entirely of siliceous phytoliths.

In a typical B horizon, the concentration of biogenic silica may be about 0.5%, while in a typical C horizon, particles of biogenic silica are usually rare. The decrease in concentration of these particles with depth in the soil profile results chiefly from leaching and dissolution by groundwaters.

Most data on the concentration and distribution of biogenic silica in soils come from studies of agricultural lands, grasslands, and forested regions, locales where one might expect the abundances of these particles, especially phytoliths, to be high. However, even in desert regions such as those in central Australia, dusts and upper soil levels may contain as much as 1.9% phytoliths (Baker, 1960). Values of this magnitude in arid regions may reflect very slow dissolution rates or the effects of airborne transport of biogenic silica from distant regions of higher productivity.

Quantitatively, wind is the most important agency for transporting particulate biogenic silica. The low specific gravity of the particles and their relatively high surface areas (14.4 $m^2\ g^{-1}$ for oat phytoliths, Jones and Milne, 1963; and up to 123 $m^2\ g^{-1}$ for diatom frustules, Lewin, 1961), together with their tendency to be concentrated in the uppermost layers of soils, render the tiny particles highly susceptible to airborne transport. Airborne siliceous particles may return to the earth's surface as dust or as inclusions in rain drops and snow. For example, a 142-ml sample of the "red rain" that fell near Melbourne, Australia, in 1903 contained 17 g of dry sediment of which 0.65% consisted of phytoliths, 0.58% of diatom frustules, and 0.06% of sponge spicules (Baker, 1959a), a total biogenic silica content of 1.54 g l^{-1}. Biogenic silica particles are also readily transported by surface waters: in fact, such particles are common constituents of ordinary tap water. In comparison with wind and water, biological transporting agencies, such as birds, grazing mammals, and man, probably are quantitatively insignificant on a worldwide basis, although biological transport may be important locally.

DIAGENESIS OF BIOGENIC SILICA IN TERRESTRIAL ENVIRONMENTS

Dissolved silica

According to Jones and Handreck (1967), silica in soil solutions is entirely in the monomeric form $Si(OH)_4$ (monosilicic acid) and is present in concentrations generally ranging from 7-80 $\mu g\ g^{-1}$, but always less than the saturation value (about 120 $\mu g\ g^{-1}$). The concentration of dissolved silica in soils depends on those factors which control dissolution rates of polymeric silica and on those which control the rate of removal of monosilicic acid from solution.

Dissolution rates of biogenic silica are dependent on a number of variables, of which the following seem most important.

(1) Crystallinity. Non-crystalline forms will dissolve more rapidly than crystalline forms; the vast majority of biogenic silica in soils is non-crystalline.

(2) Specific surface area. The higher the specific surface area (surface area: volume ratio) of silica particles, the greater their dissolution rate will be; in general, it is also true that smaller particles will dissolve more rapidly than larger particles (as much as 50-75% of biogenic silica in soils may be in the <5 μm size fraction).

(3) Contaminants. Fe and Al ions can become chemically adsorbed onto surfaces of particulate silica and inhibit dissolution (Jones and Handreck, 1967); occluded carbonaceous matter also has been suggested as possibly having an inhibitory effect on the dissolution of phytoliths (Wilding and Drees, 1974).

(4) Soil microorganisms. Many reports have documented the ability of certain soil microorganisms (especially bacteria such as *Bacillus siliceous*, which is used as a "fertilizer" in some parts of the Soviet Union (Cooper, 1959)) to depolymerize silica and convert it to the soluble monomeric form (see e.g., Lauwers and Heinen, 1974, and references therein); thus, the presence or absence of these organisms can be expected to play an important role in determining the concentration of dissolved silica in soil solutions.

The following are factors considered to be important in governing the rate of removal of monosilicic acid from soil solutions.

(1) Flushing. Soils are open systems with respect to water flow, and wherever silica-depleted groundwaters pass through soils, soluble silica is removed; this may be redeposited elsewhere or may be fed through rivers to the ocean.

(2) Adsorption. Monomeric silica can be removed from solution by adsorption onto surfaces of sesquioxide minerals (e.g., Al_2O_3, Fe_2O_3) through pH-dependent reactions that apparently involve hydrogen bonding (see Jones and Handreck, 1967, and references therein); such reactions, especially with Al_2O_3, seem to exert a major control over the concentration of dissolved silica in soil solutions.

(3) Allophane production. Allophane is a non-crystalline, hydrous aluminosilicate of highly variable composition, which forms as a common colloidal constituent of soils; it may be an intermediate in the formation of some authigenic clay minerals.

(4) Formation of clay minerals. It is generally agreed that some of the dissolved silica in soils is utilized in the formation of authigenic clay minerals; however, the quantitative significance of this process is unclear (McKeague and Cline, 1963).

(5) Formation of cements, overgrowths, and silcretes. Precipitation of silica from solution to form intergranular cements and overgrowths on primary quartz grains is regarded as an important process in soils and sedimentary rocks (Pettijohn, 1957; Breese, 1960); in addition, formation of silcretes and other heavily silicified duricrusts is a major process in certain arid regions, especially those of Australia and southern Africa (Stephens, 1971), where it occurs in near-surface horizons through massive silicification of pre-existing sediments by silica-charged groundwaters.

(6) Biological uptake. Some soil microorganisms are capable of removing dissolved silica from solution and incorporating it in their cells (Lauwers and Heinen, 1974); similarly, vascular plants, which are the major source of biogenic silica in the first place, remove silica from soil solutions and deposit it within and around their cells.

Particulate silica

Biogenic silica that is not solubilized, but persists as discrete particles in soils, is transformed ultimately to quartz. Although this transformation could proceed directly from non-crystalline silica to quartz, it is more likely that the conversion process generally involves one or more intermediate stages such as in the following reaction:

non-crystalline silica → opal-CT → quartz.

Opal-CT is a poorly ordered silica polymorph composed of interstratified cristobalite and tridymite layers (Jones and Segnit, 1971, 1972); it is the mineral which is often misidentified as cristobalite in studies of the diagenetic transformations of silica.

As is the case with other forms of non-crystalline silica, the conversion to quartz of phytoliths, diatom frustules, and sponge spicules under sedimentary conditions on land can be expected to be accelerated by elevated temperature, elevated pressure, and the presence of an aqueous phase of high pH (>9) or containing electrolytes.

Unlike biogenic silica deposited in the sea, that deposited on land apparently has seldom been observed in the various stages of transformation. However, Beavers and Stephen (1958, p. 4) note the presence in Illinois palaeosols of phytoliths showing several stages of transformation to quartz, "from unaltered opal through opal with marginal chalcedonic alteration to completely altered paramorphs of chalcedony". These observations suggest that the conversion to quartz begins at the outer margins of phytoliths and proceeds inward. In contrast to the relatively rapid conversion rates illustrated by these examples, there are two reports of opal (i.e., apparently non-crystalline) phytoliths preserved in sedimentary rocks of Tertiary age. Baker (1959b) reported the presence of opal phytoliths in sediments of

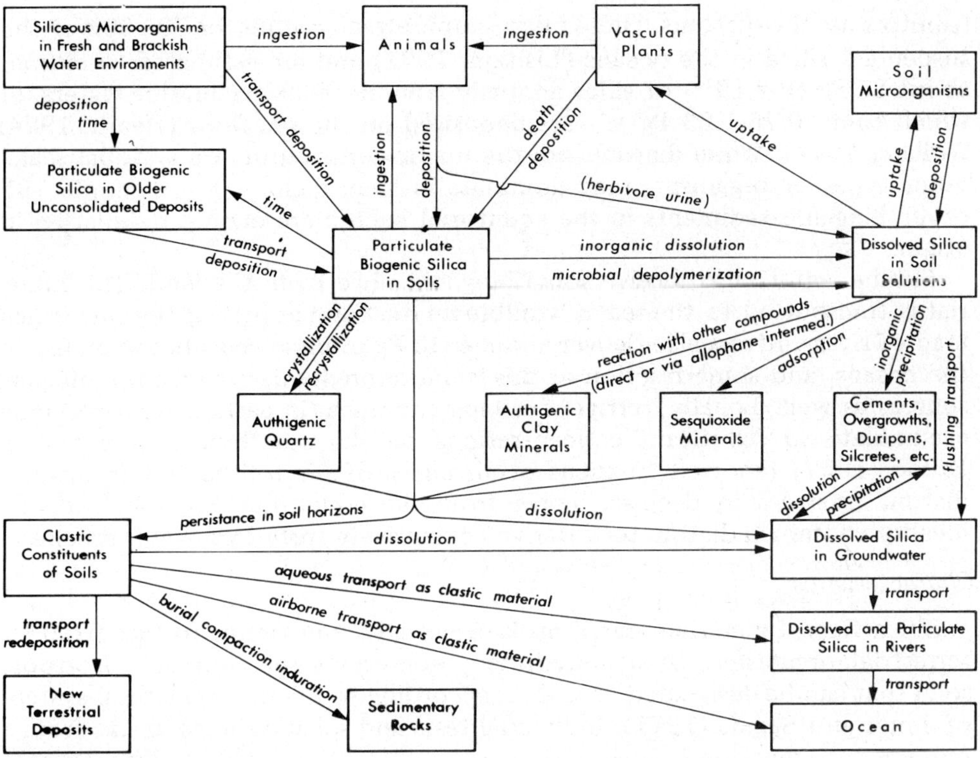

Fig. 7.3.1. Diagrammatic representation of major processes and silica reservoirs in the terrestrial silica cycle.

Holocene, Pleistocene, and Pliocene age in Victoria, Australia, and Jones (1964) reported opal phytoliths from North American sedimentary rocks as old as Palaeocene. If the phytoliths in these Tertiary samples are indeed composed of non-crystalline silica, then they provide an excellent illustration of the fact that the conversion rates of non-crystalline silica to quartz under natural conditions are variable and strongly dependent on local environmental conditions and geologic histories.

A diagrammatic summary of those aspects of the terrestrial silica cycle discussed above is shown in Fig. 7.3.1.

DEPOSITION OF BIOGENIC SILICA IN MARINE ENVIRONMENTS

Sources

The main sources of biogenic silica in marine environments are diatoms, radiolaria, silicoflagellates, and silicisponges. By far the most important contributors are the diatoms. These microscopic algae account for 70-90% of the tributors are the diatoms. These microscopic algae account for 70—90% of the suspended silica in the oceans (Lisitzin, 1971) and are estimated to extract about 25 Pg (P = 10^{15}) of silica annually from near-surface marine waters, of which some 0.75-1.23 Pg y^{-1} are deposited on the sea floor (Heath, 1974; Wollast, 1974). While diatoms are the major contributors on a global scale, other types of organisms may dominate in local regions. For example, siliceous biogenic sediments in the equatorial Pacific are mainly radiolarian in origin.

Marine waters also receive some biogenic silica from the land. This material is transported to the sea as windblown dust and as part of the suspended load of rivers. Rivers also deliver about 0.43 Pg of dissolved silica annually to the oceans, and some fraction of this is undoubtedly derived from biological sources as well. Locally, terrigenous biogenic silica (in particulate form) may accumulate to significant concentrations on the sea floor. For example, Kolbe (1957) reported frequent occurrences of phytoliths and freshwater diatom frustules in deep-sea cores from the equatorial Atlantic, and one locality contained diatom tests derived exclusively from freshwater species.

Characteristics

Silica fixed by marine organisms is essentially comparable to that fixed by terrestrial organisms. It is a hydrated, non-crystalline, optically isotropic solid that can be designated as opal-A, according to the nomenclatural system of Jones and Segnit (1971). Individual tests and spicules range in size from a few μm to a few mm in longest dimension. The silica is reportedly very pure (Lewin, 1962), although it may contain small amounts of major elements such as Al, K, and Fe, as well as variable quantities of bound water. In addition, the particles may be coated with a thin layer of protoplasmic organic material. The gross morphology of the particles varies from the rather simple, spine-like spicules of sponges to the exceedingly intricate, lace-like tests of diatoms and radiolaria. The specific surface area of the particles increases with morphological complexity and may be as high as 123 m^2 g^{-1} in some diatom frustules (Lewin, 1961). The specific gravity tends to be around 2.0 - 2.1.

Changes during settling

In the marine environment, biological fixation of silica takes place mainly in near-surface waters, in the photic zone. Following death of the organisms, their tests settle toward the ocean floor. However, recent studies indicate that only a small fraction of these siliceous particles (probably less than 5%) survive the descent to become part of the marine sedimentary pile (e.g., Calvert, 1968, 1974; Lisitzin, 1971; Hurd, 1973; Heath, 1974; Wollast, 1974). The vast majority of biologically polymerized silica in the oceans is redissolved during settling. According the Heath (1974), this dissolution during settling can be divided into two major processes, oxidative regeneration and non-oxidative dissolution. Oxidative regeneration refers to the rapid, initial dissolution of siliceous tests that occurs in the upper few hundred meters of the water column due to oxidative destruction of the protective organic coatings around the tests and the consequent direct exposure of the silica to undersaturated waters. Following destruction of the organic coatings, the dissolution rates of silica in this relatively shallow regime seem to be controlled mainly by water temperature, degree of silica saturation of the water, and available surface area of the tests (Hurd, 1972). Below the zone of oxidative regeneration, solution of settling silica particles continues through the normal process of non-oxidative dissolution. At depth, this process apparently proceeds independently of the ambient dissolved silica concentration and is governed instead by the available surface area of the particles and by the turbulence of the water (Bogoyavlenskiy, 1967; Berger, 1968; Heath, 1974). Heath (1974) has estimated that oxidative regeneration contributes about 20.3 Pg y^{-1} of dissolved silica to ocean waters, while non-oxidative dissolution (both during settling and at the sediment/water interface) contributes about 3.65 Pg y^{-1}. These figures compare with the value of 24.2 Pg y^{-1} estimated by Wollast (1974) as the amount of silica dissolved from siliceous tests prior to burial within the sedimentary pile.

From the foregoing, it can be seen that the amount of silica annually fixed by marine organisms (ca. 25 Pg y^{-1}) is approximately balanced by the amount annually dissolved from their siliceous tests (ca. 24 Pg y^{-1}). Moreover, these quantities exceed by more than an order of magnitude the yearly amounts of silica contributed to the world ocean from external sources (rivers, interstitial waters, etc.) or removed from it by burial in marine sediments (Heath, 1974; Wollast, 1974). Thus, it is apparent not only that the marine silica cycle is biologically controlled, but also that the biological subcycle of the marine silica cycle acts as a quasi-closed system.

A consequence of dissolution during settling is that large or robust tests are more likely to be preserved in deep-water sediments than are small or delicate ones. Thus, the preserved assemblage of tests in bottom sediments may not accurately reflect the biological composition of populations inhabiting overlying waters (e.g., Berger, 1968, Goll and Bjorklund, 1971). To some extent, however, this tendency to discriminate against the preservation of delicate tests is overcome by cycling of the silica through organisms at higher trophic levels. When this happens, siliceous tests are excreted with the faeces of the consuming organism, and, if the faecal pellets do not disintegrate rapidly, they can protect the contained tests from subsequent dissolution during settling (e.g., Schrader, 1971). But, as Hurd (1972) has pointed out, passage through the alimentary tracts of grazing organisms can also have adverse effects on the preservation potential of siliceous tests. Crushing during ingestion can increase the specific surface area of the particulate silica, and digestion of cell protoplasm can remove the protective organic coatings from the tests. Both processes render the silica more susceptible to dissolution.

Distribution in marine sediments

The distribution of biogenic silica in marine sediments is directly related to the distribution of silica-depositing organisms in overlying waters. These organisms are most abundant in those places where upwelling currents bring deep, nutrient-rich waters into the photic zone. Such areas include the circum-Antarctic region, the eastern equatorial Pacific, and many of the continental margins (Lisitzin et al., 1967).

The concentration of biogenic silica in marine sediments is controlled not only by biological productivity, but also by the relative proportions of other types of sediment in the same locality, especially terrigenous clastics and biogenic carbonates. For example, Heath (1974) has estimated that 85-90% of the biogenic silica deposited in marine environments is laid down in near-shore areas, where it is masked by clastic material derived from the land. Similarly, in areas where the productivity of both siliceous and calcareous plankton is high, the concentration of biogenic silica tends to be greater in sediments that lie below the calcite compensation depth than in those that lie above it. Finally, ocean currents can influence the distribution of biogenic silica in bottom sediments. This may happen during settling (e.g., Kolbe, 1957) or after deposition (e.g., Johnson, 1974). The tendency in the latter case is for tests (especially small ones) to be transported downslope to sites deeper than those from which they were eroded.

Thus, in submarine surface sediments, maximum concentrations of biogenic silica (locally exceeding 70% of the total sediment weight) are found in deep-water localities that underlie surface waters of high productivity and are situated seaward of the normal limit of terrigenous sediment influx. The largest single zone of this type is the circum-Antarctic region (Lisitzin et al., 1967), where accumulation of siliceous material is probably assisted by uninterrupted latitudinal circulation and its tendency to inhibit northerly dispersion of settling tests.

DIAGENESIS OF BIOGENIC SILICA IN MARINE ENVIRONMENTS

Dissolved silica

In addition to the considerable amount of dissolution of biogenic silica that takes place in the oceanic water column, further dissolution of siliceous tests and spicules occurs within the marine sedimentary pile. Recent estimates of the proportion of deposited particulate silica that redissolves after burial range from 50% (Heath, 1974) to 90% or more (Hurd, 1973; Wollast, 1974). Upon dissolution, this silica becomes a constituent of the interstitial waters of marine sediments. The concentration of dissolved silica in interstitial waters varies with the lithology, geographic location, and subbottom depth of the enclosing sediment. Concentrations range from a few to more than 100 $\mu g\ g^{-1}$, but are generally higher than in the overlying bottom water. Vertical profiles through marine sediments show that the dissolved silica content of pore waters increases downward through the uppermost few hundred meters of sediment, then gradually decreases at greater depths (e.g., Siever et al., 1965; Bischoff and Sayles, 1972; Gieskes, 1973). Such profiles indicate that the dissolution rate of particulate silica initially exceeds the rate of removal of $Si(OH)_4$ from solution, but that this trend reverses at depth due to diagenetic reactions that result in the incorporation of silica into solid phases.

About half of the dissolved silica produced in marine sediment pore waters escapes back into the oceanic water column (Heath, 1974; Wollast, 1974). This results from upward diffusion along the concentration gradient that exists across the sediment/water interface and from physical expulsion of interstitial water during sediment compaction. Estimates of the magnitude of this flux range from 0.06-0.79 Pg y^{-1} of silica (Fanning and Pilson, 1971; Bischoff and Sayles, 1972; Hurd, 1973; Heath, 1974; Wollast, 1974). Dissolved silica that remains within the sedimentary pile is available as a reactant in subsequent diagenetic processes. A small proportion precipitates as opaline cement in localities such as the clay fillings of siliceous tests (Heath, 1974). Some diffuses into and replaces calcareous sediment to form chert nodules and partially silicified chalks (e.g., Wise and Kelts, 1972; Wise and Weaver, 1974). Some participates in the formation and induration of deep-sea bedded cherts. And some reacts with other ions in sea water to form authigenic minerals such as sepiolite, $Mg_2Si_3O_8$ (Wollast et al., 1968). However, it appears that the majority (perhaps 50% or more) reacts with clay minerals to form either new, more silicic minerals (Wollast, 1974) or silica-rich, adsorbed surface layers on existing clay particles (Siever and Woodford, 1973). These latter two types of reactions are reversible (e.g., Mackenzie et al., 1967, Fanning and Schink, 1969; Siever and Woodford, 1973) and probably play the major role in controlling the concentration of dissolved silica in deep interstitial waters.

Particulate silica

From the preceding discussions, it is evident that only a small proportion of the silica originally fixed by marine organisms survives dissolution to be incorporated into the geologic record. Recent estimates of the fraction ultimately preserved in marine sediments range from 2 to 0.05% (Hurd, 1973; Heath, 1974, Wollast, 1974) and Heath (1974) considers that 85 to

90% of this fraction is deposited in near-shore environments, where it is masked by terrigenous debris. Thus, the thick and extensive siliceous deposits developed in deep-sea environments represent a minute percentage of the silica initially deposited on the sea floor. Siliceous tests and spicules that remain as intact, solid particles in the marine sedimentary pile are converted ultimately to microcrystalline quartz. A large body of observational and experimental data suggests that this conversion proceeds through a "cristobalitic" intermediate stage that can be identified as opal-CT (e.g., Mizutani, 1966; Ernst and Calvert, 1969; Calvert, 1971a, b; Heath and Moberly, 1971; von Rad and Rösch, 1972, 1974; Flörke et al., 1975; Oehler, 1975; Mizutani and Oehler, 1979). Although this transformation occurs in all marine sediments that receive some biogenic silica from overlying waters, it is most clearly illustrated in highly siliceous deep-sea sediments, where it leads to the formation of bedded cherts.

In deep-sea sediments, opal-CT commonly occurs as platy, blade-shaped crystals that are often arranged in spheroidal rosettes a few μm in diameter. These rosettes are known as lepispheres (Wise and Kelts, 1972). The nearly euhedral crystal habit of natural lepisphere crystals indicates that the opal-CT precipitated in situ as an authigenic mineral; this conclusion is supported by results of experimental studies (Oehler, 1973) in which completely euhedral lepisphere crystals were synthesized from amorphous silica. Thus, it appears that the conversion of biogenic silica (opal-A) to opal-CT occurs principally, if not exclusively, through a solution-reprecipitation mechanism.

The conversion of opal-CT to quartz has been studied experimentally with conflicting results. Mizutani (1966) concluded that the reaction takes place through a solution-reprecipitation mechanism. Ernst and Calvert (1969) concluded that the mechanism is by solid-solid conversion. In a re-evaluation of the data presented by Ernst and Calvert (1969), Stein and Kirkpatrick (1976) concluded that the evidence more closely supports a solution-reprecipitation mechanism. At present, it is unclear which of these two processes, if either, dominates in the marine sedimentary environment.

The transformation, biogenic opal-A → opal-CT → quartz, has been regarded widely as a simple maturation process, depending principally on ambient temperature and pressure. Recently, however, Lancelot (1973) and Greenwood (1973) have cited data which suggest that the chemical composition of the host sediment influences the transformation rates. They found that siliceous deposits in clayey sediments are composed largely of opal-CT, while those found in carbonate sediments consist mainly of quartz. One explanation of these observations is that large cations present in clayey sediments promote the formation of opal-CT by distorting the structure of growing cristobalitic crystals, while the paucity of such cations in carbonate sediments permits the development of well-ordered silica polymorphs and leads to a rapid conversion to quartz (Lancelot, 1973). Similarly, Murara and Nakata (1974) have concluded that, in siliceous strata of the Monterey Formation in California, formation of diagenetic opal-CT occurred first in layers of purest diatomite and later in layers of diatomaceous mudstone. Thus, while the maturation concept can still be regarded as generally valid, it now appears that some relationship also exists between the rates and types of silica transformations and the chemical composition of the host sediment.

A final factor influencing the rates of diagenetic reactions (solution, reprecipitation, recrystallization) in marine siliceous deposits is the porosity and permeability of the sediment. In studies of radiolarian cherts from Italy, Thurston (1972) found that the effects of diagenetic processes were least in

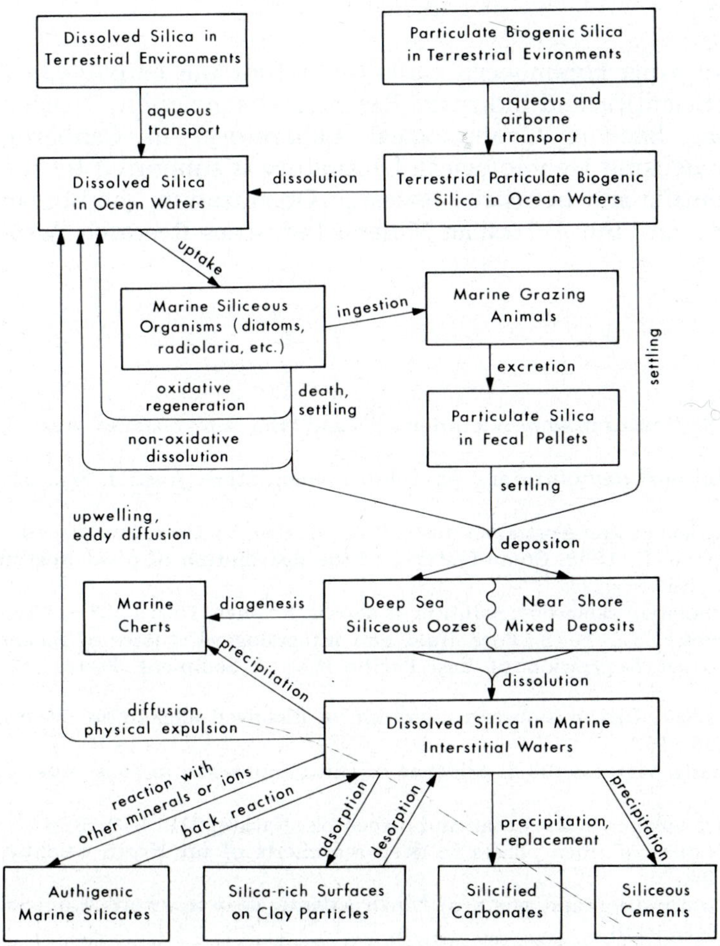

Fig. 7.3.2. Diagrammatic representation of major processes and silica reservoirs in the marine silica cycle.

those cherts where movement of pore solutions had been inhibited by the presence of fine-grained hematitic and clay material. A similar phenomenon could have contributed to the relationships found by Lancelot (1973), Greenwood (1973), and Murata and Nakata (1974) noted above. That is, the tendency to form opal-CT rather than quartz (or to form opal-CT very slowly) in clay-rich sediments could be related to inhibition of pore water movement in these sediments rather than to physicochemical effects arising from the abundance of large cations.

A diagrammatic summary of those aspects of the marine silica cycle discussed above is shown in Fig. 7.3.2.

ACKNOWLEDGMENTS

Preparation of this paper commenced while the author was employed by the Commonwealth Scientific and Industrial Research Organization, Division of Mineralogy, Baas Becking Geobiological Laboratory, in Canberra, Australia. The Baas Becking Geobiological Laboratory is supported by the Commonwealth Scientific and Industrial Research Organization, the Bureau of Mineral Resources, and the Australian Mineral Industries Research Association Limited.

REFERENCES

Baker, G., 1959a. Opal phytoliths in some Victorian soils and "red rain" residues. Aust. J. Bot., 7: 64—87.

Baker, G., 1959b. Fossil opal phytoliths and phytolith nomenclature. Aust. J. Sci., 21: 305—306.

Baker, G., 1960. Phytoliths in some Australian dusts. Proc. R. Soc. Victoria, 72: 21—40.

Beavers, A.H. and Stephen, I., 1958. Some features of the distribution of plant opal in Illinois soils. Soil Sci., 86: 1—5.

Berger, W.H., 1968. Radiolarian skeletons: solution at depths. Science, 159: 1237—1239.

Bischoff, J.L. and Sayles, F.L., 1972. Pore fluid and mineralogical studies of recent marine sediments: Bauer Depression of East Pacific Rise. J. Sediment. Petrol. 42: 711—724.

Bogoyavlenskiy, A.N., 1967. Distribution and migration of dissolved silica in the oceans. Int. Geol. Rev., 9: 133—153.

Breese, G.F., 1960. Quartz overgrowths as evidence of silica deposition in soils. Aust. J. Sci., 23: 18—20.

Calvert, S.E., 1968. Silica balance in the ocean and diagenesis. Nature, 219: 919—920.

Calvert, S.E., 1971a. Nature of silica phases in deep sea cherts of the North Atlantic. Nature, 234: 133—134.

Calvert, S.E., 1971b. Composition and origin of North Atlantic deep sea cherts. Contrib. Mineral. Petrol., 33: 273—288.

Calvert, S.E., 1974. Deposition and diagenesis of silica in marine sediments. In: K.J. Hsü and H.C. Jenkyns (Editors), Pelagic Sediments on Land and Under the Sea. Interna-

tional Association of Sedimentologists Special Publication No. 1, 1: 273—299.
Cooper, R., 1959. Bacterial fertilizers in the Soviet Union. Soils Fert., 22: 327—333.
Ernst, W.G. and Calvert, S.E., 1969. An experimental study of the recrystallization of porcelanite and its bearing on the origin of some bedded cherts. Am. J. Sci. (Schairer Vol.), 267A: 114—133.
Fanning, K.A. and Pilson, M.E.Q., 1971. Interstitial silica and pH in marine sediments: some effects of sampling procedures. Science, 173: 1228—1231.
Fanning, K.A. and Schink, D.R., 1969. Interaction of marine sediments with dissolved silica. Limnol. Oceanogr., 14: 59—68.
Flörke, O.W., Jones, J.B. and Segnit, E.R., 1975. Opal-CT crystals. Neues Jahrb. Miner., 1975: 369—377.
Gieskes, J.M., 1973. Interstitial water studies, Leg 15. In: B.C. Heezen et al. (Editors), Initial Reports of the Deep Sea Drilling Project, 20: 813—829.
Goll, R.M. and Bjorklund, K.R., 1971. Radiolaria in surface sediments of the North Atlantic Ocean. Micropaleontology, 17: 434—454.
Greenwood, R., 1973. Cristobalite: its relationship to chert formation in selected samples from the Deep Sea Drilling Project. J. Sediment. Petrol., 43: 700—708.
Heath, G.R., 1974. Dissolved silica and deep-sea sediments. In: W.H. Hay (Editor), Studies in Paleo-oceanography. Society of Economic Paleontologists and Mineralogists, Special Publication No. 20: 77—93.
Heath, G.R. and Moberly, Jr., R., 1971. Cherts from the western Pacific: Leg VII, Deep Sea Drilling Project. In: E.L. Winterer et al. (Editors), Initial Reports of the Deep Sea Drilling Project, 7: 991—1007.
Hurd, D.C., 1972. Factors affecting solution rate of biogenic opal in seawater. Earth Planet. Sci. Lett., 15: 411—417.
Hurd, D.C., 1973. Interactions of biogenic opal, sediment and seawater in the Central Equatorial Pacific. Geochim. Cosmochim. Acta, 37: 2257—2282.
Johnson, T.C., 1974. The dissolution of siliceous microfossils in surface sediments of the eastern tropical Pacific. Deep Sea Res., 21: 851—864.
Jones, J.B. and Segnit, E.R., 1971. The nature of opal. I. Nomenclature and constituent phases. J. Geol. Soc. Aust., 18: 57—68.
Jones, J.B. and Segnit, E.R., 1972. Genesis of cristobalite and tridymite at low temperatures. J. Geol. Soc. Aust., 18: 419—422.
Jones, L.H.P. and Handreck, K.A., 1967. Silica in soils, plants, and animals. Adv. Agron., 19: 107—149.
Jones, L.H.P. and Milne, A.A., 1963. Studies of silica in the oat plant. I. Chemical and physical properties of the silica. Plant Soil, 18: 207—220.
Jones, L.H.P., Milne, A.A. and Sanders, J.V., 1966. Tabashir: an opal of plant origin. Science, 151: 464—466.
Jones, R.L., 1964. Note on the occurrence of opal phytoliths in some Cenozoic sedimentary rocks. J. Paleontol., 38: 773—775.
Kolbe, R.W., 1957. Fresh water diatoms from Atlantic deep-sea sediments. Science, 126: 1053—1056.
Lancelot, Y., 1973. Chert and silica diagenesis in sediments from the central Pacific. In: E.L. Winterer et al. (Editors), Initial Reports of the Deep Sea Drilling Project, 17: 377—405.
Lauwers, A.M. and Heinen, W., 1974. Bio-degradation and utilization of silica and quartz. Arch. Microbiol., 95: 67—78.
Lewin, J.C., 1961. The dissolution of silica from diatom walls. Geochim. Cosmochim. Acta, 21: 182—198.
Lewin, J.C., 1962. Silicification. In: R.A. Lewin (Editor), Physiology and Biochemistry of Algae. Academic, London, pp. 445—455.

Lewin, J.C. and Reimann, B.E.F., 1969. Silicon and plant growth. Annu. Rev. Plant. Physiol., 20: 289—304.
Lisitzin, A.P., 1971. Distribution of siliceous microfossils in suspension and in bottom sediments. In: B.M. Funnell and W.R. Riedel (Editors), The Mircopaleontology of Oceans. Cambridge University Press, London, pp. 173—195.
Lisitzin, A.P., Belvayev, Y.I., Bogdnov, Y.A. and Bogoyavlenskiy, A.N., 1967. Distribution relationships and forms of silicon suspended in waters of the world ocean. Int. Geol. Rev., 9: 604—623.
Mackenzie, F.T. and Gees, R., 1971. Quartz: Synthesis at earth surface conditions. Science, 173: 533—535.
Mackenzie, F.T., Garrels, R.M., Bricker, O.P., and Bickley, F., 1967. Silica in sea water: control by silica minerals. Science, 155: 1404—1405.
McKeague, J.A. and Cline, M.G., 1963. Silica in soils. Adv. Agron., 15: 339—396.
Mizutani, S., 1966. Transformation of silica under hydrothermal conditions. J. Earth Sci., Nagoya University, 14: 56—88.
Mizutani, S. and Oehler, J.H., 1979. Silica diagenesis and origins of chert. in press.
Murata, K.J. and Nakata, J.K., 1974. Cristobalitic stage in the diagenesis of diatomaceous shale. Science, 184: 567—568.
Norgren, A., 1973. Opal phytoliths as indicators of soil age and vegetative history. Unpublished, Ph.D. thesis, Oregon State University, Corvallis, Oregon. Xerox University Microfilms, Ann. Arbor, Michigan, Dissertation Abstracts Number 33: 3421B.
Oehler, J.H., 1973. Tridymite-like crystals in cristobalitic "cherts." Nature, (Phys. Sci), 241: 64—65.
Oehler, J.H., 1975. Origin and distribution of silica lepispheres in porcelanite from the Monterey Formation of California. J. Sediment. Petrol., 45: 252—257.
Pettijohn, F.J., 1957. Sedimentary Rocks, 2nd edn., Harper, New York, N.Y., 781 pp.
Riquier, J., 1960. Les phytolithes de certains sols tropicaux et des podzols. 7th International Congress of Soil Science Transactions (Madison, WI), pp. 425—431.
Schrader, H.J., 1971. Faecal pellets: role in sedimentation of pelagic diatoms. Science, 174: 55—57.
Siever, R. and Woodford, N., 1973. Sorption of silica by clay minerals. Geochim. Cosmochim. Acta, 37: 1851—1880.
Siever, R., Beck, K.C. and Berner, R.A., 1965. Composition of interstitial waters of marine sediments. J. Geol., 73: 39—73.
Stein, C.L. and Kirkpatrick, R.J., 1976. Experimental porcelanite recrystallization kinetics: A nucleation and growth model. J. Sediment. Petrol., 46: 430—435.
Stephens, C.G., 1971. Laterite and silcrete in Australia: a study of the genetic relationships of laterite and silcrete and their companion materials, and their collective significance in the formation of the weathered mantle, soils, relief and drainage of the Australian continent. Geoderma, 5: 5—52.
Thurston, D.R., 1972. Studies on bedded cherts. Contribut. Mineral. Petrol,, 36: 329—334.
von Rad, U. and Rösch, H., 1972. Mineralogy and origin of clay minerals, silica and authigenic silicates in Leg 14 sediments. In: D.E. Hays et al. (Editors), Initial Reports of the Deep Sea Drilling Project, 14: 727—751.
von Rad, U. and Rösch, H., 1974. Petrography and diagenesis of deep-sea cherts from the central Atlantic. In: K.J. Hsü and H.C. Jenkyns (Editors), Pelagic Sediments on Land and Under the Sea. International Association of Sedimentologists Special Publication No. 1, 1: 327—347.
Wilding L.P. and Drees, L.R., 1974. Contributions of forest opal and associated crystalline phases to fine silt and clay fractions of soils. Clays Clay Miner., 22: 295—306.
Wilding, L.P., Smeck, N.E. and Drees, L.R., 1977. Silica in soils: Quartz, cristobalite,

tridymite and opal. In: J.B. Dixon and S.B. Weed (Editors), Minerals in Soil Environments. Soil Science Society of America Special Publication, pp. 471—552.
Wise, S.W., Jr. and Kelts, K.R., 1972. Inferred diagenetic history of a weakly silicified deep sea chalk. Trans. Gulf Coast Assoc. Geol. Soc., 22: 177—203.
Wise, S.W., Jr. and Weaver, F.M., 1974. Chertification of oceanic sediments. In: K.J. Hsü and H.C. Jenkyns (Editors), Pelagic Sediments on Land and Under the Sea. International Association of Sedimentologists Special Publication No. 1, 1: 301—326.
Wollast, R., 1974. The silica problem. In: E.D. Goldberg (Editor), The Sea, Vol. 5, John Wiley, New York, N.Y. pp. 359—392.
Wollast, R., Mackenzie, F.T. and Bricker, O.P., 1968. Experimental precipitation of sepiolite at earth surface conditions. Am. Mineral., 53: 1945—1962.

Chapter 8

BIOGEOCHEMISTRY OF URANIUM MINERALS

G.H. TAYLOR

CSIRO Fuel Geoscience Unit, P.O. Box 136, North Ryde, N.S.W. 2113 (Australia)

CONTENTS

INTRODUCTION

Uranium has become vastly more important to man during the last thirty years, and with this new importance has come increased knowledge of the chemistry and biology of uranium and its compounds. Cyclic behaviour in the earth's crust is probably easier to demonstrate for uranium than for most elements; the sections below deal with the chemical basis of that behaviour and with the roles which organisms can play during their life and — as organic residues — after their death. The way in which this behaviour has led to the redistribution of uranium in rocks (to form ore bodies in favourable cases) is an important subject, as is the related topic of biogeochemical prospecting for uranium. Many of the same considerations are relevant to the recovery of uranium by leaching from broken rock and to the way in which the cycling of uranium may affect the environment.

Topical interest in uranium in recent years has led to the publication of an immense number of papers; no more than a fraction of these can be referred to here.

CHEMISTRY OF URANIUM IN AQUEOUS, LOW-TEMPERATURE CONDITIONS

The following consideration of the chemistry of uranium is restricted to its behaviour under conditions such as commonly occur at or near the earth's surface.

Uranium is low in the list of element abundance, comprising only 0.0027% of the earth's crust (Levinson, 1974). Apart from ore deposits, uranium is most abundant in high silica igneous rocks and in shales, especially black shales. Reported contents for some common rocks and waters are listed in Table 8.1. Uranium ores are referred to on p. 497ff; here it need only be mentioned that most known ore bodies occur within sedimentary rock sequences.

With atomic number 92 and atomic weight 238.03, uranium is the heaviest naturally occurring element. There are eleven known isotopes, of which three — with atomic weights 234, 235 and 238 — occur in nature. They are all radioactive, with half-lives (in years) of 2.35×10^5, 7×10^8 and 4.5×10^9, respectively. The relative abundance of the isotopes varies depending on the age and geological history of the uranium occurrence; a typical distribution is ^{238}U: 99.28%; ^{235}U: 0.71%; ^{234}U: 0.005%. A quite exceptional deviation from this distribution has been found at Oklo in Gabon as a result of spontaneous fission chain reactions in the remote past (Anon., 1975). At Oklo ^{235}U concentrations as low as 0.29% have been found.

The radioactivity of the uranium isotopes is important in the present context since, over geological time, they become progressively converted to other elements. This means that less uranium is available in each successive geochemical cycle. Thus ^{238}U becomes ^{206}Pb and ^{235}U becomes ^{207}Pb (Stanton and Russell, 1959). The generation of other elements from the radio-

TABLE 8.1

Average abundance of uranium in various rocks, soil, river and sea water [a]

	Uranium ($\mu g\ g^{-1}$)
Ultramafic rocks	0.001
Basalt	0.6
Granodiorite	3
Granite	4.8
Shale	4
Sandstone	1
Limestone	2
Soil	1
River water	0.4
Sea water	0.002

[a] Data from Levinson (1974) and Rogers and Adams (1969).

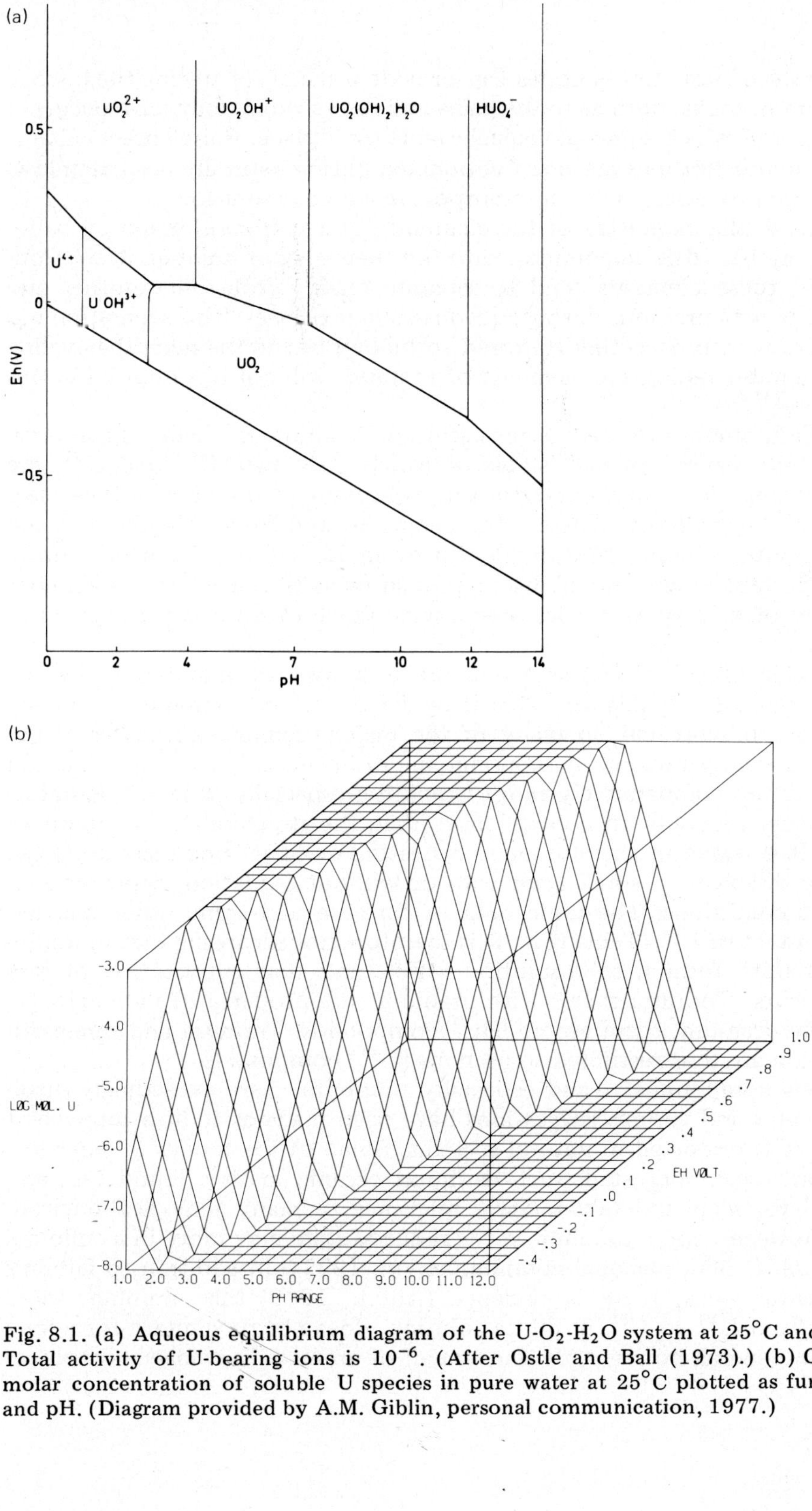

Fig. 8.1. (a) Aqueous equilibrium diagram of the $U\text{-}O_2\text{-}H_2O$ system at 25°C and 100 kPa Total activity of U-bearing ions is 10^{-6}. (After Ostle and Ball (1973).) (b) Calculated molar concentration of soluble U species in pure water at 25°C plotted as function Eh and pH. (Diagram provided by A.M. Giblin, personal communication, 1977.)

active decay of uranium isotopes can provide a means of tracing the history of uranium in rocks, both as to the gains and losses which may have occurred and the ages at which some geological events took place. For all practical purposes, uranium isotopes are not fractionated during naturally occurring low-temperature processes, including transport in aqueous solution.

In general, the chemistry of the elements formed from uranium by radioactive decay has little in common with the chemistry of uranium itself. Consequently, these elements tend to separate readily from one another and from the parent uranium during aqueous-state processes. This separation has consequences in prospecting (referred to below) but in the main the cycling of the daughter radiogenic elements of uranium will not be discussed in this chapter.

Metallic uranium does not occur naturally. Uranium in chemical combination exhibits several valence states of which only two, U^{IV} and U^{VI}, are important here. The sharply contrasting behaviour of uranium in these two states is the key to most of its cycling behaviour and forms the major theme of this chapter. Figure 8.1 (a) reproduces an Eh-pH diagram which shows some important ranges of stability; a plot showing the calculated molar concentration of soluble U species over a wide range of Eh and pH is given in Fig. 8.1 (b).

The oxide of U^{IV} (UO_2) as a mineral, is known as uraninite or by the varietal name of pitchblende. This is by far the most commonly occurring mineral of uranium and, in many of the major uranium ore bodies of the world, is the only mineral of economic importance. (Other U^{IV} minerals are known but are comparatively rare.) Uraninite, especially as it occurs naturally, appears generally to contain more than the stoichiometric amount of oxygen. It is stable in air, although on heating in air it is oxidized to U_3O_8. Uraninite dissolves in some acids with or without oxidation, depending on the exact conditions. The ionic radii of thorium and some other ions are similar to that of U^{4+} (Table 8.2). It is therefore not surprising that uraninite and thorianite form a solid-solution series, and that substitution of rare earths occurs. Thorium shares with uranium the property of radioactivity, but unlike uranium, there are no Th^{VI} compounds. Thorium and rare-earth substitution are thus restricted to uranous (U^{4+}) compounds.

Uranous compounds — and especially uraninite — are exceedingly insoluble in water and various suggestions have been made as to how tetravalent uranium is transported in natural systems. Some of the modes of transport which have been suggested, such as liquid organic matter, liquid CO_2 and gaseous UF_4, seem unlikely to have relevance to many geological environments. However, some uranium in the reduced state may move in a colloidal form (A.M. Giblin, personal communication, 1977). In addition to Giblin's experimental work, there is evidence (Anon., 1969) that uranium, when reduced from U^{VI} to U^{IV} in dilute solution, does not precipitate for a long period. In general, however, the movement of uranium as halogen complexes

TABLE 8.2

Ionic radii (nm) of uranous and some other ions [a]

Ion	Radius (nm)
Y^{3+}	8.9
Ce^{4+}	9.2
U^{4+}	9.7
Ca^{2+}	9.9
Th^{4+}	10.2

[a] Data from Weast (1970).

in aqueous systems appears to offer a reasonable explanation of the movement of uranium under conditions where only tetravalent uranium could occur (A.M. Giblin, in preparation). Such complexes as $(UCl)^{3+}$, $(UCl_2)^{2+}$ and $(UCl_3)^{+}$ are known and, granted sufficiently high chloride concentrations, are stable; dilution of such solutions or other changes in conditions leading to instability of the complex lead to precipitation of uranous compounds.

There are many minerals in which uranium occurs in the hexavalent state. In naturally occurring compounds, the uranium invariably occurs as the uranyl ion $(UO_2)^{2+}$. As is evident from Fig. 8.1, the uranyl ion is stable over a wide range of Eh-pH conditions. Of the uranyl minerals, some are unusually stable and virtually insoluble over a wide range of conditions. This is true especially of uranyl compounds such as vanadates, tantalates and titanates, whereas other compounds like sulfates tend to be short-lived species on a geological time-scale. Figure 8.2 shows the stability relations among some uranium and vanadium compounds under specified low-temperature conditions. The most striking feature of this diagram is the very large field of stability of the potassium uranyl vanadate, carnotite.

A high degree of mobility is conferred on the uranyl ion by the presence of carbon dioxide which allows the formation of stable uranyl carbonate complex ions (Tugarinov, 1975). Table 8.3 lists some complexes which have been referred to as being of great importance in the transport of uranium. Many other complex ions, including organic complexes, are known. The

TABLE 8.3

Some uranyl complexes of importance in the transport of uranium

$(UO_2)(SO_4)$
$[(UO_2)(SO_4)_2]^{2-}$
$[(UO_2)(CO_3)_2H_2O]^{2-}$
$[(UO_2)(CO_3)_3]^{4-}$

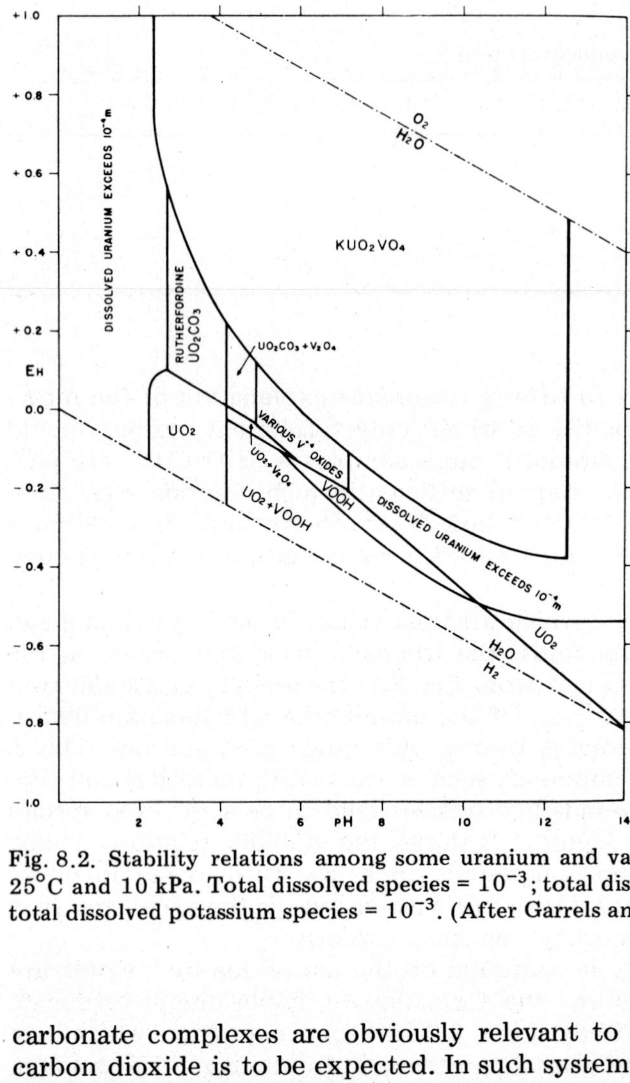

Fig. 8.2. Stability relations among some uranium and vanadium compounds in water at 25°C and 10 kPa. Total dissolved species = 10^{-3}; total dissolved carbonate species = 10^{-1}; total dissolved potassium species = 10^{-3}. (After Garrels and Christ, 1965.)

carbonate complexes are obviously relevant to biological systems in which carbon dioxide is to be expected. In such systems, a decrease in CO_2 concentration may result in uranium precipitation while an increase in alkalinity may inhibit its precipitation.

The transformation of uranous to uranyl ion obviously involves oxidation and in natural systems it is nearly always, directly or indirectly, oxygen from the earth's atmosphere which is implicated. Oxidizing conditions are not restricted to the earth's surface but, through diffusion and by the movement of water, may occur at depths hundreds of metres below the earth's surface. There have been many suggestions that oxygen was a very minor component of the atmosphere during the earlier Precambrian and Table 8.4 shows the changes which have been postulated in the earth's atmosphere over geological time. (However Dimroth and Kimberley (1976), after reviewing the subject, could not agree that an oxygen-free atmosphere had existed at any time during the span of geological history recorded in well-preserved sedimentary rocks.) The postulated increase in atmospheric oxygen has been attributed to the activity of organisms which in recent years have been recognized in fossil form in quite ancient rocks. Uraninite would thus have been stable at, or very close to, the earth's surface in earliest Precambrian times. As the atmospheric oxygen increased as a result of the activity of organisms, oxidizing conditions must have extended increasingly, although irregularly, down into the near-surface rocks. With these changes came greater opportunities for movement of uranium through the mobility of the uranyl ion.

The transformation of the uranyl to the uranous ion depends, of course, on reduction. By far the most important reductants in natural systems of the kind under discussion are the carbonaceous materials. Such carbonaceous materials include both living material such as algal mats together with organic matter derived from organisms. By far the most important components of this family are the residual materials collectively called carbonaceous matter (also known as 'kerogen', 'organic matter', and by other names). Most carbonaceous matter in sedimentary rocks is not graphite (which tends to be

TABLE 8.4

Changes in oxygen content of the atmosphere during evolution of the biosphere [a]

Era	O_2 content of atmosphere (%)	Organisms	Redox functions
Precambrian			
Archaean >2300 My	0.02—0.2	Bacteria, algae	Oxidation of abiogenic organic matter during fermentation Reduction of CO_2 during oxidation of H_2, CH_4, NH_3, H_2S
Proterozoic 2300—570 My	0.2—2.0	Blue-green algae, green algae	Reduction of CO_2 during oxidation of S, Fe^{2+} Photosynthetic reduction of CO_2
Palaeozoic 570—225 My	2—20	Brown algae, land plants	Oxidation of organic substances during respiration Development of photosynthesis in terrestrial plants
Mesozoic 225—65 My	approx. 20	As above	Adaption of land plants to reduction of CO_2 in oxidized biosphere
Cenozoic <65 My	23.01	As above	Localization of redox processes in organs and oraganelles of plants

[a] Mainly after Boichenko et al. (1975).

restricted to high-grade metamorphic rocks and some igneous rocks) but a non-crystalline solid composed essentially of carbon, oxygen and hydrogen. Common ranges of composition are carbon: 65 to 95%, oxygen: 2 to 25%, and hydrogen 3 to 6.5% (all on a dry basis) with some nitrogen and sulfur. No specific reactions for the complex chemical oxidation of organic matter accompanying the reduction of uranyl ion, can be cited, but it appears that oxalic acid may be an oxidation product formed under such circumstances. The oxalic acid itself may be precipitated as an oxalate such as the calcium oxalate, whewellite (Galimov et al., 1975). Probably some carbonate minerals associated with uranium mineralization are formed in this way.

Much carbonaceous matter bears a strong resemblance to coal and many of its properties have been inferred from coal studies (Stach et al., 1975) or by actual separation and analysis of the carbonaceous matter from the host sedimentary rocks (Saxby, 1970). Virtually all carbonaceous matter in sedimentary rocks (except perhaps some from the early Precambrian) appears to be derived from organisms, although extensive biological degradation may occur prior to deposition (Deuser, 1971). It has been suggested that some carbonaceous matter associated with uranium is the product of radiogenic degradation of gaseous or liquid hydrocarbons. In the writer's experience, the textures and properties of such carbonaceous matter are inconsistent with such an origin. While the carbonaceous matter containing least carbon commonly is most reactive, all carbonaceous matter can act as a reductant in aqueous systems. Virtually all shales contain some carbonaceous matter; black shales contain 2% or more. Some sandstones contain minor carbonaceous matter, usually in the form of thin layers or lenses, but in most comparatively coarse-grained sedimentary rocks carbonaceous matter is rare or absent.

In modern environments, reducing conditions may exist near or at the sediment-water interface where oxygen availability is small, for example in depressions in the sea floor (Kolodny and Kaplan, 1970). The removal of uranium at such interfaces is probably one of the major factors in maintaining the uranium content of sea water at a low level, while rivers continue to add 'new' uranium to the oceans.

URANIUM AND ORGANISMS

Uranium is not known to be an essential element for the life process of any organism (Rogers and Adams, 1969) and Baturin (1972) refers to uranium as "an abiogenic element" which is not concentrated in living tissues. It is claimed that small amounts (4—24 $\mu g\ g^{-1}$) of uranyl compounds stimulate growth in both bacteria and higher plants, while larger amounts are inhibitory (Updegraff and Douros, 1972). Certainly for man, uranium is a recognized carcinogen and highly toxic. Its high chemical toxicity is largely shown in kidney damage and necrotic lesions. The rapid passage of soluble uranium through the body allows relatively large amounts to be taken in (Sax, 1975). Uranium and its daughter elements also pose hazards from radioactivity.

Since the uranous ions is close in size to the Ca^{2+} ion (Table 8.2), it is not surprising that some substitution of uranium for calcium takes place in carbonate shells of invertebrates, especially since at least local reducing conditions must exist during the life and shortly after the death of the organism. However, it is claimed that the uranium in modern mollusc shells (range 2—6 $\mu g\ g^{-1}$) occurs as the uranyl compound $Ca_2[UO_2(CO_3)_3] \cdot 9H_2O$, so that reduction may not be involved in such cases (Oglobin and Khalifa-Zade, 1974). The fact that the uranium content of limestones is uniformly low (range 0.3—2.3 $\mu g\ g^{-1}$ (Rogers and Adams, 1969)) suggests that invertebrates are not efficient concentrators of uranium. Globigerina oozes are also poor in uranium. Miyake et al. (1970) measured the uranium content of phytoplankton and zooplankton from the sea water of the western North Pacific Ocean. Analyses on a dry basis varied from 17--78 $\mu g\ g^{-1}$ compared to 3.02 to 3.55 $\mu g\ l^{-1}$ uranium in sea water. They also measured the ratio of activity of ^{234}U to ^{238}U in the plankton and algae; the ratio in the plankton is in close agreement with that in the relevant sea water, confirming that there is little or no selectivity in the uptake of uranium isotopes by marine biota.

While uranyl carbonates are sparingly soluble in sea water, the phosphates have very low solubilities. This is probably one reason why phosphate minerals like apatite may contain comparatively high trace amounts of uranium. Thus Arrhenius et al. (1957) found that, while living fish have practically no uranium in their bones, fish remains which had been exposed to sea water for ten thousand years had become considerably more radioactive, probably at least partly as a result of uranium uptake. Marine phosphorites are commonly much enriched in uranium (range 50—300 $\mu g\ g^{-1}$ (Rogers and Adams, 1969)). While the phosphorus is almost certainly of organic origin *, the addition of uranium appears to occur from the small amount of uranium in the sea water; This (together with fixation in black shales) is probably another major mechanism in the depletion of uranium in sea water.

Black shales are commonly enriched in uranium with respect to other sedimentary rocks, sometimes to the extent of causing these rocks, rich in organic matter, to be considered as possible commercial sources of uranium (Vine, 1956). The uranium content of coals is highly variable and its occurrence sporadic; where higher than average, the uranium tends to be concentrated in the stratigraphically highest coals or in the vicinity of an unconformity. It is thus most probable that virtually all the uranium in coals has been extracted from solution long after the deposition of peat and that almost no uranium was associated with the original vegetation. A similar conclusion can be reached for the uranium-enriched coalified wood which occurs in some sandstones, for example in the Colorado Plateau (Breger,

* For a discussion of the origin of phosphorites, see Chapter 3.1.

1974). Gentry et al. (1976) have recently shown, on the basis of isotopic analyses, that uranium introduction may have occurred far more recently than was previously supposed. However, they find also that, in some instances, the uranium was introduced before coalification was complete since the haloes have been compressed with the coal as it increased in rank. These results are consistent with laboratory and field work by Szalay (1964) who showed that the insoluble 'humic acids' in peat are capable of concentrating uranium from very dilute solutions in natural waters. Sorption occurs as 'uranyl humate', the process following the normal kinetics of the Langmuir adsorption equation. (Where uraninite occurs in association with peat or other carbonaceous matter, the uranium may thus have been initially sorbed as a uranyl compound which was later reduced to uraninite.)

Updegraff and Douros (1972) made a detailed search for microorganisms which may be commonly associated with uranium minerals. They carried out systematic studies on 63 samples ranging from high-grade uranium ore to barren sediments from a number of localities in the U.S.A. In fact, the samples contained remarkably few microorganisms and few kinds of organisms. Seventy-two percent of the pure cultures belonged to the genus *Arthrobacter*, other bacterial genera identified being *Bacillus* and *Streptomyces*. Fungi were found in occasional samples only. There was no significant difference in flora between samples high in uranium and those low in uranium. However, Silverman and Ehrlich (1964) reported an interesting instance of bacterial action on a uranium compound. This was the reduction shown in eqn (1).

$$(UO_2)(OH)_2 + 2\,e^- + 2\,H^+ \rightarrow U(OH)_4 \qquad (1)$$

This reaction was catalysed by *Micrococcus latilyticus* which also catalyses a number of other reduction processes. It appears to be a process which could well accompany or follow the sorption of uranyl compounds by humic matter referred to above.

Magne et al. (1974) made a somewhat similar study, but their emphasis was on the enhancement of the solubility of uranium in granites through the activity of heterotrophic bacteria. In their experiments microbial activity increased the amount of uranium in solution by factors of 2 to 97. Several organisms may have been involved, *Bacillus licheniformis* being the one species definitely isolated. Species of *Thiobacillus* were absent, so that the enhancement of solubility observed was probably quite unrelated to leaching processes depending upon the oxidation of pyrite *.

There is an indirect way in which bacteria could be involved in the removal of uranium from solution in near-surface environments. Jensen (1958), on the basis of sulfur isotope measurements, suggested that sulfate waters in sedimentary rocks had been reduced by anaerobic bacteria to hydrogen sulfide. Zones rich in carbonaceous matter could have pro-

* See Chapters 4 and 6.3.

vided the environment and energy source for these bacteria. Later, uranyl ions introduced in solution would have been reduced to uraninite by the hydrogen sulfide. Apart from the fact that it satisfies the isotope data, Jensen's hypothesis does not appear to be necessary to explain the common, although not universal, association of sulfide minerals with uranium concentrations in sedimentary rocks. As is discussed below, it is possible that the sorptive capacity of carbonaceous matter is increased in the presence of hydrogen sulfide since iron and other sulfide-forming species may then tend to be desorbed from carbonaceous matter and increase the possibility of uranium sorption.

The reduction of sea-water sulfate by sulfate-reducing bacteria is well known (see Chapters 6.1 and 6.2), and certainly the hydrogen sulfide generated must have increased the rate at which reduction of uranyl ion could occur as compared with the situation where only sulfide was present. Experimental work by Viragh and Szolnoki (1970) used *Desulfovibrio desulfuricans* to reduce sulfate. The precipitation of uranium was enhanced by the presence of hydrogen sulfide and the authors concluded that H_2S was important in the formation of uranium ores. However, the experiments of Szalay (1964) referred to above and observations of uranium enrichment in the presence of carbonaceous matter where little or no sulfate has been available for reduction suggest that the intervention of bacteria such as *Thiobacillus* and *Desulfovibrio* is not essential for the reduction of uranyl ions. Also, Baturin (1972, p. 192) concluded that ". . . the final outcome of the uranium concentrating process ... generally is the same both in basins with a hydrogen sulfide type of environment (Black and Baltic Seas, productive ocean shelves) and in those with normal aeration".

Bacterial leaching of uranium ores is the subject of a subsequent section of this chapter. Here it suffices to say that the bacterial activity in such situations is that of enhancing the rate of oxidation of sulfides, through which ferric sulfate and sulfuric acid are liberated to speed up the process of solution of uraninite.

It was mentioned above that many authors conclude that the earth's atmosphere was originally deficient in, or at least of very low, oxygen content; this hypothesis is linked to the development of plants over geological time. There now seems no doubt that fossilized remains of organisms are present in rocks of most of the range of Proterozoic age (Schopf, 1975). An association of particular interest is that between the gold-uranium mineralization of the Upper Witwatersrand conglomerates in South Africa, and the variety of carbonaceous matter known as thucholite. Although he believed that the thucholite itself was the product of ionizing radiation, Schidlowski (1970, p. 4) concluded that the hydrocarbons must have originated within the sediments of the Witwatersrand System: "The $^{13}C/^{12}C$ ratio of the carbonaceous material as well as the occurrence within the latter of a suite of amino acids and monosaccharides seem to indicate a biogenic origin of the primary hydrocarbons. This would imply that photosynthetic processes were operating during the time of deposition of the Witwatersrand System, ... An electron microscopic investigation of Witwatersrand rocks has, furthermore, revealed structures which strongly resemble relics of primitive unicellular life forms". Snyman (1965) drew attention to similarities between thucholite and some more recent algal coals and although Schidlowski dismissed the similarity as coincidence, his own illustrations seem to support Snyman. The present author also reached the conclusion that thucholite from a number of sources, all 2 Gy or more old, has textures and properties which are quite compatible with an origin from algal-like plants. Recently the matter appears to have been put almost beyond doubt by Hallbauer and van Warmelo (1974) and Hallbauer et al. (1977). These authors used light and scanning electron microscopy to study what appear to be remarkably well-preserved carbonaceous remains (thucholite of a Precambrian plant of columnar habit, "shown to be a primary plant structure" (Hallbauer et al., 1977, p. 477). The internal structure of this columnar plant (for which the name *Thuchomyces lichenoides* is proposed) has many parallels in the family of modern lichens. Hallbauer et al. (1977) also described a filamentous microscopic organism which appeared to have a parasitic relationship to the lichen-like plant, and for this the name *Witwateromyces conidiophorus* has been proposed. These authors compare the latter with filamentous bacteria or primitive fungi. Recent work (Reimer, 1975) suggests that the rocks of the Witwatersrand System were deposited over a period of about 30—40 My during the time range of 2.48—2.37 Gy B.P., i.e. earlier than had been thought. Throughout the world, this type of uranium-rich, quartz-pebble conglomerate appears to be restricted to the range 2.2—2.5 Gy B.P.

If we can accept that primitive plants existed from at least as early as 2.5 Gy ago, it is necessary to ask what is the relationship between these organisms and uranium cycling. Hallbauer et al. (1974) regarded the incrustations of uranium, thorium, gold and other metals of filaments of *Thuchomyces lichenoides* as evidence of assimilation during growth of the plant.

The writer has also been impressed by the repeated occurrence of fine uraninite within thucholite bodies as illustrated on Plate 23 of Schidlowski (1970). These small crystals of uraninite give the impression of having crystallized in situ before the remains of the organism became compacted. It thus seems probable that this fine uraninite was trapped either during the life, or no later than very soon after the death, of the organisms. Certainly detrital minerals, including uraninite, are present, but a large part of the uranium appears to have been extracted from solution.

It is thus necessary to return to the controversial matter of the earth's atmosphere. Cloud (1965, p. 33) considered this in the context of fossil evidence: ". . . it is now generally agreed that the components of the present atmosphere came ultimately from within the earth, mainly by volcanism.

Free oxygen, however, is not directly available from such sources; nor, in the absence of green plant synthesis, is it formed secondarily except in trivial and readily scavenged quantities from photolytic dissociation of CO_2 and H_2O. All who have considered the problem critically, therefore, agree that at an early stage in its history the terrestrial atmosphere was essentially anoxygenic, or reducing. Only at a later date, after the appearance of a photosynthetic source of oxygen, could the atmosphere evolve towards its present oxygenic (and oxidizing) state".

Taking the evidence together, it is hard to escape the conclusion that the period 2.5—2.2 Gy B.P. was unique because of the development of plants generating sufficiently oxidizing conditions for uranium to be mobilized as the uranyl ion in surface waters and to be trapped by reduction, still at the surface, by mats of organisms in the stream beds. During this period, oxidizing conditions may have had only localized occurrence (Cloud, 1965) and would be unlikely to have extended below the surface to any significant extent. Only leter than 2.2 Gy ago, apparently, were conditions in the rocks tens or hundreds of metres below the surface sufficiently oxidizing for the uranyl ion to exist in ground and formation waters.

GENESIS OF URANIUM ORES

There have been many attempts to categorize the various types of uranium deposits (see, for example, Ruzicka (1971)). The variation between such systems reflects our uncertainty about the mode of formation of some deposits. Probably the simplest division is that currently adopted by the working groups of the International Atomic Energy Agency, namely: sedimentary basins and sandstone-type deposits; uranium in quartz-pebble conglomerates; vein- and similar-type deposits; and other deposits. However, each of the last two categories appears to contain genetically diverse deposits and, until there is a greater degree of understanding of past and present ore-forming processes, no generally acceptable classification will be possible.

It seems likely that most ore bodies of uranium were formed at or near the earth's surface by processes involving remobilization. Probably the uranium now accumulating was virtually all brought to the surface in the first instance by igneous rocks or by veins associated with these. Much known uranium ore occurs with, or close to, rocks of Precambrian age, especially rocks between 1.5 and 2.5 Gy old. In many such terrains, 'hot' granites or other igneous rocks with higher than average uranium content are present. Such rocks now present are remnants of more extensive occurrences which were weathered and leached to yield the uranium from which some of the other ore types have been formed, especially some of those which occur in sedimentary rocks, i.e. the great majority of economic uranium occurrences. It is likely that organisms, especially bacteria capable of oxidizing sulfides, played a role in the weathering of uranium source rocks, similar to their role in the leaching technology used for pyritic uranium ores (see p. 507).

Since the present context is that of recycling involving organisms, no great emphasis can be given here to processes possibly involved in the emplacement of igneous rocks and of veins related to these. It was mentioned above that the movement of uranium as halogen complexes offers a reasonable explanation for the movement of uranium under conditions (as in veins related to igneous intrusions) where only tetravalent uranium would be likely to occur. Bohse et al. (1974) discuss this point and conclude from studies of various mineral occurrences that the formation of complex halide compounds rather than increasing oxygen fugacity is responsible for the mobility of uranium in volatiles derived from magmas. The high-silica igneous rocks are amongst those richest in uranium (Table 8.1) and some of those of earlier Precambrian age are more than usually rich. Great volumes of such rocks existed and were eroded prior to the formation of ores such as the conglomeratic deposits of Witwatersrand and Blind River, and also ores such as those in northern Australia.

In the extreme case, some peralkaline, high-silica igneous rocks contain up to 1500 $\mu g\ g^{-1}$ U_3O_8 (Bowie, 1972). (Note that the actual form of uranium in such rocks, when unweathered, would be as uraninite or other minerals where uranium would be present in the tetravalent state. Uranium contents are often quoted, not in terms of the actual mode of occurrence, but as U_3O_8, the initial commercial oxide product after processing.) However, even assuming a uranium content of only a few $\mu g\ g^{-1}$, high silica rocks — granites, rhyolites and pegmatites — appear to be a more than adequate source for the deposits possibly derived from them. It seems significant that relatively few igneous rocks younger than 1.5 Gy have high uranium content. This suggests that some uranium in the earth's crust was fixed and unavailable during the processes leading to the assimilation of sedimentary material and the formation of younger generations of high-silica rocks.

True vein deposits of uranium are not very common except perhaps in Europe. Uranium in such deposits commonly occurs with "... minerals, such as tin, copper, cobalt, vanadium and arsenic ..." (Bowie, 1972, p. 3). In Europe, as in the U.S.A. (Walker and Osterwald, 1963), the assemblage commonly includes pyrite and other sulfide minerals. Moreover, there is an association of metals in veins which is significant. Walker and Adams (1963, pp. 76—77) state: "The positive correlation of certain metals — notably molybdenum, manganese, beryllium, tungsten, vanadium, niobium, yttrium, and zirconium — to uranium in veins seems to be reasonably well-established within some deposits, districts, or restricted geographic areas, but none of these metals can be shown to correlate with uranium in all or even a large percentage of vein deposits. In addition to the metals that, when present, appear to correlate intimately with uranium, many other metals such as lead, zinc, copper, silver, and cobalt are associated with uranium in many deposits only in the sense of occurring within the same favourable structure. Some uranium in veins locally occurs in economic and large deposits of other metals, principally copper, lead, zinc, and silver, as for example at Bisbee, Arizona, in many deposits in the Front Range of Colorado, in several deposits in the Coeur d'Alene district, Idaho, and in the Goodsprings district, Nevada; in many other deposits the ores are characterized by small quantities of both uranium and other metals, principally lead and zinc, or copper, or locally silver''.

It is not difficult to envisage the aqueous transport of an association of metals as complexes with deposition in sites determined by favourable rock permeabilities and, especially, by favourable chemistry. The question arises as to how uranium is transported in such mineralization processes. There seems little doubt that the uranyl ion could not exist under most conditions attributed to hydrothermal transport. While some authors have stated categorically that no complexes of U^{IV} are known to occur in geologically relevant situations, various kinds of uranous complexes have been discussed in the chemical literature (Pascal, 1960). Recent work (A.M. Giblin and Taylor, unpublished results) has confirmed that uranium in the reduced state can be transported in brines at quite appreciable rates. On this evidence and on the basis of the mineral associations and vein-margin alteration, it seems likely that the uranium moved as U^{IV} chloro-complexes. It seems probable that this is also a major process in bringing uranium to the surface in igneous rocks since uraninite has an extremely high melting point and uranium does not enter readily into most rock-forming minerals. The mode of transport of uranium as chloro- and perhaps other U^{IV} complexes is probably also important in the local remobilization of uranium which appears to have occurred during metamorphism of some ore bodies.

There is no obvious way in which biology can have entered into the processes of transport and precipitation of uranium occurring wholly under highly reducing conditions. However, organisms appear, directly and indirectly, to have been most important in succeeding parts of the uranium cycle, in which oxidation to U^{VI} and subsequent transport and precipitation occurred. There are a number of reasons for believing that many major ores formed by the following sequence of events: (1) oxidation of uranium minerals in near-surface igneous rocks or veins; (2) migration as the uranyl ion; (3) precipitation in favoured sites with or without reduction. The first pointer is that very many uranium ores occur within sedimentary rocks. Some of the pegmatites which contain enrichments of uranium are probably products of high-grade metamorphism of sedimentary rocks. When these are eliminated, direct associations of ore-grade uranium with igneous rocks are rather unimportant. Secondly, most uranium deposits in sedimentary or metasedimentary rocks are either strictly stratiform like the conglomerate deposits, or stratabound, as in the case of the sandstone deposits. Although there is much about the emplacement of uranium in sedimentary rocks which is not fully understood, it is clear that the large-scale movement of water at the surface or in permeable sedimentary rocks comparatively near the surface is extremely significant in the localization of uranium concentrations. Thirdly, there is a very common association of uranium concentrations with carbonaceous matter; this subject is developed more fully below. Fourthly, the characteristic association of elements in veins is not, in general, consistent with redox control of precipitation (as is the association in typical sedimentary rock concentrations). The association of elements in deposits within sedimentary rocks tends to be of metals which, like uranium, have two or more valence states, especially copper, vanadium, iron and gold.

The role of carbonaceous matter has been much discussed in relation to uranium mineralization. Certainly the association of uraninite ore bodies with carbonaceous matter is too common and too specific to be coincidental, although it is unwise to draw general conclusions from this for application to all particular cases. As mentioned above, "carbonaceous matter" embraces many kinds of carbon-rich material which vary as a result of differences in original organic material, differences in conditions under which it was incorporated in the sediment and differences in the changes which it has undergone since deposition. Examination by microscopic and chemical means gives some information on the history of carbonaceous matter and, in favourable cases, something can be said about the diagenetic or metamorphic stage which had been reached by the carbonaceous matter at the time of deposition of uranium minerals.

The relationship of uranium to carbonaceous matter is most likely to be apparent in the case of present-day examples but, even in these, the story is not simple. Sites where uranium is most likely to be enriched are those where reducing conditions occur locally in contact with a larger body of water where more oxidizing conditions prevail. Such conditions exist, for example, in depressions in the sea floor where water is intermittently or continuously stagnant, and reducing conditions are maintained by a steady rain of carbonaceous matter from above. Kolodny and Kaplan (1970) analysed the waters and sediments from such a depression off Vancouver Island and found that the authigenic uranium (i.e., that uranium which appeared to have precipitated on the sea floor from solution) showed good correlation with the content of carbonaceous matter. About half of the uranium in the sediment was in the form of organic complexes and the interstitial water of the sediment was also enriched in uranium. Baturin (1973a) considered the fate of uranium in marine waters and found that in streams entering the sea, uranium was present in a 1 : 1 ratio in dissolved and suspended states. The 'hydrogenic' uranium, i.e. that which reached the sea floor in solution, was preferentially fixed in layers rich in carbonaceous matter. Baturin emphasized the interesting point that uranium is not concentrated in oceanic suspensions containing up to 20% organic carbon, so that "organic components extract uranium not from water as a whole but only at the water-bottom interface" (Baturin, 1973a, p. 1034). The other half of the uranium finds its way to the sea floor in particulate matter; it remains as a minor constituent of the sediment unless there is any redistribution during diagenesis. For such redistribution to occur the uranium must be transferred from the solid phase into the interstitial waters of the sediment.

Baturin (1973b) examined the uranium cycle in the Black Sea and Azov Sea and found most transport to be in solution as a uranyl carbonate complex. The Azov Sea is a shallow-water basin in which sediments are often disturbed by storms; bottom sediments have a fairly uniform and low (0.3—2.3 $\mu g\ g^{-1}$) content of uranium. The Black Sea, by contrast, is deep, with H_2S-rich bottom waters of low Eh. The bottom water is strongly depleted in uranium, while the sediment is of variable, sometimes high (0.2—23.0 $\mu g\ g^{-1}$) uranium content. The Black Sea sediments are said to receive up to 400 Mg y^{-1} of uranium from the water. The residence time of uranium in Azov Sea waters is only 14.5 y, but in Black Sea water, is of the order of 3000—4000 y.

The examples cited above, and some others, are instances where the availability of uranium is not especially great by present-day standards. Where water reaching the sea was draining rocks highly enriched in uranium or even draining actual ore bodies, the supply to the marine sediment could have been much greater. Another factor affecting the concentration of uranium in the bottom sediments is the rate of sedimentation. Baturin (1973a) provides evidence that the ratio of uranium to carbonaceous matter in the upper sediment layer increases with decreasing rate of sedimentation, which could be interpreted as decreasing dilution by uranium-poor sediment. As mentioned above, the contribution of uranium to sediments from the shells of invertebrates must usually be quite minor.

Water associated with the sediment near the sediment-water interface may be mildly oxidizing to strongly reducing. However, with a continued supply of carbonaceous matter from above, the conditions in a particular bed will necessarily become reducing, if not so already, once further sediments have been deposited and the diffusion of oxygen hindered. Under these conditions, sulfate reduction by appropriate bacteria will proceed vigorously with the generation of H_2S and, of course, bisulfide and sulfide ions. The solubility of iron sulfide and of many metal sulfides is so small that the amount of any such metals in solution will be brought to a low level by precipitation of sulfides. When the carbonaceous matter arrives at the site of deposition it has, in general, had a long history of exposure to sea water and much adsorption will have already occurred. Some of the sorbed species will be sulfide-forming metals and, in the presence of sulfide ions, there will be a strong tendency for these to be stripped from the carbonaceous matter, leaving the formerly occupied adsorption sites available. It thus seems possible that the reduction of sulfate by bacteria may indirectly promote the sorption of uranyl ions. (As discussed above, the uranium as initially sorbed on carbonaceous matter appears to be in the uranyl and not the reduced state.) Thus there are at least two possible ways in which the precipitation of uranium may be enhanced in the presence of hydrogen sulfide. The first is by increasing the sorption capacity of carbonaceous matter and the second is by promoting the reduction of hexavalent to tetravalent uranium as suggested by Jensen (1958).

Uranyl ions or compounds present in the sediment cannot remain in this oxidation state indefinitely, although there are suggestions that reduction to uranous dioxide does not occur rapidly in carbonaceous sediments. However, with maturation of the carbonaceous matter, the adsorption sites are lost, and if reduction had not already occurred, the desorbed uranyl ion would then be reduced to UO_2. Since the maturation of carbonaceous matter is a slow, progressive process, any desorbed UO_2 would be released in very small amounts over long periods. There is thus the possibility of UO_2 entering the formation water in colloidal form and having enough mobility to move to a site of concentration.

In general, the possibility of redistribution of uranium (whether originally introduced in particulate or soluble form) in recently deposited sediments depends on the passage of uranium from the solid phase into solution. This could happen so long as the interstitial water in the sediment retains a moderately high Eh, but, when the uranyl ion is no longer stable, any uranium in solution would be precipitated as UO_2 or in the lattice of an authigenic mineral. The effect of any such movement during early diagenesis would be to increase the degree of association between uranium and carbonaceous matter. With the continuation of highly reducing conditions, any further migration would have to be in the form of uranous complexes, or locally, as colloidal uraninite.

It is not difficult to see how the various processes described lead to the observed concentrations of uranium in shales rich in organic matter. An instance which was studied in considerable detail is that of the Chattanooga Shale, which contains up to 90 $\mu g\ g^{-1}$ of uranium. A particular area of contemporaneous high land is indicated as the likely source; the percentage of organic matter is linearly related to the uranium content of the shale; and, while the uranium was probably fixed in the uranyl form in the first instance, it now exists as a phase separate from the organic and inorganic constituents, probably as uraninite (Breger and Brown, 1962). Most such concentrations are no more than geochemical enrichments but some rock units have been considered as possible low-grade sources of uranium.

It is worth bearing in mind the above considerations when seeking an understanding of the origin of conglomeratic ores. These ores are of great economic importance, especially in southern Africa and Canada, and have been much studied. There has been a great deal of controversy over their origin. Robertson (1974, pp. 507—508) gives a useful summary of their characteristics and a widely held view as to their origin:

"Pyritic, uranium-bearing, quartz-pebble conglomerates have been found in Canada, in South Africa, in Brazil and in western Australia. The conglomerates are all of similar aspect and the minerals of interest are of similar character.

The rocks in which the conglomerates lie are also similar yellow-green to grey clastics, some clean and well washed as at Elliot Lake and some being poorly sorted and dirty. They are, with one exception, only very slightly metamorphosed, they lie with profound unconformity on highly metamorphosed strata and they underlie a red rock sequence which contains no pyrite.

It is the opinion of the author that the conglomerates are all of similar age (2.2 to 2.8 My), of the same origin and that they are unique to their special part of geological time, a time which preceded the formation of the epigenetic deposits . . . . During this period of time, the earth's atmosphere was anoxygenic and syngenetic uraninite from pegmatite and gneiss areas was carried in detrital form to be deposited as a heavy mineral in what are now 'fossil' placers."

From the above, and from a recent publication of Minter (1976), there is no doubt that at least some uranium is of detrital origin and this is consistent with some thorium and rare earths being in and with the uraninite. (For example, Hiemstra (1968) reported uraninite from the South African Dominion Reef as containing 63.29% uranium oxide, 6.52% thorium oxide and 3.43% rare earths.) Nevertheless, there is a close association in some conglomeratic ores between uranium and the variety of carbonaceous matter referred to as thucholite. (While the name 'thucholite' is intended to underline an association of carbonaceous matter with thorium and uranium, Feather and Koen (1975) point out that the name is misleading since such carbonaceous matter may be quite rich in uranium, but contains very little thorium.) As mentioned above, thucholite has been shown recently to have characteristic organized forms (Hallbauer and van Warmelo, 1974), thought to be algal-like in character.

Microscopic studies have shown that much uranium is intimately associated with these organisms. Its finely divided occurrence within such organic remains suggests strongly that it arrived in soluble form and was fixed while the organism still retained a high water content; this could have been shortly after the death of the organism or even while growth was continuing, as in an algal mat. There seem thus to be several stages at which the uranium could have found its way into the thucholite. The first is during the time when the conglomerate formed the bed of a stream draining the, presumably, igneous rock terrain from which the uranium was derived. The second possible stage is soon after deposition of the conglomerate, by redistribution of the uranium from originally detrital uranium to thucholite via solution in interstitial water. The third stage is later still, by redistribution of detrital uranium or by addition of 'new' uranium if the conglomerate, now a stratiform unit within a rock sequence, acted as an aquifer.

The fact that algal-like organisms grew at the time the conglomerate was formed has often been taken to indicate that there must have been some oxygen, even if only a little, in the atmosphere * and, in the absence of a reductant, uranous minerals would have oxidized slowly, so that soluble uranyl carbonate complexes could have formed. It seems likely that the greatest opportunity for redistribution in this way must have occurred during the first of the three stages listed, and that completely reducing conditions would have been established in the formation water very soon after younger sediments had been deposited. The sequence of events which seems to the author likely to have occurred is as follows: Uranium was present in acid igneous rocks and in associated veins in a terrain which was being actively eroded, presumably by essentially physical processes. The streams draining this land mass received, in addition to quartz and other rock-forming minerals, grains of such minerals as uraninite, monazite and pyrite. Because the oxygen content of the atmosphere was low, grains could remain in stream beds for a considerable period and even form heavy mineral accumulations. However, uraninite would oxidize slowly over the long periods suggested by the extremely good sorting of the conglomerates and uranium, as the uranyl ion, would be transported to algal mats where adsorption, and eventually, reduction to UO_2, could occur.

The distributions so far discussed have been at, or quite close to, the surface. Deposits in sandstones clearly occur at somewhat greater depths, although the chemical controls of deposition are similar. Finch (1967) has described in detail the geological occurrence of uranium mineralization in sandstone in the U.S.A. More recently, Adler (1974) has reviewed the various types of sandstone, or 'roll' deposits as some varieties have been called. The host rocks are typically permeable sandstones or other clastic sediments which have been aquifers for long periods. The uranium was introduced with the ground water as the uranyl ion, from an up-gradient, usually unknown source. Various redox patterns occur in which uranium has precipitated at or near interfaces between regions characterized by oxidizing and reducing conditions. Many rocks which serve as hosts to this type of uranium mineralization contain considerable amounts of carbonaceous matter, both coal-like and petroleum-like materials. One of the most detailed organic geochemical studies of a coal-uranium association was made by Breger (1974) in the context of mineralization in sandstone. The uranium was introduced into coalified logs epigenetically along microfaults or cracks, probably as a complex alkali uranyl carbonate. After its introduction, the uranium was reduced

* *Editors' footnote*: non-oxygenic algal photosynthesis has now been demonstrated — see Chapter 6.1.

to form uraninite or coffinite. After mineralization, the coal was radiochemically degraded by demethanation and dehydrogenation.

As mentioned in the preceding section, there has been considerable speculation over the possible role of organisms in generating bacteriogenic hydrogen sulfide (Jensen, 1958). Pyrite is a common mineral with or close to most sandstone ores and its possible role as a reductant has been much discussed. It is probably simplistic to postulate any single determinant in a system where redox potential could have been affected by many factors. It seems probable to the author that the pyrite and other sulfides have been fixed by reduction of ferric to ferrous iron and of sulfate to sulfide. Thus the sulfides are considered, like the uraninite, to be precipitated as a result of the lowering of the redox potential in the formation water. The most obvious agent of reduction is the carbonaceous matter which is present in considerable quantities, although gaseous and liquid hydrocarbons could have played a role. It is possible to see banding in the sandstone around individual coaly fragments, which appears to reflect progressive lowering of the Eh as the coal fragment is approached. Such banding may be asymmetric by being extended down-gradient along the direction of hydrologic flow. It thus seems unlikely that the emplacement of uranium as it presently occurs in sandstones, owes much, if anything, to contemporaneous biological factors.

So far as is known, there is no biological component in the processes which lead to the formation of deposits such as Yeelirrie in calcrete (Dall'Aglio et al., 1974). In such deposits the uranium is present in uranyl compounds, the precipitation of which appears to depend on the solubility relationships of uranyl and other ions, including complexes containing vanadium, in waters of varying composition; carbonate concentration appears to have been especially important.

PROSPECTING

Uranium offers one very simple and direct exploration technique — radioactivity. While various field methods of detecting radioactivity have been widely used with great success, a cover of only a metre or two of soil or rock is enough to prevent any radioactivity from ores being detected at the surface. Most deposits with pronounced surface expression are likely to have been already discovered.

As is true for all mineral exploration, a knowledge of the geology is a prerequisite. This is emphasized by Bowie (1972) who points out that more than 90% of uranium reserves occur in Precambrian rocks or in Phanerozoic rocks closely underlain by Precambrian rocks. The mineral and element associations of known uranium mineralization and the environment of deposition of known ore bodies provide excellent guides in the search for further ore. For example, Cannon (1964) reported that sulfur, selenium, arsenic and molybdenum are concentrated with uranium-vanadium ores in the Yellow Cat area of Utah. Thus for at least some styles of uranium mineralization, conventional geochemical prospecting is most valuable.

Because uranium is comparatively mobile in oxidizing environments, groundwaters and waters from rivers, lakes and wells have been analysed to enable zones of mineralization to be detected (Levinson, 1974). Uranium finds its way into such waters both as a result of oxidative weathering of uraninite and of other U^{IV} minerals, and also by solution of uranyl compounds. The waters can be directly sampled and analysed, usually radiometrically, but sometimes by the very sensitive wet chemical methods which make possible determinations at levels of less than 1 ng g^{-1}. Currently available methods have been listed by Ostle and Ball (1973) in the context of the geochemistry of uranium in the supergene environment. Uranium in natural waters may be fairly rapidly removed by the formation of complexes, especially organic complexes and this may restrict the use of natural waters for uranium prospecting. However, MacDonald (1969) used lake waters successfully and was able to show that the Beaverlodge uranium district is delineated by an analytical plateau of 0.9 ng U_3O_8 g^{-1}, approximately twice the regional background value of 0.4 ng U_3O_8 g^{-1}. The lakes sample by MacDonald included some containing much organic matter and the presence of this material gave rise to erratic results.

This kind of enrichment of uranium is most pronounced where peat bogs or peaty soils lie in the path of surface water draining rocks from which uranium is being leached. Armands (1967) described a situation where water entering bogs in Northern Sweden averaged 0.1 μg U g^{-1}. He found that peat in the bogs contained up to 3% uranium on a dry basis. Szalay (1958) has shown that peat, fully saturated with uranyl ion, can contain nearly 10% uranium on a dry basis.

The strong adsorption on peat of uranium from natural waters is utilized in a prospecting technique recommended by Horvath (1960). In this procedure, washed peat samples are placed in finely woven cotton bags and anchored in a stream for a week. The peat is then removed, dried and ashed and the ash analysed for uranium. Whitehead and Brooks (1969a), who carried out such experiments in New Zealand, found that, while the uranium content of the peat ash could not be used to determine the absolute amounts of U in the river waters, it was a useful indication of the relative concentrations in such waters. At the same time, Whitehead and Brooks were able to compare adsorption on peat with the uranium content of bryophytes in the same streams (Wodzicki, 1959). Bryophytes tolerate a variety of ecological conditions and have a remarkable capacity for taking up trace elements from the substrate on which they grow. In the New Zealand study, there was broad agreement between the variation of uranium in the peat and that in the bryophytes; divergences would be expected since the mechanisms of uptake, and competition for adsorption sites must differ with the adsorbing material.

Plants in the area of northern Sweden studied by Armands (referred to above) were also found to have a high degree of enrichment in uranium, the highest figures being for willow, in which the plant ash from twigs contained almost twice as much uranium as the leaf ash: 860 versus 450 $\mu g\ g^{-1}$.

This brings up the possibility of using geobotanical prospecting methods. The subject has been comprehensively discussed by Brooks (1972) and his book has been drawn upon for the comments below. No actual indicator plants for uranium itself appear to have been reported, which is not surprising in view of its known toxicity. However, Whitehead and Brooks (1969b) reported uranium values ranging from 1.5 to 1000 $\mu g\ g^{-1}$ in the ash of the New Zealand shrub *Coprosma australis.* Since the normal background level of plants is about 1 $\mu g\ g^{-1}$, the response of *C. australis* to uranium is very good and biogeochemical prospecting with this plant may well be successful.

In some of the most successful work of its kind described, Cannon (1964) used two botanical methods of prospecting in the Yellow Cat area of Utah. The first involved the analysis for uranium of juniper needles and leaves of shrubs, and the second the mapping of the distribution of indicator plants. Cannon found two selenium indicators *Astralagus preussi* and *A. pattersoni* to be excellent indicators of mineralized ground.

The selenium indicators grew on 81% of the ground mineralized at a depth of less than 10 m and in 42% of that mineralized at a depth of 10—52 m. Here geobotanical work in this area alone led to the discovery of five uranium ore bodies. Cannon (1957) listed in all 9 indicators known to be effective in locating uranium, 25 species known to favour mineralized ground and a further 16 indicators tolerant of mineralization. Most, if not all, these indicators appear to point to concentrations of selenium which, in Utah and some other sandstone environments, is almost universally associated with uranium.

Bayer et al. (1966) reported that uranium, with some other metals, can form stable complexes with anthocyanins which are mainly responsible for flower colours in the range from orange to deep blue. It is possible that larger than usual amounts of these metals could produce a blue tint in flowers which are normally pink or red, and Brooks (1972) reports just such a case in New Zealand where the metal concerned was chromium. Effects such as these, together with toxicity and all other influences on plant growth must become increasingly important as remote sensing techniques are progressively refined.

IN-SITU LEACHING OF URANIUM: WEATHERING

Grandstaff (1976), in examining the kinetics of oxidative dissolution of uraninite concluded that the factors influencing the rate of dissolution are:

the specific surface area of the uraninite, the presence of organic matter, the proportion of cations other than uranium in the uraninite, the dissolved oxygen content of the water, the total dissolved carbonate and the temperature. These and perhaps some other factors determine if, and at what rate, uranium will dissolve in natural waters down to depths of some hundreds of metres. Much past transport of uranium appears to have been a very long-term process. For example, some concentrations of uranium which occur in sandstone have probably occupied many sites over geological time, moving down-gradient when waters become more oxygenated.

There is no sharp line between weathering and leaching. In fact, most leaching of uranium has been, in effect, an accelerated weathering process in which bacteria oxidize the pyrite in the ore to ferric sulfate and sulfuric acid; the resulting oxidizing solution dissolves uranium which is later recovered in an ion-exchange process. Under natural conditions of unbroken rock near the surface, the weathering of uranium minerals occurs slowly, usually much too slowly to be significant as a means of uranium recovery, since rock permeabilities are low and bacteria are unlikely to be able to participate very actively in the process. However, when broken rock is leached by rain or other water the rate of leaching can increase markedly and there has been considerable research over the past twenty years to establish whether leaching can be made an economic means of recovering uranium and other valuable metals, especially copper. The reactions which occur are many and complex, beginning with the wet oxidation of pyrite to ferrous sulfate and sulfuric acid (see Chapter 6.3). This, and the further oxidation of ferrous to ferric sulfate may be assisted by bacteria. At high acidities there can occur the following reaction (eqn (2)):

$$UO_2 + Fe_2(SO_4)_3 \rightarrow (UO_2)SO_4 + 2\ FeSO_4 \qquad (2)$$

in which uranium is solubilized. Lowson (1975) has recently produced an excellent review on bacterial leaching of uranium ores and this has been much used by the present author.

Leaching has been carried out both in heaps of broken ore on the surface (Cameron, 1963) and in underground mines (Duncan and Bruynesteyn, 1971), where mine waters were already acid as a result of pyrite oxidation. In both types of leaching, effective plumbing arrangements must be devised to enable leaching to continue over quite long periods — up to a year and more. For heaps, this requires careful attention to geometry and conditions which affect flow-rates.

Temperature is an important factor for bacterial activity and Pings (1968) set 35°C as the optimum operating temperature, which makes Australia's Northern Territory ideal. In Canada, however, it was found that heating of intake air was justified during winter months and, in France, bacterial leaching was abandoned on the grounds of low temperature (Mouret, 1971).

Autotrophic bacteria responsible for sulfur and iron oxidation are discussed in Chapters 4, 6.1 and 6.3. The most important with respect to leaching is *Thiobacillus ferrooxidans* which oxidizes Fe^{2+} to Fe^{3+}, and sulfides and elemental sulfur to H_2SO_4, as a source of energy. Pyrite and other sulfidic minerals may be solubilized as a result of these reactions.

In addition to an energy source, the bacteria require a carbon source which may be universally available carbon dioxide or, in some instances, simple organic compounds. Lowson (1975) quotes a number of cases which suggest that bacterial activity is not greatly hindered by lack of available carbon or of trace elements since the addition of nutrients provided little or no increase in uranium leaching rates.

Bacteria are only catalysts in the reaction sequence, and since they can be acclimatized to a particular environment, the optimum chemical conditions can be defined without reference to the bacterial action (Lowson, 1975). This means that the greatest benefit from leaching is likely to be gained when optimum grain sizes, flow-rates and temperatures can be maintained in broken rock or in ore containing abundant pyrite. Heap leaching, with heaps up to 13,000 m^2 in area and 10 m high, is now practised in several areas. The process is attractive because ores which could not be profitably mined with conventional practices can be extracted with up to 90% efficiency.

Most leaching described in the literature has followed the lines described above. Strongly basic ores, usually those containing abundant carbonate, are not well suited to bacterial leaching since a low pH is not as readily obtained as in other cases. For such ores, leaching with sodium carbonate solution has been used successfully, the uranium being transported as an anionic uranyl carbonate complex (Merritt, 1971).

The leaching of uranium-bearing rocks, whether exploited for the recovery of uranium or as an inevitable outcome of mining, has consequences for the environment. This is true whether the mining is by deep or by strip methods. Hunkin (1975) claims that, compared with other mining systems, uranium solution has a negligible effect on surface disturbance, interference with natural groundwater and aerial discharge of radionuclides. But as Lowson (1975) points out, the autotrophic bacteria involved are adaptable to extremes of environment so that whenever a pyritic ore body is exposed by mining, bacterial oxidation can be expected to occur. Leaching is retarded by keeping stockpiles dry. However, heaps of pyritic rejects are not usually accorded such attention and these, when saturated and oxidized, provide an ideal solution for leaching any heavy metal present. Such liquors may contain metals (not only uranium but also, for example, copper) at concentrations below those required for economic recovery but well above those which can cause serious pollution. The problem is, of course, not unique to uranium-bearing ores.

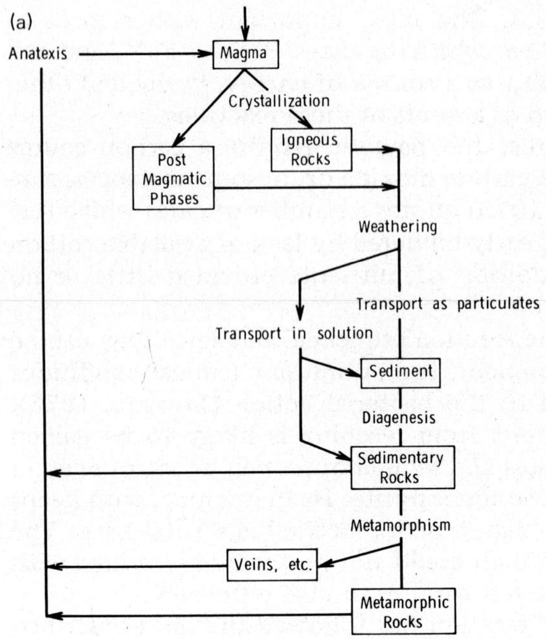

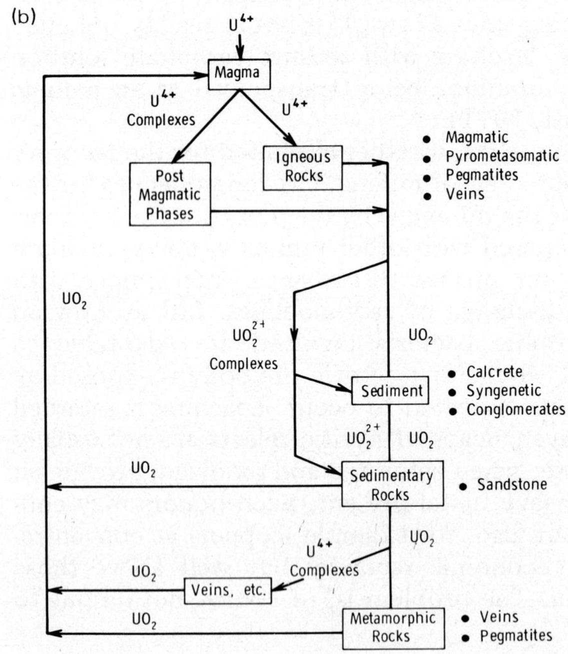

Fig. 8.3. (a). Geochemical cycle. (b). Cycle as in Fig. 8.3. (a), showing relationships of ore deposits and most important form of uranium in various parts of the cycle.

URANIUM AND THE GEOCHEMICAL CYCLE

By way of summarizing some of the conclusions reached in the preceding sections, an attempt has been made to show the geochemical cycle with special reference to uranium. For this reason, the way of presenting the cycle differs from that commonly used (e.g. Mason, 1966). Figure 8.3a shows the cycle itself while in Fig. 8.3b are shown the author's assessment of the relationship of ore deposits to the geochemical cycle, and also the most important form of the uranium at each stage of the cycle.

REFERENCES

Adler, H.H., 1974. Concepts of uranium ore formation in reducing environments in sandstones and other sediments. In: Formation of Uranium Ore Deposits. International Atomic Energy Agency, Vienna, pp. 141—168.

Anon., 1969. Research in Uranium Geochemistry. Investigations of the relationship between organic matter and uranium deposits; Part 2, Experimental Investigations. Report submitted to U.S. Atomic Energy Commission by Denver Research Institute, University of Denver.

Anon., 1975. The Oklo Phenomenon. International Atomic Energy Agency, Vienna, 647 pp.

Armands, G., 1967. Geochemical prospecting of uraniferous bog deposits at Masugnsbyn, northern Sweden. In: A. Kvalheim (Editor), Geochemical Prospecting in Fennoscandia. Interscience, New York, NY, pp. 127—154.

Arrhenius, G., Bramlette, M.N. and Picciotto, E., 1957. Localisation of radioactive and stable heavy nucleides in ocean sediments. Nature, 180: 85—86.

Baturin, G.N., 1972. Average ratios of uranium and organic matter in Holocene sea and ocean sediments. Dokl. Akad. Nauk SSSR, 207: 190—192.

Baturin, G.N., 1973a. Uranium in the modern marine sedimentary cycle. Geochem. Int., 10: 1031—1041.

Baturin, G.N., 1973b. Uranium and sedimentation in the Black and Azov Seas. Lithol. Miner. Resour. (USSR), 8: 540—549.

Bayer, E., Egeter, H., Fink, A., Nether, K. and Wegman, K., 1966. Complex formation and flower colors. Angew. Chem. Int. Ed. 5: 791—798.

Bohse, H., Rose-Hansen, J., Sørensen, H., Steenfelt, A., Løvborg, L., and Kunzendorf, H., 1974. On the behaviour of uranium during crystallization of magmas — with special emphasis on alkaline magmas. In: Formation of Uranium Ore Deposits. International Atomic Energy Agency, Vienna, pp. 49—60.

Boichenko, E.A., Saenko, B.N. and Udel'nova, T.M., 1975. Variation in metal ratios during the evolution of plants in the biosphere. In: A.I. Tugarinov (Editor), Recent Contributions to Geochemistry and Analytical Chemistry. John Wiley, New York, NY, pp. 507—512.

Bowie, S.H.U., 1972. The status of uranium prospecting. In: S.H.U. Bowie, M. Davis and D. Ostle (Editors), Uranium Prospecting Handbook. The Institution of Mining and Metallurgy, London, pp. 1—16.

Breger, I.A., 1974. The role of organic matter in the accumulation of uranium. In: Formation of Uranium Ore Deposits. International Atomic Energy Agency, Vienna, pp. 99—124.

Breger, I.A. and Brown, A., 1962. Kerogen in the Chattanooga Shale. Science, 137: 221—225.

Brooks, R.R., 1972. Geobotany and Biogeochemistry in Mineral Exploration. Harper and Row, New York, NY, 290 pp.

Cameron, J., 1963. Discussion on natural leaching of uranium ores. Trans. Inst. Min. Metall., 72: 507—517.

Cannon, H.L., 1957. Description of indicator plants and methods of botanical prospecting on the Colorado Plateau. U.S. Geol. Surv. Bull., 1030-M: 399—516.

Cannon, H.L., 1964. Geochemistry of rocks and related soils and vegetation in the Yellow Cat area, Grant County, Utah. U.S. Geol. Surv. Bull., 1176: 127 pp.

Cloud, P.E., 1965. Significance of the Gunflint (Precambrian) microflora. Science, 148: 27—35.

Dall'Aglio, M., Gragnani, R. and Locardi, E., 1974. Geochemical factors controlling the formation of the secondary minerals of uranium. In: Formation of Uranium Ore Deposits. International Atomic Energy Agency, Vienna, pp. 33—48.

Deuser, W.G., 1971. Organic-carbon budget of the Black Sea. Deep Sea Res., 18: 995—1004.

Dimroth, E. and Kimberley, M.M., 1976. Precambrian atmospheric oxygen: evidence in the sedimentary distributions of carbon, sulphur, uranium and iron. Can. J. Earth Sci., 13: 1161—1185.

Duncan, D.W. and Bruynesteyn, A., 1971. Enhancing bacterial activity in a uranium mine. Can. Min. Metall. Bull., 64: 32—36.

Feather, C.E. and Koen, G.M., 1975. The mineralogy of the Witwatersrand reefs. Miner. Sci. Eng., 7: 189—224.

Finch, W.I., 1967. Geology of epigenetic uranium deposits in sandstone in the United States. U.S. Geol. Surv. Prof. Pap., 538, 121 pp.

Galimov, E.M., Tugarinov, A.I. and Nikitin, A.A., 1975. On the origin of whewellite in a hydrothermal uranium deposit. Geochem. Int., 12: 31—37.

Garrels, R.M. and Christ, C.L., 1965. Solutions, Minerals, and Equilibria. Harper and Row, New York, NY, p. 256.

Gentry, R.V., Christie, W.H., Smith, D.H., Emery, J.F., Reynolds, S.A., Walker, R., Cristy, S.S. and Gentry, P.A., 1976. Radiohalos in coalified wood: new evidence relating to the time of uranium introduction and coalification. Science, 194: 315—318.

Grandstaff, D.E., 1976. A kinetic study of the dissolution of uraninite. Econ. Geol., 71: 1493—1506.

Hallbauer, D.K. and van Warmelo, K.T., 1974. Fossilized plants in thucholite from Precambrian rocks of the Witwatersrand, South Africa. Precambrian Res., 1: 199—212.

Hallbauer, D.K., Jahns, H.M. and Beltmann, H.A., 1977. Morphological and anatomical observations on some Precambrian plants from the Witwatersrand, South Africa. Geol. Rundsch., 66: 477—491.

Hiemstra, S.A., 1968. The mineralogy and petrology of the uraniferous conglomerate of the Dominion Reef Mine, Klerksdorp area. Trans. Geol. Soc. South Africa, 71: 1—66.

Horvath, E., 1960. Uranium adsorption on peat in natural waters containing uranium traces. Atomki Kozl., 2: 177—183 (in Hungarian).

Hunkin, G.G., 1975. The environmental impact of solution mining for uranium. Min. Congr. J., 61: 24—27.

Jensen, M.L., 1958. Sulfur isotopes and the origin of sandstone-type uranium deposits. Econ. Geol., 53: 598—616.

Kolodny, Y. and Kaplan, I.R., 1973. Deposition of uranium in the sediment and interstitial water of an anoxic fjord. Proceedings of the International Symposium on Hydrogeochemistry and Biogeochemistry, Tokyo. The Clarke Company, Washington, DC.

Levinson, A.A., 1974. Introduction to Exploration Geochemistry. Applied Publishing Ltd., Calgary, 612 pp.

Lowson, R.T., 1975. Bacterial leaching of uranium ores — a review. Australian Atomic Energy Commission Publication E356, 24 pp.

MacDonald, J.A., 1969. An orientation study of the uranium distribution in lake waters, Beaverlodge district, Saskatchewan. Colo. Sch. Mines Q., 64: 357—376.

Magne, R., Berthelin, J.R. and Dommergues, Y., 1974. Solubilisation et insolubilisation de l'uranium des granites par des bactéries hétérotrophes. In: Formation of Uranium Ore Deposits. International Atomic Energy Agency, Vienna, pp. 73—88.

Mason, B., 1966. Principles of Geochemistry, 3rd edn. John Wiley, New York, NY, 329 pp.

Merritt, R.C., 1971. The Extractive Metallurgy of Uranium. Colorado School of Mines Res. Inst., Golden, Co., 576 pp.

Minter, W.E.L., 1976. Detrital gold, uranium, and pyrite concentrations related to sedimentology in the Precambrian Vaal Reef Placer, Witwatersrand, South Africa. Econ. Geol., 71: 157—176.

Miyake, Y., Sugimura, Y. and Mayeda, M., 1970. The uranium content and the activity ratio $^{234}U/^{238}U$ in marine organisms and sea water in the western North Pacific. J. Oceanogr. Soc. Jn., 26: 123—129.

Mouret, P., 1971. In: The Recovery of Uranium. International Atomic Energy Agency, Vienna, p. 239. (This is in discussion of a paper: Mrost, M. and Lloyd, P.J. Bacterial oxidation of Witwatersrand slimes, pp. 223—239.)

Oglobin, K.F. and Khalifa-Zade, C.M., 1974. Abundance of uranium in the shells of recent and fossil molluscs. Geochem. Int., 11: 239—244.

Ostle, D. and Ball, T.K., 1973. Some aspects of geochemical surveys for uranium. In: Uranium Exploration Methods. International Atomic Energy Agency, Vienna, pp. 171—187.

Pascal, P. (Editor), 1960. Nouveau Traité de Chimie Minérale, Vol. 15, Masson, Paris, 734 pp.

Pings, W.B., 1968. Bacterial leaching of minerals. Colo. Sch. Mines, Miner. Ind. Bull., 2: 1—19.

Reimer, T.O., 1975. The age of the Witwatersrand System and other gold-uranium placers: implications on the origin of the mineralisation. Neues Jahrb. Mineral. Monetsh., No. 2: 79—98.

Robertson, D.S., 1974. Basal Proterozoic units as fossil time markers and their use in uranium prospection. In: Formation of Uranium Ore Deposits. International Atomic Energy Agency, Vienna, pp. 495—512.

Rogers, J.J.W. and Adams, J.A.S., 1969. Uranium. In: K.H. Wedepohl (Executive Editor), Handbook of Geochemistry. Springer, Berlin, Section 92.

Ruzicka, V., 1971. Geological comparisons between East European and Canadian uranium deposits. Geol. Survey of Canada, Paper 70—48, 196 pp.

Sax, N.I., 1975. Dangerous Properties of Industrial Materials, 4th edn. Van Nostrand, New York, NY, 1258 pp.

Saxby, J.D., 1970. Isolation of kerogen in sediments by chemical methods. Chem. Geol., 6: 173—184.

Schidlowski, M., 1970. Untersuchungen zur Metallogenese im südwestlichen Witwatersrand-Becken (Oranje-Freistaat-Goldfeld, Südafrika). Beihefte Geol. Jahrb., 85, 80 pp.

Schopf, J.W., 1975. Precambrian Paleobiology: Problems and perspectives. Ann. Rev. Earth Planet. Sci., 3: 213—249.

Silverman, M.P. and Ehrlich, H.L., 1964. Microbial formation and degradation of minerals. Adv. Appl. Microbiol., 6: 153—206.

Snyman, C.P., 1965. Possible biogenetic structures in Witwatersrand thucholite. Trans. Geol. Soc. South Africa, 68: 225—235.

Stach, E., Mackowsky, M-Th., Teichmüller, M., Taylor, G.H., Chandra, D. and Teichmüller, R., 1975. Stach's Textbook of Coal Petrology. Gebrüder Borntraeger, Berlin, 428 pp.

Stanton, R.L. and Russell, R.D., 1959. Anomalous leads and the emplacement of lead sulfide ores. Econ. Geol., 54: 588—607.

Szalay, A., 1958. The significance of humus in the geochemical enrichment of uranium. Geneva, Proc. Internat. Conf. Peaceful Uses Atom. Energy, 2nd, 2: 182—186.

Szalay, A., 1964. Cation-exchange properties of humic acids and their importance in the geochemical enrichment of UO_2^{2+} and other cations. Geochim. Cosmochim. Acta, 28: 1605—1614.

Tugarinov, A.I., 1975. Origin of uranium deposits. In: A.I. Tugarinov (Editor), Recent Contributions to Geochemistry and Analytical Chemistry. John Wiley, New York, NY, pp. 293—302.

Updegraff, D.M. and Douros, J.D., 1972. The relationship of microorganisms in uranium deposits. In: Developments in Industrial Microbiology. Society for Industrial Microbiology, Vol. 13, pp. 76—90.

Vine, J.D., 1956. Uranium-bearing coal in the United States. U.S. Geol. Surv. Prof. Pap., 300: 405—411.

Viragh, K. and Szolnoki, J., 1970. Role of bacteria in the formation and reaccumulation of the uranium ore of Mecsek. Fold. Kozl., 100: 43—54 (in Hungarian).

Walker, G.W. and Adams, J.W., 1963. Mineralogy, internal structural and textural characteristics, and paragenesis of uranium-bearing veins in the conterminous United States. U.S. Geol. Surv. Prof. Pap., 455D: 55—90.

Walker, G.W. and Osterwald, F.W., 1963. Introduction to the geology of uranium-bearing veins in the conterminous United States. U.S. Geol. Surv. Prof. Pap., 455A: pp. 1—28.

Weast, R.C., 1970. Handbook of Chemistry and Physics. The Chemical Rubber Co., Cleveland, OH.

Whitehead, N.E. and Brooks, R.R., 1969a. Aquatic bryophytes as indicators of uranium mineralization. Bryologist, 72: 501—507.

Whitehead, N.E. and Brooks, R.R., 1969b. Radioecological observations on plants of the Lower Buller Gorge region of New Zealand and their significance for biochemical prospecting. J. Appl. Ecol., 6: 301—310.

Wodzicki, A., 1959. Geochemical prospecting for uranium in the lower Buller Gorge, New Zealand. N.Z. J. Geol. Geophys., 2: 602—612.

Chapter 9

MINERALS AND AGRICULTURE

V.J. KILMER

National Fertilizer Development Centre, Tennessee Valley Authority, AL 35660 (U.S.A.)

CONTENTS

INTRODUCTION

Modern agriculture depends upon man's ability to cycle minerals that are nutritionally important through crops, poultry, and animals. In particular, the absorption of these minerals by green plants and their subsequent role in the formation of organic compounds are basic to almost all forms of life.

During the early stages of his existence, man nourished himself as did the animals of the wild, i.e., he was a gatherer of food. Many centuries passed before he became a producer of food. Natural forms of vegetation were slowly replaced with selected cultivated varieties that were more efficient producers of foodstuff for humans and later for the animals that he domesticated. As man struggled to insure an adequate food supply, he sought an understanding of the factors that stimulated crop yields. He found that yields gradually decreased on certain soils that were continuously cropped and reasoned that crops removed something from soils which contributed to the decline in yields. Crude attempts were made to discover the "principles of vegetation" and many theories, some containing particles of truth, were developed over many centuries.

Certain materials, such as compost, wood ashes, sewage, and animal wastes, were eventually discovered to be beneficial to crop growth. But basic research concerning the roles played by various chemical elements in the mineral nutrition of plants could not be initiated until the fundamentals of chemistry were firmly established.

Following many centuries of trial and error, scientific agriculture began to emerge in the middle of the 19th century when man began to divert and accelerate the flow of minerals through cultivated crops for his sole benefit. As population increased, so did the demand for agricultural products. This increased demand has forced man to accelerate this rate of mineral flow through the human food cycle. Concentrated mineral deposits are now processed into forms that can be utilized as fertilizers for crops, or to a much lesser extent, as supplementary feed additives for poultry and animals. As these minerals pass through the food chain, they are recycled over parts of the earth's surface on an ever-increasing scale. Man's influence over this cycle has steadily increased during the past century, and indications are that it will continue to do so in the future.

A BRIEF HISTORICAL PERSPECTIVE

Man's search for knowledge relating to the nutrition of crop plants is as ancient as his cultivation of the soil, estimated by some historians to have begun as long as 200 centuries ago. Even the ancient husbandman knew that certain soils would not continue to produce satisfactory yields when cropped continuously. This is clearly referred to in Exodus 23 : 10 : 11: "And six years thou shalt let it rest and lie still; that the poor of thy people may eat, and what they leave the beasts of the field shall eat." Thus, the resting or repose system for renewing the producing power of soils is shown to be a very old practice.

Ancient civilizations located on river floodplains were usually blessed with a fertile land. Writings dating back to 4500 B.P. mention the phenomenal yields obtained by the inhabitants of Mesopotamia, situated between the Tigris and Euphrates rivers in what is now Iraq. The naturally rich alluvium was maintained in part by annual flooding. The water was allowed to remain on the land as long as possible, so that a maximum amount of mineral-rich silt would be deposited (Tisdale and Nelson, 1966, p. 5).

The first agriculturists also learned about the value of manure by observing the effects of animal droppings on plant growth. The truck gardens and olive groves around ancient Athens were enriched by sewage from the city by using a canal system.

The idea of applying these and other organic wastes to soils became widespread. North American Indians used fish for fertilizing corn long before the white man arrived. The Incas judiciously worked the guano deposits on the coastal islands of Peru, probably as early as the 12th century, according to some historians, while others place guano collection and use as far back as 2200 B.P. or earlier.

The value of wood ashes for soil enrichment is mentioned in the Bible and by the early Greek and Roman writers. Farmers were advised to burn vines and stubble on the spot and the plow in the ashes to enrich the soil. Potassium nitrate is also referred to in the Bible in the Book of Luke and by early Greek and Roman writers as being useful for the fertilization of crops.

The use of bones as a mineral supplement to soils is a practice of great antiquity. Chinese farmers and fruit growers are said to have used calcined bones some 2,000 years ago.

The earliest geologic deposits used as mineral supplements for agricultural soils seem to be chalk and marl. The beneficial effects of these materials on crops were known to the Celts as early as 2500 B.P. The Romans, who learned this practice from the Greeks and Gauls, even classified various liming materials and recommended that one type be applied to grain and another to meadow. However, liming materials were used for nearly 20 centuries before the beneficial effects were shown to be mainly due to the neutralization of excess soil acidity.

While the value of applying various amendments to agricultural soils was recognized and practiced for over 100 centuries, serious scientific inquiry into the mineral nutrition of plants did not begin until early in the 18th century. Most of the results of these early experiments were misinterpreted, principally due to the underdeveloped state of chemistry. The famous experiment of van Helmont (1577—1644) is a case in point. He placed 90.8 kg of soil in an earthen container, moistened the soil, and planted a willow shoot weighing 2.3 kg. After 5 years the tree weighed 76.7 kg, and he could account for all but 57 g of the 90.8 kg of soil originally used. Since he had added only water, he concluded that water was the sole nutrient of the plant and attributed the loss of the 57 g to experimental error.

Many other early experiments and theories could be cited. J.R. Glauber (1604—1668), a German chemist, suggested that saltpeter (KNO_3) was the "principal of vegetation" and not water as van Helmont had suggested. About 1700, an Englishman named John Woodward grew spearmint in rainwater, sewage water, and sewage water plus garden mold. He found that the growth of spearmint was proportional to the amount of impurities in the water, and concluded that earth rather than water was the principal of vegetation. Jethro Tull (1674—1741), educated at Oxford, moved to a farm because of ill health. His experiments there led him to conclude that soil particles were actually ingested by plants through root openings.

The latter half of the 19th century was a time when great progress was made in understanding plant nutrition and basic principles of soil fertility. Among the giants of this period were Jean Boussingalt (1802—1882) and Justus von Liebig (1803—1873). Boussingalt carried out plant nutrient balance studies on crops and soils. Liebig firmly believed that, by analyzing plants, he could formulate mineral fertilizers based on the results of the analyses. He manufactured a fertilizer based on his ideas, fusing phosphate and potash salts with lime. While his ideas were sound, the plant nutrients in his fertilizer were largely insoluble (unavailable to plants) and therefore a complete failure.

Although our knowledge of plant nutrition is still incomplete, much progress has been made since 1900 in understanding the relationships between nutrient supplied in the soil and crop yields.

We now know that 20 elements are essential for plant growth; not all of these are required by all plants, but all are essential to some. An element is essential if a plant cannot complete its life cycle without it. Carbon, O, and H are obtained by plants from air and water, the remaining 17 elements — N *, P *, K *, Ca *, Mg *, S *, Fe *, Zn *, Cu *, Mn *, B *, Co, V, Cl, Na, Si, and Mo * — are obtained primarily from the soil solution. All 17 elements are important, but man's influence relates primarily to those that are followed by an asterisk, since these are commonly applied to agricultural soils. Nitrogen, P, and K are termed major nutrients, since these are required by crop plants in large amounts and are the elements that most frequently limit crop yields. Calcium, Mg, and S are the so-called secondary nutrients, although S and P requirements for many crops are similar. The remaining elements are required in relatively small amounts and are termed micronutrients. It should again be stressed that all are vital to crop growth.

Also, it has long been known that mineral* nutrients are essential to the life and continued health of poultry and livestock. The latter often require supplementary P, Ca, and NaCl. Laying hens require a diet that is very high in Ca. Much progress has been made in recent years toward understanding the nutritional needs of poultry and livestock, particularly the role played by trace elements.

NATURAL INPUTS OF MINERALS TO AGRICULTURAL SOILS

Since the age of agriculture is estimated to be about 10 ky, it follows that any discussion of natural inputs of minerals to agricultural soils from an external source will be limited to comparatively recent phenomena. Thus, rocks and minerals that have been transported in the distant geologic past by ice, wind, and water to form glacial, aeolian, alluvial, lacustrine, and marine deposits are outside the scope of this chapter, since these are generally parent materials from which modern soils were formed. Dense populations usually developed where the soils had been enriched by recent volcanic activity or by young alluvial deposits.

There is relatively little quantitative data relating to natural inputs of minerals to agricultural soils, but it is important to recognize these natural inputs, even though such recognition must be of a general nature and very limited in scope.

Atmospheric additions

Wind erosion — dust storms. Dust is injected into the upper atmosphere by various processes, but human activities probably account for much of the dust that is injected into the lower atmosphere. Although land areas are exposed to wind erosion by agricultural and forestry operations, various

---

* Webster defines the term "mineral" as "any chemical or compound occurring naturally as a product of inorganic processes." Since this chapter is primarily concerned with the use of minerals in plant and animal nutrition, terms such as "plant nutrients", "trace elements", "fertilizer nutrients", and "mineral additives" are used within the context of Webster's definition, including inorganic fertilizers derived from naturally occurring minerals.

kinds of evidence indicate that the Great Plains region of the U.S.A. was periodically subjected to tremendous dust storms long before the introduction of current methods of farming and ranching (Idso, 1976).

In recent years, photographs from satellites have revealed much about the origin and extent of dust storms. Russian scientists, A.A. Grigor'yev and V.B. Lipatov, have located five major regions that generate dust storms. Idso (1976, p. 111) has published an interesting account of the Russian work, and excerpts from his paper follow.

"The first of these regions, which is in central and western Africa, is often characterized by extremely long dust currents over the southern Sahara. They appear to result from an enormous airstream moving east-northeast over the sandy deserts of Mauritania and Niger. Giant dust bands 2,500 km long and 600 km wide often move across the area with the cold fronts. Some large systems have even carried this dust across the Atlantic to the eastern coast of South America.

The second major region where dust storms originate is the southern coast of the Mediterranean. Here the storms begin with the passage of cold fronts connected with low-pressure troughs that stretch toward North Africa from western Europe. Deep cyclones often form in the troughs.

The third major region is the northeastern Sudan . . . vast dust storms form when cold northwesterly air currents meet hot southwesterly monsoons. At such times masses of dust may be raised over large areas encompassing the region of the northeastern Sudan from the Nile River to the Red Sea. High-altitude winds often transport this dust across the Red Sea to the Arabian Peninsula.

The Arabian Peninsula is the fourth major dust-storm region . . . . storms generally develop when the wind increases on the western periphery of a pressure depression centered over southern Iran. Cone-shaped streams of airborne particulates form and move southward, often expanding. At their inception, . . . several small parallel streams, from three to five km in size, develop. They stretch for about 100 km and then merge into more powerful streams. The larger bands, swirling and expanding, stretch up to 500 km. These streams, moving with the trade current, are carried through a corridor bounded on the north by the mountains of the southern edge of Asia Minor and Iraq and on the south by the plateaus and mountains of Saudi Arabia.

The last of the five dust-storm regions studied by Grigor'yev and Lipatov encompasses the lower Volga and the northern Caucasus. The dust storms here are also of the storm-zone type. They are typically caused by an increase in the gradient of barometric pressure on the northwestern periphery of high-pressure ridges that extend to the Volga and central Asia. A gradient increase of this kind is caused by the movement of cyclones into northern Europe and the spread of low-pressure troughs to the south of the cyclones."

Large dust storms also occur in central China, originating in a vast area of barren land containing the great deserts Takla Makan, Gobi, and Ordos, and also the major loessial-soil lands of China. It is estimated that several thousand tonnes of soil are transported each year in these areas, some of it being transported as far south as Hong Kong (Idso, 1976).

In Europe, the amount of material deposited in Wesphalia during the great dust storm of 1859 averaged 30 g m^{-2}. The origin of the storm was assumed to be the Sahara.

Recent dust storms have occurred in the U.S.A. and these are fairly well characterized. A large storm occurred on May 12, 1934, extending eastward from the vicinity of the Rocky Mountains to several hundred miles over the Atlantic. It has been estimated that 180 to 270 Tg of topsoil was moved out of the Plains States as a result of this single storm (Bennett, 1939).

Bennett (1939) describes a dust storm that occurred in early February 1937 that originated in the Texas-Oklahoma panhandle. Dust from this storm was carried over 800 km and deposited on large areas of ice and snow in the midwestern U.S.A. and in Canada. Subsequent precipitation resulted in a thick layer of dust "sandwiched" between ice and snow.

TABLE 9.1

Partial chemical and physical composition of virgin soil, dune sand, and dust derived from cultivated soil as the result of a dust storm on and preceding February 6, 1937 — values are percentages (Adapted from Bennett, 1939.)

	Unplowed grassland near Dalhart, Texas	Dune sand near Dalhart [a]	Dust from Clarinda, Iowa [b]
SiO_2	87.24	91.35	66.31
TiO_2	0.24	0.15	0.63
Al_2O_3	6.07	4.37	13.93
Fe_2O_3	1.14	0.79	4.24
MnO	0.03	0.02	0.10
CaO	0.34	0.31	1.98
MgO	0.25	0.14	1.43
K_2O	2.05	1.77	2.58
Na_2O	0.62	0.25	0.92
P_2O_5	0.04	trace	0.19
SO_3	0.03	0.03	0.18
Ignition loss	1.84	0.84	7.26
Total	99.89	100.02	99.75
Nitrogen	0.06	0.02	0.19
Organic matter	1.06	0.33	3.35
Sand	79.2	91.8	0.0
Silt + clay	19.6	7.5	97.0

[a] Formed on and immediately preceding February 6, 1937.
[b] Collected from surface of snow, February 8, 1937.

TABLE 9.2

Estimated ranges in total and available plant nutrient contents in aeolian material resulting from the U.S. dust storm of February 6, 1937, and estimated amounts of available nutrients deposited at three locations (Bennett, 1939.)

Plant nutrient	Total [a] ($mg\ g^{-1}$)	Available [a] ($mg\ m^2\ y^{-1}$)		
		Iowa	Michigan	New Hampshire
N	0.2– 2	0.2– 2	0.1– 1	<0.1– 0.8
P_2O_5	0.4– 2	0.2– 1	0.1– 0.5	<0.1– 0.4
K_2O	18 –26	18 –26	9 –13	7 –10
CaO	3 –32	3 –32	1.5–16	1 –13
MgO	1 –14	1 –14	0.5– 7	0.4– 6
SO_3	8 –73	8 –73	4 –36	3 –29
Fe_2O_3	8 –42	8 –42	4 –21	3 –17
MnO	0.2– 1	0.2– 1	0.1– 0.5	<0.1– 0.4

[a] The total nutrient content of soils consists largely of nutrients that are organically and inorganically bound. Only a small fraction, perhaps 10% at most, of N, K, and S are released in plant-available form over a year's period. A 5% release rate is used for the remaining elements. However, it should be emphasized that the rate of release depends upon many factors and can only be determined for a specific material and environment.

Partial chemical and physical characteristics of original virgin soil and freshly formed dune sand in the area of storm origination are shown in Table 9.1. The sorting action of wind erosion is well illustrated here. The finer material was transported great distances, depriving the source area of material that was relatively high in plant nutrients. The windblown material contained little or no sand, but the unplowed grassland of the source area contained about 80% sand, the newly formed dune, 92%.

One can develop a rough estimate of mineral contributions to agricultural soils as a result of dust storms. Turning again to the 1937 storm, the amount of material falling at Ames, Iowa, was 10 g m^{-2}; at Marquette, Michigan, 5 g m^{-2}; and in New Hampshire, 4 g m^{-2}. The results of such estimates are given in Table 9.2. These crude estimates indicate that modern dust storms are not a highly significant source of minerals that are essential to plant nutrition. Overall, huge tonnages of soil materials are moved annually on a worldwide basis, but repeated depositions would be required to have a measurable effect on soils. That this has occurred in the geologic past is, of course, clearly demonstrated by loessial deposits aroung the world.

Precipitation — fallout. Atmospheric additions of nutrients to soils occur in precipitation and as particulate fallout. Two elements, N and S, have received the most attention in this regard, since they both can exist as gases. Combined mineral N (i.e., ammonia and N oxides) in the atmosphere is believed to arise principally from combustion and biological activity. Total annual deposition of N is reported to vary from about 0.2—2 g m^{-2} (Hoeft et al., 1972), the amounts varying with season and location. However, most of the survey data covering large areas of the globe are at least 25 y old and there is evidence that the amounts of mineral N deposited in precipitation are gradually increasing, although there is a paucity of recent information in this area. Delwiche (1970) estimates the amount of fixed N delivered to the earth by precipitation to be about 25 Tg y^{-1}.

Total annual N additions over the State of Wisconsin ranged from 1.3—3 g m^{-2}, averaging about 2 g m^{-2} (Hoeft et al., 1972). Amounts measured during the first half of this century usually did not exceed a maximum of 1 g m^{-2} annually, averaging an estimated 0.55 g m^{-2} on a worldwide basis (Buckman and Brady, 1969). Increased combustion of fossil fuels is probably the most important factor behind increased rates of N deposition in rainfall. Recent studies by Kilmer et al. (1974) showed that, in western North Carolina, the yearly N additions were 1.2 g m^{-2}, including that in precipitation and dry fallout. Other elements measured included P, 0.2 g m^{-2} and K, 0.2 g m^{-2} y^{-1}. Atmospheric depositions of S range widely over the globe, amounting to 0.5 to more than 11 g m^{-2} y^{-1}; the highest amounts are found near areas of great industrial activity (Tisdale and Nelson, 1966) *.

Volcanic activity. Volcanic activity includes atmospheric dust, ash, and lava flows. The relationship of atmospheric volcanic dust to quantitative mineral inputs is not known, but a review of the subject indicates that this mineral source is not generally significant from an agricultural standpoint (Lamb, 1970). Recent volcanic ash deposition and lava flows can make sizable contributions of minerals to soils on a localized basis. Maximum involvement in this respect concerns land areas bordering on or occurring as islands in the Pacific Ocean.

Taylor and Stoiber (1973) have recently published some excellent data pertaining to soluble material on ash from active volcanoes in Central America. It is evident from their paper that volcanic ash can provide huge tonnages of readily water-soluble nutrients that are potentially available to crop plants (Table 9.3). Taylor and Stoiber (1973) conclude that much of the soluble material was deposited on the ash surface from volcanic gas during eruptions.

Recent alluvial deposits. Soils formed from alluvium are among the most productive in the world. Floodplain and delta deposits are the most impor-

---

* See also p. 424.

TABLE 9.3

Soluble materials leached from ash (Adapted from Taylor and Stoiber, 1973.)

Volcano	Date of eruption	Quantity of ash		Chemicals leached from ash [a] (weight in Mg)									
		vol. (km^3)	weight (Tg) [b]	Cl	F	SO_4	Na	K	Ca	Mg	Zn	Cu	Mn
San Miguel	3/30—4/5/70	0.0008	0.1	711	0.9	304	256	1.7	20	228	0.7	5	14
Cerro Negro	10-12/68	0.0127	17	8,900	140	6,000	2,900	290	2,400	360	2.5	2.1	85
Paricutin	2/43—12/53	1.31	2,227	17,282 $\times 10^3$	13 $\times 10^3$	32,871 $\times 10^3$	1,926 $\times 10^3$	94 $\times 10^3$	4,387 $\times 10^3$	6,414 $\times 10^3$	27 $\times 10^3$	31 $\times 10^3$	209 $\times 10^3$
World fragmental ejecta	1 y av. based on period 1500—1914	0.8	1,080	8,380 $\times 10^3$	6.5 $\times 10^3$	15,940 $\times 10^3$	934 $\times 10^3$	45 $\times 10^3$	2,127 $\times 10^3$	3,107 $\times 10^3$	13 $\times 10^3$	15 $\times 10^3$	102 $\times 10^3$

[a] Mg of chemicals calculated. Average concentration in leachate analyses multiplied by total weight except for Cerro Negro which is a weighted average using leachate analysis for each size fraction and location.

[b] Density measured for Cerro Negro, 1.35 Mg m^{-3}, same density assumed for others except Paricutin which measured 1.7 Mg m^{-3}.

tant from the standpoint of agriculture, because of the very extensive areas of land that are involved. However, modern agriculture is likely to be more adversely than favorably affected by current alluvial deposits, because such deposition involves periodic flooding of agricultural lands. Repeated flooding can result in ruined crops, drainage problems, and the addition of coarse, infertile materials. Thus, the deposition of recent alluvium can be a negative factor as far as current agricultural production is concerned.

In a broader sense, vast amounts of soil materials are transported annually by rivers and streams, adding to deposits formed over the geologic past. Floodplain and delta deposits are developed most extensively in the U.S.A. along the lower Mississippi River. These deposits cover an area of roughly 6×10^{10} m^2 and extend from Cairo, Illinois, to the Gulf of Mexico, a distance of about 800 km. This river continues to be a formidable transporter of sediment; the load averaged over 0.45 Tg d^{-1} at St. Louis, Missouri, during the period 1948—1958 (Mack, 1967). Other alluvial areas of great agricultural importance include the deltas and floodplains of the Ganges, Po, Tigris, and Euphrates rivers.

Since the chemical composition of alluvium obviously depends upon the source of the material, one can only make general statements about plant nutrient contributions that are related to alluvial processes. Soils vary widely as to their plant nutrient contents, as shown in Table 9.4. It should be recalled that, at most, only 5 to 10% of these total amounts is likely to be solubilized in any given year. Using the figure of 0.45 Tg of sediment per day carried by the Mississippi River at St. Louis, and assuming 5% dissolution of plant nutrients contained in the sediment, an estimated 90—2000 Mg of *available* nutrients is transported daily by this river alone. The sediment plus contained nutrients is either laid over or extends existing deposits. For example, the Po River extends its delta by about 8×10^5 m^2 y^{-1}.

TABLE 9.4

Total amounts of plant nutrients ordinarily found in temperate region, mineral surface soils (Swaine, 1955; Buckman and Brady, 1969; Kilmer, 1974.)

Constituent	Range (mg g^{-1} of soil)	Constituent	Range (μg g^{-1} of soil)
N	0.2— 5.0	B	2 — 200
P	0.1— 0.2	Co	1 — 40
K	1.7—33.0	Cu	2 — 100
Ca	0.7—36.0	Mn	200 —3000
Mg	1.2—15.0	Mo	0.2— 5
S	0.1— 2.0	Zn	10 — 300

HUMAN INFLUENCES ON MINERAL INPUTS AND CYCLING IN SOILS

Man is now dependent upon mineral fertilizers to the extent that fertilizer use in developing countries accounts for an estimated 40% of the increase in crop production per unit area since 1940. Developing countries tend to be increasingly dependent upon fertilizer use, because of inadequate diets and rapidly expanding populations. Further, soils in developing countries are likely to be in a very low state of fertility. In these countries, a rough rule of thumb is that 1 Mg of plant nutrients will produce about 10 Mg of cereals, or enough to provide 40 people with 10 kJ daily for a period of 1 year (Nelson, 1972).

The diets of populations in developed countries tend to be high in animal or animal-based protein, such as meat, cheese, eggs, and milk. In the U.S.A., for example, the average per capita protein intake is 100 g d^{-1}, consisting of 30% plant protein and 70% animal protein (National Academy of Sciences, 1972). Citizens of other developed countries either have or are striving for similar diets, even though the conversion of plant protein to animal protein is a relatively inefficient process. Improved diets will generally require additional inputs of nutrients to the agricultural production cycle.

WORLD PLANT NUTRIENT CONSUMPTION

Statistics on plant nutrient consumption are generally limited to the three major nutrients N, P, and K. World consumption of these nutrients has increased six-fold over the past 25 years as shown in Table 9.5. In 1975, nearly 82 Tg of N, P, and K was used in world agriculture, with N accounting for more than 40% of this amount. By 1980, total world consumption of the three major plant nutrients is estimated to be between 105 and 123 Tg.

The intensity of fertilizer use exhibits a wide variation among nations.

TABLE 9.5

World plant nutrient consumption in Gg (FAO, 1976)

Year	N	P_2O_5	K_2O	Total
1950	3,639	5,864	3,994	13,497
1955	6,521	7,553	6,439	20,513
1960	9,626	9,532	8,109	27,267
1965	16,404	13,634	11,031	41,069
1970	28,677	18,802	15,569	63,048
1975	38,859	22,784	19,937	81,580
1980 Low (est.)	53,100	28,300	23,500	104,900
1980 High (est.)	60,800	34,100	27,800	122,700

Worldwide, the average application of N, P_2O_5, and K_2O in 1974 totaled 5.4 g m^{-2} of arable land and land under permanent crops. But in Sweden, the rate was 16; in Japan, 37.5; in Nepal, 0.6; in The Netherlands, 75.6; in the United Kingdom, 25.7; and in North America, 6.8 g m^{-2} (FAO, 1976, pp. 52—62).

Phosphorus and K reserves appear adequate for several centuries to come, but present N fertilizer technology is dependent upon a cheap and abundant supply of natural gas as a source of H and process heat.

THE USE OF MINERALS IN AGRICULTURE

Nitrogen

Mineral deposits of N are limited largely to $NaNO_3$ — one of the oldest N fertilizers used — and very small amounts of KNO_3. Natural deposits of these materials have been reported in many widely scattered locations, such as South America, Egypt, South Africa, Mexico, and the U.S.A. Thus far, the only deposits of commercial significance are those found in northern Chile, first mentioned in 1809 and initially mined by the Spaniards in 1813 (Collings, 1949). The desert in which the nitrate is found is a part of the plateau of Tarapaca. The plateau is about 724 km long, running north and south, and 24 to 145 km wide, at an elevation of 0.9—2.5 km. It is shut in on both sides by mountain ranges; thus, it seldom receives rain. The original deposits were estimated to contain 263.3 Tg of nitrate (Cuevas, 1916).

The origin of these deposits remains a mystery, many theories having been advanced to account for their existence. These include non-symbiotic and electrical fixation of N, but neither of these explains the presence of I and Br in the deposits. Other theories involve nitrification of marine, plant and animal refuse, nitrification of animal manure from huge herds of vicunas and lamas, or formation from guano deposits. A commonly accepted theory is that Chilean nitrate had its origin in the evaporation of nitrate-carrying waters reaching the plateau after passing through bedded sedimentary rocks and gravels of surrounding highlands (Collings, 1949). Possibly the most eloquent description of the Chilean nitrate beds was written about five decades ago by the Chilean Nitrate Educational Bureau (Anon, 1935):

> "When the earth was being made Mother Nature must have looked ages ahead. She must have seen then what farmers would need. She took mysterious and unknown steps. She did strange, awe-inspiring things. Up from the sea she hurled ranges of mountains, great rocky plateaus, vast arid deserts. Giant plants took form and grew. Birds filled the air. Animals roamed the land. A myriad of fishes swam. Volcanic ruptures changed the face of the earth. Terrific storms flooded its surface. Shriveling heat seared every living thing. All this took countless centuries. Out of it came the

TABLE 9.6
Average range of composition of representative samples of caliche [a] (Collings, 1949.)

	(%)
Sodium nitrate ($NaNO_3$)	12 —25
Potassium nitrate (KNO_3)	2 — 3
Sodium chloride (NaCl)	8 —25
Sodium sulfate (Na_2SO_4)	2 —12
Calcium sulfate ($CaSO_4$)	2 — 6
Magnesium sulfate ($MgSO_4$)	0 — 3
Sodium biborate ($Na_2B_2O_4$)	1 — 3
Sodium iodate ($NaIO_3$)	0.05— 0.1
Sodium perchlorate ($NaClO_4$)	0.1 — 0.5
Insoluble matter	23 —70

[a] The nitrate deposit is called *caliche*; but strictly speaking, this is incorrect, since the word is derived from the Latin *calx* meaning a lime or calcareous deposit.

world as we know it today. Out of it came the many treasures that Nature formed within the earth. Out of it came Chilean nitrate of soda, one of Nature's most valuable gifts to modern agriculture. This extraordinary plant food is entirely natural in its origin. Nature herself made it. Only Nature knows how. Only Nature knows why one area in all the world — that area in arid northern Chile — was chosen as the chest for this natural treasure."

Accordingly, we will let the theories rest at this point. An average range of the composition of the so-called caliche is shown in Table 9.6.

Up until 1914, the Chilean sodium nitrate industry practically controlled the world market for nitrates, since Chile had a natural monopoly on the world's supply. World consumption of nitrate was 203 Gg in 1880 (Weir, 1936). Chilean nitrate is estimated to have supplied about 70% of the fertilizer N during the last half of the 19th century. A process for fixing atmospheric N was developed by Haber and Bosch in Germany during the early years of the 20th century. This highly significant development, coupled with the urgent need for a reliable source of nitrates for manufacturing munitions during World War I, led to a gradual decline in the world's nitrate-dependence in Chile (Table 9.7) *.

* During this period, Wilson Dam was built on the Tennessee River at Muscle Shoals, Alabama, to furnish the power needed for the production of synthetic ammonia by the Haber process. This project was undertaken to insure a domestic source of nitrates for producing munitions and fertilizers. When the Tennessee Valley Authority was organized in 1933, the dam and facilities became a part of the Authority and thus the National Fertilizer Development Center at Muscle Shoals was born.

TABLE 9.7

World consumption (in Gg) of Chilean nitrate and synthetic nitrogen, 1914–1974 (Tim and Danz, 1964; FAO, 1976.)

Year	Chilean nitrate	Synthetic N
1914	362	4
1920	292	87
1930	326	814
1939	207	1,888
1950	245	3,285
1960	131	10,659
1970	120	30,150
1974	113	40,498

Basic synthetic processes for the fixation of atmospheric N, are three in number and include the direct oxidation of N, or *arc process*:

$$N_2 + O_2 \rightarrow 2\ NO \quad (1)$$

$$2\ NO + O_2 \rightarrow 2\ NO_2 \quad (2)$$

$$3\ NO_2 + H_2O \rightarrow 2\ HNO_3 + NO \quad (3)$$

The fixation of atmospheric N by this method required large amounts of electricity; hence it is used primarily in Norway where hydroelectric power is abundant.

The *cyanamide process* involves the reaction of N gas with calcium carbide at 1100°C to form calcium cyanamide:

$$CaC_2 + N_2 \xrightarrow{heat} CaCN_2 + C \quad (4)$$

Appreciable quantities of N fertilizers are produced in Germany by this process.

The most important method of synthetic N fixation is by the *Haber-Bosch process.* Ammonia is synthesized from a 3 : 1 volume of H_2 and N_2 at elevated temperature and pressure in the presence of a catalyst:

$$N_2 + 3\ H_2 \rightarrow 2\ NH_3 \quad (5)$$

Ammonium nitrate is produced by a simple acid-base neutralization reaction:

$$HNO_3 + NH_3 \rightarrow NH_4NO_3 \quad (6)$$

Urea is produced from ammonia and carbon dioxide:

$$CO_2 + 2\ NH_3 \rightarrow CO(NH_2)_2 + H_2O \quad (7)$$

Nitrogen fertilizers are now produced as solids (ammonium nitrate, ammonium sulfate, ammonium phosphates, and urea, for example), fluids (clear solutions and suspensions), and as a gas (anhydrous ammonia). In the U.S.A., approximately 30% of the N used in agriculture is in the form of anhydrous ammonia, 41% as solids, and 20% as fluid fertilizers (USDA, 1976a).

Nitrogen is now the principal nutrient that limits agricultural production in North America and in Europe. While it is obvious that the supply of atmospheric N is infinite, future production of N fertilizers by the Haber-Bosch method could be severely limited by the supply of economically available H that is now largely supplied by natural gas.

World consumption of N in fertilizers is now about 40 Tg y^{-1}.

Phosphorus

"There can be no civilization without population, no population without food, and no food without phosphoric acid." So wrote the United States Commissioner of Patents in his 1859 annual report to Congress.

Although the good commissioner's statement is not wholly correct, he is quite right in stressing the importance of phosphorus as an essential constituent of all life. Bones were likely among the first mineral source of phosphorus to be used in agriculture. They were used for many centuries, but peak demand in Europe and Great Britain began in the early 1800's.

As the demand for bones increased in England, the home supply became insufficient and annual imports of bones from the Continent were reported to have been 25 to 30 Gg by 1842. This activity is said to have provoked Justus von Leibig into accusing England of scouring the battlefields and ancient burial grounds of the Continent for bones. The following version of Leibig's statement is quoted from Superphosphate, Its History, Chemistry, and Manufacture (USDA and TVA, 1964).

> "England is robbing all other countries of the conditions of their fertility. Already in her eagerness for bones, she has turned up the battlegrounds of Leipzig, of Waterloo, and of the Crimea; already from the catacombs of Sicily she has carried away the skeletons of many successive generations. Annually she removes from the shores of other countries to her own the manurial equivalent of three and a half millions of men, whom she takes from us the means of supporting, and squanders it down her sewers to the sea."

In North America, buffalo bones were collected for fertilizer on the western prairies of the United States and Canada as late as 1890 (C.S. Slater, personal communication, 1956).

Although two Swedish chemists, Carle Scheele and Johann Galen, had shown in 1769 that calcium phosphate was the principal constituent of bones, the first commercial superphosphate fertilizer did not appear until 1843. John Bennet Lawes (1814—1900), an Oxford graduate, is credited with developing the process that consists of treating phosphate rock (PR) with sulfuric acid to produce ordinary superphosphate. Development of this process, i.e., treatment of phosphatic materials with sulfuric acid, was the world's first successful venture into the manufacture of chemical fertilizers and is the basis of today's phosphate fertilizer industry.

World phosphorus reserves. Phosphorus constitutes 0.10 to 0.12% of the earth's crust and lies twelfth in order of abundance among chemical elements.

The world is rich in phosphate deposits. Phillips and Webb (1971) estimate that reserves ranging upwards of 50 Pg of PR can be mined economically with present-day technology. Other estimates of total minable reserves are 117 Gg (Anon, 1971) and 1.2 Pg (Emigh, 1972). These wide variations in estimates can be explained by widely varying definitions of the term "minable reserves". There are now many factors that determine just what is or is not minable. These factors include concern for environmental and aesthetic values, the P_2O_5 content and other quality factors of PR, the market price of P, depth of overburden, etc. Most PR deposits might be termed minable if society is willing or finds it necessary to pay the price.

World production of PR in 1974 was approximately 110 Tg, having more than doubled during the past decade (FAO, 1976, p. 74). Principal world reserves are now mined in North Africa, North America, and the Kola Peninsula in the U.S.S.R. Minable reserves are scattered widely throughout the world. New deposits of PR are currently being discovered, notably in South America and in India.

Production of phosphate fertilizers. The principal phosphate deposits occur for the most part as apatite species, varying widely in their chemical and physical characteristics, as well as crystallographic properties.

The apatite bonds in PR must be broken to convert the phosphate to a form that is readily available to crop plants. This can be accomplished by heating (thermal process) or by acidulation of PR with a strong mineral acid, the latter being known as "the wet process." In theory, any strong mineral acid can be used to acidulate PR, but sulfuric and phosphoric acids are generally preferable for a variety of reasons.

Wet process phosphoric acid results from the principal reaction that takes place when PR is treated with H_2SO_4:

$$Ca_{10}(PO_4)_6F_2 + 10\ H_2SO_4 + 20\ H_2O \rightarrow 10\ CaSO_4 \cdot 2\ H_2O + 6\ H_3PO_4 + 2\ HF \qquad (8)$$

The product H_3PO_4 (about 32% P_2O_5 equivalent) is drawn off and concentrated to the desired strength.

Triple superphosphate (also termed treble, double, or concentrated superphosphate) is an impure monocalcium phosphate made by reacting phosphoric acid with PR:

$$Ca_{10}(PO_4)_6F_2 + 14\ H_3PO_4 \rightarrow 10\ Ca(H_2PO_4)_2 + 2\ HF \qquad (9)$$

The P_2O_5 equivalent ranges from 44 to 52%, almost all of the P is water-soluble, and the material has excellent physical properties.

Ammonium orthophosphates are highly concentrated sources of plant food containing as much as 18% N and 48% P_2O_5. Ammonium phosphates are generally made by reacting phosphoric acid and anhydrous ammonia. Potash salts may be added if desired and the final product is granular and free flowing.

Superphosphoric acid is made by dehydrating orthophosphoric acid to form a highly concentrated product having a P_2O_5 equivalent as high as 79% compared with 54% P_2O_5 equivalent of the unconcentrated acid. Higher-analysis materials have the obvious advantage of lower freight costs per unit of plant nutrient and also result in lower application costs. Wet process superphosphoric acid is used to prepare high-analysis liquid fertilizers and granular ammonium polyphosphates; the latter are prepared by ammoniating superphosphoric acid or by reacting NH_3 and H_3PO_4 at high temperature. Superphosphoric acid has a sequestering effect in fertilizer solutions, which is important in maintaining impurities in solution.

Nitrophosphates, made by treating PR with nitric acid, are produced in significant quantities in Europe, but to a much lesser extent in the U.S.A.

The *electric furnace method* is the most important thermal method for converting the P in PR to a plant-available form. This process provides P and H_3PO_4 of very high purity. Electric-furnace materials are now used primarily for non-fertilizer purposes, such as detergent production, since the energy input required by an electric furnace is of the order of 43.2 kJ g^{-1} of recovered P (Bixby et al., 1964). This, plus the relatively large plant investment, rules out the electric furnace method for fertilizer production alone. Accordingly, virtually all of the phosphoric acid used in fertilizer manufacture is now made by the "wet process" or "wet acid" procedure that initially involves the treatment of PR with sulfuric acid to break the apatite bonds and produce phosphoric acid. This procedure results in a production cost that is about half that of thermal plants having the same capacity.

World consumption of P_2O_5 in 1975 as a plant nutrient was approximately 22.9 Tg.

Potassium

Potassium (K), often referred to as potash (K_2O), is required by plants in large quantities, since it is essential to the complex biochemistry of plant growth. Up to the middle of the 19th century, miscellaneous materials, such as flue dust, kelp, tobacco stems, and wood ashes were used as sources of K by agriculturists. The K content of these waste materials was very low, ranging from 2 to 10% K_2O.

It is of interest to note that the first patent issued in the U.S.A., signed by

George Washington on July 31, 1790, was a process for extracting potash from furnace ashes. *

The first discovery of water-soluble K salt deposits occurred in Germany in 1856 and the potash industry was born at that time. In the absence of discoveries of similar deposits elsewhere, Germany held a nearly complete monopoly on the industry until about 1930 (Turrentine, 1946).

World potassium reserves. The world reserves of K are now estimated at over 48 Pg of recoverable K_2O. Canadian deposits represent 37% of the world's known reserves, Russian deposits 49%, and the remainder is distributed among the United States, Europe, Asia, Africa, and South America (Adams, 1968). There are unproven deposits of K salts in Tunisia and Libya, as well as subsurface brines in Niger and Nigeria. Undoubtedly there are other significant reserves as yet undiscovered in sparsely explored areas, such as Central America, Siberia, eastern Canada, and Australia.

Potassium sources and their retrieval. The average K content of the earth's crust is 2.4%, but large quantities of soluble K salts are used in the production of agricultural crops.

Potassium salt deposits are the major raw material source for K fertilizers. These undergound deposits were formed by the evaporation of sea water; hence, NaCl is always present as a major contaminant. Magnesium salts are also present and these are combined with sulfates and/or chlorides. Potassium chloride (sylvite) is the predominant K form in most deposits, but K also occurs as langbeinite (New Mexico) and as kainite (Sicily) to a limited extent. Some of the langbeinite and virtually all of the kainite is converted to K_2SO_4 (Barber et al., 1971). However, sylvinite ores are the principal reserves that are economically exploitable and these are judged to be our principal source of K for the next several decades (Adams, 1968).

Underground deposits of solid K minerals are mined by (i) *conventional shaft mining* procedures involving recovery of the ore in solid form, and (ii) *solution mining techniques* whereby the K-bearing ore is dissolved and brought to the surface in the form of a solution. Shaft mining is a very costly procedure, although the newer mines employ continuous mining machines which have largely replaced the conventional mining techniques. The function of these machines is to remove ore continuously from the K-ore vein. Some machines of the boring type have been developed that can remove up to 5 Mg of ore min^{-1}. The ore is then crushed and brought to the surface for further processing. Shaft mining is generally restricted to depths of about 1,200 m or less because of safety and cost considerations (Kapusta, 1968).

---

* Original is in Chicago Historical Society Collection.

In *solution mining*, hot brine containing NaCl and some KCl is injected into the ore vein where it dissolves more KCl than NaCl. With clarification and cooling, KCl crystallizes but the NaCl stays in solution. The brine is reheated and recirculated to continue the process. Deeper deposits may be mined by this technique and greater total recovery of K is possible than in conventional room-and-pillar shaft mining.

Use of potassium in fertilizers. The current world consumption of K_2O is approximately 20 Tg y^{-1} (Table 9.5). The amount of K applied per sq. metre of arable land varies widely among different countries of the world. In 1966, for example, Italy averaged about 1 gK m^{-2}, while Belgium applied 16 gK m^{-2}. The Netherlands and West Germany also apply high rates of K, while France, Denmark, the United Kingdom, and the U.S.A. average 6 gm^{-2} or lower. However, the U.S.A. uses about 25% of the world's present consumption (Barber et al., 1968).

Estimates indicate that KCl accounts for over 95% of the fertilizer K consumed throughout the world; the K content of this salt ranges from 50 to 52%.

Potassium salts other than KCl are used to some extent in agriculture. Agricultural K_2SO_4 is used on Cl-sensitive crops and is produced by five different methods. The production of K_2SO_4 from the double salt, langbeinite ($K_2SO_4 \cdot 2MgSO_4$), is based on the reaction:

$$K_2SO_4 \cdot 2\,MgSO_4 + 4\,KCl \rightarrow 3\,K_2SO_4 + 2\,MgCl_2 \qquad (10)$$

The Trona process involves the production of K_2SO_4 from burkeite ($Na_2CO_3 \cdot 2Na_2SO_4$) and KCl in a two-stage process.

The Hargreaves process produces K_2SO_4 directly from KCl by passing SO_2, air, and water vapor through beds of KCl briquettes.

The Mannheim furnace process is employed to produce K_2SO_4 according to the following reactions:

$$KCl + H_2SO_4 \rightarrow KHSO_4 + HCl \qquad (11)$$

$$KHSO_4 + KCl \rightarrow K_2SO_4 + HCl \qquad (12)$$

Potassium sulphate is produced in Italy from kainite, a double salt represented by $KCl \cdot MgSO_4 \cdot 3\,H_2O$. The kainite is first leached with brine to yield $MgCl_2$ solution and solid phase $K_2SO_4 \cdot MgSO_4 \cdot 3\,H_2O$. After separation, this salt is leached with fresh water to extract the more soluble $MgSO_4$.

Potassium nitrate (KNO_3) is produced by the reaction of $NaNO_3$ and KCl. The KNO_3 salt can also be prepared by reacting HNO_3 or N oxides with KCl. It is also a coproduct in Chilean sodium nitrate (Barber et al., 1971).

Sulfur

Sulfur has been used by man since remote antiquity, being referred to in the Illiad and Odyssey of 3 ky ago as a neutralizer of evil and a purifier of air. The ancients used S for bleaching linen, as a paint ingredient, for preparing gunpowder and incendiary warfare materials. The element has also played an important role in temple sacrifices and purification rits because of its unique flame and pungent odor.

Sulfur is effective in controlling various plant diseases and a limited number of insects, although its use for these purposes is rapidly declining as a result of the development of synthetic fungicides and other pesticides.

Sulfur has been recognized since the mid-1800's as an essential element in plant nutrition, crop requirements being similar to P needs. Sulfur is also used as a soil amendment in the reclamation of saline and alkali soils.

Presently, the largest use of S in agriculture is fertilizer production, H_2SO_4 being used to convert PR to superphosphate and in producing wet process phosphoric acid.

World sulfur reserves. The earth's crust contains about 0.6% S, where it occurs as elemental S (brimstone) in deposits associated with gypsum and calcite; combined S in metal sulfide ores and mineral sulfates; as a contaminant in natural gas and crude oils; as pyritic and organic compounds in coal; and as organic compounds in tar sands (Tisdale and Nelson, 1966). The elemental form commonly occurs near active or extinct volcanoes, or in association with hot mineral spings. Estimates by Holser and Kaplan (1966) of the terrestrial reservoirs of S suggest that about 50% of crustal S is present in relatively mobile reservoirs such as sea water, evaporites, and sediments. The chief deposits of S in the form of brimstone and pyrites are in Western European countries, particularly in France, Spain, Poland, Japan, Russia, U.S.A., Canada, and Mexico. World production of S in the form of brimstone and pyrites was approximately 41 Tg in 1973; other sources accounted for about 8 Tg, making a total of 49 Tg (Anon, 1973). Byproduct S from sour-gas, fossil fuel combustion, and other sources now accounts for over 50% of S used by western countries, as shown in Fig. 9.1. This percentage may increase as pollution abatement measures increase the removal of SO_2 from fossil fuel, particularly in the U.S.A. Atmospheric S, returned to the earth in rainwater, is also a very important source of S for plants.

Soils commonly contained 0.1—2.0 mg S g^{-1} (Table 9.4). Most of this S is in soil organic matter, where it is component of certain organic compounds.

The world has vast S resources, the minable fraction being governed by economics, i.e., product price and demand. Apparent identified reserves, excluding coal and gypsum, total an estimated 2 Pg (Bixby, 1977). In addition, sea water containing about 0.27% SO_4-S is estimated to hold 1,300 Eg of S (Holser and Kaplan, 1966).

Sulfur mining. The most common method of mining S is by the Frasch superheated water method, invented about 1900. Four concentric pipes are inserted into the deposit dome. Hot water is forced by compressed air into

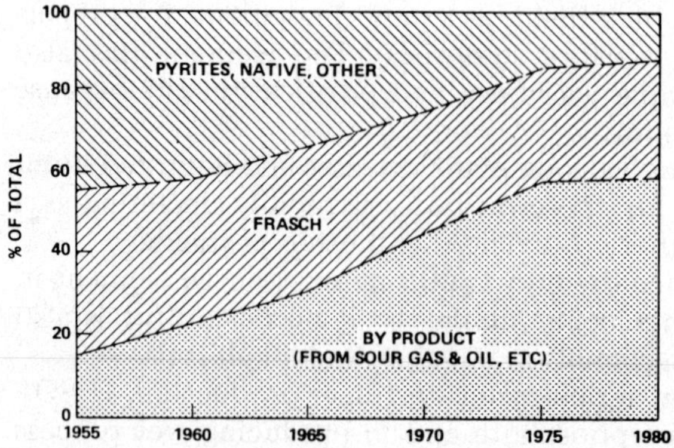

Fig. 1. Sources of sulfur in western nations, adapted from Manderson (1970).

the deposit through the two outside pipes causing it to melt. The melted S of about 99% purity is forced to the surface, where it cools and solidifies on a conveyor belt moving over shallow tanks of cold running water or is directed to large vats in which it cools and solidifies. Sulfur is also handled and shipped in the molten state in heated tank cars (Tisdale and Cunningham, 1963). More recently, nuclear energy has been suggested for mining S occurring in domes. This proposal involves the thermal energy dissipated in subterranean nuclear blasts within or just below a dome S deposit. The heat is sufficient to convert S to the liquid state and maintain it in this condition over a long period of time (Dale and Dehart, 1960).

The use of sulfur in fertilizers. Fertilizer production consumes 40 to 50% of the world's production. The use of S for this purpose is estimated to range around 30 Tg by the year 1985 (Bixby, 1977); most P fertilizer production processes require H_2SO_4. This S requirement varies, depending upon the composition of PR, the process used, and the kind of phosphate fertilizer produced (Table 9.8). Production of these fertilizers was discussed in the previous section on P.

An unknown amount of S is applied to world crops as fertilizer. Unlike the nutrients N, P, and K, statistics on S fertilizer use are not commonly available. However, the incidence of S deficiency in soils is increasing and the response of crops to S fertilization is becoming increasingly widespread, particularly in North America, Africa, and Australasia. Use of S in Australasia was recently reviewed by McLachlan (1975).

Plant roots absorb S almost entirely in the sulfate form. Thus, S applied to soils in other forms such as elemental S or as polysulfides, for example, must be converted to SO_4 before plant absorption is possible. The rate of conversion of non-sulfate compounds depends upon soil temperature, moisture, and particle-size distribution of the applied material. Under good grow-

TABLE 9.8

Sulfur requirements for phosphate fertilizers (Adapted from Hignett, 1967.)

Product	Surfur requirement, (g S g^{-1}) of available P_2O_5
Normal superphosphate	0.62
Triple superphosphate	0.67 [a]
Wet-process phosphoric acid	0.94
Ammonium phosphate	0.96 [a]
Mixed acid nitrophosphate	0.4—0.6 [b]

[a] Assuming that wet-process phosphoric acid is used.
[b] For products ranging in P_2O_5 water solubility from 0 to 40%.

ing conditions, this conversion is usually rapid when fine materials are mixed with soil, being virtually completed in a matter of a few weeks.

Sulfur fertilizers include those that exist in nature such as K_2SO_4, $K_2SO_4 \cdot MgSO_4$, and elemental S, as well as a long list of synthetic materials. Bixby and Kilmer (1975) list some 40 materials that are suitable S sources for direct application, for formulating multinutrient solid and fluid fertilizers — or for application in irrigation water. Table 9.9 lists some of the more commonly used S fertilizers and their nutrient content. Ammonium sulfate now accounts for nearly 4 Tg of plant-nutrient S, worldwide (Tisdale and Cunningham, 1963). Normal superphosphate (18 to 20% P_2O_5) contains 10 to

TABLE 9.9

Examples of some sulfur-containing fertilizer materials

Material	S content (%)
Ammonia-sulfur solution	10
Ammonium bisulfite	32.3
Ammonium polysulfide solution	40
Ammonium sulfate	24.2
Ammonium thiosulfate	43.3
Gypsum (hydrated)	18.6
Lime sulfur solution	57
Magnesium sulfate	13
Potassium sulfate	17.6
Potassium magnesium sulfate	18.3
Sulfuric acid (100%)	32.7
Elemental sulfur (brimstone)	100
Sulfur dioxide	50
Superphosphate, normal	10—14

14% "incidental" S. The use of normal superphosphate has delayed the occurrence of S deficiencies in many areas of the world, even though its role as a P source is steadily decreasing because of its relatively low analysis. Normal superphosphate is still an important fertilizer in Australia.

Gypsum is not widely used as an S source because of its low S content (13 to 19%). It is a useful fertilizer for peanuts, a crop that has high Ca requirements. Large tonnages of gypsum are used, however, in the reclamation of saline soils according to the following reaction:

$$2\ NaX + CaSO_4 \rightleftharpoons CaX_2 + Na_2SO_4 \qquad (13)$$

where X represents the soil-exchange complex. This replacement of Na by Ca tends to flocculate the soil, improving water permeability. The Na_2SO_4 is carried out in drainage water.

Recently the Tennessee Valley Authority has developed S-coated urea, a slow-release N fertilizer. The S also becomes available to plants slowly, making it particularly useful as an N-S combination fertilizer.

Calcium

Calcium is an essential element for all higher plants, but it is often overlooked as a plant nutrient because Ca levels are usually adequate for crops grown on soils above pH 5.0. The specific physiological functions of Ca are not clearly defined, but a deficiency results in the failure of terminal bud development. A paramount role played by Ca appears to be that of maintaining membranes and hence, intracellular organization (Epstein, 1972).

The Ca content of soils varies widely, ordinarily ranging from about 0.07 to 3.60%. Calcium is contained in a number of soil minerals including dolomite, calcite, Ca feldspars, apatite, amphiboles and many others. Coarse-textured soils in humid regions, particularly those formed from rocks low in Ca minerals, are generally low in Ca. In spite of this, Ca deficiencies in crops do not appear to be of widespread occurrence, although a number of other factors affect nutrition in low base status soils. Fertilizers are not generally manufactured specifically to provide Ca as a plant nutrient, since this element is economically supplied by periodic applications of agricultural lime, as discussed subsequently.

Magnesium

Magnesium is a constituent of chlorophyll and acts as an activator of numerous plant enzymes. Thus, chlorosis in plants is an early symptom of Mg deficiency, as well as weak stalks and premature leaf and fruit drop.

Magnesium constitutes about 1.9% of the earth's crust, and soils commonly contain 0.1 to 1.5%. Fine-textured and/or semi-arid soils formed from high Mg parent materials may contain as much as 3 to 4% Mg.

Soil Mg originates from the decomposition of minerals such as dolomite, biotite, olivine, serpentine, and chlorite. Soluble Mg may be held in the exchangeable form on soil clays or organic matter.

Acid and sandy soils, as well as high K soils, may exhibit Mg deficiency, the latter being the result of K-Mg antagonism, particularly on low-Mg soils.

The most commonly used Mg source other than dolomite ($CaCO_3 \cdot MgCO_3$) is $MgSO_4$, although Mg can also be added as MgO and $Mg(OH)_2$. Application rates vary with soils and crops, but 3—4 g Mg m^{-2} as $MgSO_4$ applied every 3 to 4 years has been recommended (Mortvedt and Cunningham, 1971). Like Ca, Mg is economically supplied on acid soils by periodic applications of dolomitic limestone.

Liming materials

In addition to being essential elements in plant growth, Ca and Ca-Mg compounds capable of reducing soil acidity are widely used as liming materials. Strictly speaking, lime refers only to CaO, but agricultural lime also includes $Ca(OH)_2$, $CaCO_3$, $CaCO_3 \cdot MgCO_3$ (dolomite), and calcium silicate slags. Dolomitic limestone seems to be the preferred liming material for correcting soil acidity and at the same time supplying Ca and Mg to crops.

Marl deposits were among the first geologic materials to be applied to agricultural soils, having been used by agriculturists in the pre-Christian era.

The Romans apparently learned this practice from the Gauls and the Greeks, as evidenced by early Roman writings (Tisdale and Nelson, 1966). In the U.S.A., Edmund Ruffin carried out a series of liming experiments with marl on his Virginia plantation in the early 1800's. One historian (Sitterson, 1961) described Ruffin's early experiments:

> "On a February morning in 1818 his carts began to haul the marl that puzzled Negro hands dug from pits hastily opened on his lower lands. They spread some two hundred bushels over a few acres of newly-cleared, but poor, ridge land, and in the spring he planted the entire field to corn as a testing crop. Eagerly he waited. As the season advanced, he found reason for joy. From the very start the plants on marled ground showed marked superiority, and at harvest time they yielded an advantage of fully forty percent. The carts went back to the pits. Fields took on fresh life. A new era in the agricultural history of the region had dawned."

However, modern concepts of soil acidity were not developed until the middle of the 20th century. In the U.S.A., Jenny, Stout, Coleman, Harward, and others contributed heavily to a better understanding of the nature of soil acidity and related aspects of soil chemistry and plant nutrition (Coleman and Thomas, 1967). Soil pH has a profound influence on the availability of nutrients to crops, since chemical, microbiological and, to a lesser extent, physical properties of soils are affected by the degree of soil acidity or alkalinity that prevails.

TABLE 9.10

Approximate amounts of agricultural lime used for direct application to soil in the U.S.A., 1955—1975 (Anon., 1975.)

Year	Amount (Tg)
1955	18.6
1960	20.4
1965	25.2
1970	25.9
1975	31.3

Statistics on the amounts of lime used on a worldwide basis are not available. In the U.S.A., reliable records of lime use were not available prior to 1929 (Woodruff, 1967). More recent figures set forth in Table 9.10 reveal a slow rise in the use of lime in the U.S.A. (Anon., 1975). Steadily increasing amounts of N fertilizer use tend to increase soil acidity; so the problem is becoming more acute with time. It is estimated that the amount of agricultural lime needed in the U.S.A. to bring soil pH to the desired range is three times the present annual usage. This could be accomplished by increasing lime usage by 40 to 50% for a period of 5 to 7 years, after which annual maintenance applications should slightly exceed present usage.

Micronutrients

The elements included in the class of plant nutrients known as micronutrients are Mn*, Mo*, Cu*, B*, Zn*, Fe*, Co*, Cl and V — elements marked with an asterisk being most commonly applied as fertilizers.

Since micronutrient fertilizers must be applied at very low rates to avoid plant toxicity, the prevailing practice is to incorporate them with macronutrient fertilizers. This can be accomplished by incorporation with solid fertilizers during the manufacturing process, by bulk blending with granular fertilizers, or by mixing with fluid fertilizers just prior to application. Care must be taken to select micronutrient sources and macronutrient fertilizers that are compatible to avoid the formation of reaction products that reduce availability of nutrients to plants (Mortvedt and Cunningham, 1971).

Inorganic sources include naturally occurring ores, as well as manufactured products that include oxides and metallic salts such as sulfates, chlorides, and nitrates. After grinding, the tri- and tetravalent oxides may be roasted in a kiln and reduced to divalent forms.

Organic sources include natural organic complexes that are byproducts of the wood pulp industry, as well as synthetic chelates which are generally more effective than inorganic salts. The synthetic chelates are relatively expensive to use.

TABLE 9.11

Amounts of micronutrients sold for fertilizers in the U.S.A., fiscal year 1975—1976 (USDA, 1976b.)

Micronutrient	Amount (Mg) (elemental basis)
Copper	605
Iron	1,830
Manganese	11,748
Molybdonum	110
Zinc	17,401

Fritted glasses or frits are prepared by mixing and fusing finely-ground micronutrients and silicate or glass powders. Solubility of the metallic salts in the frits is controlled by composition of the matrix and by particle size. These can contain several micronutrients to provide custom mixes for vari-

TABLE 9.12

Some commonly used micronutrient fertilizers

Micronutrient			Approx. micro-nutrient content (%)
Boron	Borax	$Na_2B_4O_7 \cdot 10\ H_2O$	11
	Sodium pentaborate	$Na_2B_{10}O_{16} \cdot 10\ H_2O$	18
	Colemanite	$Ca_2B_6O_{11} \cdot 5\ H_2O$	10
Copper	Copper sulfate	$CuSO_4 \cdot 5\ H_2O$	25
	Chelate	$Na_2CuEDTA$ [a]	13
	Polyflavonoids	—	5—7
Iron	Ferrous sulfate	$FeSO_4 \cdot 7\ H_2O$	19
	Ferrous ammonium sulfate	$(NH_4)_2SO_4 \cdot FeSO_4 \cdot 6\ H_2O$	14
	Chelates	NaFeEDTA [a]	5—14
	Polyflavonoids	—	9—10
Manganese	Manganese sulfate	$MnSO_4 \cdot 3\ H_2O$	26—28
	Manganese oxide	MnO_2	63
	Manganese chelate	MnEDTA [a]	10—12
Molybdenum	Sodium molybdate	$Na_2MoO_4 \cdot 2\ H_2O$	39
	Ammonium molybdate	$(NH_4)_6Mo_7O_{24} \cdot 4\ H_2O$	54
	Molybdenum frits		2—3
Zinc	Zinc sulfate	$ZnSO_4 \cdot H_2O$	35
	Zinc oxide	ZnO	78
	Zinc frits	(silicates)	varies
	Zinc chelate	$Na_2ZnEDTA$ [a]	14
	Polyflavonoids	—	10
	Ligninsulfonates	—	5

[a] EDTA = Ethylenediamine tetraacetic acid.

ous crops. Finely-ground metals are also used as micronutrient sources, but generally these are not as satisfactory as other sources (Mortvedt and Cunningham, 1971).

Foliar applications of micronutrients are useful for rapidly correcting deficiencies, or for plant utilization of some nutrients such as Fe, which may be immobilized when applied to calcareous soils. Foliar application may also be more practical for valuable horticultural crops such as fruit trees, vegetable crops, and ornamentals. Great care must be exercised in applying foliar sprays because of the possibility of damage to foliage.

With the possible exception of Fe — Julius Sachs having discovered its essentiality to plant growth in 1860 — much of our present knowledge concerning the role of micronutrients was acquired during the present century (Stout, 1972). These nutrients, although required by plants in relatively small amounts, are none the less essential for plant growth. In general, there is quite a narrow range between sufficiency and toxicity of micronutrients.

Consumption data for micronutrients are not available, but amounts sold for fertilizers in the U.S.A. during fiscal year 1975—1976 are shown in Table 9.11. The range of concentrations of micronutrients usually found in soils is shown in Table 9.4. Commonly used micronutrient fertilizers are shown in Table 9.12. Data for trace elements in a wide range of fertilizers have been compiled by Swaine (1962).

Boron

Commercial deposits of B include accumulations of hydrated borate minerals formed by the evaporation of sea water or salt lakes. The latter deposits are the most important, yielding borates of Na or Ca, while the marine deposits yield Mg borates. Sea water contains an average of 4.6 $\mu g\ g^{-1}$, ranking twelfth in abundance among the dissolved constituents (Goldberg, 1965).

Boron's role in plant nutrition is an essential one, but the specific role that it plays is still in question. The most persistent hypothesis is that B facilitates the transport of sugars through membranes (Gauch and Dugger, 1954).

Rates of B applied to crops vary over a narrow range; representative studies in the U.S.A. indicate optimum rates of 0.06—0.32 g m^{-2}. Sodium tetraborate ($Na_2B_4O_7 \cdot 10H_2O$) is the most commonly used B source, but colemanite ($Ca_2B_6O_{11} \cdot 5H_2O$) is often used on sandy soils where excessive leaching is a problem (Murphy and Walsh, 1972).

Copper

The stable forms of Cu in the earth's crust are chiefly sulfides: the most abundant Cu mineral is chalcopyrite, $CuFeS_2$ (Krauskopf, 1972). Copper is strongly retained by soils (retention may be increased by the application of lime) but is low in soils having a low cation-exchange capacity. The concentration of Cu in soils is reported to range from 2 to 100 $\mu g\ g^{-1}$ (Table 9.4). Copper deficiencies usually occur on organic soils, although Cu deficiencies have been found on mineral soils, particularly very sandy ones. Copper is a metal activator of several enzymes and a deficiency also causes Fe accumulation in corn plants, particularly the nodes. There is a growing list of Cu proteins such as ascorbic acid oxidase, lactase, and plastocyanin (Price et al., 1972).

Recommended rates of Cu for various crops range from 0.2 to 1.45 g m^{-2} applied to soils broadcast or banded (Murphy and Walsh, 1972). Grain crops tend to be sensitive to low levels of Cu, as are citrus, clover, tobacco, peanuts, and grasses in some areas. Copper may also be applied to crops in foliar sprays as $CuSO_4 \cdot 5H_2O$, or in chelated, polyflavonoid and ligninsulfonate froms. Residual effects of Cu applied to soils are long lasting and, under field conditions, a single application may be adequate for several years. Some commonly used Cu fertilizers are listed in Table 9.12.

Iron

Iron is often described as a ubiquitous element since it forms numerous stable compounds with S and with silicates. Iron represents about 5.0% of the earth's crust, occurring as hematite, magnetite, limonite, siderite, and pyrite.

The total amount of Fe in soils varies widely, ranging from 200 $\mu g\ g^{-1}$ to more than 10%. In soils, Fe is found as oxides, hydroxides, and phosphates, as well as in the lattice structure of primary silicate and clay minerals.

Iron is required for the maintenance of chlorophyll in plants; thus, plants that are severely Fe-deficient exhibit visual chlorotic symptoms that are often quite spectacular. Iron deficiency is most often encountered on calcareous soils, although it may occur on acid soils that are extremely low in total Fe. Soil applied inorganic Fe salts are generally not very effective in correcting deficiencies, since the Fe may be rapidly converted to highly insoluble compounds. Like Cu, organic sources of Fe are generally more effective for soil application and are useful for foliar application as well. Advantages of foliar application include economy of materials, rapid plant response and elimination of soil reactions that result in the formation of insoluble Fe compounds.

Spray or foliar applications of Fe are usually recommended over soil applications in the U.S.A. A 3 to 4% solution of $FeSO_4$ or its equivalent, including a wetting agent, is a common recommendation. There has also been some experimental work directed toward applying Fe chelates to crops in sprinkler irrigation water. Rates up to 0.45 g Fe m^{-2} have been applied in this manner (Murphy and Walsh, 1972).

Manganese

Manganese is very similar to Fe in both its behavior and occurrence, although it ranks twelfth in abundance among the elements that constitute the earth's crust. The ore minerals of Mn are chiefly oxides such as pyrolusite, hausmannite, and manganite: Mn also occurs as carbonates and silicates. Concentrations of Mn in soils usually range from 0.2—3 mg g^{-1}. Manganese deficiencies most often occur on well-drained neutral or calcareous soils, as well as organic soils or mineral soils high in organic matter content. Small grains and large-seeded legumes are especially sensitive to low levels of available Mn in soils, although Mn deficiencies have also been reported on a variety of field, vegetable, and fruit crops. Like Fe and other heavy metals, Mn functions in the activation of numerous plant enzymes, alone and with other metals. More recently, Mn has been shown to function in certain photochemical processes such as the Hill reaction (Tisdale and Nelson, 1966).

Manganese may be soil- or foliar-applied, but soil application is the most common method for correcting Mn deficiencies. Optimum rates of Mn to be applied depend on soil pH, Mn source, soil organic matter content, and method of incorporation with soil. Soluble Mn salts quickly revert to unavailable forms when applied to Mn-deficient soils, so the method of application is important. The optimum rates reported in the literature vary widely, cotton responding to as little as 0.2 g Mn M^{-2} when side-dressed. Onions required an optimum rate of only 0.9 g of Mn m^{-2} when banded, but 6.8 g m^{-2} when broadcast (Murphy and Walsh, 1972). Foliar application is a very efficient way to correct Mn deficiency; much lower rates are required than for soil application.

Manganese sulphate is the most commonly used Mn source, while chelated forms have been used successfully as a foliar application. Manganese oxide, Mn frits, and $MnCO_3$ are less widely used at present.

Molybdenum

Molybdenum is present in the lithosphere and soils in extremly small quantities, averaging about 2 $\mu g\ g^{-1}$ (Swaine, 1955). This element is required by plants in very small quantities, but Mo deficiencies are quite common, particularly in Australia, New Zealand, and North America. Deficiencies are most frequently noted on legumes, since Mo is essential for symbiotic N-fixation. However, Mo deficiency has also been observed on many other agronomic and horticultural crops. The availability of Mo to crops is greatly influenced by soil pH: availability is increased by application of lime. Vegetation grown on soils high in Mo content may be toxic to livestock, but Mo toxicity in plants appears to be very rare.

The amounts of Mo required to correct deficiencies in crop plants are very small. Hence, Mo salts are usually added with fertilizers or lime. Seed treatment has also been used with some success.

Rates of application are very low, being of the order of a few cg or less of Mo m^{-2}. This low rate often poses problems in achieving a uniform distribution over fields. Soil or seed application of Mo is common, although foliar sprays may be used in emergency situations. Sodium and ammonium molybdates are the most widely used sources of Mo. Molybdenum frits are also effective in correcting deficiencies.

Zinc

The Zn content of the earth's crust is estimated to average about 80 μg g^{-1}; soils usually contain 10 to 300 μg g^{-1} total Zn. Zinc deficiencies are quite common in many crops grown on calcareous or organic soils, or where soils have been leveled for furrow or flood irrigation. Deficiencies have also been observed on soils that are excessively high in P.

The principal function of Zn in plant nutrition is the activation of enzymes; at least nine of these have been identified thus far. Soil application of Zn is the most common and generally successful method used for crop fertilization. Foliar application is generally used as a temporary emergency measure, or as an amendment to soil application.

Relatively large amounts of Zn may be applied to soils, since residual effects from a single application may last 5 years or longer. Zinc in excessive quantities can be toxic to plants, although this appears to be a rare occurrence, particularly in alkaline soils. Recommended rates generally vary from 0.2—2.2 kg of Zn m^{-2} where inorganic sources are used. Lower rates (0.0—0.6 kg of Zn m^{-2}) can be used for chelated or other orgamic sources (Mortvedt and Cunningham, 1971).

The most commonly used Zn sources are sulfates, oxides, chelates, frits, polyflavonoids, and ligninsulfonates.

MINERALS AND ANIMAL NUTRITION

The amounts of minerals that are added as supplements to animal and poultry rations are small when compared with minerals used in fertilizers. Quantitative data relating to this use of minerals are scarce to non-existent, but it appears that this phase of use will become increasingly important in the future.

Animals respond in improved health and growth to 20 different elements, namely Ca*, Cl*, Cr, Co*, Cu*, F, I, Fe*, Mg*, Mn*, Mo, Ni, P*, K, Se*, Si, Na*, Sn, V, and Zn*. These are in addition to N and S which are required as essential amino acids and C, H, and O obtained from water and air. The minerals used most commonly in feed formulations are marked with an asterisk.

General functions of mineral elements in animal nutrition (Pfost and Swinehart, 1970) are as follows:

(1) Constituents of bone and teeth components of the skeleton.
(2) Structural constituents of soft tissue.
(3) Function in irritability of nerve and muscle.
(4) Maintenance of osmotic pressure.
(5) Absorption, transport, excretion, and metabolic mechanisms.
(6) Regulation of pH of blood and tissue.
(7) Catalysts.
(8) Cofactors in metabolism.
(9) Activation of enzyme systems.

Daily Ca, P, and NaCl requirements of some animals and poultry are given in Table 9.13. Limestone is the most common Ca source used in feed supplements, while P sources include defluorinated mono- and dicalcium phosphate, as well as steamed bone meal. Commercial salt is used as a source of Na and Cl.

Zinc and Cu, as well as Co, are also assuming increasing importance in animal nutrition. In New Zealand, Zn supplementation of farm animals has decreased cow and calf losses and has alleviated facial eczema in sheep. In the U.S.A., weight gains in beef cows and calves have been improved by Zn additions to feed (Allaway, 1975). Farm animals are often supplied with Cu in the form of dietary mineral supplements. In the U.S.A., Co is usually added to mixed feeds or salt licks where needed.

Two other elements of unusual interest and importance in animal nutrition are Mg and Se. The most urgent problem associated with Mg is grass tetany, particularly in lactating cows. This disease is caused by a "conditional Mg deficiency" in animals, due to a low intake of this element or inef-

TABLE 9.13

Daily Ca, P, and NaCl requirements of some farm animals and birds (Seiden and Pfander, 1957.)

	Ca (g)	P (g)	NaCl (g)
Swine	16—34	11—23	14—28
Dairy cattle	4—20	3—15	21—40 [a]
Beef cattle	15—20	12—20	30—38 [a]
Sheep	3— 7	2— 5	9—13
	% of diet		
Chickens	1.0—2.25	0.6	0.5
Turkeys	2.0—2.25	0.75—1.0	0.5
Laying hens	4.0—5.0	—	—

[a] These data from Morrison (1948).

fective utilization of dietary Mg within the animal. In instances where Mg fertilization of pastures prove ineffective in preventing tetany, direct supplementation of animals with Mg is required.

Extremely small amounts of Se are required by warm-blooded animals, but Se is also highly toxic in larger amounts. Two livestock diseases known as "blind staggers" and "alkali disease" were identified as manifestations of acute and chronic Se poisoning (Robinson, 1933). New Zealand scientists McLean et al. (1959) discovered that muscular dystrophy in lambs and calves could be prevented by Se therapy. Selenium deficiency areas were later identified in many other countries and it soon became apparent that these areas were of far greater extent than those affected by excessive Se levels.

Selenium is essential for growth and fertility in animals, as well as for general animal health. Grazing sheep and cattle require a dietary intake of feed containing 0.1 $\mu g\ g^{-1}$ Se. A minimum requirement of 0.06 $\mu g\ g^{-1}$ Se prevents white muscle disease in lambs. Inorganic salts such as selenates and selenites are effective sources of dietary Se. Pasture grasses and forages containing more than 0.1 $\mu g\ g^{-1}$ Se keep animals free from Se-responsive diseases (W.H. Allaway, personal communication, 1976).

Injection of Se usually as Na_2SeO_4 directly into animals is the most economical method in terms of Se cost, but this method as well as oral dosing increases labor costs. Heavy pellets containing Se as $CaSeO_4$, $BaSeO_4$, and elemental Se have been administered orally to grazing sheep, providing Se to the animal for periods of up to one year. Adding Se to Se-deficient farm soils to produce crops that contain enough Se to protect livestock is risky, because potentially toxic levels of Se may occur in plants soon after Se fertilization (Allaway, 1975).

LOSSES OF PLANT NUTRIENTS FROM AGRICULTURAL SOILS

Man's food supply depends largely upon his ability to maintain a favorable equilibrium between losses and gains of available plant nutrients in agricultural soils. As discussed in prior sections of this chapter, man's struggle to maintain this equilibrium has existed since agriculture began.

Cropping, leaching, volatilization, and erosion are the four principal pathways whereby nutrient elements are removed from soils. The harvested crop removes only those elements taken up by the crop during its period of growth. Leaching losses are governed primarily by the degree of nutrient mobility, and the amount of water passing through the soil profile. Cation leaching losses are also affected by the associated anion, the most important ones being NO_3^-, SO_4^{2-}, HCO_3^-, and, near sea-coasts or in arid regions, Cl^-. Volatilization losses are likely confined to N and S, although other elements may be volatilized when vegetation is burned.

The loss of plant nutrients through erosion is nonselective in the sense that all elements present in soils are transported by water and wind erosion, Erosion is selective in that fine organic and inorganic materials are relatively higher in available plant nutrients and are also more vulnerable to erosion than are the coarser soil fractions.

In general, it is very difficult to construct accurate nutrient balance sheets under field conditions because of the difficulties involved in monitoring the many possible fates that await plant nutrients contained in or applied to soil systems.

Cropping

The removal of nutrients by harvested crops is by far the most thoroughly and accurately documented of the four principal loss mechanisms. This is due to the relatively simple sampling and analytical procedures associated with growing plants as compared with those for field soils. Crop recovery of plant nutrients applied during the year that the crop is grown is a quite inefficient process. As a general proposition, if one applied X g m^{-2} each of N, P, K, and micronutrients at planting, the harvested crop will contain roughly X/2, X/3, X/5, and X/30-50 of N, P, K, and micronutrients, respectively. There are, of course, residual benefits to subsequent crops, particularly with P and K, but usually annual fertilization of each crop with one or more nutrients is required for maximum yields.

There are many factors that influence the recovery of applied fertilizers by crops. Assuming a fertile soil and ideal growing conditions, the principal factors that limit this recovery are the numerous reactions and transformations that occur when fertilizers are applied to soils. These events, of varying intensity, compete with the growing plant for soluble plant nutrients. These competing factors are usually very difficult to mitigate, although some improvement in nutrient utilization has occurred through an increased knowledge of soil chemistry, microbiology, as well as the use of improved soil management, cultivars and fertilizers, pesticides, and herbicides. Representative amounts of nutrients removed by crops are shown in Table 9.14. These data have been derived from a variety of sources and can vary quite widely depending upon soil, climatic, and other conditions affecting plant growth. Individual crops may vary widely in the amounts of nutrients that they remove from soils. Obviously, yields are a big factor, but for the examples shown in Table 9.14, tobacco removed a total of approximately 25 g m^{-2}, while removal by corn totaled over 62 g m^{-2}.

An attempt was made to calculate the total worldwide annual removal by crops of the three major and three secondary nutrients (Table 9.15). These crude estimates indicate that nearly 70 Tg of plant nutrients are cycled annually through the world's cultivated crops, which approaches the total world fertilizer consumption of N, P_2O_5, and K_2O in 1975, calculated to be about 82 Tg (Table 9.5). This does not suggest that there is a high

TABLE 9.14

Representative amounts of nutrients removed in crops

Crop	Yield level ($g\ m^{-2}$)	Nutrients removed annually ($dg\ m^{-2}$)								
		N	P_2O_5	K_2O	S	B	Fe	Cu	Mn	Zn
Wheat	440	134	50	101	17	0.07	0.50	0.06	0.16	0.24
Rice	500	112	45	112	11	<0.01	—	0.01	0.11	0.09
Tobacco	220	84	17	134	16	0.06	0.90	0.03	0.62	0.08
Cotton	170	224	101	179	34	0.03	0.07	0.20	0.37	1.08
Corn	940	269	101	224	28	0.13	0.17	0.07	0.10	0.17
Tomatoes	4480	112	62	235	28	0.16	1.46	0.08	0.15	0.18
Peanuts	340	168	45	168	23	0.13	1.01	0.13	0.84	—
Potatoes	4030	134	50	252	10	0.09	0.90	0.07	0.16	0.09
Soybeans	340	134	73	140	17	0.07	0.56	0.06	0.07	0.06
Sugar beets	3360	112	39	140	11	—	—	0.03	0.84	0.17
Cabbage	4480	146	39	157	45	0.10	0.22	0.04	0.11	0.09
Sugarcane	6720	168	67	280	84	—	—	—	—	—

TABLE 9.15

Estimated annual nutrient removal by major crops — harvested portion (USDA, 1967a.)

Crop	1975 World production (Gg)	Nutrients removed (Gg)					
		N	P_2O_5	K_2O	S	Ca	Mg
Wheat	338,441	7,065	3,580	2,167	471	188	848
Rice	350,774	4,692	1,876	938	312	312	391
Maize	313,215	4,016	1,607	1,160	417	447	654
Barley	144,508	2,641	1,133	754	226	75	151
Sugar (cane + beets)	82,128	165	55	138	28	91	66
Potatoes	229,226	763	286	1,432	57	32	57
Soybeans	62,965	3,938	922	1,382	105	189	189
Sorghum [a]	50,000	810	440	250	60	50	60
Oats	48,973	964	383	291	92	46	61
Rye	23,954	465	185	140	44	22	29
Peanuts	18,610	597	112	217	19	37	31
Beans, dry edible	5,828	242	80	80	16	7	7
Tobacco	5,292	186	38	299	35	189	45
Citrus fruits	2,253	5	2	7	0.7	—	—
Hay (U.S.A. only)	132,917	2,900	662	2,900	290	1,847	382
Sweet potato [a]	130,000	355	118	591	47	31	71
Cassava [a]	100,000	266	88	443	35	23	53
Cotton	12,638	1,012	506	379	51	51	101
Other (vegetables, non-citrus fruits, oil crops) [a]	210,000	725	196	725	161	123	50
Total	2,261,692	31,807	12,269	14,293	2,466.7	3,760	3,246

[a] Yield data from Harlan (1976).

recovery of nutrients applied to soils during the year that the crop is grown. Rather, plant nutrients such as P and K tend to accumulate in most cultivated soils where fertilizers are being used. In other soil areas, plant nutrients are either derived from non-commercial sources or the native soil supplies are being utilized. Per capita consumption of commercial fertilizers in developing countries, for example, was only 6.6 kg in 1974, compared with 58.2 kg for developed countries (Skinner, 1976).

Leaching

The cations and anions removed in the largest amounts from soils in humid regions by leaching are of Ca, Mg, Na, K, and NO_3, Cl, HCO_3, and SO_4, respectively. Losses of P by this mechanism are exceedingly low and indications are that only very small amounts of micronutrients move with drainage waters in most instances. Leaching losses are part of a normal, continuous cycle, that occurs to some extent in all soil systems subjected to the percolation of water through the entire profile.

The quantitative determination of plant nutrients lost by leaching under field conditions is usually a difficult problem. The difficulty arises when one attempts to relate the underground flow of water to a specific land area. As a result, leaching losses have generally been measured by various types of lysimeters or in discharge waters from tile-drained fields. Neither of these procedures is wholly satisfactory from the standpoint of well-drained, upland field soils. Lysimeters are useful for obtaining comparative data which may or may not bear a relationship to the downward movement of nutrients under field conditions. Poorly drained soils that have tile drainage are, of course, atypical since alternate reducing and oxidizing conditions occur over long periods of time relative to well-drained upland soils.

Recent reviews of fertilizer losses by leaching include those by Soileau (1969, 1971) and Doerry (1976). Also useful in gaining a perspective relating to this phase of mineral cycling is the publication of 50 years of lysimeter studies in Germany by Jürgens-Gschwind and Jung (1977). A very crude estimate resulting from literature reviews indicates that, of the nutrients applied in fertilizers, the percentages lost by leaching are generally within the following ranges: N, 10 to 30; P, <1; K, 5 to 20; Ca, 20 to 30; Mg, 10 to 15; S, 15 to 25; and micronutrients, <1. On a global scale, statistics for the three major fertilizer nutrients, N, P, and K, are quite reliable, but data relating to other nutrients are fragmentary or unavailable. Using the foregoing figures as a percentage of world plant nutrient consumption in 1975, and estimated 5 to 16 Tg of applied N + P + K could be leached from agricultural soils each year. However, leaching losses are usually very low in the presence of an actively growing crop. Such losses are lowest from forested and grassed soils and highest from fallowed soils after crops have been harvested or have ceased growing.

Volatilization

Nutrients are volatilized from soils in two principal ways: the displacement of air in soils by water, and by elevated temperatures resulting from burning vegetation. Relatively few quantitative data are to be found in the literature that relate to gaseous losses measured under field conditions.

Nitrogen volatilization is of major interest in this respect; 10 to 20% of that applied in fertilizers, or 4 to 8 Tg annually, seems to be a reasonable range of gaseous N loss according to Allison (1965). Nitrogen may be lost to the atmosphere as NH_3, N_2, NO_2, and N_2O. Ammonia losses are most prevalent from alkaline soils and from surface-applied urea which undergoes hydrolysis to form unstable $(NH_4)_2CO_3$, resulting in localized alkaline soil conditions.

Volatilization losses of S from flooded soils have not been extensively studied, but it appears certain that significant quantities of S move into the atmosphere from soils that are saturated with water either intermittently or continuously.

The shifting cultivation system of agriculture practiced in the tropics for centuries has become a rather serious problem in recent years, since it is now practiced more intensively than in the past. Vast areas of tropical forest and savannah have been transformed into less productive savannah grassland. This aspect of man's intrusion into plant nutrient cycling has received little attention from soil scientists, but sparse evidence points to a rapid decline in soil fertility as a result of burning. Nye and Greenland (1960) working in Ghana, report N losses from grass burn in high-grass savannah to be approximately 30 g Nm^{-2}. There is additonal evidence that other nutrients may be volatilized or transported in fly-ash as a result of burning. Smith and Bowes (1974) report total K, P, Ca, and Mg losses from biomass in southern Ontario to be 40 to 50% of the amounts present prior to burning. It is not possible to arrive at a quantitative estimate of the total amounts of minerals removed from surface soils as a result of the shifting cultivation system. While large quantities of minerals are released to the soil from burning vegetation, these same minerals may be subject to transport by water to an extent that is yet unknown.

Erosion

Man-induced erosion has reduced the productivity of vast areas of the world's agricultural land. Lowdermilk (1953) surveyed lands in the Middle East, China, and Europe during 1938—1939 and blames serious soil erosion and land-use problems for the decline and disappearance of many ancient cultures and civilizations. Following this study, he was invited to give a radio talk in Jerusalem in June 1939 and gave for the first time what has since been called an "Eleventh Commandment":

"Thou salt inherit the Holy Earth as a faithful steward, conserving its resources and productivity from generation to generation. Thou shalt safeguard thy fields from soil erosion, thy living waters from drying up, thy forests from desolation, and protect thy hills from overgrazing by thy herds, that thy descendants may have abundance forever. If any shall fail in this stewardship of the land thy fruitful fields shall become sterile stony ground and wasting gullies, and thy descendants shall decrease and live in poverty or perish from off the face of the earth."

While Lowdermilk gives few figures regarding land areas affected by erosion, he stated that about 2×10^{11} m^2 of U.S. farmland had been ruined in this manner.

Water erosion has been fairly well characterized in the U.S.A., but has generally received less attention in other countries. Four main factors influence the degree to which water erosion occurs: climate, soil, topography, and vegetative cover. One may assume a value for each of these factors, which by itself may indicate a soil-loss problem.

Soil erosion studies began with the work of Wollny in Germany during the latter part of the 19th century. About 1920, work began in midwestern U.S.A. by Miller and Duley, followed by a strong national soil conservation research and action program fostered and lead by H.H. Bennett. Vast amounts of research data resulted in an improved soil loss equation that was subsequently developed by Smith and Wischmeier (1962) of the U.S. Department of Agriculture.

The soil loss equation is:

$$A = RKLSCP$$

where A = the computed soil loss per unit area; R = the rainfall factor (erosive potential); K = soil erodibility; L = length of slope; S = slope gradient; C = crop management factor; and P = erosion-control practice factor.

Solutions of the equation provide alternate plans wherein the annual soil loss from a specific field can be held within the limits that can be tolerated under the soil topographic conditions involved. In spite of the fact that nearly $\$15 \times 10^9$ has been spent on soil conservation in the U.S.A. since the mid-1930's, soil erosion remains one of the biggest and most pervasive problems still facing the nation (Carter, 1977). The dust storms that occurred in early 1977 underline this statement. Nationally, erosion losses have been estimated at about 2.5 kg m^{-2} each year, but soil scientists believe that even deep soils cannot sustain a fraction of this loss within serious reductions in productivity. Discounting erosion losses, top soil possibly forms at the rate of about 0.35 kg m^{-2} y^{-1} in humid regions.

The dominant form of soil loss in the U.S.A. is caused by water runoff. Some 3.6 Pg of sediment is delivered to waterways annually in the 48 contiguous states, and 50% of the sediment originates from agricultural lands

TABLE 9.16

Estimated amounts of total plant nutrients lost annually in the U.S.A. by erosion [a]

Constituent	Loss ($Gg\ y^{-1}$)	Constituent	Loss ($Gg\ y^{-1}$)
N	360— 9,000	B	3.6 — 360
P	180— 360	Co	1.8 — 72
K	3,060—59,400	Cu	3.6 — 180
Ca	1,260—64,800	Mn	360 —5,400
Mg	2,160—27,000	Mo	0.36— 9
S	180— 3,600	Zn	18 — 54

[a] Probably not more than 10% of these amounts become available for plant growth annually.

TABLE 9.17

Annual soil loss from various crops in different regions of the U.S.A. (Adapted from Pimentel et al., 1976.)

Crop	Location	Slope (%)	Soil loss ($dag\ m^{-2}$)
Corn (continuous)	Missouri (eastern)	3.7	44
Corn (continuous)	Wisconsin (western)	16	199
Corn	Mississippi (northern)		49
Corn	Iowa (southwestern)	9	63
Corn (plow-disk-harrow)	Indiana		47
Corn (plow-disk-harrow)	Ohio (eastern)		27
Corn (conventional)	Ohio (southeastern)		6
Corn (conventional)	South Dakota (eastern)	5.8	6
Corn (continuous chem.)	Missouri (northeastern)	3	47
Corn (contour)	Iowa (southwestern)	2 to 13	48
Corn (contour)	Iowa (western)		54
Corn (contour)	Missouri (northwestern)		54
Cotton		2 to 10	43
Cotton	Georgia (northeastern)		46
Wheat	Missouri (central)	3.7	23
Wheat (black fallow)	Nebraska (western)	4	14
Wheat	Washington (southeastern)		11 to 22
Wheat-pea rotation	Washington (southeastern)		12
Wheat (following fallow)	Washington (southeastern)		15 to 22
Bermuda grass	Texas (east central)	4	0.06
Native grass	Kansas (western)	5	0.06
Forest	North Carolina (northeastern)	10	0.004
Forest	New Hampshire (central)	20	0.02

according to Wadleigh (1968). Crude estimates of losses of total available plant nutrients are shown in Table 9.16. These data were obtained by assuming that 1.8 Pg of soil is eroded annually from U.S. soils having the ranges in elemental composition given in Table 9.4. Possibly 5 to 10% of these total amounts becomes available for plant growth each year. Average annual soil losses from various crops in different regions in the U.S.A. can be observed in Table 9.17. These data represent research studies spanning 40 years. Row cropping, of course, can result in large soil losses. In most instances, two or three storm events occurring during the early part of the growing season will produce a high percentage of the annual soil loss.

Since eroded material frequently differs in composition from the original soil, the loss of plant nutrients in runoff may be expressed in terms of an enrichment ratio (E.R.). This is the ratio of the concentration of that element in the runoff to that in the original soil (Barrows and Kilmer, 1963):

$$\text{E.R.} = \frac{\text{Concentration of element in soil material in runoff}}{\text{Concentration of element in soil from which runoff originated}}$$

Massey and Jackson (1952) calculated regression equations for the E.R.'s of organic matter, N, P, and K from three midwestern United States soils. The average E.R.'s are: organic matter, 2.1; N, 2.7; available P, 3.4; and exchangeable K, 19.3.

While soil and wind erosion results in huge losses of potentially available plant nutrients, it is important to recognize the role of fertilizers in this regard. The application of fertilizers is often essential to the establishment and maintenance of strong erosion-resistant vegetative cover on soils. In addition, fertilizers make it posible for man to produce high yields on a limited land base, thus minimizing the cultivation of marginal lands. Fertilizer use also makes it possible to maintain high plant populations, particularly with corn. This in turn results in less erosion from soils planted to row crops, since the crop itself has a protective effect.

REFERENCES

Adams, S.S., 1968. Potassium reserves in the world. In: V.J. Kilmer, S.E. Younts, and N.C. Brady (Editors). The Role of Potassium in Agriculture. The American Society of Agronomy, Madison, Wisconsin, pp. 1—19.

Allaway, W.H., 1975. The effect of soils and fertilizers on human and animal nutrition. Agriculture Information Bulletin No. 378, United States Department of Agriculture, Washington, DC, 52 pp.

Allison, F.E., 1965. Evaluation of incoming and outgoing processes that affect soil nitrogen. In: W.V. Bartholomew and F.E. Clark (Editors). Soil Nitrogen. American Society of Agronomy, Madison, Wisconsin, pp. 578—606.

Anon., 1935. Nitrate Fields of Chile. The Pan American Union, Washington, DC, 21 pp.

Anon., 1971. World Survey of Phosphate Deposits. The British Sulphur Corporation Limited, London, 180 pp.

Anon., 1973. Preliminary Sulphur and Sulphuric Acid Statistics, 1973. The British Sulphur Corporation Limited, London, 20 pp.

Anon., 1975. Report of Tonnage of Agricultural Limestone Used in the United States in 1972. National Limestone Institute, Inc., Fairfax, VA, 1 p.

Barber, S.A., Munson, R.D. and Davey, W.B., 1971. Production, marketing, and use of potassium fertilizers. In: R.A. Olson, T.J. Army, J.J. Hanway and V.J. Kilmer (Editors). Fertilizer Technology and Use, 2nd edn. Soil Science Society of America, Madison, WI, pp. 303—331.

Barrows, H.L. and Kilmer, V.J., 1963. Plant nutrient losses from soils by water erosion. Adv. Agron., 15: 303—316.

Bennett, H.H., 1939. Soil Conservation. McGraw-Hill, New York and London, 993 pp.

Bixby, D.W., 1977. Sulfur requirements of the phosphate fertilizer industry. In: E.J. Kamprath, F.E. Khasawneh and E.C. Sample (Editors). The Role of Phosphorus in Agriculture. Soil Science Society of America, Inc., Madison, WI, in press.

Bixby, D.W. and Kilmer, V.J., 1975. Sulphur resources available to the fertilizer industry, the fertilizers made, their role, and the problems involved in their manufacture. In: K.D. McLachlan (Editor). Sulphur in Australasian Agriculture. Sydney University Press, Sydney, Australia, pp. 231—241.

Bixby, D.W., Rucker, D.L. and Tisdale, S.L., 1964. Phosphatic Fertilizers, Properties, and Practices. Technical Bulletin No. 8, The Sulphur Institute, Washington, DC, 85 pp.

Buckman, H.O. and Brady, N.C., 1969. The Nature and Properties of Soils. Macmillan, New York, NY, 653 pp.

Carter, L.J., 1977. Soil erosion: the problem persists despite the billions spent on it. Science, 196: 409—411.

Coleman, N.T. and Thomas, G.W., 1967. The basic chemistry of soil acidity. In: R.W. Pearson and F. Adams (Editors). Soil Acidity and Liming. American Society of Agronomy, Madison, WI, pp. 1—34.

Collings, G.H., 1949. Commercial Fertilizers, Their Sources and Use, 4th edn. The Blakiston Company, Philadelphia-Toronto, 522 pp.

Cuevas, E., 1916. The Nitrate Industry. The New Jersey State Agricultural College, 61 pp.

Dale, J.M. and Dehart, R.C., 1960. The use of underground nuclear explosions for mining sulphur. In: Proc. of Minerals and Metals Symposium, Mexico City.

Delwiche, C.C., 1970. The nitrogen cycle. Sci. Am., 223: 137—146.

Doerry, V.R., 1976. Bibliographie. Thema: Lysimeter. BASF Agricultural Experiment Station, Limburgerhof, 176 pp.

Emigh, D.G., 1972. World phosphate reserves — are there really enough? Eng. Min. J., 173: 90—95.

Epstein, E., 1972. Mineral Nutrition of Plants: Principles and Perspectives. John Wiley, New York, NY, 412 pp.

FAO (Food and Agriculture Organization of the United Nations), 1976. Annual Fertilizer Review, 1975. Food and Agriculture Organization of the United Nations, Rome, pp. 52—62, 74.

Gauch, H.G. and Dugger, W.M., 1954. The physiological action of boron in higher plants: a review and interpretation. Maryland Agricultural Experiment Station Bulletin, A-80, 43 pp.

Goldberg, E.D., 1965. Minor elements in seas. In: J.P. Riley and G. Skirrow (Editors). Chemical Oceanography, Vol. . Academic, New York, NY, pp. 163—194.

Harlan, J.R., 1976. The plants and animals that nourish man. Sci. Am., 235, 89—97.

Hignett, T.P., 1967. Characteristics of the World Fertilizer Industry — Phosphatic Fertilizers. TVA Report No. S-422. Tennessee Valley Authority, Muscle Shoals, AL, 85 pp.

Hoeft, R.G., Kenney, D.R. and Walsh, L.M., 1972. Nitrogen and sulfur in precipitation and sulfur dioxide in the atmosphere in Wisconsin. J. Environ. Qual., 1: 203—208.
Holser, W.T. and Kaplan, I.R., 1966. Isotope geochemistry of sedimentary sulfates. Chem. Geol., 1: 93—135.
Idso, S.B., 1976. Dust storms. Sci. Am., 235(4): 108—114.
Jürgens-Gschwind and Jung, J., 1977. Ergebnisse Lysimeteruntersuchungen in der Grossanlage. BASF, Limburgerhof, 177 pp.
Kapusta, E.C., 1968. Potassium fertilizer technology. In: V.J. Kilmer, S.E. Younts and N.C. Brady (Editors). The Role of Potassium in Agriculture. The American Society of Agronomy, Madison, WI, pp. 23—50.
Kilmer, V.J., 1974. Nutrient losses from grasslands through leaching and runoff. In: D.A. Mays (Editor). Forage Fertilization. The American Society of Agronomy, Madison, WI, pp. 341—360.
Kilmer, V.J., Joyce, R.T., Eklund, D.C. and Mays, D.A., 1974. Nutrient inputs and removals on a grassed watershed in western North Carolina. Agronomy Abstracts, American Society of Agronomy, Madison, WI, p. 30.
Krauskopf, K.B., 1972. Geochemistry of micronutrients. In: J.J. Mortvedt, P.M. Giordano and W.L. Lindsay (Editors). Micronutrients in Agriculture. Soil Science Society of America, Inc., Madison, WI, pp. 7—40.
Lamb, H.H., 1970. Volcanic Dust in the Atmosphere. Philos. Trans. R. Soc. London, A266: 425—533.
Lowdermilk, W.C., 1953 (revised 1975). Conquest of the land through seven thousand years. Agriculture Information Bulletin No. 99, Unites States Department of Agriculture, Soil Conservation Service, Washington, DC, 30 pp.
Mack, F.J., 1967. Sedimentation in the Upper Mississippi River Basin. In: Soils and America's Future. Proc. of the 22nd Annual Meeting, Soil Conservation Society of America, Des Moines, IA, pp. 95—102.
Manderson, M.C., 1970. World Sulfur Outlook into the Late 1970's. Presented at American Chemical Society Annual Convention, Chicago, Illinois, September 1970. Arthur D. Little, Inc., Cambridge, MA, 14 pp.
Massey, H.F. and Jackson, M.L., 1952. Selective erosion of soil fertility constituents. Soil Sci. Soc. Am. Proc., 16: 353—356.
McLachlan, K.D. (Editor), 1975. Sulphur in Australasian Agriculture. Sydney University Press, Sydney, Australia, 261 pp.
McLean, J.W., Thompson, G.G. and Claxton, J.H., 1959. Growth responses to selenium in lambs. Nature, 184: 251—252.
Morrison, F.B., 1948. Feeds and Feeding. Morrison, Ithaca, NY, 1207 pp.
Mortvedt, J.J. and Cunningham, H.G., 1971. Production, marketing, and use of other secondary and micronutrient fertilizers. In: R.A. Olson, T.J. Army, J.J. Hanway and V.J. Kilmer (Editors). Fertilizer Technology and Use, 2nd edsn. Soil Science Society of America, Inc., Madison, WI, pp. 413—447.
Murphy, L.S. and Walsh, L.M., 1972. Correction of micronutrient deficiencies with fertilizers. In: J.J. Mortvedt, P.M. Giordano and W.L. Lindsay (Editors). Micronutrients in Agriculture. Soil Science Society of America, Inc., Madison, WI, pp. 347—388.
National Academy of Sciences, 1972. Accumulation of Nitrate. National Academy of Sciences, Washington, DC, 106 pp.
Nelson, L.B., 1972. Agricultural chemicals in relation to environmental quality: chemical fertilizers, present and future. J. Environ. Qual., 1: 2—6.
Nye, P.H. and Greenland, D.J., 1960. The Soil Under Shifting Cultivation. Technical Communication No. 51, Commonwealth Bureau of Soils, Harpenden, Great Britain, 156 pp.
Pfost, H.B. and Swinehart, C.E., 1970. Feed Manufacturing Technology. American Feed Manufacturers Association, Inc., Chicago, IL, 604 pp.

Phillips, A.B. and Webb, J.R., 1971. Production, marketing, and use of phosphorus fertilizers. In: R.A. Olson, T.J. Army, J.J. Hanway and V.J. Kilmer (Editors). Fertilizer Technology and Use, 2nd edn. Soil Science Society of America, Inc., Madison, WI, pp. 271—300.

Pimentel, D., Terhune, E.C., Dyson-Hudson, R., Rochereau, S., Samis, R., Smith, E.A., Denman, D., Reifschneider, D. and Shepard, M., 1976. Land degradation: effects on food and energy resources. Science, 194: 149—155.

Price, C.A., Clark, H.E. and Funkhouser, E.A., 1972. Functions of micronutrients in plants. In: J.J. Mortvedt, P.M. Giordano and W.L. Lindsay (Editors). Micronutrients in Agriculture. Soil Science Society of America, Inc., Madison, WI, pp. 231—242.

Robinson, W.O., 1933. The determination of selenium in wheat and soils. J. Assoc. Offic. Agric. Chem., 16: 423—424.

Seiden, R. and Pfander, W.H., 1957. The Handbook of Feedstuffs. Springer, New York, NY, 591 pp.

Sitterson, J.C., 1961. An Essay on Calcareous Manures by Edmund Ruffin. The Belknap Press of Harvard University Press, Cambridge, MA, 199 pp.

Skinner, K.J., 1976. Nitrogen fixation. Chem. Eng. News, 54(41): 22—35.

Smith, D.D. and Wischmeier, W.H., 1962. Rainfall erosion. Adv. Agron., 14: 109—148.

Smith, D.W. and Bowes, G.G., 1974. Loss of some elements in fly-ash during old field burns in southern Ontario. Can. J. Soil Sci., 54: 215—223.

Soileau, J.M., 1969. Effects of Fertilizers on Water Quality. National Fertilizer Development Center, Tennessee Valley Authority, Muscle Shoals, AL, 105 pp.

Soileau, J.M., 1971. Effects of Fertilizers on Water Quality (Supplement). Bulletin Y-18, National Fertilizer Development Center, Tennessee Valley Authority, Muscle Shoals, AL, 41 pp.

Stout, P.R., 1972. Introduction. In: J.J. Mortvedt, P.M. Giordano and W.L. Lindsay (Editors). Micronutrients in Agriculture. Soil Science Society of America, Inc., Madison, WI, p. 1.

Swaine, D.J., 1955. The Trace-Element Content of Soils. Tech. Communication No. 48, Commonwealth Bureau of Soil Science. Commonwealth Agricultural Bureaux, England, 157 pp.

Swaine, D.J., 1962. The Trace-Element Content of Fertilizers. Tech. Communication No. 52, Commonwealth Bureau of Soils. Commonwealth Agricultural Bureau, England, 306 pp.

Taylor, P.S. and Stoiber, R.E., 1973. Soluble Material on Ash from Active Central American Volcanoes. Bulletin V84, Geological Society of America, pp. 1031—1042.

Tim, B. and Danz, W., 1964. History of nitrogen fixation processes. In: Vincent Sachelli (Editor). Fertilizer Nitrogen, Its Chemistry and Technology. ACS Monograph No. 161, Reinhold, New York, NY, pp. 40—57.

Tisdale, S.L. and Cunningham, H.G., 1963. Advances in manufacturing of secondary and micronutrient fertilizers. In: M.H. McVickar, G.L. Bridger and L.B. Nelson (Editors). Fertilizer Technology and Use. Soil Science Society of America, Inc., Madison, WI, pp. 269—285.

Tisdale, S.L. and Nelson, W.L., 1966. Soil Fertility and Fertilizers, 2nd edn. Macmillan, New York, NY; Collier-Macmillan, London, 694 pp.

Turrentine, J.W., 1946. Past Consumption and Future Requirements (1950) of Potash Salts in American Agriculture. American Potash Institute, Washington, DC (principal offices now located in Atlanta, GA), 31 pp.

USDA (United States Department of Agriculture), 1976a. Agricultural Statistics. United States Department of Agriculture, Washington, DC, 613 pp.

USDA (United States Department of Agriculture), 1976b. Commercial Fertilizers, Preliminary Consumption for Year Ending June 30, 1976. U.S.D.A. Crop Reporting Board, Statistical Reporting Service, Washington, DC, 12 pp.

USDA (United States Department of Agriculture) and TVA (The Tennessee Valley Authority), 1964. Superphosphate: Its History, Chemistry, and Manufacture. United States Government Printing Office, Washington, DC, 349 pp.

Wadleigh, C.H., 1968. Wastes in Relation to Agriculture and Forestry. Miscellaneous Publication No. 1065, United States Department of Agriculture, Washington, DC, 112 pp.

Weir, W.W., 1936. Soil Science, Its Principles and Practice. J.B. Lippincott, Chicago, IL, and Philadelphia, PA, 615 pp.

Woodruff, C.M., 1967. Crop response to lime in the midwestern United States. In: R.W. Pearson and F. Adams (Editors). Soil Acidity and Liming. American Society of Agronomy, Madison, WI, pp. 207—227.

Chapter 10

A SECOND IRON AGE AHEAD? *

B.J. SKINNER

Department of Geology and Geophysics, Yale University, New Haven, CT 06520 (U.S.A.)

CONTENTS

INTRODUCTION

Mention 1776 and most people think immediately of the historic Declaration of Independence by the American colonies. They would do well to enlarge their horizons and to include in their catalogue of the important events of 1776 a modest commercial transaction that took place in England. That transaction was a harbinger of changes in social structure even more far-reaching than the changes wrought by the political upheavals in the new United States of America; 1776 was the year that the newly formed engineering firm Boulton and Watt sold its first steam engine. Although this was not the first steam engine to be sold, it incorporated many of Watt's innovative ideas and thus was the first efficient example to be offered commercially.

Watt did not, of course, invent the steam engine, but prior to his time the old Newcomen engine was used exclusively as a steam pump; it was slow,

* This article was originally published in American Scientist, Vol. 64, 1976, pp. 258—269, and is reproduced with permission of the publishers, Sigma Xi, The Scientific Research Society of North America Inc. A few modifications have been made to the original, notably the deletion of Figs. 1, 2 and 5 which are photographs of a copper mine, a smelter, and three mineral assemblages.

cumbrous, and excessively wasteful of fuel. Watt transformed the old engine to a quick-working, powerful, and efficient device, then he adapted it to drive machinery of all kinds. A device made of metals dug from the ground could convert the chemical energy locked in wood and coal into useful mechanical energy. This was the key step that led to replacement of manpower by machine power. The door had been opened for the industrial revolution, and society was forever changed.

Following Watt's invention the use of mechanized devices proliferated, and demand for fuel to drive machines rose concomitantly with demand for materials to build more machines. Both demands were satisfied by a burgeoning mining industry that supplied first coal then oil and gas for fuel, together with a multitude of metals for ever more complex machines. Freed from the limitations imposed by puny human muscle power, able to travel and communicate widely and therefore no longer forced to subsist on food produced in the immediate environs of their habitations, populations grew rapidly and expanded into less and less hospitable parts of the world. Population densities rose, and as people were removed farther and farther from direct production of food, the infrastructure needed to supply and feed them became increasingly more complex. This trend continues apace, and because the infrastructure is built from and driven by mineral substances dug from the earth, every additional complexity brings demands for more fuels and more metals. Without those substances the structure must collapse and the population wither back to some small fraction of its present size.

The magnitude of the world's annual needs for newly mined substances has grown so large that it is difficult to view the volume in perspective. The annual per capita consumption of newly mined mineral products for all the peoples of the world now totals 3.75 mg. The total includes coal, oil, iron, copper, cement, and a myriad of substances used in countless different ways: the total is still rising, doubling approximately every decade, and there is no sign that it is likely to stop in the near future. If all 1975's newly mined mineral substances were somehow stacked on historic Boston Common, the result would be a column approximately 10 km high. By 1985, the annual pile would be 20 km high.

Annual-per-capita rates of consumption of mineral products are highest in the countries of the industrialized Western world (15 Mg in the U.S.A.), but the rate of increase is highest in developing countries. We can anticipate that both aspirations and populations of the developing countries will continue to grow, and that consumption of mineral products will follow suit. But populations have now become so vast that we must question the adequacy of earth's mineral supplies to satisfy the aspirations. Ours is the first generation for which this question has had to be raised, and it is embarrassing because it means that we must examine life styles, too. As is commonly the case when difficult questions are first posed, initial responses are tentative and often conflicting.

Both the questions and their answers are many-sided. Most attention has naturally been focused on energy supplies, particularly on the fossil fuels oil, gas, and coal. The controversial reports concerning the magnitude of their supplies are typical of the confusion. However, despite doubts about the exact amounts of oil, gas, and coal that will ultimately be found, the attention focused on energy means that it is now widely appreciated that each of the fossil fuel sources is limited in size, and that mankind must ultimately develop and use other energy sources. But it is equally clear, fortunately, that of energy itself there is no shortage; provided we can learn to use energy safely from sources such as the sun, earth's internal heat, and naturally occurring radioactive elements, there need never be an energy shortage.

We fool ourselves, however, if we dwell on energy alone. The uses of all natural resources are intertwined. Oil is of little use without engines built of iron, copper, zinc, and other metals. Farmlands will yield maximum crops only if they are tilled by tractors and plows and fertilized with compounds of phosphorus, nitrogen, and potassium. A failure in the supply of one resource will inevitably influence the use of others. Viewing the panoply of natural resources, we see that one group, metals, occupies a unique position. Without metals we could not build machines to replace human muscle. Without metals we could use little of the available energy. Metals are, in effect, the enzymes of industry. If supplies of metals are limited, then society must ultimately be limited too. It is my contention that the distribution of the chemical elements in nature means, inevitably, that there are natural limits to supplies of metals, and that these limits are much more important to the future of society than limits on energy. I also contend that, with sufficient work, the limits can be predicted. It is the purpose of this paper, therefore, to explore briefly the way metals occur and to attempt to place in perspective the limitations they may ultimately impose on us.

WHERE METALS ARE FOUND

With a few notable exceptions such as gold and platinum, metals do not occur in earth's crust in their elemental states. They are present instead as one of two or more elements combined into inorganic crystalline compounds called minerals. To recover the metals from their entrapping compounds, it is natural to choose the least expensive options; we mine and process only those minerals from which the desired metals can be easily obtained. In practice, this means that we choose those minerals that require the least expenditure of energy in order to effect the chemical disintegration and release of the metal.

The production of metals requires two separate and quite different kinds of steps. The first step is to mine and concentrate the desired mineral. Mining involves breaking and removing the rock containing the desired mineral from the ground; concentration involves separating it from the valueless and unwanted minerals with which it occurs, then gathering its grains into a pure aggregate. In concentration processes, the rock is crushed so that each fragment is a separate mineral. The mineral grains can then be separated on the basis of differing physical characteristics such as density, magnetism, or wettability in different liquids. By means of these processes, an ore that contains as little as 0.5% of the desired mineral can be upgraded to a concentrate containing more than 90% of the mineral.

By comparison with the second step in the metal recovery process — smelting and refining — mining and concentration are not very energy-intensive processes. During smelting and refining, the concentrated mineral is broken down and the desired metal is in turn separated from the elements with which it is combined. This is an energy-intensive step because it involves disintegration of stable chemical compounds. To keep energy demands as low as possible, the kinds of minerals from which metals are preferentially sought are sulfides, oxides, hydroxides, and carbonates — not the highly refractory silicate minerals that make up the great bulk of earth's crust. The problem of assessing metal supplies therefore reduces to determining the frequency of the desired source minerals and evaluating the problems that will attend any change from preferred to less preferred minerals.

Crustal abundance of chemical elements

The only portion of the earth accessible to mining, and therefore the only portion we can consider as a source of metals, is the crust. It has sometimes been argued that sea water should be considered a potential source of metals, but the argument does not bear close examination because the concentration of most metals in sea-water solution is vastly less than their concentration in the crust (Table 10.1). Sea water could be regarded as an ore only if a metal could be extracted at least one thousand times more efficiently from sea-water solution than from solid rocks, and the likelihood of this being achieved is small. In terms of today's technological costs, for example, if a plant capable of handling a volume as great as 4.5 Ml min^{-1} worked with a 100% extraction efficiency, the cost of the energy alone that would be needed to pump and process the water would far exceed the value of the extracted metals.

Earth's crust ranges in thickness from 10 to 50 km and contains at least trace amounts of 88 chemical elements. It can be subdivided into two distinctly different regimes: the oceanic crust that underlies the oceanic basins and the continental crust. The two differ in composition — the oceanic crust being richer in iron, magnesium, and calcium, the continental crust being richer in silicon, aluminum, and alkali elements. The 88 natural elements are all present in both crusts, though in somewhat different concentrations. Nevertheless, only 12 elements, and the same 12 elements in each case, are

TABLE 10.1

Comparison of the valuable metals in a cubic kilometer of average rock in earth's crust and in a cubic kilometer of sea water (Source: Skinner, 1976.)

Metal	Amount in crust (Mg)	Amount in sea water (Mg)
Manganese	1,809,000	1.9
Zinc	170,000	2.0
Chromium	130,000	0.2
Nickel	100,000	2.0
Copper	86,000	2.0
Cobalt	32,000	0.05
Uranium	7,800	3.3
Tin	5,700	0.8
Silver	160	0.3
Gold	5	0.01

present in amounts equal to or greater than 0.1% of the whole. Because of our knowledge of the chemistry of the continental crust, most of the subsequent discussion will refer to the continental crust.

The 12 elements most abundant in the continental crust are listed in Table 10.2. These 12 account for 99.23% of the mass of the crust, and the remaining 76 elements account for a mere 0.77% of the mass. I shall refer to the 12 most common elements as being geochemically abundant, and those present

TABLE 10.2

Major chemical elements in the continental crust

Element	Amount (weight percent)
Oxygen	45.20
Silicon	27.20
Aluminium	8.00
Iron	5.80
Calcium	5.06
Magnesium	2.77
Sodium	2.32
Potassium	1.68
Titanium	0.86
Hydrogen	0.14
Manganese	0.10
Phosphorus	0.10
Total	99.23

at levels below 0.1% by weight as geochemically scarce. The geochemically abundant metals that are widely used in industry are aluminium, iron, magnesium, titanium, and manganese. All other metals, many of which have numerous uses, are geochemically scarce.

Geochemically abundant metals

Essentially every common type of rock is composed of an aggregate of minerals, one or more of which contains a geochemically abundant metal as an essential constituent. It is apparent that this must be the case by simply considering the abundance of elements, and it means that, starting with a common rock type, it is always possible to produce a mineral concentrate in which the desired abundant element is a major constituent.

The practical miner does not, of course, select an average crustal rock as a source of metal. Instead he seeks those rocks in which the desired metal has been concentrated by some natural process and in which the metal occurs in a desired mineral form such as an oxide or a carbonate. Nevertheless, because most of the minerals he seeks are widespread in common rocks, the miner can look forward confidently to a long future with respect to geochemically abundant metals. Because many of the present ore deposits grade slowly into common rocks, as the extreme local enrichments he now

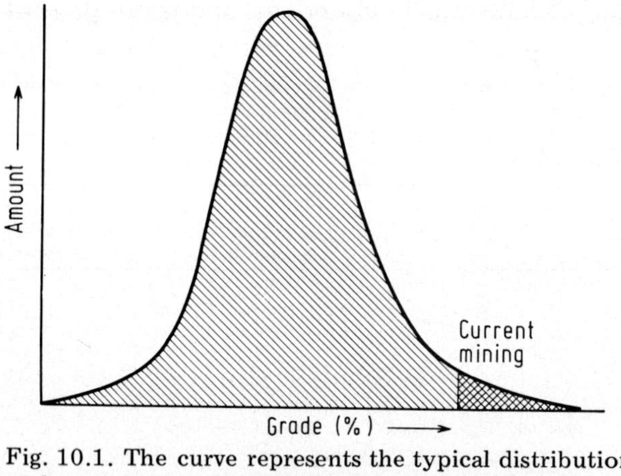

Fig. 10.1. The curve represents the typical distribution of geochemically-abundant metals such as iron, titanium, and aluminium in the earth's crust. The proportion of the total amount that has been mined up to the present is indicated at the right. The highest-grade ores have been mined first, but, because the same kinds of minerals occur regardless of grade, the concentrating techniques used in current mining processes to extract metal from high-grade ores can also be used in the future with less rich material and even common rock. As the grade, or percentage of metal in an ore, declines arithmetically, the curve indicates that the amount of metal available will increase geometrically down to a grade corresponding to the peak of the curve.

works become depleted, he will be able to continue mining material that is less and less rich, and he will always be able to prepare concentrates of desirable minerals. The situation is as depicted in Fig. 10.1. As reduction in the grade of ore mined (grade being a miner's term for the percentage of metal in an ore) declines arithmetically, the amount available increases geometrically.

One geochemically abundant metal that may seem at first glance to contradict this simple picture is aluminium. The common minerals that contain aluminium are all silicates, and the hydroxide compounds found in bauxite, which is the ore of aluminium mined at present, are not found in common rocks. Thus, once the world's resources of bauxite have all been mined out, we will have to learn to use silicate minerals such as anorthite ($CaAl_2Si_2O_6$), sillimanite (Al_2SiO_5), and kaolinite ($Al_4Si_4O_{10}(OH)_8$), as Russia, Sweden, Japan, and other countries have done at times of economic necessity. The principle of concentration of aluminium minerals will then be applicable and it will be possible to use common rocks as starting materials and to separate the anorthite, sillimanite, and kaolinite from other minerals in the rock. Carrying this line of reasoning to its conclusion, it is apparent that the entire crust can be considered a potential resource of geochemically abundant minerals and that no major technological barriers, beyond the ability to mine steadily-declining grades of ore, will have to be overcome. The smelting practices available today will also work in the future.

Geochemically scarce metals

The geochemically scarce metals pose an entirely different situation and probably have a very different grade-distribution curve, as shown in Fig. 10.2. Rarely do the *scarce* elements in common rocks form separate minerals. Instead, they are present as randomly-distributed atoms trapped by isomorphous substitution in minerals of the geochemically-abundant elements, an atom of a scarce element replacing an atom of an abundant element. For example, lead is found in most common rocks as an atomic substitute for potassium, while zinc appears as a substitute for magnesium. Thus, even though analysis of a common rock such as granite may reveal that it contains many parts per million lead and zinc, no amount of searching will disclose either a lead or a zinc mineral. The two metals are trapped in the atomic cages of common minerals such as orthoclase ($KAlSi_3O_8$) and biotite ($(K_2Mg,Fe)_6Si_6Al_2O_{20}(OH)_8$). A rock containing 10 ppm lead, in which all the lead is present in a feldspar, can yield a concentrate no richer than a pure feldspar concentrate. If the rock contains 10% feldspar, the maximum lead concentrate will contain only 100 ppm lead — hardly an encouraging result for a miner accustomed to producing concentrates with 70% lead or more. Furthermore, to release the lead from its silicate cage, the entire mineral must be broken down chemically and the lead atoms separated from all other atoms. This is a complicated and very energy-intensive process, which

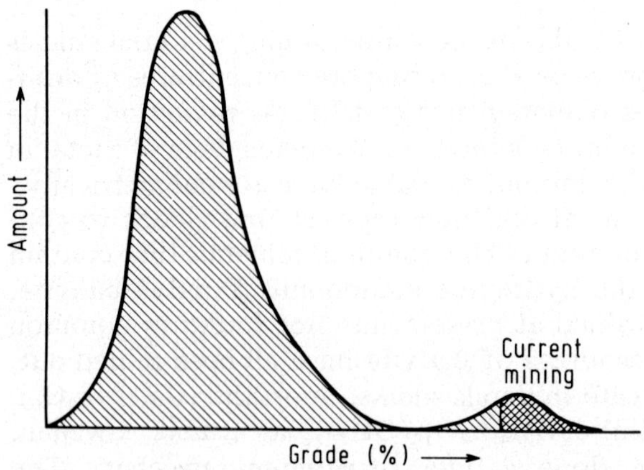

Fig. 10.2. The bimodal curve represents the probable distribution of a geochemically scarce metal in the earth's crust. The large peak is the distribution in common rocks, where scarce metals occur not as separate minerals but as atomic substitutes for abundant metals: the small peak represents deposits produced by ore-forming processes such as those resulting from the circulation of brines in the crust. Current mining has already reached the point where the distribution curve for easily accessible scarce minerals turns downward. Further reductions in the grade of ore mined will produce declining tonnages of ore and will eventually bring us to a mineralogical barrier — which lies somewhere in the area between the two humps.

at today's prices for energy could produce lead only at a cost of hundreds or thousands of dollars a pound. The present price of lead — less than a dollar a pound — is a bargain by comparison.

Miners never consider common rocks when they are seeking ores of geochemically-scarce metals. Instead they seek those rare and geologically limited volumes of the crust where special circumstances have produced marked local concentrations of geochemically-scarce elements, and have done so in a manner that leads to the scarce elements being present in compounds of their own, not as atomic substitutes in silicate minerals. These localized volumes (which we commonly call ore deposits) contain enrichments far above the average crustal abundance of an element — sometimes by as much as 100,000 times. Lead, for example, has an average abundance in the continental crust of 0.0010% but lead ores usually contain at least 2% lead, and some are known to be as rich as 20% lead. It is not surprising that such extraordinary local enrichments are limited both in number and in size. The vital question therefore is: What percentage of the scarce element content of the continental crust is concentrated into ore deposits?

Ore deposits are the result of unusual sets of circumstances. For example, most deposits of lead and zinc form as a result of reactions between brines circulating in the crust and the rocks they encounter. The brines alter and cause recrystallization of the silicate minerals, and in the process lead and zinc trapped by atomic substitution in the silicate minerals pass into aqueous solution. The concentration of lead and zinc (and other scarce metals) in solutions is low — no more than a few hundred ppm — but it is sufficient, under suitable conditions, to form ore deposits. The heated brine and its dissolved load must flow in a restricted channel and, while doing so, must undergo a further set of reactions which cause precipitation of lead and zinc minerals — usually the relatively insoluble sulfides PbS and ZnS.

Thus, three steps — solution, restricted flow, and localized precipitation — are necessary, and although brines are common in the crust, the delicate timing of events and the chemistry needed to form lead and zinc ores coincide very rarely. Similarly complex and therefore infrequent events lead to the formation of deposits of the other scarce metals. Just how the secondary concentration processes modify the distribution curve of geochemically scarce metals is not known with certainty, but the end result is, I suggest, the bimodal distribution depicted in Fig. 10.2.

Figure 10.2 tells an important story. Ore deposits of scarce metals always contain local concentrations of minerals such as PbS and ZnS. It is possible to mine and concentrate the minerals in an ore deposit and thus produce enriched aggregates ready for smelting. The area beneath the small hump in Fig. 10.2 is therefore a measure of the percentage of a scarce metal in the crust that is freed from atomic substitution and is combined as separate minerals. The large hump, by contrast, is a measure of the percentage trapped by isomorphous substitution in abundant silicate minerals. Proof of the bimodal distribution of geochemically-scarce elements in still lacking because analytical sampling of the crust is too crude. Considering the importance of the consequences, definitive sampling programs should stand high among the priorities of large government laboratories equipped to make the necessary investigations.

A bimodal distribution has two obvious consequences: first, as we mine the material under the small hump, starting with the richest grades first, there will be an initial period during which the declining grades will bring the reward of larger and larger tonnages of ore. This has been the situation faced by miners for most of this century. Eventually, however, the distribution curve turns down again, and further reductions in grade bring declining tonnages of ore. There are indications that this might already be happening in the mining of metals such as mercury, gold, and silver.

The second is less a consequence of the bimodal distribution than of mineral distribution in general: a point will eventually be reached where reduction in grade brings us to a mineralogical barrier — the point beyond which a scarce element occurs only as an isomorphous substitute and is therefore no longer amenable to concentration. The grade of the mineralogical barrier will vary for each element, each deposit, and each rock type. With a few exceptions, such as gold, uranium, and gallium, the barrier appears to lie at grades somewhere between 0.01 and 0.1%. That is, at grades below

0.01 and 0.1%, a scarce metal occurs solely by isomorphous substitution, above that grade, scarce metals form compounds of their own. Gold and uranium are exceptions on the low side, in that separate minerals are known to occur at grades well below 0.01%. Gallium is an exception on the high side. Indeed, so little gallium is concentrated into separate minerals that no large ore deposit has ever been found even though gallium is twice as abundant as lead in the crust.

Can we predict how much of each metal is present in the small "ore deposit hump" in Fig. 10.2? Until the last deposit is found and measured we cannot, of course, be sure that we are correct, but an indirect line of reasoning suggests that we can indeed estimate, at least to the correct order of magnitude, the size of the hump. A number of mass properties of ore deposits, such as the size of the largest known deposit for each scarce metal (Fig. 10.3) and the number of deposits that contain 1 Tg or more of a given metal (Skinner, 1976) are proportional to the average content (or crustal abundance) of the element in the continental crust. As first pointed out by McKelvey (1960), even the discovered reserves of the scarce metals are proportional to the crustal abundances (Fig. 10.4). These relations suggest strongly that the size of the "ore deposit hump" is directly proportional to the geochemical abundance of an element. Other lines of evidence support this suggestion, but they are no more definitive than the arguments just presented. The contention is therefore still a suggestion, albeit one that has a high probability of being correct.

Accepting the conclusion that the size of the "ore deposit hump" is proportional to crustal abundance, we need only estimate the size of one "hump" to be able to estimate how much of the other scarce metals to

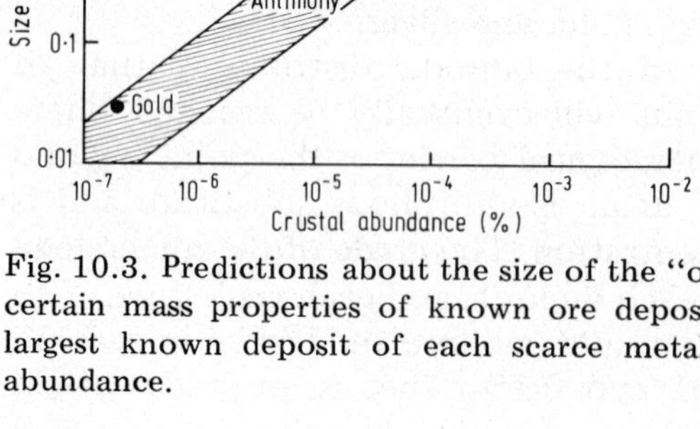

Fig. 10.3. Predictions about the size of the "ore deposits hump" in Fig. 10.2 are based on certain mass properties of known ore deposits. For example, the graph shows that the largest known deposit of each scarce metal is approximately proportional to crustal abundance.

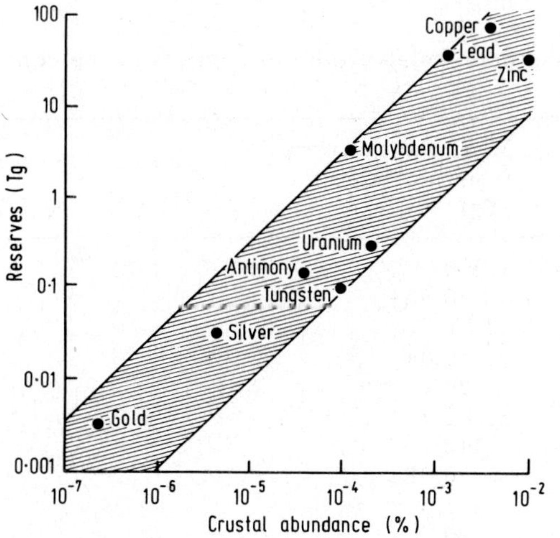

Fig. 10.4. As shown in the graph, the known reserves of scarce metals in the U.S.A. are also found to be proportional to crustal abundance. Like the relationship illustrated in Fig. 10.3, this function suggests that the amount of scarce metal ultimately available to current mining techniques is directly proportional to the geochemical abundance of the element. (Data from Brobst and Pratt 1973.)

expect. A recent report by the National Academy of Sciences' Committee on Mineral Resources and the Environment (COMRATE, 1975) has shown how this might be done. The report estimates that the mineralogical barrier for copper is reached at a grade of 0.1%. It also estimates that no more than 0.01% of total copper in the continental crust will be found concentrated in ore bodies with grades of 0.1% Cu or more. The Committee's reasoning was based on the volume percentage of mineralized rock in the most intensely mineralized regions so far discovered and on the frequency of copper deposits in the crust. Their figure of 0.01% must therefore be taken as the maximum possible yield, but it is not likely to be too large by more than a factor of ten. Thus the size of the "ore deposit hump" will probably fall between 0.001 and 0.01% of the amount of any scarce metal in the crust. The maximum estimated yields of metals from concentrated ore deposits, calculated according to be assumptions given, are listed in Table 10.3.

Using this kind of reasoning, COMRATE estimated that present reserves (in the mining sense of material from which a profit can be made) plus past production of copper already amount to 3% of the world's ultimate yield. The equivalent figure for the U.S.A. is estimated to be 16%. It does not take much arithmetic to calculate that, with use rates growing as they now are, copper will change from its present position as a metal in apparently abundant supply to a position of strategic shortage by the end of the present century.

TABLE 10.3

Estimated maximum yield of geochemically scarce metals from ore deposits in the continental crust

Element	Average abundance in continental crust (%)	Maximum recoverable from ore deposits (Tg)
Copper	0.0058	1,000
Gold	0.0000002	0.034
Lead	0.0010	170
Mercury	0.000002	0.34
Molybdenum	0.00012	20
Nickel	0.0072	1,200
Niobium	0.0020	340
Platinum	0.0000005	0.084
Silver	0.000008	1.3
Tantalum	0.00024	40
Thorium	0.00058	100
Tin	0.00015	25
Tungsten	0.00010	17
Uranium	0.00016	27

Note: The calculation assumes that mining will proceed no deeper than 10 km below the surface, and that 0.001% of all the metal in the continental crust is present in minerals available to mining and concentration. (The minerals may not be located in deposits rich enough to be considered ore by present standards). The calculation includes that part of the continental crust that lies beneath the continental shelf.

RELATIVE USE RATES OF METALS

Because supplies of scarce metals are apparently proportional to crustal abundances, we should view the use rates of metals in the same terms. Using the most widely employed geochemically abundant metal, iron, as a basis of comparison, Fig. 10.5 is an attempt to put relative use rates in a geochemical perspective. All metals falling on the dashed line are being used at the same rate, proportional to their crustal abundance, as iron. Metals that fall above the dashed line (and this includes most of the geochemically scarce metals) are being used at proportionally faster rates. For example, mercury and gold are being used at a rate about 110 times faster than iron, and lead about 40 times faster.

Assuming that we continue to use metals at rates that are not proportional to their abundances, we can read directly from Fig. 10.5 those metals which are likely to be mined out first. The farther a metal plots from the dashed line, the earlier its demise. Metals apparently in trouble include such widely-used commodities as mercury, gold, silver, copper, and lead. By contrast with geochemically-scarce metals, all of the geochemically-abundant metals

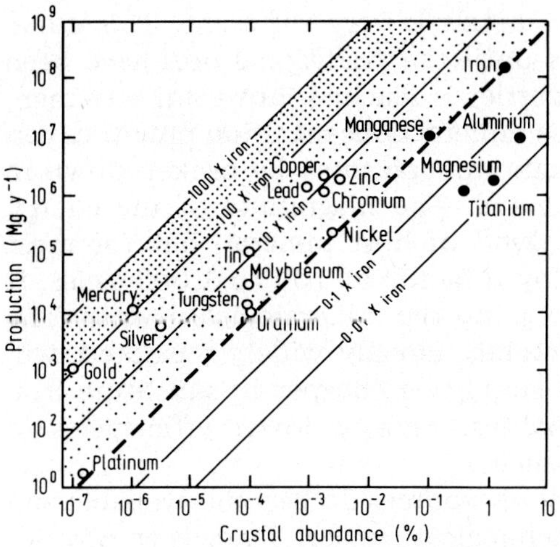

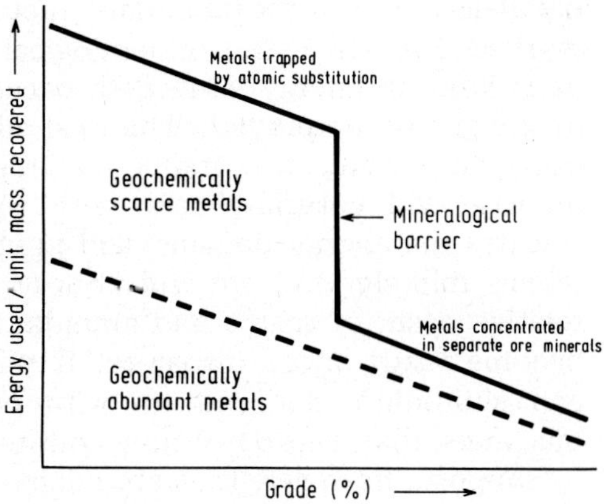

Fig. 10.5. The graph shows the relation between annual world production of newly-mined metals and their abundance in the continental crust. The dashed line drawn through iron, the most widely used geochemically abundant metal, may be considered a kind of base-line for use rates of metals: points lying on the line are produced at the same rate, relative to their crustal abundance, as iron. Metals below the line are mined proportionally slower; those above, proportionally faster. Metals farthest from the line — including many of the geochemically scarce metals — will be mined out first.

Fig. 10.6. The relationship between the grade of an ore and the energy input per unit mass of metal recovered is shown for both scarce and abundant metals. A steadily rising amount of energy will be needed to produce even geochemically-abundant metals from the leaner ores of the future, but the amount of energy needed to produce scarce metals will take a tremendous jump when the mineralogical barrier is reached. At that point, when ore deposits are worked out, mineral-concentrating processes can no longer be applied, and the silicate minerals in common rocks must be broken down chemically to separate the atoms of scarce metals from all the other atoms.

seem to be under-used. Their positions plot on or far below the dashed line. If my argument is correct, we have an unbalanced situation that cannot long continue. Clearly, we should be using abundant metals more and scarce metals less.

What happens when ore deposits of scarce metals have all been found and mined — when the mineralogical barrier is finally reached? The situation is demonstrated in Fig. 10.6. As grades decline in deposits of geochemically abundant metals, the energy input (and therefore the cost) per unit mass of metal recovered rises steadily. The smelting process remains the same because it is always possible to produce a concentrate. The steady rise in energy required is a result of the need to mine larger volumes as leaner and leaner ores are worked and to process these larger volumes by crushing and concentration.

The curve for geochemically scarce metals is very different. This curve parallels that for the abundant metals until the traditional ores have been worked out. Once the mineralogical barrier is reached, however, a tremendous jump in energy is needed, because mineral concentration processes can no longer be employed. The host silicate mineral must be broken down in order to recover the trapped scarce metal. The magnitude of the energy increase will naturally vary with the kind of host mineral, but for most silicates the energy demand will jump by a factor of 100 to 1,000 times. It seems unlikely that we will choose to jump the mineralogical barrier. The relative costs of scarce and abundant metals, already widely separated, will become vastly more disparate. It will simply be cheaper to substitute iron and aluminium and put up with penalties, such as lower efficiencies in machines, that we do not now countenance.

Suppose, however, that abundant energy sources do become available and that it is feasible to overcome the mineralogical barrier. There is an obvious reason why a future technology built largely on geochemically-abundant metals will pertain even in that event. Consider once again the relative abundance of elements. If we extract metals from silicate minerals, and therefore succeed in using average rock as an ore, all metals would be produced in amounts approximately proportional to their crustal abundances. That is, a mining industry based on average rock would have production figures for *all* metals that would be along a line parallel to the dashed line in Fig. 10.5. If we wished to use metals in the proportions in which we now use them, we would soon have vast surpluses of iron, aluminium, and other abundant metals. Instead of allowing huge unwanted stocks to accumulate, I have no doubt that we would soon find a way to reduce our demands for scarce metals and make do with the more abundant metals.

HARD TIMES AHEAD

However one views the use of geochemically scarce metals in the future, it is clear that there are very real limits to the amounts available in traditional ore deposits of the continental crust. Efficient recycling, which surely must come as existing deposits are worked out and new ones become harder and harder to find, will guarantee that even the scarcest metals will always be available in at least small amounts. But recycling can at best sustain a declining use rate. While growth or even level use rates pertain, newly mined material must continue to be added. When the traditional deposits have all been found and mined, however, our responses will inevitably be governed by geochemical abundances, and little, if any, newly mined scarce metals will be available to be added to the pool in use.

This critical time in man's future technological development cannot be pinpointed, but it cannot be too far in the future. The date depends on future use rates of metals, and since some metals will be effectively used up before others, we are likely to see an extended decline rather than a sudden cutoff. The decline has already started for gold, silver, and possibly a few other metals. The rest will follow during the next century, and by the year 2076, when the U.S.A. celebrates its tercentenary, mining of scarce metals will be increasingly a memory from the past.

The decline will be controlled, at least in part, by international politics. My estimate of the relative positions of three industrial countries, England, the U.S.A., and the U.S.S.R., is illustrated in Fig. 10.7 using relations first elucidated by Hewett (1929). The curve defining the amount of metal produced starts at zero-production when mining first commences in a country and ends again at zero when all ore deposits are worked out. The area under the curve is the total amount of metal produced in the country and corresponds to the "ore deposit hump" in Fig. 10.2. The curve defining the number of working mines is a measure of the rate at which ore deposits are discovered, and the third curve indicates the rising amounts of metals imported to supply an industry that can no longer be maintained by internal supplies. A century and a half ago, England was a major metal producer, shipping its copper, lead, tin, and other scarce metals around the world. Since that time it has become, increasingly, an importer of raw materials. Now the U.S.A. is following the same route and, as each year passes, a higher percentage of essential raw materials must be imported (Morton, 1973).

The future for production of scarce metals clearly lies with those continents where prospectors have not yet scoured every corner. These are, mainly, the less inhabited portions of Asia, such as Siberia, the central and

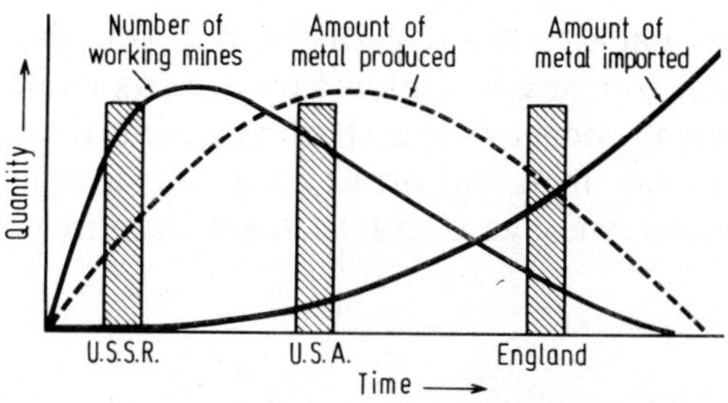

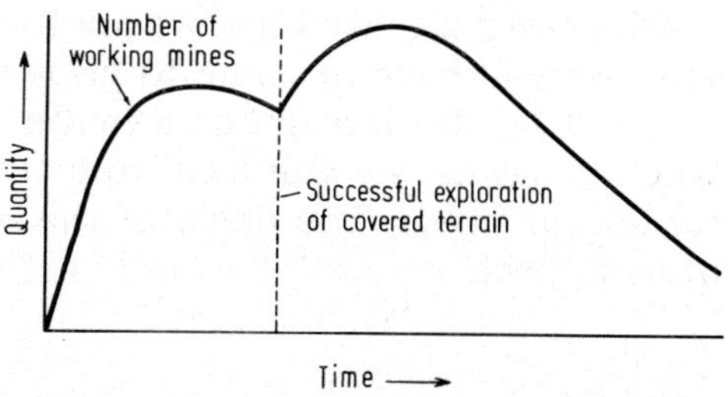

Fig. 10.7. The historical development of metal production, the number of working mines, and the amount of metal imported are shown for three industrial countries. As time passes, the position of a country moves from left to right in the graph. The U.S.A. is today where England stood early in the last century. The U.S.S.R. is today in the same relative stage development as the U.S.A. in about 1850.

Fig. 10.8. When a successful method of prospecting for ore deposits beneath covered terrain is developed, the curve defining the changing number of working mines with time will move sharply up (compare the equivalent curve in Fig. 10.7). The situation is comparable to prospecting a new country.

northern portions of Africa, much of South America, Australia, and Antarctica. Each continent still has some unexplored ground of its own — that is, areas so deeply covered by soil and by young sediments that it is impossible to use any prospecting method to sense ore deposits below. For a country such as the U.S.A., the amount of blind ground is close to 50% of the total land area. We await a breakthrough in research that will develop methods to carry out this deep prospecting. When it comes, the curve defining the number of working mines in Fig. 10.7 will have to be modified as shown in Fig. 10.8. No doubt hopes will rise when successful prospecting in covered terrain is achieved, and skepticism will then be expressed at the kind of predictions made in this paper. The end is entirely predictable, however. Figure 10.8 shows what it must be.

So far we have concentrated on the continental crust. The crust beneath the ocean floors also remains to be prospected, however, and we can consider it, in terms of Fig. 10.7, as a huge new continent open for exploration. Unfortunately, the deep ocean floor, which is about two-thirds of earth's surface, seems to offer distressingly poor prospects for most scarce metals (Skinner and Turekian, 1973). It is still much too early to draw final conclusions, but it looks as if considerably less than 0.01% of the metals in the oceanic crust is concentrated into ore deposits, even including such unusual deposits as ferromanganese nodules on the deep-sea floor. It is not too surprising that this might be so, because the ocean floor is all less than 200 My old — so young, geologically speaking, that concentrating processes have not had as much time to do their work as they have in the vastly older continental crust. Undoubtedly some deposits will be found — deposits of copper and nickel seem the most likely — but their recovery will offer a great many technological headaches.

Whichever way we turn, we are forced back to the realization that one day soon we will have to come to grips with the way in which earth offers us its riches. That day is less than a century away, perhaps less than half a century. When it dawns, we will have to learn to use iron and other abundant metals for all our needs. The dawn of the second iron age is much closer than most of us suspect.

REFERENCES

Brobst, D.A. and Pratt, W.P. 1973. United States Mineral Resources. Geological Survey Professional Paper 820. U.S. Govt. Printing Office, Washington, DC, 722 pp.

COMRATE, 1975. Mineral Resources and the Environment. Report by the Committee on Mineral Resources and the Environment, National Academy of Sciences, National Research Council, 348 pp.

Hewett, D.F., 1929. Cycles in metal production. Transactions of Am. Inst. of Mining and Metall. Engineers, Yearbook for 1929, pp. 65—98.

McKelvey, V.E., 1960. Relation of reserves of the elements to their crustal abundances. Am. J. Sci., 258-A: 234—241.

Morton, R.C.B., 1973. Mining and Minerals Policy. Second Annual Report of the Secy. of the Interior under the Mining and Minerals Policy Act of 1970. Vol. 1. U.S. Govt. Printing Office, Washington, DC, 73 pp.

Skinner, B.J., 1976. Earth Resources. 2nd edn. Prentice-Hall, Englewood Cliffs, NJ, 162 pp.

Skinner, B.J. and Turekian K.K., 1973. Man and the Ocean. Prentice-Hall, Englewood Cliffs, NJ, 149 pp.

GLOSSARY OF SELECTED TERMS

The majority of definitions have been abstracted from the following sources to which the reader is referred for more complete and, in some cases, alternate versions:

M. Gary, R. McAfee and C.L. Wolf (Editors), 1972. Glossary of Geology, American Geological Institute.

R.E. Buchanan and N.E. Gibbons (Editors), 1974. Bergey's Manual of Determinative Bacteriology, 8th edn., Williams and Wilkins, Baltimore, MD.

M.R. Walter (Editor), 1976. Stromatolites, Developments in Sedimentology 20, Elsevier, Amsterdam.

P. Gray, 1967, The Dictionary of the Biological Sciences, Rheinhold, New York, NY.

Taxonomic and trivial names of organisms are not included. Also excluded are terms for which adequate definitions are to be found in the Concise Oxford Dictionary, 6th edn, 1976, Clarendon Press, Oxford.

Ahermatypic coral: a coral which lacks algal symbionts.

Alkalinity: the number of milliequivalents of H^+ neutralized by 1 litre of sea water at 20° C.

Amphiboles: a group of dark, rock-forming ferromagnesian (q.v.) silicate minerals.

Amygdaloid: an extrusive or intrusive rock containing gas cavities filled with secondary minerals.

Amygdule: a gas cavity or vesicle in an igneous rock which is filled with secondary minerals.

Argillaceous: a term applied to all rocks or substances composed of clay or having a notable proportion of clay in their composition.

Assimilation: incorporation of simple compounds into cellular material.

ATP: adenosine triphosphate, a biological energy-transfer molecule containing "high-energy" pyrophosphate bonds.

Authigenic: formed or generated in place.

Autotroph: an organism capable of growth on purely inorganic media.

Barite: a mineral, $BaSO_4$.

Barophilic organisms: organisms requiring high pressures for growth.

Batch culture: a culture of organisms (generally microorganisms) in which the medium is not renewed (cf. continuous culture).

Biogeochemical prospecting: mineral exploration based on trace element analysis of plants and parts of plants.

herm: a circumscribed organo-sedimentary structure whose minimum width is less than or equal to 100 times its maximum thickness, embedded in rocks of different lithology.

Bioturbation: the physical disturbance of sediments by burrowing and other activities of organisms.

B.P.: before present.

Breccia: a course-grained clastic (q.v.) rock composed of large (>2 mm diam.), angular, and broken rock fragments which are cemented together in a finer-grained matrix.

Calcarenite: limestone or dolomite composed of coral or shell sand or of sand derived from the erosion of older dolomite.

Calcrete: a hard mass of surficial sand and gravel cemented by calcium carbonate.

Calvin dark cycle pathway: a pathway of biological CO_2 fixation in which early products are C_3 compounds.

Capsule (bacterial): a loose, more or less amorphous layer made up of organic polymers, which is deposited outside, and remains attached to, the cell wall.

Carbonate compensation depth: the level of an ocean at which the rate of calcium carbonate deposition equals the rate of its resolution.

Celestite: a mineral, $SrSO_4$.

Cementation: the process of precipitation of a binding material around grains or minerals in rocks.

Chemoautotroph: *see* Chemolithotroph.

Chemocline: the boundary between circulating and non-circulating water masses or layers of a lake.

Chemolithotroph: an organism that utilizes CO_2 as its principle source of carbon for growth and obtains its energy by the oxidation of inorganic compounds.

Chemostat: an apparatus in which organisms are maintained in continuous culture (q.v.) through continuous input of a growth-limiting nutrient.

Chemosynthesis: the process of dark fixation of CO_2 into cellular material coupled to oxidation of inorganic compounds.

Chemotaxis (chemotactic): the process by which motile organisms migrate to and accumulate in a part of a chemical gradient.

Chert: a hard, extremely dense or compact, dull to semivitreous, cryptocrystalline sedimentary rock, consisting dominantly of cryptocrystalline silica.

Chloragocytes: cells closely associated with the blood vessels of the gut of annelids. They contain small granules called "chloragosomes" which are released from the cells and are important in their metabolism.

Cisterna: a fluid-containing sac or cavity in an organism.

Clastic: consisting of fragments of rocks or of organic structures that have been moved individually from their places of origin.

Coccolith: very tiny calcareous plates, generally oval and perforated, borne on the surface of some marine flagellate organisms.

Coelomic fluid: fluid of the coelom, the main body cavity in which the gut is suspended, in many animals having a body made up of three parts (ectoderm, mesoderm and endoderm).

Coffinite: an important mineral in some uranium deposits, $U(SiO_4)_{1-x}(OH)_{4x}$

Colloform texture: the rounded, globular texture of a colloidal mineral deposit.

Conformable (stratigraphy): describes strata or stratification characterized by an unbroken sequence in which the layers are formed one above the other in parallel order by regular, uninterrupted deposition under the same general conditions.

Conglomerate: Similar to Breccia (q.v.) except that most of the fragments have smooth edges and worn corners.

Connate water: water trapped in the interstices of an extrusive igneous or sedimentary rock at the time of deposition.

Conodont: tiny tooth- or jaw-like fossil composed of calcium phosphate and of uncertain zoological affinity.

Constitutive (enzyme): a constitutive enzyme is one which is present in a cell at high levels under all growth conditions (cf. Induction).

Continuous culture: a culture in which populations of microorganisms can be maintained in a state of exponential growth for extended periods of time.

Coprolite: fossilized excrement of vertebrates composed mainly of calcium phosphate.

Corrinoid: the general term for compounds containing the corrin nucleus ($C_{19}H_{22}N_4$).

Cytochrome: a haem-containing protein involved in electron transport in cells.

Dehydrogenase: an enzyme catalysing the reversible transfer of hydrogen from an organic substrate(S) to a carrier (C); eg. $S—H_2 + C \rightleftharpoons S + C—H_2$.

Denitrification: process by which nitrate and nitrite are reduced to N_2.

Detritus: material produced by the disintegration and weathering of rocks that has been moved from its site of origin.

Diagenesis: process leading to changes in a sediment after deposition at low temperatures and pressures; less drastic than metamorphism (q.v.)

Diastrophism: process or processes by which the crust of the earth is deformed, producing continents, ocean basins, plateaus, mountains faults, etc.

Dissimilation: a poorly-defined term which is often applied to biochemical reactions in which the products of reaction are not used for synthetic purposes.

DNA hybridization: a method for determining the degree of similarity between two species of DNA.
Duricrust: the case-hardened crust of soil formed in semiarid climates by the precipitation of salts at the surface of the ground as the ground water evaporates.
Endergonic: consuming energy.
Endolithic: of organisms, living within rock; specifically boring organisms.
Endoplasmic reticulum: a complex intracellular membrane system.
Enterolithic: describes a sedimentary structure consisting of ribbons of intestine-like folds that resemble those produced by tectonic deformation but that originate through chemical changes involving an increase or decrease in the volume of rock.
Epeirogenesis: movement of the crust due to earth's forces which has produced the larger features of the continents and oceans eg. plateaus and basins.
Epigenesis: the changes, transformations, or processes, occurring at low temperatures and pressures, that affect sedimentary rocks after compaction, exclusive of surficial alteration; late diagenesis (q.v.)
Epilimnion: the uppermost layer of water in a lake, characterized by an essentially uniform temperature, that is generally warmer than elsewhere in the lake, and by relatively uniform mixing by wind and wave action.
Epilithic: of organisms, living on or attached to rock.
Epithelium: any tissue that lines, or covers, an organ or organism.
Eucaryotes: nucleated protists and higher organisms.
Exogonic: releasing energy.
Facies: a stratigraphic body as distinguished from other bodies of different appearance or composition.
Facultative aerobe: an organism capable of both aerobic and anaerobic growth.
Feldspar: a group of common rock-forming minerals with the general composition $MAl(AlSi_3)O_8$ where M=K, Na, Ca, Ba, Rb, Sr, or Fe.
Fermentation: an ATP (q.v.) — generating metabolic process in which organic compounds serve as both electron donors (becoming oxidized) and electron acceptors (becoming reduced). The average oxidation state of the end products is identical to that of the substrate.
Ferromagnesian: containing iron and magnesium.
Foliose: having the appearance of a leaf.
Fulvic acid: organic matter of complex composition which remains soluble when an aqueous extract of sediment or soil is acidified.
Gangue: the non-metalliferous or non-valuable metalliferous minerals associated with ore.
Gastrolith: a polished stone or pebble from the stomach of some vertebrates.
Geobotanical prospecting: mineral exploration based on the appearance and distribution of plant species.

Geodes: hollow, globular bodies varying in size, 2 to >20 cm, characteristic of certain limestone beds but rarely in shales.
Geochemical anomaly: a concentration above the natural background level (q.v.) of one or more elements in rock, soil or related material.
Geological time scale: see diagram below.
Glauconite: a green mineral, essentially a hydrous potassium iron silicate.

GEOLOGICAL TIME-SCALE

Eon	Era	Period	Epoch	Age in years
PHANEROZOIC	CAINOZOIC	Quaternary	Recent	0 - 15 000
			Pleistocene	15 000 - 1 800 000
		Tertiary	Pliocene	1·8 - 5 000 000
			Miocene	5 - 22 500 000
			Oligocene	22·5 - 37 500 000
			Eocene & Paleocene	37·5 - 65 000 000
	MESOZOIC	Cretaceous		65 - 136 000 000
		Jurassic		136 - 195 000 000
		Triassic		195 - 235 000 000
	PALAEOZOIC	Permian		235 - 280 000 000
		Carboniferous		280 - 345 000 000
		Devonian		345 - 395 000 000
		Silurian		395 - 435 000 000
		Ordovician		435 - 500 000 000
		Cambrian		500 - 570 000 000
PRECAMBRIAN	PROTEROZOIC	Upper		570 - 1 400 000 000
		Middle		1 400 - 1 800 000 000
		Lower		1 800 - 2 300 000 000
	ARCHAEAN			2 300 000 000 + (OLDEST-KNOWN ROCKS 3 700 000 000 YEARS)

Bureau of Mineral Resources, Geology and Geophysics. March, 1974

Gneiss: a foliated rock formed by regional metamorphism (q.v.) in which bands or lenticles of granular minerals alternate with bands and lenticles in which minerals having flaky or elongate prismatic habits predominate.

Goldich's sequence: the order of stability of igneous rocks towards weathering.

Haemolymph: the circulatory fluid of various invertebrates.

Halophilic: of organisms requiring high concentrations of NaCl for growth.

Hermatypic coral: coral characterized by presence of symbiotic unicellular algae.

Heterocyst: a spore-like structure produced by some cyanobacteria.

Heterotroph: an organism requiring preformed organic matter for growth.

Histoplasmosis: a disease caused by infection by a fungus, *Histoplasma capsulatum.*

Holdfast: an organelle (q.v.) by which a microorganism is attached to a surface.

Homolictic lake: one in which the entire water mass circulates at overturn periods.

Humic acid: black, acidic organic matter, soluble in alkali but insoluble in acids and organic solvents.

Humus: relatively stable dark part of soil organic matter decomposed beyond the stage of visual recognition of the original plant material.

Hydrogenase: an enzyme catalysing the reversible dissociation of molecular hydrogen into hydrogen ions and electrons.

Hypersaline: highly saline, usually with respect to sea water.

Hypolimnion: The portion of certain lakes below the thermocline (q.v.) which receives no heat from the sun and no aeration by circulation.

Induction: of enzymes, synthesis of enzyme in response to the exposure of organisms to a specific substrate.

Ionotropy: tautomerism.

Interstitial water: *see* Porewater.

JOIDES: Joint Oceanographic Institutes Deep Earth Sampling.

Kainite: a mineral, $KCl \cdot MgSO_4 \cdot 3\ H_2O$.

Karren: a surface feature resulting from differential solution of limestone and removal of residual limestone soil.

Karst: a feature resulting when limestone is dissolved by rain or rivers.

Kerogen: insoluble, organic material found in sedimentary rocks, usually shales, and sediments.

Lamella: (geological), a thin plate, scale, flake, leaf, lamina or layer.

Lamella: (biological), an organ, process or part of an organism resembling a plate.

Langbeinite: a mineral, $K_2Mg_2(SO_4)_8$.

Langmuir adsorption equation: an expression usually known as the Langmuir adsorption isotherm which relates the amount of substance adsorbed on the surface to the partial pressure of that substance in the gaseous phase.

Lithification: a complex process that converts a newly-deposited sediment into a hard rock.
Lithobionts: organisms living on surfaces of rocks.
Lithofacies: a lateral, mappable division of a designated stratigraphic unit of any kind, distinguished from other adjacent subdivisions on the basic of lithologic characters.
Lithotroph: a general term covering chemo- and photolithotrophs (q.v.).
Loessial deposit: a fine grained, slightly coherent calcareous deposit of mainly silt material.
Lysocline: the level or depth in an ocean below which there is a significant increase in the solution of calcium carbonate.
Macrophyte: a megascopic plant particularly in an aquatic environment.
Mafic: describes an igneous rock composed chiefly of one or more ferromagnesian (q.v.), dark coloured minerals in its mode.
Mantle: (geological); layer of earth between crust and core.
Mantle: (biological); an envelopment of the body usually meaning the outer soft coat of Mollusca and Brachiopoda.
Melanophore: a cell containing melanin as pigment.
Meromictic lake: one that is partly mixed and in which thermal turnover occurs only in the top layers; bottom layers are stagnant and anaerobic.
Mesophilic: of organisms, growing at moderate temperatures (ca. 15—35°C).
Metalimnion: virtually synonymous with thermocline (q.v.).
Metamorphism: process by which consolidated rocks are altered in composition, texture or internal structure by forces such as pressure, heat and introduction of new chemical substances.
Metasomatism: replacement of one mineral by another in a rock.
Meteoritic (meteoric) water: water of recent atmospheric origin.
Micrite (micritic): the semiopaque, micro-crystalline interstitial component (matrix) of limestones consisting of precipitated carbonate (calcite) mud.
Microvilli: minute finger-like projections from the cell surface about 0.1 μm in diameter.
Mixotrophic: of organisms, capable of utilizing combinations of organic and inorganic compounds as energy and carbon sources.
Monaxon: a simple uniaxial sponge spicule with a single axial filament or axial canal, or one developed by growth along a single axis.
Monimolimnion: the bottom, non-circulating water mass of a meromictic (q.v.) lake.
Monohydrocalcite: a rare mineral found in lake sediments, $CaCO_3 \cdot H_2O$.
NAD^+ (NADH): oxidized (reduced) nicotinamide — adenine dinucleotide, a hydrogen carrier in metabolic reactions.
Nannoplankton: plankton (q.v.) of the size range 5—60 μm.
Natural background: of elements, the concentration of an element in naturally-occurring material that could be regarded as "normal" as distinct from "anomalous".

Neo-Euxinian: a term applied to sediments deposited during the freshwater phase of the Black Sea.
Nepheline syenite: an alkali-rich, silica-depleted igneous rock emplaced below the earth's surface.
Nephridea: individual excretory units present in many invertebrates.
Non-axenic culture: a culture containing more than one species or strain of organisms.
Obligate: a term applied to an organism which has a strict requirement for certain growth conditions, e.g. obligate anaerobic; obligate autotroph.
Oncoid: an algal biscuit resembling the small, variously shaped (often spherical) concentrically laminated, calcareous sedimentary structures called oncolites.
Ooid: a small, round accretionary body in sedimentary rock usually formed of calcium carbonate in successive concentric layers.
Oxidative phosphorylation: generation of ATP (q.v.) by respiration (q.v.).
Palaeosol: a buried soil horizon from the past.
Palisade layer: A layer of palisade parenchyma (columnar or cylindrical cells rich in chloroplasts) in a leaf.
Pedogenesis: the process of soil formation.
Pedoscope: a more rugged adaptation of the peloscope (q.v.) for use in firm sediments and soils.
Peloscope: an array of microcapillaries (glass) inserted in water or the top layers of sediments for subsequent microscopic study of microbial development.
Pentlandite: a mineral, $(Fe, Ni)_9S_8$.
Periostracum: the thin organic layer covering the exterior or shell of brachiopods and many molluscs.
Phosphorite: a phosphate rock.
Photoautotroph: *see* Photolithotrophic.
Photolithotrophic: of organisms, able to grow with light on a strictly inorganic medium.
Photoorganotrophic: of organisms, able to grow with light at the expense of organic compounds.
Photophosphorylation: light-catalysed synthesis of ATP (q.v.).
Photoreceptors: light-trapping organelles (q.v.) or molecules in organisms.
Phototrophic: general term for photolithotrophs and photoorganotrophs (q.v.).
Phytane: a saturated hydrocarbon, $C_{20} H_{42}$.
Phytolith: a rock formed by plant activity or composed of plant remains.
Plasmalemma: the cell (cytoplasmic) membrane.
Plate tectonics: an hypothesis advanced to explain the broad features of the upper part of the earth's crust. It assumes that broad, thick plates or blocks of crust and mantle (q.v.) "float" on a viscous underlayer.
Podzol: a soil with a surface layer of organic matter overlying gray leached horizons.

Porewater: water found in the space between solid particles in soil, sediments or rocks.

Preexuvial: before exuviation, the removal of the theca of a dinoflagellate.

Procaryote: a protist in which the genetic material is never separated from the cytoplasm by a membrane.

Prograding: describes the seaward advance of shoreline resulting from the nearshore deposition of sediments brought to the sea by rivers.

pS^{2-}: negative logarithm of sulfide concentration.

Pseudomorph: a crystal formed by replacement of one mineral by another but retaining the outward form of the original mineral.

Pynocline: a layer in a water body where there is a rapid change in density with depth.

Regolith: the layer or mantle of loose incoherent rock material, of whatever origin, that nearly everywhere forms the surface of land and rests on the hard or "bed" rocks.

Repression: of enzymes, inhibition of enzyme synthesis by a product (or products) of the metabolic pathway in which the enzyme operates. A means of control on metabolism.

Respiration: an ATP (q.v.) generating process in which either inorganic or organic compounds serve as electron donors (becoming oxidized) and inorganic compounds (mostly oxygen, but also sulfate, nitrate and carbonate) serve as the ultimate electron acceptors (becoming reduced).

Resting cell: an ambiguous term generally applied to viable microorganisms which, because of nutrient limitations, cannot divide.

Rhodanese: an enzyme which catalyses the reaction $A^- + S_2O_3^{2-} \rightarrow AS^- + SO_3^{2-}$ where A^- may be CN^- or certain thiols.

Sandstone: a cemented or otherwise compacted detrital sediment composed predominantly of quartz grains, the grades of the latter being those of sand.

Sclerotized: describes the covering of an invertebrate (esp. an arthropod) hardened by substances other than chitin.

Silcrete: a conglomerate consisting of surficial sand and gravel cemented into a hard mass by silica.

Statoconia: small calcareous granules occurring in the statocyst of some animals.

Stillstand: a condition of stability, or of remaining stationary, with reference to the Earth's interior or to sea level, applicable to an area of land, as a continent or island; a period of time during which there is a stillstand.

Stratabound: of mineral deposits, confined within a single stratigraphic unit.

Stratiform ore: a layered stratabound mineral deposit generally of sedimentary origin in which the layers are conformable (q.v.) with those of the host rock.

Stromatolite: an organo-sedimentary structure produced by sediment trapping, binding and/or precipitation as the result of the growth and metabolic activity of microorganisms, principally cyanophytes.

Subaerial: occurring beneath the atmosphere or in open air.

Supergene: applied to ores or ore minerals that have been formed by generally descending water.

Syngenetic: of minerals or ore deposits, formed contemporaneously with the enclosing rocks.

Synsedimentary: of minerals; deposited at the same time as the enclosing sediments.

Syntrophic: associated or mutually dependent on one another.

Thermocline: the layer in a thermally stratified body of water within which the temperature decreases rapidly with increasing depth usually at a rate greater than 1° C per metre.

Thermophilic: of organisms, requiring high temperatures (ca. 40—90° C) for growth.

Trichome: a many-celled, frequently-branched, filament of bacterial or, less frequently algae.

Trichosperical: a term applied to a spherical microcolony (of microorganisms) consisting of filaments (or trichomes) growing radially from a common centre.

Tridymite: a mineral, SiO_2.

Trophogenic: descriptive of the trophic zone.

Tropholytic: describes the deeper part of a lake in which organic matter tends to be dissimilated (cf. trophogenic).

Tuffaceous: describes tuffs; sediments containing up to 50% of compacted pyroclastic deposits of volcanic ash and dust.

Unconformity: a substantial break or gap in the geological record where a rock unit is overlain by another that is not next in the stratigraphic succession.

Urolith: a urinary calculus.

Vaterite: a rare mineral, $CaCO_3$.

Vermiculite: a group of clay minerals of the general composition $(Mg,Fe,Al)_3$-$(Al,Si)_4O_{10}(OH)_2 \cdot 4\ H_2O$.

Vug: a small cavity in a vein or in rock, usually lined with crystals of a different mineral composition from the enclosing rock.

Zoogloea: a gelatinous or mucilaginous mass characteristic of the growth of various bacteria when growing in media rich in organic matter.

Zooxanthellae: symbiotic, unicellular algae in the endoderm of hermatypic coral (q.v.) polyps.

SUBJECT INDEX